H.-P. Blume, H. Eger, E. Fleischhauer,
A. Hebel, C. Reij, K. G. Steiner (Editors)

Towards Sustainable Land Use

Furthering Cooperation Between People And Institutions

VOLUME II

Geographisches Institut
der Universität Kiel
ausgesonderte Dublette

ADVANCES IN GEOECOLOGY 31

A Cooperating Series of the International Society of Soil Science (ISSS)

ISBN 3-923381-42-5

Die Deutsche Bibliothek - CIP Einheitsaufnahme

Towards sustainable land use : furthering cooperation between people and institutions / H.-P. Blume ... (ed.).
- Reiskirchen : Catena-Verl..
(Selected papers of the ... conference of the International Soil Conservation Organisation (ISCO) ; 9)
(Advances in geoecology ; 31
ISBN 3-923381-42-5
International Soil Conservation Organization:
Selected papers of the ... conference of the International Soil Conservation Organisation
(ISCO). - Reiskirchen : Catena-Verl.
(Advances in geoecology ; ...)
Vol. 1. - (1998)
Vol. 2. - (1998)

Managing Editor "Advances in GeoEcology":
Margot Rohdenburg

The English language editing of this volume "Advances in GeoEcology 31"
has been carried out by:
Simon Berkowicz, Jerusalem

Sponsors
German Federal Ministry for the Environment, Nature Conservation and Nuclear Safety
German Federal Ministry for Economic Cooperation and Development
Deutsche Gesellschaft für Technische Zusammenarbeit (GTZ) GmbH
Deutsche Bundesstiftung Umwelt
Technical Centre for Agriculture and Rural Co-operation (CTA)
German Foundation for International Development (DSE)
Commission of the European Union
Swiss Development Cooperation
German Scientific Foundation
Food and Agriculture Organization of the United Nations
Kunst- und Ausstellungshalle der Bundesrepublik Deutschland
Stadtwerke Bonn
Verwaltung des Deutschen Bundestages
Rheinbraun AG
Misereor
Brot für die Welt
Deutsche Lufthansa AG

© Copyright 1998 by CATENA VERLAG GMBH, 35447 Reiskirchen, Germany
All rights are reserved. No part of this publication may be reproduced, stored in a retrieval system or transmitted in any form or by any means, electronic, mechanical, photocopying, recording or otherwise, without prior permission of the publisher.
This publication has been registered with the Copyright Clearance Center, Inc.
Submission of an article for publication implies the transfer of the copyright from the author(s) to the publisher.

ISBN 3-923381-42-5

Contents Volume II

Part III: From soil and water conservation to sustainable land management   819

C. Reij, H. Eger & K. Steiner
From soil and water conservation to sustainable land management:
Some challenges for the next decade   819

Chapter 11: Sustainable land use management: policies, strategies and economics   827

Chapter 11.1: Policies and strategies   827

H. Hurni
A multi-level stakeholder approach to sustainable land management   827
J. Pretty
Furthering cooperation between people and institutions   837
A. Steer
Making development sustainable   851
M. Stocking
Conditions for enhanced cooperation between people and institutions   857
J.-C. Griesbach & D. Sanders
Soil and water conservation strategies at regional, sub-regional and national levels   867

Chapter 11.2: Economics   879

R. Clark, H. Manthrithilake, R. White & M. Stocking
Economic valuation of soil erosion and conservation - A case study of Perawella, Sri Lanka   879
P.C. Huszar
Including economics in the sustainability equation: Upland soil conservation
in Indonesia   889
Ram Babu & B.L. Dhyani
Economic evaluation of watershed management programmes in India   897
E.H. Bucher, P.C. Huszar & C.S. Toledo
Sustainable land use management in the South American Gran Chaco   905
M. Zlatić, N. Ranković & G. Vukelić
Improvement of soil management for sustainability in the hilly Rakovica Community   911

Chapter 11.3: Case studies   919

A. Arnalds
Strategies for soil conservation in Iceland   919
J. Hraško
Influence of socio-economic changes on soil productivity in Slovakia   927
F. Dosch & S. Losch

Spatial planning tasks for sustainable land use and soil conservation in Germany	933
I. Hannam	
An ecological perspective of soil conservation policy and law in Australia	945
S. Ewing	
Australia's community 'Landcare' movement: A tale of great expectations	953
J.S. Gawander	
The rise and fall of the vetiver hedge in the Fijian sugar industry	959
A.S. Kruger	
Closing the gap between farmers and support organisations in Namibia	965

Chapter 12: Institutional and planning aspects of soil and water conservation   971

M. Maarleveld
Improving participation and cooperation at the local level: Lessons from Eeonomics and psychology   971
L. van Veldhuizen
Principles and strategies of participation and cooperation challenges for the coming decade   979
A. Oomen
Political and institutional conditions for sustainable land management   985
H. Zweifel
Sustainable land use - A participatory process   991
H. Eger
Participatory land use planning: The case of West Africa   1001
T. Bekele & W. Zike
Village level approach to resource management: A project in a marginal environment in Ethiopia   1009
Lixian Wang
Mountain watershed management in China   1017
M. Schneichel & P. Asmussen
Dry forest management - Putting campesinos in charge   1023
P.S. Cornish
A farmer-researcher-adviser partnership designed to support changes in farm management needed to meet catchment goals   1029

Chapter 13: Approaches to soil and water conservation and watershed development   1037

Chapter 13.1: Case studies from Africa   1037

H. Liniger, D.B. Thomas & H. Hurni
WOCAT - World Overview of Conservation Approaches and Technologies - Preliminary results from Eastern and Southern Africa   1037
G.O. Haro, E.I. Lentoror & A. von Lossau
Participatory approaches in promotion of sustainable natural resources management experiences from South-West Marsabit, Northern Kenya   1047
M.J. Kamar
Soil conservation implementation approaches in Kenya   1057

K.C.H. Mndeme
Participatory land use planning approach in North Pare Mountains, Mwanga, Kilimanjaro Region, Tanzania — 1065

S. Minae, W.T. Bunderson, G. Kanyama-Phiri & A.-M. Izac
Integrating agroforestry technologies as a natural resource management tool for smallholder farmers — 1073

T. Defoer, S. Kantè & Th. Hilhorst
A participatory action research process to improve soil fertility management — 1083

I. Yosko & G. Bartels
Nomads and sustainable land use: An approach to strenghten the role of traditional knowledge and experiences in the management of the grazing resources of Eastern Chad — 1093

H. Paschen, D. Gomer, L. Kouri, H. Vogt, T. Vogt, M. Ouaar & W.E.H. Blum
Management of watersheds with soils on marls in the Atlas Mountains of Algeria - A proposal for a non-conventional watershed development scheme — 1099

Chapter 13.2: Case studies from Asia — 1107

B. Adolph & T. G. Kelly
What makes watershed management projects work? Experiences with farmer's participation in India — 1107

J. Mascarenhas
The participatory watershed development process
Some practical tips drawn from outreach in South India — 1117

R. Chennamaneni
Watershed management and sustainable land use in semi-arid tropics of India: Impact of the farming community — 1125

K. Mukherjee
People's participation in watershed development schemes in Karnataka - Changing perspectives — 1135

J.S. Samra & A.K. Sikka
Participatory watershed management in India — 1145

Sameer Karki & S.R. Chalise
Improving people's participation in soil conservation and sustainable land use through community forestry in Nepal — 1151

C. Setiani & A. Hermawan
Conservation farming land use on critical upland watersheds
- Social and economic evaluation study - — 1161

Chapter 14: Technologies for soil and water conservation and sustainable land management — 1167

Chapter 14.1: The role of tillage practices and ground cover — 1167

H. Liniger & D.B. Thomas
GRASS
Ground Cover for Restoration of Arid and Semi-arid Soils — 1167

R. Derpsch & K. Moriya
Implications of no-tillage versus soil preparation on sustainability of

agricultural production 1179
E. Chuma & J. Hagmann
Development of conservation tillage techniques through combined on-station and
participatory on-farm research 1187
Th. Rishirumuhirwa & E. Roose
The contribution of banana farming systems to sustainable land use in Burundi 1197
A. Calegari, M.R. Darolt & M. Ferro
Towards sustainable agriculture with a no-tillage system 1205
H. Morrás & G. Piccolo
Biological recuperation of degraded Ultisols in the province of Misiones, Argentina 1211
D.K. Malinda, R.G. Fawcett, D.Little, K. Bligh & W. Darling
The effect of grazing, surface cover and tillage on erosion and nutrient depletion 1217
Lin Kai Wang, Xiao Tian, Li Yili, Su Shuijin & Xie Fuguang
Preliminary results of the grass-tree system for rehabilitation of severely
eroded red soils 1225

Chapter 14.2: Nutrient management practices 1233

R.D. Lentz, R.E. Sojka & C.W. Robbins
Reducing soil and nutrient losses from furrow irrigated fields with
polymer applications 1233
A. Calegari & I. Alexander
The effects of tillage and cover crops on some chemical properties of an Oxisol
and summer crop yields in southwestern Paraná, Brazil 1239
T.N. Mwambazi, B. Mwakalombe, J.B. Aune & T.A. Breland
Turnover of green manure and effects on bean yield in Northern Zambia 1247
J. Lehmann, F. v. Willert, S. Wulf & W. Zech
Influence of runoff irrigation on nitrogen dynamics of *Sorghum bicolor* (L.)
in Northern Kenya 1255
M.C.S. Wopereis, C. Donovan, B. Nébié, D. Guindo, M.K. Ndiaye & S. Häfele
Nitrogen management, soil nitrogen supply and farmers' yields in
Sahelian rice based irrigation systems 1261

Chapter 14.3: Mechanical control of soil erosion 1267

Edi Purwanto & L.A. Bruijnzeel
Soil conservation on rainfed bench terraces in upland West Java, Indonesia:
Towards a new paradigm 1267
N. Sinukaban, H. Pawitan, S. Arsyad & J. Armstrong
Impactof soil and water conservation practices on stream flows in
Citere catchment, West Java, Indonesia 1275
R.N. Adhikari, M.S. Rama Mohan Rao, S. Chittaranjan, A.K. Srivastava, M. Padmalah, A. Raizada & B.S. Thippannavar
Response to conservation measures in a red soil watershed in a semi arid
region of South India 1281
K. Michels, C. Bielders, B. Mühlig-Versen & F. Mahler
Rehabilitation of degraded land in the Sahel: An example from Niger 1287

W.P. Spaan & K.J. van Dijk
Evaluation of the effectiveness of soil and water conservation measures in a closed sylvo-pastoral area in Burkina Faso — 1295
F.A. Mayer & E. Stelz
Gully reclamation in Mafeteng District, Lesotho — 1303
R.G. Barber
Linking the production and use of dry-season fodder to improved soil conservation practices in El Salvador — 1311

Chapter 15: Adoption of soil and water conservation practices — 1319

Chapter 15.1: Demographic, socio-economic and water conservation practices — 1319

B. Rerkasem & K. Rerkasem
Influence of demographic, socio-economic and cultural factors on sustainable land use — 1319
M. Tiffen
Demographic growth and sustainable land use — 1333
W. Östberg & Ch. Reij
Culture and local knowledge - Their roles in soil and water conservation — 1349
Yohannes Gebre Michael
Indigenous soil and water conservation practices in Ethiopia: New avenues for sustainable land use — 1359
H.K. Murwira & B.B. Mukamuri
Traditional views of soils and soil fertility in Zimbabwe — 1367
A.S. Langyintou & N. Karbo
Socio-economic constraints to the use of organic manure for soil fertility improvement in the Guinea savanna zone of Ghana — 1375
J. Hellin & S. Larrea
Ecological and socio-economic reasons for adoption and adaptation of live barriers in Güinope, Honduras — 1383
J. Currle
Farmers and their perception of soil conservation methods — 1389
R.J. Unwin
Farmer perception of soil protection issues in England and Wales — 1399

Chapter 15.2: The role of incentives, research, training and extension — 1405

M. Giger
Using incentives and subsidies for sustainable management of agricultural soils - A challenge for projects and policy-makers — 1405
Ch. Reij
How to increase the adoption of improved land management practices by farmers? — 1413
D. Sanders, S. Theerawong & S. Sombatpanit
Soil conservation extension: From concepts to adoption — 1421
K. Herweg
Contributions of research on soil and water conservation in developing countries — 1429

M. Douglas
Training for better land husbandry — 1435
J. Hagmann, E. Chuma & K. Murwira
Strengthening peoples capacities in soil and water conservation in Southern Zimbabwe — 1447

Chapter 15.3: Women and soil and water conservation — 1463

L.M.A. Omoro
Women's participation in soil conservation: Constraints and opportunities.
The Kenyan experience — 1463
D. Kunze, H. Waibel & A. Runge-Metzger
Sustainable land use by women as agricultural producers?
The case of Northern Burkina Faso — 1469
Swarn Lata Arya, J.S. Samra & S.P. Mittal
Rural women and conservation of natural resources: Traps and opportunities — 1479

Chapter 16: Land tenure and soil and water conservation — 1485

M. Kirk
Land tenure and land management: Lessons learnt from the past,
challenges to be met in the future? — 1485
H.W.O. Okoth-Ogendo
Tenure regimes and land use systems in Africa: The challenges of sustainability — 1493
V. Stamm
Are there land tenure constraints to the conservation of soil fertility?
A critical discussion of empirical evidence from West Africa — 1499
D. Effler
National land policy and its implications for local level land use:
The case of Mozambique — 1505
H.M. Mushala & G. Peter
Socio-economic aspects of soil conservation in developing countries:
The Swaziland case — 1511
M. Chasi
Impact of land use and tenure systems on sustainable use of resources in Zimbabwe — 1519
K. Goshu
Assessing the potential and acceptability of biological soil conservation techniques
for Maybar Area, Ethiopia — 1523
B.J. Rao, R. Chennamaneni & E. Revathi
Land tenure systems and sustainable land use in Andhra Pradesh:
Locating the influencing factors of confrontation — 1531
H. Cotler
Effects of land tenure and farming systems on soil erosion in Northwestern Peru — 1539

Conclusions and Recommendations of the 9th ISCO Conference — 1545

Personal Records Editors — 1557

Authors' Index — 1559

Part III: From Soil and Water Conservation to Sustainable Land Management

From Soil and Water Conservation to Sustainable Land Management: Some Challenges for the Next Decade

C. Reij, H. Eger & K. Steiner

Summary

Since 1980, the state-of-the-art in soil and water conservation (SWC) has developed considerably. In particular socio-economic aspects of SWC have become much more important. This paper identifies ten themes which could become new centres of interest for soil conservationists. One of these themes is farmer-innovators in SWC. Farmers who through experiments try to improve the efficiency of their traditional SWC practices or to adapt new technologies to their specific requirements, are a source of inspiration, which is often ignored by outsiders. Another theme deserving more attention is soil fertility management by farmers, because the evolution of soil fertility is the most important indicator determining the sustainability of the farming systems.

Keywords: trends in soil and water conservation; challenges for the next decade; farmer-innovators in soil and water conservation; soil fertility management

1 Introduction

The 9th ISCO Conference was attended by about 900 participants from 120 countries, so it can be assumed that the large number of papers and posters presented at the Conference reflect to a certain degree the preoccupations of soil conservationists. Since 1980 the state-of-the-art soil conservation has developed considerably as demonstrated by the papers in this volume. While only 8 out of the 42 papers published in the proceedings of the 2nd ISCO Conference in 1980 dealt with economic, social and legal aspects (Morgan, 1981), these subjects became much more important in 1996.

At the conference in Bonn, researchers and practitioners discussed six major themes represented in this book:
- **Sustainable land use management: policies, strategies and economies** (chapter 11) It is evident that the sustainability issue had not yet emerged in the 2nd ISCO Conference, but policy, strategy and economic aspects were already discussed in 1980.

- **Institutional and planning aspects of soil and water conservation** (chapter 12) Most of the papers under this heading are about how to involve land users in planning for sustainable land use management.
- **Approaches to soil and water conservation (SWC) and watershed development** (chapter 13) Many papers under this heading are about local experiences with land user participation in SWC, watershed development and community-based land use management. Given the number of papers presented on this subject it is clear that it is getting more and more attention.
- **Technologies for soil and water conservation and sustainable land use management** (chapter 14) In this chapter a number of papers on soil fertility issues are presented. This shows that soil conservationists are no longer mainly concerned with conservation engineering, but increasingly with soil fertility management, because yields cannot be sustained without adequate soil fertility management.
- **Adoption of soil and water conservation practices** (chapter 15) In the past little attention was drawn to reasons for adoption or non-adoption of SWC practices as most land users were either coerced or seduced into SWC through the use through indiscriminate use of specific incentives, often "Food-for-Work" programs
- **Land tenure and soil and water conservation** (chapter 16) Nothing on this subject was presented at the 6th ISCO Conference in Adis Ababa in 1989, less than a decade ago. Given its huge importance for land users, it can be argued that it is only slowly emerging on the agenda of soil conservationists.

2 Challenges for the next decade

Analysing the papers submitted to the 9th ISCO Conference, it is evident that certain themes still do not get the attention they deserve. The introduction to this volume is dedicated to their identification and a brief discussion of what will hopefully become new centres of interest for soil conservationists.

3 Farmer-innovators in SWC as a source of inspiration

The phenomenon of farmer-innovators in SWC has until recently largely been ignored. Individual farmers, however, try to improve the efficiency of their local conservation practices in many regions. They also adapt new technologies introduced by conservation experts or technicians to their specific requirements. A well-known example are the improved traditional planting pits, which are now widely used in parts of Burkina Faso and Niger to rehabilitate strongly degraded land. They are the result of experiments done by one farmer-innovator in the Yatenga region of Burkina Faso (Ouedraogo and Kaboré, 1996). With his work he had more impact on progress in SWC in this region than all researchers combined during the last two decades. In Southern Africa, for instance, the SADC ELMS Unit is attempting to identify farmer-innovators and to analyse their innovations (Segerros, et al. 1995). The IFAD-funded SWC and Agroforestry Project in Lesotho has organized individual innovative farmers in a network (Critchley and Mosenene, 1995). At the end of 1996 a major Dutch-funded project has started which aims to identify and organize farmer-innovators in SWC in six African countries (Burkina Faso, Cameroon, Tunisia, Ethiopia, Tanzania and Zimbabwe). The objective of this project is to link farmer-innovators, field staff and researchers in an effort to jointly generate conservation technologies which are acceptable to land users. The starting point is formed by innovations already introduced spontaneously by farmers and not by experiments proposed by researchers. This emphasis on the role of farmer-innovators is part of a trend leading to a demystification of the role of researchers and experts, to capacity building of local people in SWC (Hagmann et al 1997). It is also part of a trend to build up on local conservation practices (Stocking 1997; Yohannes 1997) and of the empowerment of local communities to manage their own natural resources (Steer 1997; Schneichel and Asmussen 1997).

In this context, it should be mentioned that participatory approaches have been taken much more into account during the last decade than ever before. While SWC has been treated as a mainly technical and scientific problem at the first ISCO Conferences, today more importance is given to the dialogue with the population. By now, Participatory Rural Appraisal (PRA) and Participatory Land Use Planning at village level have become common tools for capacity building in an effort to achieve more sustainable management of natural resources.

4   More attention should be drawn to economic aspects

Only a few papers about the economic aspects of SWC were presented at the Conference (Otzen 1997; Clark et al. 1997; Huszar, et al.1997). The case studies for Sri Lanka (Clark, et al.1997) and for the uplands of Java (Huszar et al.1997) show that the benefits derived from investment in SWC are sometimes low or even negative. This is quite in contrast to recent findings in semi-arid regions where the introduction of water harvesting techniques relieves the water constraint to crops. This produces an immediate positive impact on yields (for example, Hassan 1996). In other cases SWC measures require investments in the form of labour and capital which produce benefits to farmers only after a number of years. This could mean lower incomes to farmers in the first few years, which is not acceptable to many resource-poor farmers.

For several reasons, it is important to pay more attention to economic aspects of SWC. One reason is that donor agencies focus on cost-effectiveness. Therefore it is important to show them that SWC packages have a positive impact on plant production. In many instances conservation techniques have been promoted, which did not produce higher yields. Sometimes they even led to a decrease in yields compared to those obtained on adjacent non-treated fields. Another reason is that it is important to know whether yield increases are sustainable over longer periods.

Finally, there is a growing awareness that macro-economic changes can have considerable impact on SWC and on soil fertility management by farmers, but little data is available. The effects of macro-economic parameters on SWC and soil fertility management, such as changing prices of inputs and outputs and the use of subsidies and taxes to influence environmental behaviour, still need to be taken much more into account.

5   SWC and range management; an important but neglected theme

Three or four papers only were presented to the Conference touching on the issue of the management of pastoral lands (Liniger and Thomas 1997; Haro et al.1997; Yosko and Bartels 1997; Brummelman et al.1997), this is a theme which is largely neglected. This is understandable, because it is far easier to promote good land husbandry practices on cultivated lands than it is to improve the management of rangelands. Technically it is often easy to restore degraded rangelands to produc-tivity, but the management aspects are complex. Recent publications have argued that the conven-tional notion of „carrying capacity" of the land is inadequate and inappropriate and does not take into account the opportunistic management strategies of livestock owners in pastoral areas with a widely fluctuating availability of fodder over time and space (Scoones, 1994).

In many countries in Africa as well as in Asia, SWC projects are evolving from fairly narrow-based SWC projects with a focus on the promotion of conservation packages into projects aiming at community-based land use management. The question of appropriate management of non-cultivated land usually does arise in this context and soil conservationists are inevitably confronted with the issues of the availability of and access to fodder resources. Other issues which arise in this context are the improvement of the capacity of traditional or modern local institutions for the management of rangelands and mechanisms for conflict resolution. It is a complex and fascinating area to which soil conservationists could also make contributions.

6 How much longer will gender and equity issues be ignored ?

During the last decade an entire toolbox of Participatory Rural Appraisal (PRA) techniques has been developed (Pretty, 1997), which allows a fairly rapid, fairly low cost and fairly accurate analysis of the present situation by and with the local groups. In this way answers can be obtained to delicate questions such as "who contributes to and who benefits from SWC"? It is surprising that although many SWC and NRM projects have started using some form of PRA, they continue to pay little attention to gender and equity issues and most still seem to operate on the premise that everybody will benefit equally from SWC.

The reason why men and women do not benefit from certain activities in the same way is related to the gender-specific division of labour and the different access of men and women to natural resources, especially to land. While most women in developing countries do have access to land this does not mean that they also own it. The lack of tenure security often makes it unattractive to apply SWC measures on their fields. On the other hand, men increasingly migrate to distant cities or mining areas to earn a cash income, which means that women as household heads sometimes have to take decisions with regard to the management of the family lands.

7 Soil fertility management by farmers: a key to sustainability

The evolution of soil fertility on farmers' fields is the single most important indicator of the sustainability of the farming systems. Some researchers have argued that SWC contributes to the depletion of soil fertility and it thus may create more problems than it solves (Breman, pers. communication). The argument is that SWC leads to an increase in plant production, which means that nutrients are withdrawn and they may not be adequately replaced by the annual gift of manure and chemical fertilisers. As a result soil fertility levels are depleted each year a bit more. This process has been coined as 'soil mining' (Pol 1992). Farmers in the Yatenga region in Burkina Faso, who had invested in contour stone bunding and planting pits pointed out that they had been aware of declining soil fertility, but as soon as they detected a decline in yield levels, they added manure. Although the quantities of manure used by farmers may not be what researchers deem sufficient (at least 5 ton/ha), it is undeniable that both the quantity and the quality of manure used by farmers in the Yatenga region of Burkina Faso have increased considerably in the last decade. Moreover they use manure more rationally in the sense that they concentrate it in planting pits rather than spreading it over their entire fields. The manure which is put into the planting pits during the dry season attracts termites, which improve what Shaxson (1997) calls 'soil architecture'. A number of papers presented at the 9th ISCO Conference cover subjects, such as soil erosion and soil fertility; reducing soil and nutrient losses and the effect of grazing on erosion and nutrient depletion. Papers on soil fertility management by farmers are hard to find. An exception is the paper by Defoer, et al. 1996 on sustainable soil fertility management in Southern Mali. In the coming years more attention needs to be paid to changing soil fertility management practices of farmers, some of which takes place as a result of structural adjustment programmes removing subsidies on inorganic fertilizers.

Great hopes have been placed in green manures and in agro-forestry as a means to reduce erosion and to maintain or even improve soil fertility (Minae et al. 1997; Drechsel et al., 1996).

Success stories have been reported from central America (Bunch 1995; Mausolf and Farber 1995). The question of whether it is possible to restore soil fertility sufficiently without the use of external inputs, especially phosphorus, is still open (Pieri and Steiner 1996).

8 Incentives in natural resource management: which way forward ?

During the colonial period land users were often coerced into SWC and they were punished in one way or another if they did not contribute their labour. SWC was also often imposed in the 1960s and 1970s. Farmers were not given a choice from a menu of SWC techniques and they were „bribed" into SWC through the indiscriminate use of incentives such as Food-for- Work or Cash-for-Work. A major criticism of these incentives has always been that farmers were not interested in SWC, but in the incentives which ceased with the end of construction works. In particular SWC projects with a target-oriented approach continue to rely on the use of Food-for-Work and Cash-for-Work, but many are grappling with incentives (Critchley et al. 1992). It is surprising that only one paper in Bonn was specifically dedicated to this subject (Giger 1997).

Although it is the current paradigm that SWC should be undertaken on a voluntary basis to ensure that land users are really interested in SWC as they invest their own scarce resources into it, there also is a consensus that they need some form of incentives and support to ease the burden of work and to accelerate the rate of implementation. The question remains: "what kind of support and how much support can be provided without jeopardising maintenance of conservation works and continuation of farmer investment in improved land management practices in the post-project phase?" It is a theme which urgently needs more systematic attention.

9 SWC and cross-compliance: a tension between voluntary and top-down approaches

Cross-compliance is used in SWC programmes in the USA where it was introduced in the early 1980s. It became an important tool in the context of the Conservation Reserve Programme (CRP) proposed by the 1985 Farm Bill. The major objective of the CRP was to take erodible land out of cultivation and to introduce a number of (partially government-subsidised) conservation measures on these lands, which had to be kept out of production usually for 15 years or longer. Farmers received compensation payments for each hectare taken out of production. If they would not respect their part of the deal, they would lose the conservation payments and all other farm subsidies they were entitled to. The CRP has contributed substantially to controlling soil erosion in the USA, but because of the high costs of the programme it was scaled down by the 1995 Farm Bill.

Various forms of cross-compliance or financial support based on the fulfilment of environmental measures have now also been introduced in Europe as part of the Common Agricultural Policy (Baldock et al., 1993). It is evident that most developing countries can not afford those programs. The question is whether elements of it could be introduced in particular into externally funded SWC/NRM programmes. Credit for livestock could be, for example, given under the condition that terrace risers are planted with grass, or subsidised donkey carts could be provided on a contract basis if a number of improved land management practices, to be selected by the farmers, were implemented on the fields. The donkey carts could be withdrawn, if the contract is not respected (Reij 1997). Such arrangements acknowledge the fact that there are limits to voluntary participation and that resource-poor land users should somehow be supported. This approach is a mixture of using "carrots" and "sticks". Why not test this approach in the next few years in a variety of socio-economic and agro-ecological zones and monitor and analyse the results.

10 Land tenure: how to avoid more conflicts ?

Although papers about land tenure had been presented to previous ISCO Conferences, this topic received more attention at the 9th ISCO Conference than ever before. It is unavoidable that land tenure will appear even more firmly on the agenda of future Conferences, because population pressure on available natural resources is increasing and as a corollary, competition and conflict often arise.

Moreover, through a process of inheritance land is divided up in smaller and smaller parcels which is often a 'thorn in the eye' of soil conservationists and land use planners, who claim that this is irrational and inefficient. Land division often takes place across the slope so farmers maximise variation in their fields along the toposequence. At a recent seminar in Tunisia on participatory approaches in SWC, a forester accused the soil conservation experts of not taking into account such aspects in the implementation of conservation measures: "the fields of farmers resemble my tie and the only thing you do is constructing massive earthworks along the contours which take up one third of the farmers' fields".

11 The International Convention to Combat Desertification (CCD)

The Convention to Combat Desertification finally entered into force some months after the 9th ISCO Conference and four years after the Agenda 21 had been adopted at the Rio Conference. For the purpose of this convention desertification means "land degradation in arid, semi-arid and dry sub-humid areas resulting from various factors, including climatic variations and human activities". This definition corresponds to the wide range of topics treated at the 9th ISCO Conference, because it integrates the social/human dimension of desertification.

Combatting desertification therefore does not only include the prevention of land degradation and the rehabilitation of partly degraded land but all activities (ecological, social, economic, institutional and political measures) which are part of an integrated sustainable development of land in arid, semi-arid and dry sub-humid areas.

In accordance with the motto of ISCO IX - "furthering cooperation between people and institutions" - the convention also stresses the importance of cooperation and coordination between countries, institutions and the population. The convention correspondingly, appreciates the involvement of international organisations such as ISCO in it's implementation. The impact of this convention should be discussed at future ISCO Conferences. Experiences gained from it's implementation could provide useful lessons for the dissemination of innovative methods and concepts.

12 Monitoring and evaluation

The common methods of impact assessment of SWC activities seem to be primarily directed towards the monitoring of project performance, i.e. number of hectares treated as percentage of targets, but little or no attention is paid to monitoring their short and long-term impact. Most projects assume that a high degree of implementation is an indicator of success, but this type of quantitative information does not give any information about the impact of SWC activities, for example, on yields, groundwater recharge and on biodiversity. the sustainability of soil management. The monitoring and evaluation of SWC activities should also take into account qualitative, biophysical and socio-economic aspects.In addition to performance assessment interdisciplinary impact assessment is therefore required.

Generally, indicators and types of monitoring and evaluation which are based on local know-how are not yet sufficiently considered. Impact assessment methods rarely include experiences and criteria used by farmers. Their contribution to the identification of indicators and their active involvement in monitoring and evaluation will ensure a better fit between SWC activities and farmers objectives.

The creation of indicator-databases can facilitate the selection of indicators adapted to a specific situation and provide descriptions and standardised methods. The storage of data is important, but their accessibility should also be guaranteed.

References

Baldock, D., Beaufoy, G., Bennett, G. and Clark, J. (1993): Nature Conservation and New Directions in theEC Common Agricultural Policy. Institute for European Environmental Policy, London.

Brummelman, G., Ishaq, S.O., Ahmed, M. and Mogge, M. (1997): To root or to rot: learning experiences of the NARMAP project with community-based natural resource management in the Sahel zone of Sudan. Paper presented at the 9th ISCO Conference, Bonn 1996.

Bunch, R. (1995): Soil Recuperation in Central America: Sustaining Innovation after Intervention. IIED Gate-keeper Series **55**.

Clark, R., Manthrithilake, H., White, R.and Stocking, M. (1997): Economic valuation of soil erosion and conservation - a case study of Perawalla, Sri Lanka. Chapter 11, this book.

Critchley, W., Reij, C. and Turner, S.D. (1992): Soil and Water Conservation in sub-Saharan Africa: towards sustainable production by the rural poor. IFAD, Rome; CDCS, Amsterdam.

Critchley, W. and Mosenene, L. (1995): Individuals with initiative: network farmers in Lesotho. In: Successful Natural Resource Management in Southern Africa. Windhoek, Gamsberg MacMillan Publishers, 71-81

Defoer, T., Kanté, S. and Hilhorst, T. (1997): An action-research approach to improved soil fertility management. Chapter 13, this book.

Drechsel, P. and Steiner, K. and Hagedorn, F. (1996): A review on the potential of improved fallows and green manure in Rwanda. Agroforestry Systems **33**: 109-136.

Giger, M. (1997): Using incentives and subsidies for sustainable management of agricultural soils - a challenge for projects and policy makers. Chapter 15, this book:

Hagmann, J., Chuma, E. and Murwira, K. (1997): Strengthening people's capacities in Soil and Water Conservation in Southern Zimbabwe. Chapter 15, this book.

Haro, G.O., Lentoror, E.I. and Von Lossau, A. (1997): Participatory approaches in promotion of sustainable natural resource management: experiences from SW Marsabit, Northern Kenya. Chapter 13, this book.

Hassan, A. (1996): Improved traditional planting pits in the Tahoua Department, Niger. In: Reij, C., Toulmin, C. and Scoones, I. (eds), Sustaining the Soil: Indigenous Soil and Water Conservation in Africa. London, Earthscan.

Huszar, P.C., Pasaribu, H.S. and Ginting, S.P. (1997): Including economics in the sustainability equation: upland soil conservation in Indonesia. Chapter 11, this book.

Liniger, H.P. and Thomas, D. (1997): GRASS: grass cover for restoration of arid and semi-arid soils. Chapter 14, this book.

Mausolff, C. and Farber, S. (1995): An Economic Analysis of Ecological Agricultural Technologies among Peasant Farmers in Honduras. In: Ecological Economics **12**, 237 - 248.

Minae, S., Bunderson, W.T., Kanyama-Phiri, G. and Izac, A.M. (1997): Integrating agro-forestry technologies as a natural resource management tool for smallholder farmers. Chapter 13, this book.

Morgan, R.P.C. (ed) (1981): Soil Conservation:Problems and Prospects. John Wiley & Sons.

Otzen, U. (1997): Overall macro-economic and political conditions influencing soil management and sustain-able land use. Paper presented at the 9th ISCO Conference, Bonn 1996.

Ouedraogo, M. and Kabore, V. (1996): The zaï, a technique for the rehabilitation of degraded land in the Yatenga. In: Reij, C., Toulmin, C. and Scoones, I. (eds), Sustaining the Soil: Indigenous Soil and Water Conservation in Africa. Earthscan Publications, London.

Pieri, C. and Steiner, K. (1996): The role of soil fertility in sustainable agriculture with special reference to Sub-Saharan Africa. In: Entwicklung und Ländlicher Raum **4**, 3-6.

Pol, F.v.d. (1992): Soil mining: an unseen contributor to farm income in Southern Mali. Amsterdam, Royal Tropical Institute, Bulletin 325.

Reij, C. (1997): How to increase the adoption of improved land management practices by farmers ? Chapter 15, this book.
Segerros, M., Prasad, G. and Marake, M. (1995): Let the Farmer Speak: Innovative Rural Action Learning Areas. In: Successful Natural Resource Management in Southern Africa. Windhoek, Gamsberg MacMillan Publishers.
Scoones, I. (1994): Living with Uncertainty: new directions in pastoral development in Africa. IT Publications, London.
Schneichel, M. and Asmussen, P. (1997): Dry forest management: putting campesinos in charge. Chapter 12, this book.
Shaxson, F. (1997): Concepts and indicators for assessment of sustainable land use. Chapter 1, vol 1, this book.
Steer, A. (1997): Making development sustainable. Chapter 11, this book.
Stocking, M. (1997): Conditions for enhanced cooperation between people and institutions. Chapter 11, this book.
Yohannes, G.M. (1997): Indigenous soil and water conservation in Ethiopia: new avenues for sustainable land use. Chapter 15, this book.
Yosko, I. and Bartels, G. (1997): Nomads and sustainable land use: an approach to strengthen the role of traditional knowledge and experiences in the management of grazing resources of Eastern Tchad. Chapter 13, this book.

Addresses of authors:
C. Reij
Vrije Universiteit
De Boelelaan 1115
1081 HV Amsterdam, The Netherlands
H. Eger
K. Steiner
Deutsche Gesellschaft für Technische Zusammenarbeit (GTZ)
Postfach 5180
D-65726 Eschborn, Germany

A Multi-Level Stakeholder Approach to Sustainable Land Management

H. Hurni

Summary

Soils provide us with over 90% of all human food, livestock feed, fibre and fuel on Earth. Soils, however, have more than just productive functions. The key challenge in coming years will be to address the diverse and potentially conflicting demands now being made by human societies and other forms of life, while ensuring that future generations have the same potential to use soils and land of comparable quality. In a multi-level stakeholder approach, down-to-earth action will have to be supplemented with measures at various levels, from households to communities, and from national policies to international conventions. Knowledge systems, both indigenous and scientific, and related research and learning processes must play a central role. Ongoing action can be enhanced through a critical assessment of the impact of past achievements, and through better co-operation between people and institutions.

Keywords: Stakeholder, sustainable land management

1 The soils - a globally neglected resource

The International Soil Conservation Organisation (ISCO) invited an international group of participants to the current meeting, the 9th ISCO Conference. More than 800 participants, who are concerned with the future of this precious resource, have been registered. In my view we are here because "earth" matters: Earth in the sense of local soil, earth in the sense of land, and Earth in the sense of the whole world. At the global level, soils have been a neglected resource. On the other hand, a number of international conventions have been ratified since 1992, when the UN Conference on Environment and Development in Rio de Janeiro took place, namely on climate, biodiversity, and desertification. Furthermore, international action plans on forests and water are being implemented.

1.1 Soils - an overlooked resource of central importance

Soils as a major resource were only marginally mentioned in the various Chapters of Agenda 21, although it should be stated that the developed societies have not provided the financial means for implementing this massive global programme, as expected in 1992. Four years after Rio, merely 5% of the annual budgets foreseen in Agenda 21 are being invested in this global action plan. Soils are a neglected resource because present-day threats to the soil are much more serious and complex than in the past. More than 2.6 billion people are directly dependent on agriculture, and 85% of the world's population live in countries where agriculture is a predominant occupational sector.

ISBN 3-923381-42-5
© 1998 by CATENA VERLAG, 35447 Reiskirchen

Photo 1: Soil degradation in the form of gully erosion in the vicinity of the Aïr Mountains in the Sahara in Niger. The seemingly local damage has devastating effects on the groundwater table and on grass vegetation over a much larger area. H. Hurni, 5 March 1989.

1.2 Accentuated soil degradation

Countries where more than 50% of the population is engaged in farming are among the poorest in terms of gross domestic product, and poverty and land degradation are directly linked in many of these countries. Furthermore, degradation is accelerating at alarming rates. Some of the ancient forms of labour-intensive care for the soil, including long-term fallows after cultivation, or traditional terracing, have been replaced by mechanisation, fertilisers, chemicals, and introduced crops. New types of soil degradation resulting from inappropriate forms of agriculture, industrial development and urbanisation have been added to, and superimposed on, ancient types such as soil erosion and salinisation. Many forms of small-scale damage to soils, once perceived as local, have thus accumulated to constitute a global threat to the survival of humankind.

2 Co-operation between people and institutions

"Bringing people and institutions together" is the theme of the 9th ISCO Conference. Concerned participants at the conference, however, are only one group among many who constitute the actual stakeholders in developing a shared action programme to save our soils (Table 1).

Both direct land users and political decision makers are largely absent from this conference, while technicians and scientists are better represented. Looking at backgrounds in education and experience, and actual competence, further gaps become obvious (Table 2).

Category of stakeholder	ISCO attendance (%)
Direct land users (farmers, livestock herdsmen, miners)	1
Technicians (experts, specialists, extensionists)	48
Scientists (teachers, researchers)	50
Political decision makers	1

Table 1: Composition of participants at the 9th ISCO conference

Category of competence	ISCO attendance (%)
Environmental (soils, biology, geography)	30
Engineering (agronomy, agricultural, civil)	30
Transdisciplinary (integrative, professional experience)	30
Social sciences (sociology, political sc., anthropology)	7
Economics	3

Table 2: Components and background of participants at the 9th ISCO conference. During the keynote address, two brief participatory surveys were made among those in the conference hall. The results of these enquiries are rough estimates only.

The social and economic sciences are largely underrepresented. Considering that each discipline has its own research agenda, it appears impossible that an integrated, transdisciplinary approach, which should involve major fields of knowledge, and local as well as policy levels, can be developed within the current group. And considering that the policy environment is central in the creation of enabling conditions for sustainable land management, and that local stakeholders are central in any down-to-earth action, the potential impact of the 9th ISCO Conference will be restricted to conceptual and methodological approaches on the one hand, and to indirect effects of secondary measures on the other hand, to be initiated by participants once they return to their working environments.

3 The need for down-to-earth action

Land degradation occurs on approximately one third of the world's agricultural soils (Oldeman et al., 1990), and 83% of the soil degradation damage observed has been caused by soil erosion by water and wind, induced by inappropriate land management. If an attempt is to be made to improve degrading land use systems, it will be necessary to clarify for what purpose the concerned land and its soils should be used. The major stakeholders in down-to-earth action, however, are the direct land users, and perhaps the technical people assisting them. In principle, there are only two options to act on the ground: either to improve a current land use system, or to alter it.

3.1 Are improvements in present land use feasible?

Three princples have been put forward for improving present land use: 1. increase vegetative ground cover; 2. enhance productivity in a sustainable manner, while minimising the negative effects on soils and ecosystems; and 3. use regenerative agricultural technologies for sustainable land management (Pretty, 1995). In practical terms this means that technological adaptation is only possible in areas where land use is already more or less appropriate. Many indigenous conservation systems have a potential for improvement (cf. Reij, 1991; Critchley et al., 1994; Krueger et al., 1995).

Photo 2: Technological improvements of present land use are possible in Anjeni, Ethiopia, where farmers invest in terrace construction on moderately steep cultivated land. H. Hurni, 24 April 1986

3.2 Do alternatives to inappropriate land use systems exist?

Technological, environmental, social, economic or political constraints can hinder sustainability under present forms of land use. Sometimes, the only solution may be to live with degradation and declining productivity, particularly when food security is at stake, until resources are used up. In other cases, productivity goals have to be set at lower levels than at present, and stabilised there, using technologies which are not profitable in the short run, but which will compensate for degradation losses in the foreseeable future. Or, land users may be able to increase productivity on some farm plots or earn income in other sectors, thus compensating for losses. A final possibility is to abandon current land use and seek alternative income outside the agricultural sector. When changes in land use are required, land users may have much more difficulty accepting and inducing change.

4 A common concept of sustainable land management

Soils are often perceived as having several primary functions: agricultural production, livestock production, and mineral utilisation. However, there is also the ecological regulatory function of a soil, the habitat and living space function, and last but not least, the cultural heritage function (Figure 1).

Unfortunately, these are largely conflicting functions. A common valuation and land use designation should be shared by those members of a society who constitute the stakeholders in a particular case. At the local level, technological solutions to make present land use systems sustainable, and to keep them productive at the same time, are often lacking or not feasible.

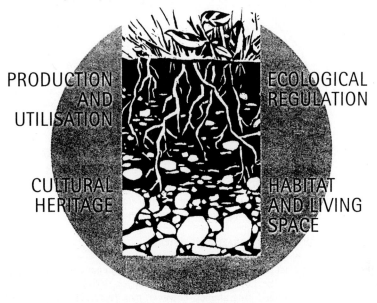

Figure 1: Soil functions as components of sustainable land management. Source: Hurni, 1996.

4.1 Soils and land

There is, furthermore, the spatial element of the soil resource: the land. "Sustainable land management" as a concept can be defined as follows (adapted from Hurni, 1996):

"Sustainable land management is a system of technologies and/or planning that aims to integrate ecological with socio-economic, and political with ethical principles in the management of land, for productive and other functions, to achieve intra- and inter-generational equity".

Hence, there is a need for more co-operation, including policy levels. Global agreements and conventions should be co-ordinated, and international action plans harmonised so that sustainable land management policies can be coordinated for the benefit of the land user. Enabling policy environments (Perich, 1993; Pierce, 1988; FAO, 1987) are an important prerequisite and an indispensable component of a "multi-level stakeholder approach in sustainable land management" (Fig. 2). This includes land security as an indispensable prerequisite for long-term investment by the land user (Anamosa, 1995; Wachter, 1996).

4.2 What are stakeholders?

Stakeholders are actor categories, and sometimes groups, defined by activity, by rights, or by organisation (Rocheleau, 1994). They can be found at all levels, from the household to the international community. Usually, stakeholder views of resources are not the same, and converging principles are difficult to find. Action decided on at local level may not be appropriate, because not all stakeholders have participated in the process of negotiation.

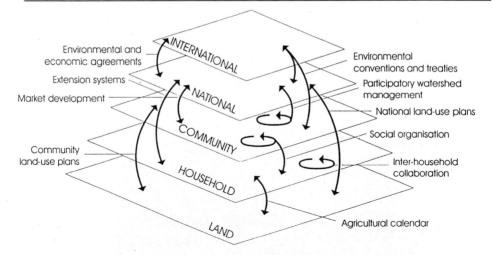

Figure 2: Intervention levels and activities in a multi-level stakeholder approach to sustainable land management. Source: Hurni, 1996

4.3 Economic feasibility of land management systems

Conceptual controversies exist at all levels. For example, costs and benefits of techniques may not be economically feasible at the household level (cf. Kappel, 1996). Direct incentives as an immediate answer in conventional programmes, however, may represent a threat to the longevity of an action. On the other hand, incentives are often considered indispensable by project staff and land users, particularly when only the initial investments are supported, and not the subsequent maintenance costs.

4.4 Prerequisites for development

The development of the national economy has often been cited as a prerequisite for sustainable land management, because poverty prevents people from taking a long-term view and from making additional investments. The need for better enabling environment calls for better horizontal and vertical communication (Samran Sombatpanit et al, 1996). This should include institutional capacity for mitigating conflicting interests at community level and at higher levels (Baechler, 1994). Careful evaluation of the land use system will improve the participatory selection and discussion of appropriate remedial measures (cf. Dumanski, 1996).

5 A world overview of sustainable land management

One may be impressed by the observation that world-wide, many specialists, land use planners and decision-makers assist local land users in improving soil productivity through soil and water conservation measures (Douglas, 1994). However, there is concern that valuable experience based on these activities is usually not available to people in other locations. In order to overcome this deficiency, the World Association of Soil and Water Conservation launched a communication and evaluation programme in 1992, called WOCAT: World Overview of Conservation Approaches and Technologies, to promote sustainable land management systems. WOCAT is organised as a consortium of international institutions such as FAO, UNEP, GTZ, ISRIC, regional organisations like OSS, RSCU, ASOCON, national institutions in many countries, and university centres such as CDCS and IRE. WOCAT is coordinated by CDE.

5.1 The WOCAT approach

WOCAT uses a standardised framework for the evaluation of soil and water conservation, consisting of a series of questionnaires to be filled out in a participatory process by groups of specialists for each technology in regional or national workshops (cf. Hurni et al, 1994). The Questionnaire on Approaches evaluates the ways and means used to implement a technology on the ground. The Questionnaire on Technologies evaluates the measures used in the field, such as agronomic, vegetative, structural and management measures. The Questionnaire on Maps assesses the areal coverage and the impact of soil and water conservation in relation to the problem of land degradation for specific physiographic area units and ecoregional zones.

5.2 The WOCAT communication system

The information collected by WOCAT contributors in different countries is later compiled in databases which will be established in regional centres and made accessible to any user. Users of the database will thus be able to select sustainable land management systems for specific conditions such as agro-ecology, land use, farming systems, and policy environment according to their requirement, or analyse and exchange collected information in cross-referenced queries. With each contribution to the database, the global availability of the information will be broadened, including use of the internet. For each continent, WOCAT is producing specific printouts on suitable technologies and approaches, overview maps, computer-based, interactive hypertext versions, and decision-support systems for participatory negotiations with land users.

5.3 Achievements and future work

To date, WOCAT has compiled information from 22 countries in Africa and Asia. Immediate interest has been shown by another 40-50 countries which are expected to join the network during the coming 2-3 years. Prototypes of outputs and cross-sectional analyses have been produced (see Liniger et al., this volume). A contribution to UNEP's second edition of the Desertification Atlas has been made for Eastern and Southern Africa. Projected activities include supporting initiatives taken by regional and national institutions which are willing to adopt the WOCAT framework and to contribute with their experience and infrastructure. Task forces and workshops are used to carry out activites within their own funding limits, with a small support component from the Secretariat. By presenting lessons learned from successful examples of soil and water management, WOCAT is expected to substantially contribute to the sustainable development of natural resources worldwide.

6 The impact of past achievements

Agricultural research and modern inputs have helped to ensure a global increase in crop yields of about 3% annually over the last 40 years, a rate which has so far kept pace with population growth.

However, this "success story" may be nearing its limits today. Although the rate of population growth is declining, production increases can no longer keep pace. At the same time, consumption levels among a growing number of people have increased. As a result, world grain production per person, which increased from 250 kg to 340 kg between 1950 and 1980, declined to 290 kg between 1980 and 1995 (Worldwatch, 1996). If these trends continue, the 1950 level will be reached before the year 2010, which is just a few years away. It is thus high time to change some of our consumption habits in the wealthier part of the world in order to adopt a global ethical code. On the other hand, natural resource management has become a crucial issue of global dimensions, and it is time to look at the impact of past achievements (El-Swaify, 1991).

6.1 Issues in soil degradation

Sometimes the need to combat soil and land degradation has been questioned in view of the multiple needs to advance agricultural production, to reduce population growth, to eradicate poverty and diseases, and to adjust the lifestyles of wealthier societies. Despite this, it is no longer necessary to justify the role of sustainable land management, particularly since clear alternatives to conventional soil and water conservation projects have been developed in the recent past (cf. Lundgren and Taylor, 1993).

Achievements in reducing soil erosion, the major degradation process, have been mixed. While a system of conservation-based land management technologies has been implemented on a large scale in the wealthier economies, much less has been achieved in poorer economies. For example, highly variable achievements were shown in the WOCAT analysis for 15 countries in Eastern and Southern Africa. While the effectiveness of the implemented measures was low to medium in general, the areal coverage of the measures was even less. Only 5-25% of the land use types in need of soil and water conservation have been covered during the past 10-20 years (cf. UNEP, forthcoming).

The term "desertification" has often been confused with soil and land degradation. It should be noted that soil degradation in desertification areas accounts for only about half of all land affected. The other half is soil degradation occurring in humid areas. Hence, there is a need for a global strategy of sustainable land management, perhaps even in the form of a convention. The Convention to Combat Desertification will hopefully be a tool to effectively reduce degradation and poverty, and to induce sustainable development in semi-arid areas and poor countries (UNSO, 1994). However, this will not suffice to institute sustainable land management at the global scale.

For example, other forms of soil degradation include chemical, physical, and biological processes. These are mostly a consequence of industrialisation, urbanisation and intensive agriculture. These processes are increasingly causing damage also in poorer countries, where the institutional, financial and technical means for effective prevention and rehabilitation are extremely small.

6.2 Societal development

Soil degradation does not discriminate between different countries, but distinctions must be made between nations when assessing their capabilities to cope with degradation. The application of sustainable land management is determined by economic status, public awareness, and educational levels (cf. Stocking, 1995). In countries where more than 50 percent of the population is employed in the agricultural sector, the gross domestic product is very small, and the capacity of these countries to cope with soil degradation is also low owing to poverty and lack of opportunity (cf. Blaikie et al, 1996). Wealthier countries, on the other hand, have the greatest financial, educational, scientific and service resources available. Here, only a fraction of the population is still employed in agriculture. Often, public awareness and willingness to engage in agriculture is lacking here, the more so since damage to the soils often originates outside the agricultural sector. This comparison stresses the need to differentiate between each country's agro-ecological, economic and social status in search of sustainable solutions, for which the approach described in this paper may be a suitable tool.

Acknowledgements
The author wishes to acknowledge the substantial inputs provided by the international group of co-authors of the pre-conference issue paper "Precious Earth", on which this introductory keynote is mainly based: Helmut Eger, Eckehard Fleischhauer, Samir El-Swaify, Wilhelm Östberg, Eric Roose, Hans W. Scharpenseel, T. Francis Shaxson, Samran Sombatpanit, Michael A. Stocking, Anneke Trux, Helen Zweifel, S.K. Choi, Eric T. Craswell, Alemneh Dejene, Malcolm Douglas, Rodney Gallacher, Jean-Claude Griesbach, Denis Sims, Jeff Tschirley, Rolf Kappel, Hanspeter Liniger, Eva Ludi, Estefan Rist, Arie Shahar, and Donald Thomas. Special thanks go to Ted Wachs of CDE for English language editing. The publication "Precious Earth" is on the Internet: <http://www.giub.unibe.ch/cde/isco/earth.htm>. The WOCAT programme can also be contacted through e-mail: <wocat@giub.unibe.ch>.

Abbreviations

ASOCON	Asia Soil Conservation Network (Jakarta)
CDCS	Centre for Development Cooperation Services (Amsterdam)
CDE	Centre for Development and Environment (Berne)
DLD	Department of Land Development (Bangkok)
FAO	Food and Agriculture Organisation of the United Nations
GLASOD	Global Assessment of Soil Degradation
GtZ	German "Gesellschaft fuer technische Zusammenarbeit"
IBSRAM	International Board for Soil Research and Management
IRE	Institute for Resources and Environment (Vancouver)
ISCO	International Soil Conservation Organisation
ISRIC	International Soil Reference and Information Centre
ISSS	International Soil Science Society
IUCN	World Conservation Union
JICA	Japanese International Cooperation Authority
OSS	Observatory of the Sahara and the Sahel (Paris)
RSCU	Regional Soil Conservation Unit of SIDA (Nairobi)
SIDA	Swedish International Development Authority
SWCS	Soil and Water Conservation Society (USA)
UNEP	United Nations Environment Programme
UNSO	United Nations Sudano-Sahelian Organisation
WASWC	World Association of Soil and Water Conservation
WOCAT	World Overview of Conservation Approaches and Technologies

References
Anamosa, F.A. (1995): African land tenure theories and policies: a literature review with particular reference to natural resources management. Dakar: IUCN
Baechler, G. (1994): Desertification and conflict. The marginalisation of poverty and of environmental conflicts. ENCOP, Swiss Federal Institute of Technology, Zurich, and Swiss Peace Foundation, Berne.
Blaikie, P., et al. (1996): Understanding local knowledge and the dynamics of technical change in developing countries. ODA Natural resources systems programme, socio-economic methodologies workshop, 25-30 April 1996.
Critchley, W.R.S., Reij, C. and Willcocks, T.J. (1994): Indigenous soil and water conservation. Land degradation and rehabilitation, Vol. 5, 293-314

Douglas, M. (1994): Sustainable use of agricultural soils. Development and Environment Reports No. 11, CDE, Berne

Dumanski, J. (1996): Guidelines for conducting case studies under the international framework for evaluation of sustainable land management (FESLM). Ottawa: ISSS Subcommission F. Newsletter, Agriculture and Agri-Food Canada

El-Swaify, S.A. (1991): Land-based limitations and threats to world food production. Outlook on Agriculture, Vol. 20, No. 4, p. 235-242

FAO (1987): Conservation Guide No. 12: Incentives for community involvement in conservation programmes. By C. Velozo. FAO, Rome.

Hurni, H., Liniger, H.P., Wachs, T. and Herweg, K. (1994): WOCAT 1993 Workshop proceedings. CDE, Berne

Hurni, H., with the assistance of an international group of contributors (1996): Precious earth: from soil and water conservation to sustainable land management. International Soil Conservation Organisation (ISCO), and Centre for Development and Environment (CDE), ISBN 3-906151-11-5, Berne, 89 pp.

Kappel, R. (1996): Economic analysis of soil conservation in Ethiopia: issues and research perspectives. Research Report, Soil Conservation Research programme, Addis Abeba, and CDE, Berne

Krueger, H.J., Berhanu Fentaw, Yohannes G/Michael, and Kefeni Kejela (1995): Inventory of indigenous soil and water conservation measures in Ethiopia. Research Report, Soil Conservation Research Programme, Addis Abeba, and CDE, Berne.

Liniger, H., Thomas, D.B. and Hurni, H. (1997): WOCAT - World Overview of Conservation Approaches and Technologies - Preliminary results from Eastern and Southern Africa. ISCO Proceedings (this volume).

Lundgren, L., and Taylor, G. (1993): From soil and water conservation to land husbandry: guidelines based on SIDA's experience. Stockholm: SIDA.

Oldeman, L.R., van Engelen, V.W.P. and Pulles, J.H.M. (1990): The extent of human-induced soil degradation, Annex 5 to the GLASOD maps. International Soil Reference and Information Centre, Wageningen, and UNEP.

Perich, I. (1993): Umweltökonomie in der entwicklungspolitischen Diskussion. Development and Environment Reports No. 8, CDE, University of Berne

Pierce, D. et al. (1988): Environmental economics and decision-making in Sub-Saharan Africa. London: Environmental Economics Centre.

Pretty, J.N. (1995): Regenerating agriculture. Policies and practice for sustainability and self-reliance. London: Earthscan Publications Ltd.

Reij, C. (1991): Indigenous soil and water conservation in Africa. Gatekeeper Series No. 27, London. International Institute for Environment and Development.

Rocheleau, D.E. (1994): Participatory research and the race to save the planet: questions, critique, and lessons from the field. Agriculture and human values. Journal of the Agriculture, Food and Human Values Society. Vol. 11, No. 2-3, p. 4-25

Samran Sombatpanit, Zoebisch, M., Sanders, D.W., and Cook, M.G. (eds.) (1996): Soil conservation extension: from concepts to adoption. Soil and Water Conservation Society of Thailand in collaboration with IBSRAM, DLD, JICA and WASWC. Bangkok, ISBN 974-7721-70-8, 488 pp.

Stocking, M. (ed.) (1995): Integrating environmental and sustainable development. Themes into agricultural education and extension programmes: a report of an expert consultation. FAO, Rome.

UNEP (forthcoming), Desertification Atlas, 2nd edition.

UNSO (1994), Poverty alleviation and land degradation in the drylands. New York: UNSO.

Wachter, D. (1996): Land tenure and sustainable management of agricultural soils. Development and Environment Report No. 15, CDE, Berne

Worldwatch Institute (1996): State of the world. New York and London: W.W. Norton

Address of author:
Hans Hurni
Centre for Development and Environment
University of Berne
Hallerstraße 12
CH-3012 Bern, Schweiz

Furthering Cooperation Between People and Institutions

J. Pretty

Summary

Agricultural development is currently facing unprecedented challenges. With some 800 million people hungry, the views on how to proceed vary hugely. There are five different schools of thought. Some are optimistic, others are darkly pessimistic. This paper focuses on the benefits of sustainable agriculture. The basic challenge for sustainable agriculture is to make better use of available biophysical and human resources, with the integration of a wide range of resource-conserving technologies to manage soil, water, nutrients and pests. Regenerative agriculture, founded on full farmer participation at all stages of development and extension, can be highly productive.

For a new era of soil and water conservation, farmers are now considered the potential solution rather than the problem, and so the value of local knowledge and skills is being put at the core of new programmes. Local organisations are strengthened through participatory processes, this participation being interactive and empowering. For widespread impact, enabling policy frameworks are still needed to encourage the spread of more sustainable practices for agriculture.

Keywords: Sustainable agriculture, erosion; indigenous technology; participation; policies; soil and water conservation; new professionalism; policies

"And the soil said to man: take good care of me or else, when I get hold of you, I will never let your soul go"
Kipsigis proverb, as told by Mr. arap Keoch, Chemorir, Kenya: 1990

1 Schools of thought for the global agricultural challenge

As this century draws to a close, agricultural development faces some unprecedented challenges. By the year 2020, the world will have some 8 to 8.5 billion people. Even though enough food is produced in aggregate to feed everyone, and that world prices have been falling in recent years, some 800 million people still do not have access to sufficient food. This includes 180 million children underweight and suffering from malnutrition.

The views on how to proceed vary hugely. Some are optimistic, even complacent; others are darkly pessimistic. Some indicate that not much needs to change; others argue for fundamental reforms to agricultural and food systems. Some indicate that a significant growth in food production will only occur if new lands are taken under the plough; others suggest that there are feasible social and technical solutions for increasing yields on existing farmland.

There are five contrasting schools of thought on how we should approach these challenges (for summaries, see McCalla, 1994; Hazell, 1995; Pretty, 1995a; Thompson 1995; Pretty and Thompson, 1996).

There are *"environmental pessimists"*, who suggest that ecological limits to growth are being approached; are soon to be passed; or have already been reached. It is said that populations are too great; yield growth has slowed, and will slow more, stop or even fall; no new technological breakthroughs are likely; and that environments have been too thoroughly degraded for recovery. Solving these problems means putting population control as the first priority (see Brown, 1994; CGIAR, 1994; Kendall and Pimentel, 1994; Brown and Kane, 1994; Ehrlich, 1968).

There are *"business-as-usual-optimists"*, who say supply will always meet increasing demand, and so recent growth in aggregate food production will continue alongside reductions in population growth. Innovations in biotechnology are expected to keep food output growth rising. But in many countries, it is also expected that the area under cultivation will expand dramatically - some estimates put the increase at extra 79 million hectares in Sub-Saharan Africa alone by 2020 (see Rosegrant and Agcaolli, 1994; Mitchell and Ingco, 1993; FAO, 1993; Crosson and Anderson, 1995).

The *"industrialised world to the rescue"* lobby believes that Third World countries will never be able to feed themselves, for all sorts of ecological, institutional and infrastructural reasons, and so the looming food gap will have to be filled by modernized agriculture in the North. Increasing production in large, mechanised operations will allow smaller and more `marginal' farmers to go out of business, so taking the pressure off natural resources, which can then be conserved in protected areas and wildernesses. These large producers will then be able to trade their food with those who need it, or have it distributed by famine relief or food aid (see Avery, 1995; Wirth, 1995; DowElanco, 1994; Carruthers, 1993; Knutson et al., 1990).

Two groups, however, believe that biological yield increases are possible on current agricultural land. But they are fundamentally divided over what is the most appropriate approach.

One group, whom we might call the *"new modernists"*, argues food growth can only come from high-external input farming, either on the existing Green Revolution lands, or expanded to the `high-potential' lands that have been missed by the past 30 years of agricultural development. This group argues that farmers simply use too few fertilizers and pesticides, which are said to be the only way to improve yields and so keep the pressure off natural habitats. It is also argued that high-input agriculture is more environmentally sustainable than low-input, as low-input agriculture can only ever be low output (see Borlaug, 1992, 1994a, b; Sasakawa Global 2000, 1993-95; SAA, 1995-96; World Bank, 1993; Paarlberg, 1994; Winrock International, 1994; Crosson and Anderson, 1995).

Another group, though, is making the case for the benefits of *"sustainable intensification"*, on the grounds that substantial growth is possible in currently unimproved or degraded areas whilst at the same time protecting or even regenerating natural resources. It is argued that empirical evidence is now indicating that low-input (but not necessarily zero-input) agriculture can be highly productive, provided farmers participate fully in all stages of technology development and extension. It is argued that evidence indicates that agricultural and pastoral lands productivity is as much a function of human capacity and ingenuity as it is of biological and physical processes (see Pretty, 1995a; Thompson, 1995, 1996; Hinchcliffe et al, 1996; NAF, 1994; Hewitt and Smith, 1995; Röling and Wagemakers, 1996).

2 Sustainable agriculture

Over the past fifty years, agricultural policies throughout the world have successfully promoted external inputs as the means to increase food production. Pesticides have replaced biological, cultural and mechanical methods for controlling pests, weeds and diseases. Inorganic fertilizers have been substituted for nitrogen-fixing crops, livestock manures, and composts. Information for management decisions comes from input suppliers, researchers and extensionists rather than building on local knowledge and practices.

The basic challenge for sustainable agriculture is to make better use of available biophysical and human resources. This can be done by minimizing the use of external inputs, by regenerating internal resources more effectively, or by combinations of both. This ensures the efficient and effective use of what is available, and ensures that any changes will persist, as dependencies on external systems are kept to a reasonable minimum. Sustainable agriculture seeks the integrated use of a wide range of pest, nutrient, agroforestry, soil and water management technologies. By-products or wastes from one component or enterprise become inputs to another. As natural processes increasingly substitute for external inputs, so the impact on the environment is reduced. A more sustainable agriculture, therefore, is any system of food or fibre production that systematically pursues the goals in Box 1.

It should be emphasised that sustainable agriculture does not represent a return to some form of low-technology, 'backward' or 'traditional' agricultural practices. Instead it implies an incorporation of recent innovations that may originate with scientists, with farmers or both. And it is not just about food production, but about increasing the capacity of rural people to be self-reliant and resilient in the face of change, and about building strong rural organisations and economies.

There is remarkable new empirical evidence to show that resource-conserving technologies can bring environmental and economic benefits for farmers, communities and nations (Pretty, 1995a, 1996; Hinchcliffe et al, 1996). The best evidence comes from resource-poor parts of Africa, Asia and Latin America, where farming has been largely untouched by the modern packages technologies. In these lands, farming communities adopting regenerative technologies have doubled to trebled crop yields, using few or no external inputs.

Box 1: A more sustainable agriculture pursues:

- A thorough integration of natural processes such as nutrient cycling, nitrogen fixation, and pest-predator relationships into agricultural production processes, so ensuring profitable and efficient food production

- A minimisation of the use of those external and non-renewable inputs with the potential to damage the environment or harm the health of farmers and consumers, and a targeted use of the remaining input used with a view to minimising costs;

- The full participation of farmers and other rural people in all processes of problem analysis, and technology development, adaptation and extension, leading to an increase in self-reliance among farmers and rural communities;

- A greater productive use of local knowledge and practices, including innovative approaches not yet fully understood by scientists or widely adopted by farmers;

- The enhancement of wildlife and other public goods of the countryside

---

But these are not the only sites for successful sustainable agriculture. In the Green Revolution lands characterised by high inputs and irrigation, farmers adopting regenerative technologies have maintained yields whilst substantially reducing their use of external inputs. In industrialised countries, farmers using sustainable technologies generally see yields fall, but dramatic cuts in input use mean profitability increase.

3 The spread and scaling up of sustainable agriculture

Despite the increasing number of successful sustainable agriculture initiatives in different parts of the world, it is clear that most of these are still only `islands of success'. There remains a huge challenge to find ways to spread or `scale up' the processes which have brought about these transitions.

Sustainability ought to mean, therefore, more than just agricultural activities that are environmentally neutral or positive; it implies the capacity for activities to spread beyond the project in both space and time. A `successful' project that leads to improvements that neither persist nor spread beyond the project boundary should not be considered sustainable.

When the recent record of development assistance is considered, it is clear that sustainability has been poor. There is a widespread perception amongst both multilaterals and bilaterals that agricultural development is difficult, that agricultural projects perform badly, and that resources may best be spent in other sectors. Reviews by the World Bank, the European Commission, Danida and DFID (formerly the ODA) have all shown that agricultural and natural resource projects both performed worse in the 1990s than in the 1970s-1980s and worse than projects from other sectors (World Bank, 1993; Pohl and Mihaljek, 1992; EC, 1994; Danida, 1994; Dyer and Bartholomew, 1995). They are also less likely to continue achievements beyond the provision of aid inputs.

A recent analysis of 95 agricultural project evaluations logged on the DAC-OECD database shows a disturbing rate of failure, with at least 27% of projects having non-sustainable structures, practices or institutions, and 10% causing significant negative environmental impact (Pretty and Thompson, 1996).

This empirical evidence of completed agricultural development projects suggest four important principles for sustainability and spread:

1. *Imposed technologies do not persist*: if coercion or financial incentives are used to encourage people to adopt sustainable agriculture technologies (such as soil conservation, alley cropping, IPM), then these are not likely to persist.
2. *Imposed institutions do not persist*: if new institutional structures are imposed, such as cooperatives or other groups at local level, or Project Management Units and other institutions at project level, then these rarely persist beyond the project.
3. *Expensive technologies do not persist*: if expensive external inputs, including subsidised inputs, machinery or high technology hardware are introduced with no thought to how they will be paid for, they too will not persist beyond the project.
4. *Sustainability does not equal fossilisation or continuation of a thing or practice forever*: rather it implies an enhanced capacity to adapt in the face of unexpected changes and emerging uncertainties.

4 The comprehensive technology package

Modernist agricultural development has begun with the notion that there are technologies that work, and it is just a matter of inducing or persuading farmers to adopt them. Yet few farmers are able to adopt whole packages of conservation technologies without considerable adjustments in their own practices and livelihood systems.

The problem is that the imposed models look good at first, and then fade away. Alley cropping, an agroforestry system comprising rows of nitrogen-fixing trees or bushes separated by rows of cereals, has long been the focus of research (Kang et al, 1984; Attah-Krah and Francis, 1987; Lal, 1989). Many productive and sustainable systems, needing few or no external inputs, have been developed. They stop erosion, produce food and wood, and can be cropped over long periods. But the problem is that very few, if any, farmers have adopted these alley cropping systems as designed. Despite millions of dollars of research expenditure over many years, systems have been produced suitable only for research stations (Carter, 1995).

There has been some success, however, where farmers have been able to take one or two components of alley cropping, and then adapt them to their own farms. In Kenya, for example, farmers planted rows of leguminous trees next to field boundaries, or single rows through their fields; and in Rwanda, alleys planted by extension workers soon became dispersed through fields (Kerkhof, 1990).

In Laos, one project used food for work to encourage shifting agriculturalists to settle and adopt contour farming with bench terraces (Fujisaka, 1989). But these fields became so infested with weeds that farmers were forced to shift to new lands, and the structures were so unstable in the face of seasonal rains that they led to worsened gully erosion. Farmers then refused to do further work when the incentives finished.

What does this mean for sustainable agriculture? How should we proceed so as to ensure farmers are fully involved in developing and adapting these sustainable and productive technologies?

5 The many interpretations of participation

There is a long history of participation in agricultural development, and a wide range of development agencies, both national and international, have attempted to involve people in some aspect of planning and implementation. Two overlapping schools of thought and practice have evolved. One views participation as a means to increase efficiency, the central notion being that if people are involved, then they are more likely to agree with and support the new development or service. The other sees participation as a fundamental right, in which the main aim is to initiate mobilization for collective action, empowerment and institution building.

In recent years, there have been an increasing number of comparative studies of development projects showing that 'participation' is one of the critical components of success. It has been associated with increased mobilization of stakeholder ownership of policies and projects; greater efficiency, understanding and social cohesion; more cost-effective services; greater transparency and accountability; increased empowering of the poor and disadvantaged; and strengthened capacity of people to learn and act (Finsterbusch and van Wicklen, 1989; Bagadion and Korten, 1991; Cernea, 1991; Pretty and Sandbrook, 1991; Uphoff, 1992; Narayan, 1993; World Bank, 1994; Pretty, 1995a, b).

As a result, the terms 'people's participation' and 'popular participation' are now part of the normal language of many development agencies, including NGOs, government departments and banks (Adnan et al, 1992; World Bank, 1994). It is such a fashion that almost everyone says that participation is part of their work. This has created many paradoxes. The term 'participation' has been used to justify the extension of control of the state as well as to build local capacity and self-reliance; it has been used to justify external decisions as well as to devolve power and decision-making away from external agencies; it has been used for data collection as well as for interactive analysis.

One of the objectives of agricultural support institutions must, therefore, be greater involvement with and empowerment of diverse groups of people, as sustainable agriculture is threatened without it. The dilemma for many authorities is they both need and fear people's participation. They need people's agreements and support, but they fear that this wider involvement is less controllable, less precise and so likely to slow down planning processes. But if this fear permits only stage-managed forms of participation, then distrust and greater alienation are the most likely outcomes. This makes it all the more crucial that judgements can be made on the type of participation in use.

In conventional rural development, participation has commonly centred on encouraging local people to sell their labour in return for food, cash or materials. Yet these material incentives distort perceptions, create dependencies, and give the misleading impression that local people are supportive of externally-driven initiatives. This paternalism undermines sustainability goals and produces impacts which rarely persist once the project ceases (Bunch, 1983; Pretty and Shah, 1994; Kerr, 1994). Despite this, development programmes continue to justify subsidies and incentives, on the grounds that they are faster, that they can win over more people, or they provide a mechanism for disbursing food to poor people. As little effort is made to build local skills, interests and capacity, local people have no stake in maintaining structures or practices once the flow of incentives stops.

The many ways that development organisations interpret and use the term participation can be resolved into seven clear types. These range from manipulative and passive participation, where people are told what is to happen and act out predetermined roles, to self-mobilization, where people take initiatives largely independent of external institutions (Table 1). This typology suggests that the term 'participation' should not be accepted without appropriate clarification.

Typology		Characteristics of Each Type
1.	*Manipulative Participation*	Participation is simply a pretence
2.	*Passive Participation*	People participate by being told what has been decided or has already happened. Information being shared belongs only to external professionals.
3.	*Participation by Consultation*	People participate by being consulted or by answering questions. Process does not concede any share in decision-making, and professionals are under no obligation to take on board people's views.
4.	*Bought Participation*	People participate in return for food, cash or other material incentives. Local people have no stake in prolonging technologies or practices when the incentives end.
5.	*Functional Participation*	Participation seen by external agencies as a means to achieve project goals, especially reduced costs. People may participate by forming groups to meet predetermined objectives related to the project.
6.	*Interactive Participation*	People participate in joint analysis, development of action plans and formation or strengthening of local groups or institutions. Learning methodologies used to seek multiple perspectives, and groups determine how available resources are used.
7.	*Self-Mobilization*	People participate by taking initiatives independently of external institutions to change systems. They develop contacts with external institutions for resources and technical advice they need, but retain control over how resources are used.

Table 1: A typology of participation for soil conservation and land husbandry (Source: Pretty, 1995b)

One study of 121 rural water supply projects in 49 countries of Africa, Asia and Latin America found that participation was the most significant factor contributing to project effectiveness and maintenance of water systems (Narayan, 1993). Most of the projects referred to community participation or made it a specific project component, but only 21% scored high on interactive participation. Clearly, intentions did not translate into practice. It was when people were involved in decision-making during all stages of the project, from design to maintenance, that the best results occurred. If they were just involved in information sharing and consultations, then results were much poorer. According to the analysis, it was quite clear that moving down the typology moved a project from a medium to highly effective category.

Great care must, therefore, be taken over both using and interpreting the term participation. It should always be qualified by reference to the type of participation, as most types will threaten rather than support the goals of sustainable agriculture. What will be important is for institutions and individuals to define better ways of shifting from the more common passive, consultative and incentive-driven participation towards the interactive end of the spectrum.

6 Systems of participatory learning and action

Recent years have seen a rapid expansion in new participatory methods and approaches to learning in the context of agricultural development (see *PLA Notes (formerly RRA Notes)*, 1988-present; Pretty et al, 1995; IDS/IIED, 1994; Chambers, 1994, a, b, c; Mascarenhas et al., 1991; Rhoades, 1990; Grandin, 1987; KKU, 1987; Conway, 1987). Many have been drawn from a wide range of non-agricultural contexts, and were adapted to new needs. Others are innovations arising out of situations where practitioners have applied the methods in a new setting, the context and people themselves giving rise to the novelty.

There are now more than 30 different terms for these systems of learning and action, some more widely used than others[1]. Participatory Rural Appraisal (PRA), for example is now practised in at least 130 countries, but Samuhik Brahman is associated just with research institutions in Nepal, and REFLECT just with adult literacy programmes. But this diversity and complexity is a strength. It is a sign of both innovation and ownership. Despite the different contexts in which these approaches are used, there are important common principles uniting most of them. These systems emphasise the following six elements (Pretty, 1995b):

- *A Defined Methodology and Systemic Learning Process* - the focus is on cumulative learning by all the participants and, given the nature of these approaches as systems of inquiry and interaction, their use has to be participative. The emphasis on visualisations democratises and deepens analysis.
- *Multiple Perspectives* - a central objective is to seek diversity, rather than characterise complexity in terms of average values. The assumption is that different individuals and groups make different evaluations of situations, which lead to different actions. All views of activity or purpose are heavy with interpretation, bias and prejudice, and this implies that there are multiple possible descriptions of any real-world activity.
- *Group Learning Process* - all involve the recognition that the complexity of the world will only be revealed through group inquiry and interaction. This implies three possible mixes of investigators, namely those from different disciplines, from different sectors, and from outsiders (professionals) and insiders (local people).
- *Context Specific* - the approaches are flexible enough to be adapted to suit each new set of conditions and actors, and so there are multiple variants.
- *Facilitating Experts and Stakeholders* - the methodology is concerned with the transformation of existing activities to try to bring about changes which people in the situation regard as improvements. The role of the 'expert' is best thought of as helping people in their situation carry out their own study and so achieve something.
- *Leading to Sustained Action* - the learning process leads to debate about change, and debate changes the perceptions of the actors and their readiness to contemplate action. Action is agreed, and implementable changes will therefore represent an accommodation between the different

---

[1] A selection of recently emerged terms alternative systems of learning and action include Agroecosystems Analysis (AEA), Beneficiary Assessment, Development Education Leadership Teams (DELTA), Diagnóstico Rurale Participativo (DRP), Farmer Participatory Research, Farming Systems Research, Groupe de Recherche et d'Appui pour l'Auto-Promotion Paysanne (GRAAP), Méthode Accélérée de Recherche Participative (MARP), Participatory Analysis and Learning Methods (PALM), Participatory Action Research (PAR), Participatory Research Methodology (PRM), Participatory Rural Appraisal (PRA), Participatory Rural Appraisal and Planning (PRAP), Participatory Technology Development (PTD), Participatory Urban Appraisal (PUA), Planning for Real, Process Documentation, Rapid Appraisal (RA), Rapid Assessment of Agricultural Knowledge Systems (RAAKS), Rapid Assessment Procedures (RAP), Rapid Assessment Techniques (RAT), Rapid Catchment Analysis (RCA), Rapid Ethnographic Assessment (REA), Rapid Food Security Assessment (RFSA), Rapid Multi-perspective Appraisal (RMA), Rapid Organisational Assessment (ROA), Rapid Rural Appraisal (RRA), Regenerated Freiréan Literacy through Empowering Community Techniques (REFLECT), Samuhik Brahman (Joint trek), Soft Systems Methodology (SSM), Theatre for Development, Training for Transformation, and Visualisation in Participatory Programmes (VIPP).

conflicting views. The debate and/or analysis both defines changes which would bring about improvement and seeks to motivate people to take action to implement the defined changes. This action includes local institution building or strengthening, so increasing the capacity of people to initiate action on their own.

The participatory methods used in these systems of learning and action can be structured into four classes: methods for group and team dynamics, for sampling, for interviewing and dialogue, and for visualisation and diagramming. It is the collection of these methods into unique approaches, or assemblages of methods, that constitute different systems of learning and action.

7 Towards a new professionalism for soil conservation and land husbandry

The central principle of sustainable agriculture is that it must enshrine new ways of learning about the world. But learning should not be confused with teaching. Teaching implies the transfer of knowledge from someone who knows to someone who does not know. Teaching is the normal mode of educational curricula, and is also central to many organisational structures (Ison, 1990; Argyris et al, 1985; Russell and Ison, 1991; Bawden, 1992, 1994; Pretty and Chambers, 1993). Universities and other professional institutions reinforce the teaching paradigm by giving the impression that they are custodians of knowledge which can be dispensed or given (usually by lecture) to a recipient (a student). Where these institutions do not include a focus on self-development and enhancing the ability to learn, then *"teaching threatens sustainable agriculture"* (Ison, 1990).

The problem for farmers is that they cannot rely on routine, calendar-based activities if they are to support sustainable agriculture. Interventions must be based on observation and anticipation. They require instruments and indicators which make more visible the ecological relationships on and between farms. Technology for sustainable farming must emphasise measurement and observation equipment, or services that help individual farmers assess their situations, such as soil analysis, manure analysis, pest identification (Röling, 1994). It also has to focus on higher system levels. Predators and parasitoids to control pests often require a larger biotope than that of a small farm. Erosion control, water harvesting, biodiversity, access to biomass, recycling waste between town and countryside and between animal and crop production, all require local cooperation and coordination.

What becomes important is the social transition, or new learning path, that farmers and communities must take to support sustainable agriculture. This is much less obvious and often remains unrecognised by extensionists. Learning for sustainable agriculture involves a transformation in the fundamental objectives, strategies, theories, skills, labour organisation, and professionalism, of farming.

In educational systems, therefore, the fundamental requirement for sustainable agriculture is for universities to evolve into communities of participatory learners. Such changes are very rare, an exception being Hawkesbury College, which is now part of the University of Western Sydney, Australia (Bawden, 1992, 1994). However, a regional consortium of NGOs in Latin America concerned with agroecology and low-input agriculture recently signed an agreement with eleven colleges of agriculture from Argentina, Bolivia, Chile, Mexico, Peru, and Uruguay to help in the joint reorientation of curriculum and research agendas towards sustainability and poverty concerns (Altieri and Yuryevic, 1992; Yuryevic, 1994). The agreement defines collaboration to develop more systemic and integrated curricula, professional training and internship programmes, collaborative research efforts and the development of training materials.

A move from a teaching to a learning style has profound implications for agricultural development institutions. The focus is less on *what* we learn, and more on *how* we learn and *with whom*. This implies new roles for development professionals, leading to a whole new professionalism with new concepts, values, methods and behaviour (Table 2).

Typically, normal professionals are single-disciplinary, work largely in ways remote from people, are insensitive to diversity of context, and are concerned with themselves generating and transferring technologies. Their beliefs about people's conditions and priorities often differ from people's own views. The new professionals, by contrast, make explicit their underlying values, select methodologies to suit needs, are more multidisciplinary and work closely with other disciplines, and are not intimidated by the complexities and uncertainties of dialogue and action with a wide range of non-scientific people (Pretty and Chambers, 1993).

Elements	Components of the new professionalism
Assumptions about reality	The assumption is that realities are socially constructed, and so participatory methodologies are required to relate these many and varied perspectives one to another.
Underlying values	Underlying values are not presupposed, but are made explicit; old dichotomies of facts and values, and knowledge and ignorance, are transcended.
Scientific method(s)	The many scientific methods are accepted as complementary; with reductionist science for well-defined problems and when system uncertainties are low; and holistic and constructivist science when problem situations are complex and uncertain.
Who sets priorities and whose criteria count?	A wide range of stakeholders and professionals set priorities together; local people's criteria and perceptions are emphasised.
Context of researching process	Investigators accept that they do not know where research will lead; it has to be an open-ended learning process; historical and spatial context of inquiry is fundamentally important.
Relationship between actors and groups in the process	Professionals shift from controlling to enabling mode; they attempt to build trust through joint analyses and negotiation; understanding arises through this interaction, resulting in deeper relationships between investigator(s), the 'objects' of research, and the wider communities of interest.
Mode of professional working	More multidisciplinary than single disciplinary when problems difficult to define; so attention is needed on the interactions between members of groups working together.
Institutional involvement	No longer just scientific or higher-level institutions involved; process inevitably comprises a broad range of societal and cultural institutions and movements at all levels.
Quality assurance and evaluation	There are no simple, objective criteria for quality assurance: criteria for trustworthiness replace internal validity, external validity, objectivity, and reliability when methods is non-reductionist; evaluation is no longer by professionals or scientists alone, but by a wide range of affected and interested parties (the extended peer community).

Source: adapted from Pretty and Chambers (1993)

Table 2: Towards a new professionalism for soil conservation and land husbandry

But it would be wrong to characterise this as a simple polarisation between old and new professionalism, implying in some way the bad and the good. True sensibility lies in the way opposites are synthesised. It is clearly time to add to the paradigm of positivism for science, and embrace the new alternatives. This will not be easy. Professionals will need to be able to select appropriate methodologies for particular tasks (Funtowicz and Ravetz, 1993). Where the problem situation is well defined, system uncertainties are low, and decision stakes are low, then positivist and reductionist science will work well. But where the problems are poorly defined and there are great uncertainties potentially involving many actors and interests, then the methodology will have to comprise these alternative methods of learning. Many existing agricultural professionals will resist such paradigmatic changes, as they will see this as a deprofessionalisation of research. But Hart (1992) has put it differently: *"I see it as a 're-professionalisation', with new roles for the researcher as a democratic participant."*

Institutions can, therefore, improve learning by encouraging systems that develop a better awareness of information. The best way to do this is to be in close touch with external environments, and to have a genuine commitment to participative decision-making, combined with participatory analysis of performance. Learning organisations will, therefore, have to be more decentralised, with an open multidisciplinarity, and heterogeneous outputs responding to the demands and needs of farmers. These multiple realities and complexities will have to be understood through multiple linkages and alliances, with regular participation between professional and public actors. It is only when some of these new professional norms and practices are in place that widespread changes in the livelihoods of farmers and their natural environments are likely to be achieved.

8 Policies for sustainability and learning

Policy reform has been underway in many countries, with some new initiatives supporting elements of a more sustainable agriculture. Most of these have focused on input reduction strategies, because of concerns over foreign exchange expenditure or environmental damage. Only a few as yet represent coherent plans and processes that clearly demonstrate the value of integrating policy goals. Nonetheless, it is clear that many policy reforms are leading to changes in the sustainability of agriculture.

The first action that governments can take is to declare a national policy for sustainable agriculture. This would help to raise the profile of these processes and needs, as well as giving explicit value to alternative societal goals. New policies must be enabling, creating the conditions for development based more on locally-available resources and local skills and knowledge. Policy makers will have to find ways of establishing dialogues and alliances with other actors, and farmers' own analyses could be facilitated and their organised needs articulated. Dialogue and interaction would give rapid feedback, allowing policies to be adapted iteratively. Agricultural policies could then focus on enabling people and professionals to learn together so as to make the most of available social and biological resources.

It is important to be clear about just how policies should be trying to address the issues of sustainability and learning. As has been suggested, precise and absolute definitions of sustainability, and therefore of sustainable agriculture, are impossible.

Sustainable land management should not, therefore, be seen as a set of practices to be fixed in time and space. It implies the capacity to adapt and change as external and internal conditions change. Yet there is a danger that policy, as it has tended to do in the past, will prescribe the practices that farmers should use rather than create the enabling conditions for locally-generated and adapted technologies.

For sustainable land husbandry to spread widely, policy formulation must not repeat these mistakes. Policies will have to arise in a new way. They must be enabling, creating the conditions for sustainable development based on locally available resources and local skills and knowledge. Achieving this will be difficult. In practice, policy is the net result of the actions of different interest groups pulling in complementary and opposing directions. It is not just the normative expression of governments. Effective policy will have to recognise this, and seek to bring together a range of actors and institutions for creative interaction and joint learning.

References

Adnan, S., Barrett, A., Nurul Alam, S.M. and Brustinow, A. (1992): People's Participation. NGOs and the Flood Action Plan. Research and Advisory Services, Dhaka

Altieri, M.A. and Yurjevic, A. (1992): Changing the agenda of the universities. ILEIA Newsletter **2/92**, 39

Argyris, C., Putnam, R. and Smith, D.M. (1985): Action Science. Jossey-Bass Publishers, San Francisco and London.

Attah-Krah, A.N. and Francis, P.A. (1987): The role of on-farm trials in the evaluation of composite technologies: the case of alley farming in Southern Nigeria. Agric. Systems **23**: 133-152

Avery, D. (1995): Saving the Planet with Pesticides and Plastic. The Hudson Institute, Indianapolis

Bagadion, B.U. and Korten, F.F. (1991): Developing irrigators' organisations; a learning process approach. In: Cernea, M. M. (ed). Putting People First. Oxford University Press, Oxford. 2nd Edition.

Bawden, R. (1992): Creating learning systems: a metaphor for institutional reform for development. Paper for joint IIED/IDS Beyond Farmer First: Rural People's Knowledge, Agricultural Research and Extension Practice Conference, 27-29 October, Institute of Development Studies, University of Sussex, UK. IIED, London.

Bawden, R. (1994): A learning approach to sustainable agriculture and rural development: reflections from Hawkesbury. Hawkesbury College, Australia, mimeo

Borlaug, N. (1992): Small-scale agriculture in Africa: the myths and realities. Feeding the Future (Newsletter of the Sasakawa Africa Association) **4**:2.

Borlaug, N. (1994a): Agricultural research for sustainable development. Testimony before US House of Representatives Committee on Agriculture, March 1, 1994.

Borlaug, N. (1994b): Chemical fertilizer 'essential'. Letter to International Agricultural Development (Nov-Dec), p 23.

Brown, L.R. (1994): The world food prospect: entering a new era. In Assisting Sustainable Food Production: Apathy or Action? Winrock International, Arlington, VA.

Brown, L.R. and Kane, H. (1994): Full House: Reassessing the Earth's Population Carrying Capacity. W W Norton and Co, New York.

Bunch, R. (1983): Two Ears of Corn. World Neighbors, Oklahoma City.

Carruthers, I. (1993): Going, going, gone! Tropical agriculture as we knew it. Tropical Agriculture Association Newsletter **13 (3)**, 1-5

Carter, J. (1995): Alley Cropping: Have Resource Poor Farmers Benefited? ODI Natural Resource Perspectives No 3, London

Cernea, M.M. (1991): Putting People First. Oxford University Press, Oxford. 2nd Edition.

CGIAR (1995): Sustainable Agriculture for a Food Secure World: A Vision for International Agricultural Research. Expert Panel of the CGIAR, Washington DC and SAREC, Stockholm

Chambers, R. (1994a): The origins and practice of participatory rural appraisal. World Development **22**, No 7, 953-969

Chambers, R. (1994b): Participatory rural appraisal (PRA): analysis of experience. World Development **22**, No 9, 1253-1268

Chambers, R. (1994c): Participatory rural appraisal (PRA): challenges, potentials and paradigm. World Development **22**, No 10, 437-454

Conway, G.R. 1987. The properties of agroecosystems. Agric. Systems 24, 94-117

Crosson, P. and Anderson, J. (1995): Achieving a Sustainable Agricultural System in Sub-Saharan Africa. Building Block for Africa Paper No 2, AFTES, The World Bank, Washington DC

Danida (1994): Agricultural Sector Evaluation. Lessons Learned. Ministry of Foreign Affairs, Copenhagen

DowElanco (1994): The Bottom Line. DowElanco, Indianapolis, IN.

Dyer, N. and Bartholomew, A. (1995): Project Completion Reports: Evaluation Synthesis Study. Evaluation Report Ev583. ODA, London

EC (1994): Evaluation des Projets de Developpement Rural Finances Durant les Conventions de Lomé I, II, et III. European Commission, Brussels

Ehrlich, P. (1968): The Population Bomb. Ballantine, New York.

FAO (1993): Strategies for Sustainable Agriculture and Rural Development (SARD): The Role of Agriculture, Forestry and Fisheries. FAO, Rome

FAO (1995): World Agriculture: Toward 2010. Edited by N. Alexandratos. United Nations Food and Agriculture Organization, Rome.

Finsterbusch, K. and van Wicklen, W.A. (1989): Beneficiary participation in development projects: empirical tests of popular theories. Econ. Development and Cultural Change **37 (3),** 573-593

Fujisaka, S. (1989): The need to build on farmer practice and knowledge: reminders from selected upland conservation projects and policies. Agroforestry Systems **9,** 141-153

Funtowicz, S.O. and Ravetz, J.R. (1993): Science for the post-normal age. Futures **25,** No 7, 739-755

Grandin, B. (1987): Wealth Ranking. IT Publications, London.

Hart, R.A. (1992): Children's Participation: From Tokenism to Citizenship, UNICEF Innocenti Essays No 4. Florence: UNICEF

Hazell, P. (1995): Managing Agricultural Intensification. IFPRI 2020 Brief 11. IFPRI, Washington DC

Hewitt, T.I. and Smith, K.R. (1995): Intensive Agriculture and Environmental Quality: Examining the Newest Agricultural Myth. Henry Wallace Institute for Alternative Agriculture, Greenbelt MD

Hinchcliffe, F., Thompson, J. and Pretty, J.N. (1996): Sustainable Agriculture and Food Security in East and Southern Africa. Report for the Committee on Food Security in East and Southern Africa, Swedish International Agency for International Cooperation, Stockholm.

IDS/IIED (1994): PRA and PM&E Annotated Bibliography. IDS, Sussex and IIED, London.

Ison, R. (1990): Teaching Threatens Sustainable Agriculture. Gatekeeper Series SA21, IIED, London

Kang, B.T., Wilson, G.F. and Lawson, T.L. (1984): Alley Cropping: A Stable Alternative to Shifting Agriculture. IITA, Ibadan.

Kendall, H.W. and Pimentel, D. (1994): Constraints on the expansion of the global food supply. Ambio **23,** 198-205.

Kerkhof, P. (1990): Agroforestry in Africa. A Survey of Project Experience. Panos Institute, London.

Kerr, J. (1994): How subsidies distort incentives and undermine watershed development projects in India. Paper for IIED Conference New Horizons: The Social, Economic and Environmental Impacts of Participatory Watershed Development, November, Bangalore, India

KKU (1987): Rapid Rural Appraisal. Proceedings of an International Conference. Rural Systems Research Projects, Khon Kaen University, Thailand.

Knutson, R.D, Taylor, J.B., Penson, J.B. and Smith, E.G. (1990): Economic Impacts of Reduced Chemical Use. Texas A&M University.

Lal, R. (1989): Agroforestry systems and soil surface management of a Tropical Alfisol. I: Soil moisture and crop yields. Agroforestry Systems **8:** 7-29.

Mascarenhas, J., Shah, P., Joseph, S., Jayakaran, R., Devavaram, J., Ramachandran, V., Fernandez, A., Chambers, R. and Pretty, J.N. (eds) (1991): Participatory Rural Appraisal. RRA Notes 13. IIED, London.

McCalla, A. (1994): Agriculture and Food Needs to 2025: Why We Should be Concerned. Sir John Crawford Memorial Lecture, October 27. CGIAR Secretariat, The World Bank, Washington, DC.

McCalla, A. (1995): Towards a strategic vision for the rural/agricultural/natural resource sector activities of the World Bank. World Bank 15th Annual Agricultural Symposium, January 5-6th, Washington DC

Mitchell, D.O. and Ingco, M.D. (1993): The World Food Outlook. International Economics Department. World Bank, Washington, DC.

N.A.F. (1994): A Better Row to Hoe: The Economic, Environmental and Social Impact of Sustainable Agriculture. St Paul, Minnesota: Northwest Area Foundation

Narayan, D. (1993): Focus on Participation: Evidence from 121 Rural Water Supply Projects. UNDP-World Bank Water Supply and Sanitation Program, World Bank, Washington DC

Paarlberg, R.L. (1994): Sustainable farming: a political geography. IFPRI 2020 Brief, No. 4. International Food Policies Research Institute, Washington, DC.

PLA Notes (formerly RRA Notes) (1988): Issues 1-25, cont. Sustainable Agriculture Programme, IIED, London.

Pohl, G. and Mihaljek, D. (1992): Project evaluation and uncertainty in practice: a statistical analysis of rate-of-return divergences of 1015 World Bank projects. World Bank Economic Review **6 (2),** 255-277

Pretty, J.N. (1995a): Regenerating Agriculture: Policies and Practice for Sustainability and Self-Reliance. Earthscan Publications Ltd, London; National Academy Press, Washington DC; Vikas Publishers and ACTIONAID, Bangalore.

Pretty, J.N. (1995b): Participatory learning for sustainable agriculture. World Development **23(8),** 1247-1263.

Pretty, J.N. (1996): Can sustainable agriculture feed the world? Biologist **43 (3):**130-133

Pretty, J.N. and Chambers, R. (1993): Towards a learning paradigm: new professionalism and institutions for sustainable agriculture. IDS Discussion Paper DP 334. IDS, Brighton

Pretty, J.N. and Sandbrook, R. (1991): Operationalising sustainable development at the community level: primary environmental care. Presented to the DAC Working Party on Development Assistance and the Environment, OECD, Paris, October 1991

Pretty, J.N. and Shah, P. (1994): Soil and Water Conservation in the 20th Century: A History of Coercion and Control. Rural History Centre Research Series No.1. University of Reading, Reading.

Pretty, J.N. and Thompson, J. (1996): Sustainable Agriculture and the Overseas Development Administration. Report for Natural Resources Policy Advisory Department, ODA, London.

Pretty, J.N., Guijt, I., Scoones, I. and Thompson, J. (1995): A Trainers' Guide to Participatory Learning and Interaction. IIED Training Materials Series No. 2. IIED, London.

Rahnema, M. (1992): Participation. In: Sachs, W. (ed) The Development Dictionary. Zed Books Ltd, London.

Rhoades, R. (1990): The Coming Revolution in Methods for Rural Development Research. Mimeo. User's Perspective Network International Potato Center, Manila, Philippines.

Röling, N. (1994): Platforms for decision making about ecosystems. In: Fresco L (ed). The Future of the Land. John Wiley and Sons, Chichester

Röling, N.R. and Wagemakers, M.A. (eds) (1996): Sustainable Agriculture and Participatory Learning. Cambridge University Press, Cambridge (in press).

Rosegrant, M.W. and Agcaolli, M. (1994): Global and regional food demand, supply and trade prospects to 2010. IFPRI, Washington, DC.

Russell, D.B. and Ison, R.L. (1991): The research-development relationship in rangelands: an opportunity for contextual science. Plenary paper for 4th International Rangelands Congress, Montpellier, France, 22-26 April 1991

SAA (1995-96): Feeding the Future. Newsletter of the Sasakawa Africa Association, Tokyo.

Sasakawa Global 2000. (1993-1995): Annual Reports. Sasakawa Africa Association, Tokyo.

Thompson, J. (1995): Participatory approaches in government bureaucracies: facilitating the process of institutional change. World Development **23 (9)**, 1521-1554

Thompson, J. (1996): Sustainable agriculture and rural development: challenges for EU Aid. EC Aid and Sustainable Development Briefing Paper, No. 8. International Institute for Environment and Development, London.

Uphoff, N. (1992): Learning from Gal Oya: Possibilities for Participatory Development and Post-Newtonian Science Cornell University Press, Ithaca

Winrock International (1994): Assisting Sustainable Food Production: Apathy or Action? Winrock International, Arlington, VA.

Wirth, T.E. (1995): US Policy, Food Security and Developing Countries. Undersecretary of State for Global Affairs, Committee on Agricultural Sustainability for Developing Countries, Washington DC

World Bank (1993): Agricultural Sector Review. Agriculture and Natural Resources Department, Washington, DC.

World Bank (1994): The World Bank and Participation. Report of the Learning Group on Participatory Development. April 1994. World Bank, Washington, DC

Yuryevic, A. (1994): Community-based sustainable development in Latin America: the experience of CLADES. Paper presented at IIED In Local Hands: Community-Based Sustainable Development Symposium, July 4-8, Brighton

Address of author:
Jules Pretty
Centre for Environment and Society
John Tabor Laboratories
University of Essex
Wivenhoe Park
Colchester CO4 3SQ, UK

Making Development Sustainable

A. Steer

Keywords: Environmental policy, people involvement, ecosystems approach, knowledge base

1 Introduction

Few issues can be more crucial to human well-being and long-term ecological sustainability than the subject of land use management. In the past half century about 2 billion of the 8.7 billion hectares of agricultural land, permanent pastures, and forests and woodlands have been degraded. Today there are more than 900 million people in 100 countries affected by desertification, which is one of the reasons why we hope to support the Desertification Convention's implementation very seriously. Increases in food production have failed to keep pace with population growth in more than 50 countries in the 1980s and the first half of the 1990s. And these were very often among the poorest countries on earth. The rate of growth of global grain production dropped from 3% in the 1970s to 1.3% in the 1983 to 1993 period and the amount of grain produced per person has fallen in the last decade. The inadequate attention to soil and water management is one of the key reasons for these declines. And the challenges ahead for food production are obviously enormous.

The IFPRI 2020 (IFPRI, 1995) study concludes that over the next 25 years developing countries are likely to increase their market demands of food grains by 75% and for livestock products by 155%. Over the next four decades, less than one working lifetime, world food production will need to almost double. Optimists of course believe that this will be achieved fairly easily. After all, the last doubling of food production occurred in just 25 years. These optimists however neglect the unique technical circumstances of the last doubling, and our current inability to replicate those growth conditions. They also neglect the 0.3 to 1.5 billion hectares being lost each year due to water logging and salinization on precisely the land that gave us the last doubling of world food production. And they neglect too the sharp decline in growth of agricultural research and technology development.

Research in low-income countries' agriculture is grossly inadequate today. The public sector expenditures on agriculture research are typically less than 0.5% of the agricultural GDP, compared to 1% in middle-income countries and 2-5% in industrial countries. In Africa agricultural research expenditures per scientists have fallen on average by 2.6 % per year since 1961. Declining foreign assistance budgets are partly responsible for this decline. Not only have real aid budgets declined over all, but the share of agriculture in the total has declined from 20% to 14% between 1980 and today.

Improved soil and land management of course is not only crucial for the long-term aggregate outlook, of equal or perhaps even more importance is its vital role in poverty reduction. Of the 1.3 billion people living in the world today living in acute poverty on less than 1 $ per day almost three quarters of these live in rural areas. Most of these are deeply dependent directly or indirectly on the productivity of the soil. When we consider aggregate figures we must never forget that the lives of individuals and families are at stake. I would like to divide the remainder of my remarks into three points. First, I would like to place the discussion into a broader context of what has been happening in the four years since the Rio Earth Summit; second, I would like to apply some of these broader themes to the subjects of sustainable soil and land management; finally, I would like to say just a few words in closing about how the World Bank is reinvigorating its efforts to support sustainable land management.

2 Four years on from the Rio Earth Summit, where do we stand?

Concern for the environment in many industrial countries has fallen from its Earth Summit peak, environmental conditions in most developing countries are worse today than at the time of Rio. What then is there to be encouraged about? I believe that there is a quiet revolution underway as environmental sustanability is slowly and gradually becoming a theme of policy making around the world. The key propositions of sustainable development laid out in the Brundtland Commission Report in 1987 and taken up in Rio's Agenda 21 in 1992 was somewhat controversial at the time. They are now widely accepted. Among such propositions that economic development and the environment must be understood as partners not enemies, that the costs of ignoring the environment can be very high indeed, that addressing environmental and natural resource problems requires that poverty has to be reduced, that economic growth must be guided by prices and policies that reflect environmental externalities, that indicators of progress including the national accounts need to include natural as well as man-made capital.

Ten years ago these propositions would not have found wide acceptance among the powerful policy-makers in the ministries of finance, planning, public works or even agriculture. In our client countries these propositions now do receive general acceptance. Broad acceptance of course does not ensure their effective implementation, but there is nonetheless a quantum shift in awareness among policy-makers and in some countries real action is underway. About 100 countries have now prepared national environmental action plans, and in about half of these numbers policies are being revised and better investment plans implemented.

It is no longer considered odd that at the biggest meeting of finance ministers and central bankers in the world, the Annual IMF World Bank Meeting, that there should be a major associated conference on environmentally sustainable development in which many of the finance ministers participate. The theme of this year's conference is sustainable rural development.I am looking forward seeing Elizabeth Dowdeswell, head of UNEP, addressing finance ministers and planning ministers and central bankers from around the world. A number of these finance ministers are beginning to put their country's money along with their stated intentions. Some 70 countries over the past five years have requested and received financial support from the World Bank to assist them in reforming their environmental and natural resource policies.

In the 1990s we have lent over ten billion Dollars which has to be repaid for this purpose. Adding domestic financing and co-financing these are investments adding to about 25 billion Dollars. These investments are targeted specifically at improving environmental conditions, addressing pollution, deforestation, soil depletion and erosion, water management, coastal zone management and the like.

3 Principles for enhancing environmental policies

We recently surveyed the programmes on natural resource management in an effort to distil common themes of this new environmental awareness. We identified ten principles that distinguish the new approach to the environment from more conventional environmental policies. They sound so obvious today, yet ten years ago they absolutely did not characterize environmental policy making around the world.

Principle 1: Set priorities carefully. That has not been done traditionally.
Principle 2: Promote cost-effectiveness vigorously.
Principle 3: Understand human behaviour and choose policies and technologies accordingly.
Principle 4: Go for win-win options first. They are of enormous scope.
Principle 5: Economize on scarce administrative and regulatory capacity.
Principle 6: Take an ecosystem-wide approach. Traditional environmental policies too often have been focused on first order symptoms rather than system-wide causes and impacts.
Principle 7: Work with, not against the private sector.
Principle 8: Involve local people thoroughly.
Principle 9: Build constituencies for change. There will always be opposition to change, there will always be losers, as well as winners.
Principle 10: Invest in the facts.

These ten principles apply quite well to the subject of land management. I would like to discuss four of those ten principles, which seem to me to be of particular importance for our deliberations here and outline the lessons for the issue of land management.

The first lesson is this: Understand human behaviour and design policies and technologies accordingly. Addressing the challenge of land management requires very close collaboration between technologists and scientists on one hand and economists and social scientists on the other. It is no use designing improved anti-erosion or nutrient-cycling techniques, if such techniques are financially, socially or culturally unattractive to farmers.

The field of soil management has suffered badly from inadequate synergy between technical and behavioural specialists. There are simply dozens of technologies that have been sound technically, but that even after repeated selling attempts to farmers have been found to be unsustainable. This has been because nobody has taken the trouble to understand what drives farmers and community behaviour.

Along with research institutions in Central America we recently completed a major analysis of costs and benefits of various soil conservation techniques in that region. The broad conclusions reinforced the findings of other recent research studies concerning the unattractiveness of large-scale terracing and other large investment options. But many of the detailed conclusions were unexpected and often highly location specific and they almost always illustrated the rationality on the part of poor farmers in making decisions based on costs and benefits.

The divorce between physical and biological science on the one hand and behavioural science on the other has been very costly to poor farmers around the world and to the world's soils. Thankfully this divorce is now ending and a new, more integrated approach is emerging and this is very good news indeed. This new approach is clear from some of the session titles in this ISCO conference and in the overall title to the conference. It is also evident from the realignment of research programmes throughout the revitalized CGIAR system. It is also evident in efforts to map out a new strategy for soil, water and nutrient management, such as that by the international board for soil research and management.

If we are to help make land management more sustainable we simply must invest more effort to understand the motivation, aspirations and calculations of the key actors, those who manage the land. We have of course learned a great deal about the behaviour of farmers over recent years. Numerous research studies have demonstrated the dynamism, rationality and efficiency of the small farm sector and its role in employment generation and in poverty reduction. Small poor farmers while smart, rational and entrepreneurial do usually have high discount rates. Recent research in Central America and South Asia suggests that 20-30% is a reasonable assumption. This means that high up-front investments with a long pay-back period are generally unacceptable to the farmers.

Expectations of high discount rates are without question an obstacle to achieving long-run sustainable development. Recognition of these high discounts of course has two implications. First, for researchers it is that lower cost investment alternatives must be the primary focus of research in low-income countries. The second, for economists and policy-makers is that efforts must be made to bring down these expectations of discount rates. We know what this requires. It requires macro-economic stability, non-distorted markets and encouragement of non-subsidized, preferably non-government rural credit. It also requires incidentally market-based land reform in many countries. This illustrates once again the importance of sound macro-economic and sectoral-economic policies as a vital part of any strategy to make development more sustainable. Concerned soil scientists, environmentalists and ministries of agricultural and natural resources need to help lobby for better economic policies in many countries today.

The second lesson for soil management from the new environmental approach is a logical extension of the first. It is to involve local people thoroughly in all efforts to improve land management. The evidence is now overwhelming that technologies and investment programmes will be more successful if developed with the beneficiaries rather than for them. Whether in the soil clubs of the state of Paraná in Brazil in the 1970s or 80s or in 100 other examples, the message is clear: participation works. This is a very hard lesson for governments to learn. It is perhaps an even harder lesson for scientists to learn. I can assure you it has been a hard lesson for the World Bank to learn. Five years ago within the Bank we set up a learning group on participation. Members of the group were taken from all grade levels and all regions of the world. Over a five-year period that group has documented both the benefits of a participatory approach and some of the best practices. It became clear to us that participation means very different things to different people. And some parts are easier to implement than others. We now think in terms of three simple levels of participation. The first level is simply information sharing. The researcher or official shares with local affected people current plans and findings. World Bank staff found no difficulty in approving this one-way communication. This level of participation is better than nothing, but not much. It is not really worthy of the name of participation.

The second level of participation involves a genuine discussion between experts and local people. Views on both sides are thoroughly aired and the expert takes what is useful from the local expertise in the design of the programme. The benefits of this second approach are considerable. This approach is now World Bank policy and increasingly our practice, too.

The third level of participation is the most difficult. It involves in addition to dialogue a transfer of a substantial part of the decision-making authority to the local people. Experts find it very hard to give up control, especially to people who they consider to be less expert. Yet it is only in this fuller meaning of participation that its benefits in terms of ownership, commitment and expertise are reaped. A growing number of Bank-supported projects follow this third approach, but it is not yet main-streamed. I hope it is not long before this form of behaviour becomes the norm.

The third lesson of the new environmental insight for land management is this: Take an ecosystem-wide approach.

Good progress is actually already being made here. The emergence, for example, of integrated nutrient management provides a conceptional and practical aid to policy-makers and extension workers with potentially extremely high long-term payment. Nevertheless, progress is partial and too much research and policy advice is undertaken with little thought of ecosystem-wide interactions. A review this year by CGIAR of 14 international research centres found good progress in reorienting research towards soil and water management, but it also found inadequate attention paid to offside interactions at the river basin and regional levels. This was of particular concern since offside costs of unsustainable management practices are often greater than their impacts on on-site productivity.

Another area of relative neglect is the flow of nutrients from rural to urban areas. Each year hundreds of millions of tons of nutrients are taken from the soils of farms in the form of food and become waste in cities. A social good is thus converted into a social bad. Some countries are leaders in the area of recycling. As Minister Merkel said, Germany already, for example, uses more than one third of its sewage sludge as fertilizer and the share will rise with tighter restrictions on dumping. For the developing world such recycling for agriculture offers great opportunities. It is also an urgent need. Of the World Bank we feel we paid inadequate attention to this issue. And we will be holding a conference on this subject entitled "Wasting waste" at the end of September 1997.

Another area of neglect at an even larger scale is the role played by land management in the global carbon cycle. A great deal of practical research is being done on the role of forests in this cycle, much less so on the role of soils. Factoring in the value to the global community of carbon sequestration in the soil may change the calculus from a global perspective of various land and nutrient management regimes. This is, for example, yet another reason to redouble efforts to restore phosphorus to the soils of Africa. In Burkina Faso for example, it is estimated that replacing phosphorus would through below and above ground biomass generation sequester almost a ton of carbon per hectare per year. Grant resources from the global environment facility through its three implementing agencies, UNEP, UNDP and the World Bank are available to finance the incremental costs of investment in the global interest. Resources could therefore be available to help finance land management including the Desertification Convention.

The fourth and final lesson of the environmental approach that I would like to highlight as of particular relevance this: invest in the facts.

The state of knowledge of trends in the quality of the world's land stock are appallingly poor. We all agree that the loss of tropical forests is a great threat to world ecosystems, yet the aggregate data that we all use still refer to the 1980s. We really have no idea in the aggregate what has been happening since 1990. We all believe that the degradation of soils in Africa is one of the greatest threats to that region's and the world's future. Yet we really have no systematic real time knowledge of trends. The figures are often old and usually of rather dubious quality. This level of ignorance would be unacceptable in other fields of endeavour. Imagine central bankers meeting together and being content with out-of-date ten year old data on GDP and money supply.

I believe we urgently need to remedy this situation. We need better knowledge for many reasons of course. One is to enable us to establish priorities for action. One example of recent careful analysis in Ethiopia found, for example, that contrary to accepted wisdom, nutrient loss was a much more serious threat to productivity than was soil erosion. And yet it has been erosion prevention that has received the great bulk of investment resources. Similarly in Mexico, when the impact of livestock overgrazing on land quality was calculated and factored into decision-making it was found that the rate of return to society of livestock investment fell in half. A major coordinated effort is needed to improve the knowledge base. And the international agencies, including FAO and UNEP, the responsible with compiling data need to be given adequate resources for this purpose. Recent joint efforts of FAO, the World Bank, UNEP, and UNDP to develop the system of land quality indicators is encouraging. This programme is built around the pressure- state -response framework which is being used in other environmental areas. I hope it will receive the support of this conference.

4 World Bank support to make land management more sustainable

At the request of the new president to the World Bank we have been rethinking our own role in the rural sector. While we have greatly strengthened our capacity on the environment over the past few years, our overall engagement in the rural sector has not been what it ought to have been. Just as our client countries paid less attention to rural development in the 1980s, so we, too, lost some of our capacity in the agricultural sector. We are currently seeking to reverse this. A major strategy paper outlining a new reinvigorated approach, the rural sector is being discussed by senior managers and the president of the World Bank at present. A key element of this approach will be deeper partnerships with many of the agencies and specialists represented here. We shall be looking forward to reviewing the findings of this ISCO conference in helping us to guide ourselves in our own future role.

Let me end with a biblical reference in the gospels. There is that wonderful story of the sower who went out to sow and some of the seeds that he spread around fell on stony ground, some of the seeds fell on paths and on roadways, some fell into shallow ground and grew up but then were choked. And some of the seeds fell on good ground. The seeds that are being sown today around the world are better than ever before, but the amount of shallow soil, the amount of roadways is growing each year. To transpose the metaphor I would like to wish you fertile soil for your deliberations and offer you deep and receptive ground for your ideas in the World Bank.

Reference

IFPRI (1995): Global Food Projections to 2020: Implications for investment. Mark W. Rosegrant, Mercedita Agcaok-Sombilla, and Nicostrato D. Perez. IFPRI Vision 2020. Food, Agriculture and the Environment. Discussion Paper 5. Washington.

Address of author:
Andrew Steer
The World Bank
1818 H Street, N.W.
Washington D.C. 20433, USA

Conditions for Enhanced Cooperation Between People and Institutions

M. Stocking

Summary
In order to enhance cooperation between people and institutions, four principal topics are addressed: the value of local knowledge; diversity within local society; different economic perspectives; and participation. Local knowledge, while not replacing formal science, enables the engagement of local people, enlisting their cooperation and interest. A typology of local knowledge is presented. Understanding the diversity and heterogeneity of local society allows the professional to appreciate choices and conflicts which might affect the uptake of technologies. Similarly, different economic perspectives affect how farmers make decisions in investing in soil improvement. Participation by local people is an essential element of any intervention, requiring techniques and approaches which are innovatory. However, enhancing cooperation means a re-analysis of thinking and practice by professionals which is best embraced by the term 'empathy' or putting oneself in the place of the person to whom you are talking.

Keywords: Institutions, local society, local knowledge, diversity, economic perspective, participation

1 Challenge

Local people often fail to solve their environmental problems - the issues are usually greater than their powers. Instead, they may just learn to cope. Institutions entrusted to solve these same problems similarly often fail. This paper examines the potential for bringing the two together to address one of the principal land use problems of the world, soil degradation. Can the combined efforts of people and institutions do better than each singly? What are the conditions for enhancing cooperation and who should take the initiative?

Consider first what local people have to offer. They have commitment and immediate responsibility, which, if they believe it to be rational, they can support with their own labour and resources. However, they also have what this paper will call 'local knowledge' (LK), sometimes referred to as 'indigenous [technical] knowledge'. Ethnoscience - "the set of concepts, propositions and theories unique to each particular culture group" (Meehan, 1980, p.385) - is similar but more restricted. The concept of knowledge encompasses the way people view and understand their worlds and how they structure, code, interpret and apply meaning to their experiences (Arce and Long, 1992). LK takes in intergenerational knowledge (that is, accumulated experience handed down from one generation to the next), new or modified knowledge which is created locally through processes of experimentation and innovation, and transferred knowledge which has been adapted and incorporated into the local way of life. It is this last aspect which makes LK different from indigenous knowledge and gives it a dynamic that has the capability of dealing with pressures such as population growth, economic crises and wars. Informal experiments and adaptations of soil and water conservation techniques are now increasingly being reported in the literature, especially from South Asia (Premkumar, 1994; Stocking, 1993), each measure being a *pot-pourri* of indigenous and externally-derived knowledges.

Institutions have formal scientific knowledge, access to new technologies and the capability, in theory at least, to develop new techniques from which major advances may derive in land management. They also have access to specialist knowledge, money and power. Through formal institutions, governments channel subsidies, control expenditures, legalise activities and attempt social order. Informal and local institutions, including NGOs, may act more discreetly but nevertheless organize society to achieve goals unattainable by individuals. The Chinese bench terraces of the Loess Plateau are a spectacular example of the role of the State and its institutions. It is no accident that as the State has loosened its economic grip on the countryside, the bench terraces now crumble. Institutions also have committed professionals who often have to labour under crushing bureaucracies and difficult socio-political conditions.

The greater power lies with institutions. Therefore, in order to enhance cooperation with people, the larger onus of responsibility attaches to them to change. Institutions do have many problems, especially inflexibilities and difficulties to cross disciplinary boundaries. However, since this paper is written from the perspective of a professional based in an institution and working for other institutions, the primary focus for cooperation between people and institutions will be taken to be how we, as professionals, can adapt or modify our thinking and approaches. It is **not** about re-educating the peasants. Education itself, if recognised as a constraint to sustainable land management, can only be achieved through enhanced cooperation. Cooperation with people signals a change in style of dialogue, mutual understanding, recognition of the value of other knowledges and humility. It particularly demands that the professional appreciates the divergent views of local people and of institutions. Although it would be quite wrong to see either side as having single sets of views, Table 1 draws some stereotypical comments that might characterise the conditions for the sustainable use of the soil. So, this paper concentrates on four main topics:

* the **value of local knowledge**; and how institutions may access and integrate local knowledge with formal scientific knowledge;
* **diversity within local society**; gender, age and ability differences; and how diversity can be utilised;
* **divergent economic perspectives**; society's interests are often different from farmers' economic rationality; and how farmer rationality may be exploited;
* **participation** in research and implementation of projects; appreciation of the various modes of participation.

2 The value of local knowledge

Local knowledge (LK) is increasingly being recognised for its value, not in replacing formal scientific knowledge, but in gaining a new, more integrative perspective on the capacity of the local environment and some tested options as to how the environment might be exploited by the people who already live there. Agrawal (1995) challenges the notion of any difference between indigenous and formal scientific knowledge, arguing that the same piece of information can be viewed from different perspectives according to who classifies it. Although the knowledge itself may be the same, its contextualisation and organisation most certainly will be different. So, typically, LK is holistic, conservative (in the sense of yielding slowly to change) and adaptive, while scientific knowledge is abstract, radical and prescriptive. These differences do not hold universally, however. The Burrungee's views, for example, on soil creation from rain with stones at the surface as evidence of soil formation are highly abstract. (Östberg, 1991). Nevertheless, the particular strength of LK is usually seen in its adaptive nature, ready acceptance and low risk. The converse epitomises to many the weaknesses of scientific knowledge (Scoones and Thompson, 1994).

Conditions for the sustainable use of soil *	Stereotypical views of:	
	Local people	**Institutions**
Biophysical Dimension:		
soil	A medium for supporting livelihoods	The loose material on the earth's surface to be prevented from degrading
role of vegetation	Better yields; more income	Protection to the soil; promotes organic cycling of nutrients; prevents surface crusting
soil degradation	Degradation - what degradation?	To be prevented at all costs
Social and Cultural Dimension:		
conformity with social and cultural norms	Essential - if it doesn't, we shall ignore you	A nuisance - why can't these people enter the real world?
secure rights to land	Essential - if you want me to invest my labour and resources	That's a political issue - nothing to do with me!
Political Dimension:		
stability and political commitment	It's nice to have support from the top	How about more money for my department? And a salary rise.
Economic & Financial Dimension:		
high rates of return to family labour	Essential - otherwise I'm better off going for wage labour in the city	Labour? - it will get them working instead of lounging
benefits in new technolo-gies must offset risks	And if this fails, who pays the price? Me!	Of course this technology is good. It comes from the research station.
Policy Conditions:		
local empowerment	You've never trusted us before	They couldn't organise a party in a brewery
minimise subsidies	If you want us to conserve the environment, then we need payment	Excellent idea! Why pay them for doing what they ought?
Institutional Conditions:		
community participation	You ask our views, then ignore us	What a bind! It just makes our job more difficult and lengthy
collaboration between institutions	One lot tells us one thing; the other contradicts	Can we trust the others? They've undermined us before
Developmental Conditions:		
respect for local knowledge	My father taught me to do it this way	Second rate knowledge; to be replaced as soon as possible
conservation integral to farming system	Of course - what else?	Conservation is a deliberate intervention to address problems with the farming system
Project Design Conditions:		
Long time horizon	My children will inherit this land in good condition	Impossible. We've only got a 3-year project funding
no technical 'fixes'	'Fixes' are fine so long as you understand my way of life	Technology is the answer to your problems

- Based on Douglas' (1994) "prerequisites for success or failure" in the sustainable use of agricultural soils

Table 1: Contesting views of the conditions for the sustainable use of soil.

Knowledge Appropriated	LK has financial value - people as 'local gatekeepers'
Knowledge Ventriloquised	LK, a language to transmit modern ideas
Knowledge Esteemed	LK as a study of culture; an entry to understanding local beliefs, attitudes and practices
Knowledge Negotiated	LK as the means for participation and mutual problem-solving
Knowledge as Empowerment	LK as the means for local people to exercise their own skills and take control of their own affairs

Table 2: Knowledge-in-Action - a typology of constructions of Local Knowledge (LK)

Clearly, there is a value in being able to understand and access LK - not least it enhances cooperation between people and institutions, and may provide insights unavailable to the normal processes of developing and applying scientific knowledge. This marriage of knowledges has led Blaikie et al (1997) to a typology of constructions of LK, or 'Knowledge-in-Action' as seen from the perspective of a professional in an institution (Table 2). Using this typology, we can see how LK may be used and negotiated with formal scientific knowledge.

'Knowledge Appropriated' typifies the exploitation of LK for commercial gain - for example, knowledge about local plants, medicinal values, insecticidal properties. It raises unresolved issues over intellectual property rights and the distribution of benefits from commercial exploitation. In soil management, this type of knowledge use is perhaps less contentious, but it may include the soil-improving benefits of individual species and the use of cover crops, intercrops and agroforestry combinations.

'Knowledge Ventriloquised' is the translation of external ideas into the local vernacular and into culturally acceptable formats. LK is in this context merely a language to transmit 'modern' inventions. The increasing use of local soil names is more about the delivery of agricultural extension messages to recognisable locations in the landscape than it is about any real professional interest in what these names represent.

'Knowledge Esteemed' is, for the present, largely confined to the academic community. Brookfield and Padoch (1994) describe how local communities of small farmers manage biological diversity, termed 'agrodiversity'. However, it is extraordinarily difficult to transfer this knowledge into practical recommendations, other than to congratulate the farmers on a job well done! Esteeming LK is the very opposite to transfer of technology approaches promoted by most agricultural research centres.

'Knowledge Negotiated' is the goal of many NGO projects where LK is mutually constructed through a negotiation between people and external agents. The examples of *nullah plugs* in India and other conservation techniques used by people to create new fields are a product of an initial local idea modified by agricultural engineers to sharpen their technical efficiency (Premkumar, 1994).

'Knowledge as Empowerment' in effect superimposes a social agenda on natural resources management, especially those resources that are communally held. Many external agencies have strong aid priorities such as poverty focus, poorest-of-the-poor, positive gender discrimination, or marginal people. The means of delivering those priorities has to include employing LK as the baseline for assistance. It will usually involve help in creating local institutions for resource management. Forest User Groups in Nepal and Community-Based Resource Management Groups in southern Africa (see Hasler, 1996, for a review of Zimbabwe's CAMPFIRE) are examples of what is increasingly, in rhetoric at least, being promoted on the grounds of social sustainability, equity and better long-term prospects of respect for environmental resources.

3 Diversity within local society

Typically, professionals perceive local society as one homogeneous group of shared interests, and common economic activities, cultural backgrounds and ways of managing natural resources. Societies are far from a united harmony. Divisions as deep as those that divide many institutions permeate local society. Conflicts are common. Inequalities of access to resources and unjust demands for labour are but two of the types of issues with which local society has to contend and to which institutionally-based professionals are often blind. The drought-prone but populous States of north-east Brazil suffer severe environmental degradation at the hands of poor, desperate people with little productive land. Yet, the water storages constructed in response to drought remain unused and controlled by the military and political elite. Access to productive resources is a privilege to gain prestige, and some sectors of society are not deemed powerful enough to worry about. In enhancing cooperation between local people and institutions, professionals need to be constantly alert to these diversities in their client populations and the reasons behind the differences. It is an intensely political arena, and uncomfortable for those with only a precarious base in their own institutions.

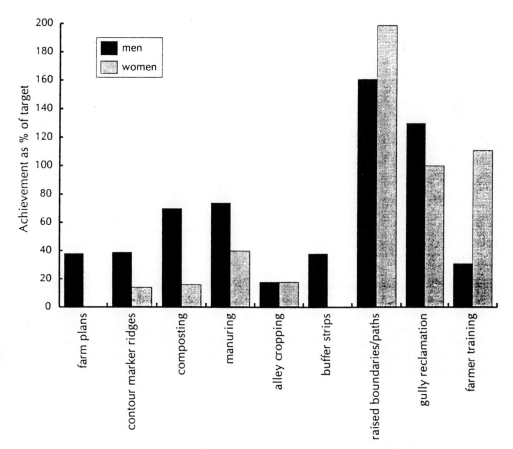

Fig. 1: Differences in adoption rate of soil and water conservation between men and women
(Source: Jackson, 1995)

Take, for example, how environmental degradation embodies gender relations. Jackson (1995) argues that soil erosion and the potential success of control programmes are better understood through the 'lens of gender concepts'. The poor are often blamed for degradation, and poverty is commonly represented as driving rural people to exploit the soil unsustainably - but which poor? Figure 1, taken from Jackson's analysis, shows the experience of one development project in which women's adoption of conservation technologies was generally lower than men's. Further division along ethnic, tribal, socioeconomic and down to family or individual lines would undoubtedly yield even greater differentiation. Taking differences down to the ultimate level is impractical. However, at the broad two-group level of the interests of men and women, a gender analysis offers frameworks which can help professionals anticipate major constraints and opportunities in the quest for encouraging development of sustainable land use. Such an analysis assists with appreciating the divisions of labour: stereotypically, African women devote more of their labour to farm activities than men, and therefore any additional burden of labour posed by a conservation measure is unlikely to be met. Similarly, it is useful to pose the question as to who suffers the cost of erosion and who gains from conservation. Yield increases could accrue either to the man or the woman of the household, representing either an encouragement or a disincentive to the excluded partner. Similar differences between young and old, such as who constructs an earth bund, also control personal attitudes within households to the management of natural resources.

4 Divergent economic perspectives

Farmers' economic rationality is now being seen as a useful indicator of the suitability of new technologies. Implicit evidence that farmers make rational land use decisions comes from the coping mechanisms against severe natural hazards employed by highland dwellers in Papua New Guinea (Sillitoe, 1993). Participation in conservation programmes usually only occurs if there is private economic benefit and personal profitability for those engaged. The perspective of society may be to protect downstream water sources. If projects ignore the farmers' perspectives, then we have a recipe for failure.

The experience in semi-arid Kenya reported by Kiome and Stocking (1995) is instructive. The study sought to identify whether farmer behaviour in natural resource management practices accords with technical efficiency and cost-effectiveness. Research and extension promotes contour tillage, tied ridging and *fanya juu* terraces as the principal means of soil and water conservation. Farmers mainly employ trashlines, a traditional system with no written technical specifications (and hence much variability in application) which gathers crop residues and weeds to form runoff barrier strips. Trashlines are somewhat scorned by the professionals. Along with a control treatment of hand tillage, the authors compared the yield performance on three soils of the four conservation treatments to find that trashlines were the top performer overall. They were most effective in the short (drought) rains. Not only were they good at conserving water but also they accumulated organic matter and led to much improved soil physical and chemical properties. Tied ridging performed marginally better when fertilizers also were used. Labour devoted to construction and maintenance is the main differentiating criterion of these practices.

Therefore, translating these yield increases into a cost-benefit analysis costing labour at the market rate and using a time horizon of 10 years and a discount rate of 15%, only trashlines gave a positive Net Present Value (NPV). For example, the NPV of trashlines on a Ferralsol at Machanga was US$1432, compared with terracing (minus $870), hand tillage (minus $974) and tied ridging (minus $1423). It would be economically irrational for farmers to engage in any conservation practice or, indeed, no conservation practice, except trashlines. These results were supported by a survey of 38 farm households which indicated a 79% adoption rate of trashlines; at least twice as high as any other single measure. Although professional conservation staff were not similarly surveyed, experience suggests that most would promote terraces as the best economic option - they look good; they are impressive to visitors; and they are nice measurable entities for the quarterly report to headquarters. Scruffy banks of dead weeds do not have quite the same *cachet*.

Contractual	Scientists contract with farmers to provide services
Consultative	Scientists consult farmers about their problems, and then develop solutions
Collaborative	Scientists and farmers collaborate as partners in the research process
Collegial	Scientists help strengthen farmers' informal research and development systems

Table 3: Modes of farmer participation in research (Source: Biggs, 1989; Blaikie et al, 1997)

5 Participation

Enhanced cooperation between people and institutions means participation by local people - there is no alternative. There has been a rapid expansion of participatory approaches which involve interactive learning between professionals and farmers. Methods now exist for (1) professionals to understand local people - e.g. ranking exercises; (2) local people to inform outsiders of their needs - e.g. community groups; and (3) local people to analyse their own conditions - e.g. mental mapping. However, participation is a variable quantity. Much that masquerades as participation is simply replacing on-station technicians and workers by off-station farmers. Such paid employment of farmers is unlikely to elicit the crucial response as to whether the techniques being researched are socially, culturally or economically acceptable; the farmer becomes a part of the experiment, rather than an objective evaluator. Such a contractual arrangement may, however, provide more realistic data on labour requirements and other resources.

It is now almost obligatory for international funding of a project to have a far greater level of participation. Table 3 gives a simple typology of modes of participation; Pretty (1995) goes into greater detail how participation may extend all the way to effective empowerment -not quite handing out white coats and eye protectors to peasants, but close!

Negotiated knowledge, a combination of collaborative and collegial participation modes, is probably the most appropriate for natural resource management. It will certainly be best for enhancing future cooperation. Take, for example, the *Projet Agro-Forestier* in Burkina Faso (Atampugre, 1993) which highlights the complex and intricate dialogue between local people and the NGO implementing the project. Many disappointments ensued for the professionals, yet distilled from all experiences is an undeniable technical impact. Local people have accepted, developed and modified *diguettes* as their preferred way of conserving soil and water. From a very different environment, the spontaneous uptake by small farmers of maize-velvetbean (*Mucuna pruriens*) intercrops on acid, nutrient-poor Oxisols in southern Brazil and Paraguay is an example of how a formal research system can sometimes, through participatory analysis, monitoring and evaluation, hit the right note. In this case, the velvetbean not only reduced fertilizer needs and increased maize yields but also created a weed-choking mulch and an ideal seedbed for the following year without cultivation.

6 Conclusion

Enhancing cooperation between people and institutions is a very real challenge, especially in developing countries where the people are poor and the institutions often weak. Barriers to cooperation abound, the most serious being what Chambers (1993) calls 'normal professionalism' or, put another way, we professionals exercise power to defend our specialisms as first priority. Against ingrained prejudice, there is no real solution. With a little innovation, however, the farmers' interest can become our specialism. This paper has been addressed to those ready and willing to make the transition from narrow science to sustainable rural development.

There is a wealth of ways to enhance cooperation, most of which mean changes in thinking, analysis and practice by professionals. Four principal ways have been identified in this paper to promote cooperation: valuing local knowledge; recognising diversity within local society; analysing the different economic perspectives; and encouraging the highest level of local participation. Of course, these ways are interlinked. One word, 'empathy' perhaps best describes the attitude needed by professionals for achieving true and lasting cooperation. Defined variously as "the power of entering another's personality", "ability to participate in another's feelings", or "entering into the spirit of somebody else's views", empathy embodies personal skills as well as technical ways of enhancing cooperation. If it were a simple kit-bag of techniques, it would be easy. The change of attitude is the challenge.

Acknowledgements

This Keynote Paper is based partly on research undertaken for the UK Department for International Development (formerly Overseas Development Administration), Natural Resources Systems Programme, and presented at a Socio-Economic Methodologies Workshop in April 1996 in London (Blaikie et al, 1997). I am grateful to my colleagues Piers Blaikie, Katrina Brown and Lisa Tang at the University of East Anglia for their stimulating ideas on how to research the role of local knowledge.

References

Agrawal, A. (1995): Dismantling the divide between indigenous and scientific knowledge. Development and Change **26**: 413-439.

Arce, A. and Long, N. (1992): The dynamics of knowledge - interfaces between bureaucrats and peasants. In: Long, N. and Long, A. (eds), Battlefields of Knowledge - the Interlocking of Theory and Practice in Social Research and Development. Routledge, London, 211-246.

Atampugre, N. (1993): Behind the Stone Lines. Oxfam, Oxford.

Biggs, S. (1989): Resource-poor farmer participation in research: a synthesis of experiences from nine national agricultural research systems. OFCOR Comparative Study Paper 3, International Service for National Agricultural Research, The Hague

Blaikie, P., Brown, K., Stocking, M., Tang, L., Sillitoe, P. and Dixon, P. (1997): Knowledge in action: local knowledge as a development resource and barriers to its incorporation in natural resource research and development. Agricultural Systems **55**: 217-237.

Brookfield, H. and Padoch, C. (1994): Appreciating agrodiversity - a look at the dynamism and diversity of indigenous farming practices. Environment **36(5)**: 6-11, 37-45.

Chambers, R. (1993): Challenging the Professions: Frontiers for Rural Development. Intermediate Technology Publications, London, 254 pp.

Douglas, M. (1994): Sustainable Use of Agricultural Soils: A Review of the Prerequisites for Success or Failure. Development and Environment Reports 11, Institute of Geography, University of Berne, Switzerland, 162 pp.

Hasler, R. (1996): Agriculture, Foraging and Wildlife Resource Use in Africa. Kegan Paul, London, 208 pp.

Jackson, C. (1995): Environmental reproduction and gender in the Third World. In: Morse, S. and Stocking, M. (eds.) People and Environment. UCL Press, London, 109-130.

Kiome, R.M. and Stocking, M. (1995): Rationality of farmer perception of soil erosion. The effectiveness of soil conservation in semi-arid Kenya. Global Environmental Change **5(4):** 281-295.

Meehan, P. (1980): Science, ethnoscience and agricultural knowledge utilization. In: Brokensha, D.W., Warren, D.M. and Werner, O. (eds), Indigenous Knowledge Systems and Development. University Press of America, 383-391

Ostberg, W. (1991): Land is Coming Up. Burrungee Thoughts on Soil Erosion and Soil Formation. Environment and Development Studies Unit, School of Geography, University of Stockholm.

Premkumar, P.D. (1994): Farmers Are Engineers - Indigenous Soil and Water Conservation Practices in a Participatory Watershed Development Programme. Pidow-Myarda, Kamalapur, India, 40 pp.

Pretty, J.N. (1995): Regenerating Agriculture: Policies and Practice for Sustainability and Self-Reliance. Earthscan, London, 320 pp.

Scoones, I. and Thompson, J. (eds.) (1994): Beyond Farmer First: Rural People's Knowledge, Agricultural Research and Extension Practice. Intermediate Technology Publications, London, 301 pp.

Sillitoe, P. (1993): Losing ground? Soil loss and erosion in the Highlands of Papua New Guinea. Land Degradation and Rehabilitation **4:** 143-166.

Stocking, M. (1993): Soil and water conservation for resource-poor farmers: designing acceptable technologies for rainfed conditions in India. In: Baum, E., Wolff, P and Zöbisch, M. (eds) Acceptance of Soil and Water Conservation: Strategies and Technologies. Topics in Applied Resource Management, Volume 3, German Institute for Tropical and Subtropical Agriculture (DITSL), Witzenhausen, 291-305.

Address of author:
Michael Stocking
School of Development Studies
University of East Anglia
Norwich NR4 7TJ, U.K.

Soil and Water Conservation Strategies at Regional, Sub-regional and National Levels

J.-C. Griesbach & D. Sanders

Summary

Soil conservation programmes have concentrated for too long on the physical problem of soil erosion which could not succeed because soil erosion is only a symptom of bad land management. The problem can only be overcome by a process of addressing the underlying causes which can be administrative, social, economic and political, as well as physical. To tackle these issues, appropriate strategies are needed at the regional, sub-regional and national levels.

Recently, FAO has been searching for new strategies and approaches, including the active participation of rural communities in conservation programmes. Communication and resolving conflicts are important tools.

FAO produced strategies for Africa, Asia and Pacific and Latin America and the Caribbean. All contain guidelines for the sub-regional and national levels and how local programmes can be developed with a new approach. This provides for the participation of rural people in conservation through the benefits that conservation can bring directly to them.

Numerous examples from different regions are now available to show us what can be done. This paper reports on progress and the way in which some soil conservation programmes are being successfully executed.

Keywords: Conservation policy, conservation strategy, participatory problem-solving, project design, institution building

1 Introduction

The problem facing FAO today is how to provide for the needs of a rapidly growing world population from a shrinking land base.

The world population now exceeds 5 billion. Although the rate at which population is increasing has slowed down since 1980, the increase in actual numbers is currently higher than at any time in history. Additions will average 97 million per year until the end of the century and 90 million per year until AD 2025. Significantly, ninety-five percent of this increase is expected to take place in developing countries (FAO, 1996a).

Although the last hundred years have seen great advances in the technology of production, such as the development of more productive crop varieties and the extension of irrigation and fertilizer use, it is becoming more difficult for technological progress to keep up with the rising demands of population growth.

Besides this, there is a finite limit to the supply of land. FAO estimates that a total area of approximately 2.5 billion ha of land in the developing world have some potential for agriculture, although two-thirds of this land is rated as having significant constraints due to topography or soil conditions. Not all this land is available for agricultural production as an estimated 94 million ha is occupied by human settlements and infrastructure, 1.7 billion is classified as forest and 0.4 billion ha as protected areas. In addition to this, land is not evenly distributed between countries. Based on an assessment of the potential from available land and projected population growth in 117 countries in the developing world, FAO concluded that 64 countries (55 per cent) would not be able to support their populations from land resources alone by the year 2000 if they continued to use low input systems.

Competition for the different uses of land is becoming acute and related conflicts are becoming more frequent and more complex. This competition is most apparent on the peri-urban fringes where the continuing pressure of city expansion, both for residential and industrial plots, competes with agricultural enterprises and with recreational demands for parks, nature reserves and golf courses.

Serious as these problems are, they are compounded by another factor - land degradation. With the excessive pressure being placed on the world's land resources, land degradation is becoming more and more widespread and serious. Huge areas of the tropics and temperate regions are now affected by water and wind erosion. Salinity and waterlogging are problems in most large-scale irrigation schemes. Over-grazing and over-cultivation are leading to soil compaction and fertility depletion and an increasingly large area is now being affected by pollution. Land degradation is a complex problem and its actual extent is difficult to assess accurately (Sanders,1994), however Table 1 gives an indication of its global extent.

Region	Water Erosion	Wind Erosion	Chemical Degradation	Physical Degradation	Total (m ha)
Africa	170	98	36	17	321
Asia	315	90	41	6	452
South America	77	16	44	1	138
North and Central America	90	37	7	5	139
Europe	93	39	18	8	158
Australasia	3	-	1	2	6
Total	748	280	147	39	1,214
Major causes	%	%	%	%	
Deforestation	43	8	26	2	384
Overgrazing	29	60	6	16	398
Mismanagement of arable land	24	16	58	80	339
Other	4	16	10	2	93
Total	100	100	100	100	1,214

Table 1: Soil degradation by type and cause (moderately to excessively affected) (million ha.) Source: In: Oldeman, L.R., Hakkeling, R.T.A. and Sombroek, W.G. 1991. World Map on the Status of Human Induced Soil Degradation (Revised Edition). ISRIC and UNEP, Nairobi.

Fortunately, the seriousness of land degradation is now appreciated and over the last sixty years large sums of money have been spent on a wide variety of soil conservation programmes (Sanders, 1989). For example, in 1986 it was estimated that the United States Government alone had spent more than $18 billion since the 1930s and was at that time spending about $1 billion per year (Napier, 1986).

There have been good programmes and success stories but, overall, in spite of the expenditure of a considerable amount of money, time and effort, the situation is worsening. If, as Hallsworth claims (1987), the basic principles of soil conservation have been understood for centuries why is land degradation still such a serious problem and why are farmers not making more use of conservation farming practices?

Can it be that our policies, strategies and general approach to soil conservation have been faulty? It is worth reviewing what these have been.

2 A framework for action

Land degradation is not solely a local problem. FAO is therefore in the process of developing and implementing strategies at three different levels: regional, sub-regional and national. This started with the launch of a programme for Africa: "The Conservation and Rehabilitation of African Lands - an International Scheme", or ISCRAL as it has become known, in 1990 (FAO, 1990). The following countries have applied to join the scheme: Botswana, Benin, central African Republic, Gambia, Ghana, Madagascar, Malawi, Morocco, Rwanda Togo and Tunisia.

A similar scheme followed for the Asia and Pacific Region, "The Conservation of Lands in Asia and the Pacific" or CLASP for short (FAO, 1996b). This was approved in 1996 and is now in the early stages of implementation. At present another scheme is being developed along the same lines for the Latin America and Caribbean Region, "Conservación y Rehabilitación de Tierras en América Latina y el Caribe" (CORTALC). A notable feature of these schemes is the use by FAO of panels of local experts. This has ensured that the schemes have been tailored as far as possible to the unique conditions and problems of the regions.

While the regional schemes do differ to some extent to reflect their particular requirements, they all outline national, regional and sub-regional strategies along the same lines and follow the same principles.

2.1 National strategies

The guidelines for national strategies are all based on the following lines of action (Sanders, 1990):

Identifying the underlying causes of land degradation
Degradation results primarily from incorrect land use and bad land management. Farmers and other land users do not deliberately degrade the land from which they have to make their living and feed their families. Therefore, incorrect land use and bad management must result from ignorance, or more likely, from economic, social and political pressures that force farmers to use the land in the way in which they do.

The first step in a strategy, therefore, should be to analyze why undesirable land uses are being practised. There may be many answers. Population pressure may be too high, agricultural pricing policies may be inappropriate, inputs could be unavailable or land tenure systems could be forcing farmers to over-exploit their land. Whatever they may be, without analysis the underlying causes of land degradation may well be overlooked and much time, effort and money spent on dealing with symptoms rather than the problem itself.

The analysis may lead to the conclusion that the problem cannot be overcome until some major constraint affecting the land users is dealt with. This may mean providing some wanted farm input, providing a market outlet or even adjusting the land tenure system. These are subjects which have rarely been considered in past conservation programmes.

For example, an analysis of a community practising shifting cultivation on steep hill sides may reveal that they have no long-term rights to the land that they are farming - circumstances which are common in many countries in the tropics. In these circumstances it would probably be a waste of time to start a programme under which the farmers are expected to invest a considerable amount of their time and energies in terracing and tree planting. On the other hand, if the farmers could first be granted some long-term rights to the land they may well become interested in its long-term protection and improvement.

Even if solutions are not immediately possible, a careful analysis of the situation can provide an understanding of the problems and prevent what so often happens - governments embarking on programmes which cannot work simply because they do not deal with the underlying causes.

Involving the land users

As pointed out earlier, too often farmers see little relevance to their needs in soil conservation programmes.

It is now realized that soil conservation programmes have little chance of success unless there is full participation of the land users in the whole process of identifying what the real problems are, developing solutions, planning what has to be done and then implementing the necessary works. The land users must be involved to the extent that they feel that the programmes belong to them.

Those who frame soil conservation strategies must recognize this fact. If conservation programmes are to be effective every effort must be made to identify, develop and promote practices which not only conserve soil but also provide short-term, tangible benefits to the farmers. Strategies must be adopted which will lead to increased yields, reduced risks or provide some other direct benefit such as taking the drudgery out of farm work, and at the same time controlling land degradation and improving the land.

This can only be done with the full participation of the land users. In recent years a range of different ways have been developed to obtain this participation. These will not be dealt with here but they are now well-documented and include such methods as Participatory Rural Appraisal and Planning, Participatory Technology Development, Agroecosystem Analysis, Beneficiary Assessment, Farming Systems Research and Extension, Diagnosis and Design and many more (Cornwall et. al, 1994).

Selecting the right technologies

In the past most soil conservation programmes have been based on introducing practices and measures aimed at slowing down and safely disposing of runoff. Heavy emphasis has been placed on diversion works, graded banks, waterways and the like. All of these are technically sound and there will always be a place for them. These types of works, however, provide little if any immediate returns to the farmer: they take up valuable space and can be costly and time-consuming to maintain. Farmers are therefore reluctant to take up these types of works and they frequently fail through lack of maintenance.

For soil conservation to succeed it must be seen as a means of attaining increased production in a sustainable way, not just as a means of controlling erosion. Soil conservation strategies should therefore concentrate on identifying, developing and promoting practices which are productive as well as erosion control effective.

This can be done in a number of ways. Practices should be promoted which make the best use of water where it falls. If this is done, the chances of healthy plant growth and better yields increases while the effects of drought and crop failure decreases. Soil management practices must be promoted which increase soil organic matter content, prevent the formation of soil crusts and compacted layers and generally improve soil structure and water holding capacity. In practice, this means making more and better use of crop residues, introducing better crop rotations, promoting relay cropping, improving pasture management and so on.

There are a number of excellent examples of how this has been done. One is the promotion of minimum tillage techniques in a number of countries, including the United States, Brazil and Australia. Benefits from minimum tillage include better yields, lower input costs, higher profits and the reduction of erosion to the extent that physical conservation measures, such as contour banks, may no longer be needed.

Obviously, not all soils or rainfall patterns allow all the rain to be used where it falls. Problems of waterlogging and mass movement of soils can occur in some circumstances if excess water is not encouraged to drain away. In many instances, however, lack of soil moisture is a major constraint to plant growth and in most circumstances better use can be made of water where it falls.

In recent years a range of other productive and conservation effective practices and systems have been developed that are attractive to farmers. These include the Sloping Agricultural Land Technology (SALT), which was developed in the Philippines. Here the technology is based on a system of agroforestry and contour cultivation, allied with a number of practices which not only control erosion, but also lead to increased production and farm incomes.

Once a technology has shown itself to be profitable, meeting the immediate needs of the land users and appropriate for the local socio-economic conditions, it is easy to promote and can be taken up over large areas relatively quickly and cheaply.

Developing institutions

With the land users themselves becoming more involved, the future role of governments will be to provide them with the back-up needed to plan and implement their own programmes.

Countries vary according to environmental, cultural, political and economic conditions, thus making it impossible to provide a blueprint which can be universally applied. In fact, there are no universal panaceas which can be applied anywhere without modification. Each country must develop its own soil conservation strategy, tailored to its own unique needs.

Nevertheless, the general principles remain the same for all countries. A fundamental requirement is that governments provide the back-up services that the land users need to overcome their land degradation problems. These include strengthening and rationalizing relevant government institutions, establishing an advisory system and attending to the legislation, training and research needs of the conservation effort.

Appointing an advisory committee

In most countries a high-level advisory committee or commission is needed to advise on the detailed formulation of a conservation strategy, to develop policy, coordinate activities and monitor progress. The committee can be made up of senior government officials and representatives from various areas of the country and from special interest groups, such as farmer associations. It could include representatives from government departments responsible for macroeconomic policy and budget allocation. The committee should be formed initially to develop strategy and policy but could remain in existence to help revise policy where necessary, ensure coordination and monitor progress.

Committees of this type already exist. The Permanent Presidential Commission on Soil Conservation and Afforestation in Kenya, for example, was established in 1981. It is made up of members representing different regions of the country. The Commission's functions include promoting conservation and afforestation, monitoring progress and coordinating the activities of the organizations involved. The Commission is supported by a small secretariat that includes a group of specialists who can advise members on technical issues. This Commission has proved very effective in creating public awareness of conservation issues.

Strengthening government services

Conservation is administered differently in different countries. In some countries, such as Malawi and Lesotho, ministries of agriculture have separate conservation units. Other countries combine conservation with forestry, land use planning or watershed development. In recent years, newly created ministries for the environment have started to take over problems of land degradation in some countries.

All these arrangements have advantages and disadvantages. Each government has to decide which system best suits its own needs. For conservation programmes to succeed, however, each country needs to have one clearly defined ministry, department or unit with the overall responsibility for conservation and with the authority to coordinate.

Continuity is important. Starting with a small number of staff, limited facilities and a modest budget - which can be maintained or slowly built up over the years - is better than embarking on ambitious programmes that cannot be maintained and that may have to be cut back or abandoned in the future.

The need for a sound legal base

Some-large scale soil conservation programmes have been imposed on land users. This was particularly so in some African countries in the colonial period, when harsh penalties could be applied and farmers who did not carry out the required works could be fined or jailed. Not surprisingly, conservation became extremely unpopular in these countries. In East Africa the conservation laws were transformed into an issue in the lead-up to independence and local politicians encouraged farmers to break these laws as a means of expressing opposition to the colonial administration.

Unfortunately, such events have led to the idea that soil conservation laws are counterproductive and best avoided. Actually, legislation can offer governments an important tool in promoting conservation. Most countries need legislation to make conservation work - to establish the necessary government institutions, to legalize their mandate and to ensure that they receive a regular budget.

A thorough review of all relevant legislation is an essential element of national conservation strategy. Particular attention should be paid to harmonizing existing legislation that may thread its way through manifold government departments and ministries.

Manpower and training

Manpower requirements, training needs and facilities have to be reviewed. Three points need emphasising:
- Technicians need training not only in conservation techniques but also in how to involve rural communities in developing their own plans and timetables for conservation.
- Farmer training courses should incorporate conservation as part of good farming; it should not be taught in isolation or as a separate subject.
- Short seminars for administrators should be held to sensitize them to conservation and to stress the important role they have to play in national programmes.

Identifying research needs

More research is needed, particularly research aimed at solving problems identified by farmers themselves, so that conservation services can produce packages of conservation measures that are appropriate to local conditions and that can easily be integrated into local farming systems. Particular attention should be paid to agroforestry, surface ridging, crop rotations, intercropping, mulching and vegetative barriers. In developing countries research is needed on traditional conservation systems as some of these systems might be modified or adapted to meet present day requirements.

Developing conservation programmes

The final stage of the process is to develop programmes. Most countries will need to develop these at three levels:

- At the local level: here the programmes must be tailored to the individual requirements of the communities (village or some lower administrative level) and developed in close collaboration with them.
- At the district or provincial level: here detailed and specific programmes should be developed, based on national policy and the requirements coming out of the local programmes, perhaps in the form of rolling five year plans that can be reviewed and updated annually. Important aspects of these programmes should be identified for specific inputs from different ministries and agencies.
- At the national level: here government policy can be combined with political, social and economic data and the programmes being developed at the district or provincial level to produce national conservation strategies for perhaps 10 to 20 years. The programmes should be published as official government documents and incorporated into national development plans where they can form the framework for subsequent legislation, administrative action and budgeting.

2.2 Regional and sub-regional strategies

Regional and sub-regional programmes provide opportunities for the transfer of experience between countries combatting land degradation. Unfortunately, there used to be little of this type of cooperation in the past but governments are realizing that many of their problems are not confined by national boundaries and can best be handled through sub-regional or regional programmes. There is now willingness for governments to cooperate and this has recently resulted in the formulation of regional and sub-regional plans to combat desertification for Africa, Asia, Latin America and Southern Europe. In addition to this, there are a number of regional and sub-regional projects and programmes which are growing in importance. A good example of these is The Asia Soil Conservation Network for the Humid Tropics (ASOCON). This network was formed with assistance from FAO and UNDP in 1989 and became independent in 1993. ASOCON aims to assist its member countries (includes China, Indonesia, Malaysia, Papua New Guinea, the Philippines, Thailand and Vietnam) through a programme of information exchange, regional workshops, expert consultations and learning activities to enhance the skills and expertise of those responsible for the development and dissemination of soil and water conservation practices to small-scale farmers. The ultimate aim is to help small-scale farmers in southeast Asia use the land that is available to them in a more productive and sustainable way. ASOCON provides a model for what could be followed in other sub-regions.

Regional and sub-regional activities provide three major benefits for participating countries:

Training in soil conservation
The senior staff of many countries have had to be sent outside their region for advanced training in conservation. This is expensive and means that fewer people than are required can be trained. It often means that the training is received in very different environmental and socio-economic conditions and therefore may not be very relevant to the needs of the trainees. The development of adequate training facilities within each region, and possibly sub-region, would overcome this problem. There are a number of institutions working to meet this need, such as ASOCON in the tropical areas of southeast Asia, but these institutions need to be strengthened and others established.

Information exchange
Many countries face similar problems, such as developing viable alternatives to shifting cultivation, and there is obviously much to be gained through the exchange of information and experience between those working on these problems. This can best be organized through the establishment and development of data bases, networks, regular sub-regional and regional meetings and newsletters.

While all of this is taking place at present, it is mostly on an *ad hoc* basis and in an unsystematic way. Not nearly enough is being done and activities need to be institutionalized, with organizations and agencies provided with the mandate and funds to carry out the necessary work. The Regional Soil Conservation Unit in Eastern Africa, which is funded by the Swedish International Development Authority (SIDA), is an excellent example of what can be done in providing the necessary links and assistance on a sub-regional basis (SIDA, 1995). Under this programme, information is shared between the participating countries through networks, regular meetings, courses and workshops. The Regional Soil Conservation Unit presently operates in Eritria, Ethiopia, Kenya, Tanzania, Uganda and Zambia.

Research on land degradation
There is an urgent need to develop soil and water conservation measures that are appropriate to the environmental conditions of the different sub-regions and also to the socio-economic conditions of the land users. Research is expensive, requiring well trained staff and good facilities. Fortunately, research results can be shared by establishing simple research networks or by building a research component into existing networks such as the Asia-Pacific Association of Agricultural Research Institutions. Several national institutions are currently participating in global research activities that have been initiated by FAO. These include three studies: the effects of erosion on soil productivity, integrated plant nutrition and sulphur deficiency in soils. These all provide good examples of what can be done.

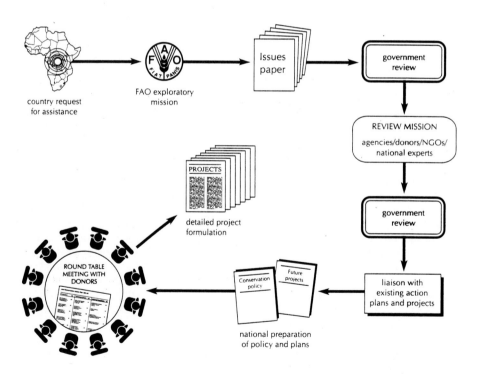

Fig. 1: *Internal scheme for the conservation and rehabilitation of African lands (ISCRAL).*
(FAO 1990)

3 Putting the schemes into action - Malawi

Once these concepts have been accepted, the question arises of how to put them into practice. Figure 1 shows the procedure which FAO is recommending in the case of the International Scheme for the Conservation and Rehabilitation of African Lands (ISCRAL). The most important part of this process, and the most difficult, is the development of the national conservation policy as it requires consultations with a large number of people and organizations. Countries are doing this in different ways but the procedure followed in Malawi can be taken as a good example to follow (Allinson, 1996).

The Government of Malawi started work on developing a national programme in November 1994. From the start it was decided that the national soil conservation policy would reflect the views of a broad spectrum of society. To achieve this, some seventy people drawn from a variety of backgrounds, occupations, income levels and organizations were involved over the following 15 months. The policy was designed so that it would fit in with other policies now being developed on such subjects as irrigation and housing. Combined together, these will make up a broad National Environment Policy.

A process called "Open Space Technology" was use. This started with a three day workshop attended by 40 people. They were first taken on a field inspection of land under a range of uses and management practices. During this inspection, a facilitator used a focused discussion technique so that by the end of the day most people had started to list issues that they or their organizations considered important enough to be included in a policy on land use.

The following two days were devoted to the creation of the agenda. This involved the individual participants briefly stating to the whole group the issues they felt were most important for policy and writing these down for all to see. Flexible groups then worked through the agenda, with the output at the end being 25 issues and option papers. All those present committed themselves to writing a first draft of policy on more than two of the issues in which they were most interested.

These first drafts were prepared over the next few months, collected and combined by an Executive Review Team, most of whom had taken part in the workshop.

A two day seminar was then convened. This was first addressed by the Minister of the Environment and the Principal Secretary for Agriculture & Livestock Development who stressed the importance of what was being done and raised points that should be considered.

The participants were then set to work on the policy. They were encouraged to adopt a visionary approach, that is to envisage what results the policy could have in 20 years' time on Malawi. This was done to assist people to think beyond the constraints of the *status quo* and to maintain a positive approach in developing strategies.

Sub-groups formed voluntarily in the seminar according to the following policy areas:
- land rehabilitation and agriculture;
- land tenure, property rights and legislation;
- public awareness, NGOs and rural communities;
- business, industry, mining and the environment; and
- waste disposal, road mitre drains and physical planning.

The facilitator asked all the groups to build into their policy areas consideration of women and gender issues.

Groups were asked to allow time for individuals to list and prioritize their views and to share these within the sub-group, choosing those that were "concrete" and "practical" and yet "the most exciting" in relation to advancing land use and management. The success of this process varied between groups but generally seemed to help them to think more positively. Sub-groups were asked during the process of amending policy to rank strategies in order of importance. Plenary sessions allowed all participants to share the progress being made and to have ownership of all the subject matter. All participants devoted much energy to their task within the sub-groups and plenary sessions. The improvements made during the process were significant.

On the second day of the seminar, two hours were devoted to starting the development of the Action Plan. The facilitator asked the sub-groups to take their one or two most highly ranked strategies and think of the actions necessary to overcome "blocks" to implementation in order to arrive at systematic and timed actions over a five year period. The output from all groups was considerable. Records from this session were given to the Executive Review Team.

This seminar was followed by a "lock-away" workshop of four days for the Executive Review Team to complete the editing of the policy component of the Action Plan. Here each strategy was taken and changed into a statement for action. Each action statement was listed as an "action programme" and assigned a time slot within the next five years when action should occur. Action which should continue beyond five years was also categorized. The major institutions regarded as responsible for future action were listed.

The resulting plan is "indicative" because, before implementation can take place, representatives from the responsible institutions need to cooperatively determine the detailed action steps necessary to achieve implementation of the action programmes.

The Executive Review Team purposely did not attempt to specify the detailed steps of action because it did not contain representatives of all the organizations that will be involved, nor did it have the necessary expertise. Furthermore, the Team recognized the importance of each responsible institution being fully involved in determining the detailed action, timing, estimation of resources needed and identifying the potential sources for resources, before negotiating for these.

At the time of writing, the Executive Review Team was completing its editing of the document. It then planned to circulate it to all concerned organizations for final comments and corrections. The document will then be placed before the National Committee on Environment (a Sub-committee of Cabinet) for final approval before it is sent out for implementation.

This has been a fairly lengthy procedure but it has allowed the participation of a large number of people and organizations. Detailed discussions have taken place and the resulting programme has been well publicized before it has started. Perhaps most importantly, the final programme is one which has been developed by those who are most concerned with the problem of land degradation in Malawi and who now have the feeling that the programme belongs to them. It is a model which could be profitably followed in other countries.

4 Conclusion

Despite the vast amount of effort, time and money that have gone into soil conservation over the last sixty years, land degradation is increasing. Past policies, strategies and programmes have not been adequate. Soil conservation will work effectively only when the land users themselves can see some immediate benefits from taking up the programmes offered to them. To treat erosion control as an end in itself is not enough. Soil conservation must be seen as a way of achieving optimum, sustainable production. Strategies for its implementation must be based on an approach which creates conditions that encourage better land use. For this to happen, the role of government must change: the problems of why land users are forced to misuse the land will have to be identified and solved and emphasis will have to be placed on conservation measures that can provide tangible, short-term benefits to the land users as well as keeping the soil in place. Strategies based on this approach offer a step forward in soil conservation. FAO is available to provide help and advice when requested.

References

Allinson, J.F. (1996): Development of a National Land Use Policy & Action Plan for Malawi by Malawians. End of Mission Report. Project Ref. No. DP/70 MLW/94/02T. FAO, Rome.

Cornwall, A., Guijt, I. & Welbourn, A. (1994): Acknowledging process: methological challenges for agricultural research and extension in Beyond Farmer First, Edited by I. Scones and J. Thompson. Intermediate Technology Publications.

Hallsworth, E.G. (1987): Anatomy, Physiology and Psychology of Erosion. A Wiley Interscience Publication. London. John Wiley and Sons.

FAO (1990): The Conservation and Rehabilitation of African Lands -an International Scheme. The Food and Agriculture Organization of the United Nations, ARC/90/4, Rome.

FAO (1996a): (In preparation). Negotiating a Sustainable Future for the Land: An Integrated Approach to the Planning and Management of Land Resources. Food and Agriculture Organization of the United Nations. Rome. DRAFT.

FAO (1996b): The Conservation of Lands in Asia and the Pacific. The Food and Agriculture Organization of the United Nations, Rome.

Napier, T.L. (1986): Socio-Economic Factors Influencing the Adoption of Soil Erosion Control Practices in the United States. In: Morgan, R.P.C. and R.J. Rickson (eds.). Agriculture - Erosion assessment and modelling. Office for official publications of the European Communities. L-2985 Luxembourg. 1988.

Sanders, D.W. (1989): Soil Conservation:Strategies and Policies. In: Tato, K. and H. Hurni (eds.) Soil Conservation for Survival. World Association of Soil and Water Conservation. Ankeny,Iowa, USA.

Sanders, D.W. (1990): New Strategies for Soil Conservation. Journal of Soil and Water Conservation, September-October 1990 Vol. **45**, No.5.

Sanders, D.W. (1994): A Global Perspective on Soil Degradation and its Socio-Economic Impact: the Problem of Assessment. A paper presented to the 8th International Soil Conservation Conference, New Delhi, India.

SIDA (1995): Annual Report, 1994/1995. Regional Soil Conservation Unit, RSCU/SIDA. Signal Press Ltd. Nairobi, Kenya.

Addresses of authors:
Jean-Claude Griesbach
Food and Agriculture Organization of the United Nations (FAO), AGLS
Via delle Terme di Caracalla
00100 Rome, Italy
David Sanders
Flat No. 1
Queen Quay
Welsh Back Bristol BS1 4SL, UK

Economic Valuation of Soil Erosion and Conservation - A Case Study of Perawella, Sri Lanka

R. Clark, H. Manthrithilake, R. White & M. Stocking

Summary

There is no single established method for valuing the on-site effects of soil erosion. Two different approaches, the resource value and production value, are employed in this study to value soil erosion and to undertake investment appraisal of soil conservation technologies in a village, Perawella, in the Hill Country of Sri Lanka.

The production value of soil erosion is the market price of the loss in crop yields that results from erosion; it represents the cost of erosion to the farmer. The resource value costs the impact of erosion on the physical attributes of the soil, irrespective of their role in current productivity. It is calculated in terms of the cumulative depletion of soil nutrient content, valued at the market prices of artificial fertilisers. For Perawella, the value of erosion calculated using both approaches represents a major cost to farmers relative to the returns to crop production.

Investment appraisal using the production and resource value is carried out for the main soil conservation technologies employed by smallholder farmers in Perawella. The results indicate that in most cases soil conservation measures are not economically viable. The farmers of Perawella have a positive attitude towards soil conservation but are unable to translate this into action. This is because the conservation technologies traditionally employed in the area are too labour intensive for them to adopt.

Keywords: Soil erosion, soil conservation, economic valuation, investment appraisal

1 Introduction

The economic impact of soil erosion and viability of soil conservation technologies are assessed through the valuation of soil erosion. Low adoption rates of conservation technologies suggest that some technologies may be unprofitable. However, there is no single established method for valuing the on-site effects of soil erosion. Two different approaches, the production value and resource value, are used to estimate the cost of soil erosion and the economic viability of soil conservation technologies for a case study in Sri Lanka.

Perawella is a remote village in the Hill Country of Sri Lanka. It is dependent on smallholder upland annual crop production, with paddy land in the valley bottom and vegetable cultivation on hill land with slopes over 100%, up to an altitude of 1500 m. Natural population expansion has increased pressure on the land and caused cultivation to expand on to the steepest hillsides. Hill land is used to cultivate potatoes during the main "maha" monsoon (November to January), a crop subject to import restrictions and so highly profitable. The small proportion of hill land with access to irrigation is cultivated for a second much less profitable vegetable crop (beans, cabbages, kholrabi) in the "yala" monsoon (May to July); hill land without irrigation is left fallow during this period. The soils in the area are red yellow podzols which are characteristically infertile, acidic and weak in structure. Average annual rainfall is 2,000 to 2,500 mm and in the absence of field data, erosion rates are estimated to range from 24 to 32 tonnes/acre/year based on expert opinion (acres are used as this is the standard unit of land area in Sri Lanka).

Land holdings range from 0.5 to 5 acres and the majority of farmers do not use credit to fund crop production; those that do are reliant on private money lenders who charge interest at 20% per month. A discount rate of 10% is used in this study, derived from the cost of capital in Sri Lanka (the average bank lending rate) which is 20% per annum, adjusted for inflation (10% per annum; Central Bank of Sri Lanka (1993,1994)), to give a real interest rate or cost of capital of 10%. This discount rate represents the most favourable circumstances for investment; at higher discount rates conservation technologies are even less economically viable. The analysis in this study is carried out from the point of view of the farmer (using financial data) and extends over a period of 20 years.

2 The production value of soil erosion

The production value represents the loss in crop yields that results from erosion and is the cost of erosion experienced by the farmer. It assumes that all other factors that influence yields remain constant (climate, pests, diseases, fertiliser rates, crop management, technology) and that farmers do not alter their cropping system in response to soil erosion. The relationship between soil erosion and decline in crop yields can be approximated by a negative exponential function (Stocking & Peake, 1986) in which crop yield decreases each year by a specified percentage of the previous year's yield (Figure 1). This incorporates the large declines in yield that usually occur as topsoil is lost and the smaller declines that occur as the lower zones of the soil profile become exposed. Yield (Y_n) in year n is represented by:

$$Y_n = Y_0 (1-r)^n$$

where Y_0 is yield in year 0 and r is the rate of yield decline. In the absence of erosion it is assumed that yields remain constant at Y_0. In any one year the loss in yield due to erosion (L_n) is the difference in yields between a crop grown in the absence of erosion and the same crop grown with erosion:

$$L_n = Y_0 - Y_0 (1-r)^n$$

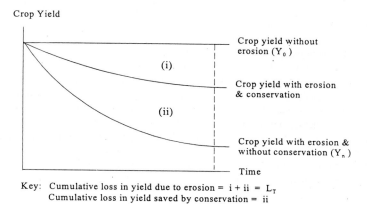

Key: Cumulative loss in yield due to erosion = i + ii = L_T
Cumulative loss in yield saved by conservation = ii

Figure 1: Model of the impacts of erosion and conservation on crop yield

The total cumulative loss in yields (L_T) over the period year 0 to N is the sum of the losses in yield experienced each year during that period (illustrated in Figure 1 as the sum of areas (i) and (ii)):

$$L_T = \sum_{n=0}^{N} Y_0 - Y_0(1-r)^n$$

This cumulative loss in yields is multiplied by the producer price of the crop (P_c) and discounted to take into account the time preference for money to give the production value of soil erosion (J_T) over the period year 0 to N:

$$J_T = \sum_{n=0}^{N} \frac{P_c(Y_0 - Y_0(1-r)^n)}{(1+x)^n}$$

where x is the discount rate. The production value of soil erosion is the discounted value of the cumulative difference in yields between crops grown with erosion and the same crops grown without erosion.

For Perawella the production value of erosion is calculated in terms of one crop of potatoes a year. Two levels of initial yield (Y_0) (2,000 kg/acre and 5,000 kg/acre) are used to reflect the range of crop management in the area and the loss in yields is valued at the producer price for potatoes (Rs 2,060/50 kg). It is assumed that at the erosion rates prevalent in the area (24 to 32 t/acre/year) crop yields decline at an annual rate (r) of 1% of the previous year's yield. The model calculates that over a period of 20 years erosion reduces crop yields by 18%. The total present value of the production value of soil erosion over the 20 year period is Rs 50,079/acre at the lower yield level and Rs 125,197/acre at the upper yield level at a 10% discount rate. The present value of net returns to potato production over the same time period is Rs 485,861/acre at a 10% discount rate (Department of Agriculture, 1993). This indicates that, under the given assumptions, the loss of crop yields due to soil erosion is a significant cost to farmers in Perawella.

3 The resource value of soil erosion

The resource value represents the physical attributes of the soil and places value on retaining them for the future irrespective of their role in current productivity. Soil is a complex resource that comprises growing medium, nutrients, fauna and capacity to store nutrients and water. The resource value recognises all of these attributes irrespective of their current contribution to soil productivity; their role in future uses of soil is unknown. This is based on a similar concept to the valuation of biodiversity of species in tropical rainforests which recognises species with identified uses as well as those for which uses are yet to be found.

Values of most attributes of the soil cannot readily be inferred from marketed goods with the exception of soil nutrients which are traded as artificial fertilisers. As a result, the resource value of soil is calculated as the value of its nutrient content, though omission of its physical and ecological qualities results in an under-estimate. Artificial fertilisers are used here as a marketed proxy for the purposes of valuation (for nutrients in both fixed and available form) and not with a view to nutrient replacement. This distinguishes the resource value from the "replacement cost" of soil erosion; the latter values the depletion of nutrients in terms of the annual marginal cost of their replacement. The resource value is based on the cumulative depletion of soil nutrient content caused by erosion. It is cumulative to account for erosion progressively impairing the quality of the soil which, over time, imposes increasing limitations on the soil's potential uses.

The depletion of soil nutrient content by erosion can be approximated by a negative exponential function, such as that shown in Figure 1. However, the estimated erosion rates of 24 to 32 t/acre/year affect only the topsoil (A horizon) over a 20 year period. The top soil is represented by the initial section of the exponential function, which can be approximated by a straight line. As a result a linear rate of soil nutrient decline can be used for the analysis (Figure 2).

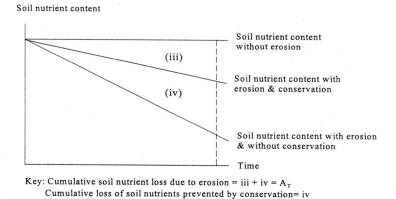

Key: Cumulative soil nutrient loss due to erosion = iii + iv = A_T
Cumulative loss of soil nutrients prevented by conservation = iv

Figure 2: Model of the impacts of erosion and conservation on soil nutrient content

Calculation of the resource value for Perawella is limited to the major soil nutrients nitrogen, potassium, phosphorous and magnesium. Soil nutrient content data for the A horizon are used from several studies carried out in the surrounding area (0.11% nitrogen, 50 ppm phosphorous, 0.59 meq/100g potassium, 0.47 meq/100g magnesium) (Rathnayake, 1994; Rezania et al, 1989; 1992). In the absence of erosion it is assumed that there is no depletion of the soil's nutrient content. The selective removal of nutrients by erosion is accounted for by the sediment enrichment ratio, estimated at 2.0 based on field data from Zimbabwe (Stocking, 1988) and Thailand (Turkelboom, 1991). For each nutrient (N, P, K, Mg) the amount eroded is converted into an equivalent quantity of the form in which it is available as artificial fertiliser (nitrate, phosphate, potash and magnesium oxide) and the value is derived from the market price of locally available fertilisers.

In year n, the total loss of soil nutrients through soil erosion (A_n) is given by:

$$A_n = D_{un} + D_{vn} + D_{wn} + ...$$

where D_{un}, D_{vn} and D_{wn} are the total losses of nutrients u, v and w that have occurred by year n;

$$D_{un} = sc_u mn$$

where s is the soil erosion rate, c_u is the soil's content of nutrient u (converted into the artificial fertiliser equivalent) and m is the sediment enrichment ratio. Over the period year 0 to year N, total cumulative depletion of the soil's nutrient content (A_T) is given by:

$$A_T = \sum_{n=0}^{N}(D_{un} + D_{vn} + D_{wn} + ...)$$

which is illustrated in Figure 2 as the sum of areas (iii) and (iv). Total cumulative depletion of each nutrient is multiplied by the market price of its artificial fertiliser equivalent (P_u, P_v and P_w respectively) and discounted (at discount rate x) to give the resource value of soil erosion (B_T) over the period year 0 to N:

$$B_T = \sum_{n=0}^{N} \frac{(P_u D_{un} + P_v D_{vn} + P_w D_{wn} + ...)}{(1+x)^n}$$

The resource value of erosion is the discounted value of the cumulative depletion in soil nutrient content.

Over a period of 20 years the total present value of the resource value of soil erosion in Perawella is Rs 104,335/acre at the lower erosion rate and Rs 139,113/acre at the upper erosion rate calculated at a 10% discount rate. Comparison with the figures calculated in the previous section indicates that the total present value of the resource value (Rs 104,335/acre) is more than twice the production value (Rs 50,079/acre) at the lower erosion rate and yield level respectively; however, the two values converge at the upper erosion rate and yield level (resource value: Rs 139,113/acre; production value: Rs 125,197/acre). The differences in values reflect the underlying concepts and assumptions of the two valuation approaches. The values calculated using both approaches do, however, represent significant costs relative to farmers' returns to agricultural production which confirms observations that the agriculture practised in Perawella cannot continue in its present form in the long term.

4 Soil conservation in Perawella

The conservation technologies employed in Perawella are constructed to a wide range of specifications and are in various states of repair. Three main forms of soil conservation technology are used by the farmers:
(1) contour drains without terracing;
(2) terraces with earth risers, scraped at the start of each cropping season to remove weeds that would shade the crops below and harbour pests;
(3) terraces with stone faced risers. Construction of these is limited to land with stone and requires skilled labourers to construct the walls (paid Rs 200/manday in comparison to Rs 100/manday for unskilled labour). There has been no recent construction of stone faced terraces in the area; farmers suggest this is due to a shortage of stone and the high cost of labour.

Table 1 summarises the labour inputs and costs for the construction and maintenance of the conservation technologies. The labour requirements are high because the land is steep with gradients often over 100%. Costs of construction and maintenance assume the use of hired labour which is what farmers specified they would use. The figures on effectiveness at reducing soil erosion are used later to calculate the benefits of the different conservation technologies.

In Perawella farmers that own the land they cultivate are more likely to conserve soil than those that rent as leases run for only one season at a time. There is no correlation between the slope of land and the conservation measures employed; the terraces that have been constructed are mostly on land that was cleared several decades ago. Farmers feel strongly that contour hedges (e.g. *Gliricidia*) are inappropriate conservation technologies as these would shade their vegetable crops. A few farmers use grass strips to conserve soil on the up-slope edges of contour drains and along the tops of earth terrace risers. Most of these strips have not been maintained and are little more than a series of clumps of grass; farmers were unable to provide data on them and for these reasons grass strips were not examined in the investment appraisal. Adoption of the strips is limited by a shortage of plants of suitable species, the timing of grass strip maintenance and damage to the strips by cows. All conservation technologies employed in the area face damage by grazing cows which are kept for milk production and are allowed to wander freely when the hill land is fallow. The number of cows kept in the village has fallen with the increase in land under cultivation; there is no stall feeding though this is common on surrounding tea estates.

	CONTOUR DRAINS	EARTH-FACED TERRACES	STONE-FACED TERRACES
COSTS			
CONSTRUCTION: Labour inputs (mandays/acre)	80 to 210[b]	560 to 1,400	skilled labour: 560 unskilled labour: 2,240
Total cost of labour (Rs/acre)	Rs 8,000 to 21,000	Rs 56,000 to 140,000	Rs 336,000
MAINTENANCE: Labour inputs (mandays/acre/year)	Clear drains twice a year: 7.5	Scrape risers: 15 Clear drains: 8[c]	Maintain risers: 5 Clear drains: 9
Total cost of labour (Rs/acre/year)	Rs 750	Rs 2,300	Rs 1,400
BENEFITS			
Effectiveness (%) at reducing soil erosion	10 to 40%	50 to 80%	80 to 95%

[a] data collected from farmers in the Perawella area
[b] reflects the additional labour required for digging into hard earth
[c] both types of terrace have drains at the bases of their risers

Table 1: Costs of the conservation technologies employed in Perawella and their effectiveness at reducing erosion[a]

The farmers in Perawella have a positive attitude towards soil conservation and have clear ideas about the technologies that they would like to use on their land. The necessary knowledge and skills are held within the community; the major barrier to the adoption of conservation technologies is the prohibitively high cost of hiring the labour required.

5 Investment appraisal of soil conservation technologies in Perawella

Cost-benefit analysis is carried out for the three main conservation technologies employed in Perawella (contour drains, earth- and stone-faced terraces) using both the production and resource value approaches. All labour is costed at market wage rates and two levels of labour inputs are used for the construction of contour drains and earth-faced terraces to reflect the ranges of inputs specified by farmers. Additional costs and benefits of soil conservation (e.g. reduced crop area, increased soil water capacity) are assumed to have no net impact and all factors other than soil depth are assumed to remain constant. The technologies are taken to be as effective at reducing crop yield loss as they are at preventing soil loss. Construction of conservation technologies is in year 0 and benefits accrue annually from year 1 for different scenarios representing ranges (specified previously) in yield, erosion rate, labour for construction and effectiveness at controlling erosion.

The production value quantifies the benefits of a conservation technology in terms of the difference in yields between a crop grown with conservation and a crop grown without conservation. The same negative exponential model of crop yield decline is used as for the production value of erosion (Figure 1). The total loss of crop yield saved by conservation is represented in Figure 1 by area (ii). The production value of a conservation technology (K_T) is given by the cumulative loss in crop yield that is saved by the technology multiplied by the producer price of the crop (P_c) and discounted over the period of the analysis (year 0 to N):

$$K_T = \sum_{n=0}^{N} \frac{P_c(Y_0(1-(1-e)r)^n - Y_0(1-r)^n)}{(1+x)^n}$$

where e is the effectiveness of the conservation technology at reducing crop yield loss and x is the discount rate. Investment appraisal employing the production value of conservation indicates that virtually all of the scenarios examined for the conservation technologies are uneconomic at a 10% discount rate (Table 2). The only exception is contour drains that are 40% effective, constructed with high labour inputs and at the upper yield level for which a net present value (NPV) of Rs 21,175 is calculated at a 10% discount rate. The other scenarios have benefit/cost ratios that range from 0.11 to 0.83 at a 10% discount rate and for most, internal rates of return cannot be calculated.

Conservation technology	Reduction in yield loss	Labour intensity of construction	Lower yield level Benefit/cost ratio[a]	IRR	Upper yield level Benefit/cost ratio[a]	IRR
CONTOUR DRAINS	10%	low	0.33	*	0.83	7%
	40%	high	0.71	6%	1.77	17%
EARTH TERRACES	50%	low	0.32	*	0.81	7%
	80%	high	0.25	*	0.62	5%
STONE TERRACES	80%	-	0.11	*	0.28	*
	95%	-	0.14	*	0.34	0%

* indicates scenarios for which an internal rate of return (IRR) can not be calculated
[a] calculated at a discount rate of 10%

Table 2: Investment appraisal of conservation technologies using the production value

To be economically viable at a 10% discount rate the scenarios for contour drains need to reduce yield loss by a range of 12 to 116%, earth terraces by 62 to 583% and stone terraces by 256 to 1053%. Figures for reduction in yield loss of over 100% indicate that yields need to increase above those obtained on uneroded land so most of the figures generated are totally unachievable. Threshold analysis indicates that some scenarios are not economically viable at any rate of yield loss: at low rates of yield loss the benefits of conservation are small and sustained over a long period but diminish due to the effects of discounting; at very high rates of yield loss the benefits are initially large but yields rapidly fall to zero and benefits then cease. To be economically viable at a 10% discount rate the scenarios for contour drains require minimum rates of yield loss in the range of 0.5 to 4%, earth terraces 1 to 5% and stone terraces 3 to 12%. All of these results suggest that at much lower discount rates the construction of contour drains and possibly even earth terraces may be economically viable, but at a cost of capital of 10% or more the adoption of conservation technologies in Perawella is economically unviable.

The resource value represents the benefits of a conservation technology in terms of the depletion of soil nutrient content (degradation of soil quality) that it prevents. The same linear model of soil nutrient decline is used as for the resource value of soil erosion (Figure 2). The total loss in soil nutrient content that is saved by conservation is illustrated in Figure 2 by area (iv). The cumulative depletion of each nutrient that is saved by conservation is multiplied by the price of the artificial fertiliser equivalent and discounted to give the resource value of soil conservation (G_T) over the period year 0 to N:

$$G_T = \sum_{n=0}^{N} \frac{i(P_u D_{un} + P_v D_{vn} + P_w D_{wn} + ...)}{(1+x)^n}$$

were i is the effectiveness of the conservation technology at reducing soil loss. Adoption of conservation technologies is slightly more economically attractive using the resource value approach than with the production value (Table 3). However, only two scenarios are found to be economically viable at a 10% discount rate: contour drains constructed with a high intensity of labour, 40% effective at reducing erosion, at both the lower erosion rate (NPV = Rs 14,349) and the upper erosion rate (NPV = Rs 28,260). For the remaining scenarios benefit/cost ratios range from 0.24 to 0.97 at a 10% discount rate and internal rates of return range from 0 to 9% (Tab. 3). To be economically viable at a 10% discount rate contour drains would need to reduce erosion by a range of 11 to 26%, earth terraces by 54 to 153% and stone terraces by 250 to 333%. These figures may be achievable for contour drains but are not for terraces.

Threshold analysis of erosion rates requires the adoption of a different model for the relationship between nutrient loss and soil erosion. The linear model (Figure 2) assumes that erosion remains within the A horizon, but erosion will proceed into lower soil horizons at higher rates of soil loss. For the whole soil profile the relationship between soil nutrient loss and soil loss is estimated by a negative exponential function, as used in the crop yield loss model (Figure 1). The resource value of soil conservation (H_T) over the period year 0 to N is then given by:

$$H_T = \sum_{n=0}^{N} \frac{i(1-z)^n (P_u D_{un} + P_v D_{vn} + P_w D_{wn} + ...)}{(1+x)^n}$$

where z is the rate of decline in soil nutrient content and x is the discount rate. Assuming a 1% rate of decline (z) in the previous year's soil nutrient content, at a 10% discount rate contour drains are economically viable at minimum erosion rates ranging from 17 to 36 t/acre/year, earth terraces at 37 to 49 t/acre/year and stone terraces at 91 to 108 t/acre/year. These results all indicate that at a 10% discount rate contour drains may be economically viable in terms of their resource value, but terraces are not.

Conservation technology	Reduction in yield loss	Labour intensity of construction	Lower erosion rate		Upper erosion rate	
			Benefit/cost ratio[a]	IRR	Benefit/cost ratio[a]	IRR
CONTOUR DRAINS	10%	low	0.73	6%	0.97	10%
	40%	high	1.52	15%	2.03	19%
EARTH TERRACES	50%	low	0.69	6%	0.92	9%
	80%	high	0.52	4%	0.70	6%
STONE TERRACES	80%	-	0.24	*	0.32	0%
	95%	-	0.28	*	0.38	1%

* indicates scenarios for which an internal rate of return (IRR) can not be calculated
[a] calculated at a discount rate of 10%

Table 3: Investment appraisal of conservation technologies using the resource value

6 Conclusions

Different scenarios are used to represent the conditions in Perawella and the ranges in results indicate that the analyses are sensitive to the data and assumptions used. The results provide only an indication of the cost of erosion and the economic viability of conservation technologies. More accurate analysis may not be warranted given the diversity of conditions faced by the farmers and the limited data available on the impacts of erosion.

The analysis indicates that the resource value of soil erosion in Perawella is the same as or up to twice that of the production value for different scenarios and that both represent major costs relative to household returns to agricultural production. Investment appraisal using the production and resource values of soil conservation indicate that some conservation technologies are economically viable at very low discount rates. However, the cost of capital for farmers in Perawella is high (20% per month in monetary terms) which renders the adoption of costly conservation technologies such as terraces unviable. This supports field evidence that there has been no recent construction of terraces. Contour drains, a low cost technology, are more economically viable and are employed by a number of farmers in the area. Information provided by the farmers suggests that there could also be potential to develop the use of contour grass strips both as a low cost measure to conserve soil and to provide fodder for stall fed cows; for reasons discussed earlier grass strips were excluded from the economic analysis.

Since this study was carried out the Mahaweli Authority has initiated a soil conservation project in Perawella based on participatory appraisal of the area. Farmers specified that they are unable to stall feed cows because of the additional labour required to gather fodder. Cows still graze freely in the area and it seems that this was not the major constraint on the use of grass strips. The project has provided farmers with plant stock of *Vetiver* and fodder grass and has successfully encouraged the widespread adoption of grass strips for soil conservation.

Acknowledgements

The research in this paper was carried out for the Overseas Development Administration (ODA)'s Forest/Land Use Mapping Project (FORLUMP) in Kandy, Sri Lanka in 1993. It would not have been possible without the time and support provided by the farmers in Perawella. The models presented in the paper were developed with assistance from Andrew Bingham, University of Cambridge. Current funding is provided by the ODA's Renewable Natural Resource Systems Programme (Hillsides).

References:

Central Bank of Sri Lanka (1993): Economic and social statistics of Sri Lanka.
Central Bank of Sri Lanka (1994): Bulletin - January. Central Bank of Sri Lanka, Colombo.
Department of Agriculture (1993): Cost of cultivation of agricultural crops. Division of Agricultural Economics and Planning, Department of Agriculture, Peradeniya.
Rathnayake (1994): Analysis of soil samples taken near Watabedda village. Unpublished.
Rezania, M., Yogarathnam, V. & Wijewardena, J.D.H. (1989): Fertiliser use on potato in relation to soil productivity in the up country areas of Sri Lanka. Department of Agriculture, Ministry of Agricultural Development & Research & FAO.
Rezania, M., Willekens, A. & Dissanayake, S.T. (1992): Status of pH, organic matter, phosphorous and potassium in soil under cultivation of field crops in Sri Lanka. Technology Transfer Division of the Department of Agriculture & FAO.
Stocking, M. (1988): Quantifying the on-site impact of soil erosion in S. Rimwanich (ed) Land Conservation for Future Generations. Department of Land Development, Bangkok pp 137-161.
Stocking, M. & Peake, L. (1986): Crop yield losses from the erosion of Alfisols. Tropical Agriculture (Trinidad) **63(1),** 41-46
Turkelboom, F. (1991): pers. comm. of results from the Jabo research site, Soil Fertility Conservation Project, Mae Jo University, Chiang Mai, Thailand

Addresses of authors:
Rebecca Clark
Overseas Development Group
University of East Anglia
Norwich NR4 7TJ, UK
Herath Manthrithilake
Environment and Forest Conservation
Division of the Mahaweli Authority of Sri Lanka
Polgolla Dam Site
Kandy, Sri Lanka
Roger White
Environmental Management and Sustained Development In the Upper Mahaweli Catchment (ENDEV)
P.O. Box 109
Kandy, Sri Lanka
Michael Stocking
School of Development Studies
University of East Anglia
Norwich NR4 7TJ, UK

Including Economics in the Sustainability Equation: Upland Soil Conservation in Indonesia

P.C. Huszar

Summary

Soil conservation programs in Indonesia have provided technical and financial assistance to upland farmers in an attempt to simultaneously reduce soil erosion and increase farmer incomes. While these programs have had short-term successes, in the longer run the gains have not been sustainable. The problem appears to be that no provision is made for the market services needed for the inputs and outputs associated with the new agricultural system introduced. The evidence suggests that by focusing on the physical requirements for soil conservation, but failing to develop the economic system necessary to support it, the upland programs are unable to sustain the gains achieved.

Keywords: Upland soil conservation, increased incomes, markets, sustainability

1 Introduction

Indonesia has aggressively invested in upland projects designed to reduce poverty and soil erosion. Their main program for increasing agricultural income and soil conservation is the National Watershed Development Program in Regreening and Reforestation, which channels about US$50 million annually to local governments for watershed activities. Over the last five years, the annual budget for this program has increased six-fold. In addition, donor countries have contributed hundreds of millions of dollars for upland development projects.

The sustainability of the income gains and reduced erosion of these programs, however, is doubtful, in large part due to the neglect of the role of markets. The purpose of this paper is to examine the sustainability of such efforts and to suggest approaches for improving the sustainability of upland development efforts. Since the problem of poverty and soil erosion is most acute on Java, the focus of this paper is on Java's uplands.

2 Background

Most of Java's watersheds are classified as "critical" in the sense that they are subject to actual or potential degradation due to erosion, though the definition of "critical" is not precise. The total area of Java's watersheds classified by the Indonesian government as "critical" is over 12 million hectares, of which over 1.9 million hectares are located in upland areas and are considered to be the most threatened. The population of the "critical" upland watersheds exceeds 10.7 million, though this figure includes upland cities such as Bandung and Malang.

Predicted soil loss from upland agriculture averages nearly 123 tons per hectare per year (t/ha/yr), with West Java having the highest predicted rate of 143 t/ha/yr, Central Java the next highest rate of 131 t/ha/yr and East Java the lowest rate of 76 t/ha/yr. The 1989 estimated annual cost of soil erosion on Java in terms of lost productivity and off-site costs was US$340.6 million to US$ 406.2 million (Magrath and Arens, 1989).

3 Upland conservation programs

The government of Indonesia embarked on a national program of land rehabilitation and soil conservation in the mid 1970's. The main purpose of the program was to increase the productivity of critical uplands and to reduce soil erosion. A series of pilot projects resulted during the 1980's, including Citanduy II, Yogya Bangun Desa, the Uplands Agriculture and Conservation Project (UACP), and the Wonogiri Project. The cost of these projects has ranged from US$ 20 million to US$ 50 million, with newer projects costing progressively more. The most recent project is the Upper Cimunuck Watershed project, with funding from the World Bank, which will cost an estimated US$ 31 million.

The basic approach of the upland soil conservation programs is to establish model farm units and to introduce a package of upland agricultural inputs and conservation practices emphasizing the construction of bench terraces and the use of new cropping patterns, seed varieties and inputs of chemical fertilizers and insecticides, on land with slopes of up to 50 percent. For land with slopes of more than 50 percent, permanent vegetation (e.g., trees) is established. Subsidies, either in cash or in kind, are provided for the construction of bench terraces and the purchase of inputs. Moreover, the projects transport the improved inputs into the project area and assist in the marketing of outputs. The basic goals of the program are to increase farm production and, therefore, incomes, while reducing soil erosion in densely populated upland areas of Java (Huszar and Cochrane, 1990).

4 Sustainability of program

Earlier studies of the program found that incomes were increased and soil erosion reduced. For example, the Citanduy II Project in West Java increased farmer incomes by approximately 25 percent, while reducing soil erosion by about 40 percent (Huszar and Cochrane, 1990). More recent studies, however, have found that while incomes are increased and erosion reduced during the implementation of the projects, these effects are not sustainable under the present approach.

The recent study by Huszar et. al. (1994) of the Uplands Agriculture and Conservation Project (UACP) has raised serious questions regarding the long-term sustainability of the benefits of the current upland conservation program in Indonesia. The UACP is located within the Jratunseluna Watershed in Central Java and the Brantas Watershed in East Java, as shown in Figure 1. UACP's goals of increasing incomes and reducing erosion were similar to those of previous upland conservation projects. These goals were to be achieved by improving farming systems, farm technologies and management (USAID/Indonesia, 1984). Subsidies were used to induce farmers to adopt improved seeds and diversify crops, to employ improved chemical fertilizers and pesticides, and to construct bench terraces and water channels. Funding for the UACP started in 1985 and came from the Government of Indonesia with assistance from the United States Agency for International Development (USAID) and The World Bank.

Figure 2 shows that project farmers one and two years after the project ended had higher gross incomes from dryland production than did the control farmers, but by the third and fourth years after the end of the project their incomes had fallen below those of the control farmers. Control farmers have similar characteristics to project farmers, but are outside of the project and were not affected by the project. These results are statistically significant at the 0.05 level. That is, while project farmers seem to initially have higher levels of dryland production, this increase is not maintained over time. Figure 2 indicates that after subsidies end, project farmers are not able to sustain the positive changes introduced by the project.

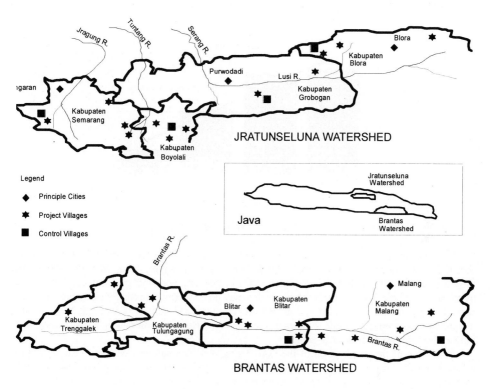

Figure 1: Map of study area

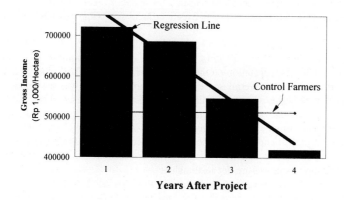

Figure 2: Average gross income

The same pattern of decline also holds for the net value of dryland production, as shown by Figure 3. That is, the surplus of revenues over costs of production initially are higher for project farmers than for farmers without the project, but the net value of production declines rapidly after the project ends, though the differences are only statistically significant at the 0.15 level. Farmers in older project villages tend to earn lower net returns from dryland crops than farmers in more recent project villages.

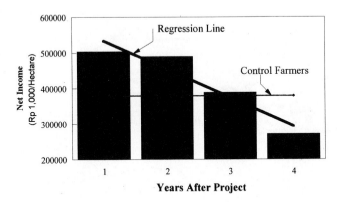

Figure 3: Average net income

The regression lines for gross and net incomes explain 95 percent and 92 percent of the variation in the observed data, respectively. In both cases, it appears that four years after the project ends in a village, farmers who were in the project are worse-off than farmers not affected by the project. One possible explanation for this may be that terracing reduces plantable land area, so that as project farmers return to conventional cropping and cultivation patterns they are at a temporary disadvantage. If this is the case, then it is unlikely that terraces will be maintained and, in fact, they may be intentionally eroded.

While one of the major goals of the UACP was to reduce soil erosion, very little data were collected by the project which measures the effects of the project on soil erosion. Data are only available for two relatively primitive measures: a rating score for terraces, which attempts to reflect the effectiveness of the terraces to reduce soil erosion, and an erosion score called Inderosi, which attempts to measure relative changes in erosion. These data were only gathered for a sample of project farmers, so that comparisons are between farmers receiving subsidies and those no longer being subsidized.

The physical quality of terracing was rated by UACP in terms of following the contour, effective planting area, slope of the planting area, main drainage channels, drop structures, drainage ditches, terrace bunds, vegetation on the terrace riser, pathways, landslides, gully erosion and conservation of topsoil. An overall score ranging from 0 to 100 was assigned to farmer terraces in the project, with higher scores reflecting more effective soil conservation.

Figure 4 shows the average terrace ratings of former project farmers compared with the average scores for project farmers still receiving subsidies. As shown by Figure 4, these scores tend to decline after the project ends, though the regression coefficient is not significant. That is, the quality of terracing appears to decline, indicating that terraces are not being maintained and that these soil conservation measures are not sustainable.

The degree to which terracing reduces soil erosion was also rated by UACP using a rating system developed by UACP called "Inderosi". The Inderosi score is based upon before and after project changes in the vegetative cover during each growing season, the amount and distribution of rainfall, and the quality of terracing. Calculation of the Inderosi score employs a modified version of the Universal Soil Loss Equation (USLE). The Inderosi rating system attempts to provide a relative score for erosion, where higher scores are associated with proportionately greater levels of soil conservation.

As shown by Figure 5 these scores are less for former project farmers than for project farmers still receiving subsidies. The regression line indicates some tendency for improvement in the score over time, but the regression coefficient is not significant. That is, soil erosion seems to be reduced less on farms after the project ends, but there is no clear tendency for the reduction in erosion to decline over time.

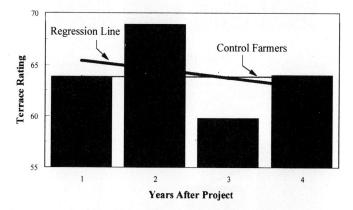

Figure 4: Average terrace rating

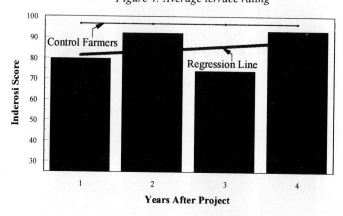

Figure 5: Average Inderosi score

The evidence suggests that maintenance of terraces declines and erosion increases after the projects end. This evidence agrees with informal observations of earlier upland conservation projects on Java, particularly Citanduy II, and with a study conducted in the Kerinci valley of Sumatra (Belsky, 1994). The study in Sumatra found that terraces were rarely maintained for longer than the 3 years in which the farmer received subsidies from the project. Moreover, erosion may be increased due to poorly maintained terraces which channel runoff more than would be the case for unterraced fields.

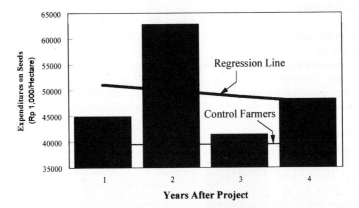

Figure 6: Average seed inputs

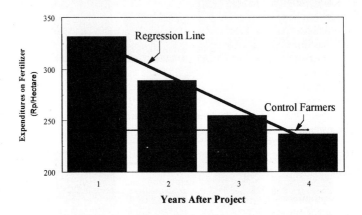

Figure 7: Average fertilizer inputs

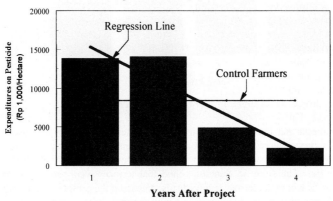

Figure 8: Average pesticide inputs

5 Causes of unsustainability

In order to overcome farmer resistance to terracing, the UACP used subsidies to reduce the initial costs of terracing and the costs of the additional inputs associated with cultivating terraces. The success rate of the program to induce farmers to terrace was high. Moreover, while the project is in operation, farmers in the area, but not receiving subsidies, also seem anxious to bench terrace and to adopt the introduced farming system (Huszar and Cochrane, 1990). The indication is that when conditions are favorable, both subsidized and non-subsidized farmers find bench terracing profitable.

However, maintenance of the terraces after the subsidies end has been a continual problem. Inputs to terraced agriculture are relatively expensive to upland farmers due to poor access to input markets. The farmer must generally travel long distances to the lowlands to purchase the improved seeds, fertilizer and pesticides needed for the cultivation practices introduced by terracing. The time and effort to obtain inputs represent a significant opportunity cost to the farmer. Without these inputs, however, the productivity of the terraces declines and, therefore, the incentive to maintain the terraces. During the project phase when the terraces are being constructed and subsidies are received, the project provides the needed inputs of seeds, fertilizer, pesticides locally. But after the project, these inputs are no longer available locally and their real cost increases. Evidence of this effect can be found in the UACP.

Figures 6, 7 and 8 show that the use of the improved seed, fertilizer and pesticide inputs introduced by the UACP declined after the project ended. While the average value of seed inputs is greater for project farmers than for the control farmers up to four years after the project ended, the tendency is one of decreased use, as can be seen from Figure 6. Fertilizer use of project farmers decreased to approximately the level of the control farmers by the fourth year after the project ended, as shown by Figure 7. Figure 8 indicates that pesticide use of project farmers fell below the use by control farmers. The more dramatic declines in the use of fertilizers and pesticides are likely related to their weight and, therefore, their transportation cost relative to seeds.

6 Conclusions and recommendations

The upland soil conservation program in Indonesia seems to be trapped in a cycle of subsidizing bench terraces and other physical agricultural improvements for short periods after which the improvements rapidly deteriorate for lack of maintenance. The question is why don't the farmers maintain these capital improvements? The answer seems to be that the introduced farming systems have higher costs and, therefore, lower net incomes over time compared with the traditional systems. If this is the case, then what can the government do to improve the long-term economic feasibility of the introduced upland farming system?

A simple solution appears to be for projects to not only provide technical and financial assistance for terracing and improved cropping patterns, but they should also provide for the establishment of local markets for the inputs necessary to make these changes profitable. In addition to subsidizing farmers, projects may also need to subsidize local businessmen to develop enterprises which supply the inputs. This may entail construction of storage facilities, the purchase of vehicles for transporting the inputs from distant distributors, and even the improvement of transportation routes.

While data are not available, it also appears that projects provide valuable marketing services for the outputs from the introduced cropping systems. As is the case on the input side, when the projects end, farmers lack local markets for their new produce and existing markets are difficult and expensive to access. If projects introduce new, higher valued crops, it seems that they must also develop the marketing system necessary for farmers to sell these crops if the change is to be sustainable.

Perhaps the most important lesson to be learned is that improved income and soil conserving measures do not exist in a vacuum. The economic system within which they operate affect and are affected by these measures. The sustainability of upland soil conservation projects is not simply a function of the technical characteristics of the measures introduced, but also depends upon the support of the economic system within which they operate.

Acknowledgments

The helpful comments of Sapta P. Ginting of the Ministry of Home Affairs and Hadi S. Pasaribu of the Ministry of Forestry, both of the Government of Indonesia, are gratefully acknowledged, though the opinions expressed in this paper are solely those of the author.

References

Belsky, Jill M. (1994): Soil Conservation and Poverty: Lessons from Upland Indonesia. Society and Natural Resources **7**, 429-443.

Huszar, Paul C. and Harold C. Cochrane (1990): Subsidisation of Upland Conservation in West Java: The Citanduy II Project. Bulletin of Indonesian Economic Studies **26(2)**, August, 121-132.

Huszar, Paul C., Hadi S. Pasaribu and Sapta Putra Ginting (1994): The Sustainability of Indonesia's Upland Conservation Projects. Bulletin of Indonesian Economic Studies **30(1)**, April, 105-122.

Magrath, William and Arens, Peter (1989): The Costs of Soil Erosion on Java: A Natural Resource Accounting Approach. Environmental Department Working Paper No. 18. Washington, D.C.: The World Bank. August.

USAID/Indonesia (1984): Upland Agriculture and Conservation Project. Project Paper, Jakarta.

Address of author:
Paul C. Huszar
Department of Agricultural and Resource Economics
Colorado State University
Fort Collins, CO 80523, USA

Economic Evaluation of
Watershed Management Programmes in India

Ram Babu & B.L. Dhyani

Summary

Soil erosion and land degradation pose major environmental and economic problems that obstruct the sustainable development of the Indian economy. Soil and water conservation programmes on a watershed basis have been implemented on large scale in the recent past for eco-friendly sustainable development. The paper highlights the productivity, employment, protection impact of 32 Watershed Management Projects representing 10 major agro-ecological regions distributed over 14 states of India. Economic evaluation studies of Watershed Management Programmes (WSM) revealed that such projects are economically sound, socially acceptable and environmentally desirable.

Keywords: watershed management, soil and water conservation, employment generation, economic evaluation, risk reduction

1 Introduction

Depletion of soil and water resources continues to be a major hazard in India. About 52 percent (173.65 m ha) of the total geographical area of India is subject to varying forms of soil erosion and yields soil loss to the tune of 5333 m t/yr (16.35 t/ha/yr). Declining productivity, under-nourishment and under-employment are direct consequences of our poor land management system. High volume of runoff, soil loss, sedimentation rates and increasing loss through natural calamities (floods, droughts, mass wasting, nutrient losses, etc.) are the indirect effects of irrational utilization of natural resources. Monetary value of indirect losses of reservoir sedimentation and nutrient losses were estimated to be 10^6 million and 4,800 million rupees (US $ 28 billion and US $ 140 million) respectively (Das, 1994).

1.1 Indian scenario

Research was intensified by establishing Regional Soil Conservation Research, Demonstration and Training Centres during the first Five Year Plan (1951-55) to conserve soil and water resources. Technology packages developed by the centres were demonstrated on farmer's fields as well as on common lands under various soil conservation and watershed management programmes. Up to 1993-94, India has spent Rs. 35915 m (US $ 1000 m) to treat 37.34 m ha. Till the early 1980s, the performances of WSM programmes were below expectation due to the fact that they were single-targeted, top-down administered and insufficiently coordinated (Dhyani and Singh, 1991).

The success of three model watersheds demonstrated by the Central Soil and Water Conservation Research and Training Institute, Dehradun during the 1970's has opened new vistas of development (Agnihotri et al., 1989; Dhyani et al., 1993 and Ram Babu et al., 1994). Consequently, WSM has become synonymous with a new developmental approach in rural India. Since then, various rural development and soil conservation programmes are under progress on a watershed basis; these include various types such as River Valley Projects, Flood Prone River Projects, Drought Prone Area Projects or other Watershed Management Programmes. During the VIIIth Plan, soil conservation and rural development programmes under various projects are being implemented on a watershed basis with an outlay of Rs. 28,000 m (US $ 800 m) to treat about 6.6 m ha area. The analysis of 32 WSM projects is presented here.

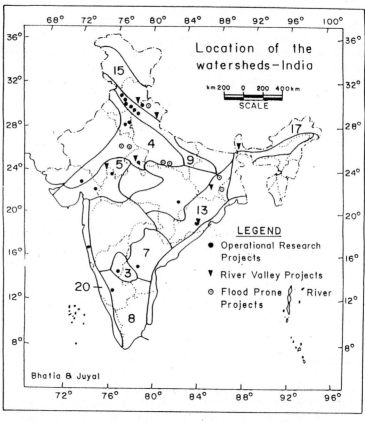

Zone	Agro-ecological zone	Zone	Agro-ecological zone
3.	Decan Plateau (Hot Arid)	9.	North Plain (Shiwalik)
4.	North Plain & Central High Land	13.	Eastern Plateau (Chhota Nagpur)
5.	Central High Land	15.	Western Himalaya
7.	Decan Plateau (Telangana)	17.	Eastern Himalaya
8	Eastern Ghats	20.	Western Ghats

Figure 1: Location of watersheds - India

2 Description of watersheds and their problems

The watersheds selected for the study include 18 Operational Research Projects (ORP), 7 River Valley Projects (RVP) and 7 Flood Prone River Projects (FPR), located across 10 agro-ecological zones (arid to humid) and 14 States of the country (Fig. 1). These watersheds vary in size from 90 ha (Nada) to 1.7 m ha (Upper Damodar Valley), have varied annual rainfall (525 to 3000 mm) and are located at elevations from 120 to 3000 m above msl. Ten soil groups represented in the study watersheds are alluvial, black cotton, red, lateritic, red yellow, silty loam, red brown, loam, black red and loamy sand. The problems faced of land degradation are unique for each region. Denudation and mass wasting in the Himalayan region; denudation, flash floods, high sedimentation rates and droughts in Shiwalik; sheet, rill, gully and ravines in the Northern Plain and Central High lands; sheet, rill and drought in the Malwa region; flood, rill and ravines in the Chota Nagpur plateau and shifting cultivation in Eastern Himalaya and part of Orissa, are the major types of land degradation problems. In each programme a nodal agency was identified and made responsible for overall planning, coordination, monitoring etc. Programmes were implemented by State Line Departments with the technical support of the Indian Council of Agricultural Research and State Agricultural Universities.

3 Watershed management plan and activities

A unique comprehensive WSM plan comprising of foundation structure and production system was developed for each watershed. The plan and activities were made compatible to physiography, hydrology, soil, land capability, vegetation, irrigability and socio-economic conditions of the region.

3.1 Foundation structures

These include small dams, tanks, water distribution system, spillways, gully plugs, check dams, silt detention basins, trenching, embankment, terracing, levelling, bunding and dug out ponds. They were constructed on the basis of plans ensuring technical feasibility and economic viability. About 60 to 80 percent of the total expenditure of watershed was utilized under this sector.

3.2 Super structure of production system

Efficient use of conserved resources was made by putting the land under most suitable productive use. For improved farm technology extension, large number of demonstrations were conducted on farmers' fields to demonstrate the efficiency and efficacy of available technologies for sustained development. These demonstrations were held in a participatory mode to assure acceptance and make them "a people's programme".

4 Watershed responses

Production with protection on a sustained basis and generation of gainful employment within the watershed constitute two of the multiple objectives of the watershed management programmes. Responses of selected watersheds on production, protection, employment generation and economics are discussed below.

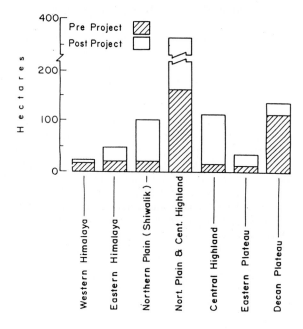

Figure 2: Impact of watershed management programme on irrigation

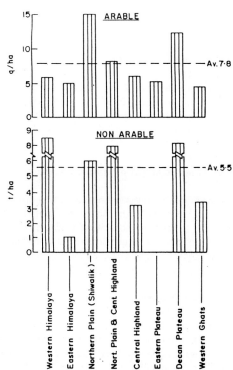

Figure 3: Increase in productivity in arable and non arable land

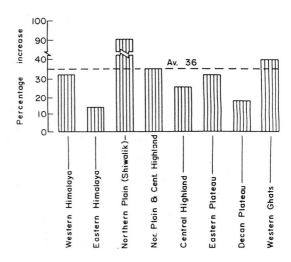

Figure 4: Percentage increase in cropping intensity

4.1 Productivity and production

Implementation of watershed management programmes enhanced irrigation potential by 40 to 333 percent with a maximum in the Northern Plain and Central Highland, improved in-situ moisture conservation and increased cropping intensity by 36 percent (maximum 90% in Northern Plain). This has increased the productivity of arable land by 4.2 to 15.4 g/ha (average - 0.8 g/ha) and of non-arable land by 1.0 to 8.5 t/ha with an overall average of 5.5 t/ha in different agroclimatic regions (Figs. 2 to 4). Milk production has also increased by 40 to 350 thousand l/yr due to substitution of low yielding local animals with high yielding animal breeds and availability of good quality fodder.

4.2 Employment generation

WSM programmes may yield productive and protective benefits in perpetuity if they are economically sound, provide gainful employment and become an integral part of the farming system. Implementation of mechanical measures generate casual (short period) employment opportunities. In our study this varied from 46 (Mandavarsa) to 506 (GK3a Gomti) man-days per hectare with an average of 215 man-days. Enhanced productive potential owing to a change in land and animal husbandary practices from extensive to intensive and traditional to improved, generated regular employment opportunities (20 man-days/ha/yr). On average, various WSM activities generated gainful employment at the rate of 2 to 54 man-days per hectare per annum (Fig.5).

4.3 Protection

The protective benefits from watersheds occur in the form of a reduction in runoff volume, in peak discharge, in sediment yield and an increase in lean period flows over a longer time and in recharge of ground water. Data collected from different watersheds indicated that WSM pro-

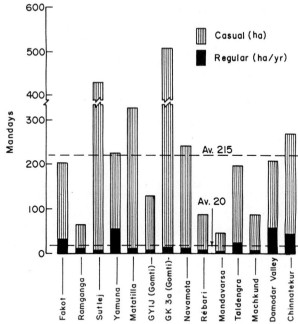

Figure 5: Employment generation through watershed management programmes

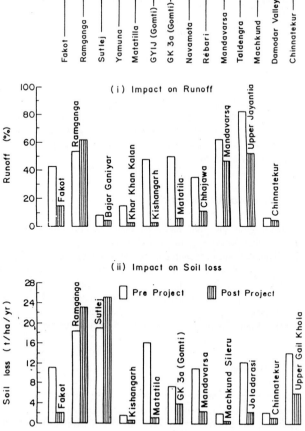

Figure 6: Impact of WSM on runoff and soil loss

Watershed	Project life (yrs.)	Discount rate (%)	Benefit cost ratio	Net present value (m Rs.)	Internal rate of return(%)
REGION : WESTERN HIMALAYA					
Fakot	25	10	1.92	0.5	24
Ram ganga	45	10	5.16*	1201.5	-
Sutlej	45	10	7.06*	4203.3	-
REGION : NORTHERN PLAIN (SHIWALIK)					
Rel Majra	20	12	1.20	0.7	-
Sukhomajri	25	12	2.06	-	19
Nada	30	15	1.07	-	12.3
Bunga	30	12	2.05	-	-
Maili	50	15	1.10	2.4	-
Chohal	50	15	1.12	1.6	16.8
REGION : NORTHERN PLAIN AND CENTRAL HIGHLAND					
Bajar-Ganiyar	20	15	1.58	-	17.0
Khar Kalan	15	15	6.07	10.9	-
Kishangarh	15	15	2.35	5.8	-
Matatila(U.P.)	12 (Agriculture)	12	3.80	9.3	41
	20 (Forestry)	12	4.50	-	-
Tejpura	10	10	3.42	-	-
GYIJ (Gomti)	21	15	3.94	12.5	50
GK 3a(Gomti)	21	15	1.97	3.7	25.8
Navamota	30	12	2.00	0.8	-
Nartora	55	15	2.25	1.3	44.6
REGION : CENTRAL HIGHLAND (MALWA)					
Rebari	20	12	2.65	0.9	37.5
Chhajawa	20	10	2.06	-	-
Mandavarsa	20	15	1.97	1264.9	66.5
REGION : EASTERN PLATEAU (CHOTANAGPUR)					
Damodar Valley (U)	15 (#)	15	2.94	-	47.8
	15 (Structures)	15	1.75	-	46.7
	25 (Forestry)	13	2.09	-	31.6
Upper Jayantia	15	15	1.28	0.3	21.4
Taldengra	25 (Agriculture)	15	1.62	15.6	41
	25 (Forestry)	15	5.23	16.9	50
Machkund Sileru	25	10	4.39*	-	-
REGION : DECEAN PLATEAU					
Joladarashi	15	15	1.45	1.7	-
Chinnatekur	15	15	1.88	18.5	-
G.R.Halli	15	15	1.48	0.9	-
REGION : WESTERN GHATS					
Khumbhave	20	15	2.10	-	N.A.

* Includes protective benefits; # (Water harvesting structures)

Table 1: Economic evaluation of waterhsed management programme, India

gramme have succeeded in achieving these objectives (Fig. 6). The reduction in runoff ranged from 2 to 42 percent, soil loss from 10 to 80 percent and peak discharge from 20 to 40 percent in all the watersheds except in Ramganga and Sutlej where soil loss increased about 30 percent and runoff by 15 percent (in Sutlej) due to heavy biotic interference in the later part of the project. This resulted in an increase in ground water table (0.8 to 2.0 meters), and volume of lean period flow and minimized the stream widening and other associated down stream environ-mental degradation. Thus, the ill effects of drought were moderated to a great extent. In addition, it also helped in changing Indian agriculture by reducing risk and uncertainty so that the farmer could develop a successful farm plan.

4.4 Economic viability

Economics of investment in WSM for management was judged by various workers and agencies/ organisations in India. Generally, budgeting techniques were employed. The range of project life varied from 10 to 55 years and the discount rate varied from 10 to 15 percent. The benefit cost ratio (BCR) of these projects varied from 1.07 in Nada, 3.42 in Tejpura and 3.94 in the GYIJ (Gomti) watershed, considering productive benefits alone (Table 1). Protection benefits like production of Oxygen, conversion to animal protein, addition to soil fertility, improving humidity, protecting birds, etc., and control of air pollution were also considered in some of the projects. This improved their BCR (7.06 and 5.16 in case of Sutlej and Ramganga) and computed @ US $ 20 per tonne of tree weight per year (ASC, 1991). The internal rate of return (IRR) were higher than 16% indicating economic soundness of the projects.

Acknowledgements

The authors are thankful to the Director of the CSWCRT Institute, Dehradun for guidance in preparation the paper, to Mrs. Nisha Singh for typing the manuscript, and to Mr. Nirmal Kumar and Roopak Tandon for computer processing work. We are also thankful to Mr. M.P. Juyal and Deepak Kaul for cartography.

References

Agnihotri, Y., Mittal, S.P. and Arya, S.L. (1989): An economic perspective of watershed management project in Shiwalik foot hill village. Ind. J. Soil Conserv. **17(2)**: 1-8.

ASC (1991): Soil Conservation Scheme, River Valley Projects, Ram Ganga (U.P.) and Sutlej (H.P.). Administrative Staff College of India, Hyderabad.

Das, D.C. (1994): Soil and water conservation for achieving goal of Panchayat Raj. Souvenier. Nat. Conf. Soil and Water Conservation for Sustainable Production and Panchayat Raj. Soil Conservation Society of India, New Delhi, 9-18.

Dhyani, B.L., Ram Babu, Sewa Ram, Katiyar, V.S., Arora, Y.K., Juyal, G.P. and Vishwanatham, M.K. (1993): Economic analysis of watershed management programme in outer Himalaya - A case study of Operational Research Project, Fakot. Ind. J. Agric. Econ. **48(2)**: 237-245.

Dhyani, B.L. and Singh, G. (1991): Socio economic aspects of environmental security and development in Himalayas. In Energy Environment and Sustainable Development in Himalayas, Pradeep Monga and P. Venkataraman (Eds.). Indus Publishing Company, New Delhi, 155-162.

Ram Babu, Dhyani, B.L. and Agarwal, M.C. (1994): Economic evaluation of soil and water conservation programmes. Ind. J. Soil Conserv. **22(1&2)**: 279-289.

Address of authors:
Ram Babu
B.L. Dhyani
Central Soil and Water Conservation Research and Training Institute
218, Kaulagarh Road, Dehradun - 248195, India

Sustainable Land Use Management in the South American Gran Chaco

E.H. Bucher, P.C. Huszar & C.S. Toledo

Summary

The Gran Chaco of South America is being rapidly altered and degraded by overgrazing, deforestation and wildlife exploitation associated with the expansion of campesino settlements. The Chaco represents the second largest biome in South America after the Amazon. The present trend of development threatens both its biodiversity and potential productivity. A promising management system has been shown to be capable of reversing the physical deterioration associated with present management methods. However, economic barriers associated with the campesinos' preferences for present versus future consumption as well as the length of their life expectancy bias the private decision process in favor of the present system. Outside assistance is likely necessary, but is unlikely to come from the governments in the region. International donors may wish to consider support for implementing the management system.

Keywords: Land management, sustainability, economic analysis

1 Introduction

The Gran Chaco is a natural region of about one million square kilometers extending over parts of Argentina, Bolivia, and Paraguay (Fig. 1). The Chaco is the second largest natural biome in South America, with only the Amazon region being larger.

The primitive landscape of the region was mostly a parkland with patches of hardwoods intermingled with grasslands. This mosaic of vegetation was kept stable by periodic fires caused by lightning or by Indians. After colonization by Europeans, fire intensity and frequency declined, particularly in the drier Western Chaco, due to the withdrawal of Indians and the reduction of fuel caused by the overgrazing of introduced domestic cattle. Consequently, the grassland patches were rapidly invaded by woody vegetation over all of the Western Chaco.

A second stage in the process of landscape alteration started in the 1880's, when railways expanded into the Argentinean Chaco. This expansion allowed intensive forest cutting and facilitated the expansion of the Chaco campesinos into the previously unoccupied region, which resulted in severe overgrazing that further deteriorated the productivity of the original vegetation.

As a result of these changes that took place in a relatively short period, nearly all of the Argentinean Chaco has gone through a process of short-lived prosperity followed by a constant impoverishment resulting from a dramatic fall in cattle and timber productivity. Poverty and lack of opportunities stimulate a steady flow of young emigrants to the main cities.

At present, the process of alteration and degradation of the Chaco's natural resources is accelerating. The expansion of campesinos into the Gran Chaco still continues and is reaching the last remaining frontiers in Paraguay, Bolivia, and the few remaining undeveloped pockets in Argentina. The recently built Trans-Chaco Highway from Asuncion, Paraguay to Santa Cruz, Bolivia is helping the expansion of large-scale ranching projects that introduce single-species pastures and eliminate all native vegetation in the relatively untouched portions of the Bolivian and Paraguayan Chaco.

Overgrazing, removal of valuable timber, charcoal production, and over exploitation of wildlife result in an increasing and dramatic negative impact on biodiversity and potential produc-tivity of the natural vegetation. Judging from the experience gained in Argentina, the most likely outcome of this rapid and uncontrolled occupation process of the Chaco is the transformation of a potentially productive and biologically diverse region into a dense and unproductive shrubland dominated by woody weeds, with resulting rural poverty and rural emigration to cities. It is very unlikely that this negative tendency will change unless a deliberate and coordinate action is taken by national and international authorities, policy makers and lending agencies.

Fig. 1: Map of Gran Chaco

2 The Salta management system

A promising model for the sustainable exploitation of the Chaco has been developed in the Argentine province of Salta, based on over 50 years of experience in the Chaco. Experience with managing both public and private enterprises has demonstrated that it is possible to restore degraded land and manage it profitably by maintaining the original forest and grassland in a productive cycle and that the model is applicable to the vast, but ecologically homogeneous area of the semi-arid Chaco.

Currently, the Fundacion para el Desarrollo del Chaco together with the Centro de Zoologia Aplicada of the University of Cordoba and the Fundación Miguel Lillo of Tucum n are refining these management techniques through research and implementation at the 10,000 hectare Los Colorados Experimental Station, located in the province of Salta, Argentina.

Crucial to the management system is the integrated management of timber, charcoal and beef production. Management units are divided into sub-units which are fenced to keep out goats and cattle. Within each sub-unit, all fallen wood is removed and almost all the trees and scrubs are harvested to provide hardwood timber and fuel for the charcoal ovens. In each sub-unit, a few mature trees of the most valuable species (especially quebracho and ironwood) are left standing to provide seeds for natural reforestation. Seeds of wild grasses are added where regrowth is poor. The area is then left undisturbed and protected from cattle until the young hardwood saplings are large enough to be immune to grazing, although limited grazing maybe allowed during this time to help disperse grass seeds spread in the cattle dung.

After about five years, when the young trees are about 2 meters tall, the saplings are thinned and all undesirable woody plants are removed to relieve competition with the young hardwoods and are used to produce charcoal. Grazing is then allowed under controlled conditions with no more than one cow per four hectares, which provides beef while allowing the grass to maintain itself and the hardwoods to mature. At intervals of about 20 to 40 years, the mature hardwoods can be harvested.

3 Evaluation

The following evaluation is based upon data collected over more than twenty years for the managed system by Carlos Saravia Toledo at the Los Colorados experimental facility and an earlier managed area know as Salta Forestal, both in the Department of Rividavia. Data on the present management system were collected with a cross-sectional survey of 194 campesinos in the northern part of the Department of Rividavia and 33 campesinos near the Los Colorados facility. All of the data are for sites with similar climatic conditions and vegetative patterns.

The progressive deterioration of the resource base results in a decline over time of the annual net income to the present management system from a high of approximately US$15 per hectare to as little as US$5 per hectare, as shown in Fig. 2. At Los Colorados, this pattern of decline has been reversed so that the productivity of the resource and the annual net returns to the managed system are increased up to a sustainable level of over US$30 per hectare (Fig. 2). However, the initial investment necessary to achieve the higher sustainable income of the managed system results in a negative net return for nearly ten years.

The net present values of the managed and present systems represent the relative economic desirability of the systems. These values depend upon both the appropriate discount rate and the expected life of the system. As shown by Figure 3, relatively high discount rates (i.e., greater than 8 percent) imply that future returns have relatively low present values, while relatively low discount rates imply that future returns have relatively high present values.

Campesinos are likely more concerned with current income and, therefore, have a time preference for income associated with relatively high discount rates. That is, for campesinos the present system has the greatest net present value. Society, on the other hand, likely places greater value on protecting the natural environment for future generations, so that relatively low discount rates are appropriate for representing the social time preference function. As can be seen from Fig. 3, discount rates greater than 8 percent favor the present system, while discount rates less than

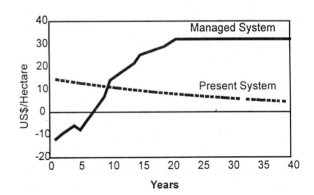

Fig. 2: Net returns per year

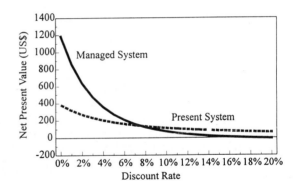

Fig. 3: Net present values for Varying discount rates

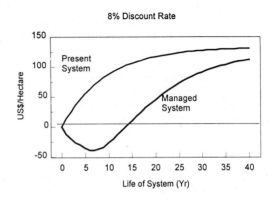

Fig. 4: Net present value of Systems for varying periods

8 percent favor the managed system. A conflict between the "best" interests of the campesinos and society likely exists due to the differing time preferences.

The relevant time period also affects the economic feasibility of the system. For example, if the appropriate discount rate is 8 percent, then Fig. 4 shows that the managed system will only have a positive net present value after about 13 years. Moreover, the present system has a greater net present value than the managed system regardless of the life of the system. That is, if the campesino's time preference is appropriately represented by a discount rate of 8 percent, then regardless of his or her relevant time horizon, it is economically rational to choose the present system.

Alternatively, if the appropriate discount rate is 4 percent, then the present system yields a higher net present value than the managed system only for periods of less than approximately 26 years, as shown in Fig. 5. If the relevant time period exceeds 26 years in this case, then the managed system yields a greater net present value and represents the preferred investment. But the age of most campesinos exceeds 40 years, their children have moved to cities for work and their operating period is likely much less than 26 years. The managed system is likely not feasible for them.

Lower discount rates and/or greater system lives will favor the managed system. But campesinos tend to have high discount rates and short time horizons. The initial negative net returns need to be overcome in order to induce campesinos to adopt the managed system.

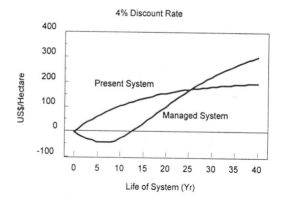

Figure 5: Net present value of systems for varying periods

4 Conclusions

The present management system is degrading the resource base and is unsustainable. The Salta Management System is capable of protecting and, in fact, enhancing the resource base and providing higher net returns in a sustainable manner. But the managed system requires an initial investment that is likely uneconomical to low income campesinos. Depending upon the accounting stance, either that of the campesinos or of society, and the associated time preferences and project life expectancies, the managed system may or may not be the most preferred. Social policy is likely necessary to overcome the rational resistance of campesinos to the managed system. Indeed, the thrust of future research must be on the socio-economics of promoting the managed, sustainable system.

One obvious measure for reducing the negative returns to the campesino is to use subsidies. Such subsidies can be justified in terms of the external benefits produced for society in general by maintaining the ecological system of the Chaco. They can also be viewed as a necessary investment in the productive capacity of the region needed to replenish depleted natural capital. However, in Argentina, as in most of the world, there is a growing tendency to abandon subsidies and government investment in favor of free market economies. As a consequence, it will likely be difficult to promote government involvement in financing the sustainable use of the Chaco.

The case of the Chaco clearly indicates a key issue in ecological economics: how can we find a balance between both the short and long-term welfare and sustainability of society and the supporting ecosystems? It does not seem that the present trend towards absolute liberalization of trade and the elimination of all subsidies provides an answer to this very crucial question. Indeed, the trend seems to run counter to the Agenda 21 document signed by nearly all of the world's leaders at the Rio Earth Summit in 1992, which recommends that effective economic and social incentives must be provided to encourage the conservation of biological diversity and the sustainable use of biological resources.

Addresses of authors:
Enrique H. Bucher
Centro de Zoologia Aplicada
Universidad Nacional de Córdoba
Casilla de Correos 122
5000 Córdoba, Argentina
Paul C. Huszar
Department of Agricultural and Resource Economics
Colorado State University
Fort collins, CO 80523, USA
Carlos Saravia Toledo
Fundación para el Desarrollo del Chaco
Del Milagro 106
4400 Salta, Argentina

Improvement of Soil Management for Sustainability in the Hilly Rakovica Community

M. Zlatić, N. Ranković & G. Vukelić

Summary

Agricultural production in the hilly community of Rakovica area has led to soil erosion problems. Sediment resulting from erosion processes can contain different harmful organic and inorganic substances and therefore presents a source of pollutants of the soil and water resources in this area.

Taking into consideration extensive erosion control works in the investigated region, structures in torrent channels, biological works and the construction of terraces for orchards in the watershed, as well as sustainable agriculture, the question is raised of ecological and economics effects of investments in erosion control. A model of soil management for sustainability is presented based on the degree of erosion hazard and the slope, as the basis for distribution of production lines. The proposal includes three types of crop rotation, self-terracing orchards, orchards with classical terraces, grassing and afforestation. In this way the soil resource can be conserved. One improvement proposes bee-keeping in the established production lines of the offered sustainable production model.

Economic efficiency of investments in sustainable soil management is satisfactory when assessed by discount methods – internal rate of return (IRR), pay back period (PBP), benefit-cost ratio and net present value (NPV). The sensitivity analysis was performed for the IRR parameter. This parameter is most sensitive to changes of benefit, thereafter to changes of regular production costs, and finally to investments in erosion control.

The offered improvements in bee-keeping production will increase economic efficiency, and are very acceptable and adaptable for small farmers. It is an additional reason for people to stay in these lands.

Keywords: Soil conservation, production models, economic efficiency

1 Introduction

From the beginning of this century to the 1950's, the hilly and mountainous regions of Serbia were characterised by a constant increase in the number of inhabitants and households. The livestock population also grew, in particular sheep and goats grazing in forests, and agricultural surfaces were established at the expense of forests on terrain unsuited for agriculture. As a consequence, soil erosion began to increase.

The period since the 1950's has experienced changes in the structure of agricultural production and large migration flows. Many plots of arable land have been abandoned leading to weed cover or return to pasture, which drastically reduced erosion. However, this led to elderly households in this region, with the worrying trend of their extinction (Zlatić, 1995).

In this paper, using the example of the hilly community of Rakovica, effects are shown of the established production model from the aspect of preservation of land resources, as well as the effects of the improvement of this model. It might provide an additional reason for people to remain in these areas.

Rakovica covers about 29 km^2, out of which forests comprise 8%, agricultural areas 54% and the rest 38%. There are 2279 landless households in total; the structure of the households in small farms is given in Table 1.

Due to the small surface areas of the land pieces, mechanised agriculture is mainly performed along the slopes, representing 81% of the agricultural area. The grasslands occupy 16% containing 8% in the form of meadows and 8% as degraded pastures. Orchards occupy 3%.

The area can be divided into four morphological units: the Rakovički Potok Watershed, the Pariguz Potok Watershed, the Kijevski Potok Watershed and the Topčiderska Reka Watershed. The percentage of arable land on steep slopes is given in Table 2.

The yield and income from agricultural products are less than should be expected according to the natural and economic conditions of these regions, especially considering the vicinity of Belgrade as a large market centre.

ha	Landless	0-0.1	0.11-0.5	0.51-1	1.01-2	2.01-3	3.01-4	4.01-5	5.01-6	6.01-8	8.01-10	10.01-15	15.01-20	>20
	2279	193	865	408	439	201	68	43	21	15	8	6	7	5

Table 1: Privately owned area

Morphological unit	Arable land participation (%)	Average slope (%)
Rakovički Potok Watershed	55	11
Pariguz Potok Watershed	85	7
Kijevski Potok Watershed	49	10
Topčiderska Reka Watershed	72	12

Table 2: Arable land on steep slopes

2 Research methods

2.1 Method of natural effects of planned models

Some data for this analysis was obtained from the Department of Erosion and Torrent Control of the Faculty of Forestry in Belgrade.

Soil losses have been estimated according to the "Universal Soil Loss Equation" - USLE (Kirkby and Morgan, 1984) for 77 sample plots of agricultural soil, under present soil utilisation, as well as under the proposed model based upon soil management for sustainability.

On the basis of field surveys, soil losses have been estimated according to USLE for the four morphological units. In this equation, the climatic erosivity factor (R) was calculated according to the total rain kinetic energy (E) and the maximum thirty-minute rain intensity (I_{30}). Rain energy (E) was calculated by the rainfall unit ($E=206+87 \cdot \log I_s$) where I_s represents the intensity of the individual rain segment. The values of the R factor were calculated at the rain gauge station in Belgrade for 1968-1992 (Wishmeir and Smith, 1978).

The values of the soil erodibility factor (K) were determined on the basis of the soil erodibility nomogramme (Wishmeir et al., 1971). Slope length (L) and slope steepness (S) factors were determined on the basis of the topographic factor (*LS*) (Wishmeir and Smith, 1978). The cover management factor (*C*) was determined on the basis of experience from the Czech Republic (1987). The supporting practices factor (*P*) was determined on the basis of Kirkby and Morgan (1984).

The precise determination of soil losses was not the objective of this work. Only an approximation is needed here to estimate the relationship between the soil losses prior to and after the performed erosion control measures.

2.2 Production models

The basic production model (model I) was developed from the aspect of soil management for sustainability, the needs of the population in this area (production lines most frequently applied in practice) and potential economic effects. Production is primarily planned in a quantitative sense, i.e. the relations are designed between the groups within arable farming (erosion-control crop rotations) and orchard production (classical orchards, orchards with self-terracing, and orchards with classical terraces), as well as pasture and forest areas. In the qualitative sense, the lines of production as per crop species are designed. Crop rotation includes cereals like wheat and oats, root crops (corn, soya beans and sunflower) and grasses. Orchard species include apple, pear, peach, apricot, cherry, sour cherry, plum, raspberry, blackberry and walnut.

On the basis of field assessments, afforestation should be considered in the markedly endangered parts of the investigated region, with economic factors involved. The land is reclaimed depending on the degree of erosion hazard, and species are selected depending on micro and macro-environmental conditions. Thus, Austrian pine is selected for the skeletal serpentine parent rock, and black locust for the regions of strong and excessive erosion. Broad-leaved species are selected for other areas, depending on site conditions, contact with the existing vegetation, and the degree of erosion hazard.

The improvements of the basic production model (models II and III) were performed by the establishment of bee-keeping in two variants: **I** variant - production of honey as the chief product, and wax, propolys, and flower powder as by-products (model II) and **II** variant - production of royal jelly as the chief product without by-products (model III). The production of honey within the frames of the existing production can be organised as an additional activity in agricultural areas (sunflower production line), in orchards (apple, pear, cherry, sour cherry, apricot, raspberry, etc.), in forest cultures (black locust and Austrian pine), as well as in meadows.

Therefore, models II and III represent an advancement of model I, aimed at the improvement of economic effects, yet preserving the protective character of the basic production model. Thereby a possibility is presented to the local population to improve their income from the existing production lines, thus enabling them to stay in these regions.

There are also other alternatives for improvement, such as medicinal herb collection, livestock raising products (such as sheep and goat breeding), etc. However, these alternatives are not characteristic of the Belgrade region and do not have such a degree of financial results and adaptability as the chosen three variants.

2.3 Quantification of costs and benefits before and after establishing models of production

On the basis of the performed calculations of unit prices of afforestation, establishing orchard terraces and development of fruit plantations, as well as regular production lines, expenses and incomes were quantified before and after performing erosion control works.

Labour costs were included in all the calculations of afforestation, orchard establishment, as well as in all production lines of the investigated models. This was calculated according to the technological norms and establishing the necessary labour quantity. Wages were determined through the price of hourly labour.

The quantification consists of the following parts:
- total revenue for the state before land reclamation,
- cost of regular production before land reclamation,
- the value of production including total revenue and residual value of investment features after land reclamation,
- financial investments including erosion control costs and costs of regular production after land reclamation.

The differences of total revenue and costs before and after land reclamation are the starting point for the evaluation of long-term economic effects of these investments.

2.4 Methods of assessment of economic efficiency

The assessment of long-term effects of the planned and the improved models were performed in terms of internal rate of return (*IRR*), pay back period (*PBP*), benefit-cost ratio and net present value (*NPV*) (Gittinger, 1982). Among the four discount methods mentioned, two (*IRR* and *NPV*) are explained here in more detail. These two methods, in our opinion, provide reliable information for the assessment of investment efficiency either for an individual (private owner) or for an investor, as well as for the wider community. This information should answer whether an investment in proposed works and production will be able to cover all project costs, and to allow for a surplus necessary for further development.

Internal rate of return is the rate of interest of invested capital. It is calculated on the basis of amount of discount rate, where the present value of all inputs is equal to the present value of all outputs in the same statement of accounts. For the investment to be economically justified, it is necessary for *IRR* to be higher than the real rate of interest. In this paper, a real rate of interest of 12% was used according to International Bank for Reconstruction and Development for Eastern European Countries.

Net present value represents the sum of total annual benefits discounted to the initial moment, reduced for the total costs discounted to the initial moment, as per the discount rate of 12% (real rate of interest). The investments are economically justified when the value of $NPV \geq 0$, i.e. when the invested value is returned or making a profit. At $NPV < 0$, the investments are not justified.

A period of 15 years was chosen for the assessment of economic efficiency according to the average production cycle of stone-fruit orchard species. Sensitivity analysis was performed for all models of production.

The prices are in German Marks (*DM*), as the most stable currency, for the period May-June 1996.

3 Results of research

3.1. Natural effects of establishing sustainable production

The average soil losses from agricultural land in this community with the present land use, according to USLE method, amount to 31 $t \cdot ha^{-1}$. Sediment, as a product of erosion, contains harmful organic and inorganic substances and has an adverse impact on the environment.

With the proposed model of production, the values of C and P in the USLE equation will be reduced several times. This will decrease soil loss below the limits of tolerance, i.e. on average 6 $t \cdot ha^{-1}$ for the whole Rakovica community.

3.2 Economic effects of the planned and improved models

3.2.1 Internal rate of return (IRR)

It can be seen that the *IRR* for the Rakovica community amounts to 17% for the planned production (model I, Fig. 1), and 18% for the improved production - variant I with honey as the chief product (model II, Fig. 2), and 22% for the improved production - variant II with royal jelly as the chief product (model III, Fig. 3). By comparing the calculated *IRR* with the real interest rate, which is 12% for reasons given above, it can be concluded that the investments in soil conservation and in the proposed production variants are cost effective. This statement is based on the fact that, after the credit commitments have been met, a certain percentage for the extended material base results from the increase of the net economic benefit. The accumulation amounts to 5% for the planned production (model I), 6% for the improved production with honey as the chief product (model II), and 10% for the improved production with royal jelly as the chief product (model III).

3.2.2 Net present value (NPV)

This parameter represents the sum of total annual benefits discounted to the initial instant, reduced by the total costs discounted to the same instant, at the discount rate of 12% (the real interest rate). According to the calculation of *NPV*, it can be seen that this parameter amounts to 3.9 million *DM* for the planned production (model I, Fig. 4), 4.4 million *DM* for the improved production variant I (model II, Fig. 5), and 7.7 million *DM* for the improved production variant II (model III, Fig. 6) models.

Since *NPV* is well above 0, one can conclude that it is cost effective to invest in the designed erosion control works and the subsequent production models in the region.

3.3 Sensitivity analysis

Long-term economic effects calculated according to the discount methods are based upon normal circumstances. However, for a multitude of reasons, perturbations relative to benefit-cost ratio are possible, either if the revenues increase or decrease, or if the same happens to the costs.

Sensitivity analysis is performed when the calculated parameters of economic efficiency are tested in order to observe what happens to these parameters if costs or benefits are modified. In this case, sensitivity analysis was performed for the *IRR* parameter. The sensitivity of *IRR* (Fig. 7) was measured with respect to the changes of the annual costs and benefits. The examined values of changes of these parameters range from 5% to 30%, both positive and negative.

One can see that *IRR* is most sensitive to benefit changes. As for the negative changes, *IRR* is on the limit of profitability (at the level of real discount rate of 12%) if the benefits are reduced by 13% with the planned production model, by 14% with the improved model (variant I) and by 22% with the improved (variant II) production models. In the same sense, cost increases by 15%, 16% and 28% are acceptable for the quoted models, respectively.

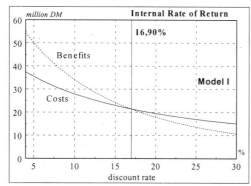

Figure 1: Internal rate of return - model I

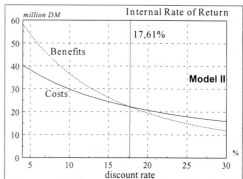

Figure 2: Internal rate of return - model II

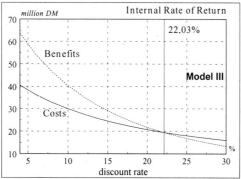

Figure 3: Internal rate of return - model III

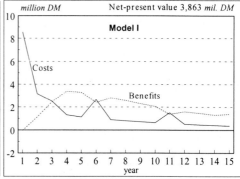

Figure 4: Net present value - model I

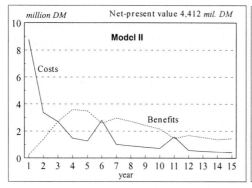

Figure 5: Net present value - model II

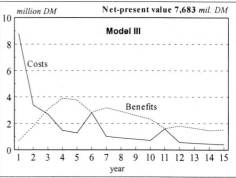

Figure 6: Net present value - model III

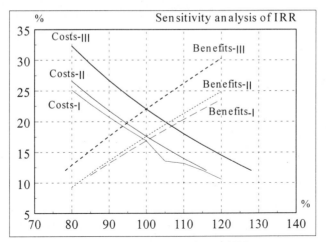

Figure 7: Sensitivity ananysis of IRR

In the case of possible changes of the quoted parameters by 20% in the positive sense, *IRR* in the case of the increased revenue amounts to 24% for the planned, 25% for the improved (variant I) and 30% for the improved (variant II) production models. In the case of decreased costs by 20%, *IRR* amounts to 25% for the planned, 27% for the improved (variant I) and 32% for the improved (variant II) production models.

4 Conclusions

The planned production model sustains the conservation of soil as one of the most important natural resources, since it reduces soil losses below the permissible limits. The soil losses are estimated to decrease from 31.06 $t \cdot ha^{-1}$ at present, down to 5.74 $t \cdot ha^{-1}$ for the conditions following the introduction of the proposed model.

The assessment of investment efficiency, carried out by the discount methods, indicates satisfactory economic efficiency. *IRR* for the offered production model amounts to 17%, 18% for the improved (variant I) and 22% for the improved (variant II) production models, in all cases higher than the real interest rate of 12%. *NPV* for all the three models is considerably above 0, i.e. 3.9 million *DM* for the planned, 4.4 million *DM* for the improved (variant I), and 7.7 million *DM* for the improved (variant II) production models.

The efficiency is enviable for all three models under consideration. The variant with honey as the chief product also gives somewhat better economic efficiency with *IRR*, and considerably higher with *NPV*. The improved production model with royal jelly as the chief product gives considerably higher efficiency for these areas.

The sensitivity analysis also indicates the significant efficiency of investments. The offered and the improved (variant I) production models can sustain the negative changes of up to 15-16%, whereas the improved model (model III) is capable of sustaining the benefit decrease of up to 22% and cost increase of as much as 28%, while still remaining within the frames of profitability. Considering the positive reserves of the production models, for changes by 20% in the positive sense, the economic efficiency is very significant for all the parameters.

According to the calculated economic efficiency parameters, sensitivity analysis of investments and their unmeasurable effects, it can be concluded that in the Rakovica community, the investments in soil management for sustainability are cost effective and beneficial for environmental control. The offered improvements of production by introducing bee-keeping have considerably increased the economic efficiency and are very acceptable and adaptable for small farmers. This can provide an incentive for people to remain in such areas.

References

Gittinger, J.P. (1982): Economic Analysis of Agricultural Projects, second edition, John Hopkins University Press, Baltimore.
Kirkby, M.J. and Morgan, R.P.C. (1984): Soil Erosion, Kolos, Moscow (in Russian).
(1987): Protierozni ochrana zemedelskych pozemku, Agropojekt Praha, Zavod OG Brno, Praha.
Wischmeier, W.H., Johnson, C.B. and Cross, B.V. (1971): A soil erodibility nomogramme for farmland and construction sites, Journal of Soil and Water Conservation **26**, Ankeny, Iowa.
Wischmeier, W.H. and Smith, D.D. (1978): Prediction rainfall erosion losses - A guide to conservation planning, Agricultural Handbook **537**, USDA, Washington.
Zlatić, M. (1995): Socio-Economic Aspects of Erosion Processes in Serbia, Proc. Scientific Conference with Participation of Foreign Specialists "90 years of Soil Control in Bulgaria", Sofia, 252-256.

Addresses of authors:
M. Zlatić
N. Ranković
Faculty of Forestry
Kneza Višeslava 1
11030 Belgrade, Yugoslavia
G. Vukelić
Faculty of Agriculture
Nemanjina 6
11000 Zemun-Belgrade, Yugoslavia

Strategies for Soil Conservation in Iceland

A. Arnalds

Summary

Through 1100 years of unrestrained land use, Iceland has lost about 60% of its vegetative cover and 96% of its tree cover. The current extent of ecosystem degradation is acknowledged, but land users have typically attributed responsibility to someone else - considering it to be a government problem. The Icelandic Soil Conservation Service is now re-directing its activities towards participatory approaches directly involving the primary and secondary land users.

Recent in-depth survey work has objectively demonstrated the extent of the degradation problem, and the media have brought the results to the attention of the nation. As a result, new strategy is evolving, with political support to provide the necessary resources and legislation for effective action.

In collaboration with farmers, special-interest groups and the general public, the Soil Conservation Service is working towards a renewed pact with the environment - development of a land ethic: acknowledging responsibility for the present state of the land, and demonstrating a practical willingness to change things and to work together towards sustainable land management for all ecosystems.

Keywords: Soil conservation, conservation policy, sustainable land use, participatory aproaches.

1 1100 Years of Unthinking Land Use

In terms of the human timescale, soil is a non-renewable resource. Once lost, it takes many generations before new soil can sustain a fertile ecosystem.

In his classic account of the conquest of land, Lowdermilk (1953) explains how degradation of soil and vegetation through unsustainable land use has been the main cause of the decline of most civilisations.

Iceland - with severe, extensive and ongoing land degradation - is, regrettably, a striking example of the consequences of long-term mismanagement of land (Fridriksson, 1978; Thorsteinsson, 1986; Arnalds, 1987). Since its settlement by Vikings in AD 874, more than half the original vegetation cover has been lost through soil erosion. The most serious environmental problems in Iceland today are considered to be the degraded state of soil and vegetation, extensive soil erosion, and continuing degradation due in part to unsustainable land use.

Prior to settlement, the vegetation had evolved in the absence of large herbivores and was extremely vulnerable to grazing and other uses (Thorsteinsson et al. 1971). Fertile land has been turned into cold and wet deserts over large areas of the country. Plant cover has decreased from about 60% to less than 25%, and the condition of much of the remaining vegetation is poor, with palatable plant species replaced by others less sought after by livestock. Woodland cover has gone from over 25% of the land area to just 1%. After accounting for the loss of plant and soil cover and the low productivity of present rangelands, Thorsteinsson (1986) estimated that the herbaceous resource had been reduced to less than 20% of its potential. The extensive ecosystem degradation - both past and present - significantly affects living conditions for Iceland's 260 000 inhabitants.

2 Organised soil conservation activities

Organised attempts to fight ecosystem degradation and desertification began in 1907 with the establishment of the Icelandic Soil Conservation Service (SCS). Much has been achieved in the battle against catastrophic erosion (Runolfsson, 1987), but despite almost 90 years of soil conservation, soil erosion is still widespread.

Conservation activities were reorganised in the 1980s, with the aim of halting degradation countrywide, increasing land user responsibility and achieving sustainable land use practices nationwide. A generally limited awareness of conservation matters needed to be countered, so a wide range of means were used to increase national awareness amongst farmers, politicians and the general public (Arnalds, 1992). The campaign has been highly successful. Widespread soil erosion and the degraded state of the vegetation resource are now recognised as the most serious environmental problems in Iceland today, and the value of restoration work is recognized. A key factor in this successful awareness raising has been close cooperation with all sectors of the mass media in focusing attention on natural resources problems and the conservation work of individuals, groups and communities.

Conservation approaches moved away from the traditional approach of using governmental machinery and personnel to halt localised, catastrophic soil erosion. A new national soil conservation strategy was developed in 1991 in cooperation with the Ministry of Agriculture. This strategy is presently under review in order to strengthen it and add supporting legislation. This policy review and reform has been greatly aided by the benefit of international experience through three consultancies[1]. One of the main aims of the new strategy is to make land conservation much more participatory, involving a wider cross-section of society, promoting a better understanding of Icelandic ecosystems and developing potentially sustainable land use systems. The focus of conservation also needs to be broadened from just soil and soil erosion to holistic ecosystem management in a multiple use context.

Although most Icelanders acknowledge soil erosion to be a major environmental problem, their actual perception of the degradation is questionable. This applies especially to the land users, where a lack of acceptance of the resource degradation has been linked with the conceptual problem of "ownership" of these problems. SCS aims to shift attitudes within Icelandic agriculture by introducing an ethic of land stewardship, with responsibility for the land resting with the land user.

Formulation of Iceland's new soil conservation strategy is still at an early stage. The review is taking a very broad view of the elements that interact in the process of developing what has to be an integrated and holistic land management strategy. Some of the key factors are considered below.

3 Resource assessment and monitoring

Assessment and monitoring of the condition and use of land are key elements in Iceland's soil conservation strategy. The long-term implications of continuing soil erosion, coupled with the degraded state of most land, have not been fully accepted by some key groups, especially in the agricultural sector.

Icelandic soils are young, friable and of volcanic origin. The profile is often characterised by many, sometimes quite thick, ash layers, leaving the soils highly susceptible to erosion by wind and water once the vegetative cover has been disturbed (Arnalds, 1984; Arnalds et al. 1987).

---

[1] Andrew Campbell, Former National Landcare Facilitator, NSCP, Australia (1993); David Sanders, Senior Soil Conservation Officer, FAO (1994); Ian Hannam, Dept. of Land and Water Conservation, NSW, Australia (1996).

The first national survey of soil erosion in Iceland (Olafur Arnalds, this conference) by the Agricultural Research Institute and SCS is almost complete. The survey is based on 1:100 000 Landsat™ satellite imagery and is available in computerised form to those interested. Survey results confirm the soil erosion problem and are already having considerable influence. The data provide a valuable tool for planning soil conservation work and for making Icelandic agriculture sustainable. Farmers in areas in good condition could advertise their products as being environmentally friendly, whereas farmers in degraded areas would need land conservation work to meet sustainability criteria.

Care has been taken to ensure that local communities have been the first to receive the relevant data and survey maps for their areas. Meetings have then been held with community representatives, usually resulting in positive dialogue concerning solution of the degradation problem. This has been a welcome change from the problem-denial syndrome so common when so-called facts are of the "We think ..." type. The soil erosion survey is also having significant influence at the political level. This objective verification of the severity of the national soil erosion problem will be fundamental to a new, government-funded, National Soil Conservation Programme planned for 1997-2001.

The degraded state of the environment - a bequest of accumulated past neglect and abuse - is not recognised by most Icelanders. Again, lack of concrete, objective information has diverted public debate into some very strange channels, distracting attention from the need to "heal the land."

Although Iceland may have lost more than half its original vegetation cover as result of 1100 years of unthinking exploitation, most people find it difficult to visualise the modern desert areas and other degraded land as ever having been vegetated. Establishing the conceptual relationship with land use, not to mention the cumulative degradation effects on the ecosystem and living conditions, requires an even more difficult mental jump. It has therefore been difficult to introduce an awareness of the need to re-vegetate land for the needs of current and future generations. Beliefs and conservation ethics usually reflect a state of knowledge. The need therefore is to re-direct the efforts of nature conservationists away from the unthinking preservation of man-made deserts and degraded vegetation associations towards restoration. However, restoration must be based on knowledge, thus a national survey of the actual growing conditions and other environmental indicators of the land's growth potential is urgently required.

4 Sustainable land use

The concepts of sustainability and ecosystem management are integral to Iceland's soil conservation strategy. In recent years, land use has been changing from predominantly free-range grazing of sheep to multiple use. The impact and needs of tourism have in particular become a significant factor in general environmental considerations. However, grazing pressure will continue to be the major determinant of the well-being of most of Iceland's ecosystems.

Agriculture in Iceland basically means grass-based livestock production as the climate is borderline for grain production. Farms and all urban areas are located in the coastal lowlands, and most farms are privately owned. The vast highlands of the interior are used for free-range summer grazing by sheep, in communal districts or "commons." The farmers of Iceland own or have grazing rights to more than 70% of Iceland, and there is an urgent need to halt erosion or restore vegetation on much of this land (Olafur Arnalds, this conference).

In general, there is a lack of stewardship towards land. Livestock owners take little responsibility for their animals during the summer grazing period, much less than in other "western" countries. Despite extensive pasture fencing in recent years, stock is often allowed to roam freely. There has been neither legal nor economic inducement for farmers to manage land in a sustainable manner. The breakdown of ecosystems is seen by most landholders to be the problem of the government.

The common grazing lands need special attention. Most of the highland areas are in a poor-to-critical condition. With the recent market-driven decrease in sheep numbers, grazing pressure on the commons has been much reduced, but SCS has stressed the urgent need for the worst degraded highland areas to be, at least temporarily, protected completely from grazing.

Horse numbers have been increasing rapidly, posing a serious management problem. Currently there are 80 000 with a doubling time of 22 years. The horses are grazed almost entirely on private land, and are owned by both farmers and city dwellers that lease or own land. The creation of conservation awareness has been particularly difficult among horse owners.

Although the Soil Conservation Act of 1965 included a range of powers to deal with overgrazing and to exclude damaged areas from grazing, these powers have not been widely used. However, given the magnitude of ecosystem damage, these powers would need to be applied over most of Iceland if land degradation is to be checked and the ecosystem improved.

The number of sheep and horses grazed in Iceland exceeds sustainable land use limits in many areas. A range of measures - from improved management and range improvements to exclusion of livestock from damaged areas - are required for ecosystem maintenance or recovery. Furthermore, consideration needs to be given to the economics of stocking on many rangeland areas (Arnalds and Rittenhouse, 1986).

The socio-economic characteristics of the sheep and horse industries have made any uniform or universal adoption of restoration practices very difficult. Implementation of improved management and restoration practices depends entirely on an individual farmer's willingness to cooperate. This has so far been spasmodic.

Several reforms are needed to attain sustainable use of Iceland's ecosystems. Development of conservation awareness and of a sustainability ethic will be key factors in the long term. Wrong land use is often the result of insufficient knowledge, and therefore a range of educational measures are needed. Property planning, especially where land holders are taught to make their own plans, have proven to be efficient means to promote and foster better land use (e.g., Anon, 1995). State support to agriculture in Iceland is huge, and needs *a priori* to be linked to goals of sustainability. Land quality considerations have to be linked with price support schemes, and a programme is required to help farmers improve the environmental quality of their farms. As a last resort, effective legal means are required to be able to intervene if land use is unsustainable.

The preparatory work for the review of soil conservation strategy has included evaluation of experience elsewhere, and their applicability to the Icelandic situation. In particular, work on conceptual aspects of sustainable land management have been consulted (e.g., Hannam, 1991; Sanders, 1989; Campbell, 1992; Mackey, no date; Gardner, 1994; ANZECC/ARMCANZ, 1996).

5 Participation

Continuing soil erosion and the degraded state of the environment cannot be shrugged off as being someone else's problem, be it the government or SCS. These problems affect the well-being of the whole nation. In order to build a strong relationship between conservation and society, SCS has been reorganising its operation and developing participatory approaches.

5.1 Farmers re-vegetate the land

With such a high proportion of Iceland being rangelands, the success of soil conservation will depend to a large extent on the land users. Participation is seen as one of the primary keys to increased conservation awareness amongst the land users. Restoration projects will be linked with consultancy, property planning and educational efforts.

Most ecosystem degradation in Iceland can be blamed on the forefathers of the current generation, and is the result of long-term land mismanagement. Only a small proportion of the degradation can be attributed to the current land users, although they may be perpetuating the degradation problems. Acknowledgement of this fact, combined with the need for ecosystem restoration for future generations, justifies increased government involvement in the restoration of rangelands and other degraded land. Government involvement is crucial, as the ability of most farmers to pay for conservation work is severely constrained. Icelandic farmers are going though tough times, and the sheep farmers in particular are facing an economic crisis.

In 1990 a project was initiated to involve farmers more in soil conservation work. Previously most projects to fight erosion and reclaim land were implemented by SCS staff and machinery. In the first year only about 40 sheep farmers participated, increasing to 120 in the second year. The results were so promising that, in 1993, the State Fertilizer Factory decided to celebrate its 40th anniversary by donating fertilisers for reclamation purposes to 250 additional farmers. SCS set itself the goal of working with and financially supporting all farmers that have land that is suitable for reclamation, i.e., over one-third of all farmers. This support will be linked with property planning, and hopefully financed in part by a government atmospheric CO_2 reduction scheme. Cooperating farmers currently receive seed and about 85% of fertilizer costs. Each farmer receives the equivalent of about $US 700/year for at least five years.

The results of the cooperative project, *Farmers re-vegetate the land,* have far exceeded expectations. The re-greening of the land is quite obvious, but, more importantly, the farmers themselves feel, and are seen to be, part of a solution instead of part of the degradation cause. The participants take pride in showing others **their** work, especially to the soil conservationist, who now is regarded as a partner. This in turn opens up positive channels for discussing and resolving other resource issues, such as overgrazing and improved management of the land. A few land-care groups have formed, but more are needed as working with individuals is very time-consuming.

5.2 Contract work

The overall battle against the widespread catastrophic soil erosion is the legal responsibility of government, and in the past the work has been carried out by SCS staff and machinery, coming into an area, fencing it and re-seeding, without much contact with the local people. This approach has been much to blame for the "problem-ownership syndrome" that has characterised the degradation issue in Iceland.

To change the ownership of these problems, most soil conservation projects are now carried out by local people, especially farmers, under the supervision of SCS staff. For the farmers this means added income in tough times. For the SCS the benefits include more efficient use of limited staff; attitude changes among the farmers, combined with high quality work; and usually a substantial cost reduction.

5.3 Public participation

It is important that all sectors of society are directly involved in soil conservation work. SCS is currently working on reclamation projects with a wide range of special-interest groups. These include municipal and rural authorities, a wide range of clubs and associations, and many individual volunteers. The President of Iceland has been an enthusiastic advocate of increased conservation, and set a public example in caring for the soil.

Public participation is a key element in the national soil conservation strategy, and SCS will both seek assistance in project implementation and support special-interest groups. Hands-on experience has proven to be an efficient tool in fostering an understanding of erosion problems and in creating a belief in the restoration work, and close cooperation with the media has led to increased public awareness, both of resource problems and the conservation work of individuals, groups and communities.

6 Improving understanding

Our actions are influenced by attitudes, and our attitudes reflect our state of knowledge. A very important element in the new soil conservation strategy will be improving understanding. This will mean increased emphasis on environmental education, research and collaboration with planning authorities and all other bodies that influence the fate of the land.

The most significant element in environmental education will be introduction and adoption of the philosophy and practice of sustainable development, i.e., the creation of a land ethic as an addition to traditional Icelandic values. SCS intends to take an active role in the formation of a strategy for environmental education. In addition to influencing environmental education through the school curriculum, positive action at community level will be encouraged through training of teachers and through continuing professional development of key groups, including farmers, land agents and planners. Soil conservation projects, with a mixture of lectures and hands-on experience, are eagerly adopted within the general school system.

Increasing knowledge is one of the main keys to success in conservation. Increased research is essential: on the causes of land degradation; on the condition of the land; on means to restore land; and the interrelationship between land use and land condition. Research has to be closely linked to the needs of the various groups active in land conservation within a framework of sustainable development and ecosystem management. In the re-design of the research effort, it is important to take into account the tremendous research and development power available at the grass-roots level, and close ties between research and application are essential at all levels.

7 Conclusion

The ecosystem degradation that has taken place in Iceland in historic times is matched in only a few countries of the world. Despite opinion polls showing that soil erosion is seen as the most serious environmental problem, most Icelanders still do not comprehend the severity of the degradation problem, nor how it affects the well-being of the nation. Understanding of natural resources – both their limits and potential – has been especially weak in the primary land user sector. Ignorance of the true nature of the problem has been compounded by a lack of systematic land resource assessment and monitoring.

The national survey results confirm that degradation problems are much more severe than realised by most Icelanders. The findings of this survey will have a major political effect and are expected to lead to an escalation in efforts to halt degradation.

A new soil conservation strategy is being developed to be followed by a new law. The strategy will promote a move from localised, single-aspect, soil conservation work towards more comprehensive and holistic ecosystem management, guided by sustainable development principles. The main aim is to restore land quality in accordance with growing conditions and the land use needs of the various user groups. SCS has set itself the aim of establishing sustainable land use practices countrywide and stemming all remaining catastrophic erosion. This called for an enhanced role and more responsibility for land users in conservation, the creation of a conservation ethic in the nation, and a quest for new methods to conserve land resources.

The current law on soil conservation (1965, with later amendments) is outdated. A new law is required to promote sustainable management of ecosystems, with effective support mechanisms and establishing increased authority to prevent and control land degradation.

Many countries are now experiencing processes of ecosystem degradation comparable to those Iceland suffered in the past. Therefore Iceland, with its well-documented history of inhabitation and a wide range of land conditions, is a unique place to study both the consequences of long-term mismanagement of land and possible means to restore land to a sustainable condition.

References

Anon (1995): Property Planning: How to produce a physical property plan. Department of Land and Water Conservation, NSW, Australia. 29 pp.

ANZECC/ARMCANZ [Australian & New Zealand Environment & Conservation Council and Agriculture & Resource Management Council of Australia & New Zealand Joint Working Group] (1996): Draft national strategy for rangeland management. Dept. Environment, Sport and Territories, Australia. 64 pp.

Arnalds, A. (1987): Ecosystem disturbance in Iceland. Arctic and Alpine Research, **19**(4): 508-513.

Arnalds, A. (1992): Conservation awareness in Iceland. In: Proc. 7th Int. Soil Conservation Conf. Sydney, Australia, September 1992, 272-275.

Arnalds, A. & Rittenhouse, L.R. (1986): Stocking rates for northern rangelands. In: Gudmundsson, O. (ed), Grazing Research at Northern Latitudes. New York, NY: Plenum Press, 335-345.

Arnalds, O. (1984): Eolian nature and erosion of some Icelandic soils. J. Agricultural Research in Iceland, **16**(1-2): 21-35.

Arnalds, O., Aradottir, A.L. & Thorsteinsson, I. (1987): The nature and restoration of denuded areas in Iceland. Arctic and Alpine Research, **19**(4): 518-525.

Campbell, A. (1992): Landcare in Australia. Taking the long view in tough times. 3rd Annual Report, National Landcare Facilitator. National Soil Conservation Program, Australia.

Fridriksson, S. (1978): The degradation of Icelandic ecosystems. In: Holdgate, M.W. & Woodman, M.J. (eds) The Breakdown and Restoration of Ecosystems. New York, NY: Plenum Press, 145-146.

Gardner, A. (1994): Developing norms of land management in Australia. Australasian J. of Natural Resources Law and Policy, **1**(1): 127-165.

Hannam, I. (1991): The concept of sustainable land management and soil conservation law and policy in Australia. In: Henriques, P. (ed), Proc. Int. Conf. on Sustainable Land Management. Napier, New Zealand, 17-23 November 1991, 153-168.

Lowdermilk, W.D. 1953): Conquest of the Land Through 7000 Years. USDA Soil Conservation Service, Agriculture Information Bulletin No. 99. 30 p.

Mackey, E.C. (ed). no date [1995?]: The natural heritage of Scotland. Scottish Natural Heritage, Edinburgh, Scotland.

Runolfsson, S. (1987): Land reclamation in Iceland. Arctic and Alpine Research, **19**(4): 514-517.

Sanders, D.W. (1989): Soil conservation: Strategies and policies. Paper presented at the 6th Int. Soil Conservation Conference. Addis Ababa, Ethiopia, 6-18 November 1989.

Thorsteinsson, I. (1986): The effect of grazing on stability and development of northern rangelands: A case study of Iceland. In: Gudmundsson, O. (ed), Grazing Research at Northern Latitudes. New York, NY: Plenum Press, 37-43

Thorsteinsson, I., Olafsson, G. & Van Dyne, G.M. (1971): Range resources of Iceland. J. of Range Management, **24**: 86-93.

Address of author:
Andres Arnalds
Soil Conservation Service
Gunnarsholt
IS-850 Hella, Iceland

Influence of Socio-Economic Changes on Soil Productivity in Slovakia

J. Hraško

Summary

Over the past 40 years a gradual increase in fertilizer consumption (mostly NPK and Ca) has occurred in Slovakia. This led to yield stabilization at an European standard and to an increase in available nutrients in soils. After 1990 the agricultural costs, including fertilizers, increased dramatically. As a consequence, farmers cut down on fertilizer application by almost eight-fold.

Keywords: soil fertility, fertilizers, available nutrients

1 Introduction

Until 1989 agriculture in Slovakia was viewed as an important branch of the national economy whose purpose was to secure enough food for the population and raw materials for the processing industry. As in the most developed countries of the world, highly productive species were introduced together with high doses of fertilizers and a good system of pest control. Farming practices, mostly on cooperative farms, were rather uniform in all production regions of Slovakia and, unfortunately, did not respect agroclimatic or geographic conditions.

Worldwide intensive farming is based on specialization and concentration of production, on adequate fertilizers and pesticides use, high technology and low-skilled labour.

When criticizing this agricultural system from the viewpoint of productivity, I have repeatedly expressed my opinion that we cannot "reproach an indisputable priority to ensure a high rate of self-sufficiency in basic food production equal to European standards of consumption" (Hraško, 1988).

2 The former and present economical situation in Slovak agriculture

Until 1989 agriculture relied on state subsidies and was a stable, relatively prosperous branch of the national economy. State protectionism in agriculture, based mainly on grants and subsidies, is characteristic not only of the former centrally-planned economy, but it is also the nation's main instrument to protect its producers. Following socio-economic changes in Slovakia, the agricultural situation became to be complicated due to a combination of a general economic recession and privatization and transition to a market economy. According to the scenario of the radical economic reforms approved in 1990 by the Government of the former Czech and Slovak Federal Republic, the prices of inputs into agriculture should have increased by 37 %, the consumer prices of food by 15%, and the purchasing price of agricultural products by 52% (Documents, 1993).

In reality, by the end of June 1993, when Slovakia was still part of the Czech and Slovak Federal Republic, the prices of inputs into agriculture increased by 214%, consumer prices of food by 203 %, and the wholesale price of agricultural products increased by 117%.

The price for one tonne of NPK fertilizers in 1989 was on average 2,180 crowns but was 5,932 crowns in 1995, while the price for 1 tonne of wheat was 3,550 crowns in 1989 and between 3,500 and 3,700 in 1996.

Under such a price disparity, agriculture cannot objectively have a chance to establish itself into market economy conditions. When the agricultural subsidies amounted to 17 billion crowns, the one year profit in agriculture was between 5 - 6 billion crowns. Present subsidies amount to only 7 billion crowns and has caused an overall loss of approximately 10 billion Slovakian crowns (Documents, 1995).

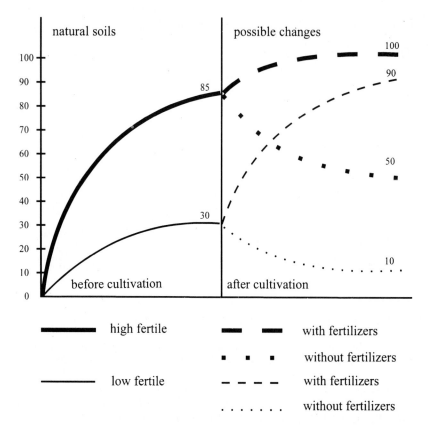

Fig. 1: Yield development on poor and rich soils (after Hraško, 1988)

3 The impact of the present economical situation on yield and on selected soils characteristics

Until 1989 profit from agriculture was dependent on inputs such as the cost of fertilizers, rather than from natural soil conditions and regeneration cycles of soil properties. Crop yield increase was higher on less fertile soils than on untreated rich soils. Fertilizer use has influenced yields in

Slovakia (ŠPALDON 1982) by almost 20% on natural rich soils, and by more than 50 % on poorer soils. This was also noted by HUEG (1977) in USA where yields of cereals increased, from 1930 to 1970, on poorer soils from 3.7 tha^{-1} to 5.9 t/ha^{-1} (i.e. 61%), but on untreated rich soils only from 6.5 t/ha^{-1} to 8.9 t/ha^{-1} (i.e. 38%).

If fertilization practices ceased, yields would decline especially on poorer soils (Fig. 1).

Both great losses in agricultural production and a decrease in farmer income levels were caused by an enormous increase in input prices while at the same time agricultural products remained stagnant. This development, together with greatly reduced state subsidies starting in 1990, meant that farmers had no means to invest in soil properties improvement. This is also why consumption of fertilizers decreased from 231 kg per hectare of agricultural land in 1989 to 41.6 kg in 1993 (Table 1).

	1989	1990	1991	1992	1993	1994	1995
N	88.6	91.6	62.8	39.5	28.4	34.8	37.8
P_2O_5	69.7	69.0	30.7	12.6	7.2	8.0	9.5
K_2O	72.9	79.1	29.6	11.8	6.0	6.6	8.3
NPK	231.2	239.7	123.1	63.9	41.6	49.4	55.6

Source: Testing and Control Institute for Agriculture, Annual Report, 1996

Table 1: Average consumption of fertilizers in kg/ha^{-1} of nutrients, 1989-1995

Monitoring of available nutrients and soil acidity testing, which was obligatory for agricultural land in Slovakia since 1956, showed an improvement of agrochemical parameters of soil up to 1990. The area of acid soils and soils with low levels of available phosphorus and potassium decreased since then (Fig. 2).

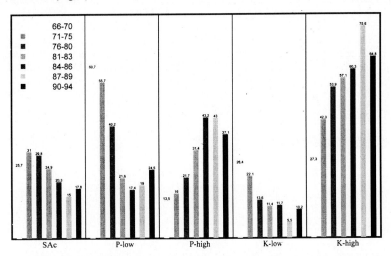

Fig. 2: Soil acidity and nutrients content(in per cent of agricultural land area)

Kobza (1995) reported that the contents of available phosphorus increased until 1990 by about 320 % and of available potassium by about 105%. The content of available phosphorus varied between 7.6 - 38.7 mg/kg^{-1} between 1961-1970, and at the end of 1990 it was between 49.2-128.2

mg/kg^{-1}. The content of potassium between 1961-1970 was between 64.4-129.1 and in 1990 137.7-306.5 mg/kg^{-1}.

Inadequate fertilization after 1992 caused a dramatic decrease in crop yield from 5.19 t to 3.76 t/ha^{-1}, i.e. 27.7 % (Table 2). There was also a decrease in the use of organic fertilizers, especially farmyard manure, because the number of cattle was reduced from 1.6 million animals in 1989 to 900,000 in 1995.

	1989	1993	1994	1995
Cereals together	5.19	3.76	4.30	4.51
of it: Wheat	5.33	3.85	4.85	4.90
Maize	5.55	4.62	4.14	4.70
Sugar beet	34.30	34.26	34.57	34.26
Potatoes	13.56	18.44	9.57	13.06
Vegetables	18.60	16.48	17.19	15.94

Source: Statistical Office of the Slovak Republic, 1995

Table 2: Main crops yields (tonnes per hectare) 1989-1995

Between 1990-1995, acid soil area increased by 2.8%, high content phosphorus soil area decreased 5.9%, and soil area with high potassium content decreased by 10.8% (Table 3).

Farmer lack of finances is also the reason why the maintenance of drainage and irrigation systems in crop farms was partly or completely mishandled. Most of the equipment is out of use and can no longer be reconstructed because the estimated cost of doing so is several billion crowns.

	Soil acidity				Phosphorus			Potassium		
Period	SAc	Ac	N	Al	L	M	H	L	M	H
1966 - 70	25.7	26.2	27.8	20.3	60.7	25.0	13.5	26.4	46.3	27.3
1971 - 75	31.0	22.2	23.6	23.2	55.7	28.3	16.0	22.1	35.6	42.3
1976 - 80	29.8	25.8	26.7	17.7	40.2	38.1	21.7	13.6	32.5	53.9
1981 - 83	24.9	27.8	259	21.4	21.6	47.0	31.4	11.4	31.3	57.1
1984 - 86	20.3	29.4	31.1	19.2	17.4	39.4	43.2	11.7	28.0	60.3
1987 - 89	15.0	28.2	32.7	24.1	19.0	38.0	43.0	5.5	18.9	75.6
1990 - 94	17.8	24.4	38.3	19.5	24.5	38.4	37.1	10.2	2.0	64.8

SAc - strong acid, Ac - acid, N - neutral, Al - alkaline   L - low, M - medium, H - high
Source: Collective (1995)

*Table 3: Evolution of soil acidity and available nutrients content
(Extension in % of agricultural land area)*

4 Conclusions

Radical economic reforms since 1990 have led to strong restrictions on soil productivity factors and consequently to the retardation of agriculture in Slovakia. Unfortunately over the last five years, some production parameters have regressed to levels experienced 20 to 30 years ago.

The deterioration of soil fertility parameters necessitates a correction in the present state agricultural policy. This policy must be compatible with the agricultural policy of EU countries as well as motivate domestic farmers to improve soil fertility.

References

Collective 1(995): Results of Agrochemical Soil Properties Testing, Cycle IX (1990 - 1994), UKSUP Bratislava (in Slovak)
Documents (1993): The Concept and Principles of the Agrarian Policy, Ministry of Agriculture of the Slovak Republic, Bratislava
Documents (1995): Report on Agriculture and the Food Industry in the Slovak Republic 1995, (Green Report), Ministry of Agriculture of the Slovak Republic, Bratislava
Hraško, J. (1988): Applied Soil Science, 467 pp. Publishing House Príroda, Bratislava (in Slovak)
Hraško, J. (1993): Agriculture, Landscape and Environment In: Proceedings of the International Conference on Evolution in Agriculture of Central and Eastern Europe and Europe - wide Cooperation, Agroinstitut Nitra
Hueg Jr., W.F. (1977): Focus on the Future with an Eye to the Past, In: ASA Special Publication Number 30, pp. 73-86, Madison
Kobza, J. (1995): Retrospective view on some soil properties changes induced by man for the period 30- 35 years, Proceedings of Soil Fertility Institute 19/I., Bratislava
Murgaš, J. (1993): Agroreforms in Transformation, Privatisation and Restructuralisation in Slovakia within the Context of European and Global Development In: Proceedings of the International Conference on Evolution in Agriculture of Central and Eastern Europe and Europe - wide Cooperation, Agroinstitut Nitra
Špaldon E. (1982): Plant production, 614 pp., Publishing House Príroda Bratislava (in Slovak)

Address of author:
Juraj Hraško
Slovak Agricultural University
Marianska 10
SK-949 01 Nitra, Slovak Republic

Spatial Planning Tasks for Sustainable Land Use and Soil Conservation in Germany

F. Dosch & S. Losch

Summary

Land sealing, contamination, devastation and the redevelopment of soils costs billions of Marks in Germany each year. Increasing living standards and a growing population are the driving forces for the impairment of natural soil functions. The continuing consumption of land, mainly on the urban fringes, intensifies urban sprawl. Spatial planning acts as the mediator between land use demands and soil conservation. The German Building Land Act prescribes a sparing *and* careful use of soil. Recent legal initiatives, such as the amended Federal Building Code and the Spatial Planning Act, also support the importance of soil conservation. Sustainable land use and soil conservation can only be realised by an economical land policy. This means that the soil must be maintained for following generations according to its ecological functions and its utilization, conservation, cultivation and development.

Key words: Settlement development, land consumption, urban sprawl, surface-sealing, sustainable soil conservation, spatial planning tasks

1 Land - an environmental medium which is forgotten and tread upon?

In the mid-1990's some 81.5 million people were living in the Federal Republic of Germany in an area of about 357,000 km^2, with over two-thirds of the population in towns. Germany is one of the most densely populated states in Europe with a population density of 228 inhabitants per km^2. As a result of dense settlement intensive use of natural resources, the sealing, contamination, devastation and redevelopment of soils entails costs of billions of Marks per year. Nevertheless environmental policy limits soil conservation mainly to end-of-the-pipe measures such as soil rehabilitation. To avoid further damage, the conservation of soil functions will be of great importance in the future.

In the old federal Länder (West Germany before reunification), the settlement area covers 13.1% 1997 and increased by 7.1 % in 1950 to 12.7 % in 1993, which equals 1.4 million hectares (fig.1a). This dramatic increase in the settlement area of almost 80 % contrasts with an increase in the population of only 28.5 % in the same period (50.8 to 65.3 million inhabitants, see fig. 1b).

The main causes are increasing mobility and increasing demands for housing space. The result is a polycentric settlement system, which leaves only few larger inhabited open areas (fig. 2).

Economic growth, increasing traffic and settlement areas are the main reasons for continous soil pollution. In the past decades, sulphur dioxide and dust were enormous air pollutants, which severely contaminated lakes and rivers. At the beginning of the 1970s environmental protection activities began emphasizing water conservation. The Law on Water Economy (Wasserhaushalts-

*Fig. 1: Land use and settlement area 1950-1993
a: Settlement area in 1950 and 1993 (Old Länder) →
b: Land use in 1950 - 1993 (Old Länder) ↘*

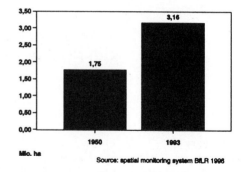

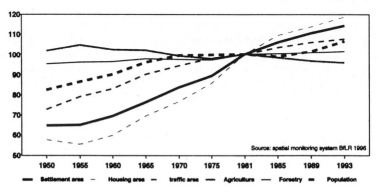

Settlement structural area types	Inhabitants in 1,000	Share of Inhabitants %	Inhabitants per km²	Share s&t[1] in %
Agglomeration areas	34,146	53.7	509	18.6
Urbanised areas	18,979	29.9	197	11.8
Rural areas	10,437	16.4	123	8.8
Old Länder total*[)	63,563	100.0	256	12.6
Agglomeration areas	9,229	51.9	352	12.3
Urbanised areas	5,255	29.6	150	8.0
Rural areas	3,291	18.5	69	6.0
New Länder total*[)	17,775	100.0	163	8.2
• Agglomeration areas	43,375	53.3	465	16.9
Central towns	20,444	25.1	2,275	51.4
Suburban counties	22,932	28.2	272	13.2
• Urbanised areas	24,234	29.8	185	10.8
Central towns	4,592	5.6	1,256	35.2
Suburban counties	19,642	24.1	154	10.1
• Rural areas	13,729	16.9	104	7.8
Germany total	81,338	100.0	228	11.3

1) s&t = Settlement and traffic areas; share of total area
*) Berlin is regarded as belonging to the new Länder
Source: Spatial Monitoring of the Federal Research Institute for Regional Geography and Regional Planning (BfLR)

Table 1: Population distribution and settlement structure

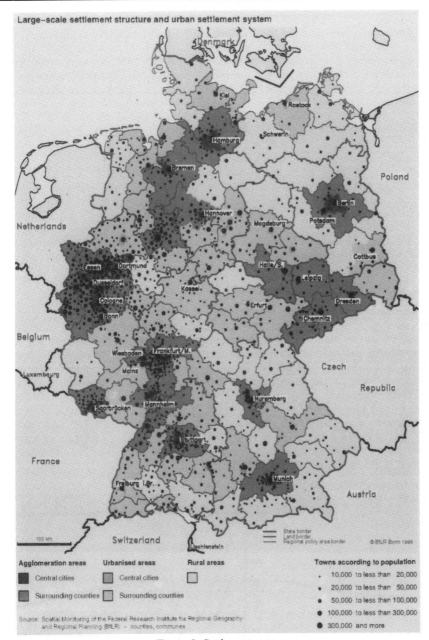

Figure 2: Settlement system

gesetz), the Detergent Law (Waschmittelgesetz) and the Sewage Tax Act (Abwasserabgabengesetz) improved water quality considerably. Although the Federal Law on Immission Control (Bundesimmissionsschutzgesetz) together with the Heating Plant Decree (Feuerungsanlagenverordnung) have paved the way for a considerable reduction of pollutants causing soil acidification, soil protection measures did not enter into environmental policy discussions until the mid-1980s when the Federal Soil Conservation Concept (Bodenschutzkonzeption) was developed. The tasks and requirements for soil conservation are unevenly distributed in the institutions of the German Länder, and they are inconsistently regulated. Nevertheless, a special Soil Conservation Act has been discussed for more than 10 years(!), and will probably be passed in 1998.

In the next century the population growth, which began in the 1990's predominantely due to a sharp increase in external migration, will continue. Increasing living standards and the growing population are the driving forces for the impairment of natural soil functions. The continuing consumption of land mainly on the urban fringes intensifies urban sprawl, surface sealing, and the segmentation of natural areas (Bundesforschungsanstalt für Landeskunde und Raumordnung 1996). As a result of the accumulation of toxic substances, contamination of soils can also be found in water riverbeds. The many functions of soil, as the section between air and water, is threatened: as a living space, filter and buffer, and as a regulator of demineralisation and metabolism.

But the surface cannot be multiplied without constraints and soils can not be burdened unlimitedly. The major traditional regulatory instruments for soil conservation are the Building Land Act, the Nature Conservation Act and the Environmental Impact Assessments (Umweltverträglichkeitsprüfungen). Spatial planning serves as the mediator between land use demands (e.g. food production, resources, building land) and soil conservation. Soil conservation by means of spatial planning has two main environmental objectives: on the one hand it is necessary to reduce consumption of land (quantity target), and on the other hand we must develop the ecological qualities of the soil (quality target).

2 On the surface - will urbanisation never end?

During the last decades, the development of settlements has been far from sustainable. During the 1970's, settlement and traffic areas in the old Länder increased daily by some 130 hectares, generally at the expense of land previously used for agriculture, which was often also meadow and pasture land valuable for nature conservation. The continuing consumption of land in the old German Länder decreased temporarily in the 1980's from 119 ha/day to 71 ha/day for settlement purposes in the period from 1989 to 1992. After that, the number of building land approvals increased dramatically (fig. 3). The daily consumption for new settlement area increased from 1953-1996 to 84 ha/day (Old Länder) or 120 ha/day in Germany. In 1993 the share of settlement and traffic areas in the total area of the federal territory was slightly greater than 11 % (fig. 4). Approximately 2 million hectares of land surface are sealed.

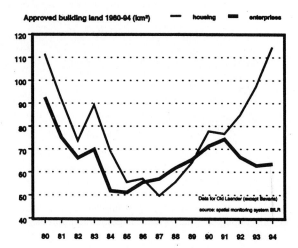

Fig. 3: Building land approvals

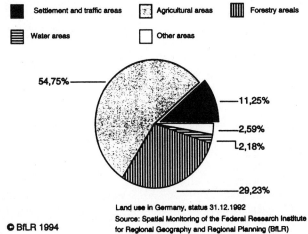

Fig. 4: Land use 1993 (Old Länder)

In the new Länder, land in the suburbs of the central towns can be made available more quickly and more cheaply and speedy building can be guaranteed. A widespread, dispersed development of settlements in the surrounding areas of the central towns is thus a fact also in the new Länder, amplified by the rapidly increased significance of cars. Between 1990 and 1995, the level of motorisation in the new Länder has increased by over 250 percent (fig. 5).

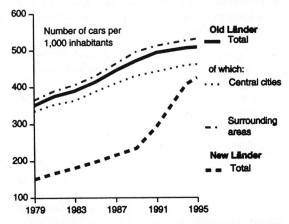

Fig. 5: Motorisation 1979-1995

Although the population declined following the German reunification, a rapid suburbanisation process began in the new Länder. The suburbanisation process in the old and the new Länder is a consequence of the rising demand for housing space and can be observed clearly in the increase of detached single-family houses and industrial enterprises. With 37 m^2 (Old Länder) and 27 m^2 (New Länder) of living space per person on average, Germany compares well internationally (fig. 6) (Federal Ministry for Regional Planning, Building and Urban Development 1996).

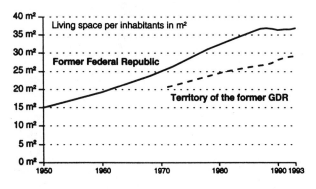

Fig. 6: Living space 1950-1993

The trends in settlement development in Germany show that it is still far from sustainable. The demands from private households for accommodation and the location demands of companies in conjunction with economic cost/benefit considerations are the primary driving forces for the continuing expansion of settlement areas. Settlement growth is shifting further and further from the central towns, the urban fringe is growing, and the building of city networks up to the merging of big metropolitan areas is induced. Specific driving forces for suburbanisation include the high land prices in the towns and the growing demands on the quality of living conditions. The trends in population and employment in the old Länder show that both are increasing more sharply outside the agglomeration areas. But many households still find less expensive possibilities for meeting their housing wishes more easily in the rural hinterland (fig. 8).

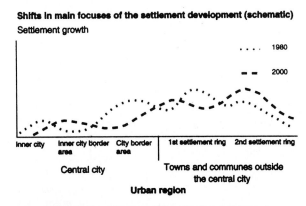

Fig. 7: Suburbanisation 1980/2000

The prognosis of the demand for land show that Germany will need approximately 100 hectares/day for new settlement purposes. With the long-distance transport of air and water pollutants, the effects of the towns are spread far beyond their boundaries into the surrounding areas. Local and regional problems increase - even to the point of becoming global risks.

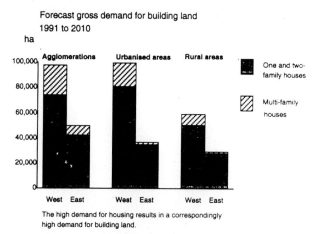

Fig. 8: Building Land 1991-2010 © BfLR 1995

The result of continuing land consumption is:
- a further land-consuming spatial expansion of urban agglomerations,
- a further increase in motor traffic with a consequent increase in environmental pollution due to immissions and noise.
- a further loss of green areas near the settlements and a further reduction in ecological compensation functions.

It leads to urban sprawl with high consumption of land for settlement purposes and, in some parts, to soil destruction.

3 Below the surface - pollution and destruction of soil functions

As a result of continuing consumption of land, open land was sealed by buildings and roads. Rainwater can no longer percolate in these sealed areas, and natural soil formation is disrupted. Almost 50 % of the settlement areas are sealed. Each year up to 18,000 hectares of newly sealed surface cover the soil. The consequence is a destruction of the soil and its natural functions.

In some parts soil devastation by mining seems irrevocable (fig. 9). Consumption of land, contamination, devastation and rehabilitation led to an economic loss of 22 to 60 billions DM annually. This includes the enormous costs for the rehabilitation of polluted former industrial sites.

Despite the reduction in air pollution, Germany is still faced with problems of soil contamination through the input of toxic substances, the large input of organic compounds, waste and sewage disposal as well as the input of radioactive material. The soil devastation from intensive agricultural land use continues regardless of the reduction in the input of fertilizers and biocides. Poaching, puddling and compaction of soils by farming (fig. 10) and soil erosion (fig. 11) are severe problems at the regional level. Tourism leads to a devastation of sensitive soils (steep slopes in the alps or marshland).

Widespread soil acidification occurs and the solution of aluminum-hydroxy-sulphate complexes (fig. 12) as well as the leaching of complex organic compounds or heavy metals from sediments and streams and produce dangers to men, animals and plants.

Fig. 9: Lignite mining (source Federal Ministry for Regional Planning 1996)

Fig. 10: Puddling and compaction (source: BUND Argumente, September 1995)

Fig. 11: Soil erosion (source: Scheffer, Lehrbuch der Bodenkunde 1989)

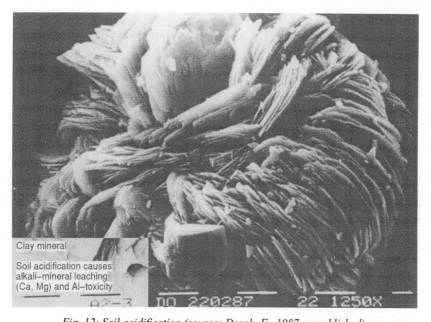

Fig. 12: Soil acidification (source: Dosch, F., 1987, unpublished)

4 Selected Challenges for soil conservation policy in Germany

Sparing *and* careful use of soils requires a reduction in consumption of land and conservation of natural soil functions in the future.

Soil must be maintained for future generations according to their ecological functions and their utilization, conservation, cultivation and development. Ecologically oriented objectives, especially at the urban level, can contribute to achieving a sustainable use of soil with the *existing* measures and building land planning instruments. Careful use of soil and land is a primary objective (Table 2).

Cause, Effects	Measures	Concepts, Instruments
• soil structure destruction caused by building, surface sealing, mining, compaction and fragmentations • loss of the soil functions buffering, filtering, regeneration of metabolism • loss of natural and cultural landscapes by the extraction of mineral resources, building and traffic • selective and widespread increase of pollution	• promotion of polycentric settlement structures by the decentralized concentration of population, jobs and infrastructure • improved management and use concepts for existing buildings, plants and areas (multiple instead of single use) • re-use of available building land (fallow and conversion areas) • a sparing and careful use of soil through concentration, mixed use and land-saving forms of settlement • measures to compensate the used areas of land and its soil functions • resource-conserving control of supply and demand	• supralocal planning and cooperation in the framework of regional and sub-regional planning according to the Spatial Planning Act, stipulations of the Federal Soil Conservation Act • improvement of soil conservation in the framework of local master planning (preparatory land-use plan, urban land-use plan as per Federal Building Code, landscape planning, nature conservation acts) • soil-conserving traffic policy: reduction of the use of automobiles by promoting local public transport, by mobility management, traffic-reducing planning, traffic avoidance • market-oriented instruments promoting a soil-conserving and land-saving behaviour (eg building land tax, sealing /land use tax, cancellation of the mileage allowance, increase of the tax on oil)

Source: own illustration

Table 2: Sustainable soil conservation planning tasks

Rules for sustainable land use:
- To re-use existing buildings, facilities and spaces more efficiently
- Distribute new buildings, facilities and spaces more sparingly and optimally within the spatial structure
- Consume less; only a reversal (change) of the continuously-expanding demand behaviour can help to maintain the remaining natural soil functions in the long-term.

In order to ensure an environmentally and socially sustainable land use in times of globalization with heavy economic competition, it is necessary to think of disciplinary legal and market-oriented instruments such as soil taxes. Spatial planning must contribute to sustainable land use. However, sustainable land protection can be achieved only if the claims for land consumption are also reduced. Perhaps sustainable soil conservation may be realised only in the long-term through additional soil taxes or surface-sealing taxes. Regarding open spaces, spatial planning should be the mediator for the integration of agriculture and nature conservation (Table 3).

Causes, Effects	Measures	Instruments
• Growing fertility losses through land cultivation (deep ploughing, compaction, clearing of crop residues, single-crop farming) • widespread soil pollution by overdressing and contamination (biocides, heavy metals) • homogenizing of agricultural production areas involving the destruction of the natural structure of species • increasing susceptibility of soil structures to wind and water erosion	• turning away from single-crop farming, crop rotations with many parts to increase biodiversity • reduction of the use of fertilizers and biocides • change of agricultural production through increased biological forms of production • reduction of the size of agricultural lots, cultivation with light machines • change of soil cultivation methods • site-adapted cultivation [Standortgerechte Bewirtschaftung]	• EU Directive 2078 • Farming Law • Fertilizer Law [Düngemittelgesetz] • Pesticide Law [Pflanzenschutzmittelgesetz] • Funding Guidelines of the EU, the Federation and the Länder • Code of good agricultural practice

Source: own illustration

Table 3: Soil protection tasks for agricultural land use

Spatial planning contributes to a sustainable use of soil if careful use of soil is implemented in land-use and local planning and if structure planning is guided by sustainable land use strategies. The Study Commission on the Protection of Mankind and the Environment (Enquete-Kommission des Deutschen Bundestages "Schutz des Menschen und der Umwelt" 1997) defined goals of environmental quality and environmental action for soils. The most important environmental goals are, apart from the protection against pollutants, a drastic reduction of surface sealing and landscape consumption. Thus additional land consumption should be reduced to 10% of the present rate by the year 2010 (Deutscher Bundestag 1997).

A further integration of agriculture and nature conservation is required in rural land (fig. 13).

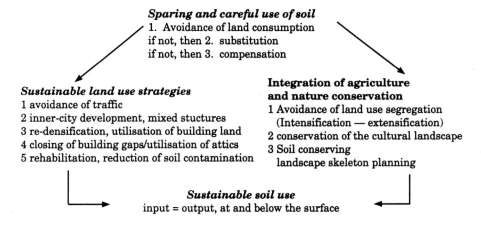

Source: own illustration

Figure 13: Spatial planning tasks for sustainable soil use

Germany is at the beginning of a long, laborious and painful path of learning to split its *economic growth and increase in the settlement area*. Soil resources are limited. Handling the ecological resource, "soil", more carefully minimize man-made natural desasters as well as creeping erosion of a necessary basis of life.

References

Bundesforschungsanstalt für Landeskunde und Raumordnung: Ausmaß der Bodenversiegelung und Potentiale zur Entsiegelung. Arbeitspapier 1/1996, Bonn 1996.

Deutscher Bundestag: Konzept Nachhaltigkeit. Zwischenbericht der Enquete-Kommission "Schutz des Menschen und der Umwelt - Ziele und Rahmenbedingungen einer nachhaltig zukunftsverträglichen Entwicklung" des 13. Bundestages. Deutscher Bundestag, Referat Öffentlichkeitsarbeit, Bonn 1997.

Federal Ministry for Regional Planning, Building and Urban Development: Human Settlements Development and Policy. National Report Germany HABITAT II, Bonn 1996.

Address of authors:
Fabian Dosch
Siegfried Losch
Federal Office for Building and Regional Planning (BBR) (ferner BfLR)
F I 6 - Soil Section
Am Michaelshof 8
D-53177 Bonn, Germany

An Ecological Perspective of Soil Conservation Policy and Law in Australia

I. Hannam

Summary

Sixty years of soil conservation in Australia has been unable to prevent the development of serious soil and land degradation. Recent land degradation surveys indicate that, ecologically, soil has received poor attention and is severely affected by a range of degradation processes. The paper covers reasons why soil conservation legislation and policy were introduced in Australia, generally, but focuses on some specific aspects of the New South Wales legislation. Comments are made on the current paradigm for soil conservation policy and law in Australia and argues its ineffectiveness in preventing degradation of soil, and inadequacy to manage soil in an ecologically sustainable manner. Characteristics of an alternative policy and legislative paradigm to manage soil in an ecologically sustainable manner are presented.

Keywords: Soil conservation, policy, soil legislation

1 Introduction

Sixty years of soil conservation legislation in south eastern Australia has failed to prevent serious soil degradation despite the legislation being introduced with powerful provisions to prevent and mitigate against land degradation (Bradsen, 1988). Surveys indicate that soil is affected by a range of degradation processes (DEHCD, 1978, Graham, 1992). It is argued that one of the main reasons for the poor ecological state of soil is the failure by successive governments and administrations to apply the legislation for the reasons it was introduced. Legislation introduced in south eastern Australia is considered inadequate to effectively assess and manage soil in an ecologically sustainable manner which is consistent with soil conservation law in general in Australia. Soil conservation policy is outdated and generally in sympathy with agricultural development objectives (Hannam, 1992). More Commonwealth and State government effort is needed to translate relevant aspects from global environmental strategies into soil conservation environmental law and policy reform (Boer, 1995).

1.1 Soil and land degradation

The Australian continent only has a small area of naturally fertile soil, and most of its soil is very ancient, shallow and low in fertility by world standards. Extreme seasonal and periodic climatic variations predispose soil to degradation through wind and water erosion, salinity, depletion of chemical and physical fertility. Woody weed invasion and mass movement, are slower to appear and have equally significant effects on the soil. Soil and land degradation accelerated in the 1960's when, under government policy, the cropping area expanded and the intensity of land use increased. Some areas of fertile soil used for intensive dryland agriculture have less than 100 years of productive life left as a result of soil erosion (McTainsh and Boughton, 1993). Conservation farming has been encouraged, but landholders are reluctant or slow to adopt these soil conservation practices (Bradsen, 1994). The severity of land degradation led to a Commonwealth-State Soil Conservation Study of 1975-1977 (DEHCD, 1978).

A systematic land degradation survey of New South Wales was carried out between 1987-88 (SCSNSW, 1989). Recent information indicates that the long term sustainability of cropping on sloping land is threatened by soil erosion (Australia, 1996a) and that only 2 per cent of Australia has soils regarded as excellent quality and the majority is not useable for agriculture (CSIRO, 1996). One of the reasons for the ineffectiveness of soil conservation policy and legislation is that the definition of soil conservation in fact encourages many of the processes which lead to soil degradation (Boer and Hannam, 1992). The current Australian definition and the global definitions emphasise productive capabilities of soil rather than their ecological sustainability. The 'productivity' definition is enshrined in Australian soil conservation legislation through their 'conservation with agriculture' theme (Hannam, 1992).

2 Soil conservation policy

The effectiveness of soil conservation policy in Australia has a mixed history, and is primarily characterised by its support for agricultural rather than ecological motives (Hannam, forthcoming). National reports recognise a general failure of legislation and land use policy in controlling land degradation. They point out that a national approach to land use policy and environmental management, and the mechanisms to plan and implement policy have been made by numerous Committees of inquiry over the years with little result (Australia, 1984). A 1989 review of policies and programs to combat land degradation also summarised the assistance given to soil conservation over the years, including the distribution of educative land use policy material, financial support for particular policies and programs, and a range of tax concessions. The review concluded that soil conservation policy had failed as evidenced by the high level of land degradation (Australia, 1989).

The 1978 Commonwealth-State soil conservation study (DEHCD,1978), was a good basis for governments to develop soil conservation and associated land and water management policies and programs, but was restricted to broad policy issues, rather than offering new policy direction and specific policies for the States to adopt. This study could have been used to initiate a comprehensive philosophical, ideological and strategic change in soil conservation policy and legislation in Australia. A significant amount of policy material has appeared in Australia in the last decade aimed at 'integrated resource management' which relies on coordination of government departments, authorities, private companies and individuals who have land management responsibilities (Blackmore et al., 1995). Some of the policies have been supported by specialist legislation (e.g., the New South Wales *Catchment Management Act 1989*). Integrated resource management is a logical way to approach and solve land degradation problems and should be supported by enabling environmental laws. In 1987 New South Wales released the first actual 'Soils Policy' in Australia as a major contribution to its integrated resource management strategy, but the effectiveness of the Soils Policy was weakened by not being linked with the *Soil Conservation Act 1938* (SCSNSW, 1987).

Landcare, Australia's most recent and ambitious land conservation initiative, is aimed specifically at encouraging the adoption of sustainable land management practices (Douglas et al., 1995). This area of policy encourages integrated management of land degradation problems through soil, water, and native vegetation conservation. State governments have reacted by amalgamating land, water and forest management agencies into 'mega departments', a move which has led to the demise of some specialist land degradation management agencies and specialist scientific expertise (Boer and Hannam, 1992). It is reasonable to say that Landcare has been partially successful as only 20-30 percent of Australian landholders have joined Landcare groups but this is considered as not sufficient to solve Australia's problems of land degradation (Bradsen, 1994). Improving the public participation capabilities of the legislation, with an appropriate duty of care on the part of government agencies, could assist with the wider adoption of integrated resource programs such as Landcare.

3 Soil conservation and environmental law

By comparison to other areas of environmental law in Australia, soil conservation law has received little attention (Bates, 1992). The common aspect emerging from soil conservation law studies is that it has not been effective (Bradsen, 1988). Land degradation alone is not a good parameter, but the effectiveness of the Acts should also be measured against an ecological or land sustainability ethic. The newest piece of State legislation is the South Australian *Soil Conservation and Land Care Act,* 1989. It moves away from some of the historic 'agricultural' concepts of soil conservation legislation in Australia, and includes a land conservation objective (Bradsen, 1991). The Commonwealth has no specific power to legislate with respect to soil conservation but has attempted to influence soil conservation policy through various Standing Committees. The main Commonwealth contribution to soil conservation was the allocation of finance to the States under a National Soil Conservation Program (DPI, 1986). Established in 1983 in recognition of the serious national land degradation problems, it stimulated increased effort by the States to control degradation. Although the position of the Constitution in relation to power over the environment has been argued in various areas, particularly the Federal governments power to exercise environmental controls over the States (Fowler, 1993), there are new mechanisms available which may achieve the same objective (Australia, 1992). Processes required to implement the *Intergovernmental Agreement on the Environment* should by their very nature bring the Commonwealth and the States closer together in natural resource legislation and policy reform (Boer, 1995).

3.1 NSW Soil conservation legislation

The New South Wales *Soil Conservation Act 1938* was introduced to protect catchment areas from soil erosion. It contains some sound eco-ethical concepts and accompanying provisions but many have not been implemented to achieve sustainable land use, or have been used in a very limited way (Hannam, 1992). Over the years, of the amendments made to this Act, only few have attempted to improve the ecological aspects of soil conservation, and most have been concerned with administrative and mechanical procedures. There have been few challenges to the philosophical basis and role of the legislation (Bradsen, 1988). The Act has provisions to prevent ecosystem damage and to restore damaged ecosystems. The 'erosion hazard' concept is based on the philosophy of penalising landholders for polluting as in the *Rylands v Fletcher (18680 LR 3 HL 330)* rule. Ecological standards could have been established from the early catchment studies to create the legal grounds to declare 'areas of erosion hazard'. Had this been done on a consistent basis, rather than resort to persuasion, the level of soil erosion would probably be less today, and other forms of land degradation may have been detected at an earlier stage of development or may not have occurred at all. Soil conservation 'projects' have treated large areas of eroded land. Their application has been limited to technical solutions without environmental impact assessment to detect unsustainable environmental consequences of land use. The complex structure of the administrative processes to implement 'projects' and to protect 'areas of erosion hazard' prevented their wider application. The Protected Land division of the Act differs from the others and uses provisions for ecologically sustainable land management based on a regulatory approach (Hannam, 1992).

Despite a consistent plea from the conservation sector for greater use of the enforcement powers in the soil conservation legislation, the landholder lobby groups have been more influential on the government. Provisions cover, 'any land', 'areas of erosion hazard', 'protected land', and 'catchment areas', so there has been at least one provision available for application to any area of land degradation. Vast areas of the State are in a highly hazardous condition today, many ecosystems have become degraded, and the public benefit of soil conservation has been reduced because of the failure of administrations to use the powers to control unsustainable land use (Boer and Hannam, 1992).

4 Legislation for ecologically sustainable land management

It is essential that soil legislative reform be approached from a sound conceptual basis with a societal goal of ecological sustainability (Harding, 1994). A change in attitude will ultimately depend on the willingness of society to accept new values in a legal system for soil conservation.

4.1 Natural rights for soil

The concept of natural rights as it relates to the environment has received quite a deal of discussion over the years (Nash, 1990). It is reasonable to take the view that all aspects of ecology should be properly considered by legislation and in the case of soil, it's fundamental role in ecosystem management should receive particular consideration. To draw on the various ecological concepts and ethics in an argument for the soil will require a major change in attitude by law makers. Legislation should reflect reasonable attitudes and responsibilities to ecological factors as well as reflecting an ecological consciousness that does not take away any rights that soil may have.

4.2 The public trust doctrine

The doctrine of public trust balances environmental concerns against the private property rights ethic. It reflects the belief that certain natural interests are so important that they must be protected in the public interest for the use of the whole population (Sax, 1989). It is relevant to soil policy and law because soil degradation is widely spread and affects the whole community. For this doctrine to be successfully adopted, the right to, and the circumstances surrounding the private ownership of soil must be reconsidered.

4.3 Sustainability

Various international and national strategies developed in response to the unsustainable use of natural resources encourage the adoption of sustainable principles in national and State environmental laws and policies (Boer, 1995). The application of sustainable concepts to soil legislation and policy will therefore involve reorganisation of soil conservation research and a change in attitude toward the development of land management guidelines which must be aimed at positive ecological benefits (Hamblin, 1991).

4.4 The precautionary principle

This principle says that where there are threats of serious or irreversible damage, a lack of scientific certainty shall not be used as a reason for postponing cost-effective measures to prevent environmental degradation (United Nations, 1992a). With soil legislation it means inclusion of specific provisions to research and study the soil, using environmental regulation and allocating adequate resources to avoid soil degradation. Where there is doubt about the consequences of an action, a decision should err on the side of caution.

4.5 Conservation of biological diversity

The conservation of biological diversity is a key element of sustainable land management, and underpins society well-being through the provision of ecological services such as those that are essential for the maintenance of soil fertility (Australia, 1996b). Provisions in soil legislation for the planning and control of land development, including environmental impact assessment, are critical to the conservation of biological diversity.

5 Reform paradigm

There is now substantial argument to move away from the legislative and policy processes which have led to unsustainable use of soil to a policy and legislative paradigm based on ecologically sustainable concepts and provisions. Globally, there is substantial argument for individual nations to make a serious commitment to the formulation of national sustainability policies and strategies and to support these with appropriate laws (United Nations, 1992b). Reforming Australian soil conservation law under the concept of land conservation would give special consideration to the sustainable land use objective. Some Australian States have already enacted legislation which specifically incorporates sustainable development concepts into its provisions and sets out the basic principles by which it might be achieved (Boer, 1995). An attitude which embraces the concept of soil as part of our natural heritage with important links to our cultural heritage is fundamental for a long term sustainability goal for soil.

5.1 Policy, legal and institutional aspects

Strategies are required to shift the commodity-oriented view of soil to a social ecological view. Legislation alone will not be sufficient, it must be supported by comprehensive on-going community education programs eg, Landcare. Existing State level policy can be redrafted with a basic duty of care to develop and implement ecologically sustainable land use systems, priorities for biodiversity maintenance, education on conservation and a reappraisal of the application of some existing important concepts, such as 'protected land' (Boer and Hannam, 1992). Secondly, ethical rules must be established for soil ecological sustainability as well as rules which recognise the need to research and establish physical and chemical limits for soil and principal land use activities (Hamblin, 1991). The evaluation of ecological soil standards for sustainable land use against existing institutional and legal standards will indicate whether philosophical and research objectives have the potential to be achieved, or are currently being achieved. This will require a reassessment of government and community attitudes to soil conservation, and must address the important areas of property rights and responsibility.

Commonwealth and State responsibilities for soil conservation/soil environmental protection will need to be examined, and a commitment to implement enforcement powers to ensure ecological standards of soil sustainability will be met is essential. Major economic and sectoral agencies of government must be made directly responsible and fully accountable for ensuring that policies, programs and budgets support development that is ecologically sustainable. The philosophy of the legislation would focus on prevention of ecological deterioration of soil, rectifying degradation and establishing decision-making systems which support sustainable land use. Basic mechanisms to achieve soil sustainability include those which prescribe behaviour and establish a *duty of care* ethic, establish rules and criteria for ecological soil standards, create cooperation in soil management, and determine the circumstances for regulatory intervention (Hannam, 1992).

5.2 Public participation in policy and legislation

Suitable soil legislation would include provisions to invite public opinion on any area of soil by: constituting community-based soil advisory bodies, preparing State soil strategies, publicly exhibiting soil management plans, providing for conservation agreements which contain ecologically sustainable techniques, including powers to control land use activities, and a standing provision so an action may be brought against an individual or corporation if an activity threatens or potentially threatens the ecological integrity of soil.

5.3 Rights and responsibilities

A national responsibility to an adequate environment for present and future generations is an important step towards the broader societal goal of ecological sustainability - as drawn from the global literature on the rights and responsibilities of individuals and States regarding sustainable development. An Australian law of this type would have considerable benefits for soil. Constitutional difficulties for soil would probably still remain until the Constitution could be amended to include an environmental power. Major rights for individuals relevant to soil include access to judicial and administrative proceedings, including redress and remedy in exercising their rights and obligations to soil legislation, the right for any person to take legal action against another person for causing soil degradation, and the right for any member of the public to participate in the planning process for soil.

6 Conclusion

There is sufficient grounds to advocate that Australia should seek an alternative paradigm for soil aimed at conserving its ecological integrity. It is difficult to see that it can be reached without significant changes within the Australian social and cultural fabric, particularly the current attitudes towards development and the preservation of 'private property rights'. Implementation of soil conservation cannot be left as a choice alone by the individual. The results of past institutions allowing this to happen, together with the promotion of soil conservation as a necessity for 'production', has not been successful. The recent 'integrated resource management policies', whilst advocating solutions to land degradation, offer little more than the past. Building these concepts into broad-based ecologically sustainable policy and legislation with specific provisions for their implementation and for on-going support (including incentives), would be a substantially better approach. At the national level, effort must be made to take up the basic responsibilities of the *Intergovernmental Agreement on the Environment* as they relate to soil. This will make clear the responsibilities of the Commonwealth, the State and local government to soil conservation policy and legislation, and it will provide a better basis for cooperative conservation programs.

References

Australia (1984): Land Use Policy in Australia, Senate Standing Committee on Science, Technology and the Environment. Australian Government Publishing Service. Canberra.
Australia (1989): The Effectiveness of Land Degradation Policies and Programs. Report of the House of Representatives Standing Committee on Environment, Recreation and the Arts. Australian Government Publishing Service. Canberra.
Australia (1992): Intergovernmental Agreement on the Environment.
Australia (1996a): State of the Environment. Australia. An Independent Report Presented to the Commonwealth Minister for the Environment by the State of the Environment Advisory Council.
Australia (1996b): The National Strategy for the Conservation of Australia's Biological Diversity.
Bates, G. M. (1992): Environmental Law in Australia. Butterworths. Sydney.
Blackmore, D.J., Keyworth, S.W., Lynn, F.L. and Powell, J.R. (1995): The Murray-Darling Basin Initiative: A case study in integrated catchment management. in Sustaining the Agricultural Resource Base. Office of the Chief Scientist. Department of the Prime Minister and Cabinet. 61-67.
Boer, B.W. (1995): Institutionalising Ecologically Sustainable Development: The Roles of National, State, and Local Governments in Translating Grand Strategy into Action. Willamette Law Review 31:2:307-358.
Boer, B.W. and Hannam, I.D. (1992): Agrarian Land Law in Australia. in, W.Brussaard, and M.Grossman. (eds). Agrarian Land Law in the Western World. C.A.B. International. Wallingford. Oxon 212-233.
Bradsen, J.R. (1988): Soil Conservation Legislation in Australia. Report for the National Soil Conservation Program. University of Adelaide.
Bradsen, J.R. (1991): Perspectives on Land Conservation. EPLJ 8:16.
Bradsen, J.R. (1994): Natural Resource Conservation in Australia: Some Fundamental Issues. In: T.L. Napier, S.M. Camboni and Samir, A. El-Swaify (eds), Adopting Conservation on the Farm, An International Perspective on the Socioeconomics of Soil and Water Conservation, Soil and Water Conservation Society. Ankeny. Iowa. USA 435-460.
(CSIRO) Commonwealth Scientific, Industrial and Research Organisation (1996): Division of Soils, figures supplied.
(DEHCD) Department of Environment, Housing and Community Development (1978): A Basis for Soil Conservation Policy in Australia, Commonwealth and State Government Collaborative Soil Conservation Study, 1975-1977. Australian Government Publishing Service. Canberra.
(DPI) Department of Primary Industry (1986): Soil Conservation Advisory Committee Annual Report.1985-1986. Australian Government Publishing Service. Canberra.
Douglas, J., Alexander, H. and Roberts, B. (1995): Sustaining the Agricultural Resource Base: Community Landcare Perspective. in Sustaining the Agricultural Resource Base, Office of the Chief Scientist. Department of the Prime Minister and Cabinet 68-76.
Fowler, R.J. (1993): A Brief Review of Federal Legislative Powers with Respect to Environment Protection. Australian Environmental Law News 1:51-64
Graham, O.P. (1992): Survey of Land Degradation in New South Wales, Australia. Environmental Management 2:205.
Hamblin, A. (1991): Sustainability: Physical and Biological Considerations for Australian Environments. Working paper No WP 19/89 (revised edition). Bureau of Rural Resources. Department of Primary Industry and Energy. Canberra.
Hannam, I.D. (1992): The Concept of Sustainable Land Management and Soil Conservation Law and Policy in Australia. in, P.Henriques, (ed). Proceedings of International Conference on Sustainable Land Management. International Pacific College. Palmerston North. New Zealand 153-168.

Hannam, I.D. (forthcoming). 'Soil Conservation Policies in Australia: Success and Failures and Requirements for Ecologically Sustainable Policy'. In: T.L. Napier and S.M. Camboni. (eds). Soil and Water Conservation Policies: Successes and Failures. Soil and Water Conservation Society. 7515 Northeast Ankeny Road, Ankeny. Iowa 50021 USA.

Harding, R. (1994): Interpretation of the Principles for the Fenner Conference on the Environment. Sustainability-Principles to Practice. Unisearch. University of New South Wales.

McTainsh, G.H. and Boughton, W.C. (eds) (1993): Land Degradation Processes in Australia. Longman. Cheshire. Melbourne.

Nash, R.F. (1990): The Rights of Nature. A History of Environmental Ethics. Primavera Press. Australia.

Sax, J. (1989): The Law of the Liveable Planet. Lawasia and National Environmental Law Association of Australia. International Conference on Environmental Law. Sydney. Australia.

(SCSNSW) Soil Conservation Service of New South Wales (1987): State Soils Policy.

(SCSNSW) Soil Conservation Service of New South Wales (1989): Land Degradation Survey New South Wales 1987-1988.

United Nations (1992a): Agenda 21.

United Nations (1992b): The Rio Declaration.

Address of author:
Ian Hannam
Department of Land and Water Conservation
10 Valentine Ave
Parramatta 2150
New South Wales, Australia

Australia's Community 'Landcare' Movement: A Tale of Great Expectations

S. Ewing

Summary
Australia's 'Landcare' program is a community-based, participatory program established, by Government, to tackle the problem of land degradation. It has been received with enthusiasm locally and has been of some influence beyond Australia (Williams et al., 1995). There have been many studies which have sought to measure the success of Landcare and those factors which constrain it. This paper draws upon the practice of Landcare in one particular region of Victoria, in south-eastern Australia: a practice which is formed out of the complex interaction of state policy, community aspirations and capabilities and the broader political and economic context.

Keywords: Community involvement, Landcare, participatory planning, policy making

1 Introduction

In 1989, on the banks of Australia's 'mighty' Murray River, Prime Minister Bob Hawke launched his environment statement *Our Country, Our Future*, and announced that the 1990s in Australia would be known as the 'Decade of Landcare' (Hawke, 1989). Land degradation poses a major challenge to the Australian community. Amongst a litany of problems we face are increasing salinity levels in soils and streams, rising groundwater levels, soil erosion, water quality decline and the loss of native flora and fauna. Most areas of cropland and improved pasture in Australia are affected. It is a bleak scenario compounded by the knowledge that much of this degradation is irreversible or, at best, expensive to rehabilitate. Set against this background of crisis, the community Landcare movement has been described as 'undoubtedly the most exciting and significant development in land conservation in Australia' (Campbell, 1992). At its core is a growing network of 'Landcare groups', made up of farmers and others in the community, working together with state support, to tackle land degradation in their local area.

Despite frequent mention in the Hansards of the State and Federal Parliaments, land degradation was, for many years, perceived by our national Governments not to be a 'sexy', vote-catching issue (Bradsen, 1988). Campbell (1994) notes, for example, that by 1988, the budget for the (then) National Soil Conservation Program was still of the same order as the landscaping budget for the nation's new Parliament House. However, this apparent apathy was to change when an alliance was formed between the Australian Conservation Foundation and the National Farmers' Federation. In partnership, these two groups lobbied the Federal Government to launch a major new national land management program (Toyne and Farley, 1989); this we now know as the National Landcare Program. In light of the substantial change that we have wrought upon the Australian landscape, expectations of Landcare and more particularly of the Landcare movement, are high.

2 A decade of Landcare

The strategic framework for Landcare is provided by the national *Decade of Landcare Plan*, made up of strategic plans drawn up by the Commonwealth and each of the States and Territories (Commonwealth of Australia, 1991). The stated goal of the Decade is to achieve 'by the year 2000, ecologically sustainable land use through the management of land degradation'. What sets Landcare apart from previous policy initiatives is its emphasis not only on community participation and a bottom-up approach, but also on a broad integrated approach to catchment management. The National Landcare Program brings together, under the one banner, a number of established programs relating to aspects of catchment management. It is instrumental in implementing not only the Decade of Landcare Plan, but also elements of other strategies such as the National Strategy for Ecologically Sustainable Development and the National Water Quality Management Strategy.

2.1 Measuring progress

In Australia, as elsewhere, public policy toward soil and water conservation issues has come under increasing scrutiny in recent years. Even Landcare, which, for much of its life has been surrounded by a good deal of euphoria, is now subject to review.

The success of the Landcare program is often said to be self-evident in that from small beginnings, there are now an estimated 2,200 Landcare groups across the country, involving one third of the nation's farmers. There is evidence that Landcare mobilises a wide cross-section of the community. In Victoria, for example, Curtis (1996) reports that a large majority of groups involve others to assist with, or study, the work of Landcare in their area (80% and 83% of groups respectively). He estimates that, during 1995, at least 40,000 visitors were involved in studying or assisting with the work of the State's 700 Landcare groups.

Impressive though they are, facts and figures such as these can be seductive and in the past three years there has been increasing concern for the long-term viability of the community Landcare movement. Alexander (1995, p. 41), for example, in a recent report to government, observes that, at the Landcare group level:

> "There is an overwhelming sense that the resources being allocated to Landcare are inadequate given the scale of the problems and the opportunities in Landcare. There is concern that Landcare can no longer be driven by the few dedicated voluntary community members alone. There is a sense that whilst they are working to maintain the social, economic and environmental fabric of rural Australia, politics and economic rationalist governments are working against it."

In order to gain some insight into the experience and concerns of Landcare groups at the *local* level, I chose recently to approach Landcare through a detailed case study of Landcare practice (Ewing, 1995). The study was based in the middle reaches of the Glenelg River catchment in south-western Victoria, one of the State's prime wool growing areas.

3 Sustaining community effort in Landcare

As indicated earlier, the extent of environmental degradation which faces many rural communities in Australia is daunting. But a still greater difficulty lies in the shortage of the requisite resources to tackle them on the scale that is needed. In Landcare groups, there is a sense that the resources being allocated to Landcare are inadequate; of those resources which are available, too few can be used for the implementation of on-ground works. The main source of federal financial support to groups is community grants. Most of these, however, can only be used for purposes of public (rather than private) benefit, such as education, planning, demonstrations and monitoring. The government's position is that it is difficult to justify spending taxpayers' money where there is evident private economic gain. This has led to widespread frustration that, as one farmer put it, the government is 'all in for community, community, community [and] if they keep pushing that way, it's not going to get trees in the ground'. At a time of belt-tightening by governments, there is a pragmatic concern for public extension agencies unable to meet a demand for their services (Cary, 1993). The government has determined that a much greater multiplier effect can be achieved through incentive support for group extension activities, than can be achieved through funding works on individual properties (Campbell, 1992).

The unfortunate reality is, that at a time of severe recession in the farm sector, the financial viability of the family farm is often a more pressing concern for farmers, than is resource degradation. Many farmers feel that more and more is being asked of them at a time when rural communities are already under stress. For example:

> "I know that eventually they want us to become independent and self-sufficient, but I think that they ... I don't know, but I think there's too much strain on country communities as it is. *If it's not Landcare*, it's the schools. If it's not the schools, it's health. If it's not health, it's the local church and ... if it's not that, it's the fire brigade. And three to four nights a week you'd often be out to meetings trying to organise and get things running."

Against these sorts of expectations, some people have identified the potential for 'burnout' amongst Landcare groups and their leaders, leading to a loss of enthusiasm and an eventual decline in membership. One of the most valued forms of support for groups, therefore, is through short-term funding for the employment of group coordinators. Coordinators assist voluntary group leaders with the tasks of planning, meeting preparation, group motivation and external liaison. For many groups, their assistance is critical:

> "You will still need people like [our coordinator] to put the enthusiasm in to make everyone keep ... thinking there's something more they can do ... If it's left just up to the farmers, [the group] will fizzle."

For convenience, many people have chosen to refer to Landcare as a movement, of which Landcare groups are a part; I have tended to use the term this way here. But this nebulous notion of a movement may, in fact, serve to mask deviations from the rhetoric which drives the program (Lockie, 1995). It implies, for example, that Landcare groups develop spontaneously, independent of government. This is not the case and the risk is that policy makers may overlook the importance of appropriate support for the voluntary groups, in a way that is inconsistent with the notion of a *partnership* between the community and the state. Funding limitations are, of course, a perpetual problem in natural resource management. Nevertheless, as Alexander (1995) and others have argued, if our expectations of Landcare are to be realised, there are important issues which the Australian community, as a whole, has yet to address; issues of public benefit, national interest, public expectations of rural land use and the allocation of responsibilities.

4 A burgeoning bureaucracy

Another frustration at the Landcare group level is an apparent burgeoning bureaucracy of Landcare indicated through a seemingly endless supply of forms of one sort or another, newsletters, restrictions, rhetoric and meetings. One respondent, for example, recalls his speech at a State Landcare conference and the effect it had on the audience:

> "What was worrying me about Landcare, was the fact that the first couple of years we were in it, it was very much a local departmental Landcare group operation. And I said, 'Mr. Minister, it's worrying me that you are just creating an employment bureau for so many people ... well for your bureaucrats'. And I said that in our little group, we only had nine people in it, we had a wool bale full of correspondence in the last twelve months. I tipped it out in our lounge room and I said 'that is ridiculous'. I said 'we want trees and pastures and Landcare work done. We don't want jobs and we don't want paperwork'. And with that, the four hundred people all stood up and clapped me like mad. I felt ten foot tall."

There is also concern at the 'bureaucratisation' of community processes, particularly in aspects of planning where the ingenuity and autonomy of community Landcare is seen to be compromised by complex bureaucratic arrangements. Regional communities in Victoria, for example, have recently been invited to participate in the development of Regional Landcare Plans. Whilst the rhetoric of Landcare refers to community *participation* in planning, in practice it often works as some form of consultation, a rather less positive level of involvement. One of my respondents, for example, reflected on his experience of participatory planning in this way:

> "We saw that with [the government planner] when he brought out the [draft Glenelg Regional Landcare] Plan ... and I was horrified about that, and you probably heard all about that. I wrote letters and all sorts of things. And he wanted to come up and talk to me, and I wrote back and said 'I don't want you to talk to me. I want you to listen to me'. And that's what worries me. If you get the wrong person in there who wants to talk to you, they're the ones with the pens."

For this Landcare member, and many others, a tension has emerged between the original idea and its enactment. The participative process has been distorted through the dominance of 'expert' discourse and through consultation upon 'beautiful, glossy' draft plans which appear as if already in operation. Through the establishment of regional advisory structures, the community members of which are appointed by government, community participation often turns out to mean the addition of another tier to the bureaucratic process which governs natural resource management (Martin et al., 1992).

5 Landcare misappropiated?

One of the other ways in which government is seen to compromise the community ownership of Landcare, is through the corporate and urban amplification of the Landcare idea. Vanclay (1994, p. 45) warns that Landcare, 'far from being the bottom-up, community-driven and empowering process that the rhetoric would have us believe, is a classic case of hegemony at work'. In the hands of the corporate sector, he argues, Landcare is expanding from its agricultural base to be everything to everybody so long as it is remotely environmental.

The formal marketing arm of Landcare is Landcare Australia Limited (LAL), a non-profit public company formed by the Federal Government in 1989, to market Landcare nationally and raise funding from the corporate sector for promotions and Landcare group projects. In order to gain the interest of national companies, LAL has found it necessary to emphasise the importance of involving the urban community in Landcare since the corporate sector is apparently not interested in a movement that only involves the rural sector (Scarsbrick, 1994). In this chase for the sponsorship dollar, it is said that LAL has hijacked Landcare's original association with better farming, shifting the direction to 'recycling rubbish and healthy kitchen sinks'. It has also been associated with corporations of dubious environmental credentials (Brown, 1994).

It may well be that Landcare does need to have broader acceptance, to be part of the way all Australians (urban and rural alike) frame their environmental practice. But it is also possible that this expansion of the Landcare idea will not be supported by an expansion in funding allocations and may result in unnecessary competition between rural and urban interests for the already stretched resources of Landcare. I question whether, in this enthusiasm for a new amplified Landcare, which embraces not only the corporate sector, but also the long-term unemployed, water supply and regional development, we will be distracted from the original task of repairing the nation's lands.

6 Conclusion

In this paper, I have drawn upon the experience of Landcare groups to help identify some of the issues which may prevent our long-term expectations of Landcare from being realised; the ways in which costs are shared between the community and the state; the way in which the structures of Landcare bureaucracy often serve to position the locus of power and ownership with the state, rather than community; and the way in which the idea of Landcare has been amplified, such that it no longer refers exclusively to Landcare groups and a voluntary, community-based approach to land management problems.

Australians have embraced Landcare with enthusiasm and there is little doubt that Landcare groups have succeeded, beyond expectation, in capturing the public interest. Indeed it may be, as argued by Bradsen (1994, p. 440), that the greatest difficulty with Landcare lies in its perceived effectiveness to date: 'that is, it will be effective *enough* to prevent a more truly effective system of land conservation being put into place...'[1].

The task which remains, however, is huge. It is critical that Landcare moves beyond demonstrations and awareness-raising to a point where it does, indeed, prompt a significant change in land management practices on the ground; to a point where we can claim with confidence that there has been an enduring shift towards more sustainable use of our agricultural lands. It is clearly a lot to expect of a voluntary 'movement' in the absence of appropriate, long-term institutional arrangements of support. The fact is, that without the voluntary efforts of community Landcare groups and of those farming men and women who, in good faith, participate in Landcare programs, the cost of implementing the national Decade of Landcare Plan would be prohibitive and its goals beyond reach.

In a recent national review of Landcare, Alexander (1995) concludes that in Landcare, Australia has the framework for sustainable development which will be the envy of the world. We continue, rightly, to hold high expectations of all that Landcare might achieve. But, if community effort is to be sustained in the long-term, there are some important issues to be addressed.

Acknowledgements

The research on which this paper is based was funded by the Rural Industries Research and Development Corporation.

---

[1] Emphasis added

References

Alexander, H. (1995): A Framework for Change: The State of the Community Landcare Movement in Australia, National Landcare Facilitator Annual Report, National Landcare Program, Canberra.

Bradsen, J. (1988): Soil Conservation Legislation in Australia: Report for the National Soil Conservation Program, University of Adelaide Press.

Bradsen, J. (1994): Natural resource conservation in Australia. In: T.L. Napier, S.M. Camboni and S.A. El-Swaify (eds.), Adopting Conservation on the Farm: An International Perspective on the Socioeconomics of Soil and Water Conservation, Soil and Water Conservation Society, Ankeny, Iowa, 435-460.

Brown, B. (1994): Landcare: recovering lost ground - from the semi-arid zone to Lake Pedder. In: D. Defenderfer (ed.), Proceedings of the 1994 Australian Landcare Conference, v. 2, Department of Primary Industries and Fisheries, Hobart, 27 - 33.

Campbell, A. (1992): Landcare in Australia: Taking the Long View in Tough Times, National Landcare Facilitator Third Annual Report, National Landcare Program, Canberra.

Campbell, A. (1994): Landcare, Allen and Unwin, Sydney.

Cary, J. (1993): Changing foundations for government support of agricultural extension in economically developed countries, Sociologica Ruralis 33 (3/4), 336 - 347.

Commonwealth of Australia (1991): Decade of Landcare Plan: Commonwealth Component, Australian Government Publishing Service, Canberra.

Curtis, A. (1996): Landcare in Victoria: A Decade of Partnerships, Johnstone Centre of Parks, Recreation and Heritage, Report No. 50, Charles Sturt University, Albury.

Ewing, S. (1995): It's in Your Hands: An Assessment of the Australian Landcare Movement, unpublished PhD Thesis, University of Melbourne.

Hawke, R. J. L. (1989): Our Country, Our Future: Statement on the Environment, Australian Government Publishing Service, Canberra.

Lockie, S. (1995): Beyond a 'good thing': political interests and the meaning of Landcare, Rural Society **5** (2/3), 3-12.

Martin, P., Tarr, S. and Lockie, S. (1992): Participatory environmental management in New South Wales: policy and practice. In: G. Lawrence, F. Vanclay and B. Furze (eds.), Agriculture, Environment and Society: Contemporary Issues for Australia, Macmillan Melbourne, 184-207.

Scarsbrick, B. (1994): Marketing Landcare. In: D. Defenderfer (ed.), Proceedings of the 1994 Australian Landcare Conference, v. 2, Department of Primary Industries and Fisheries, Hobart, 313 - 322.

Toyne, P. and Farley, R. (1989): A national land management program, Australian Journal of Soil and Water Conservation **2(2)**, 6-9.

Vanclay, F. (1994): Hegemonic Landcare: further reflections from the National Landcare Conference, Rural Society **4(3/4)**, 45-47.

Williams, M., McCarthy, M. and Pickup, G. (1995): Desertification, drought and Landcare: Australia's role in an international convention to combat desertification, Australian Geographer **26 (1)**, 23-32.

Address of author:
Sarah Ewing
Centre for Environmental Applied Hydrology
Deptartment of Civil and Environmental Engineering
University of Melbourne
Parkville, Victoria 3052, Australia

The Rise and Fall of the Vetiver Hedge in the Fijian Sugar Industry

J.S. Gawander

Summary

The Sugar industry in Fiji is a little over 100 years old. Today it is second only to tourism as a source of foreign exchange for the country. Till the 1940's sugarcane growing was confined to flat land. Population growth, high demand, attractive prices and the land tenure system necessitated expansion into less fertile, highly weathered undulating terrain.

By the early 1950's the industry launched a massive programme to plant cane on the contour, conserve trash and use vetiver as soil conservation measures to enhance sustainable development in the cane growing belt.

However, all this changed in the early seventies with localisation of the industry, unionisation of growers and with changes in the transportation mode. As a result, today, cane production on marginal land has led to higher production costs and lower productivity due to soil erosion. The growers however continue to sustain themselves on these farms mainly because of the high price of sugar due to preferential European Union markets to Africa, Caribbean and Pacific countries. This results in short-term gain for social sustainability and long term misuse of landuse.

Keywords: Sugarcane, erosion, vetiver hedge, preferential EU market, land tenure system.

1 Introduction

1.1 General background

Agriculture and the environment are inextricably linked and every agricultural development has an impact on the environment. In our bid to increase output to feed the population or to improve the living standards of our people we have often created conflict between development and the environment. In Fiji, the biggest agricultural development todate has been the cultivation of sugarcane. It was reported by Potts (1959) that crystallised sugar was produced in Fiji probably for the first time in 1862. However, sugarcane has been commercially cultivated in Fiji for a little over 100 years. Until the early 1950's, sugarcane was grown largely on flat land in moderately fertile coastal soils. The need to fulfil the country's quota under the International Sugar Agreement, attractive prices of the mid-seventies, the need for the sugar dollar in the newly independent country, and land tenure system combined to make the expansion of the cane belt inevitable. Cane cultivation spread to moderately steep and then to very steep hill lands. As soil erosion on the hill country was already a problem, this expansion to the hills inevitably caused major erosion problems. The initial technologies used to address erosion were the accepted methods of engineered contour banks and diversion banks leading to a grassed waterways. However, these structures were ineffective due to high intensity rainfall especially in the wet season and during cyclones.

The knowledge and fear of cyclones in Fiji is well documented from the early nineteenth century. Between 1925 and 1985, 62 tropical cyclones were observed in Fiji. Hurricanes are the most severe and variable of all tropical cyclones and 34 were observed between 1840 and 1985 (Thomson, 1986).

Torrential rainfall is associated with the cyclones which can exceed 350 mm in 24 hours. During cyclone Bebe in 1972, 755 mm was recorded in 48 hours over the mountains. Hurricane force winds and excessive rain were compounded by storm waves which can reach heights of 15 m. These high intensity rainfall events meant that alternative systems to control erosion had to be developed of the various systems attempted and the vegetative method of using vetiver grass (*vetiveria zizanioides*) was most successful. The sugar industry has used vetiver grass as a vegetative soil conservation measure for the past 50 years. Despite this early success, the system was abandoned in the 1970's and continues to be overlooked.

1.2 Tenant Farming System in Fiji

The existing cane farming system was developed by the Colonial Sugar Refinery Company (CSR) in the early 1920's when the industry was on the verge of collapse. The company subdivided and leased all its estates to tenant growers. The average size of farm was four hectares. With the departure of CSR in 1972 the system continues but with the difference that the land today is owned and leased by the indigenous people through the institution of the Native Land Trust Board (NLTB). This is the "Tenant System" which was mainly instrumental in saving the industry from extinction in 1924. This system worked so well that Potts (1962) quotes Professor C.Y. Shephard as saying that "to restore sugar production from less than 36,000 tons in 1923 to more than 94,000 tons in 1928, is an accomplishment of no mean order, and is not without its lessons to other parts of the cane sugar producing countries". Today, we have approximately 23,334 tenant growers with approximately 90,000 ha available for cane cultivation.

1.3 Sugarcane expansion in Fiji

The expansion of the sugarcane belt takes place from time to time in order to increase production. After World War II, the restoration of sugar production to pre-war days was the top priority. In the 1950s the high demand for sugar was relieved by cane farm expansion. This was possible due to sugar market guarantees to CSR. In the 1960s, some land reverted to native Fijian growers forced some Indian growers on to hilly lands. In the 1970s the development of some 5000 ha on undulating terrain commenced at Seaqaqa, Vanua Levu. The most rapid expansion that took place in the cane belt was between 1950 and 1970 when the area under cultivation almost doubled (Table 1).

Year	Total area (ha)
1930	31938
1940	37397
1950	38043
1960	52180
1970	66939
1980	88003
1990	99357

Table 1: Area available for cane farming with time

2 Erosion problem in cane fields

There are little quantative data available regarding soil erosion rates in sugarcane fields even though field observations indicate extensive soil erosion. Liedtke (1989) collected run-off and erosion data for a sugarcane growing area. The rate of soil was 22-80 t/ha/yr on slopes of 5 - 20°. It is worth noting that the data was recorded during the month of February 1983, which was an exceptionally dry month for that time of the year and much of the actual run-off occurred in the rains due to the cyclone that hit Fiji group at the beginning of March. In another sugarcane area Clarke and Morrison (1987) recorded soil loss of 8 - 14 cm when the first crop was harvested. Initially this land was under grass, fern and casuarina prior to cane cultivation. In some places the entire solum had been removed. This represents losses equivalent to approximately 100 t/ha/yr.

Studies were also carried out at Seaqaqa in Vanua Levu by Masilaca et al., (1986). It was originally under secondary forest prior to sugarcane cultivation. The site had a slope of 5 - 8° and was monitored from 1986 to 1992 by which time the land had been retired from cane production. Over this period, up to 50 cm of soil was lost from sections of the site corresponding to soil loss of approximately 200 - 300 t/ha/yr. There was also a marked decline in organic carbon content and increase in bulk density at this site (SQ5) as indicated by the data in Table 2.

Time	Bulk density (g/cm3)	Organic C (%)
Sept 78	0.85	4.43
Jan 80	1.12	3.75
Dec 81	1.13	3.18
Dec 83	1.17	3.00
Feb 92	1.42	1.68

Table 2: Effect of erosion on organic carbon and bulk density (SQ5)

Most of the soil losses discussed have been measured using the Universal Soil Loss Equation, thus the data should be used with caution.

3 Conservation measures: historic background

Vetiver grass was introduced to Fiji from India, most probably for the purpose of thatching material for houses. It was commonly used to stabilise embankments, terraces and to demarcate farm boundaries. The use of vetiver in the mid 1950s to reduce soil erosion on steep slopes on which cane was planted, indicated the potential of using vetiver as a conservation measure. Recognising the many advantages of the vetiver hedge it was recommended for planting in contour lines instead of graded banks and waterways. This suited the sugarcane farming system because it was more labour intensive than mechanical. This soil conservation technology was vigorously enforced over half of all sloping lands. An internal report (224N) records approximately 20% loss of top soil due to the torrential rain on 9 April 1958. The situation was described as catastrophic in the Wairuku, Penang area. It further stated that considerable contouring with vetiver hedge was hardly effective in controlling erosion. The poor control was attributed to the fact that the hedge was only recently established. The report mentions loss of faith by some growers in the erosion control measure. However, there were many who had well established hedges and they remained convinced of its effectiveness. Thus by early 1960 vetiver was well established in most of the cane belt and it was well accepted for the following reasons.
- it formed a dense permanent hedge
- it is resistant to overgrazing and fire because the crown is below the surface

- once established it is not affected by droughts or floods
- it is sterile and also does not produce rhizomes so it will not become a weed
- it is cheap and easy to establish and maintain

4 Current soil conservation constraints and implications

Inspite of clearly demonstrated usefulness of vetiver hedges in existence for over forty years it is amazing that growers are very reluctant to establish new Vetiver hedges. In many cases the well established hedges have been removed resulting in massive soil erosion and reduction in cane yields. There are several constraints in managing effective soil conservation controls and, the major ones are discussed.

4.1 Preferential price arrangement with European Union (EU)

Sugar is Fiji's largest single export and accounts for about 40% of the country's total merchandise export and about 12% of gross domestic product. The premium price received for sugar sold on quota to the EU is linked to the internal support price given to the beet farmers in the EU. It is difficult to determine what the preferential price will be, with the implementation of GATT, the reforms of EU agricultural policies and European consumer demands. As the EU reforms their Common Agricultures policy in line with GATT and in response to consumer demand there is certain to be a decrease in the preferential price. This would have a long term effect on Fiji's economy. This would also have a direct effect on some of our growers who are planting on highly erodible hilly lands.

On an average Fiji produces 450,000 tonnes of sugar from 4.1 million tons of cane. Of this amount 175,000 is sold under a preferential price to the U.K. and the rest is sold under various arrangements to other countries. As a result of the preferential markets the growers are able to sustain themselves since the price of a ton of cane is approximately F$45 and the cost of production is about F$25 per ton. However, if the 175,000 was not sold under the Sugar Protocol of the Lome Convention, then the price of cane would dramatically fall to about F$33 per ton. This would not only have an adverse effect on our national economy but would also make cane farming uneconomical particularly for farmers on marginal land. Hence, while the Lome Convention supports the nation's sugar industry it also indirectly sustains the growers who farm on the marginal lands.

4.2 Land tenure legislation

Land tenure is a major issue not only politically but also from a land management point of view. Under the existing Agricultural Landlord Tenant Act (ALTA) the grower is given a lease for a period of 30 years by the NLTB. The bulk of the leases will expire between 1997 and the year 2004. As a result for the tenant growers there is substantial anxiety regarding the renewal of the lease. Will it be renewed and for how long and on what terms etc? The uncertainty about tenure has always had an adverse effect on land management. In recent years investment on farms has decreased dramatically. Due to the anxiety about the future there is a tendency amongst growers to scale down investment and soil conservation measures. Hence, there is this real danger of further declining cane production in the future. Land is a sensitive issue and has assumed greater sensitivity in recent times. Land is an important resource and its economic utilization is vital for the industry and the country. The delay in working out an equitable arrangement only increases land degradation on the cane belt.

It is worth highlighting that high rates of erosion are occurring despite the fact that there is adequate legislation to prevent bad land husbandry. The following laws are examples of the many legal provisions that should protect against degradation: the Land conservation and improvement ordinance (1953), the Forest Act, the Fire Preventions/Rural Fire Act and the Native Lands Act.

However, it is clear that these are forgotten laws. As a result erosion is left unchecked. The instruments of title governing agricultural leases also include the following clauses to prevent land degradation:

(i) "To farm and manage the land in such a way as to preserve its fertility and keep it in good condition"
(ii) "Not to clear, burn off or cultivate any hillside having a slope more than twenty five degrees from the horizontal or the top twenty percent (measured vertically) of any hills having such slopes."
(iii) "To regularly manure the land"

Then why are these laws ignored? The apparent reason is the failure of Native Land Trust Board (NLTB) to discourage the push for more cane leases in the steep lands and to prohibit and penalise bad land husbandry. To some extent this is difficult to execute since a Certificate of Bad Land Husbandry is necessary from the appropriate authorities either to restrain or evict a grower due to land abuse.

4.3 Land Use Policy

Due to the problem of policing the soil conservation laws and in order to maximise profits under the existing land tenure system, erosion in the cane belt is increasing along with the population and rural destitution. The growers are not planting vetiver hedges anymore and are rapidly ploughing them out in order to have an extra row of cane. This is despite the fact that the soil terraces formed by the Vetiver hedges are up to 1.5 m high. The quantity of soil eroded and trapped by the Vetiver hedge clearly demonstrates the effectiveness of vetiver grass.

In the CSR era the Field Officers of the company could enforce strict land use guidelines. However this is not possible since the land no longer belongs to the miller. The vetiver hedges definitely provide effective soil conservation but it was introduced some 40 year ago when draught animals were used. Today machinery is relied upon, and the vetiver hedge is seen to cause a hindrance to farming operations. Animals can move around an overgrown hedge and terrace, whereas harvesting machinery cannot. The lorries used to remove harvested cane from steep slopes are not able to pass through the hedges nor over the terrace. Thus vetiver is ploughed out to make roadways.

4.4 Poor extension support service

There has been little support to assist growers who plant vetiver hedges because very little has been done in solving problems related to the hedges. The main concern of the growers are loss of productive land, hindrance to machinery operations and the harbouring of pests, all of which can be resolved by an effective hedge maintenance program through the extension services. In addition, the extension efforts have been negligible in the area of soil conservation. In particular no effort has been made to demonstrate the long term effects of failure to conserve soil. In the past, some growers have simply moved on to new leased land.

4.5 Economic implications of erosion

A major economic implication is the declining productivity of sugarcane in marginal areas. The sugarcane growers place the blame of low productivity of varieties and fertilizers. However, a more likely reason is the impact of soil erosion on the productive capacity of the soil. The sugarcane growers suffer from erosion induced production losses. The yields and ratooning ability in these areas have been declining rapidly.

The damage from induced erosion is also serious. Notable is the flooding and sedimentation downstream which also affect mangrove forest and coral reefs which threatens our marine productivity. It is also obvious that a minor drought appears to be a major drought mainly due to the fact that the top-soil on steep lands has been washed away, where no soil conservation measure has been adopted. As the soil profile narrows, the apparent drought effect spreads down-slope and appears more intense with much lower yields in comparison with yields ten years ago.

5 Conclusion

The land form, high intensity rainfall and cultivation practices on the cane farms are conducive to high rates of erosion. The thin line hedge is an effective system of managing soil erosion. However over forty year of experience indicates that this technology, like all other soil conservation systems, needs ongoing research, development and extension efforts to make it grower friendly. Even though the hedges are cheap to establish, permanent and easy to maintain, its acceptance by the growers cannot be guaranteed. The need to promote the concept of the green line once again is paramount if we wish to have a sustainable sugarcane industry on steep lands. In order to re-establish the hedge we need the resources and the commitment of the growers, millers and support of the European Union to educate our growers about soil conservation measures.

References

Clarke, W.C and Morrison, R.J. (1987): Land mismanagement and the development is imperative in Fiji. In Brookfield, H.C and Blaikie, P.M. Land Degradation and Society. London, Methuen, 176-185.
Liedtke, H. (1989): Soil erosion and soil removal in Fiji. Applied Geography and Development **33**: 68-92.
Masilaca, A.S., Prasad, R.A. and Morrison, R.J. (1986): The impact of sugarcane cultivation on three Oxisols from Vanua Levu, Fiji. Tropical Agriculture **64**: 325-330.
Potts, J.C. (1959): The sugar industry in Fiji - Its beginnings and developments. The Fiji Society of Transactions and Proceedings **7**: 104-130.
Potts, J.C. (1962): An outline of the successful development of the small farm system in the Fiji sugar industry. Fiji Society Transaction and Proceedings **9**: 24-38.
Thompson, R.D. (1986): Hurricanes in the Fiji area: causes and consequences. New Zealand Journal of Geography **81**: 7-12.

Address of author:
Jai S. Gawander
Sugarcane Research Centre
Fiji Sugar Corporation Limited
P.O. Box 3560
Lautoka, Fiji

Closing the Gap between Farmers and Support Organisations in Namibia

A.S. Kruger

Summary

Namibia is one of the most arid countries in the world with very low and erratic rainfall. Agricultural production is a major production area and contributes with 10% to the Gross Domestic Product. It is mainly based on extensive livestock production. Commercial farmers are mainly livestock producers, supported by efficient support structures (organised farmers unions, extension, research, credit, etc.) while the communal farmers are the majority and also mainly pastoralists. Work done by SARDEP indicated that communal farmers and community based organisations (CBO's) have very limited self-help capacities and are unable to voice their demands for sustainable development. Support organisations (governmental, non-governmental and private) are not meeting the demands and needs of the farmers and CBO's and finally that the frame conditions (e.g. land tenure) are not conducive to sustainable land use. Based on these conclusions, SARDEP embarked upon an approach to support farmers and CBO's in identifying, prioritising and finding solutions to their problems; to encourage support organisations in reorientating their services towards the needs of communal farmers and; to promote the creation of a policy framework conducive to sustainable resource management.

Keywords : Pastoralism, participation, self-help, sustainability, land-use

1 Introduction

Namibia is one of the most arid countries in the world. It has a low and variable rainfall, ranging from 20 mm in the south to 600 mm in the north-east. The high variability of rainfall (coefficient of variation close to 40 % for Windhoek), which seems to have increased during the last two decades, includes periodically years of well below-average rainfall. This significantly reduces the meaningfulness of average rainfall figures for the country (Kruger & Woehl, 1996).

Although about 34 % of the country receives more than 400 mm of rain per annum, about 95% of the soils in the area have a clay content of less than 5 %, which makes it marginal for crop production. Namibia is therefore mainly a livestock producing country where the major features are scarce productive land and fragile soils, compounded by limited water resources and an erratic rainfall regime (Kruger, 1996).

About 70 % of Namibia's total population of 1.7 million, live in the rural areas with approximately 90 % deriving their livelihood from agriculture, particularly livestock production. In relation to the overall land area (ca. 1,000,000 sq. km), Namibia is also one of the least densely populated countries in the world, although distributed highly unevenly (30% live in just 1% of the land area in the northern communal areas) with population densities of > 100 people per sq. km in relation to an average of 1.7 people per sq. km for the whole country.

2 Agriculture in Namibia

The agricultural sector is devided into a commercial farming sector where farmers operate on freehold title deed land and a communal farming sector where farmers operate under a common property regime system. In the second case land is considered state land for which no title deed can be registered. The communal sector comprises 48% of the total agricultural land and about 140,000 farm households which produce mainly livestock and subsistence food crops. The communal areas directly support 95% of the nations farming population. The main income of farmers derives from livestock production (MAWRD, 1995).

The commercial sector is made up of 6,337 freehold title deed farms belonging to about 4,200 farmers primarily oriented towards livestock production. About 90% of the red meat produced in Namibia is exported to South Africa and Europe. The agricultural sector is third after mining and fisheries as far as export earnings are concerned (MAWRD, 1995). The commercial farmers have highly productive farming practises supported by efficient and professional organisations (Namibia Agricultural Union), good infra-structure, well established information basis and adequate support structures like extension, research, veterinary services, marketing and access to credit.

In pre-colonial times, land in Namibia was considered territory of the different ethnic groups and was communally owned. Control over water points and other strategic places enabled farmers to make use of large areas of rangeland. In the north, subsistence crops were produced under a slash and burn practice since sufficient land was still available. The nomadic movement of animals tracking rainfall and better fodder resources, made it possible for farmers to keep large herds of animals. Farmers were traditionally well-organised and control over the resources was exercised by traditional leaders (Kruger & Woehl, 1996).

Today, conditions for farming in the communal areas are completely different from the past. Communal farmers lost the ability to solve their problems by themselves. They can no longer move with their animals in search of fodder and water. Traditional structures, that controlled the access to and management of natural resources, are still unclear. Support services to communal farmers are virtually non-existent, despite recent attempts by government to improve them. This results in over-utilisation of the rangeland, desertification and a further impoverishment of the rural communities (Kruger & Kressirer, 1995).

3 The Sustainable Animal and Range Development Programme (SARDEP)

It was against this background that the Namibian government initiated the Sustainable Animal and Range Development Programme (SARDEP) in the communal areas of Namibia. SARDEP started in 1991 with support from the Federal Republic of Germany through the "Gesellschaft für Technische Zusammenarbeit" (GTZ). The Ministry of Agriculture, Water and Rural Development (MAWRD) is the implementing agency (SARDEP, 1995a).

The overall goal of SARDEP is to contribute towards reducing land degradation caused by human interference in the communal grazing areas of Namibia. SARDEP started with an "Orientation Phase" in the Southern and Eastern communal areas in 1991 with the view to identify a strategy for sustainable rangeland management and improved livestock production. The programme extended activities to the North Central communal areas in June 1993 and also recently to the Western communal areas.

The Orientation phase was mainly used to
- establish the implementation structures needed;
- to analyse and document the resources, production systems, conditions and constraints for rangeland utilisation;

- to identify solutions for range management and livestock production based on the needs and demands of the communities;
- to assist the local communities in testing some of the solutions according to the identified problems;
- to monitor and assess the tested solutions for possible replications and
- to identify a strategy for sustainable rangeland management and improved livestock production.

4 Results and discussion

With the support of the CDC Consultants from Switzerland in July 1994, an analysis of the current and future situations of people in the programme areas was carried out. The purpose of this exercise was to support SARDEP in preparing the components of the future strategy in identifying and developing roles and responsibilities of all the actors at all levels of the programme. In the process consultations were held with farmers, government officials and programme staff on all levels.

4.1 The present situation in the communal areas

The current *task* of local households is to sustain a decent living for a growing population in the communal areas and to ensure subsistence for the members of the households.

The *outputs required* in order to achieve this task are however not adequate. The quantitative and qualitative performance of livestock production is not sufficient; the poorer segment of the population is increasingly impoverished and is more and more depending on off-farm income sources and migration. The over-utilisation of the rangeland reduces ist production potential resulting in degradation, loss of soil, insufficient forage and low drought tolerance.

The *environment* in which the task should be fulfilled, is also not conducive. The current land tenure system is uncertain and does not encourage initiative and responsibility for sustainable utilisation by the resource users. There is still a lack of awareness amongst politicians and some regulations remain obstacles to sustainable land use (e.g. fodder subsidies during drought periods). The population growth rate exceeds 4% in the communal areas and at the current rate, the population of the country will double every 25 years. The communal farmers also have low access to proper basic infrastructure (roads, communications, etc.) and services (extension, research, marketing, credit, veterinary services, health, education, etc.).

Despite the availability of limited *internal resources* within the communities (e.g. manpower, funds, etc.), they are not always adequately supported by external inputs from both governmental and non-governmental sources. Not all the inputs from outside are relevant to the needs of self-sustainability. An example is that the provision of food relief to sustain people is increasing.

The *internal structure and organisation* of local communities is far from adequate. Because the management capacity for new collective tasks is low, self-help capacity is low and dependency on outside support is high. High management capacities in traditional systems to cope with transhumance, rotational grazing, etc. are disrupted by outside interference. Existing formal structures for the resource management and the collective decision making are not effective. Because of land degradation, an increasing part of the poor can not be sustained by livestock production and have no alternatives for income (KEK/CDC, 1994).

4.2 The strategy components

The goal to use resources in a sustainable manner in order to reduce man-induced land degradation and to improve the welfare of the rural population, makes it necessary to bring about change in the entire system of communal land use, and not only on an individual basis. Eight strategy components have been identified that need to be addressed by the relevant actors in a well-coordinated manner in order to achieve the desired results:

- *Resettlement of large communal farmers to title deed areas*: The population in the communal areas will double within the next 25 years from the current ca. 1,000,000 people (70% of the population) to approximately 2,000,000 people (still 70% of the population). This will place tremendous pressure on the already limited resources (land, grazing, water, etc.). Government will therefore have to promote the resettlement of large communal farmers to title deed areas in such a way that it can relieve the pressure on the communal lands.
- *Improved marketing*: In a variable environment where animal numbers have to be adjusted to the variable fodder base, an efficient marketing system needs to be in place. Apart from the need for adequate marketing infra-structure (e.g. auction pens), sufficient buyers should attend auctions to encourage competition for better prices. Farmers also need timely and regular information regarding market prices for different classes and types of livestock.
- *Alternative Income Generating Activities*: The objective of this component is to reduce the dependency on livestock production. Where diversification is not possible because of a lack of natural resources, alternatives like shifting from raw material production (live animals) to semi-products (slaughtering, meat, leather, shoes, etc.) is essential, thus keeping a part of the processing of livestock in the area and thereby reducing the exclusive dependency on livestock production.
- *Land Tenure System*: The objective of this component is to promote the improvement of the frame conditions necessary for sustainable rangeland management and improved livestock production. Land in the communal areas is state owned and no title deeds are allowed. This creates the problem of open and uncontrolled access to the rangeland resources to everyone. With the uncertainty of control over and access to land and the management of the rangeland, as well as the urgently awaited Communal Land Bill, there are very little incentives for the local community to implement sustainable rangeland management practices.
- *Alternatives for Capital Accumulation*: In line with the true tradition of the African pastoralist, livestock (mainly cattle) is still considered as the major source of security and wealth. Due to the lack of alternatives, farmers will always re-invest money into cattle and put them on the already degraded rangelands.
- *Local Investment Packages*: The objective of this component is to identify, develop, test and implement packages such as saving schemes to promote local funds and schemes that will generate money from outside the community for investment in the rural areas.
- *Institution Building on Communal Land*: There is an obvious gap between the traditional authorities on communal land, that are no longer able to assume the full management responsibility, and the newly-created administrative bodies, that are not yet able to take over.
- *Sustainable Improvement of Livestock Production*: This is a clearly defined agricultural component. Farmers have to be supported to identify, test and implement sustainable rangeland utilisation and improve livestock production practices under their specific situations. Support to farmers is required in the very areas in which they are residing in terms of extension, research, veterinary services, marketing, credit, etc. Emphasis should also be put on enhancing their capacities to better track fodder resources in times of fodder scarcity (KEK/CDC, 1994).

4.3 Conclusions from the orientation phase

During a Strategy workshop in March 1995, farmers participated from the southern, eastern and northern programme areas. Some 35 governmental, non-governmental and private organisations were also invited. The purpose of the workshop was to bring communal farmers and support organisations together on the same forum to meet each other, to know each other better, to find common avenues for future co-operation and to get ideas on the future strategy for sustainable range management and improved livestock production.

Farmers were given a chance to elaborate a joint vision for the future, to identify the constraints and hindrances in getting from the current situation to the vision and to find solutions, first those that they can implement themselves and then those where they will need support from outside. Service organisations were also given a chance to illustrate current services and goods they are providing to the communal farmers, as well as what they intend to deliver in future. From this exercise, the following conclusions were made (SARDEP, 1995b):

- Communal farmers have a very good idea about their vision for the future. They however lack the ability to voice their demand for support in order to implement their solutions.
- Support Organisations (governmental, non-governmental and private) are not presently providing services matching the needs and demands of the communal livestock farmers.
- Frame conditions (e.g. land tenure) are not conducive to sustainable rangeland utilisation and improved livestock production practices.

5 The SARDEP strategy

On the basis of experiences made in the Orientation Phase as well as inputs made during the CDC Consultancy (1994) and the Strategy workshop (1995), the SARDEP Strategy for sustainable rangeland utilisation and improved livestock production practices was identified:

- Farmers and Community Based Organisations (CBO's) are supported to be in a position to identify their problems, prioritise them and find solutions to them. Farmers must be made aware of the potentials and limitations of their natural resource base, their farm economics and the overall economic situation in the country. Farmers must also be in a position to explore what problems they can solve on their own and where they need support. They must have a detailed knowledge about accessible services. They must be in a position to clearly formulate their demands to other organisations. They need to be organised in interest groups for a common task and be willing to share their experiences. Farmers must be in a position to make their own contributions into joint research, extension and other service activities and they should be encouraged to continue with animal production and rangeland utilisation independently and invest even further into their activities in a sustainable manner, should they choose to do so.
- Service Organisations (governmental, non-governmental and private) are supported to re-orientate their services towards the needs and demands of the communal farmers. These organisations must have a clear picture of the demands of the communal farmers in all fields. They must have an internal Human Resource Development System geared towards dealing with communal farmers and they must be in a position to make information on specific activities in certain regions available to other organisations and farmers.
- SARDEP has to promote the creation of a conducive policy framework where sustainable rangeland management and improved livestock production is possible. This is done through studies related to access to and management of natural resources in the communal areas, through giving direct inputs in meetings and workshops where related policy issues are

discussed, and also by exposing politicians and decision-makers to the real situation on the ground to highlight the constraints and hindrances in the current policies (SARDEP, 1995b).

6 Conclusion

It is obvious that the development of technical solutions alone might not be sufficient to support local farmers in solving their problems. What is however a prerequisite for success, is the empowerment of the local communities to resume responsibility for the management of their own resources. SARDEP moved away from the general practice of developing and offering technical packages (input approach) to the communities, to a process of enabling local communities to identify solutions to their problems and identifying those institutions which can support them in implementing solutions beyond their own capability (negotiation approach).

The success of SARDEP will at the end be measured in how far they succeeded to "close the gap" between the farmers and the service organisations. SARDEP will only be considered successful if farmers are able to solve their own problems, if they are able to get the necessary support for those problems they cannot solve by themselves, and if the service organisations are directing their services towards the needs of the communal farmers. Proof of whether SARDEP has achieved its goals, will be when farmers and CBO's do not require their services any more. This is the true meaning of sustainability.

7 References

KEK/CDC (1994): Support for the Organisational Development Process of SARDEP, Main Mission Report, Sustainable Animal and Range Development Programme, Ministry of Agriculture, Water and Rural Development, Windhoek, Namibia.

Kruger, A.S. and Kressirer, R.F. (1995): Towards Sustainable Rangeland Management and Livestock Production in Namibia. Sustainable Animal and Range Development Programme, Ministry of Agriculture, Water and Rural Development, Windhoek, Namibia.

Kruger, A.S. (1996): Environmental Threats and Opportunities Assessment for Namibia. A report on Pastoral and Agro-Pastoral Resources. Namibia Resource Consultants, Windhoek, Namibia.

Kruger, A.S. and Woehl, H. (1996): The Challenge for Namibia's Future: Sustainable Land-Use Under Arid and Semi-Arid Conditions. Agriculture and Rural Development, Technical Centre for Agricultural and Rural Co-operation, DLG-Verlags-GMbH Frankfurt am Main, Germany.

MAWRD (1995): The National Agricultural Policy, Ministry of Agriculture, Water and Rural Development, Windhoek, Namibia.

SARDEP (1995a): SARDEP Objectives and Policy. Sustainable Animal and Range Development Programme, Ministry of Agriculture, Water and Rural Development, Windhoek, Namibia.

SARDEP (1995b): Strategy Workshop for the next phase (1996-1999) for the Southern and Eastern Communal Areas, Sustainable Animal and Range Development Programme, Ministry of Agriculture, Water and Rural Development, Windhoek, Namibia.

Address of author:
Albertus S. Kruger
Chief Rangeland Researcher
and
National Co-ordinator of the Sustainable Animal and Range Development Programme (SARDEP)
Private Bag 13184
Windhoek, Republic of Namibia

Improving Participation and Cooperation at the Local Level: Lessons from Economics and Psychology

M. Maarleveld

Summary

In the area of rural development and sustainable land use, many projects try to use the principles of participation and cooperation as it has become clear that local users have the abilities and knowledge to manage natural resources in a sustainable manner themselves. However, difficulties encountered in implementing participatory strategies make clear that much needs to be learned and improved. Combining theoretical insights with practical findings is one route to understanding and fine-tuning participatory strategies. This is the path that is taken in this paper. The concept of social dilemma is used to unravel the different grounds of interdependence among stakeholders involved in resource management and the need for coordinating individual decisions and actions. Some results of experimental research in psychology and economics concerned with coping with social dilemma situations that support further development of participatory strategies are highlighted: the importance of repeated interaction, the influence of the institutional context, and the notion of evolving conditions.

Keywords: Participation, interdependence, social dilemmas, institutional context, evolving conditions

1 Introduction

In the past few decades it has been recognized that technological solutions are not sufficient to guarantee sustainable use of land and other natural resources. Many examples of projects that have failed due to difficulties with the adoption of technological solutions or because these solutions have led to new and unforeseen problems exist. Increasingly, efforts have been made to address the social aspects of both resource exploitation and technology development. Practice as well as research in these areas have witnessed, according to some, paradigmatic changes in approach. A major turning point has been the realization and acceptance that stakeholders have abilities and knowledge to sustainably manage resources themselves (Ostrom 1990, Scoones & Thompson 1994, Uphoff 1992). A stakeholder is no longer viewed as a short-sighted "homo economicus" trapped in a paradox where fulfilling self-interest inevitably leads to resource destruction. Instead, the expectation is that involving local users in the governance of natural resources will result in sustainable practices adapted to local circumstances.

Numerous approaches have evolved, both in developed and developing countries, in order to involve stakeholders in design and implementation of projects and programs to induce sustainable natural resource management through improving participation and cooperation at the local level. However, as the experiences and evaluations of participatory practices (Naraydan 1994, Hinchcliffe et al. 1995) have shown, participatory strategies are not the panacea for all problems encountered in natural resource management. The criticism should not be taken so much as failure of participatory approaches, but as instances where learning can take place and strategies can possibly be improved.

Combining theoretical insights with practical findings is one route to improve and fine-tune participatory strategies. This is the path that is taken in this paper. First, the problems of natural resource management will be characterized. Second, participatory strategies shall be discussed. Third, the concept of social dilemma is used to unravel the different grounds of interdependence among stakeholders involved in resource management and the need for coordinating individual decisions and actions. Fourth, some results of research in psychology and economics concerned with coping with social dilemma situations will be related to the use of participatory strategies. Finally, the value of using the social dilemma concept for further development of participatory strategies to induce sustainable resource management and improve cooperation on the local level will be discussed.

2 Sustainable resource management: Coping with complexity, uncertainty and change

For a number of reasons many of the problematic issues arising in the use of resources such as land and water can be characterized by complexity, uncertainty, and continuous change. The scientific knowledge regarding the complexity of resource system dynamics and extent of ecosystem resilience is limited. For example, knowledge of groundwater dynamics is still limited as researchers become aware of the delicate balances and interlinkages among processes in the ecosystem. However, policy regarding groundwater use and quality are made on the basis of these inadequate models, leading to dubious recommendations questioned by both users and environmental groups. Often, natural systems and problematic issues cross the spatial and functional boundaries of existing institutions. In the case of water pollution in rivers or desertification, a number of nations can be involved and agreements or organizations need to be created or adjusted.

As issues concerning the environment are linked to issues in agriculture, health, international relations, trade, and industry among others, attempts to change relationships in natural resource management can involve changes in many different spheres. Changes in one area can have consequences for the bargaining power of various parties in other spheres. Those dissatisfied with configurations of power will attempt to change the governance structures, while others might strive for consolidation. These changes can lead to unpredictable interactions among people and ecosystems as they evolve together. All in all, where complexity, uncertainty and continuous change dominate, no single solution exists or will last (Gunderson et al. 1995). This means that "solutions" have to be socially constructed and negotiated. Focusing on the manner in which the choices and actions of stakeholders are interdependent can lead to a better understanding of the social mechanisms involved in sustainable natural resource management.

3 Inducing participation and cooperation at the local level: Participatory strategies

Participatory strategies aim to involve local stakeholders in design and implementation of projects and programs, among others, to bring about sustainable management of natural resources. Numerous approaches have evolved, for example, participatory action research (Okali et al 1994), community development (Korten 1980), participatory appraisals (Chambers 1994, Engel 1995), participatory technology development (Reijntjes et al. 1992), co-management (Pinkerton 1989), co-production (Ostrom 1996), interactive policy making (Renn et al. 1995, Van Woerkum & Aarts 1996) and platform building (Röling 1994). The notions of joint learning processes and cooperation with local stakeholders and communities are central in these approaches. The importance of developing and strengthening local organizations and management capabilities, the necessity of institutions involved to change their attitude from expert to facilitative, and the development of strategies and procedures for a flexible dialogue are also emphasized. In general, participatory strategies are directed at the following aspects.

Empowerment:
Help people to develop self-respect, confidence and pride in their knowledge and capabilities. Develop the ability to selectively incorporate, adapt and take advantage of external technologies and ideas.

Development of a (local) collective information/knowledge base:
Develop and share information/knowledge. Creation of an information/knowledge base can result in the collective exploration of multiple learning trajectories.

Informed decision and rule making:
Develop capacities in information gathering, analysis and documentation, including the effective use of information. Address competence in planning, management, leadership and preparing proposals as well as skills for facilitating and negotiating between different interest groups.

Building of horizontal ties and vertical ties:
The existence of ties between people increases trust and facilitates credible commitments. Horizontal ties are important as they lead to greater information sharing. Vertical ties should not be neglected however, as they are crucial in linking different levels of decision-making and action.

These aspects are means to promote participation and cooperation at the local level. The degree to which these aspects are addressed differs depending on approach and context.

Although many development planners as well as local stakeholders are convinced of the value of participatory strategies to induce sustainable practices in natural resource management through improving cooperation at the local level, others remain skeptic. They believe that cooperation and the welfare of the collective are not high on the list of priorities of individuals pursuing their own interests. In addition, it has been difficult to evaluate outcomes of projects attempting to increase participation and cooperation at the local level. Projects making use of participatory strategies often continue to be heavily expert-dependent and skewed, are not always sensitive to existing local institutions not readily visible, tend to be based on a weak notion of participation, and focus on formal organization building instead of an institutional base (Fisher 1993). All in all, participation often appears to remain a rhetorical notion rather than become actual practice. Linking "islands of successful participatory management" has been difficult. Scaling-up positive results at the local level to other levels has also proven to be problematic. Difficulties in replicating and scaling-up positive results make clear that, even when taking into account that achievements are often context-specific, much remains to be learned about the conditions and factors that influence improving cooperation through participation.

4 Conceiving natural resource management problems as social dilemmas

In general, the notion of social dilemma is used to conceptualize how people are strategically interdependent, i.e., how the choice or behavior of one actor affects the set of possibilities and outcomes of other actors. Interdependence is not necessarily problematic. Individual and collective interests can be promotively interdependent, i.e., pursuing individual interests leads to the desired outcome for the collective. But intertwining of individual interests of stakeholders can lead to situations where decisions and behavior, although individually advantageous, are sub-optimal for the group as a whole. These situations are labeled social dilemmas.

The tragedy of the commons as portrayed by Hardin (1968) in which a group of herders overuse a common grazing ground illustrates this type of interdependence effectively. Each herder wants to graze as many animals as possible on the common as direct benefits are received from the animals.

The common, however, has a capacity for only a limited number of animals. Because the individual herders bear only a share of the cost of overgrazing the common, they are "locked into a system that compels each herder to increase his herd without limit in a world that is limited" (Hardin 1968, 1244). For each herder the benefits of adding one animal to the commons far outweighs the possible long-term costs of the destruction of the common grazing ground. Numerous similar interdependence structures leading to overuse and degradation can be found in natural resource management. Moreover, the intertwining of individual interests becomes even more complex when one takes into account that individuals often have divergent, multiple interests and that collective interests themselves can also conflict.

In order to cope with the sub-optimal collective outcome, a change in behavior is required. This can be achieved in two ways. Strategies can be adopted that reduce the sub-optimality of outcomes without changing the structure of the situation, i.e., agree upon a joint strategy within the set of pre-existing rules. Alternatively, those involved can decide to change the structure of the situation by changing the rules which constitute that structure (Ostrom et al. 1994). Realization of either of these alternatives requires some form of organization, monitoring and/or sanctioning. Establishing and maintaining these mechanisms often involve some degree of collective action and joint decision making. For example, if in the earlier-mentioned tragedy of the commons, herders decide to regulate the grazing of the commons, some kind of effort and investment on their part is necessary to establish rules and monitor compliance. Not everyone might be willing to make the required contributions to establish and maintain such mechanisms, especially as the nature of these mechanisms makes it extremely difficult to exclude those who have not contributed. Consequently, it is both possible and tempting to free ride on the efforts and contributions of others. As all involved are aware of this, chances are that the collective outcome is sub-optimal, i.e., individuals do not make the necessary contributions, and organization, monitoring and sanctioning are not developed or based on the efforts of a few. In this case, stakeholders again find themselves in a social dilemma.

When agreeing upon a joint strategy within a pre-existing set of rules and changing the set of pre-existing rules, actors are influenced by the institutional constraints they face as well as their individual cognitive abilities and their state of mind. Therefore, individuals must learn about the circumstances in which they find themselves and the consequences of their choices and actions. By addressing issues of empowerment, development of a collective knowledge base, informed decision and rule making, and building of horizontal and vertical ties, participatory strategies can affect structural as well as individual determinants of human choice and action. Unraveling the different social dilemmas underlying natural resource management can help to better identify and understand the processes and factors involved when stakeholders with various interests come together.

5 Coping with social dilemmas: Some research results from psychology and economics

Results of research undertaken in economics and psychology to analyze and learn more about the various ways people can cope with social dilemma situations support further development of participatory strategies to improve cooperation at the local level. Three findings are highlighted here: the importance of repeated interaction, the influence of the institutional context, and the notion of evolving conditions.

5.1 Repeated interaction

A number of factors have been found to enhance cooperative behavior when situations of strategic interdependence occur. A crucial condition is that actors are involved in a recurrent social situation, i.e., that they interact repeatedly (Raub & Voss 1986). Repeated interaction creates a future for an actor insofar as it makes visible how his behavior affects the decisions and behavior of the other actors in a situation. It also ensures a past which may provide the actor with information upon which expectations concerning future behavior of other actors is built. Repeated interaction also provides a basis for conditional strategies, "I will cooperate if you cooperate", upon which a lot of cooperation is founded. As such, repeated interactions create opportunities to reciprocate. The manner in which individuals give form and content to these interactions conveys information about the credibility of commitments, trustworthiness and reputation of the various actors involved. These can have effects beyond the specific situation as interactions are often connected through multiple relations. Thus, generalized trustworthiness can induce a person to cooperate in situations where this might not appear to be in his individual interest.

The importance of repeated interaction supports the use of participatory strategies to induce sustainable natural resource management. In order to cope with social dilemma situations, people need to be brought together and interact. In addressing issues of empowerment, development of a collective knowledge base, informed decision- and rule-making, and building of horizontal and vertical ties, participatory strategies create conditions for repeated interaction and can facilitate the learning process of cooperation.

5.2 Institutional context

The relationship between individual and structural determinants of human choice and action is also of importance in coping with social dilemmas. Although individuals tend to act on the basis of their personal preferences, to a certain extent, preferences, decisions and actions are shaped by the structural environment in which individuals find themselves. Institutional analysis in economics illuminates how structures of social interaction such as markets, governments and other institutions explain general regularities that govern the behavior of all actors (Satz & Ferejohn 1994). Not all human behavior can be explained in terms of institutional constraints and opportunities that individuals face. In situations where structural constraints on individuals are much weaker, individual profiles come to the forehand. However, in many situations the overall institutional environment determines the constraints and opportunities of individual behavior. For example, in case of a competitive market environment, the types of action patterns which will lead to survival and success are in part determined by generic market structures. Survival in such an environment is dependent on abilities to maximize profits. Consequently, motivational preferences such as maximizing self-interest in terms of cost/benefit analysis can be seen as reflections of the constraints of that environment and do not necessarily indicate the true nature of humans (Clark 1995).

This means that it is of great importance to take into account the incentive structures that individuals face and how these play a role in shaping individual preferences and problem perceptions. For this reason, letting people articulate preferences and perceptions of a problem situation is not sufficient to induce behavioral changes towards cooperation. Although individuals might be aware of the benefits of cooperation, in some situations, structural constraints and opportunities can dominate individual profiles. Therefore, further development of participatory strategies should not only focus on how differences in individual interests can inhibit cooperation but also on how the institutional environment can impede cooperation on the local level.

5.3 Evolving conditions

Changing rules in order to cope with social dilemmas, whether within the framework of a set of pre-existing rules or changing that framework itself, is not a simple matter. The complexity of changing rules and the interlinkages of these rule arenas remain difficult barriers in coping with social dilemma situations (Ostrom et al 1994). Establishing and changing rules entail historic, political and bargaining processes that are difficult to manipulate. In addition, changing the rules of the game, means changing the opportunities and constraints that an individuals face, i.e., the institutional environment. Taking the embeddedness of individual preferences and action in the structural environment into account, changes in the configuration of rules can lead to changes in preferences and behavior. As a result, new strategic interdependencies, i.e. social dilemmas, can have evolved. Moreover, the unraveling of social dilemmas makes clear that "solving" one dilemma can involve new social dilemmas. Consequently, there is a need to be able to adapt conditions as they evolve.

The actuality of sustainable resource management is characterized by complexity, uncertainty and change. In order to cope with the dynamics of evolving conditions, flexible response and problem-solving capacities are desirable. This means not only "solving" the issues at hand, but also realizing that problem situations will arise continuously. In theory, participatory strategies aim to strengthen problem-solving abilities. However, in practice, participatory strategies tend to be used as intervention tools directed to solve a problem at hand (often pre-defined by a project or program), and the notion that a flexible response and continuous problem solving are required tends to disappear to the background.

6 Conclusion: Improving participation and cooperation

Linking the problems of sustainable natural resource management to the concept of social dilemma can provide a framework to analyze strengths and weaknesses of participatory strategies that are currently in vogue to induce sustainable natural resource use. Participatory approaches are often successful in getting together the different stakeholders to express their divergent interests and points of view regarding strategies to manage a natural resource. Problematic however, remains setting up the necessary institutional arrangements for organization, monitoring, and sanctioning in order to change social dilemmas to more promotive interdependencies. Greater understanding of how social dilemmas are nested in one another and the embeddedness of individual decisions and actions in the institutional environment can not only improve the establishment and maintenance of the above mentioned mechanisms, but also aid in the replication and scaling-up of positive results of participatory management at the local level. After all, the nestedness of social dilemmas and structural embeddedness expose already existing horizontal and vertical linkages. These linkages could provide possible anchor points for replicating and scaling-up outcomes and processes of participatory strategies.

Since participatory strategies aim to improve cooperation at the local level through involving stakeholders in project design, implementation, and evaluation emphasis is often on individual perceptions and interactions. To induce cooperation within a specific setting, it is necessary to establish conditions for conditional cooperation, reciprocity, and trust. This can be achieved through stimulating interactions among individuals. This is what participatory strategies strive to facilitate. However, when focusing on the behavior of particular actors, the general regularities that govern behavior of all actors should not be forgotten. In many situations, not so much an individual's psychology but the environmental constraints and opportunities they face explains their behavior. As a strong assumption of "malleability of society" and interventionist attitude underlies current application of participatory strategies, the tendency exists to underestimate the power and influence of the institutional embeddedness of individual preferences and choices.

Thus, participatory strategies only partly succeed in coping with problems of natural resource management. As long as participatory strategies are used from an interventionist point of view as a tool to solve natural resource problems in a specific setting, without taking the broader dynamics of social processes into account, improving the quality and scope of participation and cooperation at the local level will remain difficult. Recognition of the influence of structural constraints and opportunities, and that these and individual preferences are in a continuous flux, makes it important to detect and adapt to conditions as they evolve. By unraveling the grounds for strategic interdependence among stakeholders involved in resource management and the need of coordinating individual decisions and actions, the social dilemma concept helps to expose how individual choice and action are linked to the structural constraints and opportunities facing individuals. In this manner the causes of which resource overexploitation and degradation are symptoms can be laid bare. Insights thus gained lay open pathways to further develop participatory strategies as approaches that induce sustainable resource management through adaptive management and self-governance.

Acknowledgments

The author thanks the Netherlands Organization for Scientific Research (NWO) and the Department of Communication and Innovation Studies for financial support to visit the Workshop of Political Theory and Policy Analysis in Bloomington, Indiana, U.S.A. where the ideas for this paper were developed. Thanks also to the reviewers for their insightful comments.

References

Chambers, R. (1994): The origins and practice of participatory rural appraisal, World Development **22**, 953-969.
Clark, A. (1995): Economic Reason: The interplay of individual learning and external structure, Paper in honor of Douglas North, St. Louis, Department of Philosophy, Washington University.
Engel, P. (1995): Facilitating Innovation. An action-oriented approach and participatory methodology to improve innovative social practice in agriculture, Dissertation, Wageningen.
Fisher, B. (1993): Creating space: Development agencies and local institutions in natural resource management, Forests, Trees and People Newsletter **11**, November.
Gunderson, L.H., Holling, C.S. & Light, St.S. (eds.) (1995): Barriers and Bridges to the renewal of ecosystems and institutions, New York, Columbia University Press.
Hardin, G. (1968): Tragedy of the commons. Science **162**, 1243-1248.
Hinchcliffe, F., Guijt, I., Pretty, J. & Parmesh Shah (1995): New horizons: The economic, social and environmental impacts of participatory watershed development, London, IIED, Gatekeeper Series no. **50**.
Korten, D.C. (1980): Community organization and rural development: A learning process approach, Public Administration Review **40**, 480-510.
Narayadan, D. (1994): The contribution of people's participation: Evidence from 121 rural water supply projects, UNDP-World Bank Water Supply and Sanitation Program, World Bank, Washington D.C.
Okali, Ch., Sundberg, J. & Farrington, J. (1994): Farmer Participatory Research. Rhetoric and Reality, London, Intermediate Technology Publications.
Ostrom, E. (1990): Governing the Commons, Cambridge, Cambridge University Press.
Ostrom, E. (1996): Crossing the great divide: Coproduction, synergy, and development, World Development.
Ostrom, E., Gardner, R. & Walker, J. (1994): Rules, games and common-pool resources, Ann Arbor, University of Michigan Press.
Pinkerton, E. (1989): Co-operative management of local fisheries. New directions for improved management and community development, Vancouver, University of British Columbia Press.
Raub, W. & Voss, T. (1986): Conditions for cooperation in problematic social situations. In: A. Diekmann & P. Mitter (eds.), Paradoxical Effects of Social Behavior: Essays in Honor of Anatol Rapoport, Heidelberg, Physical.
Reijntjes, C., Haverkort, B. and Water-Bayers, A. (1992): Farming for the future: An introduction to Low-External-Input and Sustainable Agriculture, London/Basingstoke, MacMillan Press Ltd.
Renn, O., Webler, Th. & Wiedemann, P. (1995): Fairness and competence in citizen participation. Evaluating models from environmental discourse, Dordrecht, Kluwer Academic Publications.
Röling, N. (1994): Platforms for decision-making about ecosystems. In: L.O. Fresco, L. Stroosnijder, J. Bouma & H. van Keulen (eds.), The future of the land: Mobilizing and integrating knowledge for land use options, New York, John Wiley & Sons Ltd.

Satz, D. & Ferejohn, J. (1994): Rational Choice and Social Theory, Journal of Philosophy **91**, 71-87.
Scoones, I. & Thompson J. (eds.) (1994): Beyond Farmer First: Rural People's Knowledge, Agricultural Research and Extension Practice, London, IT Publications.
Uphoff, N. (1992): Learning from Gal Oya. Possibilities for participatory development and post-Newtonian Social Science, Ithaca, Cornell University Press
Van Woerkum, C. & Aarts, N. (1996): Communication and policy processes: Reflections from the Netherlands, Available at the Department of Communication and Innovation Studies, Wageningen Agricultural University.

Address of author:
Marleen Maarleveld
Department of Communication and Innovation Studies
Wageningen Agricultural University
Hollandseweg 1
6706 KN Wageningen, The Netherlands

Principles and Strategies of Participation and Cooperation Challenges for the Coming Decade

L. van Veldhuizen

Summary

Although interest in participatory approaches to natural resource management has increased considerably, a number of very important challenges present themselves to ensure that participatory approaches remain on the agenda in future and that their effectiveness is maintained and further improved. This paper argues that the basic concepts used in the analysis of farmer participation still need to be better understood. More efforts are required to develop effective participatory approaches and methods that go beyond planning and ensure farmer participation in subsequent stages. Other challenges identified include the scaling-up of participatory programmes for which a number of directions are given, the move required from giving training in participatory approaches towards providing support to institutional change, improved understanding of costs and benefits, and the final impact of participatory approaches. Clearly this latter analysis needs to differentiate between the relevant actors involved, including between socio-economic or gender groups within the farmer community. The final challenge is to ensure that the answers generated in the field to these issues are being documented, analyzed, and disseminated.

Keywords: agricultural development, cost-benefit analysis, gender, institutional development, participation.

1 Introduction

Because of the efforts of a great number of organizations, interest in participatory agricultural development approaches has increased dramatically worldwide over the past ten years. Publications such as Okali et al. (1994) give excellent overviews of the relevant literature while periodicals such as the PTD (Participatory Technology Development) Circular from ETC (Educational Training Consultants) in the Netherlands and the PLA (Participatory Learning & Action) Notes from IIED (International Institute for Environment and Development in the UK continue to spread information on current experiences.

Rather than presenting another overview of developments over the past years and the progress made in promoting participatory approaches, I present here what I see as important challenges if we are to ensure that principles of and approaches to participation and cooperation not only remain on the agenda in the next decade but also become increasingly understood and embedded effectively in mainstream development programmes and organizations. These challenges are formulated from the perspective of a professional working in a support organization based in Europe.

2 Understanding the basic concepts

There continues to be an urgency in clarifying the basic concepts used in the discussion of farmer participation. In the context of the present ISCO workshop, two areas of concern present themselves very clearly. The use of the participation concept itself causes considerable confusion. There is the danger that the concept is referred to in so many different ways that its fundamental meaning is lost. If this continues, all development projects and programmes will be participatory on paper but without a fundamental change in the way they interact with farmers and other stakeholders. In the past, several publications have tried to define levels of participation in terms of the influence of the target group in setting the agenda and implementing main activities (Biggs 1989, Pretty 1994). But the challenge to practitioners and scholars is what participation entails, especially the level of influence of the various players involved. Factors influencing the choice for a particular form in a particular situation can than be addressed more systematically.

Second, links between participation advocates and scientists in psychology, political sciences, and sociology will need to be strengthened to bring the analysis of participatory development beyond the continued reviewing of practical experience. The Beyond Farmer First project (Scoones and Thomson 1994) took a step in that direction but also showed the difficulties in linking these two worlds. Papers presented at this conference (Maarleveld 1996, Zweifel 1996) indicate that such linkages continue to be sought and how this leads to increased understanding of what happens in the field.

3 Beyond participatory planning

A great number of participatory agricultural development approaches have emerged, each with its own acronym and its specific features depending on its history and the context in which it was developed (e.g. van Veldhuizen 1993). But a close study of these approaches reveals that the emphasis is generally on interaction of local people and support organizations during planning stages; studying local situations, analysing problems and opportunities, and agreeing on future action. Although participatory planning is extremely important, the challenge is to develop and document approaches that continue a participatory mode systematically in subsequent stages of interaction: during first experimental activities, monitoring and evaluation, re-planning, spreading of results to others, and strengthening local managment capacities. The recent shift of PRA (Participatory Rural Appraisal) to PLA (Participatory Learning and Action) seems to reflect this concern.

Often, natural resource management programmes will not best be served if one single participatory approach, out of the many, is adopted directly. An understanding of the respective strengths of the different approaches in their different contexts allows practioners to select and use (parts of) them when appropriate, such as the case study of the Soil Conservation Research Project in Madagascar (Madrid et al. 1996).

4 Scaling-up

Most cases reported on farmer participation in natural resource management operate at a relatively small scale, in a few villages or a district. Successful cases reported at international fora such as the ISCO conference often do not cover more than 1 or 2 growing seasons. Only gradually are larger, formal, government organizations responding to the challenge to take farmer participation seriously within their national country-wide mandates (Scheuermeier and Sen 1994, Backhaus and Wagachchi 1995). The implications of this are only just starting to be understood. Assessment of the experiences of the national Landcare programme in Australia (Ewing 1996) points to the danger of funds and resources being monopolized in the bureaucracy, preventing real decentralization to the local level.

In other cases, larger scale coverage is sought outside the government structure, as in the example of the "Campesino a Campesino" movement in Central America (Holt-Gimenez 1993). A network of farmer trainers, farmer organizations and non-governmental organizations has emerged which implements farmer-led extension approaches to sustainable agriculture development. The implications of this, including the institutional arrangements required to ensure longer-term sustainability of the system, are now slowly becoming clear.

5 The process of institutional change

After more than a decade of staff training in farmer participatory research and extension methods, it has become clear that much more than training is required if organizations are to change towards more participatory approaches. A recent analysis of efforts to transform large, agriculturally-oriented, government agencies in Kenya, the Philippines and Sri lanka (Thompson 1995) identified at least 10 core elements that influence the outcome of such efforts. The challenge is to aim for change processes that:
- allow field staf and their coordinators to develop the locally most appropriate participatory approaches and methods; small-scale pilot activities, well documented, may play an important role in this;
- stimulate adjustments of institutional arrangements and structures to make these work; and
- provide for training and support over a longer period of time, involving key players at different levels.

Van Campen et al. (1996) provide is a rare example documenting such a support process.

6 Who benefits?

This is not at all a new challenge (see Scoones and Thompson 1994) yet one that needs to be stressed over and over again. Struggling to develop locally effective participatory approaches, practitioners continue to find it difficult to pay attention to differentiation among members of the target group, mostly farmers or pastoralists. Frequently, this leads to involvement of only a limited number of often better-off farmers, frequently men, in programme activities. How can participatory programmes really take local socio-economic and gender differentation seriously, go beyond the concept of "the farmers" and ensure that people who need support are involved, whether this be the poor, the ethnic minorities, the less sedentary, or (certain groups of) women? Realizing the importance of this is only one step. Identifying the relevant groups in a particular situation as part of the participatory programme is an important second step. Finding mechanisms to take their interests seriously and involve them in various activities is the ultimate step. There are no general solutions for this, and locally effective ways need to be found.

7 Costs and benefits

Advocates of participatory approaches never stop praying the many advantages as compared to more top-down approaches: increased acceptance of proposed programme activities, rapid adoption of innovations promoted, development of more appropriate solutions, greater self-confidence of participating farmers, and improved self-management capacities of farming communities. Critics, however, stress the large amount of contact time between field agents and farmers, the high costs involved in terms of staff and transport, and the high quality of field staff required to make participatory approaches work.

In spite of the methodological difficulties and potential pitfalls, analysis of costs and benefits of participatory approaches urgently needs to be addressed, if alone to be able to provide well-founded arguments to policy makers and managers of larger development institutions. Important considerations in such analyses include:
- The framework: what are important elements on both sides of the equation? Which costs and benefits need to be taken into consideration?
- Assessment of learning processes: on the benefit side, does this focus of participatory approaches need to be considered in the analysis alongside the physical outputs?
- Quantification of the quantifiable: much more can be done to make the costs and benefits encountered visible.
- Multiple stakeholders and multiple objectives: a benefit in the eyes of one group may be less so in the eyes of others.
- Participatory monitoring and evaluation methods: interactive assessment methods need to be used with the relevant stakeholders to ensure that relevant criteria and data are obtained and that the exercise contributes to learning at the various levels.

8 Capacity to self-manage change?

Participatory approaches claim that the aim not only at reaching physical outputs, e.g. of improved natural resource management practices, but also at developing people's capacity to face other challenges, solve other problems, and seek further improvements in the future. This simple and widely accepted statement in itself leads to a whole series of challenging questions that few have started to addressed (see Gubbels, 1997). For example:
- What does this self-management capacity entail? A trainers' guide for Participatory Technology Development (van Veldhuizen et al., 1997) distinguishes between capacities of individuals (analytical skills, experimental skills, self confidence), and collective capacities (development of groups and institutions, linkages with other groups, direct contact with sources of information and services).
- What is wrong with present problem-solving capacities of farmers and their communties?
- What are effective mechanisms to help strengthen this capacity?
- How can progress made in this respect be monitored? By the people involved themselves? By others?

9 Continue to learn

There are no easy answers to most of the above questions. Yet they need to be addressed if the present momentum supporting the promotion of participatory approaches is to be continued and expanded. If they are not addressed, participatory approaches may either turn into standard, routine practices in which participation does not go beyond some oblique form of consultation, and in which creativity and committment are lost; or the frustration of not reaching the often very high objectives may lead to gradual loss of credibility of and interest in these approaches.

Field practitioners, rural people and their support organizations play a crucial role in finding answers to the issues raised by experimenting with new methods and institutional arrangements. A final challenge facing organizations such as ours is to ensure that those (parts of) answers are being documented, analyzed, shared and made more widely known so that all involved (and those hitherto not involved) continue to learn.

References

Backhaus, C. and Wagachchi, R. (1995): Only playing with beans? Participatory approaches in large-scale government programmes, PLA (Participatory Learning & Action) Notes **24**: 62-65.

Biggs, S.D. (1989): Resource-poor farmer pariciption in research: A synthesis of experiences from nine NAR systems, OFCOR Comparative Study Paper 3, ISNAR, the Hague, Netherlands.

Campen, W.A.M. van, Mgomba, S.S. and Mhina Mngube, F. (1996): Coming together: Transformation of district induced soil conservation activities into population-based land and water management programme in Mbulu District, Tanzania, paper prepared for the 9th ISCO Conference.

ETC (Educational Training Consultants), The PTD Circular: Six monthly update on Participatory Technology Development, ETC-NL, P.O. Box 64, 3830 AB Leusden, Netherlands.

Ewing, S. (1996): Australia's Community Landcare Movement: a tale of expectations, paper prepared for the 9th ISCO Conference.

Gubbels, P. (1997): Strengthening community capacity: The missing link in sustainable agriculture. In: L.R. van Veldhuizen, D. Johnson, R. Ramirez, J. Thompson and A. Waters-Bayer (eds.), Farmer research in practice: Lessons from the field. London: IT Publications.

Holt-Gimenez, E. (1993): Farmer-to-farmer: The Ometepe Project, Nicaragua. In: C. Alders C, B. Haverkort and L.R. van Veldhuizen (eds.), Linking with farmers: Networking for Low-External-Input and Sustainable Agriculture, London: IT Publications.

IIED (International Institute for Environment and Development), PLA Notes: Notes on Participatory Learning and Action, The Sustainable Agriculture Programme IIED, 3 Endsleigh Street, London WC1H 0DD, UK.

Maarleveld, M. (1996): Improving participation and co-operation at the local level: Lessons from economics and psychology, paper prepared for the 9th ISCO Conference.

Madrid, K., Madrid, R. and Tampe, M. (1996): Potentials and limitations of participatory technology development in soil conservation: An example in the highlands of Madagascar, paper prepared for the 9th ISCO Conference.

Okali, C., Sumberg, J. and Farrington, J. (1994): Farmer participatory research: Rhetoric and reality, London: IT Publications.

Pretty, J. (1994): Alternatives systems for inquiry, IDS Bulletin **25** (2).

Scheurmeier, U. and Sen (1994): Starting-up Participatory Technology Development for animal husbandry in Andhra Pradesh, LBL, CH-8315 Lindau, Switzerland

Scoones, I. and Thompson, J. (1994): Knowledge, power and agriculture - towards a theoretical understanding In: I. Scoones & J. Thompson (eds.), Beyond farmer first: Rural people's knowledge, agricultural research and extension practice, London: IT Publications.

Thompson, J. (1995): Participatory approaches in government bureaucracies: Facilitating the process of institutional change. World Development **23(9)**:1521-5

Veldhuizen, L.R. van (1993): Many ways, one perspective? Paper prepared for the workshop on PTD training and development, St. Ulrich, September 1993, ETC-NL, P.O. Box 64, 3830 AB Leusden, the Netherlands.

Veldhuizen, L.R. van, Waters-Bayer, A. and de Zeeuw, H. (1997): Developing technologies with farmers: A trainer's guide for participatory learning, London: ZED Publications.

Zweifel, H. (1996): Sustainable use of land: A participatory process. paper prepared for the 9th ISCO Conference.

Addresses of authors:
Laurens van Veldhuizen
ETC Netherlands
P.O. Box 64
3830 AB Leusden
The Netherlands

Political and Institutional Conditions for Sustainable Land Management

A. Oomen

Summary

For sustainable land management, an adequate political, policy, and institutional environment is a prerequisite. Land, soil fertility and water resources, are both public and private goods (multi-uncionality). Management of natural resources (land, water, soil) does not happen by itself, nor will it be sustained without a conducive environment.

Thus, sustainability will not be generated in the free market; it will only take place in a synergy between government, and the private sector/civil society. Sustainable development requires develop-ent of sustainable technology, and considerable private and public investment. Long term invest-ents in technology development, however, require a stable and predictable policy environment, supported by transparent decision-making and a reliable judiciary system. After all, a sustainable environment requires first and foremost clear political choices from which effective policies and measures result. In order to achieve this, a country requires a set of appropriate institutions at the level of the Government and of the private sector/civil society. These should be capable of developing the policy formulation including research and consultation, the implementation and the enforcement of the policy, the monitoring and the evaluation. They refer not only to appropriate legislation, rules and regulations and procedures, but also to specific organisations, institutions and consultation mechanisms.

The division of management tasks between the private and the public sectors can be different from country to country. Where the principle of subsidiarity is applied, management usually lies in the hands of actors of society. The Government as the custodian of public goods, has a supervisory role to play. More and more conflicting interests in society are related to disputed access to natural resources. These can lead to violent conflicts. Conflict prevention requires anticipating political decision-making, consultation, legislation and law enforcement. In many cases regional cooperation is needed.

1 Introduction

The title of this paper refers to sustainable land management as a form of sustainable natural resources management. It refers also to the external conditions which constitute the enabling environment or framework for sustainable management. Natural resource management encompasses the protection and sustainable utilisation and, where necessary, the preservation of natural resources.

The management level that can be addressed realistically for its responsibility for Natural Resource Management (NRM) is the nation and, at the level of the nation, the government.

ISBN 3-923381-42-5
© 1998 by CATENA VERLAG, 35447 Reiskirchen

The national level includes however all national and sub-national actors, ranging from government to the private citizen, the individual 'manager' and utilizer of natural resources. It does not refer only to Government.

In the context of the European Union as a matter of fact we have in some areas managed to arrive at inter-governmental regional arrangements of the same strength as national legislation. But, usually, at the regional and the international level, arrangements are not as strongly binding as national legislation or can be escaped from, due to lack of effective control.

Thus, if we speak of political and institutional conditions for sustainable land management, we focus primarily on the national level.

2 Political and institutional conditions at the national level.

As a rule, at the level of the Constitution of a nation, natural resources are defined as both a public good and a private good, because, for both the society as a whole and for the individual citizen, they constitute the basis for economic and social life. Thus, management of natural resources is both a public and a private responsibility. Only in close collaboration between the public sector and the private sector and civil society is sustainable natural resource management feasible. The division of tasks between the private and the public sectors can be different from country to country. Where the principle of subsidiarity is applied, management usually lies in the hands of civil society. However, the Government, as the custodian of the public goods, has a supervisory role to play. From the private sector, responsible stewardship is required and expected. When the private sector fails to fulfil its responsibility, the Government will intervene in order to protect the public good.

In order to arrive at a well functioning working relationship between the Government and a responsible civil society, decisions at the political and policy levels and a variety of institutional provisions are required.

When dealing with legislation and policy design, the political and the policy analysis levels in respect of natural resources need to take into consideration a great number of aspects, such as, administrative, (other) institutional, technological, financial (public and private investment), economic, social, cultural and physical aspects. Here we focus on the political, administrative and other institutional aspects which together constitute the enabling environment or framework for sustainable natural resource management, in particular land management. If the framework which the Government provides proves to be stable and trustworthy, private investments in technology for sustainable land management will become sufficiently remunerative to be attractive.

At the level of the interface between Government and civil society a number of institutional provisions are required to involve the 'managers of land' in civil society in the decision-making process and to enhance the legitimacy and acceptability of the legislation and regulations which are envisaged.

We have come to realize though that policy analysis, decision-making and implementation require the capacity to manage these processes in order to arrive at sustainable natural resources management at the national level.

One striking example of a natural resource which is both a public good and a private good is soil fertility. Soil fertility is the basis for agricultural and food production. As part of its national food security strategy, a nation will see to it that its base for food and agricultural production is maintained properly. At the level of the individual farmer however, proper maintenance of soil fertility can become difficult or even impossible when farm gate commodity prices go down and the cost of the inputs like fertilizer rises above the marginal value of the product. Presently this is the case in Africa. Structural Adjustment Programmes, imposed by the World Bank and the other donors, not only suppressed the so-called subsidy on fertilizers and the dismantling of the public inputs distribution system but also the introduction of a 'liberalised' market of inputs. No credit system was put in place of the subsidy. As a result of factual and manipulated scarcity (market monopolies), prices of fertilizer rose sky high and, in Sub-Sahara Africa, fertilizer use went down to 5 kgs/ha. As an immediate consequence large scale soil nutrient depletion takes place up to 75 kgs/ha/annum. Serious soil degradation, sometimes irreversible, follows. This however is totally contradictory to the World Bank's policy regarding 'environmentally sustainable agriculture' (World Bank, 1996). As the farmers, especially the small and poor farmers which are the majority, are not able to reverse the trend, the Government as the custodian of the public good of soil fertility has to intervene and use public funds to redress the current process. If soil fertility is indeed considered a public good, fertilizer should rather be looked upon as an investment than as a consumers' good. World Bank loans and other public international assistance to African countries should under certain conditions allow for reduced fertilizer prices in order to restore and maintain soil fertility.

3 Political conditions

At the political level, decisions need to be taken as to which direction the society should follow in approaching the environmental space. Difficult decisions need to be taken. Political courage is needed. Choices are to be made on the orientation of the proper use of natural resources. As a rule these choices have to be made in the dark. Science does not have the answers ready. Politicians cannot hide behind scientists.

The following quote from a report of the Netherlands Scientific Council for Government Policy illustrates the position in which politicians find themselves and their relation with the scientific community:

'Policy is characterized by factual and normative uncertainty; this already applies in the current situation and even more so with respect to the future. Factual uncertainties are characteristic not just of the ecological but also of the social domain. While it is true that the lack of correspondence between the desired and the expected ecological situation provides ground for talking about non-sustainability, not just ecological but also economic and other social risks play a role in formulating possible solutions to this problem. The assessment of such information and the weighing of the risks is the essence of politics.'

'Political conditions' refers to decisions at the political level, in this case at the highest political level, at the level of the Constitution.

3.1 Framework

These political decisions lead to a political environment that does or does not allow individual citizens and organisations in civil society to pursue their interests within a framework. In the Plan of Action of the World Food Summit one finds the wording 'political, economic, social and environmental framework'. In principle it sets the rules and the rights of the individual citizen. Citizens know what they can do and which rights they can exercise. These political decisions either favour or prohibit legislation with regard to equitable distribution of assets, to access to land and water, to credit, to inputs, to markets. Absence of political decisions and absence of legislation and institutional provisions are in fact political decisions as well.

3.2 Conflict management

An issue that returns continuously is the fact that conflicting interests can lead to conflicts if the highest political level, the government and the administration do not take their reponsibility properly.

So, when political and policy aspects of natural resource management (NRM) are at stake, the issue of absence or presence of pre-empting political decision-making with regard to potential conflicts that are related to access to natural resources cannot be overlooked neither in developing countries nor in the so-called developed world. If political decision making is felt to be weak or absent, parties in conflicts may feel free to go ahead and to let the conflict erupt.

If political decision making is weak or absent, no administration or administrator will start developing strong regulations or legislation and risk his or her career.

4 Administrative provisions

Institutional provisions regarding the administration refer to legislation, rules and regulations, procedures, organisations, institutions, systems and mechanisms that a Government may deem suitable to govern a sector, a cross-sectoral issue, or a region. They also deal with the human capacity in each of the provisions.

First, institutional provisions at the level of the administration relate to legislation, to the application of the law by the judicial system, to civil rights such as property rights and entitlement to assets and services, and to provisions such as investment protection. Secondly, institutional provisions relate to consultation mechanisms between the administration and the political level, between the government and local governments, civil society, private sector such as the farmers in the agricultural sector, and professional organisations. Thirdly, institutional provisions at the level of the administration refer to information systems, statistical services, all other services as well as to all institutions related to research and development.

Through maintaining a link with the research community, the administration is able to translate new knowledge into policy options and policy proposals.

We have seen in the last decade that neo-liberalism was in the winning mode. Government interventions were considered to be harmful for the full creativity of the private sector. Hardly any recognition of public goods remained. Multifunctionality of sectors like the agricultural sector was rejected for the sake of the so-called free market.

The view which is now emerging is that we should be very careful in accepting the non-proven positions promoted by the neo-liberalists, especially in the area of NRM. NRM cannot be left to the free market. Natural resources are both a public good and at the same time a private good. Natural resource management can take place only in a synergy between the private sector/civil society and the public sector. The public sector has the responsibility and the means to intervene where necessary. The private sector will indicate the limitations related to economic performance.

Agriculture is a multi-functional sector. On the one hand, for private entrepreneurs, agriculture is a private sector in which private interests play an important role while on the other hand, the sector takes up the responsibility for public functions in full collaboration with the administration.

In addition to specific services and supplies to the individual user of natural resources i.e. the small farmer, such as credit, protection of land ownership and water rights, the Government also is responsible for the supply of basic services such as formal and non-formal education and health services.

This approach to the role of Government sounds logical. But, at present, precisely in this area, a major political discussion continues to take place world wide on the size of the administration, on its role and functions, on the way the administration deals with civil society.

5 Other institutional provisions

In Germany and The Netherlands we are accustomed to a great number of institutional arrangements at the interface between Government and the private sector and civil society. They are an indispensable part of our democratic system. The formal and informal networks and interactions between the private and the public sector are referred to as the Rheinland Model. Apart from formal parliamentary democracy, a multitude of institutional arrangements involve actors in civil society, also at sectoral level, in decision making.

We also know a great number of institutional arrangements within the private sector and civil society to deal effectively with the interface, from the financial and industrial sector, employer's organisations, trade unions, consumer's unions, to smaller organisations such as associations to protect nature and environment.

In developing countries on the other hand we see that many of the institutions we are so used to, are weak or do not exist. Many interactions between parties do not take place or are part of a small in-crowd. Institutions that are accessible to a much larger spread of stakeholders are important in the context of policy formulation and implementation and in conflict prevention and resolution.

As in so many other areas of governance, natural resources management needs transparency and accountability. Too many conflicting interests are involved and - in view of employment, agriculture, food production, shelter, etc. many people depend on the decisions taken.

6 Technological development and investment

Once a stable, trustworthy enabling political and institutional environment is in place, the public and the private sector will increase their investments in research and technological development. The entrepreneurs in the private sector will be able to take care of the necessary growth in a sustainable manner. Research agendas will be developed in consultation with the end users. Research will be able to develop specific opportunities for a specific setting of assets and constraints, in other words, develop specific comparative advantages within the limits of the natural resources available.

7 The challenge ahead

According to several scenarios that have been developed recently (IFPRI, 1995; FAO, 1990) the world faces the challenge to double the food production in the next 30 years. Export crop production must increase. All this will have to be done in a sustainable way.

It calls both for a green revolution as we have witnessed, and for a green revolution in the sense that we need to take account of the limits of sustainability. That is what French researchers call a 'Doubly Green Revolution' (Griffon and Weber, 1995).

Without continuous political support and without the necessary institutional framework at the national level this revolution that has to be carried forward by the private sector, researchers and so many other actors, will not come about in due time.

We could ask the Heads of State and Government who participate in the World Food Summit in Rome in November to take their commitments with regard to sustainable development of the food and agriculture production serious and, this time, take the necessary steps for their implementation.

References

IFPRI (1995): A 2020 Vision for Food, Agriculture, and the Environment, Washington D.C.
FAO (1990): Agriculture towards 2000, Rome

Griffon, M.and Weber, J. (1995): Les aspects économiques et institutionnels de la Révolution Doublement Verte, contribution at the International Seminar 'Vers une Révolution Doublement Verte', CIRAD, Montpellier

Netherlands Scientific Council for Government Policy (1995): Sustainable Risks: a Lasting Phenomenon, The Hague.

World Bank (1996): From Vision to Action in the Rural Sector, Washington D.C.

Addresses of authors:
Ad Oomen
European Centre for Development Policy Management
Onze Lieve Vrouweplein 21
6211 HE Maastricht, The Netherlands

Sustainable Land Use - A Participatory Process

H. Zweifel

Summary
The sustainable land use has a central role in the sustainable use of renewable natural resources. This paper places sustainable land use in a wider context, by presenting a conceptual approach to natural resource management developed by the Centre for Development and Environment at the University of Berne (GDE 1995). It is argued here that natural resources should be understood as components of nature and that sustainability is primarily a question of evaluation. A definition of sustainable use of natural resources must be formulated as a broad participatory process, in which the "internal perspective" - the perspective of the local land users - and the "external perspective" - the perspective of development experts or researchers - are both involved. The negotiation and evaluation process necessary for defining sustainable resource use and solving conflicts over resources therefore becomes the core of any strategy for sustainable land use. Two cases of conflict resolution through negotiation are presented - one from India and one from Switzerland - to illustrate preconditions and mechanisms involved in negotiation processes. The role of development agencies in the process of negotiation to find consensus on sustainable resource use.

1 Natural Resources - Components of Nature

What are natural resources? The term "natural resource" implies that people utilise nature and attach significance and value to it. Natural resources therefore constitute only one part of an ecosystem that is used by people, in contrast to nature as an all-encompassing entity. In other words, natural resources as components of nature are defined from a human perspective; they are not a given entity defined by nature (GDE 1995).

Perception and use of natural resources varies in space and time. People in different societies at different times and places can have other perspectives. As an illustration of the dimension of time, we might consider the Arabian peninsula, which was regarded as an area very poor in natural resources in the 19th century. Today, with petroleum an important source of energy, the Arabian Peninsula is considered one of the richest areas in the world. Every society has its own view of resources. For example, local people in dry areas might not consider aquifers deep beneath the surface as a valuable resource, although trained hydrologists certainly would.

In a given context, at least two different perspectives on natural resources are normally encountered. One is the perspective of local land users - a view which is based on the ways in which nature and soil have traditionally been perceived, and which varies from place to place and region to region. This has been called the **internal perspective on natural resources**. The other perspective - which is found throughout the world and is represented by researchers and scientists, environment and development experts, and politicians and administrators - usually reflects an economic world view and is characterised by its scientific approach. This view has been called the **external perspective on natural resources**. Both the internal and the external perspective - and the values they attach to a given component of nature - are bound to change in relation to changes in the ecological, economic, cultural or technical context (GDE 1995).

The dimension of space is important because the ecological impacts of a certain resource use may differ, depending on the special features of the ecosystem affected. People may perceive slow changes resulting from the ecological impacts of a particular land use, and record them within a known, definable space. But many types of degradation are too slow to be perceived, and many types of land use may be characterised by transregional impacts and feedback which may also need to be considered. This makes it impossible to formulate a universally valid definition of the space in which natural resources are valued and used.

2 Sustainability - in search of a definition

Since the Earth Summit in Rio in 1992, the term "sustainability" has become fashionable in politics, science and industry as well as in relation to development and the environment. Reference will be made here to sustainability as defined in the Brundtland Report (WCED 1987, p.43), although this is a very vague definition which constitutes little more than a declaration of intent to insure that development meets the requirements of the present without compromising the ability of future generations to meet their needs. In order to clarify the concept of sustainability, a frame of reference is needed. We have already defined resources as components of nature which are of use to people, and which fulfil different functions in serving people's needs. We take these functions as a standard of measurement.

From a human perspective, natural resources in general - and soils in particular[1]- are characterised by three broad categories of interrelated functions:
1. The **productive functions** of natural resources involve those components and attributes of nature that are of importance to a particular society in connection with processes of production or reproduction. Productive functions such as soil fertility (through provision of rooting conditions for production of biomass) are the basis of agricultural production in its widest sense.
2. The **social or cultural functions** of natural resources are associated with components of nature that are of some sociocultural value to the people in a particular society. The soil as earth, which many societies call "Mother Earth", has this function. Many people - either as individuals or as groups - have a strong sense of belonging to a particular place, whether as sedentary agriculturalists, nomadic pastoralists or even city dwellers. Communities reserve parts of a landscape as "cultural space", exemplified by graveyards, mountain areas, and places for worship and celebration, etc.
3. The **ecological functions** of natural resources are linked to components of nature which contribute to the sustainability of ecosystems in the view of a particular society. These functions include such aspects as humus formation in topsoil, nutrient mobilisation in the subsoil, the buffering of soil systems, and preservation of soil resilience against sudden alterations.

To achieve sustainable use of soils, all three of these functions should be sustainable in practice. Sustainability relates to future conditions and uses, and is therefore linked with uncertainty. Moreover, sustainability is directly related to the ways people assess future conditions and events. The meaning of sustainability can only be determined by a society in a participatory process of political evaluation and assessment. It has little meaning in absolute terms. A society will attach the label "sustainable" to certain land uses only when it has determined that, at the very least, these types of land use will continue in the long term and all three natural resource functions will be sustained. It is important to stress that sustainability does not mean that things never change, or that given practices or uses must be conserved. Certain types of land use may be replaced by new ones, while others are modified or abandoned.

---

[1] See Douglas (1994) for the functions and the multidimensional nature of the sustainable use of soils.

Defining the sustainability of land uses in a given situation must be a societal process, whereby the principles of sustainability are evaluated in the light of both the internal and the external perspective. Ideally, common ground will be found on the basis of evaluations by external actors as well as evaluations by local land users, i.e. all actors will agree that a certain land use fulfils productive, ecological and cultural functions in a way that will meet the long-term needs of the people and ecosystems affected. (GDE 1995)

This definition reflects the fact that sustainability can only be defined in approximate terms. It can be of help in the search for more effective and more sustainable forms of land management that take into account both the internal and the external perspective on natural resources. The following three points are of central importance in open negotiation processes over sustainable land use:

1. The three functions of natural resources (productive, social, ecological) should be considered
2. All actors involved in or affected by a given land use should be able to participate and articulate their views
3. There should be a chance to assess potential options and their consequences as far as possible, while incorporating both the internal and external perspective.

This approach focuses on negotiation and mediation involving different (often conflicting) perspectives and interests. Differences and conflicts are always present in the management of natural resources, since they are the natural expression of the needs and desires of different groups. The central position of the participatory negotiation process implies that strategies and approaches for sustainable land management and soil conservation measures can only be effective if they involve the perspectives of the different stakeholders, insiders as well as outsiders, at the local, national and international level.

3 The principle of participation

Defining sustainable land management in terms of a process involving both the internal and the external perspective on natural resources means that all people involved in and affected by a given land use and its changes need to be involved in the negotiation process to determine the meaning of sustainability. People's participation becomes a guiding principle.

In the history of soil conservation and sustainable land management, both national and international development agencies have attempted to involve people in certain phases of project intervention. The need for people's participation has been widely acknowledged and featured in many strategy papers produced by international development organisations, including NGOs, government offices, and banks. A whole new generation of "participatory projects" has emerged. As a by-product of its popularisation, the terms "people's participation" and "popular participation" have became catchwords, meaning different things to different people[2]. With regard to involvement of local people in development and environmental projects, two main schools of thought have evolved (Pretty 1995, 173):

1. **Instrumental participation**: Soil and water conservation projects are characterised by "top-down" methods of planning and implementation. Typically, an outside conservation "expert" identifies the "problem" in the field and comes up with a technical solution. People's participation - for example food or cash for work - was seen as a means to increase efficiency, because, if people are involved they may be more likely to agree with and support a new technology or service.
2. **Functional or interactive participation**: In this concept, people's participation is seen as a right and a necessity. The main aim is to initiate collective action, empowerment and institu-

---

[2] See Östberg and Christiansson (1993) for the changing rhetoric over participation and soil conservation.

tion building. This may involve the development and/or promotion of externally initiated social organisation, or people participating in joint analysis, which leads to plans of action and the formation and strengthening of local institutions. Some groups tend to be dependent on external institutions, while others take control over local decisions. The main objective of this bottom-up approach is to empower farmers to learn, adapt, and do better.

A study of 230 rural development institutions revealed that in the great majority of projects, participation was almost always understood in the instrumental way described above (Pretty 1995, 172). In marked contrast to this, there is considerable evidence that long-term economic and environmental success is achieved when people, as well as their ideas and their knowledge, are valued, and when they have the power to make decisions independent of external agencies. The use and misuse of the term "participation" make it necessary to define the term when using and interpreting it.

The principles of **internal and external participation** are central to the concept of sustainable management of land (GDE 1995):

- The principle of **internal participation** - i.e. the involvement of local land users - is based on recognition of the aims, knowledge and experience of local people, as well as assessments of the options open to them.
- Internal participation should be complemented by **external participation** - the participation of development organisations, researchers and others in the local negotiation process - in order to achieve the widest possible perspective on resources.

The principle of internal participation recognises the fact that women and men who manage the land and its resources are the real experts on their respective ecosystems, with ideas, aims, technical insights, and organisational capabilities that are needed for development (Zweifel 1996; Scoones and Thompson 1994). "Traditional" land use systems are not static, but dynamic, because local communities have always had to cope with new problems as they arise, especially on fragile lands (where most of the poor live). Variations in seasonal climate, processes of cultural and economic interaction, demographic change and other phenomena constantly demand adjustment to new situations. If local land users perceive that particular recommendations or technical innovations for improving their land are beneficial and are also in accordance with their own goals and aspirations, they are likely to adapt, adopt and maintain them. The uniqueness and value of local perspectives on land and resources are rooted in long experience and in the complexity of local land use systems. Western science and research have so far given little attention to these factors. Any strategy for sustainable land management, therefore, will aim to explore different alternatives in close collaboration with local people[3].

In view of the wide spectrum of local knowledge and experience, why is external intervention necessary? Is external intervention by development organisations, researchers and soil experts justified? We have already seen that societies attach different values to resources over time. As ecological or economic conditions change, a certain type of resource use which is sustainable today might be harmful tomorrow. Because sustainable management of land and its resources is linked with future conditions, the widest possible perspective on sustainability must be adopted - one which combines the internal and external perspectives on natural resources.

Processes like rapid modernisation, concentration, migration, and social breakdown, which threaten sustainable resource use, may also undermine and destroy local competence. This makes it important to consider all relevant environmental knowledge, both local and scientific. External participation broadens the perspective, offers new insights, and has the potential to stimulate a common search for ways to support sustainable systems of resource use. Resource use and ecological changes whose impacts transcend the local level and have negative impacts on land use

---

[3] Talking of the "internal perspective" should not hide the fact that communities are not homogenous entities but consist of diverse groups differentiated by class, caste, tribe religion and/or ethnicity, and within each of these groups, by gender and age (see Sarin 1996, Rocheleau 1987)

in neighbouring areas provide another argument for external participation. Decision-making processes at the local level may ignore the needs of neighbouring regions, which can lead to severe conflicts over resource use. In such cases, national and regional decision-makers, research institutions, and development agencies can contribute their expertise and their knowledge in the debate over resource use. Development agencies, with their own long-term environmental commitments and know-how, may also be able to supplement local perspectives and broaden the criteria used in evaluating sustainable land use and assessing potential consequences.

Joint participation involving local people and outsiders implies that a basis of trust and shared goals needs to be established, and also calls for transmitting and pooling knowledge. The aim of the process is not to fashion a rapid, ill-conceived compromise in order to reconcile conflicting objectives, but to attain consensus and achieve shared goals. We need to realise that dialogue and negotiation processes must take place within communities as well as between local populations and outsiders. With this in mind, we can go beyond the bipolar perception of participation as consisting of interactions between outsiders and local people, as implied by concepts of "local participation" (Gould 1996). These interactions are actually part of an ongoing process of negotiation, adoption and transfer of meaning. It is important to understand and analyse struggles and differences in power between actors and the dynamics of the interaction.

Some of the complexity of the **negotiation process** between governments, other outside agencies and local groups are explored in the following two examples, one from India (Joint Forest Management) and one from Switzerland (National road construction, Wallis). These two cases illustrate how the involvement of all actors concerned is important and here the understanding of different (and often conflicting) perspectives, knowledge and ideologies is central. They also illustrate mechanisms, forms and degrees of institutionalisation of negotiation (and mediation) processes in conflicts over resources.

4 Joint forest management - An institutionalised participatory process

The concept and practice of "joint forest management" (JFM) in India[4], or more generally, "co-operative management" or "shared control of resources" (since it might include not only forests but also watersheds, water courses etc.) is based on the idea that forests may be saved or rehabilitated by sharing both responsibilities and rights between the state (the formal owner of land) and villagers. JFM is concerned with participation and empowerment (since communities are being given access to and control over resources previously not officially theirs), with sustainable development, in terms of saving the physical environment, and with social and institutional sustainability. Rights and obligations are specified in legal contracts between the two parties involved - the villagers and the state. The villagers' side is represented by village forest management committees (VFMCs), which are elected bodies. To allow the forest to recover, the villagers have to agree to enforce limited and specified annual take-off quotas. In turn, they are entitled to a significant part of the timber value of the forest (25-50%). The Forest Department (FD) has to ensure that the forest tract concerned is protected from any third party such as dealers from outside. Quotas are negotiated and stringently set in the first year by the FD and VFMCs. As the health of the forests starts to recover, measures are jointly assessed and new rules negotiated. For example, a complete ban on collecting living matter may be lifted and collection of firewood may be allowed on certain, specific days, or controlled hand cutting of fodder grasses and leaves may be permitted. The specifics of the situation determine the nature of the negotiations and the contract, and every case is different. Experience shows that JFM offers local people a modest

---

[4] Sources: Skutsch (1996), Sarin (1996)

opportunity to improve living standards, but more important, it is a local institutionalised forum that allows them to participate as organised partners with the government and its representatives.

Conflicts over resources and the need to re-negotiate sustainable use emerged in cases in which the forest area to be protected was "owned" by a number of different villages with competing claims to benefit sharing (conflicts with an external party involved). Conflicts may be essentially internal, concerned with the sharing of resources within the community. As Sarin (1996) notes, women are by and large excluded from VFMCs or are silent members (members are generally "heads of households" acting on behalf of the family). However, it is acknowledged that gathering forest products such as firewood, fodder, leaves, flowers, nuts, wild fruits etc for home consumption and for sale is primarily the responsibility of women and is important for the food security of the family. In a number of cases, the VFMCs perceived active forest management and conservation primarily in terms of increased timber production (productive function). As local resources are under "conservation", women had to travel further to gather firewood, which increased the workload, led to overuse in certain places, or even forced women to break the ban.

To negotiate and settle conflicts, JFM normally relies internally on local or traditional means of settling disputes. These are usually uncodified practices that have developed over centuries, based on norms accepted by the community, and they are normally rooted in structures of authority. Moreover, whenever required, the new VFMC can call upon the formal (codified) legal system. This usually happens when parties which do not recognise the authority of traditional decision-making systems enter into the arena of conflict, as in the case of invasion by other communities, timber contractors etc. In some cases, mediation by officers of the technical institute helped in a dispute to diffuse the conflict. In situations of open conflict, individuals may call on "outside law" if they feel it helps their case. Skutsch (1996, p.10) points out that "much of the strength of the JFM system is to be found in this pluralistic legal framework". The traditional systems of conflict management, which may have been satisfactory in the past, are insufficient under modern conditions, because modern forces are different and the scale of disputes is larger.

5 Solving environmental problems through negotiation processes

The second case of conflict resolution, which concerns construction of National Highway N9 in Switzerland, exhibits institutionalised as well as informal negotiation processes (see Knoepfel 1995). This case involves conflict over the use of land, and even more over *Lebensraum* (living space), a resource which is becoming very rare in Northern countries. A recent study (Mosimann 1996) discloses the fact that one third of the soils in Switzerland are already destroyed or threatened by degradation and contamination. Urbanisation and traffic infrastructure, which consumes and pollutes vast areas of mostly highly valuable land, are the main causes of soil degradation. The planned construction and enlargement of a national highway through the fertile Rhone valley of the mountainous canton of Wallis was opposed by some local communities. The opponents, especially environmental protection activists, filed a "petition" to the federal government, in which they demanded revision and re-dimensioning of the projected national highway. The national administration mandated Professor Philippe Bovy of the Federal Institute of Technology in Lausanne to evaluate the plans, assess the requirements of the industry, traffic and tourism, and determine the impact of the new road on the environment, landscape and agriculture. Professor Bovy initiated a broad participatory process, in which local groups of people concerned and affected, such as land users (farmers) and other stakeholders (environmentalists and regional administration) were invited to participate, to evaluate the plans and propose corrections and improvements. The local groups developed a number of alternatives, which were then assessed in terms of technical feasibility. The plans were then evaluated and re-evaluated by the local groups, in the light of the productive, cultural and physical function of the land (and other criteria). As a result of this process, the thrust of the proposal was changed, and the proposition was ultimately accepted by and large, even by hard-core opponents (except for one part leading through a valuable natural forest known as the Pfynwald). As a result, most of the propositions passed through official planning channels without difficulty. Bovy, the "mediator", never knew if a consensus would be achieved through this negotiation process, nor did he know if such a consensus would be accepted by the authorities in charge.

This and other Swiss cases illustrate that many cases of informal negotiation and mediation processes can be found, even in Switzerland, which has a wide range of legal and institutionalised decision-making fora at national, regional and local level. These processes run parallel or even replace formal administrative procedures. Public or semi-public fora are thus created and used to resolve conflicts. These mediation processes are equally important wherever legislature decisions are made. Anyone who wants to have a stake in *Lebensraum* as a resource must realise that he not only needs a formal democratic mandate from the national parliament, but the tacit consensus of the people affected by such a decision.

- In both cases we find a double strategy of using formal and informal channels or fora of mediation and decision-making to bring the internal and external perspectives together. A certain degree of **institutionalisation** for open and transparent debates on sustainable use of natural resources seems to be a necessary condition, but at the same time people can use it in a flexible way, according to their needs.
- All stakeholders need to be involved. Both examples show that, as a rule, local communities, organisations or households are not homogenous units but heterogeneous groups of people with different and often conflicting perspectives and interests. Within a group loosely defined as "land users", there may be several types of actors who neither own nor manage the land as such but have a stake in it. Not only is there a multiplicity of land user groups; the users may even be divided into distinct groups according to various criteria, namely by tenure (ownership/terms of access) and by unit of organisation (Rocheleau 1987).
- In addition to a multiplicity of users, there is an equally wide range of **knowledge** (see Scoones and Thompson 1994). There are many ways of experiencing, perceiving, understanding and defining reality. Western science is characterised by conflicting interpretations and schools, and within local communities there are also differences of knowledge and interpretation: the knowledge of the elite is different from that of peasants, the knowledge of men different from that of women, and so on.
- A crucial feature of the negotiation processes is that they are characterised by dialogue. Administrators, scientists and farmers experience and conceptualise their lives and knowledge in terms of rather distinct life experiences. It is important to note that knowledge is not easily portable from one context to another - because "knowledge about agricultural practices is embedded in the performance of those practices and in the lining of these practices into an overall farming and livelihood system" (Drinkwater 1994, p.38). The task of bringing together the knowledge, perspectives and skills of rural land users, the representatives of the state, aid agencies and/or researchers must be an ongoing process in which there is continuing dialogue.

6 The role of development agencies: "Facilitating a participatory process"

Defining sustainable management of natural resources in a given context has to be a broad participatory process among partners on an equal footing. As Dahlberg (1994, p.46) points out, "even when we strongly believe that we know what we are doing, recommendations and regulations concerning land use promoted in the name of sustainability have to be reached through negotiations involving all concerned". The challenge ahead for development co-operation and local communities is to establish and/or support institutional and structural mechanisms that encourage processes of negotiation, mutual learning, and local participation in the long run. To meet this challenge, the programmes concerned will need to develop mechanisms for broad, participatory negotiation processes in which the different functions of natural resources can be assessed and the various interests of the different stakeholders can be negotiated and addressed. Genuine collaboration will be needed in the search for viable solutions.

1. The role of development organisations in this process is primarily a supportive one. Development agencies might play the role of a **mediator or facilitator**, to ensure that all social groups are equally represented and have a chance to express their views and concerns. In addition, they should attempt to support and strengthen socially or economically powerless actors and groups whose options for participation are severely limited. They have to ensure that women participate not only in the negotiation process but also in the design and implementation of projects. Some of the questions raised for discussion are: Can outside mediators or conflict management programmes identify and address conflicts related to marginalised groups and women without a clear understanding of the complex power dynamics within and between communities? Do they have the long term commitment to empower the least powerful?

2. Development agencies **participate as outsiders** in the local process. They have expertise on ecological systems and their interrelationships, as well as a wider perspective on global problems such as climate change. As engineers they may provide information and technical advice, if desired. Development agencies themselves are stakeholders in the negotiation process with their own agenda. The line between "facilitator" and "participant" becomes very thin - an issue that needs to be discussed.

3. A certain degree of **institutionalisation** is necessary for open and transparent debates on sustainable use of natural resources. Development agencies, together with their local partners, can promote, support and strengthen local decision-making institutions or support them in creating new and more democratic ones. An enabling environment at the national and international level is necessary, that allows diverse social groups to express their concerns and become involved in the negotiation process for sustainable land use. Development agencies will have to reflect the potentials of the policy dialogue for creating an enabling environment.

References

Dahlberg, A. (1994): Contesting Views and Changing Paradigms. The Land Degradation Debate in Southern Africa. Nordiska Afrikainstitutet, Uppsala

Douglas, M. (1994): Sustainable Use of Agricultural Soils. A Review of the Prerequisites for Success or Failure. Development and Environment Report No.11, Institute of Geography, University of Berne, Berne

Drinkwater, M. (1994): Developing Interaction and Understanding: RRA and Farmers Research Groups in Zambia. In: Scoones I. and Thompson J. (eds.) 1994. Beyond Farmer First. Intermediate Technology Publications, London, 133-38

Gould, J. (1996): Beyond 'Negotiation? Challenges of Participatory Projects for the Anthropology of Development. Working Paper 9/96, Institute of Development Studies, University of Helsinki

Group for Development and Environment (GDE). (1995): Sustainable Use of Natural Resources. Development and Environment Reports No. 14. Institute of Geography, University of Berne, Berne

Knoepfel, P. (1995): Von der konstitutionellen Konkordanz über administrative Konsenslösungen zum demokratischen Dezisionismus - zur Vielfalt von Verhandlungsarrangements in Konfliktlösungsverfahren in der Schweiz. In: Knoepfel, P. (ed.) Lösung von Umweltkonflikten durch Verhandlung. Helbling & Lichterhahn, Basel und Frankfurt

Mosimann, T. (1996): Die Gefährdung der Böden in der Schweiz. Syndrome der Bodengefährdung, Abschätzung der zerstörten und belasteten Bodenflächen, Trends. Eine Synthese im Auftrag des WWF Schweiz, Bubendorf

Östberg, W. and Christiansson, C. (1993): Of Lands and People. Working Paper No. 25, Environment and Development Studies Unit. Stockholm University. Stockholm

Pretty, J.N. (1995): Regenerating Agriculture: Policies and Practice for Sustainability and Self-Reliance. Earthscan Publications, London
Rocheleau, D.E. (1987), The User Perspective and the Agroforestry Research and Action Agenda. In: Gholz. H.L. (ed.) Agroforestry: Realities, Possibilities and Potentials. Nartinus Nijhoff Publishers, Dordrecht, pp. 59-87
Sarin, M. (1996): Actions of the Voiceless: The Challenge of Addressing Subterranean Conflicts Related to Marginalised Groups and Women in Community Forestry. Paper prepared for the FAO Email Conference on Addressing Natural Resource Conflicts through Community Forestry
Scoones I. and Thompson J. (1994): Beyond Farmer First. Rural People's Knowledge, Agricultural Research and Extension Practice. Intermediate Technology Publications, London
Skutsch, M.M. (1996): Conflict Analysis in Joint Management: Incorporating Gender Components. Paper prepared for the CERES Seminar on Gender, Use and Management of Natural Recourses, Leiden, 9-10 May 1996
World Commission on Environment and Development (WCED). (1987): Our Common Future. Oxford University Press, Oxford and New York
Zweifel, H. (1996): Gender, Biodiversität und lokales Wissen. Paper presented at the FemWiss Conference on "Geschlecht in Frage", University of Zurich, 9 March 1996

Address of author:
Helen Zweifel
Centre for Development and Environment
Institute of Geography
University of Berne
Hallerstraße 12
CH-3012 Berne. Switzerland

Participatory Land Use Planning:
The Case of West Africa

H. Eger

Summary
"Cooperation" between people and institutions in the West African rural context centred for decades around the paradigm that extension agents of state or para-governmental organisations "advised" the farmers what to do, where and when. Traditional knowledge and organisational behaviour were rarely considered worthwhile of being integrated in the production of, for example, cash crops. Land care and proper land husbandry were not achieved under these conditions. With the introduction of a village level **P**articipatory **L**and **U**se **P**lanning (PLUP), this top-down "cooperation" slowly changed into a cooperation between equal partners. Flexible planning with a minimum legal framework, the absence of planning hierarchies and an iterative procedure based on dialogue are the characteristics of PLUP. PLUP is therefore regarded as an important tool to further cooperation between people and institutions and as a basis for sustainable land management.

Keywords: Land care, proper land husbandry, Participatory **L**and **U**se **P**lanning (PLUP), iterative procedure, sustainable land management, **S**oil and **W**ater Conservation (SWC), **N**atural **R**esources **M**anagement (NRM), cash-crop production, land degradation, natural resource management participation, project cycle, target groups, gender, vertical and horizontal links, bottom-up, Participatory Rural Appraisal (PRA)

1 Introduction

In reviewing literature on the Sahel published during the last 30 years it can be seen that many problems of this drought stricken area were studied in detail and many solutions were proposed. Although remarkable results in some Sahel regions were obtained through Soil and Water Conservation (SWC) and Natural Resources Management (NRM) projects, unabated degradation of the environment continues in many places (Williams & Balling, 1996).

For decades, "cooperation" between people and institutions in the West African Sahel centred around the paradigm that extension agents of state or para-governmental organisations "advised" the farmers on what to do and where and when to act. This advise would then "qualify" the farmers for input supply, credits and/or the commercialisation of their cotton harvest. Traditional knowledge and organisational behaviour were rarely integrated in any of the cash-crop production oriented enterprises (Reij et al., 1996).

Under these conditions land care and proper land husbandry were not achieved and land degradation has meanwhile become a widespread phenomenon in most parts of this region (Rochette, 1997). Large, internationally funded SWC projects were started in order to improve the situation. The major objective of these projects was usually to solve erosion problems and to improve living conditions of the rural population.

The SWC measures were often initiated by foreign experts without local "cooperation" or even interest; ignored customary land tenure systems and land management rules and prevention was not even taken into account.

Accordingly the failure of these measures was foreseeable.

Cash crop production, which had been set up by foreign experts, soon disappeared as a source of income in many regions, but the established links between the farmers and institutions offering services continued. However, the extension agents were not able to give any applicable advise neither on traditional or subsistence agriculture, nor on sustainable land use practices. Their advice became thus irrelevant to the farmers and in most cases, the cooperation even came to a complete halt, i.e. the extension agents "concentrated" on office work or had no means to visit farmers. The latter were thus forced to form Self Help Organisations (SHO) or seek support from Non-Governmental Organisations (NGO).

For the reasons above, the large majority of small rural producers are not reached by the service system, nor can they attain it. This service system consists, amongst other things, of marets for inputs or means of production and services such as loans and information (GTZ, 1993).

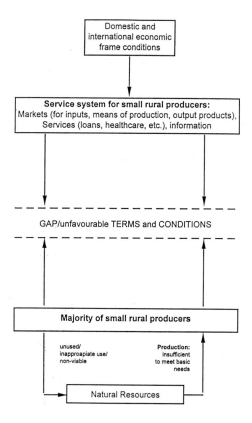

Figure 1: Scenario of problems: Small rural producer poverty (GTZ, 1993)

One of the prime reasons for the problems described above is the top-down approach applied by institutions at all levels, from the ministries down to the extension agents. In the case of SWC, for example, the ministry negotiated a project and handed down a so called "offer" of measures that the villages had to accept. The top-down approach is also used in sectors such as training and infrastructure where the needs of the target groups are always defined by the next higher level.

2 Planning: Tool or target

As the failure of such planning-strategies became more and more obvious, an increasing number of national, regional and international activities were carried out to tackle these problems. One of the major steps taken was that the Sahelian governments created a regional organisation "*Comité permanent Inter-Etats de Lutte contre la Sécheresse dans le Sahel*" (CILSS), the "Permanent Interstate Committee for Drought Control" in the Sahel. This was in the 1970's, particularly as a response to the major droughts of previous years. However, in the beginning of the 1980's the above described situation had not as yet significantly improved.

The CILSS-member countries continued with their scattered and uncoordinated efforts to deal with this regional problem. The situation was to a change when from 1984 onwards a regional strategy for desertification control for the Sahel as well as national plans for all CILSS member states were elaborated by CILLS and its member states (Winckler et al., 1995).

The international donor community appreciated the new approach. The National Resource Conservation or Desertification Control Plans were soon sponsored by the CILSS, *International Union for Conservation of Nature and Natural Resources* (IUCN), *Food and Agriculture Organisation* (FAO), *United Nations Sudano - Sahelian Office* (UNSO) and *United Nations Environment Programme (UNEP)*.

In general, the national desertification control plans and their recommendations were not translated into any activity at the field level. However, lessons were drawn from this experience and a new approach was developed that aimed at transferring responsibilities for Natural Resource Management from central levels to village or inter-village levels (Rochette, 1996).

Participation of the stakeholders in all phases of the project cycle was considered the key to success and the concept of community-based land use management or 'Gestion de terroirs' was tested by many projects.

The implementation of the resource conservation planning approaches was one of the topics of a CILSS conference, which was held in 1989 in Segou (Mali). One important result was that from now on, regional and national strategies were to be translated into action on the field level through a village land management concept. This approach has also been taken up by the "Convention to combat desertification" (Winckler et al., 1995).

3 Participatory Land Use Planning at the local and regional level

During the past decade a PLUP approach has evolved to induce local-level sustainable land management.

Participatory Land Use Planning can be defined as an iterative process based on dialogue among all concerned. Its goal is to reach decisions on sustainable forms of land utilisation in rural areas and to implement the initiation and perpetuation of relevant measures (GTZ, 1995).

PLUP should be oriented not only methodologically but also contextually to local conditions and should be built on local environmental knowledge and traditional strategies for problem and conflict management. PLUP is based on the assumption that development is a process brought about "from below" and based on self-help and responsibility. It is an interdisciplinary task and is intended to improve the participants capacity to plan and act. This requires transparency which means free access to information for all participants in the process.

Differentiation among target groups and a gender approach are further basic principles for PLUP since they are due to their economically and socially determined roles and tasks. Men and women often have unequal access to resources, different possibilities for articulating their desires, varying interests and different requirements, vis-à-vis the configuration of the planning process.

The elements of the process of Participatory Land Use Planning are:
- Preparation
- Gathering of information, analysis
- Creation of institutions and organisational structures
- Drawing up a plan
- Applying pilot measures
- Coordination and decision-making
- Plan implementation.

These elements can be applied according to the local conditions at the same time, in sequence or in a reduced combination (GTZ, 1995).

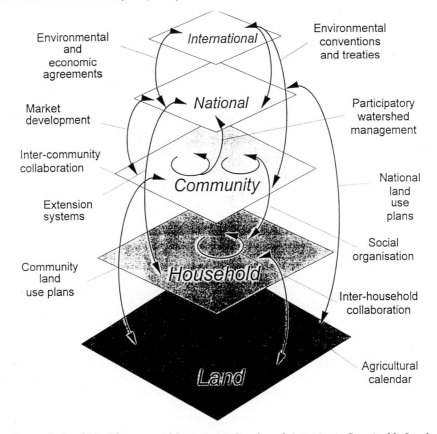

Figure 2: Land Use Planning and Intervention Levels and Activities in Sustainable Land Management (Hurni, 1997).

PLUP can be linked with other activities at different levels and can be used as a tool to bring about further cooperation between people and institutions. In this context, the organic built-up of vertical and horizontal links is important.

The horizontal links are of great importance at the household and community level, and especially influence Community Land Use Plans and their implementation. The vertical links play an important role when community land use plans are linked with national land use plans and when adequate support from extension and other support systems must be mobilised (Hurni et al., 1997).

It is important for the vertical and horizontal relations to create a favourable planning environment. The emergence of incentives marks a paradigm shift away from command and control approaches. This forces land users to contribute to the creation of a planning environment in which they are free to determine their own course of action (Hurni et al., 1997).

4 Limitations of the Participatory Land Use Planning approaches

In West Africa as well as in other countries, experience regarding the effective scaling-up of PLUP from the village level to the inter-village and district or provincial levels is very limited, despite the successes of the bottom-up and participatory approaches.

The village land use plans are mostly elaborated in isolation and the participatory approaches rarely examine relationships between neighbouring villages. This approach is adequate as long as the development activities in one village have no effect on neighbouring villages. Problems arise, for example, when one village in a watershed area restricts the water flow to another village because of its decision to construct a small dam (OECD, 1995).

Furthermore, the distribution of limited resources at the district level requires a scaling-up of the village level planning to the next higher level, as well as a development vision at this higher level.

Even though Participatory Rural Appraisal (PRA) is used effectively in many villages, some villages still show a tendency to simply come up with "lists of requests" instead of a development strategy supported by prioritised projects. In addition, not all villages in a district participate in the PRA exercises, which results in isolated pockets of development taking place. Such a piecemeal-approach to development will not produce any synergetic effect.

Many studies have revealed that participatory planning processes demand a considerable amount of time from the local population. Many land users simply cannot afford to participate because they are involved in their daily struggle for survival.

The initiative for "participation" usually comes from foreign donors and might be a condition for support rather than an open non-binding offer. This is because the ultimate rationale is not only to serve the perceived needs of local communities, but also to meet the "challenge of participation". This problem has become one of the top-issues in the world-wide discussion on development-strategies, although it is not openly acknowledged.

Policy makers and project staff in the field transfer the responsibility for success or failure to the land users themselves by giving farmers freedom of choice. However, participatory exercises are not always performed professionally and do not always reflect local conditions and aspirations correctly. They might, in fact, be a form of coercion under the guise of democracy. Many cases have been reported in which quick and sloppy work by "consultants" is disguised by the rhetoric of participatory learning and action.

5 Participatory Land Use Planning as a tool in rural development projects

Despite all these limitations PLUP has meanwhile become an important tool to further cooperation between people and institutions. It has been observed that development cooperation is shifting away from top-down methods of planning and implementation. This consists of the following steps: the "conservation expert" from outside identifies the "problem" in the field and presents technical "solutions" which are developed at the research station and are then implemented by using incentives such as food or money for work. This is a marked improvement in the approach to move away from conservation works towards land husbandry.

This improvement implies a redefinition of the role of projects as well as the knowledge needed in promoting the shift from SWC to land husbandry and to community based land use management. It has certainly to be an iterative process with PLUP as an instrument to enhance cooperation between people and institutions. Projects will have to promote sustainable land use and should be designed for a 10 to 15 years period. At least three major phases should mark the path of implementation.

The Orientation and Innovation Phase starts with the elaboration of the
1- Project implementation strategy,
2- Identification of technical, institutional and organisational solutions,
3- Creation of concertation mechanisms.
This should last about 3 years.

Stone bunds and filtering water retention walls in the semiarid zones of, for example, Burkina Faso have been developed within the context of a project pilot activity and seem to be promising SWC measures. The quality and the site-appropriate design of the measures created a huge demand from the rural population. In this way stone bunds served as an entry point for the first stage of the project implementation strategy.

In order to use the momentum of mobilisation which was initiated by this "highly demanded product", concertation mechanisms for all the intervening institutions in the region were established which helped to satisfy the growing demand for these SWC measures. In order to guarantee that all strata of the target population profited from the measures, and that the technical knowledge on all levels was adequately used, PLUP was introduced. The backbone of PLUP was the traditional land classification that built upon the local population's perception of their environment. The knowledge needed to successfully design and carry out an orientation phase is primarily technical, however with a strong view to the elaboration of a development strategy for the region concerned.

The Dissemination Phase includes the
1 - Widespread implementation of innovations,
2 - Scaling-up of village level interventions (lasts up to 6 years)

In the dissemination phase of SWC measures, widespread implementation is gained especially when traditional knowledge is sufficiently incorporated in the dissemination approaches.

The providers and users of services, information and markets have to be brought to a level which enables them to reach each other in order to guarantee widespread implementation of innovations.

It was possible to scale up village level interventions to intervillage and provincial levels in many regions of the Sahel by aggregating the results of PLUP exercises at the village level. The knowledge, which is needed to implement the dissemination phase successfully, concentrates around extension organisation, conflict moderation, monitoring and evaluation as well as widespread implementation management.

The Consolidation Phase goes beyond the widespread implementation of sectoral activities and tries to
1 - Link village approaches of sustainable land use,
2 - Establish mechanisms to mediate between competing land use at the village and higher levels.
This phase should last about 3 years.

The scaling-up of village level interventions goes beyond techniques and concentrates on approaches of sustainable land use. The establishment of mechanisms to mediate between competing land uses on inter-village and higher levels is necessary since interests and aspirations vary between villages. The mechanism that monitoring and evaluation results are fully used in management decisions during the consolidation phase should be operational. The evaluation and readaptation of technical and organisational solutions within a system approach requires a high technical and social expertise. Experience in the West African Sahelian context shows that the above outlined results are obtained during the planned phases when they are matched with the needed knowledge.

Phase	Major results	Know-how needed
Orientation- and Innovation phase (3-6 years)	• Project implementation strategies worked out • Creation of concertation mechanisms • Village land use plans • Socially, economically and economic and ecologically compatible technical and institutional/organisational solutions identified in close collaboration with the agencies in charge, and tested under conditions wich are representative for the region	• Technical know-how to evaluate appropriate situation specific innovations (technical, institutional, organisational) • Vision to elaborate development strategy for the region • Skill for the elaboration of on approach for land-use planning on village level
Dissemination Phase (3-6 years)	• Wide spread implementation of innovations • Scaling up of village level interventions (NAP's) • Rural population in a position to reach necessary services, information and markets	• Extension organisation • Conflict moderation • Planning- implementation- Monitoring & evaluation • Organisation of wide spread implementation
Consolidation Phase (3 years)	• Scaling up of village level interventions • Linking village approaches of sustainable land use • Establishment of mechanisms to mediate between competing land use on village and higher level • Functioning M&E	• Planning and organisational development expertise • Conflict moderation • Technical know-how

Figure 3: The role of "Projects" in promoting the process

6 Conclusion

Participatory Land Use Planning should be oriented not only methodologically but also contextually to local conditions and should be built on local environmental knowledge and traditional strategies for problem and conflict solution. PLUP is based on the assumption that development is a process brought about "from below" and that it is based on self-help and responsibility. It is an interdisciplinary task and is intended to improve the participant's capacity to plan and act. This requires transparency which means free access to information for all stakeholders/participants. Differentiation among target groups and a gender approach are further basic principles for PLUP. A piecemeal-approach to development will not produce any synergetic effect. As many villages as possible have to participate actively in the planning-processes as well as to express their demands.

Although the participatory approach is mobilised through the external agent, it should nevertheless be effectively promoted in the projects work. Participation should not remain a catchword. Participatory exercises should be performed professionally and should correctly reflect local conditions and aspirations. This implies a redefinition of the role of projects and the knowledge needed in promoting the shift from SWC to land husbandry and to community based land use management. PLUP has to be an iterative process in order to be an instrument for enhancing the cooperation between people and institutions and in the long run to reverse degradation and improving the living conditions for the people of West Africa.

References

GTZ (1993): Ländliche Regionalentwicklung - LRE Aktuell. Abt. 425, Eschborn.
GTZ (1995): Landnutzungsplanung. Strategien, Instrumente, Methoden. Arbeitsgruppe LNP, Eschborn.
Hurni, H. (1997): Precious Earth. From Soil and Water Conservation to Sustainable Land Management. ISCO, Berne.
OECD - Club du Sahel/CILSS (1995): Atelier de Restitution sur "La Gestion des Terroirs" et le Développement Local au Sahel. Rapport de Mission, Niamey, 30 mai - 2 juin 1995.
Reij, C., Scoones, I. and Toulmin, C. (eds.) (1996): Sustaining the Soil. Indigenous Soil and Water Conservation in Africa. Earthscan, London.
Rochette, R.M. - Club du Sahel (1996): Cooperation 21 - Cooperation 1964-1994 avec le Sahel: Optimisme et Désillusions. Rapport de Mission, Paris.
Rochette, R.M. - Club du Sahel/CILSS (1997): Contribution a Sahel 21. Rapport de Mission, Grenoble.
Williams, M.A.J. and Balling Jr., R.C. (1996): Interactions of Desertification and Climate. WMO/UNEP, New York.
Winckler, G. et al - OECD/Club du Sahel (1995): Approche "Gestion de Terroirs" au Sahel Analyse et Évolution. Mission de Dialogue avec le Projets GT/GR du Club du Sahel au Burkina Faso, au Niger et au Mali. Rapport de Mission, Paris.

Address of author:
Helmut Eger
Deutsche Gesellschaft für Technische Zusammenarbeit (GTZ) GmbH
Postfach 5180
D-65726 Eschborn, Germany

Village Level Approach to Resource Management: A Project in a Marginal Environment in Ethiopia

T. Bekele & W. Zike

Summary

A Village Level Resource Management Project (VLRMP) was initiated within the framework of development cooperation between the governments of Germany and Ethiopia to alleviate the accelerating resource degradation and stabilize the living condition of the population of Were-Jarso Woreda. The Project began in 1994 by CPAR (Canadian Physicians for Aid and Relief)-Ethiopia, an NGO involved in integrated rural development, with technical support from GTZ (Deutsche Gesellschaft für Technische Zusammenarbeit), the German Agency for Technical Cooperation. Its components include strengthening grassroots organizations, natural resources conservation, agricultural intensification and diversification, promotion of non-farm income generating activities and improving community services. This paper presents some experiences of the Project.

Keywords: Resource management, indigenous knowledge, people's participation, attitude of people, institutions, land use planning, multi-sectoral, gender specific

1 Introduction

With the objective of alleviating the accelerating resource degradation and stabilizing the living condition of the population of Were-Jarso Woreda, a Village Level Resource Management Project (VLRMP) was initiated within the framework of development cooperation between the governments of Germany and Ethiopia. The Project is being implemented starting March 1994 by CPAR-Ethiopia, an NGO involved in integrated rural development, with technical support from GTZ. Its components include strengthening grassroots organizations, natural resources conservation, agricultural intensification and diversification, promotion of non-farm income generating activities and improving community services.

2 Characteristics of the project area

The Project is located in one of the most fragile and disaster prone areas in Ethiopia, in the Blue Nile Gorge, Oromia Regional State. It lies between 9^0 58' N to 10^0 09' N latitude and 38^0 08' E to 38^0 26' E longitude with a total area of about 35,000 ha. The altitude ranges from 2500 m.a.s.l to about 1050 m.a.s.l down at the Blue Nile River. The landscape is characterized by steep slope and rugged topography where a considerable proportion of the land is unfit for cultivation. The area receives an annual average rainfall of about 900 mm with more than 50 per cent occurring in July and August. The total population of the area is about 40,000 and the primary source of livelihood is agricultural production.

3 The problem context

Rapid population growth and the age-old subsistence cereal based farming system are important factors responsible for widespread poverty and environmental degradation. Rapid population growth in the past few decades has increased the demand for food and energy, overtaxing the local resources especially the land resource. The severe land shortage and absence of gainful employment outside agriculture has forced many households to cultivate marginal lands unsuitable for cultivation exacerbating soil erosion. Moreover, flooding, due to the rugged topography and short erosive rainfall, is the other growing threat contributing to land degradation (Kefeni, 1995).

The growing landless and near-landless population generates its livelihood from non-farm income sources that do not demand advanced skills and virtually no capital investment. Unfortunately, in most cases activities such as charcoal making, fuel wood sale and lumber production are by their very nature detrimental to the environment (Fig. 1).

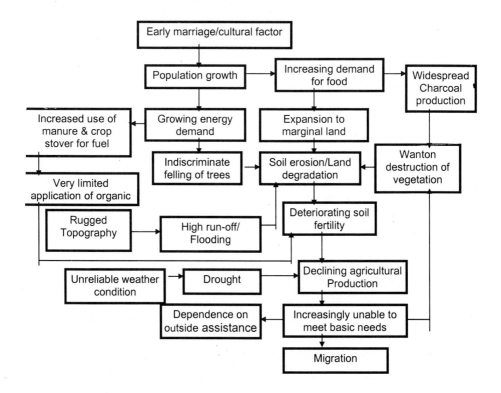

Figure 1: The Causal Relationship of Resource Degradation

4 Project objective, approach and strategy

4.1 Objective

The overall objective of the project is to improve or at least stabilize living conditions by using natural resources on a sustainable basis. Searching and identification of compatible solutions with the participation of the local people, the ultimate beneficiaries, is the immediate objective.

4.2 Approach

The Project approach is problem and target group oriented, strongly emphasizing the use of local resources and self-sufficiency, helping people to help themselves. Given the many casual factors contributing to poverty and resource degradation, the project addresses the problems using a **multi-sectoral** approach involving soil conservation, agriculture, forestry, family planning, health, education and non-farm income generating schemes. The VLRMP approach and methodology was conceived with the very principle of **people's participation** through the various committees, traditional institutions and interest groups so that interventions address priority problems as perceived by villagers. Since the single most important agent for resource degradation is man, the VLRMP focuses on changing the **attitude of people** through environmental awareness creation that includes training, orientation and visits. Undertaking **gender specific** target groups analysis, identifying problems and solutions that are gender specific, and strengthening institutions that promote women participation are elements of the approach.

The formation of participatory groups bringing the affected individuals together, working with innovative individuals who have accumulated life-long experience, and working with traditional institutions are ways of utilizing **indigenous knowledge** and developing sustainable local solutions (Courado, undated). In order to address sustainable resource management having a long-term benefit, the VLRMP **tries to address some of the immediate needs** of the community such as food, clear water and a range of social services.

4.3 Project strategy

To meet both resource conserving and productivity raising objectives, the Project's strategy has three elements. First, promotion of conservation based farming system development through appropriate land use and direct implementation of different conservation measures and farm intensification and diversification.

The second is the promotion of non-farm income generating activities as part of an effort of relieving the pressure on land. Credit and training are the two important tools in this regard. The third involves improving community services that include health, infrastructure, education and family planning through village-based contraceptive distribution.

5 Experience in implementing project components and use of indigenous knowledge

5.1 Formation of village level structures

The VLRMP utilizes the already existing rural institutions such as Peasant Association (PA) and Service Cooperative, without excluding creation of new ones. The PA, the lowest administrative structure recently renamed to the Kebele Administration[1] (KA) with its defined boundary and resource, is the most important grassroots institution. Figure 2 shows a village level structure created by initiation of the Project and accepted to be used by local government as well as community bodies.

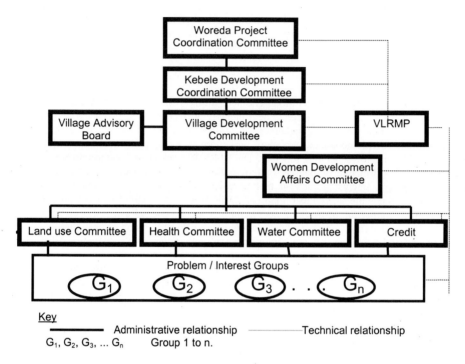

Figure 2: Village level structure for resource management

The problems facing the VLRMP in running these different development-oriented institutions include frequent turnover of members and lack of incentives. The frequent change of line department staff and government officials is also of concern.

5.2 Promotion of appropriate land use planning

As a first step in developing a village land use plan, the project developed an educational campaign to raise awareness of the community regarding conservation of natural resources and the need for a better land use. With the formation of the Land Use Committee that represents villagers in land use planning, resources have been delineated on maps and aerial photographs, enlarged to a scale of 1:12,000, forming a mosaic of the resources. Based on the mosaic, implementation of planned problem solving measures primarily focused to a shift in the land use from cereals production towards horticulture and agroforestry.

---

[1] Before the recent restructuring that resulted in the merging of two or more Kas, the area of a KA was roughly about 800 ha with up to 600 households and about half a dozen of villages.

5.3 Soil and water conservation

Emphasizing existing indigenous knowledge, the soil and water conservation practices promoted by the Project are physical and biological (Kefeni, 1995).

Physical conservation practices: Using diverse indigenous soil and water conservation know-how, realignment of waterways and construction of checkdams and cut-off drains were implemented to control the visible threat of flooding. The activities have received full-hearted support as the impact is immediately apparent and addresses food insecurity where landless and near-landless individuals participate on a food-for-work basis.

The project has also tried to implement conventional terraces that keep contour alignment. However, acceptance by farmers is very limited because, in their view, they are unsuitable for ox plowing. Conventional terraces take too much land to construct, competing with the already limited land available. For this reason the Project has relied on indigenous practices where the community has expertise, making the project interventions more sustainable and locally based.

Biological conservation measures/agroforestry: Stabilizing physical structures by introduction of multipurpose tree species such as *Sesbania sesban, Leucaena leucocephala*, phalaris, vetiver and pigeon pea has been tried in the orientation phase. There is, however, an early indication that success in this regard is very limited due to free grazing practice, competition with limited land available and hindrance to ox plowing.

5.4 Tree planting and reforestation

Private planting, community wood lots and protection of natural woodlands are activities under consideration. The Project raises both exotic and indigenous tree seedlings based on farmer demand and for testing purposes. To ensure the long-term sustainability of tree planting and reforestation, private and group nurseries are encouraged to raise seedlings that have market value while the project concentrates on seedlings that are unlikely to have a market value and are new to the area. The long dry spell and termites are the natural factors that limits peasants' tree planting efforts. Land shortage and tenure insecurity are among the socio-economic factors contributing to the slowing down of forestry development (Berehanu, 1995). As a result private planting is confined to homesteads and takes place on a very limited scale.

The protection of natural woodland that involves restricting livestock and human intervention was found to be very difficult due to a shortage of land for cultivation, grazing, and charcoal production. However, it was tried in one village and the regeneration of indigenous trees in less than two years was impressive.

5.5 Agricultural productivity enhancing measures

Apart from conservation activities, increasing the productivity of the agricultural sector by introducing new and more appropriate crops and varieties is one of the priority areas of the project. The following are some of the possible solutions being tested and promoted on a village basis.

Introduction of improved crop varieties: With the objective of increasing the yield of the two important crops, sorghum and teff, the VLRMP in collaboration with the local Ministry of Agriculture, promoted improved extension packages that have contributed in increasing the yield by up to 50 per cent.

Introduction of non-cereal crops: In order to improve the local farming system, dominated by cereals that require each season, criss-cross plowing accelerating soil erosion, the introduction of drought tolerant and perennial crops such as Enset, Cassava, pigeon pea, Sweet Potato and fruit trees is being conducted. So far their performance is encouraging.

Improving irrigation: Construction of micro-dams to extend the irrigation period, diversion of untapped streams and minimizing water loss by improving the existing traditional irrigation structures, are some of the project activities welcomed by the local people.

Conservation of landrace seeds: Resulting from recurrent drought, pests and diseases, farmers argue that selected varieties of indigenous cereals and pulses are quickly disappearing from the area. In order to support the conservation and multiplication of endangered local varieties the project runs seed collection, multiplication and distribution with interested farmers.

Goat rearing: Goat rearing is an important component of the farming system identified by the local people for consideration under the credit scheme. The Project, however, adopted a cautious approach since such an intervention would result in an increase to the stocking rate which would be hostile to the already degraded environment and contrary to the very objective of the project. The initial plan was to test dairy goats production under controlled feeding. However, other experience in the country indicated that introduction of crossbred dairy goats should be preceded by the rearing of local goats on a controlled basis. On the other hand, in the project area the culture of tethering is unknown and the community argues that goats can browse on the most steep slopes, not accessible by any other kind of livestock, thus defending their current production system. As a result the project is still unable to identify needy community members who can undertake rearing in an environmentally friendly way.

Apiculture: Apiculture is a common traditional activity but productivity is very low, about 5kg/hive/year, while it is possible to produce up to 30 kg/hive/year using improved beekeeping technology. With the objective of realizing the potential for honey production, the project has started testing different beekeeping technologies of Kenyan Top Bar Hive and Modern (Longstroth) Hive with interested farmers. However, there is an early indication that bees can not stay long due to teh hot micro-climate within the hives.

5.6 Promotion of non-farm activities.

The VLRMP tries to diversify and increase the income of the local people by encouraging and supporting non-farm income generating activities in the form of training, credit on a revolving basis and market information. Village saving and credit groups in handicrafts and petty trade/ village shop have been established and their credit repayment performance so far is very good. Identification and organizing of interest groups on small-scale mining (sand and gypsum) and incense (*Boswellia papyrifera*) collection, primarily with landless and near landless individuals is undertaken. However, implementation is lagging far behind due to the lack of a clearly defined policy both at national and regional levels as well as the long bureaucratic process in legalizing the groups.

5.7 Improving access to community services

Health and family planning: In order to improve the health service of the community, the Project assists in the construction of health posts, training of Community Health Agents (CHAs) and Trained Traditional Birth Attendants (TTBAs) and provision of start up drugs. CHAs and TTBAs undertake first aid, contraceptive distribution, safe birth attendance and health education.

Rising awareness of population pressure and depletion of natural resources with women, men and mixed groups has been the initial family planning activity undertaken. Through continuous discussion, interested women requested contraceptives mainly pills and condoms and currently there are about 144 clients.

In order to ensure the long-term sustainability of the health service, a revolving system for drug funding was established by including a marginal profit on the cost of drugs that would be enough to cover the cost of running the system. However, the low level of income of the local people and the long bureaucratic channel involved in securing drugs are impediments.

Safe water supply: So far, by capping springs and installation of hand pumps, about 20 per cent of the total population has access to a safe water supply (Wondimu, 1995). The local people were enthusiastic about the water supply component and they contributed free labor in collecting stones and transporting materials. It has also been instrumental in winning community confidence and helped the project to smoothly run other resource management activities.

Improved stoves and alternative construction materials: In order to reduce fuel wood demand, the project provides and demonstrates cooking stoves developed in the country. However, promotion is very limited where, unlike the traditional stove, the improved stoves require the use of fuel wood instead of cow dung, twigs and crop residue which are the current common source of fuel. The promotion of alternative construction materials such as stone and mud bricks has just started in selected villages and, initially, farmers have appreciated the technologies.

6 Conclusion

Although the VLRMP experience is drawn from a relatively brief orientation phase, it still provides some grassroots evidence regarding resource management in a low income and marginal environment. Considering the existing socio-economic condition of the population, the project first tried to integrate resource management, which may not necessarily yield immediate benefits given the primary problem of food insecurity. This was done on a food-for-work basis.

The Project can be considered the first of its kind in the country that created village level structure for realizing participation of the local people and government agencies in developing compatible solutions using existing diverse indigenous knowledge. Implementation of project components of soil and water conservation, agroforestry, reforestation, farm intensification and diversification are steps in the right direction towards improved land use planning. Non-farm activities are promising complementary measures for increasing income. The promotion of family planning, though with a long-term impact, is encouraging as people are trying different birth control methods.

Given the magnitude of the problem and the prevailing level of poverty, most of the solution measures being tested yield benefits in the long-term; however, the local people cannot afford to wait. Among the socio-economic problems that critically hinders farmers from adopting more resource conserving farming practices, is a land shortage which forces many households to concentrate on short-term gains at the expense of long-term and sustained resource use. The level of investment required in the form of land and labor is the challenge the Project is facing at its early stage. This raises the need for subsidies, at least in the short-term. The other lesson is that while implementation of conservation and productivity enhancing measures is important, the long-term solution to resource degradation should also include the control of population growth through promotion of family planning.

References

Berehanu, D. (1995): CPAR Were-Jarso Project Area Rural Development Plan.
Courado, S. N. (not dated): The Philosophy, Principles and Practice of Rural Reconstruction (IIRR Staff Discussion Paper)

Kefeni, K. (1995): Soil Erosion Problems Prevailing in Ten Peasant Associations of Jarso Lowland and Appropriate Soil and Water Conservation Techniques for Village Level Resource Management Project.

Deutsche Gesellschaft fur Technische Zusammenarbeit (GTZ) GmbH (1993): Regional Rural Development: RRD Update; elements of a strategy for implementing the RRD concepts in a changed operational context. Deutsche Gesellschaft fur Technische Zusammenarbeit (GTZ) and Bundesministerium fur wirtschaftliche Zusammenarbeit und Entwicklung.

Wondimu, Z. (1995): Indigenous and Acquired Knowledge Systems within Village Level Resources Management Project Area. (unpublished VLRMP survey paper).

Address of authors:
Tilaye Bekele
Wondimu Zike
CPAR
P.O. Box 2555
Addis Ababa, Ethiopia

Mountain Watershed Management in China

Lixian Wang

Summary
This paper outlines mountain watershed management in China. Such watershed management includes land use planning, appraisal of soil erosion, establishment of a protective system which closely integrates biology with engineering, supervision and management. Through a 40 year effort, China has achieved remarkable successes in watershed management such as higher productivity, improved structure of land utilization and expansion of the basic farmland area.

Keywords: watershed management, erosion control, soil and water couservation, land use planning

1 Introduction

The total area of China's territory amounts to 9.6 million km^2 which is about 1/15 of the land area of the world. However, owing to a large population, the area per capita is only 0.8 ha. At present, China has 95.3 million ha of arable land. The per capita arable land area is 0.087 ha, which is 25% of the per capita arable land in the world. China has 7% of the world's arable land but feeds 22% of the world's population.
   The features of China's land resources are as follows:
1) a shortage of reserve resources of arable land, with 33.33 million ha of existing waste land suitable for farming, of which 11.33 million ha can be reclaimed as arable land.
2) a sharp decrease in arable land by 40.67 million ha between 1957 to 1986. A national decrease of 625,000 ha of arable land occurred in 1993.
3) the pollution of industrial wastes and pesticides, deterioration of land quality, shortage of P and K, thin arable layer and soil compaction impairing land productivity.
   China is a "great country of resources" from the viewpoint of water, but a "great country poor in water" in terms of per capita amounts. The annual amount of river runoff in China is 2711.5 billion m^3, ranking sixth in the world. However the annual per capita water resource is only 2474 m^3 (88th in the world), which is 25% of the global annual per capita water resources (9360 m^3).
   Water resources are unevenly distributed in terms of time and season. In the south, the rainy season is longer, and the rainfall from March to June or from April to July is about 50% - 60% of the annual rainfall occurring in the north. There the rainy season is shorter and the rainfall from June to September accounts for about 60% - 70% of the annual rainfall, often in the form of rainstorms. As a result of the concentrated rainfall, flood and waterlogging occur frequently, and drought is liable to occur in a season short of rain.
   Water pollution further aggravates the shortage of water resources. About 80% of unprocessed waste water is directly drained into rivers, lakes, streams and reservoirs, polluting more than 1/3 of river courses and water resources of water supplies in some cities and towns (Duan, 1994).

2 Damage caused by soil erosion

China is not only in a difficult situation regarding land and water resources, but also suffers from some of the most serious soil erosion in the world. According to a remote sensing survey in 1990, the area of China's soil erosion amounts to 3.67 million km^2, i.e. 38.2% of the total area of the national territory (Guo, 1991). About 1.79 million km^2 of the area is eroded by water and 1.88 million km^2 of the area is eroded by wind. Water erosion occurs in varying degrees in different provinces and cities throughout the country. At least 5 billion tons of lost soil is caused by water erosion each year(Sun, 1992). According to incomplete statistics, arable land decreased by 2.667 million ha over the past 40 years, causing 10 billion *yuan* of economic losses each year (Wang, 1994). National per capita arable land has decreased by 50% of that in the early 1950's. At present, per capita arable land of one-third of the provinces of the country dropped to less than 0.067 ha.

Soil erosion not only causes damage to land resources, such as environmental deterioration of agricultural production and ecological imbalance, but also impairs the development of the national economy and society. The main consequences of soil erosion are as follows:
1) Disasters of flood and drought are aggravated owing to soil erosion as well as land reclamation by deforestation in mountain areas.
2) Soil erosion causes arable slopes to become increasingly poor, compounds the development of drought or desertification, and leads to low and unstable production of grain or even no production at all.
3) Soil erosion causes the deposits of silt in channels and ports, resulting in sharp decreases of shipping mileage and in-berth damage.
4) Soil erosion and poverty interact as both cause and effect become a vicious circle. Year after year soil erosion causes the limited land resources to be seriously damaged leading to desertification.

3 Achievements of mountain watershed management

In China the concept of mountain watershed management relies on protection, improvement and rational use of water, land and other renewable natural resources of a mountain watershed, in order to make the most of the potential ecological, economic and social benefits. By taking a watershed as a unit, planning permits rational use of the land agriculture, forestry, animal husbandry and sidelines, taking into account comprehensive control measures for local conditions.

Over 40 years, China has achieved remarkable successes in watershed management. Watershed management has played an important role in improving the basic conditions of agricultural production, increasing agricultural yield, promoting the development of the rural economy, accelerating the shaking off of poverty and improving people's living standard. Watershed management reduces sediment in rivers and protects and improves the ecological environment, etc. Over the past 40 years, comprehensive control of 610,000 km^2 of areas experiencing soil erosion has been completed, of which 33.3 million ha of soil and water conservation forest and 3.7 million ha of economic forest were planted. About 3.4 million ha were preserved by planting grass, 10 million ha of basic farmland were established, facilities for soil and water conservation were built, thus effectively checking the development of soil erosion. A total of 35.5 billion tons of sediment have been blocked by dam land, 1.56 million ha of land have been deposited, and more than 63 billion yuan of output value have been generated. The existing facilities for soil and water conservation can increase their capacity for conserving more than 18 billion m^3 of water and for reducing soil erosion each year by more than 1.1 billion tons.

4 Measures of mountain watershed management

According to practical experience of watershed management in China, a number of measures for comprehensive management of watershed must be adopted in order to attain the object of sustainable development. This includes:
1) Land use planning of soil and water conservation in the watershed.
   On the basis of detailed surveys of land resources, land types should be classified: Local social and economic conditions as well as state policies should be included in determining the direction of land utilization of each plot in the watershed, the proportion and specific position of lands to be used for the productive undertaking of agriculture, forestry, animal husbandry and fishery in the watershed, and the place and time to carry out different measures of soil and water conservation. In order to raise the quality and efficiency of land utilization planning in the watershed, remote sensing and a geographical information system should be applied to the soil and water conservation planning.
2) Appraisal of the dangerous nature of soil erosion.
   In order to prevent and control soil erosion, it is necessary that the soil erosion ha zard of each plot in the watershed should be surveyed. In zones of hilly slopes, the intensity of surface erosion (including sheet erosion and scaly surface erosion of forest land, grassland and waste land) and the distribution and dangerous nature of gravity erosion (including slip-slope and collapse, crumble, etc.) should be surveyed. In gully channels, the stage of gully development and the impacts of the mountain torrents and mud rock flow should be surveyed. In zones of alluvial cones, a hazard map of mountain torrents or debris flow should be drawn to determine hazard classes for different positions. In this way, the safety of people's life and property will be ensured.
3) Establishment of a protective system which closely integrates biological and engineering measures.
   Biological measures mainly refer to forest-grass establishment, while engineering measures refer to the slope surface, the gully channels and engineering of soil and water conservation in the zone of the alluvial cone.
4) Measures of supervision and management.
   In addition to soil erosion caused by natural factors, steps should be taken to counter Man's irrational productive management activities, such as destructive cutting of forests, reclaiming unused steep slopes, and ignoring soil and water conservation during mining and road repairs. In the face of illegal actions, the legal system of supervision and management must be strengthened.

5 Models and benefits of mountain watershed management

Over the past 40 years watershed management has proved that all management carried out in accordance with the technical specification of the state have achieved remarkable successes and that many successful models have emerged. The main benefits are as follows:
1) Improved condition and extent
   Watershed management promotes the development of productive undertakings of agriculture, forestry, animal husbandry and fishponds, improves the structure of land utilization and raises the rate of land utilization. According to survey data of 20 representative mountain watersheds (5 in the Yellow River valley, 6 in the Yangtze River valley, 1 in the Zhujiang valley, 1 in the Haihe valley, 3 in the Huaihe valley and 4 in the Songliao valley), changes have been positive during period 1982-1986 (Table1).

	year before control area		year after control area	
	(km²)	(%)	(km²)	(%)
total area	1012.91	100	1012.91	100
farmland	347.76	34.3	302.31	29.8
forest-grass land	315.84	31.2	553.95	54.7
waste land	222.54	22.0	47.22	4.7
non-productive land	126.77	12.5	109.43	10.8
rate of land utilization		65.5		84.5

Table1: Changes in land utilization and the rate of land utilization of 20 watersheds (Wang, 1994)

2) **High production rate of land and increase of income**
   In the course of watershed management, measures to obtain quick economic benefits were adopted. The changes of land output value, total income and per capita income of 20 watersheds are shown in Table 2.

3) **Expansion of basic farmland area and rapid increase of grain yield**
   The 20 watersheds have expanded their basic farmlands and increased grain yield. The changes before and after management are shown in Table 3.

	1982 output value		1986 output value		difference
	(10^4 yuan)	(%)	(10^4 yuan)	(%)	(%)
Land total output	4845.92	100	10121.31	100	209
Agricultural output	2872.79	59.3	5025.52	49.7	175
Forestry output	533.92	11.0	1487.97	14.7	279
Animal husbandry	964.81	19.9	1925.14	19.0	199
Sidelines output	474.0	9.8	1682.68	16.6	355
Output value per hectare	546 (yuan)		1122 (yuan)		205

Table 2: Land output value and per hectare output value of the 20 watersheds before and after control (Wang, 1995)

	1982	1986	Difference (%)
Arable land area (ha)	31467	29993	-4.7
Basic farmland (ha)	16347	23727	45.1
Per capita arable land (ha)	0.136	0.124	- 8.8
Per capita basic farmland (ha)	0.07	0.098	38.7
Total grain yield (10^4 kg)	8896.91	11762.93	23.2
Per capita grain (kg)	384.2	468.1	26.5

Table 3: Changes in basic farmland and grain yield before and after control (Wang, 1995)

4) **Improved production conditions and developed commodity production**
   Watersheds are distributed in vast hilly areas rich in natural resources. The establishment of comprehensive management and development systems of the watersheds has created favorable conditions for the development and utilization of natural resources and for established a production base of commodities such as grain, cotton, oil, livestock, fowl, fish, melon, fruit, vegetable, timber and medicinal herbs.

5) Improved ecological environment

In the above-mentioned 20 watersheds, the total area of controlled soil erosion amounts to 569.55 km2 . Soil erosion have been put under control. Forest/grass coverage increased from 31.2% to 54.7%. Once bare mountains and hills are now covered with dense vegetation. In addition rainwater is being retained, soil erosion has decreased, micro-climate, temperature, evaporation, humidity, etc., have been altered and animal and plant communities are increasing.

6 Conclusions

In order to take mountain watershed management further, the Chinese government has placed watershed management in to a plan entitled "Agenda of the 21th century in China". A national plan of watershed management for the period 1996-2010 has been adopted and put into effect. Mountain watershed management is expected to contribute more benefits to sustainable development in China.

References

Duan Qiaofu (1994): Economy of small watersheds, Harbin Press, Harbin, 18-20.
Guo Tingfu (1991): Soil erossion and their comprehensive control, Jilin Science and Technology Press, Jilin, 25-28.
Sun Lida (1992): Theory and Practice of Small watershed control, China Science and Technology Press, Beijing, 212-216.
Wang Lixian (1994): China Encyclopedia , Vol. Water Conservancy, Branch of Soil and Water Conservation, China Encyclopedia Press, Beijing, 15-18.
Wang Lixian (1995): Soil and water conservation, China Forestry Press, Beijing, 890-892.

Address of author:
Lixian Wang
College of Soil and Water Conservation
Beijing Forestry University
100083 Beijing, China

Dry Forest Management - Putting Campesinos in Charge

M. Schneichel & P. Asmussen

Summary

The dry forest of the southwestern Dominican Republic has been degenerating at a fast pace because of indiscriminate exploitation, first for precious wood, and later for charcoal, fence posts and railway sleepers. The campesinos depending on the forest are among the poorest in the country. A Dominican - German project developed a strategy based on their potential to help themselves to improve their living and working conditions, while at the same time assuring fast regeneration of the forest. The main elements of success were promotion of independently functioning campesino organisation, integration of women in all activities, marketing without intermediaries, devellopment of an easily to understand forest management system and the obtention of title deeds for the campesino organizations.

Keywords: dry forest, participation, forest management, Dominican Republic, GTZ, marketing, title deeds

1 Introduction

In the southwestern region of the Dominican Republic lies the "dry forest", a secondary thornbush vegetation comprised of small leguminous trees and cactus and which covers approximately 500,000 ha This type of vegetation developed following the virtual elimination of the original forest. As a consequence of the original forest's disappearance and the subsequent exploitation of the dry forest, rivers and small creeks have also dried up or considerably diminished. Consequently, the "campesino" dependent on irrigation for his home gardening as well as on water for his daily household consumption, has been seriously affected.

Working and producing in the dry forest is not only hard labour, but is generally regarded as economically and socially unappealing. Charcoal is therefore produced mainly by campesinos who have no other practical economic alternatives. There are few job opportunities in this area to match the level of education and skills possessed by a majority of the campesinos. The 16, 000 families who make their living from the dry forest reside primarily in villages without the benefits of electricity, drinking water, public health or educational facilities.

2 History of the project

In 1987 when the Dominican-German project "Rational Management of the Dry Forest" was initiated, the dry forest had come under threat of extinction. Due to continued pressure of meeting fuel demands by a growing urban population that relied heavily on charcoal for domestic cooking, and a large number of industrial consumers such as bakeries, restaurants and sugar mills, the forest was indiscriminately harvested at an alarming rate. The project, therefore, began with two equally important objectives: (1) Regeneration of the natural resources of the dry forest, and (2) Improvement of working and living conditions of the forest dwellers economically dependent on its products.

Formally the project is being executed in co-ordination with a government institution for regional planning, INDESUR (Instituto para el Desarrollo del Suroeste). However, prevailing conditions there made it impossible to establish a structure assuring sustainable management of the dry forest.

3 Participative procedures

The project created a basic structure relying directly on the campesino population earning a living by selling charcoal, railway sleepers and fence-posts. Therefor the campesinos were to become the main protagonists of the project. The actual daily planning, decision-making, execution and evaluation of activities is largely in the hands of the campesino organization FEPROBOSUR (Federación de Productores del Bosque Seco del Suroeste). The project team limits itself to support functions, providing elements of hands-on education, direction and advice in planning and decision-making. The involvement of the GTZ staff in these processes has gradually been reduced as the campesinos progressed in their technical knowledge and skills of leadership and organizational abilities.

4 Development of rational management procedures

At first the project team, together with some campesinos, carried out an inventory of the types of vegetation of the southwest region. A classification system was developed, that refers only to the density of the vegetation and the ease of forest management.

The following procedures of Rational Management were defined: extraction of only dry wood in the first years, integrated use of all wood and non-wood resources, use of distinct tree species according to their economic potential, and the repeated use of the same spots for charcoal burning .

These guidelines resulted in a management of individual trees as opposed to the traditional cutting of all trees in one area. As a consequence, labor input substantially increased. To convince the campesinos to accept this, other conditions had to be changed. Gradually, and according to the needs and abilities of the campesinos, the following four basic elements were developed.

4.1 Organisation of the campesinos

It soon became clear that the team would not be able to continue working directly with a growing number of community groups. The necessity of a campesino-based organization composed of village associations of "forest users" led to the formation of the federation, FEPROBOSUR, representing the campesino's interests to outsiders. FEPROBOSUR has become the true counterpart of the GTZ team, being prepared to ensure the sustainability of Rational Management.

In a social setting where paternalism, traditionalism, and "machismo" are dominant cultural factors, a main concern was to encourage women to get involve in the decision-making processes and execution of all activities in their communities. In order to improve their economic and social standing, women first needed to gain greater access to available natural resources as well as to the financial benefits of their labour, which was previously denied to them. The campesino organisations, dominated by men, had to be convinced to not only allow this to happen but actually to take part in ensuring the integration of women at every level of village affairs.

4.2 Rational and integrated management of natural resources

The process of involving the campesinos continues as Rational Management procedures are revised and adapted depending on the progress observed in the regeneration of the forest. The project encourages the integrated use of all natural resources available, e.g. goat- and beekeeping, and supports the campesino with technical assistance and small loans managed by the federation. It also provides small investments in infrastructure required to obtain water for home consumption and irrigation of kitchen-gardens.

After 4-5 years of extracting only dry wood, the dry forest had considerably regenerated itself. It became necessary to manage live trees in order to promote the proper growth of straight thick stems and to reduce competition for space by less desirable trees. A management system simple enough to be understood and executed by the campesinos and forest technicians was developed. The objective is to ensure growth of the forest towards its original diversified composition and density and, at the same time, to facilitate the acquisition of higher incomes from products with higher economic value than charcoal.

This system consists of two models, based on the forest inventory. One emphasises improving growth conditions for the more desirable and more valuable species where they still exist. The other is designed to extract a certain amount of marketable products while maintaining a dense coverage of the soil where the precious species have disappeared.

To ensure the non-production functions of the forest, like sufficient shading for the soil, limits for the extraction of wood for thinning purposes were set at either 20% of the volume available or 50 - 70% of the anual growth, depending on the density of the forest. The more restrictive limit should be applied.

4.3 Direct marketing of wood products

To compensate for the increase in labour input resulting from the sustainable management, additional income had to be generated. Since the major part of the consumer price charged for charcoal remained in the hands of the "middlemen", direct marketing was promoted by the federation to the charcoal dealer in the markets of Santo Domingo, effectively bypassing the local middlemen.

Although charcoal is still an important source of fuel, mainly for poor families, charcoal burning generally is an illegal activity in the Dominican Republic. Before the existence of FEPROBOSUR, the project had convinced the military-run forest authority of its ecologically sound manner of charcoal making and had obtained legal permission to produce, transport and sell charcoal. Today, a control post established and maintained cooperatively by the military and the campesino federation works to suppress illegal competition in the charcoal trade, as well as that of other wood products.

4.4 Title deeds

In the past, most campesinos lived and worked on state-owned land without any guaranteed right to remain there. Virtually anyone could enter the forest and cut dry wood wherever they chose. To make sure that the campesinos would reap the benefits of their efforts to promote the regeneration of the dry forest, the project has applied for title deeds in the name of the associations for those areas traditionally considered to belong to their communities and where today Rational Management techniques and procedures are being used.

5 Results

The consequent application of the above-mentioned elements over several years, and the struggle of the target group supported by the project team and local NGOs, along with respective government institutions, has yielded the following results:

5.1 Organization of the campesinos

The target group is organised into approximately 60 associations of men and women in 35 communities, all belonging to FEPROBOSUR, now an economically independent representative of the majority of people making their living from dry forest products. The federation is self-reliant in financing its staff, communication and transportation costs for all routine activities. This is achieved by charging a fee to its members on every bag of charcoal, fence post, railway sleeper or any other wood product being produced in and sold out of the dry forest. FEPROBOSUR also effectively deals with state and non-governmental institutions defending the campesinos' interests and manages its own affairs.

5.2 Rational and integrated management of natural resources

The combined efforts of the project team and the target group has resulted in a visible change of appearance and improvement of density in the forest, as well as a measurable growth in the diameter of trees. For example, the volume of wood available in the areas defined as dense increased from 14 m^3 in 1992 to 32 m^3 in 1996. To a certain extent, FEPROBOSUR controls violation of the rules of Rational Management by any of its members and, if unable to prevent the occurrence, it notifies the authorities if outsiders cut wood in their communities.

5.3 Direct marketing of wood products

The target group has completely displaced the middlemen and organizes the marketing of charcoal and other wood products by itself, obtaining the necessary permits from the forest authority, contracting transportation, paying taxes in advance, and selling and distributing benefits amongst its members. Today FEPROBOSUR practically has a regional monopoly authorised by the forest authority. Together the two institutions converted the illegal local markets into better organised, legally functioning, tax producing and supervised activities. Together they maintain a checkpoint on each of the only two access roads out of the southwest region to the main market in Santo Domingo, controlling the transport of wood products.

5.4 Title deeds

In accordance with the Dominican Land Reform the associations which represent their communities received title deeds to more than 100,000 ha of land traditionally considered to be the rightful property of their communities and where they now apply the procedures of Rational Management. Further titles are in preparation in order to ensure that large parts of the area covered by dry forests will soon be managed by campesinos in a sustainable way for the benefit of its inhabitants.

6 Conclusion

These results have not been easily obtained. Neither were the campesinos quickly convinced of the necessity of all the steps involved nor did the established commercial and bureaucratic structures simply disappear or change for the sake of logical argument. It was after the first successful actions had been executed by the project team with participation from the campesinos that the leaders of FEPROBOSUR showed more initiative and took responsibility in a process they now understand as being beneficial to themselves and their communities. A crucial element seems to have been that campesino participation in planning, actions and evaluations were not elements artificially disconnected from one another or restricted to certain formal events but were part of an interrelated and natural process which occurred continually on a daily basis. Also the objectives outlined by the project were based on potentials rather than problems, simple, clear to everybody involved, economically beneficial and personally meaningful to the campesinos who felt responsible for reaching these objectives.

Address of authors:
Martin Schneichel
Peter Asmussen
Deutsche Gesellschaft für Technische Zusammenarbeit (GTZ)
Apartado Postal 3960
Santo Domingo, Dominician Republic

A Partnership Between Farmers Researchers Advisers Designed to Support Changes in Farm Management Needed to Meet Catchment Goals

P.S. Cornish

Summary

Participatory methods are a key element in the approach to agricultural development known as Farming Systems Research (FSR). The approach has evolved in developing countries over 20 years, but until recently, FSR has received only token support in developed countries. Recent interest in FSR in Australia has focused mainly on technology adoption, whilst participatory approaches have also been embraced by the "Landcare" movement. In neither case has there been any consistent emphasis on research to advance scientific understanding or develop new technology.

This paper describes a new approach to FSR in Australia, prompted by the need to develop and employ farming practices which are demonstrably sustainable.

Farming systems research is conceived of as a learning or innovation system, in which researchers, farmers and farm advisers, collaborate in the continuous improvement of farming systems to achieve twin goals of profitability and sustainability.

Farmers in the northern grain-growing region of Australia, and the Australian Government, have each committed over $4 million to an FSR program based on the principles outlined here. The paper describes the theoretical basis for the program and its operational management, as well as outlining the difficulties encountered in implementing the plan.

Keywords: Action-learning, Farming System Research, learning, innovation

1 Introduction - The need for new approaches

Agriculture has been associated with degradation of the Australian landscape. At times, the decline of soil and water quality has been severe. Rural industry leaders and governments, alike, have recognised the need for land-use practices to change and become more sustainable. There is also strong community support for change, in part because of the very successful Landcare program.

Landcare is a community-based program which has brought groups of people together to plan and work on land restoration projects (Junor, 1991). Although a very successful program, there have not been strong links between this community program and scientific research on land and water quality management. Nor have there been strong links between the conservation aspects of "landcare" and the need to maintain or improve farm productivity and profitability.

In rural areas, the members of Landcare groups will often also be members of farmer discussion groups, formed in the wake of changing approaches to agricultural extension and reduced government funding for extension. These discussion groups are based on adult-learning principles. They focus on continuous productivity improvement through monitoring key indicators of farm enterprise performance. There is scope to extend enterprise monitoring to include monitoring of the resource base, and to establish what should be a natural linkage with Landcare and its focus on community action projects. This offers the potential to link improvements at the paddock and farm scale, to improvements at the community and regional scale. This approach recognises that changes at catchment scale can only be achieved through changes by *individual* manager-farmers.

Whilst extension in Australia is rapidly adopting a participatory approach, there has been little change in the approach to research. This paper is especially concerned with the crop and soil management research which is expected to deliver *sustainable* increases in productivity.

Australia, like all developed countries, has not widely embraced participatory approaches to research. Yet this seems to be a sensible way to go about focusing research on issues of significance to all parties: farmers who need to be both sustainable and profitable; governments who set policy (including sustainable land-use), have the power to regulate, and provide some research and extension funds; and researchers who can provide the understanding needed to underpin changes in land-use and land management, towards a sustainable future.

This paper describes a major new research and extension initiative which is designed to support the development and implementation of sustainable practices at the local scale, towards meeting desired outcomes at regional and catchment scales. The participatory project is based on a partnership between farmers, researchers and farm advisers, and is funded by the Grains Research and Development Corporation.

2 Theoretical background

Although reductionist and discipline-based approaches to science have delivered enormous technological change, there is growing recognition that more holistic approaches are also needed to make improvements to the more complex problem situations confronting modern society. Quite clearly, catchment issues such as eutrophication of surface water and dryland or irrigation salinity require changes in social values and individual attitudes, as much as they require new government policies and new farming technology.

In the program described here, a systems approach has been adopted to help define issues that require action and to define the appropriate responses to the issues.

In the first instance, biological systems are conceived of as layered hierarchical systems, in an extension to the thinking of Conway (1986) and Passioura (1979) (Figure 1). A search for understanding of phenomena at any level drives the enquirer down, to lower levels in the hierarchy. This is the traditional reductionist approach to science, and it applies to physical as well as biological systems, and has also been applied in psycho-social systems as well. Whilst reductionist research results in understanding, the phenomenon or process understood is not necessarily a significant one.

Enquirers concerned about significance look to higher levels in the hierarchy. Actions at any level in the hierarchy provide the understanding that is used to improve operations at higher levels; or to identify processes or phenomena *of significance*, for study at lower levels (Passioura, 1979).

Unless an upward search for meaning is balanced by a downward search for understanding, there can be a frenzy of well-intentioned activity, focused on significant problems, but a frenzy that delivers no beneficial and lasting improvement. Understanding provides the power to change.

Systems analysis at higher levels can focus research on significant phenomena, so that research at lower levels delivers improvement at higher levels. Crop and farm system models are often said to be useful tools for this (e.g. Bowden, 1992).

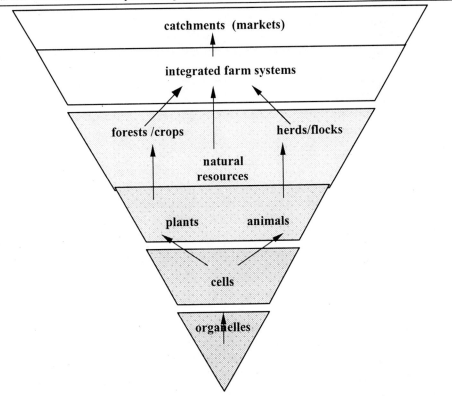

Fig. 1: *Agroecosystems are conceived of biophysically as layered hierarchies*

Agriculture is a human activity system (Bawden et al., 1985). Therefore, any model of the farm is incomplete without the farmer and, for that matter, consideration of the off-farm "system environment", including government policy, markets etc. Approaches are available to work with systems which include people, the so-called "soft systems methodologies". There are therefore, systems methodologies which can be used to *analyse* the farming system as both a biophysical and socioeconomic system, and to *focus* attention on the system components which will provide the best chance of improving both the productivity and sustainability of the system. Sustainability is here taken to include the welfare of the people who manage and derive benefit from the system.

Although these methodologies are available, few programs in developed nations employ them in an integrated or systemic way in major research (R), development (D) and extension (E) programs. The basic premise behind the program described here is that a systemic approach will deliver the answers to the complex question of raising agricultural productivity whilst maintaining and enhancing the natural resource base and sustaining rural communities.

In the program described here, the farmer, the researcher, and the farm adviser, are all seen as "learners" or "co-researchers". In the past, the researcher and adviser have behaved as observers of the biophysical and socioeconomic farm system (Squires, 1991). This new approach recognises and values the farmer's knowledge and experience. Because the emphasis in the program is on learning, the participants can be viewed as components of a learning system. The program is an extension to the innovative work of McCown and colleagues (e.g. Foale et al., 1996, McCown, 1989), who have placed heavy emphasis on improving crop management through the use of crop models in facilitated, participatory learning groups.

The learning system is intended to lead to sustainable innovation; that is, the development and adoption of sustainable technology for profitable farming systems. Therefore, the learning system can also be seen as an innovation system.

The system is illustrated in Figure 2. Note that each of the boxes is a subsystem. These are linked functionally, in some cases by flow of information which is obtained through continuous monitoring of key system attributes. Definition of the system boundary is important, as it leads to serious consideration of *who* needs to be included in the program (the stakeholders), and *what* needs to be included. Exclusions are also important, as unnecessary inclusions will dilute effort. In systems terminology, each subsystem will have inputs and outputs (of information, energy or materials). Subsystems are included if their outputs directly affect inputs to another relevant subsystem.

The fundamental concept underpinning this program is that it is a *process* of learning, and that all of the participants are learners. Farmers, researchers and advisers pass together through recurrent cycles of monitoring, evaluation, planning and action. Actions may occur at any level of the hierarchy in Fig. 1, as appropriate.

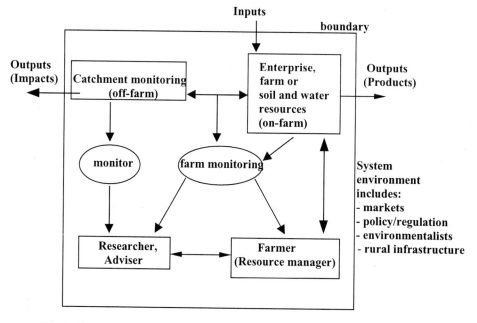

Fig. 2: The conceptual innovation or learning system and its component subsystems.

3 Operational organisation

To address the challenges of having many stakeholders involved with developing productive agriculture with minimum spill over (off-site) effects, we developed the operational structure shown in Figure 3 (Martin et al., 1996). The theoretical considerations outlined briefly in Figure 2 provide the foundations for the program. Like the foundations of any structure, they are important, but out of sight. The structure is designed for administrative efficiency *and* to provide the linkages, shown in Figure 2, that are vital for proper functioning of the innovation system.

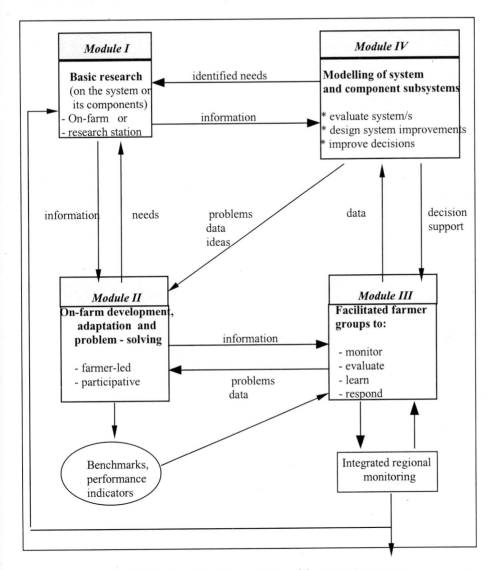

Fig. 3: Operational organisation of the innovation system

The innovation system is organised into four operational sub-systems or modules for funding and management purposes. Modules I and II provide for integrated research at multiple layers in the hierarchy (Fig. 1); Module III supports continuous system improvement by the resource managers, using a process of monitoring and group evaluation, and Module IV directly integrates all Modules. A communication system has been established to ensure flow of information between Modules and stakeholders.

Note that activities are carried out within modules. These activities are designed to support the learning process outlined previously. It is important to note that this structure supports a *process* of learning and innovation. This process determines the precise activities to be undertaken in each of the modules.

4 Appraisal of the "Innovation System"

Systems can be easily modelled, but we are concerned with real change, not theoretical abstractions. The models are useful only if they help us to design and manage better programs of research, development and extension.

The Australian Government and the graingrowers of Australia will jointly contribute about $A9 million over a five year period, towards three pilot projects in the northern grains region of Australia. The first of these is the program described in this paper. This substantial level of funding, and the very large scale of the "pilot" projects, is the first success of the approach described in this paper. All stakeholders have participated in identifying key issues and deciding in broad terms on what is needed to address the issues.

Experience over 18 months in the first of the three programs has revealed the importance of:
1. participative approaches based on adult learning principles;
2. investment in group development, leadership and ongoing support;
3. researchers and advisers **participating** as part of a **learning** system, rather than **observers** of an independent farming system;
4. addressing both on- and off-site impacts of agriculture on natural resources;
5. the system environment (e.g. policy, markets);
6. the need to contextualise the basic research;
7. good communication within the program, as well as between program participants and the wider community;
8. formal program evaluation which includes both program processes and outcomes.

5 Concluding remarks

Genuine participation can pose a threat to all stakeholders. Farmers, researchers and advisers, as well as research organisations and other government agencies, all need to learn how to operate and learn effectively in a collaborative environment. Experience so far suggests that many researchers have difficulty with collaborative approaches that recognise and value the knowledge of other stakeholders, in particular the farmers. But farmers have also found it difficult to accept new responsibilities. The formation of effective farmer groups has presented a significant challenge. Government bodies also have difficulty with the idea that R D and E funds are allocated to support a *learning-process*, rather than a set of R, D and E *activities* prescribed in an application for funding. The emphasis is on a process of learning, which often requires flexibility in the way R, D and E funds are spent. This flexibility does not sit easily with the traditional need to account for expenditure against stated purposes.

References

Bawden, R. J., Ison, R. L., Macadam, R. D., Packam, R. G. and Valentine, I. (1985): A Research Paradigm for Systems Agriculture. In: J. V. Remenyi (ed.), Agricultural Systems Research for Developing Countries, Proc. of an International Workshop held at Hawkesbury Agricultural College Richmond, N. S.W. Australia, 12-15 May 1985. ACIAR Proceedings No 11, 31-42.

Bowden, J. W. (1992): Predicting and Shaping the Future, Proc 6th Australian Agronomy Conf, Armidale N.S.W., Australia, 125-92.

Conway, Gordon R. (1986): Agroecosystem Analysis for Research and Development, Bangkok: Winrock International.

Foale, M. A., Carberry, P. S., McCown, R. L., Probert, M. E., Dimes, J. P., Dalgliesh, N. P. and Lack, D. (1996): Farmers, Advisers and Researchers Learning Together Better Management of Crops and Crop Lands, Proc. 8th Australian Agronomy Conf., Toowoomba, Qld., Australia, 258-261.

Junor, R. (1991): The Significance of the Landcare Movement for Organisations and Communities, Proc. Soil and Water Conservation Assoc. Conf., Australia, 1-7 Nov., 7-14.

Martin, R.J.; Cornish, P.S. and Verrell, A.G. (1966): An integrated approach to link farming systems research, extension, co-learning and simulation modelling. Proc. 8th Australian Agronomy Conference, Toowoomba, Qld., Aust. 413-416.

McCown, R. L. (1989): Adapting Farming Systems Research Concepts to Australian Research Needs, Proc. 5th Australian Agronomy Conf., Perth, W. A., Western Australia, 221-234.

Passioura, J. B. (1979): Making Plant Physiology Useful, Search **10,** 347-50.

Squires, V. R. (1991): A Systems Approach to Agriculture. In: Squires and Tow (eds.), Dryland Farming - A Systems Approach, Sydney and Oxford University Press, 3-13.

Address of author:
Peter S. Cornish
University of Western Sydney (Hawkesbury)
Richmond, New South Wales, Australia, 2753

WOCAT - World Overview of Conservation Approaches and Technologies - Preliminary Results from Eastern and Southern Africa

H. Liniger, D.B. Thomas & H. Hurni

Summary

The **W**orld **O**verview of **C**onservation **A**pproaches and **T**echnologies (WOCAT) is a program of the World Association of Soil and Water Conservation (WASWC), organized as a consortium of several international institutions. The overall goal of WOCAT is to contribute to sustainable utilization of soil and water. WOCAT collects and analyzes information on soil and water conservation (SWC) technologies and approaches world-wide, and presents the collected information in computer databases and decision support systems, and in the form of handbooks, reports and maps readily accessible to SWC specialists and policy-makers world-wide. WOCAT has prepared a framework for the evaluation of soil and water conservation and has started data collection.

The paper presents preliminary results with promising SWC technologies and approaches used in Eastern and Southern Africa. The first finding is that hardly any promising SWC activities could be found on common grazing lands. Analysis of the cropland shows some of the bio-physical and socio-economic conditions under which certain SWC technologies and approaches are used, including land use types, climatic zones and land tenure, and looks at issues such as participation and costs. Furthermore, classification criteria for SWC technologies and approaches are discussed.

Keywords: SWC conservation technologies, approaches, database, methodology, Eastern and Southern Africa.

1 Introduction

The **W**orld **O**verview of **C**onservation **A**pproaches and **T**echnologies, launched in 1992, is a program of the World Association of Soil and Water Conservation (WASWC), in collaboration with several international institutions e.g. UNEP, FAO, ISRIC. It is being coordinated by the Centre for Development and Environment (CDE), Institute of Geography, Berne, Switzerland (WOCAT 1997).

The **overall goal** of WOCAT is to contribute to sustainable utilization of soil and water. WOCAT collects and analyzes information on soil and water conservation (SWC) technologies and approaches world-wide, and presents the collected information in computer databases and decision support systems, and in the form of handbooks, reports and maps readily accessible to SWC specialists and policy-makers world-wide. WOCAT has prepared a framework for the evaluation of soil and water conservation. Through questionnaires and regional workshops, SWC specialists assist in the collection of promising SWC technologies and approaches world-wide. The developed database system allows easy access to the data and the analysis according to the users' needs. The aim of this paper is (1) to present preliminary results so far, mainly for the African continent and (2) to show how the data can be analyzed and used to find solutions for sustainable soil and water management.

WOCAT started data collection with two **regional workshops** in Eastern and Southern Africa in 1995 [1]. The purpose of the workshops was to initiate data collection for the African continent on widely applied and promising technologies and approaches for soil and water conservation and to allow a regional exchange of experience from implementation. Some 56 SWC specialists from 15 Eastern and Southern African countries compiled 52 SWC technologies and 36 approaches [2].

In the context of WOCAT, the following **definitions** were used:

SWC: Activities at the local level which maintain or enhance the productive capacity of the soil in erosion-prone areas through: prevention or reduction of erosion, conservation of moisture, and maintenance or improvement of soil fertility

SWC Technology: Measures used in the field (agronomic, vegetative, structural and management)

SWC Approach: The ways and means used to implement a SWC technology on the ground

2 Results on SWC technologies and approaches [3]

Of the reported technologies, 24% were introduced and new to the area, 27% were mainly new but based on previously introduced technologies, 29% were mainly new but with indigenous / traditional elements and 20% were mainly indigenous including new elements. The distinction is not clear cut. Certain practices such as ridging in Malawi have been extensively used for at least 50 years and are effectively inculturated into the farming system. The younger generation would probably consider ridging a traditional practice although its wide adoption belongs to the colonial era.

The distribution of technologies according to **land use type** is shown in Figure 1. The largest number of technologies are those for conservation of land under annual crops. Technologies for conservation of grazing land were recorded much less frequently. This no doubt reflects the continuing bias towards the conservation of cropland and the difficulties of conserving soil and water on degraded grassland.

The distribution of technologies by **climatic zone** is shown in Figure 2. This shows the predominance of technologies reported from semi-arid areas and, as before, the relative importance attached to soil and water conservation on cropland. Land users in dry areas are much more concerned about soil and water conservation because of the significant gains in production associated with the reduction or prevention of runoff. Such large gains rarely arise from conservation in humid areas.

Figure 3 shows the high proportion of **farm sizes** between 1 and 5 ha. These farms are dependent almost entirely on hand labour or ox-drawn equipment. Of all SWC technologies, 62% were implemented by manual labour, 22 % by animal traction and 16% by mechanized equipment. Further analysis will show which of the technologies are more suited for small- or large-scale land use and for different levels of inputs.

Figure 4 shows the distribution of technologies according to **land use rights**. The question of tenure is very complex but it is notable that there are more technologies recorded for situations with individual rights to land. This includes land under individual ownership in the modern legal sense but also land held under usufruct right. The options for conservation measures under individual tenure are clearly greater under this system than where there are communal rights to land but without long-term security. The least satisfactory situation is where land is under "open access" and free for all to use or misuse without restraint.

Figure 5 illustrates that on cropland farmers have mostly individual rights whereas grazing land is mostly under communal or even open access management. This may indicate why it has been difficult to find viable solutions for grazing land, as it involves communities and not individuals. Forests which are less susceptible to erosion have an almost even share of the different land use rights.

WOCAT - World Overview of Conservation Approaches and Technologies

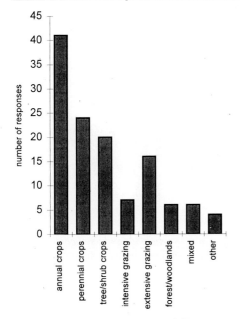

Fig. 1: Technologies used on different land use types

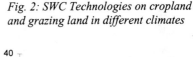

Fig. 2: SWC Technologies on cropland and grazing land in different climates

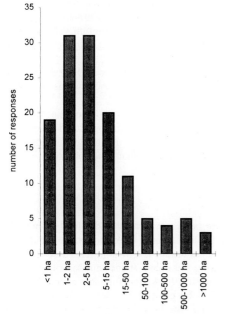

Fig. 3: Distribution of SWC Technologies according to farm size

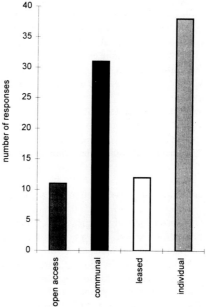

Fig. 4: Distribution of SWC Technologies according to land use rights

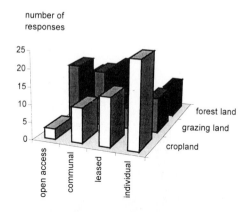
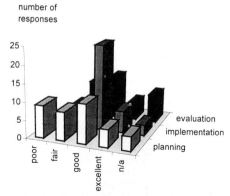

Fig. 5: Land use rights for different land use types within the area where SWC is applied

Fig. 6: Participation of local community during different project phases

Figure 6 shows **local community participation** during different phases. About half of the cases reported had good to excellent involvement during the planning and evaluation phase, and during the implementation the majority was closely involved. Bearing in mind that WOCAT collects promising examples, the results indicate that good participation of the local community during all stages is a prerequisite for success. However, more detailed analysis from the database is needed for specific conclusions.

The **establishment costs** for SWC technologies (Figure 7) show a wide range from 10 to more than 1000 US$ per ha. For the majority of the reported cases, the annual **maintenance costs** are below 50 US$ per ha. This can be compared with the average annual **production values**, which for Eastern and Southern Africa were mostly between 200 to 500 US$ per ha, although there was difficulty obtaining reliable estimates on SWC costs and production value.

When **labour input** was substantial (Figure 8): about half of the work was done voluntarily, and the rest to about the same extent with food-for-work, payment in cash, and rewards with other incentives.

The results presented above are only examples of possible analyses from the WOCAT database. In order to come up with a clearer picture of why certain technologies and approaches are successes in one place and failures in other situations, a more detailed analysis needs to be done, the current data quality has to be improved, and more examples should be collected.

3 Classification of SWC technologies and approaches

The terms used for technologies vary widely and are not always easily understood outside the region or country where they are used. For example the terms bund, bank and terrace are used in different places for the same structure. The term *Fanya juu* terrace is a Swahili term describing the way the terrace is made by digging a trench and throwing soil uphill. It does not describe the final conformation. In other situations it would be called an earth bank. One outcome of the WOCAT project will be a clearer definition of the terms used and a glossary indicating which terms are equivalent. This will be a very valuable reference tool for those trying to benefit from the experience gained in different parts of the world.

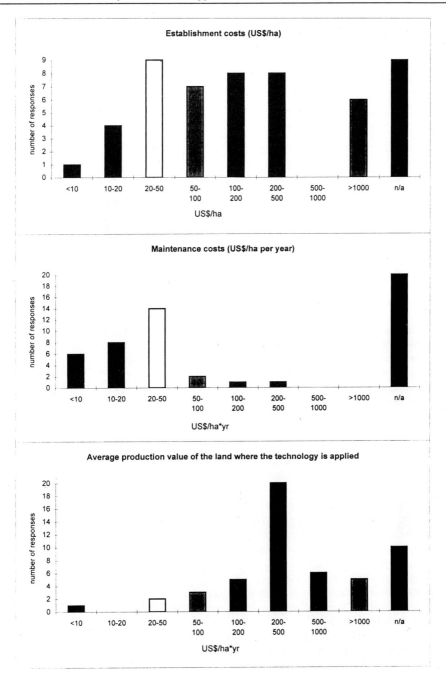

Fig. 7: Costs for establishment and maintenance of SWC technologies and average production value of land

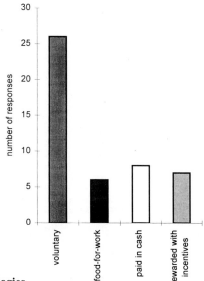

Fig. 8: Labour input

3.1 Agronomic, vegetative and structural technologies

The technologies have been classified in 4 major groups as agronomic, vegetative, structural and management (including combinations) with the following meaning attached to each:

Agronomic measures such as mixed cropping, contour cultivation, mulching, etc. which:
- are usually associated with annual crops
- are repeated routinely each season or in a rotational sequence
- are of short duration and not permanent
- do not lead to changes in slope profile
- are not zoned
- are independent of slope

Vegetative measures such as grass strips, hedge barriers, tree planting, etc. which:
- involve the use of perennial grasses, shrubs or trees
- are of long duration
- often lead to a change in land profile
- are often zoned on the contour or at right angles to wind direction
- are often spaced according to slope

Structural measures such as terraces, banks or bunds, etc. which:
- lead to a change in land profile
- are of long duration or permanent
- are carried out primarily to control runoff and erosion
- require substantial inputs of labour or money when first installed
- are zoned on the contour
- are spaced according to slope

Management measures such as land use change, area closure, rotational grazing, etc. which:
- involve a fundamental change in land use
- involve no agronomic and structural measures
- often result in improved vegetative cover
- often reduce the intensity of use

IMPACT / TECHNOLOGY	Type	Control of splash	Control of dispersed runoff	Control of concentrated runoff	Method of runoff control	Infiltration	Plant water use	Soil structure	Soil fertility	Improve outputs / income	Reduction of slope	Remarks / commonly used
Intercropping /planting	A	2	2	2	R	2	1	1	1	2	0	common maize/beans
Mulching and minimum till	A	3	2	1	R/I	2	3	2	2	2	0	high potential
Tied ridging	A	0	3	1	R	3	2	0	0	2	0	in Malawi and Zimbabwe
Trash lines	A	1	2	1	I	2	1	1	1	1	1	widespread
Matengo pits	A	0	3	2	R	3	3	1	2	2	0	Central Tanzania
Diagonal drainage ditches	A	0	2	2	D	0	1	0	0	1	0	Ethiopian highlands
Grass strip	V	1	2	2	I	2	1	1	1	2	1	in Kenya and Swaziland
Contour hedge N-fixing	V	2	2	2	I	2	1	1	2	2	1	experimental Malawi/Kenya
Stone bunds	S	0	1	2	R	2	1	0	0	-1 to 0	1	in Ethiopia
Fanya Juu, terrace level veg.	SV	1	1	2	R	2	1	0	0	-1 to +1	2	most effective in dry areas
Soil bunds level vegetated	SV	1	1	2	R	2	1	0	0	-1 to +1	1	suited for dry areas
Soil bunds graded vegetated	SV	1	1	3	R/D	1	0	0	0	-1 to +3	3	effective in dry areas
Bench terrace level vegetated	SV	1	3	3	R	3	3	0	0	2 to 3	0	Kenya dry and subhumid
Retention ditch vegetated	SV	1	2	3	R	2	2	0	0	0.	0	common in Kenya
Cutoff drain vegetated	SV	1	1	3	D	0	0	0	0		0	
Banana trench	SV	1	2	3	R	2	2	1	0	2 to 3	0	in W. Uganda

A: Agronomic, V: Vegetative, S: Structural. -1: negative impact, 0: no impact, 1: minor benefit, 2: moderate benefit, 3: Major benefit
Method of runoff control: R: Retain/trap, I: Impede/retard, D: Drain/divert

Tab. 1: Classification and impact of selected SWC technologies on cropland in Eastern and Southern Africa

3.2 Function and impact of technologies

One outcome of the WOCAT project will be a clearer understanding of the function and impact of different technologies, which is important in the search for better solutions. Table 1 shows how each technology has been assessed according to eleven different criteria, namely:
- control of water erosion caused by raindrop splash
- control of water erosion caused by dispersed runoff (i.e. inter-rill erosion)
- control of water erosion caused by concentrated runoff (i.e. rill erosion)
- method of controlling runoff
- control of wind erosion
- improvement of infiltration
- improvement of water use by plants
- improvement of soil structure
- improvement of soil fertility
- improvement of land productivity
- reduction of slope

There is some overlapping of criteria but they are all significant in different ways. The rating system applied indicates whether a given technology has a negative effect (-1), no effect (0), minor benefit (1), moderate benefit (2) or major benefit (3). These are subjective assessments provided by SWC specialists. The criterion "method of handling runoff" can be divided into four categories with one or another of the following functions: a) drain/divert, e.g. cutoff drains and diversion ditches, b) retain/trap, e.g. retention ditches, c) impede/retard e.g. grass strips and d) supplement, e.g. through water harvesting or spreading technologies. By clarifying the role of different technologies it becomes clear that an effective SWC system requires a combination of technologies which collectively have a broad impact on the problems.

3.3 Classification of SWC Approaches

WOCAT only collects information on approaches which have resulted in SWC technologies that are realized in the field. Thus technologies are linked to approaches.

In the attempt to classify SWC approaches, it was realized that each approach is unique and that there is no simple categorization like: food for work, catchment approach, training and visit, participative, etc. As a preliminary result WOCAT proposes that the following aspects have to be described in order to give a short characterization of an approach:

1)	**Objective**	• soil conservation / water conservation / drainage • production increase/maintenance • holistic	
2)	**Focus**	• social • geographic	farm, groups (gender, interest, ...), village, district, country, region, catchment / basin
3)	**Scale**	• funding / costs, • aerial extent	
4)	**Duration**	• no of years, ongoing • incremental / gradual • "one-hit"	
5)	**Process**	• initiation: by client, donor, government, NGO • management: participative, directive, compulsive	

6)	**Purpose**	• preventive (treat cause) • facilitative (treat constraint) • restorative (treat effect)
7)	**Method**	• SWC technology / management / land use change • incentives, payment, voluntary, subsidized • tenure / rights, legislation: enabling/punitive • education: interpersonal / impersonal

4 Conclusions

WOCAT has collected valuable information so far and the regional workshops had the added benefit of allowing SWC specialists from different countries to share their experience. Ways of improving the quality of the data and ensuring comprehensive coverage have been planned. The questionnaires were revised for clarity. Plans are underway to search for additional funding to allow data collection for the other regions and countries in Africa and eventually for the globe. The major outputs will include handbooks on SWC technologies and SWC approaches, a SWC map and a database and a decision support system - first for Africa and later on for other continents - and will help in clarifying options for SWC for different bio-physical and socio-economic conditions. As the database expands it will provide a unique source of information. This information will be made easily available to all contributors and interested institutions [4].

Acknowledgments:

Many SWC specialists have contributed to this presentation. We thank all the participants in the regional workshops in Africa and the core collaborators of WOCAT that participated in the last planning workshop and the first Steering Committee meeting.

Endnotes

[1] Each workshop lasted one week during which participants filled in detailed questionnaires (Liniger (ed.), 1998a, 1998b) on the particular Technologies and Approaches with which they were familiar. Participants were advised to choose those SWC activities which were important and already widely used or appeared to be promising, including indigenous as well as modern technologies and approaches. One participant for each country filled in a third questionnaire for the SWC map (Liniger and van Lynden eds., 1998) for which results will be presented in a separate publication. Each workshop was attended by several experienced facilitators who assisted with the compilation of data. Data on technologies and approaches were taken to CDE for computerization and analysis.

[2] **Funding** for the Eastern Africa Workshop came from the Swiss Agency for Development and Cooperation (SDC), Regional Soil Conservation Unit (RSCU/SIDA), FAO and Deutsche Gesellschaft für Technische Zusammenarbeit (GTZ) through the Observertoire du Sahara et du Sahel (OSS). Funding for the Southern Africa Workshop was provided mainly by UNEP with additional support from FAO and SDC.

[3] The results presented here are preliminary because there is more information to be collected and that which is already in the data base has not been fully analyzed. Even with a comprehensive data collection system, there will always be some technologies and approaches which have not been described. However, the data base has been designed in such a way that it can be continually expanded and updated.

[4] WOCAT is a consortium and is open to collaborating institutions and individuals. If you are interested in collaborating with WOCAT or in receiving further information, please contact the secretariat at CDE (see address of first author).

References:

Liniger, H.P. (ed.) (1998a): Questionnaire on SWC Technologies. A Framework for the Evaluation of Soil and Water Conservation (revised). Centre for Development and Environment, Institute of Geography, University of Berne, Lang Druck AG, Berne.

Liniger, H.P. (ed.) (1998b): Questionnaire on SWC Approaches. A Framework for the Evaluation of Soil and Water Conservation (revised). Centre for Development and Environment, Institute of Geography, University of Berne, Lang Druck AG, Berne.

Liniger, H.P. and van Lynden (eds.) (1998): Questionnaire on the SWC Map. A Framework for the Evaluation of Soil and Water Conservation. Centre for Development and Environment, Institute of Geography, University of Berne, Lang Druck AG, Berne.

WOCAT (1997): World Overview of Conservation Approaches and Technologies - A programme profile. Centre for Development and Environment, Institute of Geography, University of Berne.

Addresses of authors:
Hanspeter Liniger
Hans Hurni
Centre for Development and Environment
Institute of Geography
University of Berne
Hallerstraße 12
CH-3012 Berne, Switzerland
Donald B. Thomas
Consultant
P.O. Box 14893
Nairobi, Kenia

Participatory Approaches in Promotion of Sustainable Natural Resources Management Experiences from South-West Marsabit, Northern Kenya

G.O. Haro, E.I. Lentoror & A. von Lossau

Summary

Natural resource management schemes in the arid and semi-arid areas of Kenya have been generally unsuccessful. This is because they were based on conventional range management models that put little or no emphasis on traditional resource management practices. As a result of these failures it became a generally held believe among the development agents that pastoralists are disinterested in development activities and have no ability to solve their own problems. At present, the development agencies have come to realize the mistakes they have made in their approaches.

Experiences gained within the participatory natural resource management programme in Marsabit, Kenya through the Marsabit Development Programme GTZ is a possible model for slowing down present land degradation processes. Existing local institutions and decision making structures of the target communities were used as an entry point. The process of community mobilization for improved utilization of natural resources started to generate awareness campaigns at the neighbourhood level (the smallest recognizable management unit) and later at the territorial level (largest management unit) for conflict negotiation where influential traditional leaders and elders were brought together to discuss issues that could not be tackled at the neighbourhood level. Ways and means to include women and youth in this process not represented in the traditional resource management forum, are described.

It is concluded that sustainable utilization of natural resources could only be achieved through a process of dialogue that builds upon the local knowledge and increases the understanding of the user group at the grassroots level on the causes and consequences of a degraded environment.

Keywords: Natural resources, traditional management, participatory approaches, conflict negotiation, local institutions, Kenya

1 Introduction

The geographical location of South West Marsabit District is shown in Figure 1. Despite its aridity, the region has diverse vegetation communities whose distribution is influenced by soil moisture regimes, geology, rainfall variability and human activities. The pattern of land use is opportunistic and aimed at taking advantage of unpredictable changes in vegetation productivity.

The main inhabitants of this region are Rendille and Ariaal (Figure 2). The latter represent a community in transition between the Rendille and Samburu production systems. The Ariaal mainly keep cattle and small stock while the Rendille keep camels and small stock. The traditional grazing ranges of Ariaal are in the areas of Olturot, Ilaut, Ngurnit, Merrile, Laisamis, Loglogo and

Karare while Rendille occupy Kargi, Korr and Hedad. Despite their economic differences, the Ariaal identify more with the Rendille than with the Samburu.

The land degradation or desertification process in South West Marsabit is caused by a combination of factors that are mainly related to human activities and recurrent droughts. Over the last three decades, the traditional grazing patterns of Rendille and Ariaal pastoralists have changed because of a number of factors including; loss of access to traditional pastures due to conservation, agricultural encroachment, insecurity in some areas, population pressure and inappropriate water development.

As a result of these changes, there has been a general decline in nomadic movements and consequently over-grazing in particular localities. For instance, more than 80% of the Rendille households are living within 10 km of any major settlements like Loglogo, Korr, or Kargi because of easy access to water, health centres, educational facilities and food relief distribution. This imbalanced use of the area has led to:
- weakening of traditional resource management strategies
- mismanagement of common property resources
- accelerated land degradation
- intensified resource use conflict between user groups

At present, traditional resource managers are loosing grounds as a result of the many external changes that threaten their very basis of survival. The GTZ programme's strategy for adapting and revitalizing traditional resource use practices to address land degradation processes will be addressed below.

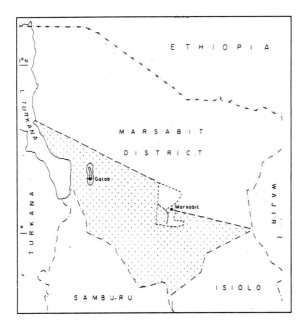
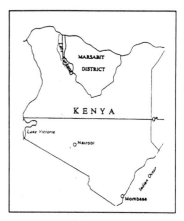

Figure 1: Location of South West Marsabit

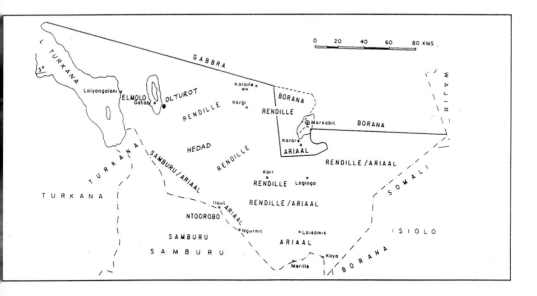

Figure 2: The peoples of South West Marsabit and their neighbours

2 Traditional decision making structure of Rendille and Ariaal

2.1 Social organization

The Rendille are divided into nine autonomous clans: Saale, Dibsahai, Galdeelan, Tupcha, Nahagan, Gobanai, Matarba, Nebei and Rengumo. The Ariaal are relatively mixed and are divided into five groups: Mosola, Lukumai, Ilturiya, Lorokushu and Ongelli. Their settlements are based on kinship and each clan moves in a large camp called *gob* (40 - 80 households). Each *gob* has two central meeting places, the *Naabo* (an enclosure in the centre of the *gob*) and *gei-makhabale* (elders' tree) where elders (married men) meet in the evening and during the day respectively. Each *gob* has an elder leader referred to as *makhabal-i-gob-iwen* or father of the *gob*. *Naabo* is their "parliament". It is here that issues of common concern that affect the clan such as:
- livestock and *gob* movement
- pasture conditions
- water availability
- security
- internal disputes, and
- communication from development agents or the administration

are all discussed. The *naabo* is central to the life of the Rendille and Ariaal and is therefore an important entry point to facilitate better dialogue on natural resource management. Figure 3 represents the programme's adaption and modification on the traditional approach to decision making to increase community participation at *gob* level.

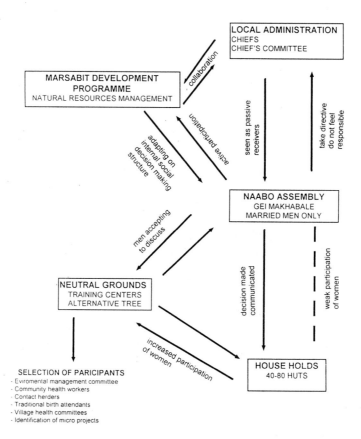

Figure 3: Channels of participatory discussions for awareness creation through gob leaders: Circumventing the Naabo and Gei Makhabale to meet on neutral ground in order to include women and youth (source: Haro 1996)

2.2 Spatial organization

The settlement pattern of the Rendille and Ariaal are not restricted; each gob has access to a certain area it is usually associated with, especially during the dry season (Oba, 1992). These areas or neigh-bourhoods called *latia* (Samburu language) or *olloh* (Rendille language) form the smallest resource management unit. A neighbourhood has well defined geographical features and covers about 500 sq km (Swift and Omar, 1991). It is inhabited by a group of adjacent camps between 5 to 10 gobs with a popula-tion of about 4,000 people. Within the neighbourhood the users are in a position to:
- coordinate annual grazing patterns
- develop area-specific resource management plans
- select elders deemed best to represent their interests
- set aside an area as dry season pasture reserve
- close an area for rehabilitation by enforcing by-laws

Neighbourhood members by virtue of using a common resource are in constant dialogue as shown by Figure 4 which illustrates channels of communication within the neighbourhood to the households through *gob* leaders.

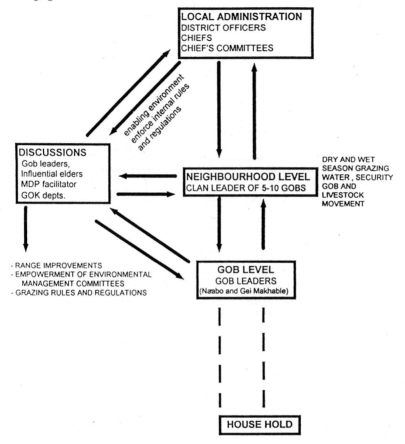

Figure 4: Discussions on improvement of natural resources management with the environment of the neighbourhood (Latio/Olloh) (source: Haro 1996)

2.3 Territorial level

This is the largest unit of natural resource management. It consists of 10 - 15 neighbourhoods. At this level clan leaders and influential clan elders are responsible for regional resource management matters that could not be handled at the neighbourhood level. The clan leaders are capable of:
- negotiating conflict between neighbourhood user groups
- endorsing customary laws and area-specific recommendations
- developing a mechanism for improved resource management at territorial level

Figure 5 represents channels of communication within the tribal territory to the clan elders of the neighbourhood.

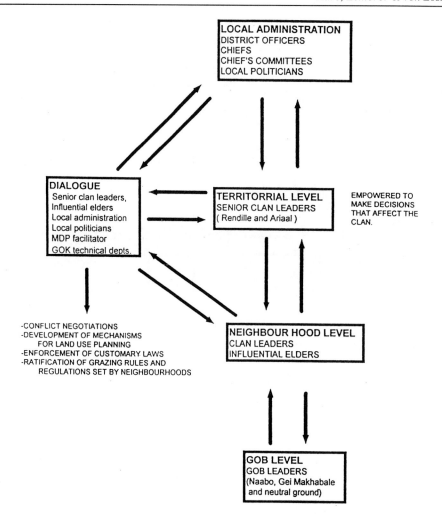

Figure 5: Process of dialogue for resource use conflict management at territorial level with clan leaders and MDP facilitators (source: Haro 1996)

3 Marsabit development programme

Marsabit Development Programme (MDP) GTZ is a community based programme with the aim of supporting communities to increase their self-help capacities in managing their social, economic and physical environment. The programme has three main sectors, livestock development, natural resource management and human resource development. MDP works in close co-operation with a number of Government departments, Non Governmental Organizations, Church Organizations and communities.

3.1 Objective of natural recources sector

The projects aim to develop and promote strategies for sustainable utilization of natural resources through:
- sensitization of the community to the problems of desertification
- revitalization of traditional resource use strategies
- promotion of a process of dialogue to create social responsibility for sustained management of common property resources

3.2 Community mobilization approach

In realising this objective, the programme collaborates with the Ministry of Agriculture Livestock Development and Marketing (MALD&M) Range Management section. It recognizes of the traditional social and spatial organization of the target group as a means to facilitate a process of dialogue for sustained management of communal resource.

Through a series of participatory natural resource management workshops at the neighbourhood level, environmental degradation processes were discussed with the help of posters depicting the various stages of land degradation caused by human activities. The discussions were also combined with visits to actual problem sites, where the participants were given an opportunity to come up with practical solutions. The consensus reached in the workshop and accepted in the naabo was then used for drafting neighbourhood natural resources management by-laws and for selection of members of the Environmental Management Committee (EMC).

3.3 Integration of women and youth

Neither the *naabo* nor *gei makhabale* is accessible to women and youth. In the traditional set up, elders are responsible for management of natural resources and women are rarely involved in activities geared towards environmental management. Traditional leaders only see them as users and not managers. Thus women and youth who are the primary users of natural resources do not feel responsible for the land degradation processes that are going on around them.

To enhance the participation of women and youth in the management of the environment, MDP has circumvented these traditional barriers and brought together men and women into a dialogue for participatory natural resource management on neutral grounds such as training centres or alternative trees to *gei-makhabale*, where all can sit for discussion (see Fig. 3).

4 Results

During our follow-up activities we found that EMC have been established in most neighbourhoods and have attempted to implement their area-specific recommendations. For the Ariaal, the proposed and adapted measures agreed upon in specific neighbourhoods during the series of environmental management workshops were nothing new but rather a reinforcement of their own traditional methods for control of common resources. In contrast, their Rendille counterparts had no strict traditional rules and regulations for the management of their neighbourhood range resources. For them all natural resources are *open access* and nobody has any right to refuse or control.

The successful implementation of these action plans was however, hampered by conflicting concepts of resource use and management between two user groups sharing a common area. By-laws initiated in an Ariaal neighbourhood were not respected by Rendille. Mobile camps and herds still settle where they wish regardless of whether the area has been reserved by camps within a neighbourhood. In the words of *mzee* Lengima an Ariaal elder, "the Rendille herders from other areas invade an area, cut down trees as though we did not know what to do with the trees".

4.1 Process of conflict management between neighbourhoods

Implementation of neighbourhood specific resource management plans brought up contentious issues that could not be handled within the neighbourhood. Noting that it was difficult to continue with the plans further, the members of the EMC requested MDP to facilitate a regional workshop that would initiate a dialogue with different neighbourhood user groups for managing these conflicts.

This workshop was convened in Korr and among the participants were administrative leaders (District Officers, Chiefs) political leaders (councillors), technical officers, MDP and traditional leaders (*makhabal-i-gob-iwen*, influential EMC elders) and also women opinion leaders. The major purposes of the workshop were to:
- discuss existing local level resource use conflict
- reach a common consensus and develop a joint plan of action
- strengthen traditional resource management in each neighbourhood

The participants were given an opportunity to share their experiences in terms of successes and problems encountered in the implementation of neighbourhood specific recommendations (Tables 1 & 2).

Activities initiated for proper use of neighbourhood recources	Number of centres
Formation of environmental management committee	8
Areas kept aside as grazing reserve for milk animals	6
Endorsement of neighbourhood rules and regulation	6
Controlled cutting of important trees	4
Awareness creation at gob level	3
Enforcement and collection of fines	3
On compound tree planting	1

Table 1: Successful activities in 14 Neighbourhoods

Resource use conflict areas between neighbourhood groups	Number of centres
Disregard for locality specific rules and regulations	10
Exclusion of dry season pastures from other herders	9
Indiscriminate cutting of important trees by fora herders	7
Lack of cooperation between the administrators and elders	5
Influx of livestock from other areas into neighbourhood	4
High population concentration	3
Insecurity	3
Cultural ceremonies	2

Table 2: Summary of conflict areas

4.2 Dialogue for conflict management

To mitigate local level conflict and increase community participation in use of neighbourhood resources within the wider context of region, the leaders came up with the following suggestions:
- Increased awareness at gob level of the outcome of the leaders workshop through meetings, field days and naabo assemblies to harmonize the implementation of natural resource management in the South West region
- Establish EMCs in all neighbourhoods with representation from all gobs
- Empower EMCs and gob leaders to take charge of natural resource management matters within their neighbourhoods through:
  - enforcement of neighbourhood by-laws
  - briefing others herders on neighbourhood specific by-laws
  - closing and/or opening dry season pasture reserves
  - setting aside degraded areas for rehabilitation
  - identifying important tree species for protection
  - management of surface water sources
  - use fines collected as incentives

4.3 Recommendations

The leaders also recommended that the development agents involved in natural resource management should:
- Encourage discussion between neighbouring social and ethnic groups in the district to enhance more balanced use of range resources
- Intensify environmental awareness at gob level to reach the majority of resource users
- Follow up the dialogue processes initiated
- Issue EMC members with environmental logos for easy identification

5 Lessons learnt

(i) Environmental awareness workshops at gob level have served as real eye openers for the majority of the user groups who traditionally did not feel responsible for the management of the environment. Women now feel duty-bound in environmental conservation as the traditional leaders have appreciated them as managers and accepted them in EMCs.

(ii) EMCs with the support of the local administration have reactivated moribund traditional rules and regulations for management of key resources like forest and water sources.

(iii) Cooperation between neighbourhoods with respect to conservation of dry season pasture reserves has increased. The Ariaal management system of reserving the pasture in the back and using the forward pastures was revived in the Ariaal / Rendille Neighbourhood.

(iv) EMCs identified with Environmental management logos are now recognized by local administration and development agents for delivery of services to the community:
  - Handling micro-development activities in the neighbourhood
  - Selection of participants for specialised community based training programmes like Community Health Worker, Contact Herders, Traditional Birth Attendants, Environmental management committees
  - Handling of improved breeding bucks, rams and bulls
  - EMCs within the neighbourhood are now being considered for representation in sub-locational development committee meetings

Acknowledgement

The report was based on a series of participatory natural resources management workshops supported by Marsabit Development Programme / GTZ in collaboration with department of livestock production. The authors wish to acknowledge Isaak Learamo and Haret Gudere for assisting in the facilitation of the workshop and James Impwi for computer layout of the document. The views expressed are those of the authors and not that of MDP/GTZ or individuals acknowledged.

References

Oba, G. (1992): Traditional grazing systems of the Rendille and Ariaal pastoralists: changing strategies and options in the Southern District of Marsabit Northern Kenya. Consultaney report for Marsabit Development Programme.

Swift, J. and Omar, A.N. (1991): Participatory pastoral development in Isiolo District: socio-economic research in the Isiolo Livestock Development Project: final report.

Addresses of authors:
G.O. Haro
Marsabit Development Programme / GTZ
P.O. Box 204
Marsabit, Kenya
E.I. Lentoror
Ministry of Agriculture
Livestock Development and Marketing
P.O. Box 204
Marsabit, Kenya
A. Von Lossau
GTZ
Division 424
P.O. Box 5180
D-65726 Eschborn, Germany

Soil Conservation Implementation Approaches in Kenya

M.J. Kamar

Summary

Soil conservation activities in Kenya started in the 1930's and can be divided into the pre-independence period, the period between independence and the United Nations Conference on the Human Environment held in Stockholm in 1972, and the post-conference period. During implementation, different soil conservation approaches were used at different times yielding varied results.

In this paper the different approaches were carefully researched and for each period and approach, the historical background, implementation strategies, and their strengths and weaknesses are discussed.

The study revealed the main approaches used since the inception of soil conservation in Kenya to be; forced conservation works, use of incentives and subsidies, paid work by the government and to some extent, farmer's mobilization. While forced conservation achieved the best implementation, the structures were soon abandoned and even destroyed due to farmers' bitterness toward the colonial government. The use of incentives and subsidies was characterised by success whenever the incentives were supplied. The government-paid soil conservation was also well implemented as long as people were paid but farmers failed to do maintenance. The present "Catchment Approach" uses total community mobilization and participation during planning and implementation of the conservation plan. Its merits and limitations are discussed.

Keywords: Soil conservation, implementation strategies, approaches, subsidies, incentives, catchment-approach, community-mobilization.

1 Introduction

Soil conservation activities in Kenya started in the 1930's and can be divided into: the pre-independence Soil Conservation Service (1930-1963), a low activity post-independence period up to the United Nations Conference on the Human Environment held in Stockholm, Sweden in 1972 (1963-1974) and the post-conference period of the Kenya National Soil Conservation Project (1974 to date). During these periods, different approaches were used in implementation and yielded varied results. In 1991 a new approach known as the Catchment Approach (CA) to conservation was introduced.

The objective of this paper is to present an analysis of the different approaches used and the reasons why most of them had limited success.

2 Methodology

The different soil conservation implementation approaches used were carefully researched. For each period and approach, historical background, conservation activities implementation strategies and their strengths and weaknesses are discussed. Farmers and extension workers who participated in the different approaches were interviewed to get an input of their experience and perceptions to these approaches.

3 Pre-independence Soil Conservation Service (1930--1962)

3.1 Historical background

Soil Conservation was introduced to the Department of Agriculture during the colonial government in the 1930's, and in 1938 a Soil Conservation Service was established under an agricultural officer, Colin Maher (Erikson, 1992; Maher, 1943). The concern over soil erosion in Kenya in the 1930s has been identified by Anderson (1984) as chiefly influenced by; the alarm caused by the dust bowl in the United States, which reached its height in 1935, a noticeable increase in the African population in the mid 1920s, the economic consequences of the Depression which made the government avoid famine relief by increasing agricultural output, and the increase in drought incidence. In addition, during the World Wars, the heavy demands of food to feed troops led to over-exploitation of land resulting in land degradation and specifically erosion (Cone and Lipscomb, 1972). At the very beginning, the Soil Conservation Service in Kenya emphasized the need for good land and crop husbandry for maintenance and improvement of soil fertility (Annual Report, 1937). In the two decades that followed, terracing took place in the large farms in what was called the 'White Highlands' as well as in the native reserves. Colonization significantly reduced the use of old indigenous practices such as shifting cultivation, slash and burn, indigenous irrigation systems and even old agroforestry systems, due to more permanent settlements. This period could further be sub-divided (Tiffen et al. 1994) into 1930-1945, a time of increasing concern of erosion which resulted in forceful interventions prevailing over persuasion, external funding over local resources and external advice over consultations with local people. In addition, the period 1945-1962 experienced large scale investment and the use of force with little local consultation. Since both periods faced the same implementation strategies the two have been lumped together.

3.2 Conservation activities

The soil conservation activities which were advanced for and undertaken by the Soil Conservation Service can be summarised as follows:
(a) Terracing - This involved the setting out of terrace lines, measuring slopes and setting vertical intervals, determining terrace capacity and calculating dimensions, selecting the appropriate terraces and the actual construction of terraces.
(b) Design and construction of waterways protected with grass.
(c) Contour planting and cultivation.
(d) The making and use of boma manure and compost, including the use of crop residue.
(e) Paddocking and the use of hedges for rotational grazing and prevention of overstocking.
(f) Planting and use of Nippier grass for cattle feed.
(g) Planting fruit trees, shade trees, fuel trees and wind breaks.
(h) Protection of streambanks, catchment areas and water supplies.
(i) Prevention and healing of gullies.
(j) Protection of roads and paths from erosion.

3.3 Implementation strategies

The Soil Conservation Service implemented its activities by using terracers who operated terracer machines like tractors, caterpillars and hammer graders. It also employed soil conservation boys who worked manually under very close supervision by levellers. The levellers were themselves answerable to the European Soil Conservation Officers. This implementation was characterized by a chain of command where orders were taken from the Soil Conservation Officers by levellers whose job was to ensure proper implementation at the farm level. The success in passing the message and supervising the work was then used for or against the levellers in promotions or salary increments. This resulted in very good supervision at the farm level.

Terracing took place in both the European and native reserves under the supervision of the levellers. While the European farms were being terraced using terracer machines or soil conservation boys at a government subsidized rate, the native farmers were forced by the colonial administrators to hand-dig terraces on their farms as group work. Anybody who refused was punished or prosecuted using the Chief's or Agricultural Act. Schools were also ordered to practise terracing on their land. This resulted in resentment towards soil conservation by the native communities which influenced their attitude for many years.

3.4 Strengths and weaknesses of the approach

According to most farmers interviewed, the Soil Conservation Service was able to undertake soil conservation work to a reasonable level of success (>80%) at the point of implementation. This was mainly attributed to the efforts of the colonial government including enforcement of the law, use of force as well as the use of trained personnel (levellers and terracers). The use of trained levellers and terracers ensured proper implementation while the incentive of promotion for best performers created competition and enhanced both quality and quantity of work done. The use of force was, however, to later become the greatest setback to the approach as native farmers perceived soil conservation as part of colonialism. Most farmers regretted that farmer training sessions were lacking and farmers were only treated with orders from levellers who dictated what should be done. The farmers, therefore, remained ignorant of the benefits of soil conservation except for farmers who had been previously affected by flooding. This major setback featured more strongly soon after independence (1963) when farmers abandoned soil conservation. Most of them stopped maintaining ready structures while others even destroyed existing structures, all in the name of freedom. Most interviewed farmers now regret the act but blame it on the forced adoption of programmes without any explanation to the farmers.

4 Post-independence low conservation period (1962-1974)

Just before and soon after Kenya obtained independence in 1963, soil conservation activities dropped rapidly. In the Machakos district, the area conserved by terracing had fallen to 2700 ha in 1961 compared to 4200 ha in 1958 (Tiffen et. al. 1994) and all other districts faced the same trend. As mentioned earlier, this was mainly because of the reaction by farmers who had believed that soil conservation activities were part of colonialism. At independence, the use of force was stopped with devastating effects. Most activities stopped, structures were not maintained and many were even destroyed. Closed grazing areas were re-opened and erosion features soon re-appeared. At the same time, there was no supervision from the new government, since, during the transition of governments, more attention was paid to the political changes and stability. During this period, therefore, conservation was left to farmer personal initiatives, and even agricultural staff were reduced (Mbogoh, 1991).

5 The Kenyan National Soil Conservation Project

5.1 Historical background

In 1972, when Kenya participated during the United Nations Conference on the Human Environment held in Stockholm, Sweden, attention focused on Soil Conservation. During the conference, Kenya identified land degradation as its most severe environmental problem. After the conference, the Kenya government approached the Swedish government for assistance to tackle the problem of soil erosion. This resulted in the starting of the Kenya National Soil Conservation Project (KNSCP) in 1974 under the Ministry of Agriculture with financial and Technical Assistance from the Swedish International Development Agency (SIDA).

5.2 Conservation activities

The project initially limited itself to soil conservation on cultivated land but over time broadened to a more integrated approach of sustainable landuse. The activities were mainly:
(a) Design and construction of cut-off drains to prevent overland flow from upper uncultivated catchments, from reaching and eroding cultivated lower parts of catchments.
(b) Terracing in cultivated land, including the use of "fanya-juu" terraces which serve both soil and water conservation.
(c) Establishing artificial grassed waterways to deliver flows from terraces and cut-off drains safely.
(d) Application of stone and trash on terrace lines as terrace starters.
(e) Planting suitable and useful trees and grasses along terraces.
(f) Establishing nurseries for raising tree seedlings and bulking grass cuttings.
(g) Stream bank protection.
(h) Gully control.
(i) Training of technical and extension staff, farmers, school children and the community at large.

5.3 Implementation strategies

The implementation of soil and water conservation activities involved the preparation of a national work-plan by the Soil and Water Conservation Branch of the Ministry of Agriculture. The work plan was based on requirements from the field which were submitted annually by every District involved. The field requirements included the number of farms to be terraced, total length of terraces to be constructed (or laid and planted with grass), total length of cut-off drains to be constructed, number of staff and/or farmers to be trained, number of gullies to be controlled including sizes, and the estimated costs of all activities including supervision. Requirements and national workplans including budgets were prepared annually. It is worth noting at this point that the decisions on which activities should be done where, were decided entirely by the field extension officers without involvement of the farmers. It was only when a farmer's farm had been identified for some activity or when a farmer was selected for training that they started being involved. The training sessions, however, played a key role in exposing the needs and importance of soil conservation as well as encouraging farmers to be involved in these activities. Subsidies and incentives were paid for by the project to encourage farmer participation and two approaches were use. Initially, implements or tools were provided as incentives and cash payments were made to farmers as subsidies for the construction of the more labour intensive common measures such as cut-off drains (CODs), fanya juu terracing, artificial waterways and gully control. Farmers were then expected to do farm conservation work such as terracing, applying trash or stones on terrace lines, planting grasses and trees along terraces as well as maintaining the constructed structures. This strategy was soon abandoned when it was discovered that, to earn more from work, farmers initiated many constructions but there was very little or no maintenance. It was also observed that most farmers did not fulfil their obligation of doing on-farm conservation even where terracing was recommended below constructed CODs. This led to the second approach where extension was directed towards the individual farmer.

Farmer training workshops were emphasized in this approach. The incentives for farmer participation changed to certificates of attendance and the sustained yield resulting from good conservation and management after training. It was assumed that following the training workshops, farmers would adopt conservation measures as they appreciated the need. Only in few instances did the project pay money as subsidies in public or common areas. This second approach, resulted in the slowing down of soil conservation activities and in some areas, the extension messages were not even being adopted. As a result, in 1987 a new extension approach, called the "Catchment Approach" was introduced (Section 6).

5.4 Strengths and weaknesses of the approach

The success of this approach can be attributed mainly to the paid subsidies by the project, resulting in donation of tools and the use of allowances as incentives to both staff and farmers. The training workshops also played a key role since it improved the technical knowledge of the staff and exposed farmers to erosion problems and the opportunities and techniques of soil and water conservation. Changing from material to non-material incentives, however, resulted in a decline in adoption of conservation work. This was a weakness in itself because it is difficult to stop doing what people are used to. In any case, they had not been involved in deciding what the best incentives and approach should be. It became clear during interviews that farmers were after incentives more than the conservation benefits.

Except for the training workshops, the project had poor farmers participation. This meant that only trained farmers understood the benefits while the rest did not have the opportunity and thus appeared reluctant to invest in the long-term benefits of conservation. Some farmers said they felt they could not invest their time and money on conservation structures which may never pay back, meaning they were still sceptical about conservation benefits. Some even felt the government should continue to pay for all conservation work since it does not pay under subsistence farming.

Another weakness in the approach was poor follow-up by extension workers as this became too expensive for the project.

The approach also gave blanket recommendation, without taking in account the socio-economic and ecological status of specific areas. The construction of Fanya Juu terraces and CODs cannot be done easily by a farmer on his own and yet it was too expensive for subsistence farmers who rightfully complained as the project had already set high rates. The recom-mended conservation structures were also insensitive to ecological or soil differences. For example, gully control was done using standard gabions in all soil types resulting in most of them failing because of inapplicability. Most conservation measures were never tested and adapted to varying climatic and soil conditions.

6 The catchment approach

6.1 Historical background

The catchment approach is a soil conservation extension strategy whereby field technical assistants and soil conservation officers work with farmers in a given catchment area (Mbegera et al. 1991). It was started in 1987 mainly to facilitate faster adoption of soil conservation measures. Its goals are improved planning and implementation of soil conservation activities through a participatory and interdisciplinary approach. It was felt there was a need for total farmer participation right from problem identification in order to take care of catchment needs and socio-economic factors. The involvement of other sectors with an interest in the catchment, such as the departments of Livestock, Water, Forestry, as well as non-governmental organizations, was also found important as it could lead to concerted efforts and sharing of resources. This approach aims at giving farmers a complete package of soil and water conservation.

6.2 Implementation strategies

To achieve its goals, the catchment approach consists of three main components which must be observed:

(a)   Total community mobilization and participation.
This will be achieved through:
(a.1) Formation of catchment committees by the farmers themselves.
(a.2) Intensified publicity through field-days, demonstrations and tours.
(a.3) Interviews of the farmers by the planning teams.

This will help spread information widely to the catchment inhabitants, to develop better understanding of the conservation problems specific to each area and to cultivate closer collaboration between the farmers, Ministry of Agriculture extension staff and other agencies.

(b)   Catchment planning
This comprises:
(b.1) Identification of problems by extension staff and farmers through interviews.
(b.2) Discussing these problems and opportunities with other government ministries and other agencies, NGOs, Presidential Commission on Soil Conservation and Afforestation (PCSCA).
(b.3) Production of physical plans with details of best measures already discussed by the affected individuals.

(c)   Implementation of plan
This will be achieved through:
(c.1) Allocation of duties to the catchment planning team.
(c.2) Actual lay out of appropriate conservation measures.
(c.3) Technical and organizational supervision by the extension staff and local Soil and Water Conservation Committees to ensure adoption by the farmers in the whole catchment.
(c.4) Disciplinary action against those not willing to implement the plan.

6.3 Strengths and weaknesses of the approach

The catchment approach has shown indications of progressive success since its inception and this can be attributed to:
(a) Community participation from problem identification to implementation.
(b) Working in a watershed rather than individual farms. This resulted in systematic collection of information and analysis, specific to the catchment, and thus use appropriate information for planning.
(c) Concentration in small areas led to faster implementation and easy supervision.

(d) Formation and use of catchment committees during implementation made farmers supervise each other and exert pressure on those lagging behind. The farmers take this responsibility seriously.
(e) Good staff - farmer relationship.
(f) Provision of tools for communal ownership which ensures care by the community.

This approach has generally resulted in better success in the implementation of conservation work and facilitated the farmer's full participation as compared to previous approaches. The weaknesses reported so far are the lack of tools in some watershed and poor or no follow-up in watersheds which were left to the committee too early. The approach should design a method of smoothly phasing out staff supervision of a catchment while maintaining their relationship through visits once in a while. This approach appears more promising in the sustainability of soil conservation activities.

7 Conclusion and recommendations

(a) The pre-independence conservation activities were successful because of the use of force and law but did not educate farmers to appreciate what they were doing. Separation of the enforcement of the law and soil conservation as a management tool, could have helped minimize the destruction of the many structures that had been put in place.
(b) The use of subsidies and incentives did assist in speeding up conservation. However, one must be aware of the incentives and how they are administered to avoid being misunderstood. It was clear from farmer interviews that it was not clear what the incentives were meant for. Some thought it was a salary to do conservation and so they expected more pay for maintaining them.
(c) Subsidies, whether money or tools, should only be used when it can be justified. Farmers should be involved in reaching the decision that there is a need. Money incentives should not be used except when it is earned competitively. This means there is a need to create farmer competition in Soil Conservation where the winners are well rewarded. Farm competitions have been used in Kenya successfully to enhance good farm planning and management but there is a need to create competition in Soil Conservation activities.
(d) Educating and training farmers is vital for the success for any conservation activities. This also better motivates them into doing conservation work. Also, training ensures the farmer understands what is going on in any land use practice that he undertakes, the consequences and the remedies.
(e) The participatory approach offers the best results. Most farmers feel they have been tilling their land for many years and they should be consulted rather than ignored and forced in the name of law. It is also clear that farmers will implement activities better if they have developed it themselves.
(f) Soil and Water Conservation activities are more successfully handled by catchment. This is because a catchment is a unit in which every area affects and is affected by another area.
(g) The use of farmer committees, as in the case of catchment committees, is vital for successful implementation and supervision. Farmers should also be allowed to develop penalties for defaulters and collect them if possible.
(h) It is necessary to create farmer competition between different catchments with incentives to motivate further conservation work.

References

Anderson, D.M. (1984): Depression, dustbowl, demographya nd drought: the colonial state and soil conservation in East Africa during the 1930s. African Affairs, **83**(332), 321-343.

Annual Report (1937): Department of Agriculture. Colony and Protectorate of Kenya. Government Printer, Nairobi.

Cone Wiston, L. and Lipscomb, J.F. (1972): The History of Kenya Agriculture. University Press of Africa. Nairobi, Kenya

Erikson, Arne (ed.) (1992): The Revival of Soil Conservation in Kenya. Carl Gosta Wenner's personal notes (1974-1981). Regional Soil Conservation Unit/SIDA -Publication. Nairobi. Kenya.

Maher, Colin (1943): Soil Conservation Service: Notes on Procedure. Colony and Protectorate of Kenya. Government printer, Nairobi.

Mbegera, J. M. (1991): Soil and Water Conservation Training and Extension - Kenyan Experience. RSCU Publication. Nairobi. Kenya.

Mbogoh, S.G. (1991): Crop Production; In: M. Tiffen (Editor), Production Profile: Environmental Change and Dryland Management in Machakos District, Kenya 1930-90: ODI Working Paper No. 55, Overseas Development Institute, London.

Tiffen, M.., Mortimore, M. and Gichuki, F. (1994): More people less Erosion. John Wiley and Sons. Chichester.

Address of author:
Margaret J. Kamar
Department of Forestry
Moi University, Chepkiolel Campus
P.O. Box 1125
Eldoret, Kenya

Participatory Land Use Planning Approach in North Pare Mountains, Mwanga, Kilimanjaro Region, Tanzania

K.C.H. Mndeme

Summary

The North Pare Mountains are experiencing high population pressure, land shortage, shortage of fuelwood and mismanagement of natural resources (e.g. forests and soil).

The Project described here operates in the North Pare Mountains, part of the Eastern Arc Mountains in the Kilimanjaro and Tange Regions in Tanzania and covers two Divisions (Usangi and Ugweno) with an total area of 420 km^2. The region is characterized by a high population density as compared to the lowlands.

The Tanzania Forestry Action Plan Project was launched in May, 1992. It adopted a participatory land use planning approach in order to assist farmers in improving the production per unit area through proper land husbandry, sustainable management and utilization of the natural resources.

Three strategies, namely village-based participatory land use planning around national forest reserves, neighborhood land use planning including agroforestry, and afforestation and forest management are presented and discussed.

Keywords: Participatory land use planning, land tenure, Tanzania forestry Action Plan

1 Introduction

The North Pare Mountains form part of the Eastern Arc mountain ranges in the Kilimanjaro and Tanga Regions in Tanzania.

The Tanzania Forestry Action Plan (TFAP) North Pare Project is one of several TFAP projects in Tanzania and was launched in May 1992 by the Ministry of Natural Resources and Tourism with technical support from the German Federal Ministry for Economic Cooperation and Development through the Germany Foundation for technical Cooeration (GTZ).

The North Pare mountains slopes form a major part of the agricultural land in the Mwanga District. This area receives adequate but unreliable rainfall. It is characterised by exhausted soils. It is rich in forests including both forest reserves (national forests) and traditional clan forests. Several water sources exist.

For the past two decades, there has been severe destruction and over-utilization of the land resources and in particular the national forests, catchment forests and clan forests, due to land scarcity.

It has been observed that, together with continuous changes in social, the economic, political and environmental conditions, farmers have adopted various cropping patterns, land use systems and land management practices. These practices do not actually meet their multiple socio-economic goals such as increased food production, availability of fuelwood, cash income etc. However, they have expanded the acreage of land as the population grows for agriculture and settlement through various practices such as forest clearing, river bank and water source encroachment, and cultivation of marginal lands. Certainly some of these practices provide short-term increased productivity although most of them have accelerated the degradation of the land resource base as illustrated in Figure 1.

2 Location, physical environment and population

The Project is situated in the Mwanga District, Kilimanjaro Region and confined to the highlands with an altitude of 1000-2000 m a.s.l. The project area covers 420 km^2. The population is about 60,000 inhabitants and growing at an average of 3.7% per annum (National Census, 1988). The population density in the highlands is relatively high (120-200 people per km^2) compared to the lowlands (15-25 people per km^2.). The average farm size has declined from 2 ha. in 1950's (Kimambo, 1969) to less than 1 ha in 1994 (Socio Economic Survey 1994).

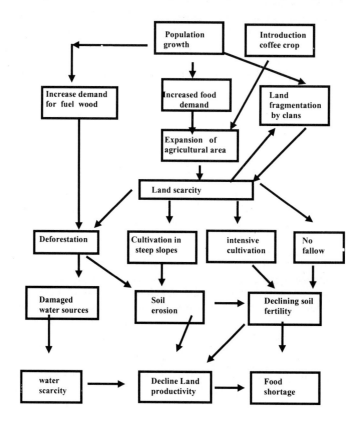

Figure 1: Land degradation in North Pare mountains slopes: Problem analysis

The mean annual rainfall ranges from 850-1200mm. There are two rainy seasons, the short rains (*'Vuli'*) that fall from November to December and the long rains (*'Masika'*) that fall from March to May. There is a dry spell in February.

The dominant soil types are Nitisols, Cambisols and Fluvisols (FAO/UNESCO, 1977). The last two occur at the pediments of the hill slopes and in the river valleys, respectively. The main rock types are granulates and granulitic gneisses (GST, 1962).

2.1 Past and present land tenure in North Pare Mountains

The North Pare mountains have a long history, even during the colonial era, of external efforts to reverse land degradation and improve productivity through soil/land management practices. This attempt did not take into consideration issues related to land use planning at village level despite the fact that the forest areas and water sources were given due respect. Large forests were gazetted as national forest reserves while small forests were under the control of different clans. The later was a sort of informal land use planning where by Chiefs and clan leaders protected the forests and water sources. Due to socio-economic changes and political instability the latter was given less priority. The land tenure system did not favour land use planning practices.

The existing land tenure structure and legal regime governing land in Tanzania have their origin in the colonial political economy. The colonial land regime established by the Germans (1885-1916) and the British (1918-1961) assumed that indigenous occupants had no ownership rights over land. This was a convenient assumption. The colonial state was not only a political sovereign installed by conquest, but was also interested in exploiting the resources of the colonized peoples. The property regime established by the colonial state was intended to facilitate such exploitation under close supervision. The central mechanism to achieve this was to place all lands in the hands of the state. Property and sovereignty merged in one entity. The vesting of radical title (ultimate ownership and control) to all lands in the state by the relevant legal instruments was the fundamental premise of the colonial land regime (Shivji, 1996)

The land ordinance passed by the British in 1923 is the principal piece of legislation which to this day governs land tenure system in the country (Shivji, 1996). Thus, both statutory and customary land tenure are applied in Tanzania, the North Pare mountains being no exception. In this case, customary land tenure considered customary holder where legislation was regulated by 'native law and custom'. Customary regime had no legal security of tenure. The statutory land tenure is considered under the state and the rights of the statutory holder were created, defined and protected by the statutory and contracted law. The holders of granted rights, largely foreign immigrants and companies, had well-defined and legally entrenched security of tenure (Shivji, 1996).

At present, land tenure and land policy reform states that all land in the country is under the state. At village level, the village council has overall mandate over land. The problem is that the boundaries of the villages are not clearly demarcated. The same applies to individual land holders land is marked by planted trees or shrubs. This has resulted in land conflicts and loss of security. Further-more, there is no legal right of occupancy or title deeds issued to individuals despite the fact that there may be verbal agreement by the former chiefs and the villagers. The control over land was passed over to village councils immediately after independence.

The land use planning at household level is arranged in such a way that within a piece of land, the upper-most portion is set aside for settlement, the middle part is for mixed cropping (coffee, banana, shade trees) while the lower part is for growing of maize, beans, cassava, sweet potatoes and sugar canes. Other land further away, but within the village, was given to individual families for large scale cultivation and tree planting (woodlots).

Due to population pressure, no land was left unoccupied in the North Pare mountains. For this reason there is no land available for communal cultivation including tree planting. Nevertheless, the land in general has undergone fragmentation as clan members increase as a result of high demand for cultivation area, and settlement. Continuous cultivation without fallowing has resulted in declining soil fertility and consequently low crop production which cannot support the fast-growing population (growth rate of 3.7% per annum, Bureau of Statistics, 1988).

Basing on these grounds, and due to the fact that land is a limited resource and that farmers have to optimize production of both food and cash crops from their small plots of land, the problem has to be handled properly. This required proper land use planning and land husbandry aimed at sustainable use and management of natural resources.

At present the revised land tenure and land policy reforms in Tanzania (Ministry of Lands 1995), provides a right of occupancy to a villager. The owner of the land, whether an individual, a family or a clan head as the case may be under the local customary law, would be issued with simple certificate called *'Hati ya Ardhi ya Mila'* (HAM). The boundary of the customary owner's land is demarcated in the presence of members of the Village Council appointed for that purpose by the village assembly, and entered on the certificate and countersigned by neighbors who share boundaries with the owner.

Any dispute will be resolved by the Elder Land Council. Owners of HAM would be required under Village Assembly rules to identify their boundaries by growing prescribed trees which will be *prima facie* evidence of the boundary. Unlike in the past where women had no right over the land, HAM will carry the name of the owner, and it will be mandatory to include the name of the spouse or spouses on the certificate. In this way the Lands Commission hope to make an inroad into the problem of gender inequality which is often observed in customary land regime. This further guarantees and gives confidence over the right of occupancy thus helping to the land managed properly (Ministry of Lands 1995).

Areas within the village which need to be conserved (water catchments, sacred sites and village forests etc.) would be under the direct control and supervision of the Village Assemblies.

Due to high population pressure, land scarcity, declining soil fertility and the effect of the past land tenure system, the TFAP North Pare project had to adopt a participatory land use planning (PLUP) approach towards sustainable management and utilization of the natural resources.

Participatory means the involvement of the target group (villager) in the whole process of land use planning of their natural resources from the assessment of the potential of the natural resources through a Participatory Rural Appraisal (PRA) exercise where problems are identified, followed by planning, implementation of proposed solutions and evaluation. In this approach, the farmer is viewed as a 'partner' and not as a client in the whole process of land use planning (Douglas, 1989).

3 Participatory land use planning approach applied by TFAP North Pare project

This approach is applied at ward, divisional and village levels and the main partners are the village communities in the highland. The major role of the Project is to facilitate the establishment of a permanent dialogue between the village communities and the Government authorities concerned on the use and management of natural resources in the North Pare mountains. The participatory land use planning is implemented under three strategies, namely Village Based Participatory Land Use Planning (VBPLUP) around national forest reserves, Neighborhood Land Use Planning (NLUPA) and Agroforestry, and Afforestation and Forest Management.

3.1 Village based participatory land use planning

The VBPLUP strategy is currently being implemented in 7 villages situated in the near vicinity of national forest reserves. The idea behind this strategy is the 'bufferzone' principle: reducing the pressure on these forests by supporting sustainable land-use practices by the villagers bordering the forests.

As a first step, together with the villagers and collaborating staff from relevant District departments, a PRA is being conducted in selected villages.

The project assists the village community concerned in the establishment of a Village Land Use Planning Committee (VLUPC). These committees initiate public debates at village level which focus on the villager's perception of problems or constraints related to natural resources in the village area and potential solutions to overcome them. This important process eventually results in a 'proposal for action'.

For every committee a three dimensional model is constructed, depicting the village environment including all physical features. Furthermore, the project in cooperation with the villagers has produced a video film recorded in the North Pare mountains, which explains the way villagers may 'plan the use of their land'. Both the model and video film have proved to be effective tools in the discussions ('public debates') at village level.

3.1.1 The roles of the working tools (village model and video film)

The village land use model is a tool prepared by an artist with the guidance of the villagers. It is a true presentation of the respective village showing the present environmental situation. Thus, it serves as a useful tool for planning since it familiarizes the whole village community on the actual situation existing in their environment.

The video film is an important tool for awareness creation to the villagers. This was shot in the North Pare mountains. It brings about a change of attitude and perception among the local community on issues related to environmental degradation and helps them to plan the sustainable use of their land.

3.1.2 Cooperating agents for implementation of the approach

The role of the VLUPC is to establish close cooperation with village government authorities as an in-charge and legal supervisor of activities at village level. It has the role of conducting public debates with other villagers and interested groups by using the model in order to identify problems in land use and come up with possible solutions. With the problems at hand the VLUPC facilitates ranking and comes up with specific activities to be carried out in one year (work plan). The VLUPC is charged to make a close follow - up of implementation of those activities and also to conduct mid-year evaluation of the work plan, and finally has to advise the village government on issues related to land use planning.

The main agents during the PLUP process are *the villagers* who are in fact the implementors as well as beneficiaries. Therefore, in order to facilitate the process, the VLUPCs are supposed not only to incorporate and link the agents but also to ensure a good relationship among the members.

On the other hand, *facilitators* have been identified at the Divisional level. Their role is through VLUPC to facilitate the participation of the target group in participatory land use planning.

The role of the project is not to guide but facilitate the process and to promote linkages with different agents from District to village level.

The role of the village government, is to be in charge of all activities in the village concerning land use planning and to approve the work plan together with the VLUPC supervise implementation. It is also authorized to link and cooperate technical agents at village, ward and divisional levels. All meetings conducted in the village are organized by the village government.

3.1.3 The implementation of the work plans

As mentioned earlier, different agents are involved in the whole process of PLUP at village level. There has been some deficiencies and weaknesses during the implementation of the work plans. These weaknesses may be categorized as follows:

The work plans have to be approved by the respective village governments before being implemented. In some cases, village governments have given inadequate support to the VLUPC for smooth implementation of the activities.

Another weakness was at the level of the VLUPC when it fails to effectively facilitate contacts between the villagers and higher authorities at Ward, Divisional and District levels, hence causing unnecessary delays on the implementation of activities.

The linkage between the TFAP North Pare project and the District partners is not strong enough to provide the required support for implementation. The reason for this is that the TFAP project lags behind the District instead of the District considering itself as a 'partner' in implementation. For example when the District does not inform and involve the ward and divisional authorities on the activities to be done for the work plan, then frustrations arise in the target group and consider land use planning as something outside their daily village development program. However, the degree of the problems differ from one village to another.

3.2 Neighborhood land use planning and agroforestry

Farmers in villages that are far away from the national forest reserves are assisted by the project to implement soil and water conservation measures, agroforestry and tree planting activities. Farmers in such areas form groups in which members help each other (a practice called '*kikwa*' in the Kipare language) in digging bank over ditches (*Fanya juu*) and bench terraces. These groups are mostly water user groups which are well organized and work according to a time table.

Up to June 1996, a total of 23 catchments have been selected with over 700 farmers participating in soil and water conservation and agroforestry. About 50 ha have been constructed with *'fanya juu'* , 53 ha. have been put under bench terraces while 15 ha. have been established with contour hedges. The first two structures are stabilized with fodder grasses, shrubs and trees.

3.3 Afforestation and forest management

The Project is assisting farmers to plant trees as part of VBPLUP. Formerly the project was assisting the District Natural Resources Office (DNRO) to run 7 central nurseries whereby seedlings produced were distributed free of charge to interested villagers. The project revised its policy of producing and distributing seedlings. Now, under the umbrella of the DNRO, the project is assisting individuals, schools, and groups to produce seedlings on their own for woodlots, and on farm planting. If there is any surplus it is sold at a price set by themselves to interested parties.

Interested groups or schools are contracted by the project to produce seedlings for village afforestation. For the period 1995/96, a total of approximately 200,000 seedlings were produced.

Afforestation sites have been established by certain villages around national forest reserves. They are 5 in number and in each afforestation site, plots of 0.25 ha have been allocated to individuals. The individuals sign an agreement with the Village government that the land shall remain the property of the Village government, while planted trees shall be tended and utilized by the plot owners as shall be agreed between owners and the Divisional forester. Individuals are not interested in planting trees on a communal basis for reasons related to land tenure (as mentioned in section 2.1 above).

Regarding forest management, the project is involving the local community in the formulation of simple joint management plans for village forests and sacred clan forests through discussions and resources assessment. Currently, 4 management plans for village forests have been prepared with villager participation. These simple joint management plans are based on ideas from the villagers. In the case of village forests, protection measures are proposed while penalties and fines are given based on the utilization of the forest products. For example, products may be collected by the villagers freely, while forest products which require permission from the village authorities would be prohibited. The fine rates are set by themselves. For sacred clan forests, traditional protection measures and fines are applied. For both categories, approval is given by the village government, ward and divisional secretaries, and at the District level by the District Executive Director.

4 Conclusions

- The introduction of village based participatory land use planning has made it possible to the villagers to practice proper land husbandry on their farms. Improved production at farm level is now realised as a result of practicing soil and water conservation measures coupled with application of required inputs (organic and inorganic fertilizers) which improves soil structure as well as soil fertility.
- Water run-off and soil erosion are expected to be reduced considerably, thus increasing infiltration and moisture content in the soil resulting in a better crop performance.
- Although the project has been operational for 4 years, the degree of change of attitude by the villagers towards adopting participatory land use planning is considered satisfactory. It has given farmers the ability to more easily practice tree planting and soil and water conservation measures through group formation.
- The current land tenure reforms in Tanzania will encourage the local community to the use of the natural resources in their area.

List of acronyms:
TFAP	Tanzania Forestry Action Plan
VLUPC	Village Land Use Planning Committee
VBPLUP	Village Based Participatory Land Use Planning
NLUPA	Neighborhood Land Use Planning and Agroforestry
PLUP	Participatory Land Use Planning
PRA	Participatory Rural Appraisal
DNRO	District Natural Resources Office

References

Bureau of Statistics (1988): Population census. Preliminary Report. Bureau of Statistics, Dar-es-salaam.

Douglas, M.G. (1989): Integrating conservation into farming system: Land use planning for small holder farmers. Concepts and Procedures. Commonwealth secretariat.

FAO/UNESCO (1977): Soil map of the World, UNESCO, Paris.

Geological Survey of Tanganyika (GST) (1962): Quarter Degree Sheet No.73. North Pare mountains. Geological Survey of Tanganyika, Dodoma, Tanzania.

Kimambo, I.N. (1969): A Political History of the Pare of Tanzania 1500-1900. East African Publishing House, Nairobi.

Ministry of Lands (1995): National Land Policy. The United Republic of Tanzania, Ministry of Lands, Housing and Urban Development, Dar-es-salaam, Tanzania.

Shivji, I.G. (1996): Land tenure problems and reforms. A paper presented for Sub- regional workshop for East Africa on land tenure issues in Natural resources management, March 11-15, 1996. Addis Ababa, Ethiopia. University of Dar-es-salaam, Tanzania.

TFAP (1994): Social Economic Survey in Usangi and Ugweno Divisions, North Pare Mountains, Mwanga.

Address of author:
K.C.H. Mndeme
TFAP North Pare Project
P.O. Box 195
Mwanga, Kilimanjaro, Tanzania

Integrating Agroforestry Technologies as a Natural Resource Management Tool for Smallholder Farmers

S. Minae, W.T. Bunderson, G. Kanyama-Phiri & A.-M. Izac

Summary

Africa faces an enormous challenge to increase agricultural productivity under a sustainable natural resource base to address its chronic food shortages. An action research and development platform that aims at increased soil fertility under improved natural resource management, through use of agroforestry technologies, is being tested in Malawi with promising results. A community-based approach was used to identify factors which impact on natural resource management at the household and community levels. This information was used to facilitate a participatory on-farm trial to evaluate the potential of agroforestry technologies to enhance soil fertility and increase crop productivity.

The first phase focused on characterisation of the land use systems at the farm and micro-watershed levels using a battery of participatory methods including; resource mapping, linear transect walks, village group discussions and individual farmer interviews. The second phase consisted of an extensive farmer participatory on-farm trial, to evaluate performance of a 'basket' of agroforestry technologies aimed at enhancing soil fertility. Technologies included were; improved fallows, relay cropping and strip cropping, of multipurpose tree species and grain legume intercrops. Group discussions, individual farmer interviews, field visits and tours were used to elicit farmers assessment criteria.

Farmers gave first priority to soil fertility improvement at the household level, while natural resource management concerns requiring communal action ranked second. Farmers assessment revealed that, crop diversification and risk aversion were as important as increased crop yield in the evaluation criteria. Though improved fallows showed promise and have the highest potential to enhance soil fertility, intercropping seemed best for the land-constrained farmers in Malawi, in terms of the attributes they are currently are seeking in natural resource management.

Keywords: Agroforestry, micro-watersheds, participatory methods, farmer evaluation, natural resource management

1 Introduction

Low soil fertility is a major limiting factor to smallholder agricultural production in Malawi. High input prices and poor access to credit from recent fiscal structural adjustments have exacerbated the situation, resulting in an economic environment where inorganic fertilisers are out of reach for resource poor farmers (Conroy and Kumwenda, 1995). Unprecedented pressure on land has resulted in diminishing and fragmented uneconomic farm holdings, estimated to average 0.4, 0.7 and 1 hectare in the Southern, Central and Northern Regions, respectively. There is an urgent need to enhance soil fertility through locally available organic matter (Dallard et al, 1993; Giller et al, 1996). Agroforestry can contribute to sustainable resource management through nutrient cycling, organic nitrogen build up, and soil and water conservation, while providing wood products to reduce deforestation (Izac, 1995).

2 Rationale and objectives

The overall goal of the project was to improve crop productivity under sustainable natural resource management. The main focus of the study was to identify agroforestry technologies that could be targeted for soil fertility improvement of the resource poor farmers in Malawi. The research approach used in this ongoing study emphasised participatory methods to ensure integration of indigenous knowledge and local experiences, in planning and implementation of agroforestry technologies development. In order to facilitate adoption of potential agroforestry technologies which address soil fertility concerns of different household categories, farmers were presented with a 'menu' of options to select from and determine which technologies to test through participatory on-farm trials. Natural resource management issues were also incorporated by implementing field activities through a community-based approach that focused on micro-watersheds and households as basic implementation units (Minae, 1996). Careful selection of the micro-watersheds used in the trial, ensured that this approach provided an opportunity to extrapolate these results to other areas in Malawi through the use of GIS (Geographical Information Systems) analysis. The research project also piggy-backed on existing agroforestry extension projects with a community-based approach to share experiences and resources. Participatory methods were used to avoid the common pitfalls of traditional top-down, supply-driven approaches (Chambers, 1994). Specific objectives of the study were to:
- Evaluate the performance of different agroforestry technologies in selected micro-watersheds to identify key bio-physical and socio-economic characteristics that influence their effectiveness to improve soil fertility.
- Identify potential adopter categories for different agroforesty technologies by evaluating how farm characteristics and farmer attributes such as farm size, available household labour, gender of the household head and household income correlate to the performance of the different technologies used in the trial.
- Integrate natural resource management in agroforestry technology development through a community-based approach.
- Facilitate adoption of agroforestry technologies by setting up a farmer participatory research process which incorporated use of indigenous knowledge in design of field protocols.

3 Methodology

The study was conducted in Lilongwe Plateau of Central Region, Malawi, which is characterised by an average precipitation between 800 to1000 mm and altitude between 600 and 1200 m above sea level. Average farm size is about 0.7 hectare and the population density is 112 persons per km^2 (Government of Malawi, 1993). The land use system is maize-based, putting high pressure on soil nutrient requirements especially nitrogen. Seven micro-watersheds were systematically selected to typify the overall study area while allowing researchers to capture effects of localised features. Two micro-watersheds located in the Shire Highlands in the Southern Region, which has a slightly higher precipitation and population density were included in the characterization exercise but had to dropped from on-farm trials, for logistical reasons. Sites were contrasted in terms of bio-physical and socio-economic characteristics such as soil type, slope and access markets and roads hypothesised to have the potential to influence soil fertility management requirements at the micro-watershed level. All sites face land shortages with cultivation encroaching into marginal areas (Kamangira et al, 1995).

A micro-watershed was defined in the study 'as a land resource which can be delineated by a water catchment area that is utilised in common by a culturally bound community. It comprises crop land, woodland, grazing land, homesteads and low lying wetlands (*dambos*)'. In Malawi, agricultural land cultivated by smallholder farmers is communally owned and held in trust by a village headman, under customary tenure. A micro-watershed therefore consists of land under crop production covered by individual user rights, while the rest is held as a common pool resource and used for livestock grazing and provision of fuelwood, and other wood and non-wood products. Land area under selected micro-watersheds consisted of 1 to 4 villages and averaged about 750 ha, supporting about 680 households.

The first phase of the study focused on characterisation of the land use systems at the household and community levels using a battery of participatory methods. This baseline information was used to describe the farming systems and production constraints at the farm and micro-watershed by establishing the current levels of the natural resource base, such as mean farm size and distribution, available household labour. Farmers' strategies to manage these household resources and exploit common pool resources were also established. Community level characterisation was used to prioritize problems, map out micro-watershed resources and identify landscape features attributed to land degradation and which can influence land-use allocation. Individual farmer interviews were used to identify different farmer categories in each micro-watershed, such as small group discussions, resource mapping and transect walks were used to characterise the micro-watersheds.

The second phase consisted of an extensive participatory on-farm trial, to evaluate a 'basket' of potential agroforestry technologies aimed at enhancing soil fertility, as this had been identified as a priority problem in the planning phase. Field implementation was carried out on individual farm basis, on the assumption that it was much easier to address natural resource management on land that is under individual user rights and later introduce communal activities to address degradation of common pool resources, once farmers had appreciated the benefits from individual efforts (Minae, 1994). Target households constituted farmers with limited capital, land and labour. Individual farmers interested in participating in the trial were encouraged to volunteer themselves and to organise themselves into groups and select local leadership through committees. As an integral component of the design and development process, aimed at incorporation of local knowledge in the field protocols, these groups were used in planning, implementation and evaluation. This approach facilitated participation of large numbers of farmers (about 60 per micro-watershed) which ensured there was a reasonable sample size per site for data analysis purposes and improved cost efficiency in training, dissemination of technical information and establishment of group activities (Atta-Krah and Francis, 1987). A collaborative process involving farmers, development agencies (including non-governmental organisations) and researchers was set up. Training and demonstrations were organised to provide farmers with technical information on the potential benefits and investment costs of the different agroforestry technologies. Field materials, such as seed and other inputs necessary to implement the trial were identified with farmers participation and supplied by the research team through the extension system. Farmers were presented with a 'menu' of potential agroforestry technologies of which they were facilitated to assess and select four of the ones they considered to have highest potential to address their particular needs and interests (Lightfoot and Noble, 1993; Ashby, 1991; Raintree, 1993). As a community they were encouraged to discuss and determine their long and short term natural resource strategies and identify which activities would be carried out on individual basis and those that would be implemented by the whole group. They were also assisted in setting up local tree nurseries and establishing over 400 individual on-farm trial plots in the seven micro-watersheds. Demonstration plots were also set up at each site and used for training and evaluation.

Each farmer had 4 plots (10 m × 10 m) of unreplicated treatments of different organic matter practices selected from the 'menu'. The combination of treatments varied in each farm depending on the technologies each farmer selected, although they were encouraged to combine treatments based on traditional legume intercrops and agroforestry technologies to facilitate the evaluation process. All the labour, management decisions and adaptation of the various technologies depended solely on farmers. The 'basket' of options presented to farmers and included as treatments in the on-farm trial consisted of agroforestry technologies which have been demonstrated to improve soil fertility and hence increase crop production under on-station conditions, and are currently under on-farm review (Saka et al., 1994; Kwesiga and Coe, 1994; Bunderson et al., 1994, Bunderson et al., 1995). They consisted of maize-based systems of:

- Mixed (relay) cropping using leguminous multi-purpose species such as *Sesbania sesban, Tephrosia vogelii* and *Cajanus cajan* (pigeon pea).
- One to two year improved fallows using the same species. (Maize [*Zea mays*] was planted after the fallow period).
- Strip cropping (one row of *Sesbania sesban,* or *Tephrosia vogelii* or *Cajanus cajan* planted between two rows of maize)
- Mixed intercropping of *Gliricidia sepium*.
- Traditional intercrops of grain legumes like cowpeas (*Vigna unguiculata*), groundnuts (*Arachis hypogaeana*), soybeans (*Glycine max*), common beans (*Phaseolus vugalis*), and velvet beans (*Mucuna ssp*).

Regular farm visits were carried out to monitor growth performance, farmer attitudes, management levels and labour utilisation in for various farm operations. A participatory on farm evaluation programme was set up by the research/extension staff and farmers, consisting of individual farm visits, group discussions and community based workshops accompanied by field tours. Two group evaluation exercises were conducted during each growing season. Individual farmer interviews using a semi-structured questionnaire on a sample of randomly selected farmers from all the watersheds was conducted at the end of the growing season. The evaluation process consisted of comparison of the organic matter technologies using farmers' criteria. Information collected pertained to, farmer assessment of the potential of each technology to increase productivity per unit of land in terms of maize and legume grain yield. Leafy biomass from legume residue was also considered as an output in the form of organic matter and hence nitrogen. Assessment of labour requirements as well as overall management was also carried out. The sites are in their third year of the trial.

The agroforestry extension component of the project had a much wider scope of activities and area of coverage. They, in addition, introduced systematic integration of *Faidherbia albida*, hedgerow intercropping, rotational fallows and green manure banks of *Gliricidia sepium, Senna spectabilis* and *Tephrosia vogelii*. Hedges of *Tephrosia vogelii* and contour planting of vetiver grass (*Vetiveria zizanioides*) were integrated in sites prone to soil erosion. Physical soil conservation measures to align ridges along the contour and to establish marker ridges were also carried out.

4 Initial Findings

Individual and group participatory methods complemented each other in terms of the overall information collated and provided the opportunity for farmers, researchers and extension workers to confer on the description of land use systems, diagnosis of production constraints and identification of potential agroforestry technologies in a manner that cemented partnership (Minae, 1994). Farmers involvement in the project, from the planning stage to implementation and evaluation, gave them a high sense of ownership. The battery of methods, while confirming and cross-checking information from different sources, for example, between village group discussions and key informants, also provided additional information that would normally not be elicited through traditional formal individual farmer interviews. For instance, mini group discussions with women ensured that gender balance in problem analysis was integrated while resource mapping and line transect walks gave an overall picture of distribution of land resources and current land use patterns within a village. Line transect walks demonstrated farmers' understanding of the landscape attributes and implications in terms of resource management requirements. In general, group methods were used to establish community characteristics and concerns while individual farmer interviews identified variations between farmer categories. The fairly intensive interaction between farmers and project staff through others group meetings, field visits, cross-farm visits and on-farm trials meant that farmers were highly sensitized to natural resource management concerns.

Land limitation was identified as a major constraint in all the micro-watersheds. Farm sizes varied within micro-watersheds, ranging from 0.2 ha to 4 ha per household. Chilobwe which is located in a high altitude area and is near the main road, had predominantly smaller unit holdings which were intensively intercropped while Chikwete had larger farm sizes and practised less intercropping (Table 1). In general, farmers with smaller farm sizes had more interest in organic matter practices such as incorporation of crop residues and intercropping. Cropping on erodable slopes (more than 15%), was more prevalent in micro-watersheds located in high population density areas. Maize as the main food crop, was grown extensively in all micro-watersheds but varied in the cropping intensity and intercropping combinations of minor crops such as grain legumes and root crops, reflecting micro-climatic differences and cultural preferences. It is not possible to give details of the characteristics of each watershed in this paper, suffice to note that variations (such as farm size, availability of household labour, gender of the household head, livestock ownership) within micro-watersheds, were reflected in the prevailing natural resource management practices of the household, in terms of use of fertiliser and manure, intensity of intercropping and crop rotation and incorporation of crop residue (Kanyama-Phiri et al., 1995).

Declining soil fertility was prevalent in all micro-watershed as pointed out by farmers during group discussions. Frustrations and limitations of inorganic fertilisers were also expressed. It is interesting to note that soil fertility was mainly described in terms of crop yield though expressions such as 'soil has become *hard*', 'it has lost *natural fertility*', ' the fertility has been *diluted*', to reflect the current decline in soil fertility. Local vegetation associated with low soil fertility, such as love grass, were also mentioned. Though farmers recognised soil erosion as a serious problem, technologies which enhance soil fertility got higher priority ranking. In general, they showed a preference to implement on-farm trials on individual basis as opposed to communal action. The only group activity implemented was in the establishment tree nurseries.

Examination of the initial selection of the agroforestry technologies used in the on-farm trial, showed farmers had more interest in species that could provide food as well as improve soil fertility. For example, out of the 420 farmers involved in this trial, 44% choose *Cajanus cajan* as their first choice, as opposed to 25% in *S. sesban* and 20% in *T. vogelii*. Though *Sesbania* and *Tephrosia* are known locally, they are not traditionally used in soil fertility improvement, which may have influenced farmer reluctance to test these species. There were minor but insignificant differences between the micro-watersheds, in farmer selection of the species and spatial arrangements used in the trial. The most popular spatial arrangement was intercropping. Only a few farmers chose improved fallow, possibly as a reflection of land pressure. Mode of propagation was an important consideration in species selection. Species such as *S. sesban,* that require propagation through seedlings were less popular than those which germinate easily by direct seeding.

Since soil fertility status was low in most farms, increments in maize yield were small in this early phase of the trial. Results from individual farmer assessment indicated that there were no discernible differences in the performance of agroforestry technologies. Mean maize output in year two of the trial was about 1.8 tons per hectare. Grain yield from different legumes showed similar trends with a mean output of one half to 1 ton (Fig. 1). Figure 2 shows the estimated mean total nitrogen input from leafy biomass (crop residue). The best output was obtained from *S. Sesban.*

	Chilobwe	Mpingu	Mganja	Chikwete	Nachisaka	Gowa	Nathenje
Total area (ha)	450	515	445	1100	1240	1450	725
Project area (ha)	40	60	85	400	207	95	24
Total Households (No).	932	286	473	450	580	1988	955
Participating HH	62	62	35	173	52	78	60
Mean farm size (ha)	0.5	1.8	0.9	2.6	2.1	1.2	1.1
Slope / area (%)							
<5%	16	72	67	34	7	19	72
6-15%	49	27	31	42	75	22	20
>16%	35	1	2	24	18	59	8
% do conservation	7	38	38	12	18	5	21
% have intercrops	100	42	76	19	29	53	43
% incorporate residue	70	34	51	0	0	11	18
Main crops	Maize Beans, vegetables	Maize Tobacco	Maize G'nuts, Cow peas	Maize Tobacco	Maize G'nuts, Tobacco	Maize G'nuts, Tobacco	Maize G'nuts
Minor Crops	Fruits, Irish Potatoes	G'nuts, Pumpkins, Sweet potatoes	Sorghum, Soy beans, Cotton	G'nuts, Soy beans, Sweet potatoes	Cassava, Beans	Beans, Cow peas, Soy beans	Tobacco, beans, Cassava
Main soils	Sandy-shallow (partially waterlogged)	Sandy-loam	Clay-loam	Sandy-clay loam	Sandy loam	Sandy-loam	Sandy-loam

Table 1. Farming Systems Descriptors of the Micro-watersheds

NB: HH -- household

Agroforestry technologies for natural resource management

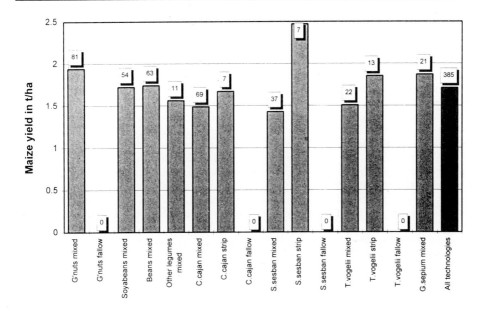

Figure 1: Estimated mean maize yield comparing different organic matter technologies, for all micro-watersheds combined, Malawi 1996. Boxes indicate the number of plot samples used in the computation of the mean.

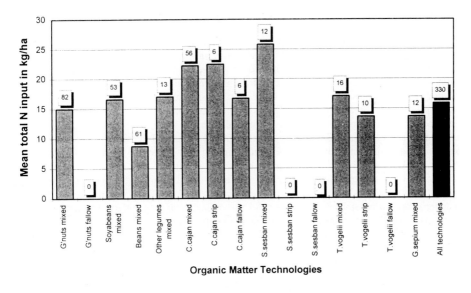

Figure 2: Estimated mean total N input from legume residues (leafy biomass) comparing different organic matter technologies, for all micro-watershed combined, Malawi, 1996

There was a notable variation in farmer evaluation criteria of different technologies. Highlights from group discussions indicated that, the criteria used to assess the pros and cons of improving soil fertility through agroforestry technologies depended on farmers priorities, objectives and farm resources. The weight given to costs and benefits, seemed to be correlated to whether a farmer's production objective was to diversify crop production, as opposed to increased maize production. For instance, though the main focus of the trial was to improve soil fertility, crop diversification was used as the main criteria to assess performance by farmers whose priority was risk aversion. As farmers indicated, 'at least one is able to harvest something even if the maize crop fails'. Intercropping was considered as a means of coping with land pressure. Intercrops, especially of grain legumes were ranked highly from this perspective. Accordingly, farmers pointed out the benefit of harvesting two crops from the same plot. For these farmers, it was acceptable to have lower maize yields and a slower soil organic matter build up, in comparison to additional benefits gained from crop diversification and risk aversion. Intercropping was also seen as a labour saving mechanism in terms of high returns per unit of land as a result of improved land productivity. Farmers who preferred improved fallow pointed out the benefits of potentially higher biomass production and hence speeded up organic matter build up, leading to increased yields. Legume-maize interactions were observed from intercropped plots. The general consensus from group discussions was that pigeon pea caused the minimum negative interaction while the most serious was observed from soybeans/maize intercrops. Late planting of either maize or soybean caused noticeable reduction in the yield of the other crop. Shading of *S. sesban* and to a lesser extent *T. vogelli* in the early stages of growth were quite noticeable especially if the agroforestry species were planted late relative to maize. The choice of whether to go for soil fertility improvement through agroforestry technologies such as relay cropping (intercropping) vs improved fallows (pure plots) depended on whether a farmer could afford to put land aside to fallow.

Though there was a wide range of alternative organic matter technologies, their impact on improved soil fertility and hence crop yield were low in the short run. There were conflicts of interests between household short term requirements (immediate food security) and long term benefits of improved natural resource management. Technologies which had relatively high potential to improve soil fertility such as improved fallows and green manures (Sanchez, 1995), also had high investment cost in term of land and labour. Intercropping systems of grain legumes, though popular with farmers had a limited potential to increase soil fertility since they have a lower capacity to fix nitrogen and/or recycle nitrogen. A significant proportion of nitrogen was removed with the grain (Giller et al., 1996). The amount of nitrogen that can be expected from intercrops ranged from 10 - 40 kg per hectare. Much higher nitrogen levels are likely to occur only if moisture and Phosphorus were sufficient to create substantial legume growth. Undersown improved fallows (the agroforestry species is intercropped in the year of fallow establishment and then left to develop in the second year) was considered as a possible option for land-limited farmers, because it seems to provide a technical balance in production constraints, (between intercropping and improved fallows) by reducing the length of the fallow period while increasing leaf biomass production.

Farmers did not perceive any problem with the additional labour requirements for establishing agroforestry technologies. However the picture may change when they expand and increase the area under agroforestry, since many farmers did also indicate that they preferred to plant trees after banking maize because it reduced labour competition with other cropping operations. The implications from this inconsistent response on labour requirements points to the need to pay more attention to this issue especially in relation to gender balance in adoption analysis. Female headed houses tend to have poor access to labour, and this may act as a major constraint to adoption.

Field observations and group discussions evidenced an increase in the awareness of resource management concerns in the micro-watersheds. For example, soil incorporation of crop residue increased even in sites where they are traditionally burnt. The combination of extension and research efforts complemented each other. For instance, as a result of the trial, there was a high demand for pigeon pea seed which was communicated to extension workers, who are now doing promotions in areas where farmers do not grow this legume. Establishment of vetiver grass was dramatic with visible barriers to soil erosion as early as the second year, leading to demand for planting material. Assessment of the state of the art of various agroforestry technologies and their potential for adoption suggested a broad range of factors that influenced decision making. Farm-level concerns hinged around tree tenure. There were reported cases of accidental destruction of agroforestry plots from uncontrolled livestock herding and field fires in the dry season. It was also observed that farmers sometimes failed to carry out timely field activities while awaiting instructions from the extension field assistants. This lack of initiative was partly a product of a top-down extension approach which has been in place for many years. The participatory methods promoted in this study should alleviate this problem. Matching of agroforestry interventions to priority problems rather than a blanket recommendation was preferred as indicated in group discussions. Farmers also wanted to see illustrative results from neighboring farms before trying new technologies. Unfortunately benefits from agroforestry take longer than conventional technologies to show results. Agroforestry needs a well established technical support in the following areas;

- production and dissemination of effective, user-friendly training materials to transfer skills and simple tools to end-users;
- support services to communities to facilitate provision of initial seed and other inputs, training and on-farm demonstrations;
- participatory farmer experimentation and extension to identify appropriate agroforestry systems and species to suit different agroecologies ;
- merit promotion of agroforestry technologies rather than using non-sustainable incentives like issuing free inputs.

5 Concluding remarks

The results reported in this paper are preliminary as they are based on field observation from the early phase of the on-farm trial which is still ongoing. This is a long-term trial which requires at least 5 years before conclusive observations on performance of the different agroforestry technologies can be made. The emphasis of the paper is therefore on the effectiveness and utility of the methodological approach. Conducting characterisation at the household and community levels ensured integration of concerns and constraints of improved natural resource management in the development process. Incorporation of indigenous knowledge systems in the agroforestry evaluation process was ensured by farmers participation in all the phases of the project. Farmer enthusiasm in the trial has been overwhelming because as they indicated 'the inputs required in the trial were affordable and are easily available locally'. Benefits accrued to farmers included: increased crop productivity, improved soil organic matter, reduced soil erosion and provision of fuelwood.

References

Ashby, J.A. (1991): Adopters and Adapters: The Participation of Farmers in On-farm Research. In: R. Trip (ed.), Planned Change in Farming Systems: Progress in On-farm Research, Chichester UK, John Wiley and Sons Ltd.

Atta-Krah, A.N. and Francis, P.A. (1987): The Role of On-farm Trials in the Evaluation of Composite Technologies: The Case of Alley Farming in Southern Nigeria, Agrofrestry Systems **23**: 169-183

Bunderson, W.T., Saka, A.R., Itimu, O.A., Phombeya, H.S.K. and Ntupanyama, Y. (1994): Comparative Maize Yields under Alley Cropping with Different Hedge Species in the Lilongwe Plains, Malawi. In Proceedings of the Second National Agroforestry Symposium, Bvumbwe, Malawi.

Bunderson, W.T., Bodnar, F., Bromley, W.A. and Nanthambwe, S.J. (1995): A Field Manual for Agroforestry Practices in Malawi, MAFE Publications; Series No. 6, Lilongwe, Malawi.

Conroy, A. C. and Kumwenda, J.D.T. (1995): Risks Associated with the Adoption of Hybrid Seed and Fertiliser by Smallholder Farmers in Malawi. In: Jewell D., S.R. Waddington, J.K. Ransom and K.V. Pixley (eds), Maize Research for Stress Environments, CIMMYT, Harare, Zimbabwe.

Chambers, R. (1994): The Origins and Practices of Participatory Rural Appraisal. Institute of Development Studies, University of Sussex, Sussex.

Dalland A., Vaje, P.I., Mathews, R.B. and Singh, B.R. (1993): The Potential of Alley Cropping in Improvement of Cultivation Systems in High Rainfall Areas of Zambia. Effects on Soil Chemical and Physical Properties. Agroforestry Systems **21**: 117-132

Giller, K.E., Itimu, O. and Masambu, M. (1996): Nutrient Sourcing by Agroforestry Species and Soil Fertility Maintenance in Maize Associations. In: S.R. Waddington (ed.), Soil Fertility Network for Maize-based Farming Systems in 1994 and 1995, Harare.

Government of Malawi (1993): Poverty Situation Analysis, Government Printer, Zomba.

Izac, A.M.N and Swift, M.J. (1994): On Agricultural Sustainability and its Measurements in Small Scale Sub-Saharan Africa, Ecological Economics **11**.

Kamangira, J.B. (1996): Assessment of Soil Fertility Status for Agroforestry Interventions Using Conventional and Participatory Methods. Unpublished Thesis, University of Malawi, Malawi.

Kamangira, J., Kanyama-Phiri, G. and Wellard, K. (1995): Assessing Soil Characteristics Using Participatory Methods: A case Study of Zomba, Malawi. Paper Presented at the Second International Crop Science Conference for Eastern and Southern Africa, Blantyre, Malawi.

Kanyama-Phiri G.Y, Wellard, K., Minae, S. and Kamangira, J.B. (1995): Preliminary Findings on the Potential Adoption of Agroforestry Technologies by Small-holder farmers in Zomba, Malawi. Paper Presented at Phase II of Farmer Participatory Research Planning Workshop, Zomba, Malawi.

Kwesiga, F. and Coe, R. (1994): The Effect of Short Rotation Sesbania sesban Planted Fallows on Maize Yield, Forest Ecology and Management **64**, 199-208.

Lightfoot, C. and Tuan, N.A. (1990): Drawing Pictures of Integrated Arms Helps Everyone: An example from Vietnam, Aquabyte **32**, 5-6.

Lightfoot, C. and Noble, R. (1993): A Participatory Experiment in Sustainable Agriculture, Journal for Farming Systems Research-Extension, Vol. 4, No.1.

Maghembe, J.A., Chirwa, P.A. and Kooi, G. (1997): Relay Cropping of Sesbania sesban with Maize in Southern Malawi. Paper in Proceedings of the International Symposium on The Science and Practice of Short-term Improved Fallows. Lilongwe, Malawi.

Mascarenhas, J. (1991): Participatory Rural Appraisal and Participatory Learning Methods: Recent Experiences from Mypada and South India, Forest Trees and People Newsletter No. 15/16.

Minae, S. (1994): Role of Local Organisations in Technology Design and Testing in Agroforestry. In: Proceedings of International Symposium on Systems-Oriented Research in Agriculture and Rural Development, Montpellier, France.

Minae, S. (1996): Participatory Farmer Evaluation of Organic Matter Technologies to Enhance Land Productivity and Sustainability based on Micro-watershed Systems. In: S.R. Waddington (ed.), Soil Fertility in Selected Countries of Southern Africa, CIMMYT Research Results and Network Outputs, 1994 and 1995, 143-149, Harare, Zimbabwe.

Raintree, J.B. (1993): Farmer Participation in On-farm Agroforestry Research, Winrock International, F/FRED.

Saka, A.R., Bunderson, W.T., Itimu, O.A., Phombeya, H.S.K. and Mbekeani, Y. (1994): Effects of Acacia albida on Soils and Maize Grain Yields under Small-holder Farm Conditions in Malawi, Forest Ecology & Management **64**, 217-230.

Sanchez, P.A. (1995): Science in Agroforestry: Agroforestry Systems **30**: 5-15.

Addresses of authors:
S. Minae
A.-M. Izac
Intrenational Centre for Research in Agroforestry, ICRAF, Nairobi, Kenya
W.T. Bunderson
Malawi Agroforstry Extension Project, Lilongwe, Malawi
G. Kanyama-Phiri
Bunde College of Agriculture, University of Malawi, Lilongwe, Malawi

A Participatory Action Research Process to Improve Soil Fertility Management

T. Defoer, S. Kantè & Th. Hilhorst

Summary

The Farming Systems Research team (ESPGRN) of the Agricultural Research Institute (IER) in Mali is developing a participatory action-research approach to enable farmers together with researchers, to analyse and understand farmer strategies and practices of soil fertility management and to develop integrated technologies for sustainable soil fertility management. The approach encompasses four phases: diagnosis/analysis, planning, implementation and evaluation. First a village territory map is made to analyse the management of the natural resources of the village. Then, the diversity of soil fertility management practices between farms and their underlying causes are diagnosed. A classification of farms is made by the villagers themselves according to these key criteria for distinguishing levels of fertility management. Subsequently, the actual management practices are visualised through resources flow models, drawn by "test" farmers, representing each of the different farm classes. The combination of analysis and regular feed-back of farmer results, combined with the exposure to information on new technologies, motivates farmers to plan actions. New flow models (planning maps) are made for next year management practices. Evaluation of the planned activities is done by visualising the resource flows effectively implemented on the planning map. Planning and evaluation of activities is done on a yearly basis, which allow farmers and researchers to evaluate changes and progress in soil fertility management. There are clear indications that this approach has improved soil fertility management practices. Farmers have started recycling considerable amounts of crop residues as litter and fodder. Also contour farming is tested, fodder crops in association with cereals are installed and many new compost pits are actually made. Moreover, the resource flow models made by farmers allow for quantification without loss of participation. Sustainability parameters are selected, allowing for assessment of the evolution over-time and comparison between classes and with research/extension recommendations.

Keywords: Participatory action-research, resource flow modelling, soil fertility management, sustainability parameters

1 Introduction

The introduction of cotton and animal traction, combined with a substantial population growth, has changed the farming systems of southern Mali. In the oldest cotton producing area most of the arable land is now cultivated and farmers are now bringing shallow, erosion-sensitive upland under cultivation. Depletion of the organic matter and nutrient reserves of the soil has been reported due to insufficient fertiliser application, both organic and mineral (van der Pol, 1992; Pieri, 1989). Maintenance of soil fertility demands more intensive management strategies to guarantee sustainable agricultural production. At the same time, the farming systems and soil fertility management practices are becoming increasingly diverse, which presents a challenge for research and extension activities. Technologies proposed as recipes for the "average" farmer become less and less effective. Therefore, there is a need for **actor oriented** approaches, based on **interactive and collaborative learning**, **systems analysis** and **action research** (Scoones and Thompson, 1994; Campbell, 1996; Röling, 1996).

A participatory action research (PAR) process has been developed by the farming systems research team (*ESPGRN: Equipe Systèmes de Production et Gestion de Ressources Naturelles*) of the Malian agricultural research institute (*IER: Institut d'Economie Rurale*). The aim is to assist farmers in improving their soil fertility management practices (Defoer et al., 1995). The PAR heavily draws on visual techniques developed in the context of Participatory Rural Appraisal and resource flow modelling as described by Lightfoot et al. (1991). Also the quantification of these resource flows is a point of attention in order to measure the effect of improved management practices on the sustainability of the farming systems and to use the results in the collaborative PAR process with farmers.

2 The process

The participatory action research process to soil fertility management encompasses *four phases*: (1) diagnosis and analysis; (2) planning; (3) implementation and (4) evaluation. The diagnostic phase is unique, while the planning-implementation-evaluation cycle is done on a yearly basis.

2.1 Diagnosing and analyzing farmers' strategies

The analytical phase in the field is done by a multi-disciplinary team of researchers and extensionists, using Participatory Rural Appraisal (PRA) tools, consists of *four steps* and takes three days. Step one, two and four take place in village meetings, while the third step is implemented at the farm level.

Step one: Mapping natural resources of the village
The village resource map is made by a small group of experienced villagers of both sexes. Indigenous soil types, catena and land use, including agricultural land (of men and women), fallow, pastures and woodlands are demarcated. The relative importance of the different land uses and a short historical profile of land occupation is made. All farms are then placed on the map. Areas of land degradation are also visualized, followed by a discussion on the causes of the degradation. Finally, the map is analyzed to identify potentials and constraints of land use and to decide if communal actions for natural resource management should be planned.

Step two: Analyzing the diversity of soil fertility management among farms
Farmer criteria for identifying and explaining the diversity of fertility management practices are assessed separately by three groups of farmers: older men, women and younger men [1].

In each group, the issue of diversity among farms is discussed and each group prepares a list of criteria that reflect this diversity. The criteria are divided into two types: (1) indicators for good

---

[1] The three groups of farmers represent the different age-groups and sexes, dealing with agriculture: (i) older men are household heads with decision making power; they generally have a better knowledge on the history of land use and management that (ii) younger men, who are more involved in the day to day agricultural and livestock practices; (iii) given their responsibilities, women have a more detailed knowledge on management of wood, water and lowlands, used for cultivation of rice or vegetables, crops generally cultivated by women.

soil fertility management, such as organic fertilizer production, erosion control and litter use, and (2) socio-economic farm characteristics influencing these practices, such as the availability of labour, cattle and transport facilities. The lists of the three groups of farmers are brought together to make one common list of criteria. Then, each group prioritises these criteria. The outcomes are pooled together in a list of key criteria.

The specific values of the key criteria are obtained for each farm, through a rapid assessment done by a number of well informed farmers. First, the name of each household head is written on a separate card. The value of all key criteria is asked and then noted on the back of the card. Thereafter, another group of farmers, again selected by the village assembly is invited to classify all farming households according to their level of soil fertility management. The farmers first decide on the number of classes and their definition. At least three classes are generally considered: *Class I (good), Class II (average)* and *Class III (poor)*. The cards are now taken one by one. The names of the household heads are read and farmers commonly decide in which class each card is to be placed. For each card, the reason of the decision is asked and noted, while the card is placed in the appropriate stack, name upward. When a large number of cards accumulates in some of the piles, farmers may decide to further re-classify these piles.

From each of the classes, at least two farms are chosen for next day's farm-level discussions, preferably with clear differences in key criteria, specifically soil type (see their position on the village map). Selection of the 'test' farmers is done by the team in consultation with the farmers.

Step three: Visualizing farmer soil fertility management practices
Test farmers draw Resource Flow Models to analyze their soil fertility management practices. Since many farmers are not literate, symbols are preferred, of which a standard list has been developed with farmers. This enables them to easily understand the maps made by their colleagues, and to compare their own maps over time.

After a walk around the farm fields, the members of the household are invited to draw on a large sheet of paper the different farm components such as fields, grain and fodder stores, animal pens, compost heaps, etc. Types of soils, acreage, erosion spots and erosion control works are also marked. On each field, both present and preceding crops are noted. Afterwards, farmers draw arrows to represent resource flows between fields and other farm components. The utilization of last year's crop residues of each field is depicted and estimated, using pie diagrams. Crop residues can be used as fodder or litter, for grazing or burning. Then, organic and inorganic fertilizer application on the actual crops is visualized as well as other resource flows entering the farm. The arrows are labelled with the amount of material (and percentage in case of crop residues), given in local terms and units (cart loads, bundles, baskets, etc.) or in conventional units. Two examples of Resources Flows Models are presented in Figure 1.

This visualization of actual soil fertility management enables farmers, together with the team, to discuss the present situation and to identify improvements adapted to farmers' conditions and strategies. Most of the improvements relate to increased crop residue recycling, improved crop-livestock, and soil and water conservation. Generally, the proposed technologies, such as litter use in cattle pens, composting, contour ploughing, supplemental feeding of livestock, fodder production and storage, have already been tested by ESPGRN in on-farm trials.

Step four: Motivating other farmers through presentation of findings
The test farmers from the different fertility classes present their Resource Flow Models during a village meeting. Farmers are invited to exchange ideas on differences in soil fertility management and on possible improvements. Thereafter, the team gives some feedback on concepts and technical implications of the suggested improvements for each of the classes. This presentation aims at stimulating other farmers of the same classes to consider similar improvements, while taking into account their possibilities and limitations.

Class I (good) soil fertility management:

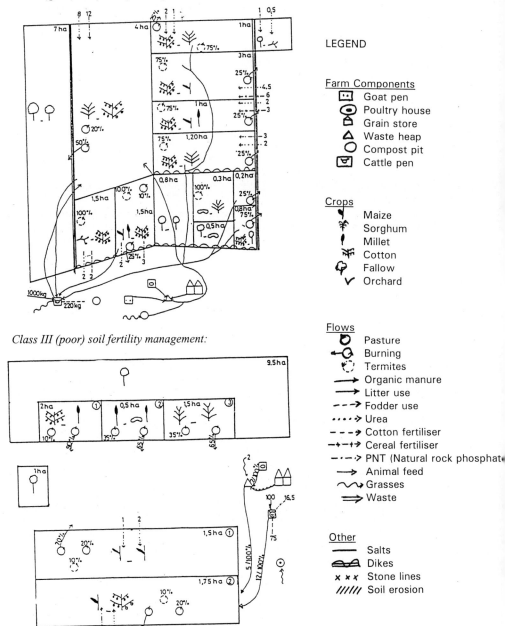

Class III (poor) soil fertility management:

Figure 1: Resource flow models of "Good" (Class I) and "Poor" (Class III) soil fertility management classes

2.2 Planning and implementing system adapted improvements

After the analytical phase, the test and other interested farmers, together with the researcher/extensionist, start planning activities.

First, farmer workshop and exchange visits are organized to expose farmers to new technologies and the experiences of other farmers. A visit is made to villages that have been using improved practices for several years. During the exchange, experienced farmers demonstrate new techniques or tools. Organizational matters for obtaining necessary inputs are also discussed.

The test farmers then visualize their plans for the next season by drawing a new resource flow model of their farm, a so-called "planning map". First, new farm components to be installed such as compost pits, fodder storages, cattle pens and new fields to be cleared, are visualized. Then, the planned activities for the next season are indicated using arrows. These include the proposed crop rotations, the use of last year's crop residues, fertilizer application, organic matter use and feed sources for animals. Proposed resource flows are quantified and marked on the arrows. Planned erosion control works are also noted.

Intended improvements are based on farmer production objectives and available resources. They are discussed in terms of types of flows and quantities recycled, leaving or entering the farm between farmers and researchers/extensionists. The plans made by farmers individually are presented to the other farmers during a village meeting, followed by a discussion on the technical implications.

2.3 Evaluation of the planned activities

The planned activities are evaluated a year after the initial analytical phase. Each test farmer evaluates together with the team, the implementation of their planned activities. The resource flows effectively implemented are visualized on the planning map in order to compare planning with execution and discuss reasons for discrepancies. Comparison with the initial Resource Flow Model is made to assess improvements of management practices, compared to the previous year.

After the individual evaluations with the test farmers, the findings are discussed with all other interested villagers, divided into the 3 classes of farmers (see classification). The discussions in separate classes tend to better highlight the specific socio-economic constraints in implementing improved soil fertility management. The evaluation exercise is concluded by a general village meeting in which farmers of the 3 classes report their findings.

3 Some results

The analysis of the diversity between farms is quick, participatory and relatively easy, while resulting in a reliable farm classification based on farmer knowledge.

Farmers are generally very positive on the utility of the Resource Flow Models as a tool for soil fertility management analysis and planning. Visualization enables them to analyze the intensity of residue recycling, the input and losses of resources and the differences between fields. It helps them to identify and to plan priorities, as well as keeping records of the changes. They also mentioned that planning and visualising activities motivated them to execute these. Self-evaluation of the planned activities allows for comparison of planning with execution.

Some of the farmers show the Resource Flow Models to other household members and use them to discuss farm activities among them. Farmers recognize the risk of having only one household member trained in the mapping technique. They feel that more members need to be trained, so that no one is indispensable. Moreover, this would stimulate discussion between the farm members. Further, more and more farmers are now joining the presentation meetings at the village level. The regular feedback during group and village level sessions allow for a comparison between farmers with similar resources and objectives, and stimulate them farmers to exchange experiences. Increasingly, farmers conduct the discussions and use this forum to exchange results and stimulate others farmers.

The tools also stimulate a dialogue between farmers and researchers/extensionists and allow for stepwise improvement of management practices. The proposed actions directly link to the analysis and problems diagnosed. Farmers are motivated to undertake actions through a combination of analysis, empowerment and exposure to information on new technologies. The comparison of the test farmer planning maps and the actual implementation shows that many improvements in soil fertility management are made. Farmers have started recycling considerable amounts of crop residues as litter and fodder and are burning less. Also contour farming is tested, fodder crops in association with cereals are installed and many new compost pits are actually made (Table 1).

CLASS OF FARMS (Level of soil fertility management)	Class I (Good)	Class II (Average)	Class III (Poor)
Composting (Nb of farmers who implemented/Nb of farmers)			
. near house	5/5	2/2	3/5
. near field	3/5	1/2	1/5
'new' fodder storage (Nb of farmers who implemented/Nb of farmers)	5/5	2/2	1/5
Fodder (average per class)			
. stored (carts) [1]	38	25	12
. chopped (carts)	1,5	1	0,8
Maize/dolichos (average per farm in Ha)	0,5	0,5	0,35
Contour ridges (Nb of farmers who implemented/Nb of farmers)	2/5	0/2	1/5
Farmers constructed a new type of fodder storage facility, using local materials and wire fencing, to protect the stalks from straying animals. The wire fence was provided on credit by the *Association Villageoise*. However, only class I and class II farmers constructed such storage facilities. Although some of the class III farmers also planned a new storage facility, time constraints and lack of creditworthiness did not enable them to construct the store or to obtain wire netting. The amounts of cereal stalks that were stored for fodder during the dry season range from 1,500 kg to more than 8,000 kg per household. They also experimented with enriching the cereal stacks using a chaff cutter and a mixture of urea and molasses. They also started using salt blocks.			
Due to livestock mortality during the dry season, however, it turned out that most of the farmers were only in need of small amounts of the stored fodder. Surplus cereal stalks has been used as cattle litter or was transferred to the compost pit. Some farmers used rock phosphate to improve the quality of the organic fertiliser. While all farmers have tried maize/dolichos association, the area planted is about double for class I and class II farmers as compared to class III farmers.			
Only three farmers thus far have experimented with contour farming in the framework of a test executed in collaboration with ICRISAT/Mali. The considerable effect on soil and water conservation, however, attracted many other farmers. Several of them have decided to try contour farming during the next cropping cycle.			

1: 1 cart load = 120 kg Dry matter of cereal stalks

Table 1: Example of activities implemented in one of the research villages (1995)

4 Assessing the impact on the sustainability of the farming systems

To assess the impact of the participatory action research and thus of improved soil fertility management practices, parameters have to be selected, which can be easily measured and which are objective, widely accepted and based on time series and existing data.

Soil fertility management

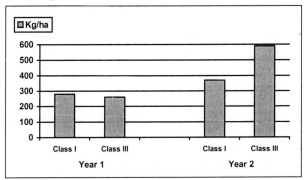

Figure 2a: Organic Fertiliser utilisation (Kg/ha)

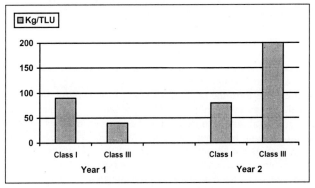

Figure 2b: Manure production (Kg/TLU)

TLU: Tropical Livestock Unit

In the first year, organic fertiliser application per hectare is quite similar for the 2 classes (Figure 3a). In the second year, however, class III (poor) farms applied considerably more organic fertiliser per hectare (about 600 kg/ha), than class I (good) farms (about 400 kg/ha). Hence, class III farms (which cultivate smaller areas seem to be more sustainable in terms of maintenance of organic matter content of the soil. The quantities are, however, still considerably below the recommended rate of 2,5 tonne/ha/year, to ensure a minimum level of organic matter in the soil (Pieri, 1989; van der Pol, 1992).
Manure production per TLU (Tropical Livestock Unit) shows the same tendency (Figure 3b). Although in the first year, class I farms produced more manure per TLU than class III farms, the increase after one cycle of action-research in considerably higher for the class III farms. The figures (about 80 kg/TLU for class I farms and 200 kg/TLU for class III farms) are still far below the optimal production of 1000 kg/TLU (with the use of 750 kg of litter per TLU), for the north Guinean zone, calculated by Bosma et al., 1995. . Hence, there is considerable scope to increase manure production, through increased litter use in pens.

Figure 2: Exampes of parameters for Class I and Class III farms, at the initial diagnostic stage (year 1) and after one year of participatory action research (year 2).

Selecting sustainability parameters is difficult since different actors are involved: the farmer, the researcher, the extensionist and the policy maker. However, given the scope and objectives of the action-research, priority has been given to parameters which allow farmers to improve their soil fertility management practices. For this reason, the selected sustainability parameters are based on farmer key criteria, determining differences in soil fertility management. Since farmers have identified these criteria themselves, the selected sustainability parameters may be more useful in adding value to farmer perception of good soil fertility management and hence to use these parameters to improve farmer planning and evaluation of improvements.

Farmer key criteria for good soil fertility management concern basically a better *residue recycling* or crop-livestock integration: *litter use* and *use of organic fertiliser*, and *mineral fertiliser use* (see diagnostic phase). Farmers classify their peers on the basis of total quantities per farm. According to the farmers, the underlying factors influencing good management mainly relate to *area cultivated, labour* and *cattle* availability.

The selected parameters we have chosen are a combination of farmers' criteria for good soil fertility management and the underlying causes. Thus, the selected parameters are not the total quantities per farm (as farmers use to judge soil fertility performance), but **ratios** of *residue use, organic fertiliser production/use* and *mineral fertiliser use*, per labour, animal or land unit.

The Resource Flow Models made by farmers allow the generation of many useful data. They are first transferred onto monitoring forms, which are subsequently transferred into a computer data base for further reorganisation, aggregating and merging of the data base. The ratio's for each of the test farms can be easily calculated from the aggregated/merged data file and subsequently, averages per class of farms can be calculated (for more details, see Defoer et al., 1996).

These ratios allow one to **compare farmer performance with research findings and extension recommendations**, to **compare the different classes of farms**, and to **monitor changes of soil fertility management over time.** Some examples are given in Figure 2.

Although farmer criteria form the basis of assessing sustainability of the production systems, other parameters will also be taken into account. Sustainability parameters in relation to flows between village resources and farm resources such as grazing on pastures, grass for fodder or litter gathered from communal land, have not yet been determined. Sustainability in terms of productivity and stability as well as the economic efficiency of the various practices have not yet been calculated. Also parameters which approach more qualitative aspects of the farm and its members, such as training, knowledge and internal household organization, have not yet been determined.

5 Perspectives and challenges

The methodology is quick and provides effective tools for analysing and understanding diversity between farming households. Farms are classified by villagers themselves on the basis of their criteria. Farmers appreciate resource flow analysis as a tool for soil fertility analysis. Visualisation enables them to analyse their strategies and practices of soil fertility management. It is participatory and action oriented and thus creates common ground for a discussion and exchange between farmers and researchers/extensionists and enables farmers to identify measures for improvement. The methodology has been developed to serve both researchers and extensionists.

Extension
The process has been tested in an extension setting with encouraging results. The extensionists appreciate especially the Resource Flow Models, which proved to be a valuable input for the final village meeting. Discussions are underway on how to simplify the identification of criteria which explain diversity, the classification process, and to develop a clearer procedure for the selection of the 'test' farmers.

The extension service has to realise that the overall introduction of this approach in the extension service will have important consequences. Capacities of their agents which will have to be stressed are animation, analysis, linking problems with solutions and identifying appropriate measures and management practices. The extensionist will no longer just transfer messages conceived by the organisation, but become a farm counsellor. The approach will require more flexibility from the extensionist and extension service.

Experience made clear that the more experienced farmers can assist their peers in the elaboration of Resource Flow Models, as well as planning and evaluation maps. The role of the extension worker would then be more confined to assistance in organising workshops, demonstrations and inter-village visits.

Research

The paper shows that a Resource Flow Model allows for quantification. Parameters that are based on farmer perception (key criteria) of soil fertility management, permit comparisons (1) between farm classes, (2) with extension/research recommendations and findings, and (3) over time. This assessment confirms the effectiveness of the participatory tools for diversity analysis, farm classification, and soil fertility strategy analysis. Moreover, the implemented 'new/improved' technologies, that are an outcome of the participatory analysis, indeed, contribute to a more ecologically-sustainable way of farming, in terms of resource recycling and crop-livestock integration. At the same time, this quantification makes clear that resource recycling (litter use and organic production/use) in terms of ratio's per animal and hectare are still quite below optimal levels. Hence this quantification is useful for the researcher and helps him/her to better target advice for the different farm classes.

However, to obtain results that allow comparison between farmers (or farm classes) and over time, the data obtained from the Resource Flow Models need to be fairly standardised. For this reason, interview guides that are used by the researcher to assist the farmer in making a RFM, have to be well structured to ensure systematic collection of all essential data.

The importance of this type of quantification of participatory action research may also be seen in the light of keeping researchers interested in using PRA tools and methodologies in research. It should be noted that researchers are evaluated on scientific results, which are still often based on quantification of findings. When the upgrading of PRA methods is ignored, they are likely to return soon to more 'classical' research methods, once the PRA wave passes.

Further work is also needed to test the usefulness of the sustainability parameters in assisting farmers in better planning activities. Feedback from farmers will therefore be essential. The idea is to "add value" to farmer criteria (and thus perception) and to "use" these parameters to improve farmer planning and evaluation of management improvements. Linking farmer participation and quantitative analysis with a natural resource focus may change not only the way researchers and extensionists generate new technologies for sustainable farming systems, but also the way farmers manage their natural resources.

References

Bosma, R., Bengaly, M. and Defoer, T. (1995): Pour un système durable de production: plus de bétail. Le role des ruminants au Mali Sud, dans le maintien du taux de matière organique dans les sols. In: J.M. Powell, S. Fernandez-Rivièra, T.O. Williams and C. Renard (Editors), Livestock and sustainable nutrients cycling in mixed farming systems of Sub-Saharan Africa, Vol. II. ILCA, Addis Ababa, Ethiopia.

Campbell, A. (1996): Landcare in Australia: Landcare and other new approaches of inquiry and learning for sustainability. In: A. Budelman (Editor), Agriculture r&d at the crossroads. Merging systems research and social actor approaches. KIT, Amsterdam, Netherlands.

Defoer, T., Hilhorst, T., Kanté, S. & Diarra, S. (1995): Analysing the diversity of farmers' strategies. ILEIA Newsletter vol. 11 (N 2).

Defoer, T., Kanté, S., Hilhorst, T & De Groote, H. (1996): Towards more sustainable soil fertility management. AGREN network paper N 63. Overseas Development Institute, ODI, London, England.

Lightfoot, C., Noble, R. and Morales, R. (1991): Training resource book on a participatory method for modeling bioresource flows. International Center for Living Aquatic Resource Management (ICLARM), Educational Series 14, 30pp.

Pieri, C. (1989): Fertilité des terres de savanne. Bilan de trente ans de recherche et de développement agricole au sud du Sahara. Ministère de la coopération, CIRAD/IRAT, Paris, France.

Röling, N. (1996): Creating Human Platforms to Manage Natural Resources: First Results of a Research Program. In: A. Budelman (Editor), Agriculture r&d at the crossroads. Merging systems research and social actor approaches. KIT, Amsterdam, Netherlands.

Scoones, I. and Thompson, J. (1994): Beyond Farmer First: Rural People's Knowledge, Agricultural Research and extension practice. Intermediate Technology Publications Ltd., London, England.

Van der Pol, F. (1992): Soil mining: an unseen contribution to farm income in southern Mali. KIT Bulletin N 325, KIT, Amsterdam, Netherlands.

Addresses of authors:
Toon Defoer
Royal Tropical Institute (KIT)
Mauritskade 63
1092 AD MSTERDAM; The Netherlands
Salif Kanté
Equipe Systèmes de Production et Gestione Resources Naturelles (ESPGRN Team)
Institut d'Economie Rurale (IER)
PO Box 186
Sikasso, Mali
Thea Hilhorst
The International Institute for Environment and Development (IIED)
3 Endsleigh Street
London WC 1 H0DD, UK

Nomads and Sustainable Land Use:
An Approach to Strenghten the Role of Traditional Knowledge and Experiences in the Management of the Grazing Resources of Eastern Chad

I. Yosko & G. Bartels

Summary

Undesirable changes in arid environments are often attributed to overgrazing by domestic livestock. Despite heavy grazing, however, the rangelands in eastern Chad are still in relatively good condition, especially if compared to those in other Sahelian regions. It is believed that this is primarily due to the existence of a viable form of transhumance livestock production. This paper describes an approach by which the knowledge and experience of transhumance pastoralists may be documented and eventually utilized to improve land use planning. On the basis of interviews with 15 different pastoral groups, it is concluded that there is a real need for an appreciation of the opportunistic strategies of the itinerant livestock producers and that development plans should accommodate for the traditional practices of these people.

Keywords: Dryland degradation, transhumance, pastoral organisation, Chad

1 Introduction

It is now generally accepted that dryland degradation is caused by a combination of climatic and anthropogenic factors. However, much disagreement remains with respect to the relative contributions of these two factors (Hulme and Kelly, 1993). For instance, overgrazing by domestic livestock is seen by many (e.g. Lamprey, 1983; Sinclair and Fryxell, 1985; Le Houérou, 1989) as a major cause of land degradation in the Sahel. This belief is based on the assumption that degradation occurs when stocking rates surpass some 'calculated' carrying capacity: high grazing pressures tend to push rangelands towards lower condition classes, a trend that can be countered only by a reduction in animal numbers.

This equilibrium concept of grazing systems is generally correct where conditons for plant growth are fairly stable between years but its validity for arid and semi-arid ecosystems has been seriously questioned (Ellis and Swift, 1988; Mace, 1991; Behnke et al., 1993). Here non-equilibrium dynamics appear to dominate, i.e., abiotic agents, rainfall in particular, have a far more important impact on plant growth and community change than grazing pressure. These dynamics may be particularly in evidence in regions such as the Sahel, where herbage consists mainly of annual grasses and ephemeral dicotyledons whose growth and development are little influenced by grazing animals, but respond primarily to yearly rainfall patterns (Pitt and Heady, 1979).

In a review of the literature, Dodd (1994) found no scientific evidence that livestock in Sub-Saharan Africa causes irreversible changes in range vegetation away from watering points and settlements, supporting our belief that, if degradation in the Sahel occurs as a result of animal impact, it is due to unbalanced animal distribution rather than to overall stocking rates being too high. The view that domesticated animals impact their environment in a negative way frequently fails to distinguish between different forms of livestock rearing. Spatial heterogeneity and intra-seasonal and interannual variability in rainfall make forage availability in arid regions poorly predictable. To optimize secondary production, livestock should track the pulses in productivity that are characteristic for these environments. Livestock systems differ greatly in their degree of mobility and thus in their ability to capture resources. Sedentary livestock production, therefore, seems an ill-adapted form of land use in arid rangelands (Danckwerts et al., 1993).

The indigenous knowledge and experience of mobile pastoral societies have often been neglected by development programs. The aim of our work is to make an inventory of the practices and fortunes of a number of pastoral groups in eastern Chad and, once armed with this information, to make more appropriate recommendations to the authorities responsible for land use and regional development. In this paper, we describe our approach to penetrate the insular world of the Chadian pastoralist and indicate to what extent his experience is in agreement with current thinking on the functioning of arid grazing ecosystems.

2 Description of the area

The Ouaddaï-Biltine region in eastern Chad borders on Sudan and is located between 11 and 16 latitude N. Mean annual rainfall ranges from 100 mm in the north to 700 mm in the south. Rain falls during a single short season of two to four months and peaks in August; the rest of the year is dry. Rainfall has decreased significantly during the last 30 years. At Abéché, the only station with a long-term record, mean annual rainfall was 492 mm during the years 1932-1964 with a coefficient of variation of 28% and 370 mm for the period 1965-1994 with a coefficient of variation of 33%. Temperatures are high with maxima reaching values of 40°C or higher at the end of the dry season (April-May). Potential evaporation exceeds 2300 mm/yr. As a result, rangelands in the region are dominated by annual grasses.

The northern Sahelo-Saharian zone is used exclusively by nomadic pastoralists; rainfall is too low for crop production (<200 mm). Main land use in the central Sahelian zone is livestock grazing but the cultivation of millet has become increasingly important. The southern Sahelo-Sudanian zone is an area of crop cultivation with a fast expanding population. It is also the zone to which many of the transhumance livestock producers retreat during the dry part of the year (November-June).

Geologically the Ouaddaï-Biltine region is a granitic massif (up to 1500 m above sea level) with sandy or skeletal soils surrounding inselbergs in the east and piedmont plains in the west. Due to the geological features of the region, groundwater resources are scarce and aquifers difficult to locate. There are no permanent watercourses in the region; the area is dissected and drained by a large number of wadis that run from east to west and whose beds often are the only source of drinking water. With the exception of the vertisols in the south and the alluvial soils along the wadis, soils in the region are low in organic matter, nitrogen and phosphorus (GAF, 1990).

The main vegetation form is an open grassland made up of annual species with varying numbers of bushes and trees. Tree canopy cover varies from virtually zero per cent in the north to about 20 percent in the south. Most perennial grasses of the Sahelian zone proper, such as *Cymbopogon proximus* and especially *Andropogon gayanus*, have disappeared during the drought of the early seventies (Monnier, 1990). Other important vegetation types recognized in the region are the forest galleries along the main wadis, and the thornbush savannas on the heavy vertisols with *Acacia seyal* as the dominant species. A detailed description of the vegetation can be found in Pias (1970).

There are essentially three nomadic peoples in the region, the Rizegat Arabs, the Goranes, and the Zaghawa. We have worked with 15 encampments, 12 Arab, 2 Gorane and 1 mixed Arab-Gorane. The Arabs tend to be more herders of camels than of cattle. During the rainy season they assemble around temporary waterpoints between 15° and 16° N where excellent pastures of annual species are found on the sandy soils of the so-called Mortcha. After the rainy season most of them move south again to the grand wadis Batha, Kadja and Bahr-Azoum. The Goranes of eastern Chad, traditionally important camel producers, have gradually descended from the Ennedi to the center of Biltine Prefecture in response to the drought of 1984. Since then they have started to take an interest in cattle production as well. During the rainy season they disperse to the sandy plateaux north of the 16th parallel.

3 Methodology

It is difficult to have access to the complex organizational world of the pastoralist with its fragmented and sometimes dispersed residential units, the pastoralist's mobility and his opportunistic lifestyle. He does not distinguish between projects, NGOs or state services and his initial attitude is always marked by mistrust and doubt. Our approach is based on recognition of the fact that for the pastoralist his clan is the most important social institution, and on the conviction that innovations must be adapted to the pastoralist's way of life and not the other way around. At the same time, we are convinced that the pastoralist himself must play a more active role in preserving his lifestyle.

Our approach then centers around two concepts, the encampment as the basic pastoral unit and the resource person. For both Goranes and Arabs, the clan, or lineage depending on the context, is the most important social unit. Members have a common ancestry and a strong feeling of solidarity based on blood relationships (Bonfiglioli, 1990). Geographically, however the lineage is not an entity; its members tend to disperse and only fractions stay together in mobile encampments composed of several households. The residents of these encampments utilize the same resources at the same time. They have a leader, whose influence is founded more on personal qualities and prestige than on vested authority: his task is not so much to decide as to try to reach a consensus among members. As intermediary between the chief of the tribe and the heads of families, he is an indispensable contact person for anyone trying to get in contact with the nomads. His engagement implies that of the community around him.

Contrary to what is generally believed, the pastoral world is not really closed to all innovations. In fact, the pastoralist constantly updates his information concerning relevant resources, and economic and political aspects. When dealing with agents of public services or projects he will always use euphemisms and never compromise the foundations of his opportunistic strategies. To evade this obstacle, we used a local resource person, a pastoralist with profound experience in livestock production and who was respected for his moral and professional qualities as well as someone open to innovations. Better than any outsider, he can introduce innovations to his fellow pastoralists and lead them to some kind of consensus on actions to take. In general, they were close to the clan but able to maintain a certain independence. The reason for not using tribal chiefs or clan leaders as contact persons is that these are often mistrusted by their own people because of their close relationship with the authorities. In most instances these leaders also lack the kind of professionalism that comes with being in almost daily contact with the pastoralist's world. Due to their traditional and administrative obligations, they are not as involved as the resource persons whom we selected. Nonetheless, these leaders were consulted when we sought access to the encampments.

4 The nomad's perception of the environment and his role in it

The mainfocus of the pastoralist's management strategy, in fact his life, is his animals' well-being. His perception of the environment and his management of the resources all revolve around this central theme. In this respect we have not found any difference between the Arabs and the Goranes. All pastoralists classify rangelands -and other resources, such as water and salt- according to their livestock production potential. Pastoralists in the region recognize valleys, depressions, sandy plateaux, regs, clay loam plains, dunes, massifs and each range site is evaluated according to this criterion during each of the five pastoral seasons. During the rainy season the clay loam plains, the regs, the sandy plateaux and the dunes are used; in the dry season the valleys and depressions with their woody species and perennial grasses, and the massif.

The pastoralist grazes and browses these sites with rigorous time management in response to the spatially variable availability of forage. Phytomass availability and quality are evaluated continuously using animal condition and behavior as important indicators. In this respect they differ from 'western' range managers who do not program their grazing as a function of the stochastic nature of rainfall and who are more concerned with ecosystem characteristics such as soil stability and vegetation composition. Much of the information on the resources' condition is acquired in discussions with other livestock producers belonging to his own or other encampments.

The choice of daily pasture, the rhythm of watering, the choice of campsite, however, remains an individual decision without an obligation to the collective group. This does not mean that use of the resources is anarchic; everybody is well aware of his rights to exploit the resources and to control their use by others. This control is exercised by forging a consensus, made possible by the importance that pastoralists attach to belonging to the same kinship group. This reglementation of access to and use of resources does not have the power of law but is subject to alteration when the group considers it necessary. As far as land use is concerned, the pastoralist prefers to act within a social structure in which one's rights and obligations are established by consensus.

The droughts of the early seventies and mid-eighties led to considerable environmental deterioration in Chad (Monnier, 1990, Gaston, 1991). The pastoralists agree that a degradation of their environment has taken place during the last 30 years, but do not feel that livestock have been a contributing factor. They attribute vegetational changes in the northern part of the region to a reduction in rainfall. Here trees have suffered the most due to insufficient rainfall and a general drop in the watertable. Among ligneous species, they note a strong reduction in the presence of *Cordia sinensis*, *Salvadora persica* and *Acacia ehrenbergiana*, an important species for camels. These species are still found in depressions and other sites with higher moisture availability. The dominant tree species now are *Acacia raddiana*, *Balanites aegyptiaca* and *Capparis decidua*. In this zone with mainly sandy soils and dunes they have also observed a diminution of annual leguminous plants (*Tephrosia* spp., *Indigofera* spp.), but they do not view this as degradation because the seed bank of these species is still intact. As our monitoring has shown, in high rainfall years these species may become important components of the herbaceous layer, Other herbaceous species reportedly to have greatly diminished are *Blepharis linearifolia* and *Monsonia senegalensis*.

They state that the northern region used to support more livestock than at present. In fact, before the drought of the early 1970's, all 2 Gorane groups and all 12 Arab groups interviewed by us used to stay yearlong in this zone, called the Mortcha. After the droughts several of them decided to move south each year during the dry season; only one of the Gorane and four of the Arab groups still remain yearlong in the Mortcha, except during extremely dry years when forage becomes a limiting factor. In wet years, on the other hand, some groups may decide to stay in the Mortcha during the dry season and not make the journey back south.

Whereas degradation in the north is attributed to climate change, further south it is regarded as mainly man-induced. In the central Sahelian zone, which for most pastoralists has become a transit area, they equate degradation with the conversion of rangelands into cropland, as this involves almost complete destruction of ligneous species. This expansion of cultivated land has reduced the area available for grazing and browsing and has deprived the pastoralist of what he considers to be an essential forage resource. To him, who is careful to avoid overexploitation of this resource, this loss of trees is extremely wasteful. Surprisingly, the pastoralists we talked to did not link the removal of trees and reduction of vegetative cover with the phenomenon of large scale erosion. To them the impact is more direct: the loss of pastoral space and of a strategic forage resource for their animals. Further south, in the Bahr-Azoum area, forage quality rather than forage availability is their main concern, but the extension of cropland and the occurrence of wild bushfires are increasingly viewed as endangering livestock rearing.

5 Conclusions

As a result of the dessication over the past 30 years, pastoral life has become more difficult. Due to an overall decrease in rangeland productivity and greater spatial variability in the availability of water and forage, the risks for pastoralists have increased. Average herd size has decreased during this period as pastoralists have responded to the deterioration of their environment with an increase in herd mobility. Many of them have also gradually retreated to less arid regions, not because they feel that the annual pastures in the north have lost their productive capacity, but because years with abundant rainfall have become more rare.

Since meteorological events are considered acts of God and the most conspicuous land degradation is caused by cultivation, pastoralists cannot be expected to enthusiastically embrace often untested technical solutions to problems they feel they did not create. We tend to agree strongly with their sentiment that in eastern Chad rainfed crop production is by far the most destructive force.

The pastoralists' interpretation of the vegetation changes in the northern zone of Ouaddaï-Biltine concurs with our own observations and measurements and lends credibility to the non-equilibrium model of range dynamics suggested by Ellis and Swift (1988). It also tends to support the doubts expressed recently, based on remotely-sensed data, on the so-called encroachment of the Saharan desert (e.g. Tucker et al. 1991, Helldén 1991). At the same time, however, their observations warn against making too sweeping generalizations based on such data -especially NDVI (normalized difference vegetation index) data- as these tend to obscure the fact that the term degradation implies more than just a reduction in green phytomass. Only by systematic long-term monitoring of plant composition, biomass production and soil surface condition under grazing and protection from grazing, will we be able to gain a more complete understanding of livestock's role and its relative importance in the process of arid land degradation.

The nomads we work with are aware of the precariousness of their current situation but do not appear to have a clear vision of the future of their production system ("with the help of God and the government we are going to survive"). The main task for pastoral development programs should be to reconcile the empirical knowledge of the nomads with the theoretical insights and methodologies of the ecological sciences. Maybe then, governments and donor agencies will be convinced that, in the short run, traditional pastoral production may not only be the most productive or profitable economic activity in arid lands but that it is also the most sustainable one when a high degree of mobility is preserved.

References

Behnke, R.H., Scoones, I. and Kerven, Carol (1993): Range ecology at disequilibrium. New models of natural variability and pastoral adaptation in African savannas. ODI, London.

Bonfiglioli, A.M. (1990): Eleveurs du Tchad oriental. Rapport Projet Elevage Adapté, Abéché (Chad).
Danckwerts, J.E., O'Reagain, P.J. and O'Connor, T.G. (1993): Range management in a changing environment: a southern African perspective. The Rangeland Journal **15**: 133-144.
Dodd, J.L. (1994): Desertification and degradation of Africa's rangelands. Rangelands **16**: 180-183.
Ellis, J.E., and Swift, D.M. (1988): Stability of African pastoral ecosystems: alternative paradigms and implications for development. Journal of Range Management **41**: 450-459.
GAF (1990): Morphopedologie et aptitudes des terres des prefectures Ouaddaï et Biltine. Munich.
Gaston, A. (1991): Rapport de mission d'appui auprès du réseau d'observation des pâturages naturels ROPANAT. IEMVT, Maisons-Alfort (France).
Helldén, U. (1991): Desertification - Time for an assessment? Ambio **20**: 372-383.
Hulme, M. and Kelly, M. (1993): Exploring the links between desertification and climate change. Environment **35**: 5-45.
Lamprey, H.F.(1983): Pastoralism yesterday and today: the overgrazing problem. In: F.Bourliere (ed.), Tropical savannas: Ecosystems of the World. Elsevier, Amsterdam.
Le Houérou, H.N. (1989): The grazing land ecosystems of the African Sahel. Springer, Berlin.
Mace, R.(1991): Overgrazing overstated. Nature **349**: 280-281.
Monnier, J.P. (1990): Le cadre pastoral au Tchad et les stratégies des éleveurs. Manuscript Projet ASETO, Abéché (Chad).
Pias, J. (1970): La végétation du Tchad. ORSTOM, Paris.
Pitt, M.D. and Heady, H.F. (1979): The effects of grazing intensity on annual vegetation. Journal of Range Management **32**: 109-114.
Sinclair, A.R.E. and Fryxell, J.M. (1985): The Sahel of Africa: ecology of a disaster. Canadian Journal of Zoology **63**: 987-994.
Tucker, C.J., Dregne, H.E. and Newcomb, W.W. (1991): Expansion and contraction of the Sahara desert from 1980 to 1990. Science **253**: 299-301.

Address of authors:
Idriss Yosko
Gerrit Bartels
Deutsche Gesellschaft für Technische Zusammenarbeit (GTZ)
Projet Elevage Adapté, Abéché
B.P. 123
N´Djamena, Tschad

Management of Watersheds with Soils on Marls in the Atlas Mountains of Algeria - A Proposal for a Non-Conventional Watershed Development Scheme

H. Paschen, D. Gomer, L. Kouri, H. Vogt, T. Vogt, M. Ouaar & W.E.H. Blum

Summary

In semi-arid Mediterranean regions the pressure on natural resources such as water and soil is increasing. In the Maghreb soil degradation and reservoir sedimentation are serious problems, particularly in watersheds with a high proportion of marls.

Investigations in watersheds are mostly oriented towards the exclusive profit of the population downstream of reservoirs, due to water supply of cities, industries and irrigation of the plains for intense agriculture. The catchment area upstream is mostly only considered as a water supplier, which must be treated to limit the transport of sediments to the reservoir.

The principal supplier of sediments in the Atlas are the marls.

The watershed of the Oued Mina, lying in the "Tell occidental algérien", has a semiarid Mediterranean climat, an annual rainfall up to 400 mm, traditional agro-pastoral land use in catchment areas and with a high proportion of marls. It was choosen by the Algerian authorities and the Deutsche Gesellschaft für Technische Zusammenarbeit (GTZ) GmbH for a Technical Cooperation Project.

It is shown that traditional land use by the local population is the most efficient way to limit the production of sediment in this area.

Keywords: Mountain watershed, erosion control, socio-economic development, Tell Atlas Algeria, Oued Mina catchment.

1 Introduction

Water supply is a basic requirement for planning and development in semi-arid Mediterranean environments. It is usually done by implementing reservoirs within the framework of national or regional planning schemes, with the only aim of developing the areas downstream through the water supply of towns, industries and intensive agricultural irrigation in plains. In contrast, the upstream catchment is considered only as a water supplier, and must be treated so as to favour the delivery of water and the minimisation of sedimentation in the reservoir, in order to optimise its cost-benefit ratio. This point of view is increasingly inadequate in Maghreb countries for two reasons:
- from a political point of view, it is an obstacle to harmonious development in especially if there is an agro-pastoral tradition, a relatively dense rural population in the upper catchment, as is often the case in the Tell Atlas mountains. This gives rise to an economic imbalance, so that downstream territories attract the upstream catchment population.

- from a technical point of view, it will be shown here that in semi-arid marly territories, agriculture limits erosion, while forest planting is not efficient and limits water availability by evapotranspiration.

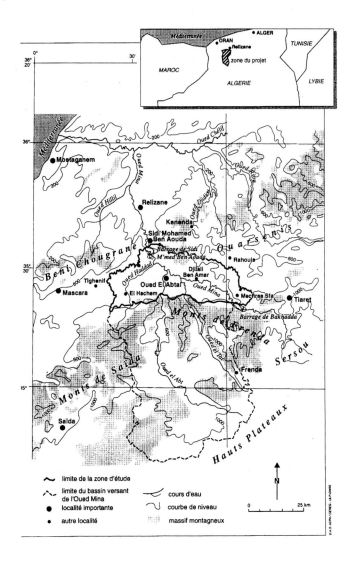

Fig. 1: Situation of the Oued Mina catchment

The example of the Oued Mina catchment upstream of the Es-Saada reservoir in the Western Algerian Tell mountains was studied within the framework of an Algerian-German technical cooperation programme which led to a non-traditional development scheme (Blum, 1989, Paschen, 1989, Vogt et al., 1996).

2 The Oued Mina catchment

The Oued Mina is a 130 km long affluent of the Chelif River (Fig. 1). Ist catchment has two parts, the upper one being calcareous and marly and the lower one being only marly; the latter part delivers the bulk of sediments to the reservoir (Meddi, 1992) and will be considered here. It is representative of regions with a semi-arid Mediterranean climate, with an average rainfall of less than 400 mm, marls, and an agro-pastoral tradition. In 1979, the Oued Mina was dammed up by the Es-Saada reservoir, just before entering the Relizane plain. The long-term average rainfall at Oued-el-Abtal is 350 mm, but can be as low as 250 mm on slopes facing the East.

The marly, northern part of the catchment extends to the south-eastern part of the Ouarsenis massif and the south-western part of the Beni-Chougrone mountains. Like most of the Tell Mountains, these two have probably been created by the Middle and Upper Pleistocene uplift. An unequal development of deep gullies and bad-lands is characteristic for the marly part of the catchment. The gullies and bad-lands concentrate in the area lying downstream, nearest to the reservoir ES-SAADA, whereas the area lying more upstream shows extended plateaus where deep vertisols still exist. This distribution has geomorphologic reasons: gullies developed in response to the progressive upstream movement of the Oued Mina incision, whereas the deep soils in the upper part are remnants of the time before the uplifting.

3 Ecological character

3.1 Sediment fluxes

Gullies supply nearly the total sediment input of the reservoir (Gomer, 1994). Water and sediment fluxes were measured on slopes by means of 30 plots, under natural and simulated rain, in the outlets of 6 gully-catchments, during three years (1989/90 - 1991/92). The results show that there is only very slight inter-gully erosion: 0,6-3 t/ha/year; values > 2 t are exceptional. At least 95 % of the sediment flux comes from gully activity (Kouri, 1993).

In contrast, if inter-gully areas do not produce sediment, they supply water for the flow that triggers gully-erosion. On more clayey soils, the runoff coefficient is slight, due to numerous soil cracks; It does not exceed 50 %, even with intense rainfall and high previous humidity. But on more silty and sandy-silty soils, which are prone to surface sealing, the coefficient is 70 %, even for low previous humidity, and can reach up to 100%. So the problem to be solved is to limit rain wash.

To allow spatial extrapolation of the runoff-coefficient and of sediment supply, a quasi-physical distributed process-based model for runoff and erosion was established (Gomer, 1994). The input of terrain conditions was provided by a GIS procedure with SPANS software, combining a digital terrain model. Two maps were produced by remote sensing (Landsat TM) processing: one of the physical characteristics of soils and one of soil humidity (Vogt and Gomer, 1992).

3.2 Afforestation is ecologically illusive

Provincial, national and international agencies usually foster afforestation to reduce sediment fluxes. But with a low average rainfall of 250-300 mm, as in the western part of the sub-catchment, and edaphic dryness of marly soils, this region is unfit for afforestation. Botanical studies showed that potential natural vegetation does not even seem to be a real forest but steppe. In the rainier eastern part (350-550 mm, depending on the altitude), it is probably a clear perennial forest characterised by thuya (Ogryssek, 1994), which is not suitable for planting a protective tree cover, either. This has been proven by some afforestation attempts (Vogt et al., 1996).

3.3 Tillage notably limits the runoff coefficient

Afforestation is an unsuitable protection measure. Nevertheless, experiments showed that tillage greatly reduces runoff, provided that the fallow is short(Gomer, 1994). Mellowing fosters porosity, hence infiltration and vegetal productivity. In addition, enhanced roughness facilitates water ponding in micro hollows. Thus, a tilled, deep clayey soil gives riuse to runoff only if the water content is about 50 % for a 10 mm/h storm or 40 % for a 20 mm/h storm, and this threshold is not easily reached in this porous medium. Tillage reduces the runoff coefficient by 90 % (Fig. 2).

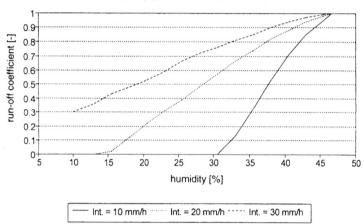

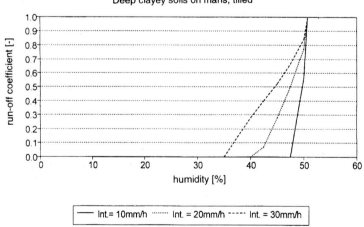

Fig. 2: Run-off coefficients of non tilled and tilled soils on marls

4 Proposal for a development scheme

It is advisable to promote a development scheme that requires the soil to be tilled, and that provides the necessary ecological and socio-economic conditions.

4.1 Ecological conditions are fit for the cultivation of cereals

The vegetation is very much degraded in this area, due to overgrazing, scarce rains and unfavourable edaphic conditions. Nevertheless, experiments in areas where grazing was forbidden showed extremely good results, indicating that the potential of seeds is not destroyed by overgrazing (Meierich, 1992).

Pastures can be regenerated. Yet peasants prefer the cultivation of cereals, which yield more profit because they are used as fodder as well as for human consumption. Cereals are decended from gramineae, indigenous in the Near East, which has a similar climate. The culture of cereals enhances the soil structure and diminishes soil sealing, and therefore run-off (Gomer, 1994).

4.2 There is an agro-pastoral tradition

There is pedological as well as sociological evidence of an ancient agricultural tradition in the marl mountains of the Oued Mina catchment and more generally in the Western Tell Atlas (Paschen, 1996). This agro-pastoral civilisation was very well adapted to the conditions of the marly hills and was founded on the cultivation of cereals for fodder and human consumption. The quantity of biomass produced is much higher than that of maquis pastures and cultivation of leguminous plants; gramineae are the appropriate fodder plants, and the cultivation of cereals is ecologically optimal on marly soils.

In the more humid years, breeds are fattened before sale and excess cereals stored in underground silos ("matmars"); in dry years, straw is pastured on the fields and stocked cereals are used for human consumption.

Landscape degradation started when this civilisation was interrupted.

In the second half of the 19th and in the 20th century, first the French colonial policy and then the socialist centralised policy were more interested in developing lowlands and valley bottoms. Recently, only the mechanised culture of cereals was extended to the wider marly interfluves.

Although some government subsidies were available, there has not been any significant progress in the rural development of the marly regions in the western Tell Atlas. Its population lives in scattered villages ("douars"), in the middle of the their farmland in extended families. The supply of drinking water and the access to amenities such as electricity, stores, medical care and schools are difficult. The population tends to move down to the nearby valleys. This tendency has recently been reinforced for two reasons:
- the dry period from 1980 - 1993
- the reduced importance of socialised agriculture, which had been initiated in 1988 with the liberalisation of the cultivation of market-garden produce, allowing better access to monetary income.

Due to this population movement, large areas, especially pastures in the marly hills, are subject to extensive grazing or are abandoned. Because of the low vegetation recovery, they are prone to run-off and gully extension and foster sediment production.

Hence a development scheme combining socio-economic development and anti-erosive measures should be implemented, combining socio-economic development and technical measures to fight erosion (Vogt et al., 1996).

4.3 Erosion control through socio-economic development and technical measures

4.3.1 Maintenance of agriculture on the interfluxes

The proposed development model aims at keeping the landscape stable, allowing a modern way of life through:
- Cultivation of cereals in dry farming, as a basis for cattle breeding, taking into account the natural and socio-economic constraints, the main condition for limiting gully development and sediment delivery.
- Allowing peasants a modern way of life. Scattered douars must be left to temporary settlement because permanent dwellings are down at the bottoms of the valleys, where modern, little townships developed. Thus it is necessary to develop a net of local roads in the catchment, in order to allow the connection of cultivated areas and permanent settlements. They must be planned so as to allow the transit of tractors with an axle load of 5.0 t, at a speed of 30 km/h. There should be one-way roads which are enlarged in the area of crossings and which follow divides, thus allowing for the lowest risk of erosion (Paschen, 1996).

4.3.2 Technical anti-erosive measures

However, the land use described above is constrained by a series of limiting factors (Paschen, 1996):
- areas where soil degradation is still too high and where there are too shallow soils or no soils at all.
- steep slopes (40 %)
- slopes with southern exposure, subject to high solar radiation and consequently high evapotranspiration.
- saline soils (20 % of the total area)
- heavily gullied areas (10 % of the total area)

The distribution of areas which cannot be tilled is variable: there are bigger areas in the western, dissected part of the catchment's marly zone and much smaller ones in the eastern, less dissected part.

Two types of management must be combined in the degraded areas:
- on the interfluves, the access of grazing animals must be prevented, in order to allow a progressive recovery of the vegetation.
- the gullies are treated by combining physical and biological measures: retention dams are established, beginning upstream and progressing downstream, to allow a gradual stabilisation.

Such a system interrupts sediment transit; when these little reservoirs are filled with sediments, enough water is retained to allow the planting of sowing a hygrophyte vegetation, which will have two advantages: It stabilises slope foots and provides peasants with wood for building and heating, and it provides a grazing area and fodder (e.g. atriplex halimus).

Moreover, dams and biological measures modify the hydrological behaviour of the gullies: they lead to reduction of water flux velocity and flood reduction.

Peasants must have a part in the implementation of these measures and mainly in the setting up and management of the vegetation.

Implementations or applications of the proposed measures of integrated watershed management in Algeria had to be postponed.

5 Aspects of public policy

Usually, administrative districts do not match the boundaries of natural resources or eco-systems. Measures of watershed management within different administrative districts inside the same watershed might be contradictory. For this reason, the spatial units of planning must have flexible limits and this should lead to the setting up of spatial planning authorities which are in charge of coordinating the various skills necessary to implement the development model. Such planning units could be based on catchment areas.

Recent developments in Algeria (Décret sur la création des Agences de Bassin-Versants) and Morocco (Loi sur l'Eau N°. 10-1995) favours the management of water resources within its natural limits of watersheds.

6 Conclusions

The proposed development model, which combines tilling with technical anti-erosion measures in gullies and on interflux areas with heavy soil degradation, is fostered for areas with an agro-pastoral tradition, but which are ecologically unfit for afforestation. Most certainly, the attempt to keep farmers in the mountainous areas will somewhat diminish the economic efficiency of the main reservoir, due to water consumption for domestic use, cattle breeding and irrigation in the valleys.

The major benefits would be the increase of the lifespan of main reservoirs due to reduced soil erosion. Another benefit mentioned is the reduction of migration to economically attractive regions and major towns, which leads to a reduced demand for housing, infrastructure and employment in overcrowded centers.

For this reason, one has to question if the model could also be of political advantage in areas that are ecologically suitable for afforestation, provided that they have an agro-pastoral tradition. The economic advantages of the model cannot be expressed in financial terms at the moment, it can only be expressed by means of long-term scenarios of about 30 years. In addition, it must be taken into account that not all favourable features of this model can be expressed in economic values.

References

Blum, W.E.H. (1989): Bodenerosion und Bodenschutz im Mergel-Einzugsgebiet des Oued Mina-Algerien. Z.f. Kulturtechnik u. Landentwicklung, Berlin **30**, 324-330.

Gomer, D. (1994): Oberflächenabfluß und Bodenerosion in Kleineinzugsgebieten mit Mergelböden unter einem semiariden mediterranen Klima. Mitteilungen. Institut für Wasserbau und Kulturtechnik, Universität Karlsruhe. Heft 191. 293 p; also available in French language by: Bureau du projet ANRH-GTZ, Ob der Eichhälden 7, D-76228 Karlsruhe, 308p.

Kouri, L. (1993): L'érosion hydrique des sols dans le bassin-versant de l'Oued Mina (Algérie). Étude des processus et types fonctionnels de ravins dans la zone des marnes tertiaires. Thèse de doctorat; Université Louis-Pasteur, Strasbourg. 238 p.

Meddi, M. (1992): Hydropluviométrie et transport solide dans le bassin-versant de l'Oued Mina. Thèse de doctorat, Université Louis-Pasteur, Strasbourg. 103 p.

Meierich, S. (1992): Vegetationsaufnahme des Wassereinzugsgebiets des Stausees Sidi M'hamed ben Aouda. Unpublished report; GTZ, Eschborn.

Ogryssek, H. (1994): Vegetationskundliche Studien im algerischen Tell-Atlas (bassin-versant de l'Oued Mina). Diplomarbeit, Universität Bonn, Geographisches Institut. 186 p.

Paschen, H. (1989): Vom integrierten Erosionsschutz zur Bewirtschaftung eines Wassereinzugsgebietes am Beispiel des Oued Mina - Entwicklung einer algerisch-deutschen Zusammenarbeit, Zeitschrift für Kulturtechnik und Landentwicklung, Berlin **30**, 310-314.

Paschen, H. (1996): Conception d'ensemble, in Vogt et al.(1996): L'aménagement des zones marneuses dans les bassins-versants des montagnes de l'Atlas tellien semi-aride. Schriftenreihe der GTZ, no. 256. GTZ, Eschborn. 5-20.

Vogt, H, Blum, W.E.H., Chader, A., Gomer, D., Kouri, L., Ouaar, M., Paschen, H., Ramdane, M., Vogt, T. (1996): L'aménagement des zones marneueses dans les bassins-versants es montagnes de l'Atlas tellien semi-aride. Schriftenreihe der GTZ, no. 256. GTZ, Eschborn. 142 p.

Vogt, T. & Gomer, D. (1992): Estimation du ruissellement et des matières en suspension par l'utilisation d'un SIG intégrant MNT., données Landsat TM et donnés hydrologiques. Bulletin de la Societé française de photogrammétrie et télédétection **128**, 7-17.

Addresses of authors:
H. Paschen
Deutsche Gesellschaft für Technische Zusammenarbeit (GTZ) GmbH
27, Rue des Cimatiéres
F-67240 Bischwiller, France
D. Gomer
Gomer Ingenieur Consultant
Ob der Eichhälden 7
D-76228 Karlsruhe, Germany
L. Kouri
Department of Soil Science, INFSA
BP 300
Mostaganem, Algeria
H. & T. Vogt
Université Louis Pasteur
3, rue de l'Argonne
F-67083 Strasbourg, France
M. Ouaar
Agence Nationale des Ressources Hydrauliques
Av. Mohammedi
Bir Mourad Rais, DZ-16000 Alger
W.E.H. Blum
Institute of Soil Research
University of Agriculture and Natural Resources
Gregor-Mendel-Straße 33
A-1180 Vienna, Austria

What Makes Watershed Management Projects Work? Experiences with Farmer's Participation in India

B. Adolph & T. G. Kelly

Summary

The results of watershed management (WSM) projects in India have been disappointing, being mostly short-lived and scattered. This study seeks to identify more effective approaches to WSM resulting in sustainable project impact in terms of bio-physical, economic and social aspects. The issues addressed in this study are: (1) What key factors are responsible for project success, such that differences in these parameters lead to differences in impact; and (2) What are the mechanisms associated with these key factors, i.e., how do these factors make a difference? A two-phased case study approach was utilized, starting with an exploratory survey of 13 WSM projects in South India. This led to the identification of three key factors related to success; all are characteristic of participatory approaches:

- farmers' involvement in the choice of both location and design of soil and water conservation (SWC) technologies, as well as the incorporation of indigenous practices into project design;
- farmers' contribution (in cash or labour) to the implementation and maintenance costs;
- existence and functioning of local institutions.

During the second phase of the study, an in-depth analysis of two WSM projects in Andhra Pradesh, representing participatory and non-participatory approaches, was undertaken to determine how participation leads to a greater impact. Both Rapid Rural Appraisal (RRA) and formal questionnaire surveys were used for eliciting information. Two mechanisms were identified. First, farmers select and implement technologies that contribute to achieving their multiple objectives and that respect existing ownership patterns; thus conservation measures are better adapted to local production conditions and as a result are maintained by farmers. Second, the existence of local institutions enables farmers to organise group action for the use and protection of community resources and for conflict resolution.

1 Introduction

Watershed management, an integrated approach to combine conservation of soil, water and vegetative resources with increased productivity and reduced risks in dryland farming, has been promoted in India for the past 15 years[1]. WSM programs, managed by government departments, research institutions, or non-governmental organizations (NGOs), typically consist of three main components: water harvesting and percolation structures, soil conservation, and tree plantations. Results, however, have been disappointing, being mostly short-lived (due to maintenance problems) and scattered (limited to project areas only)[2]. There are a few examples of "successful" WSM projects, those which have had a sustainable impact on the natural resource base and have resulted in higher agricultural productivity and rural incomes[3]. Evidence from these projects suggests a close relationship between project approach and impact.

The purpose of this study is to identify a more effective approach to WSM, i.e., the strategies for planning and implementation most likely to result in sustainable impacts. Specifically, it seeks to answer the following questions in relation to WSM:
- What are critical factors associated with project approach that result in project success, such that differences in these parameters lead ultimately to differences in the program impact?
- What is the mechanism in the more effective approach that leads to differences in impact?

Answers to these questions are important to project implementing agencies for identifying weaknesses in project design and for achieving greater impact on target groups.

2 Clarifying the terminology

"Project approach" refers to the process of planning, implementation and monitoring of all project-related activities. "Project impact" refers to the changes, both positive and negative, that result from implementation of the project activities within and beyond the project area. Both need to be defined in greater detail, however, to be useful operational concepts. For the purpose of this study, the following components or characteristics have been used in defining "project approach"[4]:
- farmers' involvement throughout all phases of the project;
- source of technology (e.g., farmers' or project's?);
- social organization of the target community initiated by the project (e.g., self-help or credit groups);
- farmers' contribution to project cost;
- project flexibility regarding farmers' time schedule;
- existence or non-existence of non-WSM components;
- accountability to the farmers;
- involving all interest groups in the planning process.

For each characteristic two extremes, completely "top-down" and completely "bottom-up", i.e., participatory, are possible. These extremes, however, are in reality hardly ever found, as most projects use a strategy that is somewhere in between. Pretty (1994: 41) distinguishes various levels or degrees of participation: passive participation, participation in information giving, participation by consultation, participation for material incentives, functional participation, interactive participation, and self-mobilisation. For the purpose of this project and in the context of WSM, participation is defined as follows:

> Participation is the active involvement of all user groups of a watershed in the identification of problems and solutions, the planning and implementation of these solutions, and the monitoring and evaluation of their performance. Full participation includes joint decision making, based on mutual agreement, of the project implementing agency and the people in the project area, and incorporates an element of accountability on behalf of the implementing agency to the people concerned.

Like "participation", "impact" requires clarification. Essentially, how impact is assessed depends on the indicators used[5]. But whatever indicators are used, sustainable impact looks beyond short-term improvements to the long-term physical, economic and social changes. Accordingly, the following definition has been used in this study:

> A project's impact are the changes - short and long term - that result as a direct or indirect consequence of the project's interventions. These include bio-physical, economic and social changes.

Since "objective" measurements of these bio-physical, economic and social changes are difficult to obtain, in this study, farmers' perceptions of these changes were used as the primary measure for project impact. This is consistent with the "emic approach" described by Uphoff (1992), using the beneficiaries own frames of reference for project evaluation. Indeed, farmers' perceptions of impact will ultimately determine their willingness to adopt and maintain a conservation practice. The information obtained from farmers was complemented by observations of the study team and secondary sources (project reports, etc.). Several components or indicators of impact were used in this study to cover key aspects of the SWC measures:

- changes in the resource base and agricultural productivity as a result of project interventions; specifically, yields of the main crops and extent of soil erosion, well water levels, irrigated area, availability of fodder grasses and fire wood, and shifts in cropping pattern, e.g., towards crops with higher water and nutrient demand;
- state of repair and maintenance of SWC measures (on-farm measures as well as community structures such as check dams and percolation tanks);
- state of exploitation and protection of common property resources (CPR);
- changes in the status of women, landless and members of low castes as a result of project interventions.

3 Methodology

The study was organised in two stages, corresponding to two village surveys. The objective of the first (exploratory) Village Survey 1 (VS1), was to identify specific characteristics of project approaches associated with project success. After key factors associated with successful projects had been identified through VS1, Village Survey 2 (VS2) tested for two projects under almost *ceteris paribus* conditions whether differences in project approach are indeed responsible for differences in project impact. Equally important, it permitted an in-depth assessment of the mechanisms leading to differences in impact.

3.1 Part I. Establishing the linkage between approach and impact

For VS1, 13 WSM projects in four South Indian states were purposively selected using the following criteria: the project is at least two years old; a minimum of 50 farms participate in the project; high, medium and low rainfall zones in red and black soil areas are represented; villages are socially heterogeneous; a variety of institutional approaches are represented; and the implementing institution and the villagers are willing and able to host the study team for a week.

The methodological steps used are as follows:

a) Defining/ identifying specific approach and impact indicators.

Impact and approach indicators were selected on the grounds that they characterize the specific approach of each project, reflect the bio-physical, economic and social dimension of the project, are reasonably quantifiable within the time and resources available for the study, and are considered relevant by the target population. Keeping these considerations in mind, 13 impact indicators and 15 approach indicators were identified (see Table 1).

b) Gathering information about these indicators in each project.

In order to measure these indicators, three sources of information were used: 1. farmers' perception (as identified through Rapid Rural Appraisal - RRA[6]), 2. observations of the study team (consisting of a researcher and two field assistants/ interpreters), and 3. secondary sources (project reports) and interviews with project staff.

c) Transcribing this information into cardinal rankings.

In order to compare the information obtained from each village, three independent coders (i.e., the members of the study team who were involved in the RRA exercises) transcribed the qualitative information obtained on each indicator into cardinal rankings on a scale from 0 to 3.[7] The rankings of the three coders were combined by calculating the arithmetic mean to obtain one single ranking per project and indicator.

Indicators	All 13 projects			Top 8 projects[a]			Bottom 5 projects[b]		
	MIN	MAX	AVG	MIN	MAX	AVG	MIN	MAX	AVG
Overall impact	3.66	30.50	19.11	20.50	30.50	24.71	3.66	15.14	10.16
1. Biophysical impact	2.66	12.33	8.33	7.00	12.33	10.12	2.66	7.00	5.46
- Yields of major crops	0.33	3.00	1.73	1.50	3.00	2.21	0.33	1.50	0.97
- Area under irrigation	0.33	3.00	1.77	1.00	3.00	2.08	0.33	2.00	1.27
- Cropping pattern	0.00	2.50	1.47	1.00	2.50	1.79	0.00	1.50	0.97
- Bio mass availability	0.50	2.50	1.54	1.00	2.50	1.79	0.50	1.67	1.13
- State of repair of SWC structures	1.00	3.64	1.82	2.00	3.64	2.25	1.00	1.64	1.13
2. Economic impact	1.00	9.00	6.00	6.00	9.00	7.75	1.00	5.50	3.20
- Income from agriculture	0.33	3.00	1.78	1.50	3.00	2.21	0.33	2.00	1.10
- Income distribution	0.00	2.50	1.44	1.00	2.50	1.92	0.00	1.00	0.67
- Access to credit	0.00	2.50	1.41	1.00	2.50	1.83	0.00	2.00	0.73
- Extent of migration	0.00	2.50	1.37	1.00	2.50	1.79	0.00	1.50	0.70
3. Social indicators	0.00	9.17	4.78	5.00	9.17	6.83	0.00	5.00	1.50
- Women's status	0.00	2.50	1.18	1.00	2.33	1.48	0.00	2.50	0.70
- Low caste and landless status	0.00	2.00	1.27	1.50	2.00	1.69	0.00	1.50	0.60
- Co-operation between farmers	0.00	2.67	1.12	1.00	2.67	1.77	0.00	0.33	0.07
- Self reliance	0.00	2.67	1.27	1.00	2.67	1.90	0.00	1.00	0.27
Approach									
Participation	0.50	8.00	4.78	5.00	8.00	6.40	0.50	5.00	2.20
- P. in problem identification	0.00	2.50	1.41	1.50	2.50	1.87	0.00	1.50	0.67
- P. in planning/solution finding	0.00	2.50	1.56	2.00	2.50	2.06	0.00	1.50	0.77
- P. in implementation	0.00	3.00	1.81	1.50	3.00	2.46	0.00	2.00	0.77
Contribution	0.00	8.00	3.38	1.00	8.00	4.58	0.00	3.50	1.46
- C. for soil conservation	0.00	3.00	1.22	1.00	3.00	1.69	0.00	1.50	0.47
- C. for water harvesting	0.00	3.00	1.10	0.50	3.00	1.44	0.00	1.50	0.57
- C. for tree plantation	0.00	2.50	1.22	0.67	2.50	1.71	0.00	1.33	0.43
Institutions	0.00	5.00	2.90	2.50	5.00	3.94	0.00	2.50	1.23
- Natural resource management group	0.00	2.00	0.88	0.00	2.00	1.38	0.00	0.50	0.10
- Women's group	0.00	2.00	1.55	1.50	2.00	1.94	0.00	2.00	0.93
- Other village group	0.00	1.00	0.46	0.00	1.00	0.63	0.00	1.00	0.20
Other project activities	0.00	6.00	2.95	0.00	6.00	3.94	0.00	3.00	1.37
- Income generation programs	0.00	1.00	0.64	0.00	1.00	0.81	0.00	1.00	0.37
- Training programs	0.00	3.00	1.31	0.00	3.00	1.63	0.00	1.00	0.80
- Infrastructure development	0.00	3.00	1.08	1.00	3.00	1.63	0.00	1.00	0.20
Agency's involvement	1.00	8.00	4.31	3.00	8.00	5.88	1.00	3.00	1.80
- Frequency of interaction with villagers	1.00	3.00	2.08	1.00	3.00	2.50	1.00	2.00	1.40
- Quality of interaction with villagers	0.00	3.00	1.38	1.00	3.00	2.00	0.00	1.00	0.40
- Accountability to villagers	0.00	2.00	0.85	0.00	2.00	1.38	0.00	0.00	0.00

[a] These are the eight projects with overall impact scores above the average of all 13 projects.
[b] These are the five projects with overall impact scores below the average score for all 13 projects.

Table 1. Impact and approach indicators for VS1 and summary statistics for all projects and project sub-groups (n=13)

d) Combining these rankings to complex impact and approach ranks.

The project-wise rankings of the 13 impact and 15 approach indicators were then compiled into three mayor impact indicators and five mayor approach indicators by adding up the rankings for each project. The three mayor impact indicators (i.e., bio-physical, economic, and social impact) were then compiled by adding up the ranks to one single overall impact indicator for each project. Table 1 shows the scores of impact and approach indicators for all 13 projects together, and separately for the eight "best" and the five "worse" projects.

e) Cross-tabulating these ranks by using scatter plots and fitting a linear function to them.

The resulting aggregated indicators for approach and impact were cross-tabulated by using scatter plots (with x-axis for the overall impact indicator and y-axis for the approach indicators). Where a linear relationship between approach indicator and impact seemed to emerge, curve estimation was used to fit a linear model to the data. The approach indicators that produced the best fit with a linear model (highest R^2) were identified. This means that, the higher the scores for this approach indicator, the higher is the project impact. These project characteristics are therefore considered to be key factors for successful WSM.

3.2 Part II. Identifying the mechanism at work

To test whether the factors identified in VS1 do result in differences in project impact, and in order to understand the mechanisms leading to those differences, a case study of two WSM projects was undertaken during VS2. These are the MYRADA Kadiri project (sub-watersheds D, E and F), an NGO-managed project, and the Drought Prone Area Program (DPAP) in Jesta watershed, a government-managed project under the Ministry of Rural Development. Both projects are located in the same agro-ecological zone (Kadiri Mandal in Anantapur District of Andhra Pradesh, approxi-mately 20 km apart) and both commenced in 1990. These projects were selected because they represent the two ends of the spectrum with respect to participation, i.e., the former being characterized as a more participatory, bottom-up approach. Other than that, however, they are quite comparable in most bio-physical and socio-economic aspects. The methodology consisted of the following steps:

a) Gathering qualitative information on the study area and the projects.

Background information on the study village was collected from the project implementing agencies, government offices, and the inhabitants of the villages. Qualitative information on farmers' perception of approach and impact, as well as farmers' opinion on the relationship between the two, were gathered from the population in both case study projects, using PRA.

b) Analysis of records from local institutions.

In the more participatory project, the minute books of all 23 sanghas (project initiated self-help groups) were analyzed according to frequency and attendance of meetings, issues discussed and decisions taken in meetings, and frequency and degree of involvement of staff members from the implementing agencies in meetings.

c) Questionnaire survey and statistical analysis.

In order to test whether the information obtained from a) is confirmed by quantitative data, a questionnaire survey was carried out to capture the individual farmer's perception of approach and impact. It consisted of 4 parts: Part 1 = all households, part 2 = only land owners (plot wise information), part 3 = only well or tank users (well wise information), and part 4 = only owners of large numbers of livestock. A 20% random sample of all households in the two projects was interviewed.

Two sets of tests are conducted with this data. First, to test whether differences between the two projects in the distribution of impact and approach indicators are significant, non-parametric tests (Mann-Whitney test and others[8]) were used. Second, to test for relationships between approach and impact at the micro-level, i.e., household and plot basis, correlation coefficients between a range of approach and impact variables will be calculated to find out whether linear associations exist between them. Regression analysis will be used to examine the relationship between selected impact indicators (e.g., changes in productivity) and a set of approach indicators, so as to identify the approach component that seems to have the largest influence on impact. This part of the analysis of the questionnaire data has not yet been completed.

d) Farmer Evaluation Workshop.

During a 3-day farmers' workshop, farmers from both case study watersheds visited each others project site and critically observed and discussed the SWC measures implemented by the respective project agencies. On the last day, all farmers met for an evaluation session to present their findings to each other and to the project implementing agencies[9].

4 Key factors for project success

As a result of VS1, three main factors associated with project success were identified.

Factor	Underlying Principle	Result
Functioning local institutions	Local institutions are essential to enforce commonly agreed rules and regulations relating to SWC (such as social fencing of grazing land and forests) as well as to resolve conflicts within the community	Local institutions take over management responsibilities, once the external project support is withdrawn
Incorporation of indigenous knowledge & ideas into project design	Local technologies are generally more suitable to meet farmers' multiple objectives and are more cost-effective	Farmers maintain structures that fit into their environment and that can be maintained locally at low cost
Contribution (in cash or kind/ labour) from farmers for SWC measures	Farmers invest only if they are convinced that the investment contributes to achieving their objectives (e.g. income stabilisation)	Farmers are interested in maintaining structures in which they invested

Table 2: Factors associated with project "success"

While all three key factors deserve attention, we are going to concentrate here on the first two of them, because hardly any SWC project can be found in India that is not subsidized in one way or another[10].

4.1 Local farmers' institutions

Local farmers' institutions that were established as part of the WSM program were found in 9 out of 13 project. However, there were large differences in their organization, activities, and membership composition. Only in six projects the local institutions are actively involved in the watershed programs for planning (e.g. preparation of a treatment map) and organizing labor for the implementation of SWC works. Among these six, one project (MYRADA/ PIDOW) used a micro-watershed approach. The rationale behind this lies in the easier manageability of small, homogenous group with similar interests ("stake holders"), who can later send representatives to general meetings, when issues are discussed that extend beyond the micro-watershed.

MYRADAs experience over the past years also showed that watershed activities need to be linked to income generating activities in order for people to be motivated enough to invest time for group meetings and to participate in the planning of conservation activities. All the watershed groups that were found in the 13 villages are involved in such income generating activities, mostly related to thrift and credit groups, training activities and bulk purchase of agricultural inputs.

4.2 Indigenous ideas and technologies

A wide range of local technologies for SWC were observed in the study villages. The most common ones are:
- farm bunds made of earth or stones to reduce erosion, keep water, fertilizer and manure in the field and demarcate property line, this can eventually lead to terracing of fields;
- agave planted along streams beds (to stabilize the stream bank) and field boundaries (to strengthen bunds and keep cattle out of the field),
- deposition fields: gullies are blocked with stones to trap silt; the stone wall is gradually increased in height (see Kerr and Sanghi 1993 for details);
- tanks for irrigation and percolation (the latter being small scale).

The difference between farmers' and project's technologies are:
- Farmers invest selectively in SWC; fertile land is given priority; leveling is more important than soil conservation.
- Farmers invest gradually, increasing the size of structures every year.
- Farmers combine SWC measures with other objectives (agave for robe making, bunds for fodder cultivation, stone bunds for boundary demarcation).
- Farmers use water harvesting structures for several purposes (not only groundwater recharge, but also irrigation, cattle, household purposes).

5 Why participation works

The case study of a participatory and a top-down project shows that the former obtains better results in almost all aspects that were included as impact indicators. The qualitative evidence obtained from the PRA exercises is confirmed statistically when comparing the distribution of impact variables[11] between the two projects. Similarly, the differences between the project approaches that were recorded during the PRA exercises manifest themselves when analyzing related variables[12] that are clearly distributed differently. The hypothesis that a participatory project approach leads to higher project impact can thus be accepted.

As an outcome of the findings during VS1, the underlying reasons for the success of participatory approaches were assumed to be the following (see also Singh 1991):
- If farmers decide themselves what type of SWC measures to implement, where, when, and how, the measures are likely to be better adapted to local conditions and needs than blueprint recommendations from outside.
- WSM requires collaboration between farmers, as some of the required measures transcend beyond the property of an individual. Therefore, local farmers' institutions are needed that can motivate and organize the community for collective action, enforce commonly agreed upon rules and regulations, resolve conflicts and that represent the community in matters involving external agencies[13]

These mechanisms are confirmed when analyzing the data from VS2. There is a strong relationship between farmers' involvement in the planning and design of the SWC measures and the perceived impact. Wherever farmers were consulted and informed about the SWC measures carried out, and wherever they where involved in deciding about the design and location of the structures, they observed a positive impact and, as a result, maintained them.

Local institutions play an important role in organizing and monitoring collective activities such as NFR (natural forest regeneration), nurseries, construction of check dam and roads, and in a few cases for conflict resolution. However, the latter aspect played a minor role, and conflicts between "winners" and "losers" of conservation measures seemed to occur less frequently than expected. One reason for this is probably the fact that farmers select technologies that require less cooperation. The results of the research team have been confirmed by farmers' findings during the "Farmer Evaluation Workshop".

Endnotes:

[1] See von Oppen and Knobloch (1990) for the emergence of the WSM concept in India, and Tejwani (1981) for technical aspects of its implementation.
[2] See Rajagopalan (1991) for problems related to WSM in India.
[3] Several such successful cases were presented at a workshop (New Horizons: The economic, environmental and social impacts of participatory watershed development) in Bangalore, in 1994. The workshop summarised the findings of a collaborative research project, coordinated by IIED (International Institute for Environment & Development), London. All the case study projects utilised participatory approaches to WSM. The results of the workshop indicate that increased involvement of farmers in project planning and implementation results in a positive project impact. See Hinchcliffe et al. (1995) and Pretty (1995).
[4] See also Uphoff (1992) for a concept of project approach evaluation.
[5] See Gregerson et al. (1988) on methods and indicators for economic evaluation of WSM programs.
[6] This consisted of focused discussions, mapping and diagramming exercises with different groups of farmers (small and large landowners, landless, well owners, women farmers, etc.), transect walks with farmers, semi-structured individual interviews, using visualisation aids, and final village meetings to present the findings to the villagers.
[7] This method was also used by Isham et al. (1995) for a comparative study of water supply projects.
[8] These tests can be used to test the hypothesis that two independent samples come from populations that have the same distribution.
[9] See Adolph and von Oppen (forthcoming) for details of this workshop.
[10] For a detailed analysis of the impact of subsidies on SWC programs in India see Kerr et al. 1996.
[11] These include farmers' perception of soil quality, degree of erosion yields of major dryland crops, input use, area under irrigation, fuel and fodder availability, et al.
[12] E.g., contact with project staff, consultations about design and location of SWC structures, origin of work force for SWC works, etc.
[13] See Fernandes (1993) for the role of local institutions in WSM.

References

Adolph, B. and von Oppen, M. (forthcoming): Farmers' workshops as a tool to evaluate watershed management projects: An account of experiences and problems in South India. In: Simpson, B. (ed.), Proceedings of the Working Session: "Heuristic Tools for Understanding and Working with Local Knowledge, Creativity and Communication", International Conference on Creativity and Innovation at the Grassroots, Ahmedabad, India, 11-14 January 1997.

Fernandes, A.P. (1993): The MYRADA experience. The interventions of a voluntary agency in the emergence and growth of peoples' institutions for sustained and equitable management of micro-watersheds, MYRADA, Bangalore (India).

Gregersen, H.M., Brooks, K.N., Dixon, J.A. and Hamilton, L.S. (1988): Guidelines for economic appraisal of watershed management projects. FAO Conservation Guide 16, FAO, Rome.

Hinchcliffe, F., Guijt, I., Pretty, J.N. et al. (eds.) (1995): The economic, social and environmental impacts of participatory watershed development. Gatekeeper Series No. 50, IIED, London.

Isham, J., Narayan, D. and Pritchett, L (1995): Does participation improve performance? Establishing causality with subjective data, The World Bank Economic Review **9**(2), 175-200.

Kerr, J.M. and Sanghi, N.K. (1993): Indigenous soil and water conservation in India's semi-arid tropics. In: E. Baum, P. Wolff and M.A. Zöbisch (eds.), Acceptance of soil and water conservation. Strategies and technologies. Topics in Applied Resource Management in the Tropics, Volume 3, DITSL (Deutsches Institut für Tropische und Subtropische Landwirtschaft), Witzenhausen (Germany), 255-289.

Kerr, J.M., Sanghi, N.K. and Sriramappa, G. (1996): Subsidies in watershed development projects in India: Distortions and opportunities. Gatekeeper Series No. 61, IIED, London.

von Oppen, M. and Knobloch, C. (1990): Composite watershed management: A land and water use system for sustaining agriculture on alfisols in the semiarid tropics, J. of Farming Syst. Research-Extension **1**(1), 37-54.

Pretty, J.N. (1994): Alternative systems of inquiry for sustainable agriculture, IDS Bulletin **25**(2), 37-48.

Pretty, J.N. (1995): Regenerating agriculture. Politics and practice for sustainability and self-reliance, Earthscan, London.

Rajagopalan, V. (1991): Integrated watershed development in India: Some problems and perspectives, Indian Journal of Agricultural Economics **46**(3), 241-250.

Singh, K. (1991): Determinants of people's participation in watershed development and management: An Exploratory Case Study, Indian Journal of Agricultural Economics **46**(3), 278-286.

Tejwani, K.G. (1981): Watershed management as a basis for land development and management in India. In: R. Lal and E.W. Russell (eds.), Tropical Agricultural Hydrology. John Wiley & Sons Ltd., Chichester, 239-255.

Uphoff, N.T. (1992): Approaches and methods for monitoring and evaluation of popular participation in World Bank-assisted projects. World Bank Workshop on Popular Participation. Washington D.C., Feb 26-27, 1992.

Address of authors:
Barbara Adolph
KFSR/E, Mashare Agricultural Research and Development Institute, Private Bag 2078, Randu, Namibia
Timothy G. Kelley
ICRISAT (SEPD), Patancheru 502324, India

The Participatory Watershed Development Process
Some Practical Tips Drawn from Outreach in South India

J. Mascarenhas

Summary
The urgent issue of sustainable natural resources management is closely linked to the participation of local communities. Such approaches involve complexity in social, economical, political, administration, technological managerial, and environmental terms, all of which need to be taken into account. Target oriented or 'blue print' approaches would never work in such a context. A community oriented approach stands a better chance of success.

This paper gives a detailed illustration of the approach developed and being followed by OUTREACH in South India. The context is that of micro watershed development. It identifies 4 definite stages in the process:
  I - Preparatory process;
  II - Planning process;
  III - Implementation process;
  IV - Withdrawal process.

This is not a rigid format and different stages may overlap or run concurrently. However the most important points made here are:
a) The need for a preparatory activity before program implementation.
b) The use of Participatory Rural Appraisal (PRA) approaches in the planning, implementation and management process.
c) The need for developing the stake of the client community in the program.
d) The need for the development of local community based institutions which will continue and sustain the development process.

Keywords: Participation, preparatory process, natural resources management, Participatory Rural Appraisal (PRA), local institutions, sustainable.

Introduction

As the issue of sustainable natural resources management (NRM) becomes more and more urgent, it is also becoming clear that sustainability is closely linked to the participation of communities who are living in close association with these natural resources. Several approaches today focus on the need to involve communities in the planning, implementation and management of natural resources. Several experiments have been initiated to try and enhance this process. One of the most significant efforts in this regard has been that of NGO's in different parts of the 3rd World. They have brought with them the skill of community mobilisation, as a very valuable new input into the development process. It is this that holds the key to meaningful involvement of client communities in NRM and consequent sustainability. This paper puts together the experiences of OUTREACH in South India, mainly derived from its own direct experience and from its association with Government and other NGOs who are involved in NRM projects. The context that is described is that of watershed development (WSD) in semiarid and drought prone areas in the states of Karnataka, Andhra Pradesh and Tamil Nadu in South India. These are areas in which the staff from OUTREACH have been involved for over the last 10 years, in trying to develop approaches to NRM that are relevant, participatory and sustainable. This paper attempts to offer some guidelines for practitioners who are engaged in this complex task.

Processes in participatory watershed development: This involves several stages and activities as described below.

Stage I - Preparatory process: (Preparing village / MWS Communities)

In well-established NGO projects, communities are invariably organised into functioning self help groups (SHG's) and are aware of development activities which are going on in the area. They are able to place demands on the 'delivery system'. Hence it is only necessary to orient communities towards new programs which are related to WSD activities. This is done through a combination of exposure trips and field interactions with other well established watershed communities. The start up period from which MSD activities (actual physical works as well as institutional and training aspects) can take place is about 5-6 months.

Areas where there are no NGO's working would need a longer startup period, of about 12 to 18 months. The process will start with an awareness of general development needs and activities generating community participation, initiation of SHGs, savings and credit activities and lead to environmental and watershed awareness. Preliminary 'PRA's to understand more about the watershed community and various interest groups within the community, existing livelihood systems, (especially those that are based on the land) community needs and motivations as well as the history of the village/community are needed (PRA I). This leads to the identification of entry point activities, which are used by the project to mobilise local communities as well as organise them into SHG's. It is found that communities are not unaware of the causes and manifestations of natural resource degradation. The evidence of communities making efforts in terms of soil conservation, moisture conservation, tree planting and maintenance of soil productivity, is plentiful. But these efforts are limited due to lack of resources (mainly financial and organisational). What is new to the community is the scientific basis of the watershed as an ecological system or a micro environment which links one farmer with others, in his own village and beyond, in more ways than he earlier was aware of.

Stage II - Planning process

As soon as communities have been brought to a stage where they have been introduced to the watershed concept and can relate to it, the project moves to the second stage consisting of the following:

1) Demarcation/delineation of the /Micro-Watershed (MWS). Here the drainage lines and the ridge line of the MWS are identified. In this part of the process the two important items to consider are:-

a) **The size of the MWS** - It should not be too large as the size of the MWS group becomes unmanageable. Individual group members will not be able to identify and relate his/her land with the rest of the watershed. The MWS should not be too small either because community organisation will lack the minimum size needed for a dynamic group interaction to take place. Experiences indicate that an ideal MWS size is between 100 to 200 ha, depending on the location, population densities, slopes, land holdings, etc. Once the MWS is delineated, the drainage map is drawn. This is a magnification and elaboration of the toposheet and also includes additional details of the drainage lines as seen from the ground.

b) **The structure of the MWS** relates to the portions of land which belong to various villages and hamlets which are located nearby. This has implications in terms of community organisation and management of the MWS program and hence the area itself should be finalised in consultation with community.

2) Identifying stake holders: This consists of farmers owning land in the watershed who live in and around the MWS as well as landless and other families who are using the watershed resources. In addition to Government records, such information can also be obtained from the community.

3) Study of MWS resources: Here the local community is exposed to the complexities and relationships that exist in their watershed in a comprehensive way. This includes the delineation of the MWS itself, its boundaries and drainage lines, its contours, slopes, ravines, erosion and degradation, water bodies, land use farming systems, trees, indigenous technologies and local management systems. The group looks at their surroundings differently. Familiarisation with the new context takes place. For the first time, they meet and relate to farmers from adjoining areas and villages and begin to consider how they are linked and how they can work together. This is done in a general sense (PRA II) and is featured by numerous familiarisation transect walks and group meetings to identify problems and opportunities.

4) Participatory planning: Based on the motivation of the community to initiate work in their MWS, dates are set for a participatory planning exercise (PRA III). This is usually done over 3 to 4 days and involves the Gramsabhas and Panchayats (local civic bodies). During this exercise a detailed analysis of the watershed is done in terms of its agro-climatic and socio-economic conditions. The PRA exercise is generally divided into two main parts:

a) **Exploratory**: This exercise generates trends (historical transects and analysis) in resource use, land-based and non land based livelihood systems, traditional ways in which the resources of the WS have been managed and used (eg. fuel, fodder, trees grown, water availability and use, grazing systems, indigenous technologies etc.). Problems and opportunities, relationships of the MWS with the main village and the other neighbouring villages, study of local formal and informal institutions and their relevance, seasonal patterns of activity such as agricultural and domestic operations, migration, diseases of humans and cattle, fodder availability credit needs, difficult times of the year, slack periods and so on are addressed in a more detailed manner. Of particular importance are the study of the management and use of resources and the wealth ranking exercises which are carried out to identify who the poorest members of the community are. This is done by means of participatory social and resource mapping exercises. These, exercises also indicate occupational and land holding status and other information of a socio economic nature. The study of the management and use of natural resources in the MWS includes the status of land use - (private & public, cultivated, non-cultivated, grazing, fallow, single/multiple cropping, irrigation, wastelands, ravines, problem areas, different soil types, fertility, productivity and so on.)

b) **Concluding:** Where the micro watershed group facilitated by the NGO arrives at a comprehensive plan for treatment and management of the watershed. This has two parts:

1) INITIAL - where the watershed community transects the watershed (sweeping transects) and identifies problems and opportunities which are specifically connected with the land. These include individual field/farm development plans, soil conservation bunds, tree species required, appropriate land use, treatment of gullies, and other eroded patches, nullah treatments, location and construction of water harvesting structures and so on. Some issues that are likely to arise during this process are a discussion on the use and management of private fallow lands and of the upper reaches of the watersheds whether commonly or privately used and whether owned by the revenue department, forestry department or members of the community. This includes grazing land and its management. Members of the micro watershed community are willing to start discussing and planning on their own land, where the title and benefits are clear. They are able to relate better to investment decisions here rather than on the commons where they are not exactly sure about various aspects such as investments, usufruct rights, procedure for joint management and so on (this usually happens towards the end of Stage III).

The transect observations (problems and opportunities) result in the identification of treatment activities in the micro watershed. These are represented visually on a map of the micro watershed prepared by the community (participatory mapping). This map indicates and illustrates the treatment plan as proposed by the micro watershed community in physical terms such as bunds, plots for horticulture, diversion drains, placement of checkdams etc. Reconciliation of the topo maps with the revenue map and the treatment map prepared by the community takes place. Discussions are held between the various parties involved about the treatments to be undertaken on different survey numbers in the MWS whether on individual or common lands.

2) Final: encompasses the following elements:

1. Treatment Plans (Activity)	Who does? What? Where?	This has been discussed in detail above. Finalisation of the treatment plan is agreed upon by the community.
2. Budget Plan (Financial Plan)	Unit cost & total costs - Who contributes? How much?	Here the costing for the different activities is worked out, as well as the contribution from the community, NGOs, Government and others. It is during this stage that agreements are negotiated with the community vis a vis cost sharing, CPR management, gender, landless etc. This exercise also deals with the aspects of zero cost, low cost, medium and high cost activities in order to place emphasis and focus on the mobilisation of resources from different sources including the community.
3. Time Plan (PERT Plan)	What is to be done? When?	The time plan done by means of seasonality exercise indicates the peak periods and slack periods of the village/watershed communities. It indicates the timings for execution of the various watershed development activities according to preferences of the community.
4. Implementation Plan	Who does? What?	The implementation plan is purely operational in nature. Basically it deals with which group agency or individual is responsible for each activity or sub activity. For example, if the activity is tree planting, then who will select the species to be planted? Who

		will raise the nursery? Who will transport the plants to the field? Who will dig the pits and trenches? What layout and type of pits/trenches? Who will do the planting, etc.

5. Management Plan — Who does? What? — Similar to the implementation plan, the management plan will come to terms with who will manage various aspects of the complete work. To continue with the above illustration, who will protect the plants? Who will water them? Who will manure and maintain them?

It is also at this time that decisions on the Watershed Management Committees (WMC) are taken, such as their composition, functions, frequency of meetings and so on. Agreements on the plan are entered into and signed by all parties on a Memorandum of Understanding.

Stage III   Implementation process

Assuming that the lead up to the beginning of this phase takes 6 to 18 months, this period will take up to 2 - 2 1/2 years and has the following stages:-

1) Strengthening of community organisations within the watershed: Community organisation and establishment of the WMC & MWS group is not a 'one time' activity. It continues throughout this period, basically through group strengthening activities such as training, discussions, exposures etc. The groups have to be supported and facilitated in their attempts to evolve, and have to be enabled to carry out their various activities particularly in the NRM and Credit Management fields. **This is CRITICAL to the development process. Groups must be given opportunities to learn and grow, even if some of these have to be deliberately created.** For example in most cases the entire experience of book-keeping is new. Similarly the opening of a joint bank account, of planning jointly, of monitoring quality, of negotiating rebates on bulk purchases of fertilisers, etc., are all new. But the groups quickly get used to these things and are able to evolve their own self-regulatory mechanisms, such as rules, norms, fines, penalties, rewards and sanctions.

2) Implementation of Watershed Management Activities on Private Lands: During this period it is quite natural for farmers to start with what they can see and relate to very closely, such as their own and neighbouring farmers' land. Each farmer has a choice of doing work (agreed upon in the plan) on his own farm either on his own or with the help of the Watershed Management Communities. Usually farmers make their own arrangements, by negotiating reciprocal arrangements with their neighbours and are likely to fall back on the WMC when they run into a conflict which they cannot resolve themselves, eg. boundary disputes where soil conservation structures encroach on the neighbouring farmers field or water from one farmers fields spills over into anothers. It may also happen that some additional work is needed to be done, which was not foreseen earlier for which the approval of the WMC is needed.

Works in the MWS are carried out according to the plan at times which are convenient to the MWS group. Earth works, bunds, gully checks & plugs, nallah bunds, forestry and horticulture, pits and trenches, etc, are usually done during the period from November to June, after the harvests are over and before the new sowing season starts. Tree planting takes place from the middle of August onwards, after the crops are well established and all cultural operations are complete. This means advance of nursery preparations. Similarly for crop cultivation, decisions on cropping patterns take place well in advance. Farmers usually have a general plan by the end of the harvest, but firm it up around March/April. They also need to get their inputs (seeds, chemical fertiliser,) ready early i.e., by end April or mid May.

3) Development and management of CPR's: Once a rhythm of work is established in the watershed programme and the micro watershed group's confidence is enhanced, the issue of treatment and management of the common lands, particularly those in the upper reaches is addressed. Demonstrations of these aspects are taken up on a sample basis during the first phase of the watershed treatment, to give the community an exposure to the benefits of such an activity and to enable them to develop a greater stake in it. The treatment plan for the commons may need to be emphasised through another PRA exercise (PRA IV) in which technical staff NGOs and GOs also participate to make it more definite. While the system of contribution towards costs is true of privately owned lands, in common lands the system is somewhat different. MWS communities may agree to contribute towards the costs of development, provided their level of confidence in the system is high and they are sure of the benefits. This contribution may be in the form of labour, prevention of grazing and theft or even in terms of agreeing to rights of usufruct in favour of landless and women's groups in the watershed. However situations vary, and each common property agreement would have to be negotiated separately.

4) Non-farm Income Generating Programs (NFIGPS) gender & equity: These are targeted at the marginalilzed groups in the watershed, such as the women, landless, rural artisans and marginal farmers. The Wealth Ranking exercise carried out earlier in Stage II helps in identifying these categories of people. NFIGPS are closely linked to the regeneration of natural resources in the watershed as they will offset some of the pressure on these resources, particularly the perennial biomass such as grasses, shrubs, trees and other vegetation in the upper reaches which serve the function of providing soil cover against the impact of wind and rain erosion. These are usually sources of fuel, fodder and timber and are collected and sold by the more marginalized groups. Commonly, NFIGPs consist of small businesses, services and industries. The individuals concerned initially get their working capital from their respective Savings & Credit funds, but provisions need to be made to supplement this fund.

NFIGPs also serve to address the issue of **equity** in the watershed where a large portion of investment is taking place on the land by apportioning resources to enable those without land or with small portions of land to stabilise their livelihood systems. The issue of equity is also linked to the management of the CPR's in the watershed and the sharing of usufruct as described earlier.

Gender: Throughout the program sufficient attention has to be paid to the practical and strategic needs of women. A strong emphasis must be placed on the development of women in the watershed, particularly in terms of their economic condition but not at the cost of their quality of life, i.e. they should not be burdened more than they already are with additional jobs and responsibilities **even if these are of an income generating nature.** A few important 'Gender Actions' in the program are:- 1) Organising & Strengthening of Women's Groups separately, 2) Savings & Credit Programs for Women 3) Initiating Income Generating Programs (of their choice), 4) Access and Control over portions of CPR especially fodder and fuel lots, 5) Including suggestions of the women in the watershed plans (e.g. selection of tree species etc., location of fuel lots, location of check dams, etc.), 6) Giving adequate representation to women in the MWS.

It is important that the domain in which the PRAs are held or the planning takes place is not male-dominated. Or, even if they are, ways should be found to give access to women and include their ideas and suggestions in the MWS plan.

5) Monitoring of Work: Monitoring addresses those works which have to be measured physically such as soil and moisture conservation and forestry activities. The purpose of monitoring is to have a look at the quality and quantity of work, whether it is according to the agreed plan, changes and the basis for these, additional works needed and also for the purpose of payments. Monitoring is usually done by a group nominated by the WMC or by the MWS group themselves. In addition there may be one person each from the NGO and the Government to complete representing all partners at the MWS level. When the work is on farmers land, where each individual is expected to contribute according to the agreement reached at the end of PRA III, the WMC members take the lead in terms of assessing quality, measuring, making payments and recovering contributions. Contributions are usually credited to the MWS groups and will form part of the MWS capital which is used for income generating activities of a non-farm nature and for maintenance of the assets created.

Other aspects connected to the monitoring of work relate to looking at the way the crop trials and demonstrations are progressing, sharing of biomass, maintenance of horticulture & forestry plants, soil conservation works and repairs, including damage which may occur due to heavy rains, etc. This is an initiation towards future 'maintenance of assets' and long term sustainability. One means by which this is done is by weekly 'Watershed Walks' or 'Sweeping Transects' where members from the WMC, MWS members and the external agents walk around together in the watersheds and look at what is happening, particularly in relation to the treatments that have taken place and their effect on soil and moisture conservation and revegetation. Observations are discussed in the field itself, and at the WMC meetings. Any issues which cannot be resolved at the individual, MWS group or WMC levels are brought to the Gram Sabha and if they cannot be sorted out here are likely to move to a higher level and placed before the Panchayat. There might also be some issues mainly of a policy nature which would need interventions at higher levels either at the District or State levels.

Stage IV - Withdrawal from the MWS
This lasts for around 2 to 2 1/2 years after Stage III. This period is characterised by continued group strengthening activities such as training of the groups in systems and procedures, book-keeping, technology generation, and other managerial functions. This is also the time when linkages are established with other institutions inside and outside the watershed area. By this time, if the community organisational activities have been effective, the groups should be confident of managing their affairs on their own, with very limited support. This includes capability and confidence in placing legitimate demands on the Government system and mobilising Government schemes and programs which are meant for them. It is also at this stage, once again, if community organisation has been successful, that the established MWS groups act as promoters for the emergence of new groups in the adjoining areas of the watershed. During this phase, beginnings are made towards the establishment of Apex watershed institutions. These emerge out of the MWS Groups and WMC's and will gradually take over the management of the watershed activities. This also is the time when the facilitating NGO should withdraw from direct interaction with the group, and hand over its role to the apex institution.

Conclusion

Participatory watershed development does not just happen on its own - it has to be made to happen. What is described above gives an idea of the complexity that exists if we are to address participatory NRM. A target or blue print approach would never work in a such a context, and a process approach is essential. However in order to arrive at this, a policy and institutional environment needs to be established. The role of NGO's in the preparation of watershed communities needs to be emphasised and enhanced. Appropriate HRD programs need to be urgently developed and implemented on a large scale. Armed with this preparation, we can be sure of better results.

Address of author:
James Mascarenhas
109 Coles Road
Fraser Town
Bangalore - 560 005, India

Watershed Management and Sustainable Land Use in Semi-Arid Tropics of India: Impact of the Farming Community

R. Chennamaneni

Summary

A watershed management approach combining improved farming practices with soil/water conservation and appropriate land use offers scope for sustainable development of agriculture. The analysis shows a positive impact on land use intensity, cropping pattern, crop yields, human labour utilisation, creation of assets etc. Combined with favourable institutional support, such as credit, subsidies/incentives, and input supply, these measures could substantially contribute towards income objectives of the farmers. However, problems such as a sectoral approach in implementation, lack of different package options in tune with farmer household strategies such as income diversification/stabilisation, and neglect of common property resources hamper the overall gains of watershed management. The successful replication of this capital-intensive technology to other areas necessitates the provision of adequate infrastructural and institutional prerequisites. Farmer participation is a key element for the adoption of management measures.

Keywords: Watershed management, technology adoption, sustainable land use, impact assessment, institutional response

1 Introduction

Semi-arid land accounts for about 170 million ha of land in India (54% of total) and supports over 400 million people (CRIDA, 1994). The average productivity here is estimated at about 1/3 that of irrigated lands and per capita income of farmers is about 1/4 of the national average (Walker, 1990). Poverty and micro food security problems are significantly related to problems of increasing the crop yields and incomes in this region. On average, about 10-40% of total rainfall is unutilized and through soil erosion and sedimentation about 8.4 million tonnes of nutrients are lost (Govt. of India, 1991a). This poses a serious limitation on moisture storage and sustainability of crop production. As several ad-hoc efforts have been unsuccessful, a major policy response at the national level is to follow a watershed approach for developing agriculture in areas with little access to assured irrigation (Govt. of India, 1991b).

Watershed management is the integration of a range of sectoral activities, such as soil and water conservation, minor irrigation, animal husbandry and other rural development activities. The micro-watershed is usually the most appropriate level of intervention. The main components of a pilot project at Maheswaram (1987-92) under study are

a) **conservation measures**, such as contour bunds/vegetative hedges to filter runoff and control soil erosion, contour cultivation for in-situ moisture conservation throughout the field, opening contour dead furrows at appropriate intervals to trap moisture for soil recharge and treatment of drainage lines,

b) **Production systems**, such as diversified cropping systems including mixed cropping, intercropping, crop sequences, introduction of new crops, dryland horticulture and cultivation of fodder,

c) **Treatment of non-arable lands**, such as reforestation of forest lands, silvipasture in degraded forest lands, pasture development, subsidiary occupations such as dairying and other agro-based industries.

Several problems concerning adoption and maintenance, imposing standardised and expensive technologies, lack of farmer involvement, etc., have been reported from implemented watershed programmes (Whitaker et al., 1991). However, less is known about the impact of a watershed management approach on the farmer and his farming. This paper assesses the impact of watershed management on the farmer and his production systems. This analysis is based on field studies undertaken between 1992 and 1995.

2 Material and methods

Primary and secondary data were used for comparing two villages under watershed and two under non-watershed conditions of same agro-ecological and socio-economic region. There were 152 farmers of small (below 2 ha), medium (2 to 4 ha) and large farm groups (4 ha and above). The analysis focuses on the impact of watershed management on sustainable land use of farmers taking productivity, employment, income etc., as examples.

3 Impact of watershed management

3.1 Soil conservation, land development and irrigation

The important components of the conservation programme were engineering and mechanical measures, vegetative barriers, agronomic practices and land development. Farmers of all size groups were covered under this programme (Fig.1). The most important measure of the soil conservation programme was khus (vetivaria) grass plantation which acts as a vegetative barrier and has been tested as a quick and cheap method. Agronomic measures such as deep ploughing, cultivation across the slope and ridge, dead furrow, etc., were followed as non-monetary inputs.

Providing irrigation facilities and creating permanent assets through deepening of old wells, digging of new wells and supply of electric motors was one of the most important measures. The capital for this investment was made available by project staff through government, cooperative and commercial banks. The analysis reveals the fact that all large farmers invested in deepening of old wells, whereas digging of new wells was the preferred investment by small and medium farmers.

Conservation measures were government implemented programmes and people treated them as such rather than their own. Concerning agronomic practices, in contrast, farmers were convinced of the real benefits and there was awareness about technologies created by the project staff.

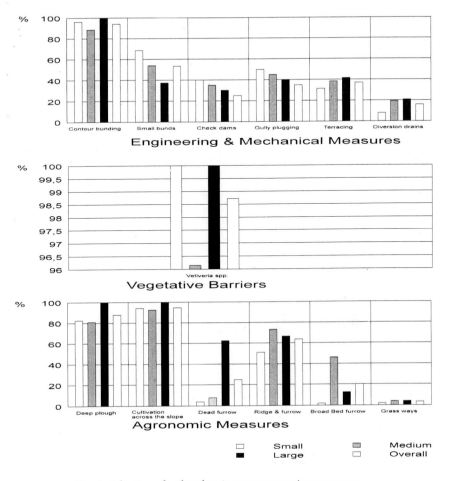

Fig. 1: Adoption of soil and moisture conservation measures
(Percentage of adopters)

3.2 Land use intensity

During the study, land use intensity was 135.2, 135.1 and 139.8 % in the watershed area which decreased with increase in size of holding for both years. These results may be attributed to the fact that small farms use their lands intensively and also to the fact that medium and large farms keep comparatively large areas under fallow and permanent pastures. The land use intensity in non-watershed areas with 105.9, 106.0 and 108.5% for the years 1992-93 to 1994-95, respectively, was lower than in watershed areas. Contrary to the trend observed in watershed areas, land use intensity has not shown any remarkable decreases with increase in size of holding.

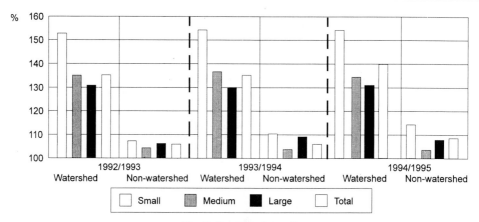

Fig. 2: Land use intensity

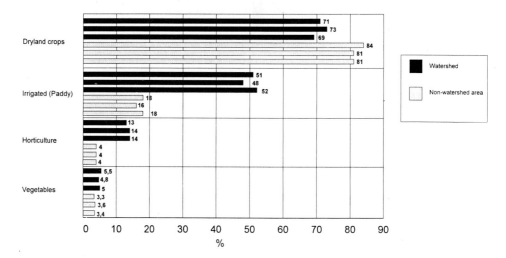

*Fig. 3: Cropping pattern in watershed and non-watershed areas
(in % of gross cropped area for 1992/93, 1993/94 and 1994/95)*

3.3 Cropping pattern

Implementation of the watershed management project witnessed a change in cropping pattern. There was substantially more area brought under paddy, hybrid jowar and jowar + redgram intercrop in the watershed area. For the period of study, the overall level of gross cropped area under dryland crops was 71, 73 and 69%, respectively, for irrigated area, 51, 48 and 52% of the gross cropped area was devoted to paddy, 13, 14 and 14 % was under horticultural crops, in 5.5, 4.8 and 5.0% of gross cropped area vegetables were cultivated (see Fig.3).

In the non-watershed areas, for the years 1992-93 to 1994-95 respectively, as much as 84 and 81 % of gross cropped area was under dryland conditions, only between 16 and 18% under paddy, about 4 % under horticultural crops and around 3.3 to 3.6 % under vegetable cultivation.

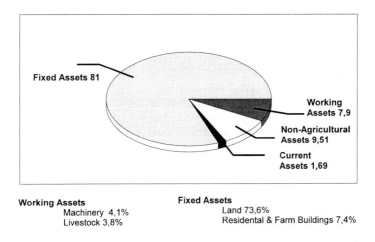

Fig. 4: Total assets in watershed area (in %)

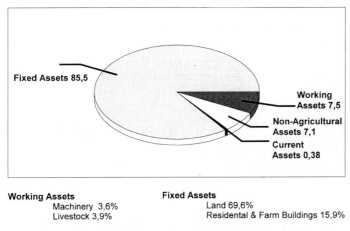

Fig. 5: Total assets in non-watershed area (in %)

3.4 Assets and liabilities

Farm productivity and economic efficiency of the farmers depend largely upon the quality and type of farm assets. Assets are calculated as total assets per household and per acre in different farm groups. At the overall level, about Rs. 411 000 and Rs. 301 000 was the value of assets in watershed and non-watershed areas. The largest assets possessed by the farmers, lands and buildings constituted about 81 and 85 %, respectively, of total assets in watershed and non-watershed areas (Figs. 4 & 5). More clearly, land was the single largest asset in both the areas with about 73.6 and 69.6 %, respectively. The availability of liquid assets was very low with 0.4 % of total value of assets in non-watershed area versus 1.7 % in the watershed area. In the watershed area the share of land in total value of assets increased with the increase in size of holding whereas this trend was not observed in the non-watershed area. Total assets per acre was more for small farmers of the watershed group. Total value of assets per acre increased with a decrease in the size of holdings in the watershed area, whereas it was the opposite in the non-watershed area.

As is well known, the ability of farmers to create capital assets depends to a large extent on their accessibility to sources of finance. Taking short, medium and long-term loans in different farm groups, it was found that farmers in the watershed area secured greater amounts when compared with the non-watershed area. It was observed that the percentage share of short-term loans decreased with increase in farm size whereas in the case of medium-term loans it was vice-versa in both areas. However, the percentage share of long term loans was found about equal in small and medium farms and higher in large farmgroups of the watershed area, whereas it increased with farm size in the non-watershed area.

The results indicate significant variations in the possession of total assets and liabilities of the farmers in watershed and non-watershed areas as well as between different farm size groups. Farmers in the watershed area are clearly in a favourable position for investments and in running their farm enterprises economically.

3.5 Cost of cultivation and productivity of crops

The operational, fixed and total costs of cultivation of paddy, jowar, jowar+redgram in watershed and non-watershed areas is presented in Fig. 6. The analysis reveals that for all crops the total costs of cultivation were higher in the watershed area than in the non-watershed area. Operational costs are greater in the watershed area due to the higher use of different recommended inputs. Fixed costs were higher in the watershed area than in the non-watershed area due to the higher level of rental value of owned land.

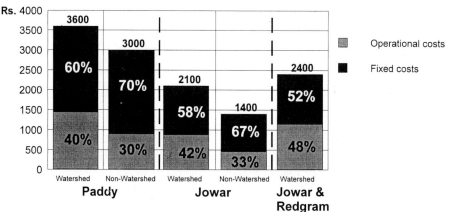

Fig. 6: Cost of cultivation and structure of different crops overall in watershed and non-watershed areas

Productivity is one of the important indicators of agricultural development. The productivity levels of different crops in watershed and non-watershed areas are compiled in Fig. 7. The variations in yield levels of paddy was low among the three different categories of farmers in watershed compared with non-watershed areas. The average yield of paddy in non-watershed areas was comparatively low with variations among farm groups. Average yields of jowar were high in watershed areas than in non-watershed areas and showed slight variations among farm groups. Though with substantially lower yields, variations in non-watershed areas were also present.

However, in both these crops, a positive relationship between the size of holdings and average yield in non-watershed area could be seen, whereas the opposite trend was relevant for watershed areas. Yields of intercrop jowar+redgram, cultivated only in the watershed areas, showed the same trend of direct relationship with farm size.

Detailed analysis of yield levels shows that paddy growing small farmers have obtained more than a 100% increase, medium farmers about 57% and large farmers about 50% in yield over the same farm size groups of non-watershed areas, which was significant at the 5% level. Overall there was an yield increase of about 69% over the non-watershed group. In jowar, overall there was a yield increase of 129% over the non-watershed group, which was significant at the 5% level. Small, medium and large farmers of the watershed group obtained 176, 106, and 109 % increases over the same farms of non-watershed groups, which was significant at the 5% level. This clearly shows that watershed farms have increased their crop productivity.

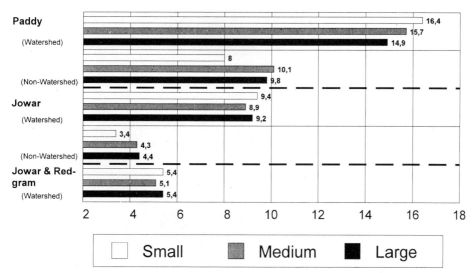

Fig. 7: Productivity of different crops (Qtl/acre)

3.6 Employment

Analysis of human labour utilisation shows that the total number of days employed by an adult worker was 235 and the number of idle days per worker to be about 130 in the watershed area (Fig. 8). 126 days were utilized for own farm work. Hiring out of labour on other farms, an important aspect of labour utilisation, was 44 days per year. About 33 and 22 days were utilised for off-farm work and self-employment, respectively.

Among different size groups, the number of days employed on own-farm varied from 70 in small farms to 169 in large farms, with 139 days in the case of medium farms. The number of days utilized on own-farm increased with the size of holdings and vice-versa with regard to off-farm work. It was observed that small farmers hired out more (98 days) and could find self employment for more number of days, where as idle number of days were at a minimum in the case of large farmers.

In the non-watershed area, the total number of days utilized for human labour employment was much less when compared with the total number of idle days. An adult worker could secure employment only for about 174 days per year as against 191 days which were idle. Own-farm provided employment for about 101 days; off-farm work for 12 days only. Hiring out of labour accounted for 49 days and self-employment could absorb the human labour for only about 9 days. The number of days utilized for own-farm work by adult workers ranged from 65 days to 127 days in small and large farmers while it was about 111 days in medium farmers. As in the case of the watershed, the number of days employed on off-farm work decreased with size of holdings, while the number of days employed on own farm work increased with the size of holdings in the non-watershed area.

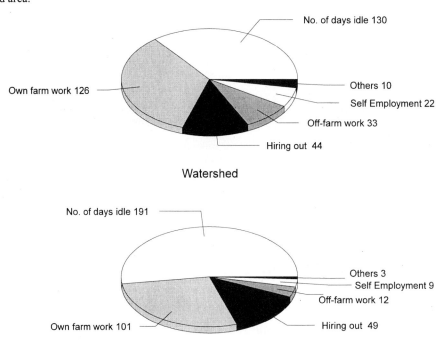

Fig. 8: *Labour utilisation in watershed and non-watershed areas (average for all farms; in number of days*

The analysis shows that own-farm provided more employment when compared to all other activities in both watershed and non-watershed areas. The number of days utilized for own-farm work were more in medium and large farms in the watershed area than in the non-watershed area. The variation in labour utilization between watershed and non-watershed areas was statistically significant in off-farm, self-employment, hiring out and other activities.

3.7 Income

Income of farm holdings includes the income obtained by all members of the family from different sources. Figure 9 clearly shows the large difference in income levels of farm holdings between watershed and non-watershed areas. The average income per farm holding was Rs. 25237, Rs. 46201 and Rs. 99765 for small, medium and large farms in watershed and Rs. 9642, Rs. 23441 and Rs. 50277 for the same categories in non-watershed area. The analysis shows that the difference in the income levels of farm holdings between watershed and non-watershed areas was large and was statistically significant at the 5% level.

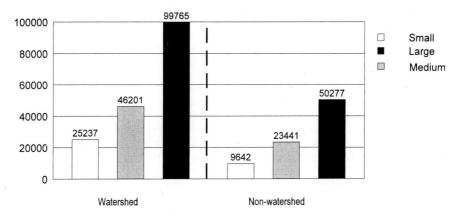

Fig. 9: Average gross income of different farms in watershed and non-watershed areas (in Rupees; 1 US$ = 37 Rupees)

The results of quantification of measures of distribution of income showed the following results. For watershed areas, higher inequalities were associated with large farmers compared with small and medium farmers whereas in non-watershed areas these inequalities were associated with small farmers when compared with medium and large farmers. On the whole lesser inequalities were associated with the watershed group when compared to the non-watershed group.

3.8 Creation of permanent assets

Agro-forestry is an important technological component of the watershed management programme. The fruit trees such as mango, guava, pomegranate and sitaphal and also forest trees were included in agroforestry. The total area brought under forest trees was about 62 acres. A number of fruit plants and forestry plants were planted around the farm sheds, houses etc. In addition to fruit plants, fodder trees, such as subabul, timber plants such as causurina were also planted.

To facilitate additional income through animal husbandry activities in watershed areas, fodder component received due attention in livestock. Pasture development was encouraged both in private and government lands and incentives such as seed supply was practiced. Cultivation of important fodder crops and pastures, such as stylosanthus humata, cyrarto, fodder jowar, paragrass, etc., was sucessfully introduced.

4 Conclusion:

1. A Watershed Management approach can contribute towards sustainable land use and income objectives of the farmers. However, without provision of the necessary infrastructure and institutional prerequisites as well as initial subsidies/incentives, this capital intensive

technology may not be accepted by the farmers of other regions. More research on incentives/ constraints faced by the farmers and villages is needed. If not we run the risk that governments, NGO's and researchers will try to push innovations on farmers who find them too costly/not sufficiently profitable and out of sync with food security, income diversification/ stabilisation objectives of their households.
2. Initially, problems concerning adoption and maintenance of watershed management technologies have come up due to several reasons, most important of which is related to scale. Institutional solutions such as watershed committees with project back-up at village panchayat level proved successful. The best way would be to integrate watershed development programmes with village development plans to ensure greater acceptability.
3. Imposing standardised technologies, following a sectoral approach without coordination has been a major shortcoming. Implementing agencies remain sometimes blind to farmers rejection of some aspects of technology such as huge capital outlays, low fodder, labour constraints, off-farm activity, competition of organic manure as fuel, of water between trees and annuals in agro-forestry. A multidisciplinary approach should be taken up by researchers and policy makers to ensure that complexities of traditional farming systems and farmers choices are addressed and different package options in tune with their farming strategies are offered.
4. Concerning common property resources, only agro-forestry and pasture development was accorded priority, while wasteland utilisation through silvipastoral system did not get much attention. Due to bureaucracy in collective management in the past, maintenance of village tanks is in a deplorable state. Though motivation of villagers for collective action prevails, the lack of technical and financial back-up by the irrigation department and clearly-defined abilities could not bring about the desired change.

References

Central Research Institute for Dryland Agriculture (CRIDA): Annual Report 1993-94, Hyderabad.
Government of India (1991a): National Watershed Development Project for Rainfed Areas. Guidelines. Ministry of Agriculture, New Delhi.
Government of India (1991b): Eighth Five Year Plan. Planning Commission, New Delhi.
Walker, S. & Ryan, J. G. (1990): Village and Household Economics in Indias Semi-arid Tropics. Baltimore, M.D: Johns Hopkins University Press.
Whitaker, M., Shenoy, P.V. & Kerr, J. (1991): Agricultural sustainability, growth and poverty alleviation in the indian semi-arid tropics. Paper presented to DSE-IFPRI Conference on "Agricultural Sustainability, Growth and Poverty Alleviation: Issues and Policies" Sep.23-27. Feldafing.

Address of author:
Ramesh Chennamaneni
Faculty of Agriculture and Horticulture
Department of Agricultural Economics and Social Sciences
Humboldt University
Luisenstraße 56
D-10099 Berlin, Germany

People's Participation in Watershed Development Schemes in Karnataka - Changing Perspectives

K. Mukherjee

Summary

Karnataka is a large state in southern India where farming is predominantly rainfed and watershed development has considerable potential. Centuries ago, maintenance of the catchments of irrigation tanks with popular participation brought all components of watershed development into play. Later, erosion of people's initiatives placed an excellent traditional system into disuse. Repeated famines in the region at the beginning of this century led to research-initiatives in dry-farming. These, however, did not yield dramatic results and much of it was not new, but traditional wisdom packaged in scientific garb. Then came massive river-valley projects where watershed development was nothing but a technology applied to protect catchments of reservoirs. Irrigation brought problems of damaged soils and small farmers who were bypassed by the green revolution. The doctrine of dryland farming was revived, this time on a watershed basis and with an eye for the participation of stake-holders. Karnataka's watersheds saw diverse stategies for sustaining agriculture as well as the interest of the people. NGOs entered the field and have come to stay. Successes have been notched up in areas of management of common property, participative research and motivating people to share costs of development, but challenges still remain.

Keywords: Soil, water, sustainable, participation, groups, watershed development, India.

1 Not lands but peoples' minds

Watershed Development schemes are designed to touch the lives of people and improve the environment they live in. The multi-disciplinary thrust of such schemes has the ultimate aim of improving the productivity of the limited natural resource available, so that living off such a resource can be sustainable. Needless to say, watershed development cannot be sustained without the participation of the people. It is fascinating to see how watershed development which was essentially conceived as an "area development program" to conserve moisture and soil in a catchment, metamorphosed into a highly "people-oriented" program. Today it would not be an exaggeration to assert that it is not a means to treat land but the minds of the people.

India has a long history of people's involvement with management of natural resources. In the state of Karnataka, located in southern India, watershed development and conservation of soil and moisture has always been important. Deprived of the luxury of the snow-fed rivers as in northern India, Karnataka's rivers carry only rain water and most of it trickling down from thousands of streams from evergreen forests (the Western Ghats). The copious rain showered on this dense canopy is soaked up like a sponge and released year-round. Whatever rain could not be captured by the forests was (and continues to be) trapped in thousands of irrigation tanks which pock-mark the map of Karnataka. All this however does not add up to beyond 22% of lands having the facility of assured irrigation. About 8 million hectares of dry lands remain a challenge of watershed development in Karnataka. In the following pages we will examine the changing flavour of people's participation in watershed development with the passage of time.

2 Think-tanks

The earliest involvement of the rural community in watershed development was in the upkeep of irrigation tanks. Some of these tanks are more than 500 years old. Many of these old tanks have been integrated within modern river-irrigation projects. Epigraphic evidence shows a prominent role played by the villagers, especially the landed gentry in preserving these tanks. The construction of a tank called for celebration and thanks-giving and the village temple was erected on its bank. Temple institutions in those days owned sizeable tracts of land and thus maintenance of tanks took on religious fervour. The Chalukyan empire (973-1336 AD) was the golden age of tanks when popular participation was at its zenith. This tradition continued in the subsequent centuries and became further refined.

Of all the maladies affecting tanks, damage to the sluice-gates and waste-weirs were the commonest. The siltation of tanks presented the biggest challenge to the rural community. The history of Karnataka is replete with evidence of voluntary effort to remove the fertile silt from tank-beds. However, at the same time, being confronted with the back-breaking task of removal of silt, people also realised the importance of preventing erosion of the catchment of a tank and all classical ingredients of watershed development came into play. Epigraphic evidence points to some cases of self-imposed restraint exercised by villagers in grazing livestock in steep and vulnerable areas of a catchment. Afforestation of the upper catchment by village groups was not unknown.

War and troubled times in the 17th and 18th centuries caused the degeneration of many well-entrenched traditions and maintenance of tanks was not an exception. Added to this was the attempt of the state to institutionalise a machinery to collect "water rates". The farmers paid the price for the state's military adventures. This is not to say that such a system of pricing water was unknown in the past, but earlier systems were localized, friendlier and more accurate in gauging whether the farmer was being coerced for paying for water that he had not received. A rigid machinery of administration reinforced the belief that "tanks belonged to the government and not to the people". By paying for water (a maintenance cess was added later) the landed gentry absolved itself of the responsibility of maintenance. A silted up tank generated a demand for digging a new one! Attempts were made to undo the damage. Sir K.Seshadri Iyer, the Dewan of the princely state of Mysore (one of the constituents of Karnataka) observed (in his address to the Representative Assembly) with some desperation in 1884:

> *"Any reform to our tank system must start with a clear recognition of the fact that it is beyond the ability of government to undertake repairs and maintenance of all tanks in the province, nor will it be equitable to throw the burden on the ryots (farmers) after the village system or what little remained of it has been disorganised after the ryot has tacitly been relieved of his responsibility by the imposition of special cesses for repair of tanks."*

Attempts to involve the people in maintenance of tanks continued well into the middle of the 20th century, but the results were indifferent. When independence came in 1947 from the British, 70% of minor tanks had fallen into disuse in the Mysore state (now the southern part of Karnataka).

3 A famine of ideas

The later part of the 19th century saw the "agricultural policy" of the colonial British government conditioned by recurring famines. Between 1933 and 1935, dry-farming research stations were set up in Hagari, Raichur and Bijapur in Karnataka. The work in these research stations culminated in the "dry farming systems". The recommendations of these systems were almost trivial like contour bunds, deep ploughing once in 3 years, application of farmyard manure, inter-plantation and use of a low seed- rate. The obvious thread in all this is in-situ conservation and efficient use of moisture. Coming to the crux of people's participation, these packages designed by scientists in controlled conditions failed to enthuse the farmers. After all these efforts, the increase in yields was not significant (about 15-20%). The main constraint was the lack of genetic material capable of withstanding moisture stress.

There is also ample evidence that our agricultural scientists were re-inventing the wheel. Records show that in 1917, A.K.Y.N. Aiyer (an agricultural expert) listed the important farming practices followed by farmers in southern India and found them to be in conformity with Hardy.W.Campbell's "Soil Culture Manual" of 1904 which was a highly respected treatise in the U.S., which too at that time was coming to grips with dryland farming. Field bunding, fall ploughing, frequent inter-cultivation, drill-sowing and the use of drought resistant crops were some of the prevailing practices listed by Aiyer which pointed to native wisdom in operation, though not backed by formal research. In summary, the first half of the 20th century was a lack-lustre period for the dryland farmer who stood alienated by the decay of old informal institutions on one hand and the inability of agricultural science to come to his rescue on the other.

4 Holding back the silt

The problem of accumulation of silt in the reservoirs of major river-irrigation projects was recognised as far back as the first quarter of this century. The setting up of the Damodar Valley Corporation to manage the truant river Damodar was the first organised effort at treating a major catchment in India, with the twin objects of controlling floods and lengthening the life of the reservoirs in the system. The Damodar Valley Corporation was organized on the lines of the Tennessee Valley Authority in the U.S.A. and it gave way to a national program of treating major river catchments. In Karnataka, the "River Valley Project" was started in 1.57 million ha of land in 515 watersheds located in the catchments of the reservoirs of Tungabhadra, Nizamsagar and Nagarjunasagar. Since the financial resources required for this task was massive, watersheds were queued up for treatment based on their vulnerability to erosion.

The importance of popular participation in this scheme came to light in the second decade of its implementation. While the upper catchments of all rivers are forest lands in the custody of government, the lower reaches are invariably privately owned farmlands. While small and marginal farmers did not pay for soil conservation operations on their lands, for large farmers this was treated as credit (to be recovered in 15 equal annual instalments). An amount of 2% of the investment in the watersheds was set apart for the maintenance of structures created. The intention was to involve the rural community in this task. Incentives were made available in the form of "demonstration plots" to those farmers who agreed to effect changes in land-use as dictated by the working plans. In spite of good intentions, with the passage of time, the following points became clear to planners and implementers of the project:

(a) It is extremely difficult to motivate the farmers upstream of a reservoir to adopt soil-conservation measures or a change in the land-use to save a reservoir from whose waters they do not benefit! The message that such measures also increase the productivity of the treated lands needed to be driven home. No state can cause a permanent change in land-use by pure economic incentives. In the Tennessee Valley this was attempted supported by a close interaction with farmers and close monitoring of the health of farms and of farmers in a few "pilot watersheds" but later abandoned as it was too expensive to replicate elsewhere.

(b) To preserve vegetative cover on community lands like pastures, prior consultation with village communities is necessary. In Karnataka, the pastures developed and preserved by rural communities have been demonstrated to be sustainable although many of them have succumbed to agriculture dictated by the pressures of an increasing population in recent years.

(c) Home-steads and residential areas with their uncovered backyards and unpaved roads sometimes add more to the flow of silt than agricultural lands. This has been demonstrated by a few studies in the catchment of the Tungabhadra reservoir. This is an area where watershed development is integrated with the entire gamut of rural habitat development.

The River Valley Project provided an excellent opportunity to planners and administrators in Karnataka to exploit the participative strength of the village communities, but the prognosis has not been entirely satisfactory. The scheme was implemented with a classical "blue-print" approach and the villagers were "handed down" treatment plans for their micro-catchments. The results varied widely depending on whether people enthusiastically accepted these plans or merely tolerated them. The state too was more comfortable with legislation than with participation. The flavour of those days is aptly captured in one of the recommendations of a national level seminar on the river valley project: *"Proper machinery and ways and means be developed for handling requirements of maintenance, which if ignored, means total loss of investment made. The system should have provisions for recoveries of the cost of maintenance from individuals or the community as the case may be."*

5 The dryland farmer- waiting for a better deal

The famines became a thing of the past in India (and Karnataka) by the time of the late 60's. Sizeable addition to the irrigated area coupled with the use of high yielding hybrids ushered in the "green revolution". The first half of this century saw the construction of the Krishnarajasagar dam on the Cauvery river and the Tungabhadra dam. Later came the Kabini, Malaprabha and Ghataprabha projects. Then came a phase of stagnant productivity and problems. The cost of new irrigation projects (both in fiscal and environmental terms) became prohibitive. Undisciplined use of water rendered fertile land unfit for cultivation. In Karnataka, roughly, 100,000 ha of land are affected by salinity and alkalinity. Above all, small and marginal farmers remained "back benchers" during the unfolding of the green revolution. The idea of watershed development in dry tracts once again appeared to be attractive.

Schemes for the dryland farmer and the small and marginal farmer mushroomed. There were the area development programs like the Drought Prone Areas Program (DPAP), the Desert Development Program (DDP), the Hill Areas Development Program (HADP) and the Western Ghats Development Program (WGDP). The Indian Council for Agricultural Research (ICAR) started a program in 1983 wherein 47 model watersheds were selected all over the country for treatment in accordance with scientific principles. The activities in each watershed were backed up by "on farm" research. The results achieved in terms of creation of biomass and increase in agricultural productivity in these watersheds were impressive. In the Mittemari and G.R.Halli model watersheds in Karnataka, the crop productivity increased in excess of 50%. A visit to most of these watersheds today shows that these changes have not been sustained. Certain benefits derived by individual farmers like mango trees and field bunds have been preserved. Water-harvesting structures have been maintained by such farmers who benefit from them directly. However, many improved practices coerced by the projects (though not assimilated by farmers) have vanished (Krishnaiah, 1991) and so have most of the common property resources. The level

of community participation in most of these schemes remained low because of the following reasons:
(a) The schemes depended very heavily on the extension machinery of the agriculture department which is characterised by a hierarchical authority. They spent more time looking inwards than outwards towards client farmers. There was no mechanism of making the extension machinery accountable to the farmers who actually bore the risk of "farming decisions".
(b) Scientists and agricultural professionals had their "reward systems" pegged in such a manner that they received greater recognition in publishing papers than in working for extension. Specialist professionals tend to have a higher status than those working closely with farmers (R. Chambers, 1985).Genetic engineers and bio-technologists command far greater respect than operations researchers!

c) All of these schemes were based on "give aways" and did not have any provision for people's contribution. Little effort was made to build local skills, interests and capacity and thus the rural community perceived no stake in maintaining structures or practices once the flow of incentives stopped.
(d) All guidelines talked of participation but assumed that it would come easily. The technical booklets were in the nature of sermons to an "ignorant" farmer. The tone struck in a 1986 FAO booklet (Ostberg & Christiansson, 1993) Protect and Produce: Soil Conservation for Development is typical: *"One thoughtless action by one human being can remove tens of tonnes of soil from each hectare. In a few days, the legacy of thousands of years of patient natural recycling can vanish for good."* Participation meant coaxing the farmers to follow a pre-determined action plan.

The entire effort of research was directed to improve the productivity of the land and available moisture. The results which were impressive would have been more so had the human resource been tackled with the same sense of purpose. Kartar Singh, professor, Indian Institute of Rural Management has in his study of the Mittemari model watershed (Singh, 1991) found that groups had been formed hurriedly and that most farmers did not understand the meaning of a watershed and the implication of the proposed practices. Crop yields which went up due to close supervision of the project staff could not be sustained once the project withdrew from the area. Interestingly in some other watersheds, the farmers resisted development in the fear that once the government treated their lands it would acquire them!

6 Kabbalanala - experiments with water and people

Kabbalanala is the name of a 30,000 hectare watershed taken up for treatment in 1984 in the dry tracts of Bangalore district. This was assisted by the World Bank and the project has been a torch-bearer for the southern part of the country. Its innovative thrust in developing orchard horticulture on marginal lands is well recognised, and so is its pioneering effort in the use of a dedicated multi-disciplinary team. The lack of coordination between line departments which was the Achilles heel of so many watershed projects was unknown in the Kabbalanala project as the entire multi-disciplinary team worked under an exclusive project director. The project staff were freed from procedural shackles and encouraged to experiment.

For all its technical appropriateness and innovation, Kabbalanala would not have been a "pilgrimage" for watershed connoisseurs had it not experimented with popular participation. Interestingly, the component of participation came into the project as a "mid-course correction" rather than a pre-conceived parameter. Groups were formed at Kabbalanala by bringing together farmers of a micro-watershed. The main objectives of forming these groups (or sanghas as they were called) were (D. Satyamurthy, 1992):

(i) To provide the local input to micro-watershed planning
(ii) To manage common property resources developed by the project.
(iii) To act as thrift and credit groups outside the clutches of the exploitative money lender and the cumbersome banking system.
(iv) To help in the continued adoption of improved techniques and to draw out non-cooperative farmers from their indifference.

To keep the groups from becoming unwieldy, the membership was restricted to 30 in each. Each member contributed Rs.100, while the project contributed Rs.300 for each member, and these amounts formed the initial seed capital of the group. These sanghas were deliberately kept informal to free them from statutory red-tape and the project coordinators tried not to spoon-feed them. The groups were engaged in participatory preparation of treatment plans. The treatment plans emerged after repeated iterative interaction between the villagers' groups and the project specialists. The most important contribution of Kabbalanala was the hypothesis that a less than perfect solution which has its roots in participatory methods is more sustainable than a technically perfect fix.

It is revealing as to how popular participation changed the emphasis of the project. Orchard horticulture which was just another component became a movement. Water harvesting structures far in excess of designed numbers had to be provided. An eye-opener was the general rejection of vegetative barriers of vetiver. In spite of severe pressure from the World Bank and the project staff, the popular attitude to vetiver was indifferent at best and outright hostility was not unknown. Hindsight suggests the following reasons:

(i) Planting of vetiver on contour (key) lines may have minimised soil loss, but it also fell in the way of routine agricultural operations like maneuvering bullocks.
(ii) Vetiver did not make good fodder like some local grasses did.
(iii) A vetiver line with gaps caused formation of gullies because of fast flow of water through the gaps.
(iv) The repeated re-planting in ill-maintained vetiver lines on farmers' fields by the project staff convinced the farmers that the government was duty bound to look after these vegetative strips.

At that time every one except the farmers was convinced that vetiver was the panacea for all kinds of erosion. There were conferences on vetiver and many studies to show that it was highly coveted by the farmers, but the potential of self-deception by questionnaire surveys is well known (Gerry Gill, 1993)! There were those die-hard vetiver adherents who even produced syrup out of it. The truth is that the voluntary adoption and enthusiastic preservation of vetiver as a vegetative soil barrier by the farmer never exceeded 30% Today it is no wonder that respectable (and somewhat red-faced!) scientists are having a second look at vetiver.

Now that the watershed project has withdrawn from Kabbalanala, a visit to Dombaradoddi, one of the micro-watersheds is interesting. The mango trees are tall and fruit laden. The check-dam nearby has water. The sangha had to repair it once. The seed capital of the sangha has grown though the president castigates a few defaulters. The fields of farmers do not exhibit cultivation along precise mathematical contours, but it is in general across the slope. Inter-cultivation of pulses has come to stay. The farmers have even preserved a few blemishless lines of vetiver for the occasional specialist visitor!

7 Self-Help Groups - the beacons of sustainability

Self Help Groups (SHG) were created by MYRADA to get over the inadequacies caused by an unresponsive and indifferent rural banking system. The idea of using such groups to plan and preserve structures in micro-watersheds came later. The groups functioned by exerting cohesive moral pressure on its members to keep them from straying. So important was the necessity of cohesion, that initially NGOs went about forming socially homogenous groups to ensure their stability. Needless to say, the recovery of loans advanced by SHGs is far better than the recovery in the conventional banking system. Typically, in MYRADA's Huthur watershed project a mere Rs.6921 is outstanding against an advance of Rs.1438355 (source: MYRADA Huthur project statement of accounts ending Dec.95) by all the SHGs. In almost all the watershed development projects in Karnataka spearheaded by NGOs, the following could be observed:

(a) The farmers who are members of SHGs are willing to contribute anywhere from 10% to 33% of the amount required for agreed items of work involving soil-conservation. The groups also agreed to maintain the structures out of their own resources.
(b) Substantial reduction in unit costs (up to 50%) have been achieved for structures due to the initiative of these groups.
(c) The groups take up activities which have the potential of quick pay-back.
(d) Groups of women and landless were formed in addition to the regular groups.
(e) The sphere of credit activity of any group went much beyond the technicalities of watershed development. A high priority was given to the need for consumption related credit.

NGOs gained confidence of the rural communities by contributing to their development initially by traditional means (so called entry-point activities). When the people turned receptive, group formation was attempted. Studies have shown that the long-term survival of a self-help group is heavily dependent on its initial motivation. Strong self-help groups have been shown to be the key to sustainability and project designers were willing to bear the extra cost and time required to set them up. It became possible to expect a contribution from the beneficiaries of a watershed project even at the stage of design.

Another bonus was the tendency of SHGs to minimise the regime of inequity inherent in a watershed development project. The SHGs of landless villagers are encouraged to take up productive non land-based activities (NLBAs). It was realised that the biggest challenge of watershed development is to create in it a stake for the landless and marginally placed farmers. While the classical technical ingredients of watershed development remained relevant, the rigid "ridge to valley" was abandoned. It was realised that rural communities cannot be equally receptive or simultaneously prepared. There was also more willingness (on the part of project facilitators) to accept a "less than optimal" technical solution for soil conservation with the guarantee of greater sustainability.

The current picture that has emerged shows that the partnership of NGOs and the government has come to stay. In Karnataka there is no bilateral watershed project without the involvement of NGOs. In the Federal government funded programs funded by the departments of Agriculture and Rural Areas, an in-built package exists for NGOs. The scheme floated by the department of Rural Areas, even integrates the locally elected bodies with the development of micro-watersheds.

8 Common properties - uncommon management.

In most watersheds in Karnataka, the common property resources (CPR) hold the key to sustainabilty as well as suffer the most vulnerability to degradation. The conventional method of treating common lands was to afforest it or make it a pasture. Very little thought went into the productive capacity of the CPR in relation to the population that had access to it. The result was severe depletion of biomass on these lands making them useless to the community which in turn became more tolerant to encroachment on common lands by adjacent farmers. Though on paper, the CPRs belonged to the local institutions (panchayats), in actual fact they were exploited by all and managed by none. There were no "social fences" to protect the thousands of small patches of forests created as CPRs under various programs. The upper reaches of many treated watersheds denuded denuded within less than 5 years.

A beginning was made in the management of CPRs when the self-help groups obtained control over them. Since the CPRs figured in the participative plan of the community, the yield of common lands was sustained consciously. The community decided the species of trees to be planted on these lands. The grazing blocks were rotated appropriately and fines were levied for violation of collective decisions. The pressure on the CPRs exerted by the landless was deflected by providing them with small business and income from non land-based activities (NLBAs). Where the CPR was inadequate the SHGs exhibited ingenuity by taking on lease the lands of absentees on mutually beneficial terms. In many watersheds of north Karnataka, the non-arable upper reaches are privately owned, and some arrangement with absentee landlords is the only way to create biomass on such lands.

In a far-reaching decision, on 12th April 1994, the government of Karnataka operationalised a scheme of Joint Forest Planning and Management (JFPM). Besides common lands, the community was given control of degraded forest areas (with a canopy cover of less than 25%) outside sanctuaries and national parks. Village Forest Committees (VFCs) were set up and these committees channelled 25% of the produce from the forest to the beneficiaries. 50% of the output went to the government and the remaining 25% to the Village Forest Development Fund operated by the VFC.

9 The "P" of PIDOW

The Participative Integrated Development of Watersheds (PIDOW) program operates in Karnataka with the collaboration of the Swiss Agency for Development (SDC) in the semi-arid district of Gulbarga (average annual rainfall of 650 mm). This is perhaps the only program of its kind where peoples' participation is as important an objective as the treatment of the watershed. The statement of objectives of the project has a familiar ring to it : *"To evolve and test a replicable strategy for participative, integrated, sustainable development and rehabilitation of small watersheds in semi-arid areas with special consideration of the interest of the weaker sections of the rural communities such as small and marginal farmers, landless and women."* This would have been quite a boastful mouthful but for the fact most of what has been stated has been achieved in the last 12 years. The PIDOW program which started as a community development scheme in 1984, later blossomed into one of the finest watersheds projects known. The project is a joint partnership between the Swiss and Karnataka governments and MYRADA which put into use its experience with self-help groups.

The self-help groups were federated into apex-level "watershed development associations" (WDA) at the level of the sub-watershed. The WDAs play an extremely important role in macro-planning and would act as management engines when the project withdraws. The WDAs have already taken over from the project where MYRADA has withdrawn. The ingredients of sustainability are clearly seen in the following table.

Number of functioning self-help groups:	125
Total common fund:	Rs.5,856,300
Total savings:	Rs.1,532,774
Loans:	Rs.11,133,176
Total number of members:	2759

Table 1: (source: MYRADA)

Studies have shown that crop-yields have gone up by 20-50% in the watershed. The SHGs in the project have substantially added to the common property resources by leasing lands from absentee landlords. Interestingly, the farmers stuck to indigenous soil and water conservation practices (source: Farmers are engineers) and refused to toe the expert-line. Some of these examples are stone-bunds, diversion drains and silt harvesting structure. Not that these technologies were unknown to the project specialists, but acceptance of the farmers' practices (with slight refinements) paved the way for better community participation. In some cases it sowed the seeds of "participative technology development". The success of PIDOW has enabled the replication of this model to 4 other districts of Karnataka.

10 Towards a more participative future

It is certain that compulsions of financial stringency and of replicating models sustainably, would demand a high quality of participation from all the stake-holders. The future holds many challenges. How the stake-holders tackle these challenges will determine the future of a successful watershed development effort in Karnataka. Consider the following:

(a) **Relation of SHGs and elected institutions**: In Karnataka, watershed development is the statutory responsibility of local elected bodies (panchayats). More often than not, the perspective of these bodies is somewhat coloured by short-term gains, whereas watershed development requires a distant vision. Panchayats are concerned about issues like roads, drinking water and sanitation and not unduly perturbed by ecological degradation or increasing biomass. It is however, not possible for the SHGs to flourish without the tacit approval of the panchayats, and indeed, a supportive panchayat can really kick-start self-help groups. Very little work has been done by NGOs too in bridging the gap between panchayats and SHGs. Using the synergy of a panchayat having a statutory mandate for watershed development and that of a self-help group which is motivated is perhaps the biggest challenge of human resource development.

(b) **Managing common property resources**: Whatever ingenuity is used, the increasing population would put pressure on the common property resources of a watershed. On one hand the landless would live off it and on the other, land-holders would try to encroach upon it. Not only would the CPRs would have to become more productive but the rural community would have to develop a visible stake in protecting it. It would not be exaggeration to state that the management of CPRs by the watershed community is an indicator of sustainability of the watershed.

(c) **High-Technology, a tool but not an end in itself**: Use of remote sensing technologies has increased the options of the scientist and the project administrators but not necessarily that of the watershed farmers. In some extreme cases, technology has increased alienation rather than participation. A satellite photograph suggests solutions which are obvious at the macro-level to the specialist, but these solutions may not be sustainable ones without acceptance by rural communities.

(d) **Cost sharing**: Classical watershed development projects were 100% state planned, funded and implemented. The rising costs of treatment coupled with the sheer size of the task ahead has shown the states the wisdom of mobilising peoples' contribution. Raising peoples' contributions has not been easy. If the planners give a visibly high priority to watershed development and speed up the treatment of watersheds, people would sit back and watch expecting the project machinery to do everything. On the other hand if the implementation of projects is guided spatially by the preparedness of the farmers to contribute, the treatment would proceed in patches diluting the synergy of watershed development. To this problem there are no easy answers and each project must find its own way out of this dichotomy. The present levels of contribution have not exceeded 30% (in the case of orchard horticulture on farmers' fields) but this is known to be far below the perceived potential in Karnataka. In certain villages, the author found farmers unwilling to pay 10% of the cost of bunds on their own land but more than ready to contribute towards building a temple in the village.

(e) Participative Technology Development: PTD is extremely difficult to sow in an environment of agricultural extension! PTD does not advocate thrusting technologies down the throat of the farmer, but nor does it envisage passive approval of native technologies. To document useful native technologies is a task that has yet to be completed. The lack of communication skills of the farmer is as much a hindrance to PTD as the rigidly formal specialisation of the scientist. One area of immediate concern is the slow but sure loss of bio-diversity in agricultural crops. High productivity taken as an over-riding criterion, does not justify the propagation of a bulk of available genetic material. The loss of genetic diversity is even more glaring in advanced countries (Jules N. Pretty 1995). USA has lost nearly 88% of its vegetable and fruit varieties. Greece has lost 95% of its local wheat varieties. In the Netherlands, 80% of the potato land is confined to one variety. India too which at one time had 30,000 varieties of rice would soon have 10 varieties cover 75% of the land. Many of the varieties discarded by the scientist may well have possessed resistance to drought and pests, and undoubtedly genetic stability. Documenting the local genetic resources and evaluating them in consultation with the rural communities is perhaps the biggest challenge of participatory technology development.

References:

Chambers, R. (1985): Normal professionalism, new paradigms and development: Paper for Seminar on Poverty, Development and Food: Towards the 21st Century, IDS,Sussex.
Farmers are Engineers, Swiss Agency for Development & Cooperation.
Fernandez, Aloysius (1994), The MYRADA experience, The Interventions of a Voluntary Agency.
Gerry, Gill (1993): "O.K., the data's lousy, but its all we got." Sustainable Agriculture Gatekeeper Series, SA38 IIED, London
Krishnaiah, K.R. (1991): PhD thesis to the University of Agricultural Sciences, Bangalore.
Ostberg & Christiansson (1993): Of Lands and People, Environment & Dev Studies Univ. of Stockholm.
Pretty, Jules N. (1995): Regenerating Agriculture, Earthscan Publications Ltd London,
Satyamurthy, D. (1992): Micro-watershed Sangha - A model for peoples' participation.
Singh, Kartar (1991): Indian Journal of Agricultural Economics Vol.46 No.2, April-June,1991

Address of author:
Kaushik Mukherjee
Agriculture & Horticulture
Government of Karnataka
4th floor, M.S. Building
Vidhana Veedhi
Bangalore-560001, India

Participatory Watershed Management in India

J.S. Samra & A.K. Sikka

Summary

Sustainability, replicability and equity in the management of environmental resources through people's participation has engaged attention of researchers, NGOs and international donor agencies for quite some time. Limited initial financial investments and community organization inputs were made to create common use resources and to improve common as well as private property resources of a village. Participatory watershed management programmes through local level institutions increased income from the common as well as private property resources and provided livelihood gathering opportunities for the rural people. Voluntary social fencing against uncontrolled grazing and felling of trees improved biomass production and biodiversity of the community owned forest land. Harvesting of runoff rain from community land, its storage and equitable use increased productivity of private farm lands several-fold. Many environmentally-positive benefits associated with reduced soil erosion in the catchment and storage of runoff were also realized. Encouraging results of the watershed management programme suggests the adoption of an incentive-based and community driven "bottom up approach" for managing degraded watersheds on a sustainable basis. The payback period of financial investments made in participatory approaches varies from 5 to 7 years.

Keywords: Resource conservation, equity, empowerment, participation, local institution, common property resources

1 Introduction

An agrarian economy, ever-increasing demographic pressure, land-based livelihood and environmental security are the main issues of resource management in India. Integration of bio-physics, socio-economics, gender, policy, equity, community participation and institutionalized management of common property resources is being harmonized for sustainable development on a watershed basis. Experiments on community participation in watershed management were initiated by Central Soil & Water Conservation Research and Training Institute (CSWCRTI), Dehradun since 1974 at four places under a Ford Foundation Grant. Since 1987, an NGO (MYRADA) took up participatory watershed management in the Gulbarga district of Karnataka in collaboration with the Swiss Development Corporation. After 1990, NGOs such as the Aga Khan Rural Support Programme, Development Support Centre, Satguru Water and Development Foun-dation and about 190 NGOs became involved in rural development programmes.

Dynamic group processes are being invoked to minimize social conflicts by ensuring community participation for rural resource appraisal, planning, implementation, equitable sharing of benefits, management and maintanance of watershed resources. A set of direct and indirect measures of rainwater harvesting, voluntary activities, attainments, investments, improved productivity of common property resources, amelioration of land and environmental externalities were analysed to highlight stakeholder's participation (Arya and Samra, 1995; Singh, 1994).

2 Watershed management models

CSWCRTI, Dehradun, and its Regional Research Centres have been playing a pioneering role in popularizing the watershed approach through research, demonstration, extension and training. An integrated watershed management approach was first demonstrated in the mid 1970´s through the model Operational Research Projects (ORPs) on watersheds at: i) Sukhomajri (Haryana) representing Shiwalik foothills, ii) Fakot (U.P.Hills) representing middle Himalayas and iii) G.R. Halli (Karnataka) representing red soils of low rainfall region. People's participation was the key to the success of these projects. As a result of the achievements and benefits of these model ORP's, 47 watersheds were established in different agro-ecological regions of the country in 1983 (Dhruvanarayana et al., 1987). Subsequently, the schemes such as National Watershed Development Project for Rainfed Agriculture (NWDPRA), Integrated Watershed Development Project (IWDP), Drought Prone Area Programme (DPAP) and Desert Development Programme (DDP) were launched for development of watersheds in the country. A policy decision has been taken in India that many productive employment generating schemes during 1997-2002 will also be implemented on a participatory watershed management basis.

3 Management of common property resources

Ownership of degraded lands generally belongs to the Forest Department, undivided or jointly owned/used by the community, religious bodies and private individuals. Successful restoration of such watersheds involving community-driven cooperative efforts was first demonstrated in the Shiwalik foothills through integrated watershed management at Sukhomajri village in Haryana state (Grewal et al., 1995). Similar concepts were subsequently replicated in village Nada and Bunga of Haryana, Relmajra in Punjab in the Shiwaliks and many other locations by NGOs. Institutionalization of the arrangements in the form of "Hill Resource Management Society" (HRMS) for managing stored rainwater, common land resources and catchment protection was the key to the success of many projects.

It has been established from the results of such experiences that development and management of common property resources is important to lay a solid foundation for the livelihood support system of rural people and encourage their active participation. People are made to realize that they have to manage the resources not only as a common property but also as common responsibility for delivering and sharing benefits and costs.

4 Resource Development and Management Societies

Peoples in the watersheds are the major resources and also the stakeholders of the development programme. Involvement of each and every farmer of the watershed including the landless is considered as essential component of the programmes for sustainable development of both common and private property resources. Their active involvement in managing the most vulnerable and degraded part of watershed, called CPRs is of utmost importance. A local institution at the watershed level is created in the form of Resource Management Society and entrusted with the responsibility to protect natural resources, ensure equidistribution of benefits from CPRs and their maintenance from their own resources. The society has a set of bylaws that can be amended, modified or changed by the general body as and when necessary. Members of every family, residing permanently in the village are eligible for the membership of the Society, if he/she pays for membership and abides by the rules and by-laws of the Society.

The Government contributes towards initial development of the forest land and water storage reservoir which are handed over to the society by evolving its rules and mechanisms of generating revenue and sharing the resources. The centralized control of common property resources through government officials is thus substituted by the end users themselves in this "bottom up" approach. Awarding of contracts for fodder and industrial grasses in the forest land as well as ensuring stall feeding through community regulated activities clearly demonstrates the effectiveness of a participatory approach. Most of the resources generated are invested in the maintenance of watersheds and creation of common use facilities such as paving of village streets with bricks, establishment of dispensaries or hospitals for animals, and construction of school buildings, etc.

In the Bunga watershed, the Society generated a total income of Rs.3,68,578 (US $ 10,531 @ 1$ = Rs. 35) over the period 1984 to Feb., 1993 from the sale of stored water for irrigation, lease of catchment area for grass (155 ha) and lease of reservoir for fish culture (Table 1). The society has so far spent Rs.1,79,889 in various activities related to maintenance of watershed infrastructure and other common use development works. The sediment yield to the reservoir has reduced from a record high of 528 t/ha (average between 1984-92) to 91 t/ha/year (average between 1991-95) as a result of people's driven catchment protection, popularly called 'social fencing'. Very nominal or token fines were imposed on those who did not respect the rules. Interestingly, there was no defaulter after four years of participatory development.

Year	Sale of grass water	Sale of irrigation	Fish culture	Fines	Total
1984	565	2157	3000	-	5722
1985	3050	6404	3500	158	13112
1986	2100	3478	3500	15	9093
1987	30000	16384	-	15	46399
1988	2367	18546	3300	-	24213
1989	11000	10915	9894	-	31809
1990	19450	16800	4000	-	40250
1991	26000	10551	6500	-	43051
1992	36000	85235	33694	-	154929
	130532	170470	67388	188	368578

Source : Arya and Samra (1985).

Table 1: Income (Rs.) of society from various common property resources in Bunga village (1984 - Feb. 1993)

	Product	Average level of attributes during		
		Pre project 1974-75	During external interventions, 1975-86	After withdrawal of external interventions, 1987-95
A.	Food crops (qtl.)	882	4015	5843
B.	Fruit (tons)	Neg.	62	1962
C.	Milk ('000 lit.)	57	185	237
D.	Floriculture ('000 Rs.)	-	-	120* (1994 & 95)
E.	Cash crops ('000 Rs.)	6.5	24.8	202.5
F.	Animal rearing method	Heavily grazing	Partially grazing	Stall feeding
G.	Dependency on forest (%)	60	46	18
H.	Runoff (%)	42	18.3	13.7
I	Soil loss (t/ha/annum)	11	4.5	2.0

* Community diversified into floriculture in 1994.

Table 2 : Impact of participation in watershed management program at different stages of development (Fakot village, U.P., India)

The results of a similar watershed management project at Fakot (U.P.Hills) were also quite encouraging in providing protection and enhancing production. Impact of participation in this watershed on various activities especially after withdrawal of the interventions of CSWCRTI, Dehradun is evident from Table 2. Production of food grains, fruits, milk, flowers and cash crops continued to increase after withdrawal of active support due to participatory management of the watershed. Reduced dependency on forest for fodder, lower runoff and soil loss from the entire watershed after its participatory management indicated positive environmental benefits.

5 Integrated Watershed Management at Relmajra: A Case Study

Resource regenerative and sustainable watershed development strategies were taken up during 1991-92 in the village of Relmajra, Punjab. Watershed restoration through integrated resource management, involving an appropriate mix of structural and low-cost vegetative measures with the help of community participation was evolved and demonstrated in a 627 ha Shiwalik foothill (Appendix I) (Samra et. al., 1995).

6 Extent of participation and securities generated

The results have been encouraging in generating food, forage, fuel, ecological and social securities with a high level of participation both on common and private property resources. In a period of three years, the members of the society increased from 4 in 1993 to 30 in 1995. In this way the level of voluntary participation rapidly increased.

Sources	Income in Rupees			
	1993	1994	1995	Total
One time membership fee	200	750	550	1500
(No. of members)	(4)	(15)	(15)	(30)
Sale of Napier grass (common land)	2060	4405	3825	10290
Sale of *Bhabbar* grass (common land)	-	500	1980	2480
Sale of *Saccharum munja* (Common land & Road sides)	-	700	3700	4400
Lease of reservoirs for fish	-	1000	5800	6800
Sale of irrigation water	120	5040	13572	18732
Total	2380	12395	29437	44202

Table 3: *Income of society from different CPRs at Relmajra (up to Dec., 1995).*

7 Participation on CPRs

The forest land in the catchment, the water storage reservoir and arable common lands constitute the major common property resources being managed by the village society. Involvement of the people in the form of manpower supply and voluntary contribution of labour was an important input in creating common property assets. As a result of community driven voluntary catchment protection, forest vegetation improved and grass yield increased many fold. Income to the society from the sale of grasses, leasing of reservoir for fish and sale of water for irrigation (Table 3) created much interest in the community. A progressive increase in income of all grasses from a low of Rs. 2,060 in 1993 to Rs. 9,505 in 1995, indicated sustained participation. A total income of Rs. 44,202 in just two years to the society apart from reduced soil erosion and moderation of flash floods, was encouraging from the erstwhile degraded lands. The society opened a joint bank account and is maintaining records of income and expenditure.

The project generated an employment of 11,780 man-days in the construction of water harvesting infrastructure, catchment treatment and rehabilitation of command area. Another 1750 mandays were generated for clearance of area, planting of grasses and other operations in the agri-horti and agroforestry systems. They were paid from the funds provided by CSWCRTI, Dehradun to initiate the process on common use activities only.

Over the past year, after partial withdrawal of external support by the centre, the society has incurred an expenditure of about Rs. 10,000 in various development and maintenance works for sustaining resource generation. Items of work on which expenditure was incurred included bunding of commonland plots to protect from flash flood, maintenance of a dam body during monsoon, construction of brush wood check dams and plantation in the main channel upstream of dams, clearance of roads, and labour for grass plantation in a hortipastoral plot. In accomplishing all these tasks, villagers have also been voluntarily contributing their time and labour. The villagers and society is estimated to have contributed about 680 mandays towards this activity in the past year.

Biodiversity indices such as relative value index (RVI) of trees and shrubs increased by three and two times, respectively. RVI of grasses decreased by 37% indicating that succession took place in favour of more economical species. These indicators showed effectiveness of watershed protection by people's participation in promoting more tree and brush growth, for ensuring ecological security. Also the flow of sediments into the reservoir was substantially reduced.

8 Participation on private lands

Participation on the private lands was gauged from their involvement and contribution in land development, adopting agri-horti and horti-pastoral systems on their farm lands and following recommended crop as well as water management practices. Out of 25 ha potential command area of the dam, over 15 ha has been levelled by the farmers by investing from their resources. Availability of limited irrigation water and land levelling motivated the farmers to improve productivity of their farm lands by adopting improved agronomical and water management practices. The area under the monsoon crops during 1994-95 increased to about 26 ha against 10.8 ha in 1992-93 (prior to dam construction), as was the case with the area under winter crops. The farmers have now started using fertilizer, farm yard manure (FYM) and other inputs. They are also introducing vegetable farming. The production of fodder crops has gone up from 76 tons to 156 tons and wheat yield increased by over two fold of the benchmark level.

The local people were also fully involved in developing horticultural systems in their fields. The Institute provided grafts of improved fruit trees and rooted slips of Napier and Saccharum munja grass for planting in their fields. The farmers dug the pits, arranged fertilizer, FYM and transplanted the grafts or rooted slips of grasses. More and more farmers are coming forward to establish hortipastoral and agri-horti systems on their farm lands. This model is being replicated extensively in the Shiwalik belt of the lower Himalayas in North India.

References

Arya, S.L. and Samra, J.S. (1995): Socio-economic Implications and Participatory Appraisal of Integrated Watershed Management Project at Bunga, Bull. No. T-27/C-6, CSWCRTI, Research Centre, 27-A Madhya Marg, Chandigarh, India.

Dhruvanarayan, V.V., Bhardwaj, S.P., Sikka, A.K., Singh R.P., Sharma, S.N., Vittal, K.P.R. and Das, S.K. (1987): Watershed Management for Drought Mitigation, Bull., ICAR, New Delhi.

Grewal, S.S., Samra, J.S., Mittal, S.P. and Agnihotri, Y. (1995): Sukhomajri Concept of Integrated Watershed Management, Bull. No. T-26/C-5, CSWCRTI, Research Centre, 27-A, Madhya Marg, Chandigarh, India.

Samra, J.S., Bansal, R.C., Sikka, A.K., Mittal, S.P. and Agnihotri, Y. (1995): Resource Conservation Through Watershed Management in Shiwalik Foothills-Relmajra, Bull. No. T-28/C-7, CSWCRTI, Research Centre, 27-A, Madhya Marg, Chandigarh, India.

Singh, Katar (1994): People's participation in micro watershed management - a case study of an NGO in Gujarat. Indian J. Soil Conservation, **22(1-2),** 271-278.

Appendix I

Major components of the integrated watershed management project included:

1. Community participation
- Involving local people at planning, designing, implementation and subsequent management stages.
- Formulation of a registered Water Users Society for maintenance, management and distribution of common property resources
- Voluntary contributions (Shramdan) by the people on community and private lands.

2. Catchment treatment
- Staggered contour trenches (2 x 0.45 x 0.45 m)
- Plantation with resource conserving & economical trees (*Acacia modesta, Acacia catechu* and *Melia azadirach*) and grasses (*Eulaliopsis binata*).

3. Channel treatment (Bio-engineering measures)
- Brushwood check dams
- Vegetative spurs (single and double lines) of *Arundo donax* (Nara)
- Vegetative filters and cross barriers of *Ipomea cornea* and *A. donax* above submergence area
- Gully plugs of *A. donax, Sacchurum munja* (kana) and *Ipomea*
- Loose boulder check dams at few critical points

4. Water resource development
- Runoff water was harvested and stored in suitably designed structures and recycled for giving protective irrigation during post-rains stress periods.

4. Diversification of land use system
- Common lands productivity was improved through horti-pastoral, silvi-pastoral and agroforestry systems.
- Private lands were also put to more economical and sustainable alternate land use sytems.

5. Arable lands
- Bush clearing and land levelling by farmers
- Agroforestry
- Demonstration of drip irrigation in horti-pastoral system

Address of authors:
J.S. Samra
A.K. Sikka
Central Soil and Water Conservation Research and Training Institute
218 Kaulagarh Road
Dehradun - 248 195, India

Improving People's Participation in Soil Conservation and Sustainable Land Use Through Community Forestry in Nepal

Sameer Karki & S.R. Chalise

Summary

This paper presents some issues in rehabilitation of degraded lands in Nepal through the use of a community forestry approach. The paper is based on the authors' experiences in implementing the ICIMOD[1]/ IDRC project on 'Rehabilitation of Degraded Lands in Mountain Ecosystems Project' in a mid-hills district of Nepal. It examines some strengths and weaknesses of the current practices in community forestry in Nepal in relation to rehabilitation of degraded lands, but the paper also illustrates that it provides a very good legal framework for undertaking such activities.

Keywords: Community forestry, rehabilitation, degraded lands, community participation

1 Introduction

Soil erosion and deforestation are the major land degradation issues in Nepal. Though most forest loss is reported to have occurred before the middle of the last century (Mahat et al., 1986), degradation of forests is still recognised as having significant adverse effects on biodiversity and on the local inhabitants. Water induced soil erosion from degraded common lands is a major concern in many Nepali mid-hills watersheds as they contribute the most sediments in rivers (Shah et al., 1993). Erosion gullies threaten settlements and infrastructures as well as valuable agricultural lands. Common lands, particularly forests, are important for maintenance of agricultural soil fertility for application of composted leaf litter or manure resulting from fodder fed to domestic animals (Mahat, 1987, Wyatt Smith 1982). Forests also provide construction materials, food, medicine and other sources of other income generation (Edwards, 1996). Rehabilitation of degraded common lands is very important to support agrarian societies and to protect biodiversity.

A Regional Project on 'Rehabilitation of Degraded Lands in Mountain Ecosystems' has been implemented by the International Centre for Integrated Mountain Development (ICIMOD)[1] in China, India, Pakistan and Nepal to develop a better understanding of the extent and processes underlying land degradation, and to identify measures for restoring and developing degraded lands by using options that are field tested and found to be economically, environmentally and socially viable (ICIMOD, 1993). This paper describes the experiences of the Project implemented in Kabhre Palanchok District, Nepal for the rehabilitation of common degraded lands since 1993 under Nepal's Community Forestry framework.

2 Community forestry in Nepal

The Community Forestry programme emerged in Nepal after reflection on the failure of government to manage forests after forest nationalisation in 1957 and the increased recognition of people's right and capabilities to manage their forests. In the 1970s, the focus of community forestry had been on reforestation of degraded lands, but more recently, it has embraced a broader participatory forest management and rural development. Community Forestry is a major focus in the Forestry Act of 1993. Fisher (1995) highlights the significant features of Community Forestry in Nepal as:

- Forest management agreements (operational plans) are negotiated between the forest department and forest user groups (ie groups of people with direct interest in use of a particular forest and claiming usufruct rights, forest user groups) rather than political or geographic units. Forest user groups are extensively involved in designing the operational plans. There is the potential for considerable flexibility in management and for a high level of local control, subject to the ultimate authority of the District Forest Office. Substantial forest use and harvesting are possible.
- Forest user groups are entitled to use all products raised through management and may use all income raised for development purposes and there is no benefit sharing by the Forest Department.

The Forest Rules and Schedules (1995) and the Operational Guidelines for Community Forestry Development Programme (1995) have strengthened the role for the forest user groups to work as community-based organisations. Forest user groups have their own written constitutions, can operate bank accounts and can also approach other organisations for support. Many forest user groups have started generating income and investing into community development. For example, Jackson and Ingles (1994) estimated that in 1994, 227 forest user groups in Sindhu Palchok and Kabhre Palanchok districts accumulated about US$ 20,000, after spending an equivalent amount in development and afforestation activities.

3 Project sites of rehabilitation of degraded lands in mountain ecosystems project in Kabhre Palanchok, Nepal

The Project selected two sites in Kabhre Palanchok in April 1993 in collaboration with the District Forest Office. Both sites are located on degraded common lands on 'red soil' which Shah et al. (1993) describe as the most degraded lands in the watershed. This soil has been described by Carson (1992) as Rhodudult (an Ultisol), which has a tendency to crust on surface, has low phosphorous fixation, and is subject to severe gullying. Characteristics of the sites are summarised in Tab.1.

3.2 Project supported activities

All land in Nepal other than private lands fall under the jurisdiction of the Ministry of Forests and Soil Conservation. Under the Forest Act, forests can be managed as religious, community, leasehold, private or government forests. As their management as government forests had been unsuccessful in the past at both sites, this option was not considered, and the sites were neither religious forests, nor could they be privatised. Leasing the forests was less acceptable to the communities as only a few of the households could potentially lease and benefit from the land. Thus, the local communities decided to manage the lands as community forests and forest user groups were formed at both sites. Training and visits were organised for forest user group members to enhance their capacity. The sites have been protected from domestic animals through social fencing. This was achieved firstly through consensus in forest user groups meeting involving the majority of stakeholders, and reinforced by forest watchers employed by the forest user groups. A modified Sloping Agricultural Land Technology (SALT) (Partap and Watson, 1994) combining nitrogen fixing contour hedgerows, with tree seedlings planted between the rows, was used at the sites. It is probably the first time in Nepal that this technology has been applied for rehabilitation efforts at such a scale with community participation. Details on plantations at both sites are presented in Table 2 and constraints, solutions identified, and the associated activities carried out are listed in Table 3.

Name of the forest land	Bajrapare Ward 1, Rabi-Opi VDC	Dhaireni Ward 9, Panchkhal VDC
Total households (1994)	18	259
Population (1994)	130	1667
Number of FUG settlements	1	10
Ethnic composition	Brahman and Chhetri	Brahman, Chhetri, Newar, Danuwar, Sarki, Kami, Damai, Tamang, Magar
Average land holding	0.5 to 1.5 ha per household	0.5 to 1.5 ha per household
Accessibility	Seasonal dirt road, far from market	All weather road, close to market
Area	6.76 ha	15.93 ha
Altitude	925-1150 m	900-1000 m
Climate	Sub-tropical (sub humid)	Sub-tropical (sub humid)
Mean annual rainfall	1200mm	1200mm
Temperature extremes	0 °C to 35 °C	0 °C, to 35 °C
Slope	15-25 %	10-25 %
Aspect	South facing	South facing
Soil	Degraded red clay loam	Degraded red clay loam.
Dominant vegetation in 1993	Some *Shorea robusta*, *Themada triandra*, *Rhus parvifolia*	Scattered *Pinus roxburghii* planted in 1973. *Rhus parvifolia*, *Themada triandra*

Table 1: Social and Biophysical characteristics of Bajrapare and Dhaireni

Site	Total plants planted in pits	hedgerows length (m)	Total area under plantation (ha)
Bajrapare	6800	1908	2.5 (of 6.76 ha)
Dhaireni	7286	3318	5.05 (of 15.93 ha)

Source: Chalise, S.R., Karki, S., Shrestha, B.G., 1995

Table 2.: Details on plantation activities.

3.3 Summary of results to date

People's participation in Project supported activities have been very good. The forest user groups organise regular meetings and they are generally well attended. The area under vegetation from natural regeneration has increased on average by 15% to 100% at various plots. Data from erosion monitoring have been collected for two rainy seasons and are yet to be analysed. However, results from China (from a site with hedgerows on degraded common lands under the same Regional Project implemented earlier) suggest that contour plantation reduces soil erosion better than just pitted plantation (ICIMOD, 1996).

Challenges	Options followed	Research component
Social constraints		
• Lack of mechanisms to organise people	•. Formation of Forest User Groups (forest user groups) as per Nepal's Forest Act 1993	• Monitoring participation and
• Free grazing	•. Social fencing to stop free grazing	• Impact on natural regeneration
• Uncontrolled resource uses (soil, grass, firewood)	•. Production of resources on private lands, particularly fodder	• Adoption of on-farm tree planting
	•.Management Constraints	•
• No indigenous management as newly settled area (after eradication of malaria in mid-50s)	•. Formation of forest user groups. Regular meetings	• Monitoring participation and
• Lack of information on options for rehabilitation	•. Training and cross visits to other demonstration sites	• Farmer-to-farmer transfer of
Natural constraints		
• Eroded soils with poor nutrient status	• Plantation of species that fix nitrogen and tolerate low nutrient status; contour hedgerows (modified Sloping Agriculture Land Technology); promotion of indigenous species	• Performances of species and biomass production monitored • Soil erosion monitoring • Impact of project activities on soil fertility status
• Warmer south aspect, long dry season, thus periodic shortages of water	• Mulching; planting of introduced and indigenous drought resistant species, micro catchments around individual plants; constructions of water harvesting ponds, supply of water from close-by streams through provision of pipes.	• Rainfall and soil erosion monitoring • Amount and efficiency of rainwater harvesting • Effect of mulching on moisture and survival of plants
• Low vegetation cover	• Social fencing to encourage natural regeneration, planting of grass, shrubs and trees	• Natural regeneration monitoring, survival and performance of planted species
• Gully erosion	• Brushwood check dams constructed, planting of bamboo and other species. Water diversion channels	• Erosion pins to record advancing of gullies

Table 3: A summary of project supported activities

4 Issues affecting community participation

For people's participation in Project supported activities, Karki et al. (1995) identified a collaborative mode of project implementation, achieving consensus of local communities on protection and contribution, investment in providing material and in capacity building of the local communities through training, and field visits to demonstration sites as being important. The formation of forest user groups and support to a water supply scheme, which the forest user groups planned and implemented, were important in generating and sustaining the interest of the local people. In addition to these issues, related directly to Project involvement, there are other factors that influenced people's participation. In the following section these issues are examined however no attempt is made to quantify their relative importance.

4.1 Issues that have strengthened participation

Five issues seem to have had the major impact on strengthening participation of local communities. They are:
- The degradation of the sites were relevant to local communities (not just the perceptions of 'outsiders'). Gullies at the sites threatened houses and agricultural lands. The communities also realised that production of biomass was decreasing every year. However, mechanisms did not exist for the people to come together as the settlements were comparatively new, having been established after eradication of malaria in the late Fifties.
- People's perception of environmental degradation has also changed due to changes in these areas from improved access to education, mass media and increasing contacts with the capital Kathmandu in the last fifteen years. Planting of trees has been promoted by the government and the media as being a 'noble' deed and any vocal opposition to such activities are liable to be seen by local communities as being 'un-progressive' and 'bad'.
- In the past, cattle and goats which are normally grazed in forests, were the main animals raised in these areas. Buffaloes (which produce more milk with higher fat contents) have become more important recently due to a thriving trade of fresh milk to Kathmandu. As buffaloes are normally stall fed, the need for open grazing has reduced. For example, a part of Dhaireni had not been planted in 1973 as the local people wanted to retain it for grazing. When pine was planted forcibly, they were quickly destroyed by them. At the same time, the demand for fodder has also increased in the areas. As people have started moving away from cereals to tomatoes and potatoes, the availability of fodder from agricultural residues (such maize stem and rice and wheat stem) has decreased. Therefore, there is now less resistance to closing areas for rehabilitation activities.
- In many districts in Nepal, farmers have responded to fodder and fuelwood shortages from common lands by increasing trees on private lands (Carter et al., 1989, Gilmour et al., 1991). This holds true for both Project sites. Farmers with more land, who tend to be the leaders and are more vocal in forest user groups Committees and in enforcing the rules, have more trees on their lands. Thus, strong support from the leaders for closing off the areas to uncontrolled resource extraction can be easily understood, as their dependence on the common resources is low. The poorer farmers did not benefit from this in the short term. However, the natural growth of grass was so impressive after protection that the forest user groups allowed cutting of grasses and they have benefited.
- The Project emphasised biological engineering. The communities could contribute more with their knowledge of indigenous species and such techniques also tend to be cheaper to implement, 'repair' and to replicate on the farms by farmers. The focus on plants has the advantage that if they are unsuitable, they can be easily taken out at little cost. Such techniques have also motivated farmers to take interest. Replication of species planted on degraded sites to private lands is already in evidence with farmers planting *Dalbergia sissoo* trees and NB21 (a hybrid of *Pennisetum purpureum*) grasses on farms.

Thus, a number of historic and socioeconomic processes, changes in people's perceptions, resource decline and project activities contributed towards the participation of people in rehabilitation activities.

4.2 Challenges to community participation

Despite considerable achievements in community participation, constraints to fuller participation were also encountered. The major challenges include the following three points:

- The villagers' general view of 'development' tends to be associated with supply of goods or services by an outside agency or the government. The project attempted to adopt a more process-oriented rather than a target-oriented approach. The Project's objective to also organise people to enable them to tap resources was only partially met as the `process based' approach was not easily understood by the communities. There was some feeling that as the Project had targets, it would be up to the Project staff to meet them. Thus in many instances, it was difficult to achieve the level of participation by local communities.
- As plantation activities are carried out during the Monsoon, also the busiest agricultural season, participation of people was lower. Most of the planting work was done by students. At other times of the year, too, the farmers at both sites are very busy in agriculture, and there is hardly any free season, thus many farmers could not participate despite their interest.
- In the small and tight knit community of Bajrapare, more women participate in meetings than at Dhaireni where the forest user group consists of many settlements. At Dhaireni, the forest user groups organised their meetings at a local school or at the Project's field office. Women appeared to be discourage from attending such meetings outside their settlements. A solution to this may have been to organise meetings in different settlements, but this suggestion from the Project has not yet been taken up by the forest user groups.

As with the issues that strengthened community participation, the challenges are also a complex interaction of human (including historical and institutional issues) and biophysical aspects of the area and the people.

5 The Community Forestry approach and rehabilitation of degraded lands

For rehabilitation activities, a framework that guarantees the right of the users to products produced and also their role in decision making and management is crucial. The legislation focuses on people's participation, which is important for changing the attitude and practices of government agencies. The influences of the Community Forestry Policy on the Project's activities are identified as the most important issue. Thus the Community Forestry legislation for rehabilitation of degraded lands warrants critical examination as this is increasingly becoming the most important framework for such activities in Nepal.

5.1 Strengths of Community Forestry for rehabilitation of degraded lands

The concept and practice of community forestry in Nepal has a number of features that make it an ideal medium to organise and support community efforts at rehabilitation of degraded communal lands. Some of these points are highlighted below.
- Subedi (1996, *in preparation*) has noted that forest user groups are compatible with indigenous decision making units in rural communities of Nepal. Formation of forest user groups brings the real stake holders together and can mobilise labour and other resources. Work within geographical boundaries (catchments) or political boundaries (such as Village Development Committees) may not bring real stake holders, (who maybe outside such boundaries) and attempts in mobilising people may not work.
- Forest user groups also have 'legal' backing, can work in other spheres of community development and thus are likely to be more sustainable in the long run than any other form of sponsored community organisation.
- As forest user groups do not have to share income generated from a community forest with the government, there is no incentive for government staff to hand over only the productive

forests. If benefit sharing was instituted, it would probably make it more attractive to District Forest Office staff to concentrate only on productive forests.
- Networks of forest user groups are emerging as an important lobby in Nepal, through national networking organisation the Federation of Community Forestry User Groups of Nepal (FECOFUN) and forest user groups networks within districts. They can also be important for dissemination of innovations and can have influence on the policy level. Thus, too, the Project's activities were undertaken through the mechanism of community forestry. The Project sites have been used by many institutions for demonstration purposes, and there are indications that the technology and the processes have been emulated by the others.

Income generation from forests is encouraged and forest user groups could invest income generated from a less degraded part of a forest for rehabilitation activities on a more degraded part. If the forest user groups do not have any access productive forests then they also have the option of planting crops such as ginger, turmeric and cardamom. However, if the sites are too degraded for production, simply protection of an area has shown to have increased the production of grass has raised interest of local people. However, 'outsiders' may act as a catalyst for such activities and in supplying innovative technologies such as SALT. One of the reasons for the introduction of SALT in community forests by the Project was so that the farmers could 'test' the technology on a common land with little risk to themselves. Afterwards it is hoped that such technologies will be adopted by farmers on their own lands, thus serve the dual purpose of being a technology that is useful for common and private lands. Some farmers have already indicated interest in applying such technology on their land, albeit at a smaller scale.

5.2 Weaknesses of Community Forestry for rehabilitation of degraded common lands

Though the strengths of community forestry have been highlighted above, there still needs to be an improvement in the way activities are targeted and implemented. For example,
- As requests for handover are increasingly coming from users themselves, handover of degraded sites is probably becoming sidelined. Local people are unlikely to request handover of degraded sites as, if there is no forest at present, the local people may be unaware that they could establish and manage plantations as community forests. If no support is provided for plantation establishment, local people may not have resources to do rehabilitation work themselves. In many instances, a place has degraded due to lack of indigenous management and in such areas, organising communities may be more difficult. On many degraded areas, people graze animals and any attempts to stop this could result in conflict. Thus, for community forestry to be effective in rehabilitation of degraded lands, the District Forest Office will need to keep on being pro-active. They may have to spend more time in organising communities, and with the recent cuts in their staff, this may be difficult to achieve.
- Under community forestry, the focus of rehabilitation activities is through afforestation. In many instances, sites may be too degraded for afforestation, and the forest user groups may wish to, or need to, manage such lands as grassland or shrubland. Thus, options such as the use of innovative approaches like SALT for rehabilitation of degraded areas need to be promoted. District Forest Office does not generally support soil conservation activities such as checkdam construction, which is considered to be the 'turf' of the Department of Soil Conservation and Watershed Management. A closer working relationship between the two departments needs to be fostered. Management of water (such as harvesting as well as channelling of excess water) is an important aspect of soil conservation and rehabilitation, and these have not received adequate attention in Nepal.
- Forest user groups can be strong viable institutions with external assistance but they may need to be fairly organised to approach other institutions The support role government agencies or

other NGOs could play are extremely important to enhance their organisational capacities. Forest user groups are often not even capable enough to approach other organisations for support and few organisations are involved in strengthening forest user groups.

6 Conclusions

Experience of ICIMOD's Rehabilitation of Degraded lands in Mountain Ecosystems Project in the Kabhre Palanchok district shows that local communities' participation in rehabilitation of degraded community lands is possible. For this Project, the innovation in approach lies in the use of community forestry concept in rehabilitation of degraded lands, and the introduction of grass, shrub and tree species contour hedgerows (modified Sloping Agriculture land Technology, SALT) by the Project as the major rehabilitation technology on community forests in Nepal. The experience of the Project shows that community participation is possible and effective in managing land resources and it can be sustainable if benefits expected are achievable and within a reasonable time. Experiences also suggest that participation is dependent upon a collaborative mode of project implementation, the incentives provided by the project, and because land degradation threatens local communities. Socioeconomic changes, such as increases in number of trees on private lands, have been conducive to participation. Appropriate technological inputs, combined with a suitable framework for bringing stakeholders together, such as through community forestry, are identified as being particularly important.

Acknowledgments

We are grateful to Steve Hunt, David Hinchley, P.B. Chhetri of NACFP for their helpful comments on this paper. Sameer would like to thank ICIMOD for its support.

Endnote

[1] ICIMOD is an international centre, established in 1983, which focuses its activities in the Hindu-Kush Himalayan region. Its member countries are Afghanistan, Bangladesh, Bhutan, China, India, Nepal, Myanmar and Pakistan.

References

Carson, B (1992): The Land, the Farmer and the Future, ICIMOD Occasional Paper No. **21**, ICIMOD, Kathmandu, Nepal.

Carter A.S. & Gilmour, D.A. (1989): Increase in tree cover on private farm land in Central Nepal, in Mountain Research and Development, Vol. 9 Issue 4.

Chalise, S.R., Karki, S., Shrestha & B.G. (1995): Rehabilitation of Degraded Lands In Mountain Ecosystems Project: Nepal Site II, Kabhre Palanchok, Final Report: Phase I (April 1993-October 1995), unpublished, ICIMOD, Kathmandu.

Edwards, D.M. (1996): Non-Timber forest Products from Nepal: Aspects of the trade in medicinal and aromatic plants. FORESC Monograph 1/96. Forest Research and Survey Centre, Ministry of Forests and Soil Conservation, Babar Mahal, Kathmandu, Nepal.

Fisher, R.J. (1995): Collaborative management of Forest for Conservation and Development, IUCN and WWI, SADAG, Bellegarde/Valserine, France.

Gilmour D.A. & Nurse. M.C. (1991): Farmer initiative in increasing in tree cover in central Nepal, Mountain Research and Development, Vol. 11 Issue 4.

ICIMOD (1993): ICIMOD Methodology Workshop on Rehabilitation of Degraded Lands In Mountain Ecosystems in Hindu Kush Himalayan region, May 29 to June 3, 1993, Kathmandu: ICIMOD.

ICIMOD (1996): Rehabilitation of Degraded Lands In Mountain Ecosystems in Hindu Kush Himalayan region (January 17, 1992- December 31, 1995), Final Report submitted to International Development Research Centre, Ottawa, Canada. Unpublished, ICIMOD, Kathmandu.

Jackson, W.J. and Ingles, A.W. (1994): Developing Rural Communities and conserving biodiversity of Nepal's Forest through Community Forestry. Seminar on Community Development and Conservation of Forest Biodiversity through Community Forestry, Bangkok, Thailand, 26-28 October 1994.

Karki, S. & Chalise, S.R (1995): Local Forest User Groups and Rehabilitation of Degraded Forest Lands, in Challenges in Mountain Resource Management in Nepal Processes, Trends and Dynamics in middle mountain watersheds, Proceedings of a Workshop held in Kathmandu Nepal 10-12 April 1995, ICIMOD.

Mahat T.B.S., Griffin, D.M. & Shepard, K.R. (1986): Human Impacts on Some Forests of the mid-hills of Nepal 2: Some major human impact before 1950 on the forest of Sindhu Palchok and Kabhre Palanchok, in Mountain Research and Development, Vol. 6, No.4.

Mahat, T.B.S. (1987): Forestry-Farm Linkages in the Mountains, ICIMOD Occasional Paper No. 7, Kathmandu.

Partap, T. & Watson, H. J. (1994): Sloping Agriculture Land Technology, A Regenerative Option for Sustainable Mountain Farming, ICIMOD Occasional Paper No. 23, Kathmandu.

Shah, P.B., Shrestha, B. & Nakrmi, G. (1993): Resource Dynamics, Production and Degradation in a Middle Mountain Watershed : Integrating Bio-physical and Socio-economic Conditions in the Jhikhu Khola Watershed with GIS and Simulation Models. First Annual report submitted to International Development Research Centre (IDRC), Singapore and Ottawa. Unpublished. Kathmandu: ICIMOD.

Subedi, B. (1996): Participatory Forest Management: Principles and practices of Community Forestry in Nepal, Unpublished a report submitted to Nepal Australia Community Forestry Project, Kathmandu.

Wyatt-Smith, J. (1982): The Agricultural System in the Hills of Nepal: The ratio of Agriculture to forest land and the problem of animal fodder, APROSC Occasional paper No. 1, Kathmandu, Nepal.

Addresses of authors:
Sameer Karki
Nepal Australia Community Forestry Project
PO Box 208
Kathmandu, Nepal
S.R. Chalise
ICIMOD
PO Box 3226
Kathmandu, Nepal

Conservation Farming Land Use on Critical Upland Watersheds
- Social and Economic Evaluation Study -

C. Setiani & A. Hermawan

Summary

Population growth in Indonesia has led farmers to cultivate hilly areas. Economic viability has been reduced as farmers tend to cultivate land without proper land conservation measures which increase soil erosion. Research regarding land use with conservation farming orientation has been conducted in Pasekan Village, Semarang District, Central Java-Indonesia since 1991. Land was managed by taking into account soil slope, land suitability and farmers' preferences. Results showed that conservation farming implementation decreased erosion rates by 58 to 73%. A conservation farming technologies strategy, developed for critical upland watersheds, was accepted by farmers, because it gave direct economic benefit to them and was in accordance with local cultural values. Another benefit of conservation farming implementation was the increase in labour productivity. In addition, farmers have a better chance to develop productive economically viable off-farm activities. Introduced technologies should be more sensitive to socio economic conditions favorable to the farmer.

Keywords: Upland watersheds, conservation, erosion socio economic, cultural value

1 Introduction

Although Java island conprises only 7% of Indonesia's area, 63% of all Indonesians live there. The population presure has led farmers in Java towards the highly erosible hilly mountainous areas. Cultivation on such sloping upland areas, without proper soil conservation efforts, rapidly decreases land productivity to a critical level. These at risk critical areas are a crucial problem in Indonesia. Karama and Irawan (1995) report that such land in Indonesia covers around 18.4 million hectares. A further 6 million hectares are non forest and cultivated by farmers.

In response, Government of Indonesia has developed rehabilitation and conservation programs through related institutions. One of these is the Upland Agriculture and Conservation Project (UACP). UACP started in 1984/1985 and was funded by the Government of Indonesia supported by the United States Agency for International Development (USAID) and International Bank for Reconstruction and Depelopment IBRD. UACP managed two priority watersheds on Java Island, Jratunseluna and Brantas.

The UACP consisted of five components; Farming Systems Research (FSR), Sustainable Upland Farming Systems (SUFS), Project Innovation Fund (PIF), Human Resource Development (HRD), and Access Road (AR). UACP formally ended in 1993. Because of the strategic function of the FSR component, the Agency for Agricultural Research and Development (AARD) that is responsible for FSR management, sustains FSR activities by Government of Indonesia funding.

FSR aims to generate information on farming conservation technologies that will be able to (1) increase farm production and farmers' income, (2) improve water and soil conservation, (3) support the changes from low quality to high quality commodities and (4), to be effective for the sustainability of environment, and acceptable by farmers (USAID, 1984). Several promising components of technologies have been developed by FSR. To increase their utility, components of technologies should be designed into a package of conservation farming systems suitable for agroecological zone and farmers' socio economic circumstances. FSR designed models are based on several criteria. including soil erodibility, solum depth, and soil slope.

This paper presents conservation farming research results conducted by FSR at Pasekan Village, Semarang Regency, Central Java. Pasekan Village represents an upper upland of the Jratunseluna Watershed.

2 Research methods

Farming system research activities in Pasekan Village began at the end of 1991. Before implementation, a diagnostic survey was carried out in order to obtain in-depth knowledge about the social situations. Basically, land was divided into three categories, based on land slope, land suitability, and farmers' preferences.

The following stages were followed during implementation:
a) Field arrangement and management include assisting and encouraging farmers to improve their soil conservation practices
b) Farmers extension and guidance in groups and provided with extension activities continuously.
c) Data collection was conducted weekly by questionnaire on technical, economic and social aspects.
d) Data analysis

3 Results and discussion

3.1 Soil conservation aspect

Farmers have used traditional bench terraces to conserve their land. As these were too vertical and unstable then during 17 months of implementation, most of the participating farmers refined their terraces' risers, made waterways and drop structures. Soil coverage was better when shaded by perennial trees (fruit and leguminous trees) As they helped to protect the soil from the effect of rainfall.

Soil conservation development was analyzed by an 'Indicative Erosion Factor' that compared "before" and "after" conditions respectively. Relative erosion is calculated from the change of the coverage factor of soil (C factor) and soil conservation treatment factor (P Factor). Index erosion calculations showed that the erosion rate decreased by 58-73% (Table 1). Full participation of farmers, extension agents as well as formal and informal leaders were the key factors.

	Year of implementation		
	1991/92	1993/94	1994/95
Inderosi value	3.91	5.26	6.26
Erosion decrease (%)	73.0	67.50	58.13
Target achieved	99.9	87.58	78.64

Source : P3HTA, 1994

Table 1: Value of index of erosion at Pasekan Village, Semarang District

3.2 The impact of conservation farming technology development

3.2.1 Intensification of ruminant farming

Grasses and legumes planted in ridges and risers of terraces increased the carrying capacities for ruminant livestock. The population of fattening cattle increased 38% during 5 years.

Total additional income earned from the forage plantation and changes in feed supply, was 16%, while the labour used decreased by 5%.

3.2.2 Labour allocation impact

The impact of conservation farming implementation on labour allocation can be seen in Table 2. The implementation of conservation technologies appear to have decreased labour time. Labour decreases mainly happened in upland activities and livestock raising. Even though the number of kept livestock increased, time consumed for forage collection was lower because of fodder cultivation. Labour used in upland activities decreased as a result of better land. Terrace maintenance was no longer needed.

	Year of implementation				
	March 1992[a]	March 1993[b]	March 1994[b]	March 1995[b]	March 1996[b]
Kinds of activities (%):					
- Lowland	27	27	24	31	30
- Upland	26	28	29	23	20
- Livestock	28	19	14	15	17
- Off-farm	25	26	35	31	33
Total (man days)	289.6	288	241.7	255	255

Sources: [a] Hermawan and Soelaeman (1992); [b] Primary data

Table 2: Farmers' labour allocation to economically productive activities in Pasekan Village, Semarang District.

The other impact was that less fertilizer was needed (manure and organic fertilizers) for food crop production since top soils were no longer lost. This meant that less labour was required to carry fertilizer to the field. Also, tree canopy growth reduced the area of land cultivated with food crops. Farmers were encouraged to engage labour farmers in off-farm activities (Table 2). From a land utilization view, the above phenomena were positive, since exploitation and pressure on the land were reduced.

3.2.3 Income generating impact

Research results showed that conservation farming development has changed farmer's income structures. At the initial phase, farmers' income was dominated by farming activities. Income from livestock and off-farm activities increased over thes years of conservation implementation (Table 3).

From the Pullin (1993) and Ortiz (1979) schemes, it seems that conservation farming will sustainably be practiced by farmers. Arguments of this are describe below.

	Year of implementation				
	March 1992[a]	March 1993[b]	March 1994[b]	March 1995[b]	March 1996[b]
Kinds of activities (%):					
- Lowland	23	.23	23	18	17
- Upland	25	16	16	16	8
- Homeyard	24	31	21	23	17
- Livestock	13	18	21	22	35
- Off-farm	10	12	19	21	23
Total Income (000 Rp./household)	1,970.4	2,050.6	2,160.2	2,860.0	2,740.2

Sources: [a] Hermawan and Soelaeman (1992); [b] Primary data from 18 households

Table 3: Farmer's income sources based on economic productive activities at Pasekan Village, Semarang District

4 Conservation farming technologies for sustainability in Pasekan Village

4.1 Farmers' perception of conservation farming technologies

In general, the technologies introduced were complex, since they contained many aspects, such as soil conservation, farm production aspects, and group management. However, for most farmers it could be very complicated. Pullin (1993) argued that complicated/sophisticated technologies will be abandoned by farmers. To overcome this, technologies were introduced in several phases. The first phase was terrace refinement and other conservation technologies. The second phase was the introduction of industrial crops and advice farmers on how to improve their farming practices by using high yielding varieties, appropriate crop spacing and recommended fertilizing. This strategy seemed to work properly. Farmers did not feel that the conservation technologies were complicated. Five years of implementation seem to be enough for farmers to realize the essence of the introduced technologies. Some farmers no longer cultivated annual crops since their uplands were shaded by the industrial tree crops. They kept the trees and intensified their livestock raising.

4.2 Orientation and motivation of farmers in farming activities

Upland farming was usually the second priority of farmers. As a consequence, farmers started upland cultivation oncelowland activities were completed. The lack of labour changed farmers' orientation of upland farming from cash crops cultivation to grass planting. At least five farmers planted grasses in the cultivated areas as they felt that it give then more advantages. Setiani and Hermawan (1995) report that grass cultivation gave higher income as compare to corn cultivation, by about 1100%.

4.3 Farmers resources and capabilities

Availability of financial means influenced farmers' participation in conservation farming activities, as such activitis were relatively costly.

Their income was above the poverty limits (minimum physical requirement for rural areas according to the Central Java Regional Planning Bureau were Rps. 19,278.00 per capita per month i.e., around US $ 9). Thus they were more like it to continue practicing the introduced technologies.

4.4 Decision making pattern related to risk aversion

Womens' role in farming activities was equal to men. However, guidance given by the project which was mainly directed towards men has caused an unbalanced perception between wives and husbands. There could be potential problems after the end of the project activities (Hermawan, 1995). Furthermore, the cultural value to share with others has encouraged participating farmers to share with their neighbours. For example some of their grasses and legumes were given to their neighbors to be planted. It hastened the spread of individual technologies. In this case, more innovative farmers that less risk averse, were needed to encourage other farmers to practice the improved technologies.

4.5 Infrastructure support

Research findings show that the farming community enthusiastically responds to introduced technologies if infrastructure, such as access roads and markets, are made available. Existence of extension institutions and source of information as well as problem solving agents were also very important (Hermawan, 1996). In many cases, technologies were no longer implemented by farmers due to a lack of institutional support. In Pasekan Villtage, some contact farmers have been prepared to act as extension agents in case official agents were not available. This strategy was adapted to guarantee the sustainability of the implemented conservation farming practices.

5 Conclusions

a) The implemented conservation farming technologies implemented were able to decrease erosion rates from 58 to 73%.
b) Most of the farmers felt the advantages of conservation farming technologies implementation. 'Demonstration effect', 'trickle-down effect' and formal local leaders support were the main factors.
c) Conservation farming technologies development could increase livestock carrying- capacity as well as decrease labour. It increased farmers' opportunity to find other economically advantageous jobs (such as off-farm jobs) which has increased farmers' income. The technology introduction process should be done in stages to ensure that farmers will be able to fully understand the essentials of the technologies to be introduced.
d) It appears that conservation farming technologies will continue to be practiced by farmers. Social and cultural factors. As well as infrastructure, are the ket reasons for this.

References

Hermawan, A. (1995):Masalah keberlanjutan penerapan teknologi usahatani konservasi di lahan kering DAS (Telaahan hasil- hasil kegiatan P2LK2T di lahan kering DAS Jratunseluna dan Brantas). Dalam Prawiradiputra et al. (Eds). Prosiding Lokakarya dan Ekspose Teknologi Sistem Usahatani Konservasi dan Alat Mesin Pertanian. Jogjakarta 17 -19 Januari 1995. Pusat Penelitian Tanah dan Agroklimat. Bogor, 235-252

Hermawan, A. (1996): Pengembangan rumput vetiver dalam sistem usahatani konservasi lahan kering. Dalam: Setiani et al. (Eds). Prosiding Seminar Hasil Penelitian Pola Usahatani Konservasi Tanah di Lahan Kering. BPTP. Badan Litbang Pertanian, 69-81

Karama, S. and Irawan (1995): Penguasaan lahan untuk penerapan teknologi dan peningkatan pendapatan keluarga tani. Makalah pada Konggres HITI ke VI, Jakarta 12-15 Desember 1995.

Ortiz, S. (1979): The effect of risk averson strategies on subsistence and cash crop decisions. In Roumasset et al. (Eds). Risk, Uncertainly and Agricultural Development SEARCA College, Laguna, Philippines

Prasetyo, T., Setiani, C., dan Triastono, J. (1993): Kiprah kelembagaan dalam penelitian pengembangan di DAS Jratunseluna hulu. Dalam Abdurachman et al. (Eds), Risalah Lokakarya Pelembagaan Penelitian dan Pengembangan Sistem Usahatani Konservasi di Lahan Kering Hulu DAS Jratunseluna dan Brantas, Tawangmangu, 7-8 Desember 1992. P3HTA. Badan Litbang Pertanian. Salatiga

Pullin, R.S.C. (1993): An overview of environmental issues in developing country aquaculture. In Pullin et al. (Eds) Environment and Aquaculture in developing Countries. ICLARM Conf. Proc., 1 -19

P3HTA (1994): Laporan Tahunan 1993/1994. Badan Litbang Pertanian. Deptan

Setiani, C., Hermawan, A., dan Prasetyo, T. (1995): Evaluasi sosial ekonomi pengembangan hijauan pakan di lahan kering DAS hulu. Dalam Subiharta et al(Eds). Prosiding Pertemuan Ilmiah Komunikasi dan Penyaluran Hasil - hasil penelitian dan rencana Penelitian untuk Mendukung dan Mendorong Pengembangan Peternakan di Jawa Tengah. Temanggung 22 maret 1995. Sub balittank. Puslitbangnak. Badan Litbang Pertanian. Klepu.

Setiani, C., dan Hermawan, A. (1995): Keuntungan komperatif penanamana rumput pada bidang olah di lahan kering DAS Jratrunseluna bagian hulu. Dalam Subiharta et al. (Eds). Prosiding Pertemuan Ilmiah Komunikasi dan Penyaluran Hasil -hasil penelitian dan rencana Penelitian untuk Mendukung dan Mendorong Pengembangan Peternakan di Jawa Tengah. Temanggung 22 maret 1995. Sub balittank. Puslitbangnak. Litbang pertanian. Klepu.

Soelaeman, Y. Hermawan, A., Rachman, A. Triastono, J., Suparno, Sugito, F.K. Andadari, Sarjito, dan J.H. French (1991): Laporan Survei Diagnostik Penelitian Pengembangan Usahatani Konservasi, Integrasi SUFS - FSR, Desa pasekan, Kecamatan Ambarawa, Kabupaten Semarang, P3HTA. Badan Litbang Pertanian. Salatiga

Soelaeman, Y., Juanda, D., dan Sudadiyono (1995): Penampilan dan perkembangan rakitan teknologi sistem usahatani konservasi di lahan volkanik DAS Jratunseluna. Dalam Prawiradiputra et al. (Eds). Prosiding Lokakarya dan Ekspose Teknologi Sistem Usahatani Konservasi dan Alat Mesin Pertanian. Jogjakarta 17 -19 Januari 1995. Pusat Penelitian Tanah dan Agroklimat. Bogor, 77-84

USAID (1984): Project Paper. Upland Agriculture and Conservation Project. USAID. Jakarta.

Addresses of authors:
Cahyati Setiani
Agus Hermawan
Institute of Assessment for Agriculture Technologies
Agency for Agricultural Research and Development (IAAT-AARD)
PO Box 101
Ungaran 50501, Indonesia

GRASS
Ground Cover for Restoration of Arid and Semi-arid Soils

H. Liniger & D.B. Thomas

Summary

The Global Assessment of Soil Degradation (GLASOD) map shows that half of the degradation on over 200 million ha of degraded soil in Africa is caused by overgrazing. The paper underlines and illustrates that grass cover is the key factor for improving the soil structure, increasing infiltration rates and primary productivity, and reducing surface runoff and gully erosion damage on semi-arid and arid grazing lands in Africa. A review of research and development activities in Eastern and Southern Africa illustrates that although the importance of grass cover is commonly known and has been reported in scientific reports, there has been little success in improving grass cover and primary productivity on grazing lands. Reasons for the slow progress are highlighted and a search for improved management solutions is recommended. For the sustainable use of grassland an international initiative, called GRASS, is urgently needed: Ground cover for the Restoration of Arid and Semi-arid Soils.

Keywords: Grazing land, cover, productivity, management, Africa

1 Introduction

"The degradation of rangelands is an important problem, but there are very few examples of successful rangeland improvement. It is therefore a lower priority for SDC" (SDC, 1994). This statement from the Swiss Agency for Development and Cooperation (SDC) is typical of the general pessimism about the possibility of controlling degradation on semi-arid and arid rangelands. This paper aims to initiate more support for restoring grass cover on denuded grazing lands.

The lack of cover on grazing land in semi-arid and arid areas of Africa is a problem of great magnitude. The Global Assessment of Soil Degradation (GLASOD) map gives a figure of 243 million ha of soil degradation caused by overgrazing in Africa, which is half of the land affected by human-induced soil degradation in Africa (UNEP/ISRIC, 1991). Since the UN Conference on Desertification in 1977, there has been a considerable effort to understand the causes of desertification and to differentiate climate from human activity as causative agents (Warren and Agnew, 1988; UNSO, 1992). The following definition of desertification was agreed to at UNCED in 1992 and adopted by UNEP: *"Desertification is land degradation in arid, semi-arid and dry sub-humid areas resulting from climatic variations and human activities"*. The most striking symptom of desertification is the denudation of the herbaceous cover and the loss of perennial grasses (Liniger, 1995). Lack of cover can be both a cause and a result of degradation. Figure 1 illustrates the vicious cycle of degradation due to overgrazing: the loss of perennial grass cover, followed by loss of water and soil. All of these consequences reduce the potential of the land to support pasture and livestock and thus increase the grazing pressure. The decline in potential is particularly serious because it comes at a time when population growth is causing an influx of people from high-potential areas in search of land for cultivation and grazing.

Fig. 1: Effect of grazing pressure on grass cover and land degradation for the same natural environment (same land form, geology, climate) in Mukogodo, Kenya: (a) With little grazing: good perennial grass cover; (b) Under heavy grazing: perennial grasses disappear and between the remaining grasses the soil is bare and erosion starts; (c) Under continued heavy grazing: perennial grasses that are not protected by trees / bushes disappear, topsoil is eroded and the surface has a crust; water cannot infiltrate and thus there is no production: Under trees the soil is protected, and there is no surface crusting, resulting in higher infiltration, less evaporation and thus higher grass production; (d) Even though bush cover is still considerable, deep gullies develop in the lower parts of slopes where perennial grasses disappeared, where the energy of surface runoff is strong enough to cut through the crust and remove the subsoil.

Progress in restoring cover on denuded grazing land in semi-arid and arid areas has been sporadic and generally poor. The World Overview of Conservation Approaches and Technologies (WOCAT) project requested information about successful soil and water conservation measures in Eastern and Southern Africa and found that 76% of the techniques recorded were for cropland and only 19% for grazing land (Liniger et al., 1996). Progress in conserving cropland has been quite marked in certain areas (e.g. Tiffen et al., 1994) and contrasts sharply with the continuing degradation over extensive areas of grazing land.

The slow progress in reversing degradation on grazing land is usually attributed to socio-economic factors such as overgrazing associated with communal land tenure, but it may also be due to failure to understand the grazing land ecosystem, and in particular, the role of cover in determining the movement of water within the system.

The grazing land ecosystem in semi-arid and arid areas is subject to change due to seasonal and long-term fluctuations in rainfall. However, these areas will normally support a cover of perennial grasses provided that there is some control of grazing and some reduction of livestock numbers during droughts, either by sale or transfer. Prolonged droughts lasting several years can lead to the death of perennial grasses, partly due to the pressure of livestock and wild animals but also due to the depredations of termites, whose biomass per hectare may equal that of livestock (Leparge, 1977).

When a prolonged drought ends and rains return, the land may remain bare even though stock numbers have been greatly reduced. Superficially the land may appear to be overgrazed, but the real problem is lack of infiltration. If land is bare because of overgrazing, the solution would appear to be reduction in stock numbers, but if it is bare because of high runoff, the first step would be to improve infiltration, which will lead to the restoration of ground cover.

This paper summarizes some of the experience gained by research and development projects in Eastern Africa, some of it under the ongoing Natural Resource Monitoring, Modelling and Management (NRM3) project in Kenya funded by SDC, the Swiss National Science Foundation and the Rockefeller Foundation.

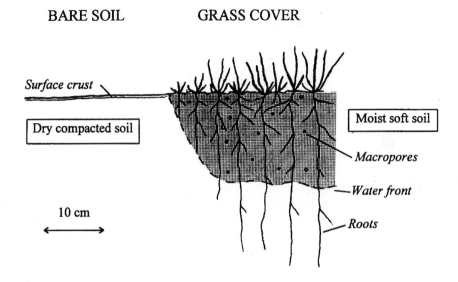

Figure 2: Difference in infiltration and soil characteristics due to grass cover on a semi-arid Lixisol (Liniger et al. 1993)

2 Importance of ground cover

Various research workers studying patterns of runoff and soil loss on various soils have shown the relationship between ground cover, runoff, soil erosion and soil productivity in Eastern Africa.

Infiltration experiments on a wide variety of soils from shallow clay loams to deep clays showed that soil cover conditions are more important in determining infiltration rates than soil types (Liniger, 1992). Infiltration under good bush and grass cover is about three to ten times the rate under overgrazed land with sparse cover. Kironchi et al. (1993) illustrated how reduced cover due to overgrazing reduces infiltration and organic matter, and increases bulk density. Figure 2 shows a soil profile dug after a rainfall of 20 mm where there was a good cover of grass adjacent to bare ground. Under the grass cover, water infiltrated to a depth of about 25 cm. The soil was soft with lots of macropores and biological activity. Where the soil was bare, the surface was sealed, the topsoil was hard and sterile, and no water infiltrated. Similar results were obtained by

Chepkwony (1980) with infiltrometer tests on severely sealed basement complex soils at Kamweleni in Machakos District in Kenya. He found that removing a few millimeters of surface seal with a penknife made a very significant difference in the rate of infiltration.

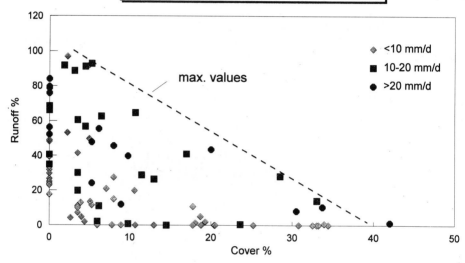

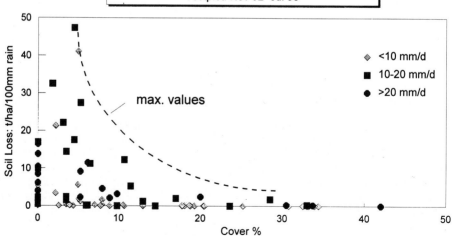

Figure 3: Effect of land-use/cover on runoff and soil loss from small plots (Mutunga et al. forthcoming)

Figure 3 illustrates the effect of cover on runoff and soil loss for soil derived from metamorphic gneisses. The straight line in (a) encloses the maximum percentage of runoff that occurred. It indicates that with a herbaceous cover of less than 5 %, over 90 % of the rain could be runoff; whereas at a cover of 40 %, runoff was reduced to zero. For the soil loss, cover is even more important. The enclosing curve shows that with a cover of more than 20 %, soil loss is reduced to almost zero (Mutunga et al., forthcoming).

Dunne (1977), working with a rainfall simulator in an arid area in Amboseli, Kenya, found a relationship between ground cover and soil loss which indicated that a minimum value of 20% cover was essential to promote infiltration and control soil loss. Moore et al. (1979) at Iiuni, Machakos, in a semi-humid to semi-arid area, reached a similar conclusion using simulated rainfall. Zöbisch (1986), using small runoff plots with Gerlach troughs in a similar zone in Machakos, concluded that a minimum cover of 40% was essential.

Thomas and Barber (1978) found that runoff and soil loss were linked with the surface conditions and that where there was a residual layer of quartz stones on the surface, infiltration was much better than in the absence of quartz. Plant cover was always more abundant in the quartz-strewn areas than in other areas where the soil was deeper but the surface was firmly sealed. Sealed ground without a vegetative cover lost up to 80% of the rainfall as runoff during simulated rainfall of 70 mm per hour.

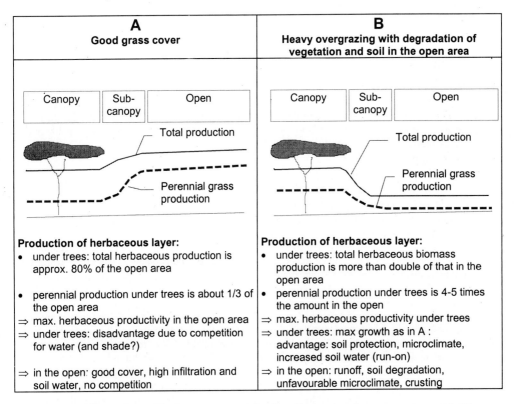

Figure 4: Effects of trees / bushes on grass production: Comparison between an area with (A) good grass cover and (B) with heavy overgrazing and degradation of vegetation and soils.

Whereas these exceptional rates of runoff can be found on small areas, most catchments have variations in land surface; some parts are bare and sealed, some have stones, some are covered with shrubs. There are areas of erosion and areas of deposition. Measured catchment water and soil loss are usually much lower than the extrapolated values calculated from small representative testplots because areas of deposition are not properly accounted for in small testplots. However, in experiments at Mukogodo, Laikipia District, Kenya, the rate of sediment in tons per hectare produced by the overgrazed catchment was roughly equal to the rate from bare, heavily grazed runoff plots. Due to surface sealing, small runoff plots produced high runoff but relatively little soil loss, whereas runoff from the whole catchment resulted in heavy gully erosion.

The same experiments showed how runoff can accumulate under bushes where the grass cover has remained intact even with heavy grazing and trampling, and infiltration rates are high. These run-on areas are important because they allow grasses to produce seeds which can facilitate the natural process of revegetation. Where land is in good condition, the presence of bush can lower the overall production of grasses, but where it is denuded and degraded, the presence of bushes can enhance production (Fig. 4) (Kinyua et al., forthcoming; Mutunga et al., forthcoming).

Maintenance of cover is clearly of prime importance in minimizing losses of soil and water and restoring the productivity of land. Restoration of cover should therefore be a top priority for land which has become denuded.

3 Restoration of cover

3.1 Research findings

The restoration of cover can take place either (a) by resting alone or (b) by resting combined with measures to promote infiltration and, if necessary, reseeding. Resting alone involves less immediate cost but may be slow, and the closure to livestock may create problems elsewhere if prolonged. The evidence presented below suggests that recovery by resting alone is most likely to succeed where the following conditions prevail:
- the soil is relatively fertile
- the soil is relatively deep
- the A horizon has not been completely eroded
- the soil surface is partly covered with stones
- there are still some perennial grasses present
- the environment is semi-humid or semi-arid but not too arid

Pereira and Beckley (1952) at Makavete (a sub-humid to semi-arid area in the Machakos District), using a randomised block experiment with half acre plots, showed that cover could be restored on denuded land by resting alone, but this took at least one year longer than active measures of land preparation and reseeding. Although the land had been severely denuded, the soil (probably a Luvisol) was in relatively good condition.

In Mukogodo (a semi-arid part of Laikipia District) an experiment showed the importance of initial conditions in determining whether resting alone would be effective (see Figure 5). In a first situation, although the land appeared to be bare, there were still a few traces of perennial grasses. On closure to grazing, a dense cover of annuals and perennials developed rapidly during the rainy season. In the second situation, less than 100 m away, there were no traces of perennials and there was no significant regeneration of cover after over 2 years of closure (Mutunga et al., forthcoming). The explanation appeared to lie partly in the fact that in the former situation the soil was less degraded and still relatively fertile, whereas in the latter it had been eroded down to the illuvial B horizon which is less fertile, has a higher clay content than the A horizon, greater propensity for surface sealing, and reduced infiltration.

The above experiments show the importance of understanding the soil conditions and weighing the potential for natural regeneration of cover. If this is slow, e.g. one or two years, there are costs associated with loss of production which might be reduced if active steps were taken to promote rapid recovery, even though initial costs of land treatment and reseeding may appear to be high.

Fig. 5: Comparison of different situations within the same natural setting: 4% slope on heavily grazed Lixisol in the semi-arid rangeland of Mukodogo, Kenya: From top left to bottom right:
(a):Before resting of an area with some remaining perennial grasses and topsoil not eroded (7.11.92)
(b): After resting of (a): fast recovery after the first rainy season (29.1.93)
(c): Before resting of an area with no perennial grasses left and topsoil eroded (7.11.92)
(d): After resting of (c) for 1 season: no recovery even after 2 years (29.1.93).

Bogdan and Pratt (1967) and Pratt and Gwynne (1977) have given detailed recommendations on methods of reseeding. Selection of suitable drought-resistant grasses such as *Eragrostis superba* and *Cenchrus ciliaris* is important and the procedures for harvesting seeds, storing and sowing need to be well understood. *Chloris gayana* has sometimes been used in semi-arid areas and can be useful in providing a quick cover, but it is not persistent and must be accompanied with drought-resistant species. There are many other species which can be used.

3.2 Field experience

The HADO Project (Mndeme, 1992) involved the forcible removal of livestock from extensive areas of mixed farm land with denuded pastures in a semi-arid area of central Tanzania. The socio-economic consequences for the population were severe. Closure of grazing land did allow recovery but bush encroachment created new problems. The measures were drastic and the results mixed, which indicates that more attention should be given to methods of rapid reclamation which mini-mize social disruption and pave the way for better management.

Mututho (1989), working on denuded rangeland in Kitui District, Kenya, showed the effectiveness of pitting coupled with reseeding for restoring cover. The pits, referred to as Matengo pits because of some similarity with the pitting system used by the Wamatengo people on cropland in Tanzania, were circular, about 2 m in diameter, and laid out in a staggered arrangement.

The pitting system was further developed by researchers at Katumani, Machakos (Simiyu et al., 1992), who recommended semi-circular pits in an interlocking system, which was effective but costly. In order to defray the costs they planted cowpeas around the pits during the first season. Although pitting systems are effective, they are also labour-intensive and unlikely to be adopted voluntarily by pastoralists.

The Baringo Fuel and Fodder project carried out intensive measures of reclamation and revegetation of denuded land in the Lake Baringo basin, which is grazed by agropastoralists. (Meyerhoff, 1991 and de Groot et al., 1992). The land is part of the Rift valley system and volcanic in origin. The soils are mainly derived from alluvial and colluvial deposits. Reclamation involved enclosure with a solar-powered electric fence and the construction of large contour bunds spaced at about 3 m to harvest runoff. The bunds were made by a motorized grader and cross ties were made by hand to prevent water breaking through at low points. Grass seed was used to restore ground cover and some planting of local trees was carried out. Technically the project was successful, but the costs were high and the commitment of the local people to maintain the communally owned land, which was reclaimed, is still uncertain. Though useful in showing the potential for revegetation when runoff is prevented, the high costs limit replication in the absence of external aid.

The Baringo Pilot Semi-Arid and Arid Project supported by the World Bank carried out reseeding with the aid of water harvesting on the alluvial soils adjacent to Lake Baringo. Hand-dug semi-circular bunds were used to harvest water and grass seeds. The grasses established included *Cenchrus ciliaris* (Buffel grass), *Eragrostis superba* (Maasai Love Grass) and *Chloris gayana* (Rhodes grass). The area was fenced and livestock excluded to allow the grasses to grow and seed. From a technical angle the project was quite successful; the grasses grew well. But from a socio-economic point of view the project was a failure. When the experimental phase was over, no effective community management had evolved, and even before the World Bank support came to an end, the fences and pasture were destroyed and the land reverted to its former state.

The pasture rehabilitation accomplished by changing the grazing management was successful in a few areas in Kenya in the 1950s. In the Esageri grazing scheme in South Baringo and the Riwa grazing scheme in West Pokot, a four block grazing scheme was established for the rehabilitation of the grass cover. Each of the four large blocks was grazed in turn for a three month period, after which there was a nine month rest before it was grazed again. All these reclamation schemes succeeded from a technical viewpoint because of the strong Government control. But as soon as control was withdrawn with Independence, the situation reverted to what it had been before.

In the semi-arid and arid areas of Turkana in Kenya, water harvesting in micro-catchments was used for tree planting and regrowth of grasses. Again, this method worked well under the control of a project, which strongly supported the initial establishment and controlled the grazing.

Neither before nor after Independence has there been much success in promoting voluntary adoption of pasture reclamation on communal rangelands. What is interesting, however, is how much progress there has been in reclaiming formerly denuded pasture lands in Machakos which are under individual tenure. Mutiso and Mutiso (1995) have given an account of the reclamation of denuded Acacia/Commiphora bushland at Wamunyu in Machakos District, Kenya. The land is semi-arid but is now held under individual ownership. The work of reclamation has been carried out by the owner and his family over a long period of time with virtually no outside help. Measures include enclosure, terracing to trap rainfall, selective clearing of unwanted bush, pruning of trees to encourage grass cover below, and rotational grazing. The farmer family who initiated this development has shown that, with good management, it is possible to keep high-grade Friesian cows economically and successfully on previously degraded land under semi-arid conditions.

Although the above example has shown what can be done by individuals in a semi-arid area without external aid or mechanization, the problem of reclaiming land in communal areas and under arid conditions is much more difficult. Little attention seems to have been given to low-cost methods of reclamation such as scratch ploughing or laying brush in lines on the contour. Prior to Independence in Kenya, extensive areas were reclaimed in Kitui District by means of scratch ploughing with an ox-drawn mouldboard plough, from which the mouldboard had been removed, and seeding with drought-tolerant grasses such as *Cenchrus ciliaris* and *Eragrostis superba* (Jordan, 1957). Other cheap methods such as laying slashed brushwood in rows along the contour could be used to protect the soil and increase infiltration, protect seedlings during emergence, and hasten the process of revegetation.

The WOCAT program has collected information on soil and water conservation in Eastern and Southern Africa, where the biggest portion of the land surface is used for grazing and browsing. Preliminary results show that where land is under a communal or open access regime, there is little progress towards improved systems of management, but where land is privately owned or managed, improvements are being made (Liniger et al., 1996).

The wide fluctuations in rainfall and the cyclical nature of droughts have led to the conclusion that concepts of carrying capacity and equilibrium in relation to semi-arid and arid rangelands are unreal (Behnke and Scoones, 1991; Behnke et al., 1993). Disequilibrium is a permanent feature of pastoralism and resource management. Development programs and Government Departments working with annual budgets lack the flexibility needed to make a quick and effective response to opportunities which arise for the restoration of cover when there is a spell of good rains.

4 General conclusions

Even though the importance of ground cover to reduce soil degradation and to improve productivity is generally known, little has been achieved to improve vastly degraded semi-arid and arid rangelands in Africa. Reasons for slow progress in revegetation of denuded land can be due to many factors but the following are common problems:
- Insufficient international concern: The main international focus has been on trees, whereas the grass layer has been neglected. While the loss of tree species and the reduction of forest area

has been of great international concern, the massive loss of valuable grass species and the loss of vast areas of grassland and its biodiversity have been greatly neglected.
- Insufficient awareness of the basic ecological and hydrological dynamics in semi-arid and arid grazing lands: The effects of vegetation, stones and other ground cover on infiltration and runoff, and the consequences for plant production and erosion have been given insufficient attention.
- Breakdown of traditional management systems on communal lands and failure to develop alternative management systems where people accept responsibility for using the land in a sustainable manner.
- Lack of adequate knowledge and training in cost-effective, locally adapted and appropriate methods of revegetation and management of grazing lands.
- Insecurity of tenure.

Based on the little experience gained so far, the following key factors in restoring cover need to be mentioned:
- Sustainable restoration of the ground cover is only possible if the current management is changed. There is no need to invest in expensive rehabilitation programmes if the management problems that have caused the degradation are not addressed and solved. The key to restoring grass cover and productivity in semi-arid and arid areas lies in the search for a flexible and locally fine-tuned management system that does not lapse into the vicious cycle of overuse, water loss, erosion and loss of grassland productivity, followed by increased pressure on the land and further degradation. This search for improved management is the main challenge on semi-arid and arid grasslands, and involves local people and rangeland specialists.
- Restoration of cover by resting alone on land which is firmly sealed is only likely to occur if there is a minimum ground cover with perennial grasses of 10-20% to facilitate infiltration. If there is less than 20% cover, the A horizon has gone, and the surface is firmly sealed, recovery is likely to be very slow with resting alone because of high rates of runoff. In this situation measures to improve infiltration are essential. Reseeding may also be necessary to ensure rapid recovery.
- Where there is a residual cover of stone on the surface, infiltration is likely to be adequate for the restoration of ground cover, but a stony cover may indicate that the topsoil has been stripped off and fertility is low.
- Measures to improve cover should be timed to take advantage of periods of average or above-average rainfall. Grasses used for reseeding should be locally adapted and drought-tolerant.
- Individual tenure can greatly facilitate the gradual reclamation of land by allowing progressive individuals to implement grazing control and reclamation measures. However, this should not lead to the recommendation to subdivide pastoral areas into small individual plots. Individual tenure is appropriate in semi-humid to semi-arid areas, but not generally in the semi-arid to arid areas, where livestock need to move over vast areas to make the best use of fluctuations in rainfall and vegetative production.

5 Recommendations

The following broad recommendations are intended to encourage renewed efforts to promote GRASS: Ground cover for the Rehabilitation of Arid and Semi-arid Soils:
- Improve knowledge and strengthen training in rangeland ecology and hydrology and, in particular, cost-effective methods for reducing runoff, promoting infiltration, and improving cover and biomass production.

- improve existing or develop new management systems, especially on communal and open access land, that allow for climatic fluctuations, maximize the grass production, reduce environmental degradation, and ensure the restoration of ground cover after drought periods.
- involve local people in making and implementing decisions and in the search for better management systems.
- train local technicians and national and international rangeland specialists in the participatory approach with pastoral communities.
- reclaim small areas initially to demonstrate appropriate measures, and assist pastoralists and farmers to visit places where such methods of reclamation have been adopted.
- establish security of tenure and land use rights, either on communal or individual basis.
- launch an international initiative to increase awareness and support for GRASS.

Acknowledgments

Several postgraduate researchers and research assistants of the Natural Resource Monitoring, Modelling and Management Project in Kenya have contributed to this paper. Special thanks go to Christopher C. Ondieki, Geoffrey Kironchi, David M. Kinyua, Bell D. Okello, Joseph N. Ngeru and Joseph Mitugo.

References

Behnke, R.H. and Scoones, I. (1991): Rethinking Range Ecology: Implications for Rangeland Management in Africa. London, Commonwealth Secretariat.

Behnke, R.H., Scoones, I., and Kerven, C. (eds.) (199): Range Ecology at Disequilibrium. London, Overseas Development Institute, IIED and Commonwealth Secretariat.

Bogdan, A.V., and Pratt, D.J. (1967): Reseeding Denuded Pastoral Land in Kenya. Government Printer, Nairobi.

Chepkwony, P.K. (1980): Restoration of vegetation cover on degraded grazing land in a low rainfall area of Machakos District. Postgraduate Diploma Project Report, Department of Agricultural Engineering, University of Nairobi.

Dunne, T. (1977): Studying patterns of soil erosion in Kenya. FAO Soils Bulletin, no. 33.

Groot, P. de., Field-Juma, A. and Hall, D.O. (1992): Taking Root - Revegetation in Semi-Arid Kenya. ACTS Press, Nairobi.

Jordan, S.M. (1957): Reclamation and pasture management in the semi-arid areas of Kitui District, Kenya. East African Agric. Journal **23**: 84-88.

Kinyua, D.M., Liniger, H.P., and Njoka, J.T. (forthcoming): The micro-environmental influence of the canopy of Acacia etbaica and Acacia tortilis in Mukogodo Rangelands, Laikipia District, Laikipia Mt. Kenya Papers, Nanyuki, Kenya.

Kironchi, G., Liniger, H.P., and Mbuvi, J.P. (1993): Degradation of soil physical properties of overgrazed rangelands in Laikipia District. Proceedings of a national workshop on Land and Water Management, Nairobi, 15-19 February, 1993, Department of Agricultural Engineering, University of Nairobi.

Leparge, M.G. (1977): Influence of grass feeding termites in a tropical savanna. Paper submitted to ICIPE/UNEP Group Training Course, 30-22 July, 1977, Nairobi.

Liniger H.P. (1992): Soil cover and management - Attractive water and soil conservation for the drylands in Kenya. 7th International Soil Conservation Conference (ISCO) Proceedings, volume 1, Sydney, 27th - 30th September 1992; 130-139.

Liniger, H. P., Ondieki, C.N. and Kironchi, G. (1993): Soil cover for improved productivity (SCIP) - Attractive water and soil conservation for the drylands of Kenya. Proceedings of a national workshop on Land and Water Management, Nairobi, 15-19 February, 1993, Department of Agricultural Engineering, University of Nairobi.

Liniger H.P. (1995): Endangered water - A global overview of degradation, conflicts and strategies for improvements. CDE, Development and Environment Reports, University of Bern.

Liniger H.P. Thomas D.B. and Hurni H. (1996): WOCAT - World Overview of Conservation Approaches and Technologies - Results from Africa. Paper presented at ISCO, Bonn, August 1996.

Meyerhoff, E. (1991): Taking Stock - Changing Livelihoods in an Agropastoral Community. ACTS Press, Nairobi.

Mndeme, K.C.H. (1992): The HADO experience in Tanzania. In: Kebede Tato and Hans Hurni, eds. Soil Conservation for Survival, Soil and Water Conservation Society, Ankeny, Iowa.

Moore, T.R., Thomas, D.B. and Barber, R.G. (1979): The influence of grass cover on runoff and soil erosion from soils in the Machakos areas, Kenya. Tropical Agriculture, Trinidad, **56**, 339-344.

Mutiso, G.C.M., and Mutiso, S.M. (1995): Kambiti Farm - The Role of Water in Capitalising Drylands. Lectern Publications, Nairobi

Mutunga, C.N., Liniger, H.P., and Thomas, D.B. (forthcoming): The influence of vegetative cover on runoff and soil loss - a study in Mukogodo, Laikipia District, Laikipia Mount Kenya Papers, Nanyuki, Kenya.

Mututho, J.M. (1989): Some aspects of soil conservation on grazing land. In: D.B.Thomas et al. (eds.), Soil and Water Conservation in Kenya, Proceedings of the Third National Workshop, 16-19 September, 1986, Department of Agricultural Engineering, University of Nairobi.

Pereira, H.C., and Beckley, V.R.S. (1952): Grass establishment on an eroded soil in a semi-arid African reserve. Empire Journal of Experimental Agriculture, **21**, 1-14.

Pratt, D.J. and Gwynne, M.D. (1977): Range Management and Ecology in East Africa. Hodder and Stoughton, London.

SDC (1994): Sustainable Management of Agricultural Soils. Swiss Development Cooperation, Sectoral Policy Document, Bern.

Simiyu, S.C., Gichangi, E.M., Simpson, J.R., and Jones, R.K. (1992): Rehabilitation of degraded grazing lands using the Katumani pitting technique. In: M.E. Probert (ed.), A Search for Strategies for Sustainable Dryland Cropping in Semi-arid Eastern Kenya. Proceedings of a symposium 10-11 December 1990. ACIAR Proceedings no. **41**, 83-88.

Thomas, D.B. and Barber, R.G. (1978): Report on rainfall simulator trials in Iiuni, Machakos, Kenya. Unpublished report. Faculty of Agriculture, University of Nairobi.

Tiffen, M., Maltimore, M. and Gichuki, F. (1994): More people, less erosion - Environmental recovery in Kenya. ACTS Press, Nairobi.

UNEP/ISRIC (1991): Global Assessment of Soil Degradation (GLASOD), Map of the status of human induced degradation. Den Haag.

UNSO (1992): Assessment of Desertification and Drought in the Sudano-Sahelian Region 1985-1991. New York, UN Sudano-Sahelian Office.

Warren, A. and Agnew, C. (1988): An Assessment of Desertification and Land Degradation in Arid and Semi-arid Areas. London, IIED.

Zöbisch, M. (1986): Erfassung und Bewertung von Bodenerosionsprozessen auf Weideflächen in Machakos-Distrikt von Kenia, Der Tropenlandwirt, **27**.

Addresses of authors:
Hanspeter Liniger
Centre for Development and Environment
Institute of Geography
University of Bern
Hallerstraße 12
CH-3012 Bern, Switzerland
Donald B. Thomas
Consultant
P.O. Box 14893
Nairobi, Kenya

Implications of No-Tillage Versus Soil Preparation on Sustainability of Agricultural Production

R. Derpsch & K. Moriya

Summary

The key problem of tropical agriculture is the steady decline in soil fertility, which is closely correlated to duration of soil use. This is due primarily to soil erosion and the loss of organic matter associated with conventional tillage practices, that leave the soil bare and unprotected in times of heavy rainfall and heat. The implications of soil preparation on soil erosion and the sustainability of agricultural production was studied with special reference to experience and projects carried out in Paraguay, Brazil and Argentina under tropical and subtropical conditions.

Scientific data show that under tropical and subtropical conditions, tillage generally has a detrimental effect on chemical, physical and biological soil properties. Investigations also show that erosion damage is enhanced when the soil is bare. Water infiltration rates are increased and consequently erosion is reduced when mulch covers the soil in a no-tillage system. Tillage also releases considerable amounts of CO_2 into the atmosphere contributing to global warming. In order to achieve sustainable agriculture in the tropics and avoid global warming, soil tillage has to be reduced to a minimum or avoided completely and the soil has to stay as long as possible covered with mulches, sod and growing crops. No-tillage in mulches of previous crops or green manures in combination with adequate crop rotations is the production system of the future if sustainable agriculture is to be achieved.

Keywords: No-tillage, tillage and soil degradation, soil quality, agricultural sustainability, erosion, organic matter, tropical agriculture.

1 Introduction

One of the main factors to be considered in relation to agro-ecological sustainability is the soil, as it is the basis for food production for humanity. Therefore, an effort has to be made to minimise soil erosion so that soil is not transported by runoff to rivers, lakes or to the sea, and to ensure sustainability of food production.

In this paper, sustainable agriculture is defined as establishing high, lasting and economic soil productivity, without damaging the soil and the environment, improving quality of life. Definitions of sustainability that consider only one dimension (i.e. soil fertility) are insufficient. Ecological, social and economic dimensions must always be considered (Hailu and Runge-Metzger, 1993).

The results of exploiting agricultural systems are evident in those regions where the soil is cultivated intensively and continuously, without considering soil degradation caused by soil preparation under hot/humid conditions. In Central Paraguay, the regions which used to be the granaries of the country and where food used to be produced and exported to Argentina, many soils are so degraded and depleted that it is not possible to obtain economic production of basic products such as maize, cassava and cotton, and are gradually being abandoned. In southern Chile a hilly region close to the city of Concepción named "Cordillera de la Costa", also a granary of the country some 40 or 50 years ago, has suffered such catastrophic erosion that some areas cannot be used even for forestry. In the Andean region of Bolivia and Peru deep erosion gullies are destroying entire landscapes.

Such examples can be found not only in Latin America but world-wide. Rapid depletion of soil fertility and non-sustainable land use particularly in developing countries is both the cause and the consequence of widespread poverty. It is therefore necessary to change actual soil-degrading agricultural systems based on intensive soil preparation which leave the soil bare and unprotected, to sustainable production systems based on permanent soil cover with plant residues and mulches.

Soil is a <u>non-renewable</u> resource and it is available only in limited quantities. Conventional soil tillage that leaves the surface of the soil bare, is one of the major causes of the occurrence of erosion on agricultural land. Highest sediment amounts as well as phosphorus and nitrogen content in the water of the Itaipú dam (shared by Paraguay and Brazil), was measured in times of soil preparation for winter and summer crops (Derpsch et al., 1991).

2 The problem of soil degradation

The key problem of conventional agriculture in the tropics is the steady decline in soil fertility, which is closely correlated to duration of soil use (Fig. 1). This is due primarily to soil erosion and the loss of organic matter associated with conventional tillage practices, that leave the soil bare and unprotected in times of heavy rainfall and heat.

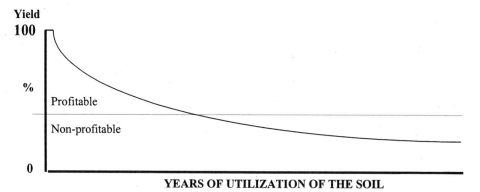

Figure 1: Soil degradation through time in conventional agriculture (Derpsch, R., unpublished)

Despite progress in genetics and breeding, fertilisation, plant protection and management, there is a clear tendency of diminishing yield over time. FAO predicts, that if soil losses continue unchecked the potential rainfed crop production will decline by about 15% in two decades in Africa, about 19% in Southeast Asia, and by more than 41% in Southwest Asia (Kelly, 1983)

The result of soil degradation is not only that farm land has to go out of production, but also that there is an increasing need for more inputs and investments to maintain high levels of productivity. In the United States, 50% of fertiliser needs is applied only to compensate for the losses in soil fertility due to soil degradation. In Zimbabwe, soil nutrient losses by erosion are three times higher than the total quantity of fertilisers applied (Stocking, 1986).

3 Erosion

Occurrence of erosion can be considered the most important factor causing soil degradation. Under the concept of sustainability, the first negative factor in relation to productivity and profitability, and the major aggressor of the environment is soil erosion. Consequently, sustainability can only be achieved if soil erosion is stopped completely.

When agriculture is practised on slopes in undulating topography, and rains of a certain intensity occur, soil preparation especially with disc implements results in bare soil, and this results in water erosion, or in regions of heavy winds in wind erosion.

It is estimated that soil losses in cropland in Latin America reach 10 to 60 t/ha/year (Steiner, 1996; Derpsch et al., 1991). Average soil losses in the State of Paraná, Brazil, where good soil conservation is practised, are as high as 16 t/ha/year. In Paraguay, on 4000 m^2 plots with 6% and 8% slope on high clay content Oxisols, average soil losses of 21.4 t/ha were measured in conventional soil preparation, while only 633 kg/ha of soil loss were measured in No-tillage (Venialgo, 1996). For the same experiment after extreme precipitations of 186 mm on June 9 and 18, 1995, soil losses of 46.5 t/ha were measured under conventional tillage, as compared to soil losses of only 99 kg/ha under No-tillage (both plots on 8% slopes). This resulted in 470 times higher soil losses when soil was prepared. (Venialgo 1996)

The high losses from agricultural soils have to be compared against the annual rates of soil regeneration that are estimated to be not more than 250 to 500 kg/ha/year. When soil losses are higher than natural soil regeneration rates, sustainable agriculture is not possible.

Recent studies show that soil erosion is a selective process, with the most fertile soil particles taken away. Eroded soil sediments usually contain several times more nutrients than the soils they originated from (Stocking, 1988).

Applied fertilisers are also transported by erosion to streams, rivers, lakes and to the sea, and therefore lost forever. Considering that world phosphate reserves are going to be exhausted in 40 to 50 years (Hoffman et al., 1983), present generations are acting irresponsibly when allowing soil management practices that produce high erosion rates. Even under the assumption that phosphate reserves are going to last much longer, it has to be kept in mind that reserves are finite.

Research has shown, that soil cover is the most important factor that influences water infiltration into the soil, thus reducing runoff and erosion (Mannering and Meyer, 1963).

4 Organic matter

In the tropics and subtropics organic matter content of the soil has an overriding importance in relation to soil fertility. According to Cannel and Hawes (1994), organic matter content of the soil is probably one of the most important characteristics in relation to soil quality, due to its influence on soil physical, chemical and biological properties.

Due to the fact that cation exchange capacity of most tropical soils is very low (Sánchez, 1976), organic matter has a much higher importance to store nutrients in the tropics than in temperate regions. Therefore the efficiency of mineral fertilisers is greatly reduced if at the same time organic matter is not added. On the other hand it is necessary to consider that organic matter is mineralised about five times more rapidly in the tropics than in temperate regions.

Therefore we can state that any agricultural production system that does not add sufficient organic matter and/or gradually reduces organic matter content of the soil below an adequate level, is not site appropriate, will result in soil degradation and is not sustainable.

4.1 Influence of soil preparation on soil organic matter content and yield

Soil tillage results in rapid mineralisation of organic matter stored in the soil, liberating nitrogen that will be available for plants. This can lead during a few years to an increase in yield. However, when soil tillage is performed under favourable conditions for mineralisation of organic matter (heat, humidity, good aeration) leaving the soil under fallow (bare), valuable nitrate reserves are lost by lixiviation (washed into deeper soil layers), without crops being able to utilise them.

Once organic matter has been consumed, more nitrogen cannot be liberated and yields of crops remain low. The result is a depleted soil, where the indispensable organic matter is missing.

Many depleted soils of Paraguay and other countries of Latin America are an example of bad land management, with excessive soil tillage resulting in organic matter exhaustion. The long term influence (100 years) of soil preparation on the organic matter content in northeastern United States (temperate climate) is described by Rasmussen and Smiley (1989). In that period a reduction in the organic matter content of the soil from 2.7 to 1.5% could be observed when plant residues were not burned. When 22 t/ha/year of manure was applied from 1930 to 1981 only a small increase in the organic matter content of 1.9% to 2.1% was measured. This shows how difficult it is to raise organic matter content of the soil once it has fallen.

Here it is necessary to remember that in tropical climates organic matter reduction is processed much more quickly, and reductions below 1%, sometimes as low as 0.2% can be reached in only one or two decades of intensive soil preparation.

The influence of 20 years of different soil preparation on the organic matter content of the soil in Kentucky, USA, is reported by Thomas (1990) (Table 1).

Nitrogen appl. /year kg/ha	No-tillage	Conventional tillage
	% Organic Matter	
0	4,10	2,40
84	4,93	2,53
168	4,28	2,45
336	5,40	2,73

Table 1: Organic matter content of the soil after 20 years of maize

These organic matter contents were also reflected on maize yields after 20 years in the same experiment (Thomas, 1990).

Yields of maize without nitrogen were initially much lower in no-tillage than in conventional tillage. The situation changed after 13 years due to organic matter depletion in conventional tillage, and since then yields under no-tillage without nitrogen have been always higher (G. Thomas, 1996, pers. commun.).

5 Influence of no-tillage (NT) on different soil properties

There is enough scientific evidence from warmer areas that shows, that no-tillage has positive effects on chemical, physical and biological soil properties compared to conventional soil preparation (Kochhann, 1996). First, because erosion is drastically reduced, and second, because organic matter levels in the soil are not only maintained, but are increased in this system, and third, because soil temperatures are kept low.

5.1 Influence of NT on chemical soil properties

Compared to conventional tillage, no-tillage has positive effects on the most important chemical properties of the soil. Under no-tillage, higher values of organic matter, nitrogen, phosphorus, potassium, calcium, magnesium and also a higher pH and cation exchange capacity, but lower Al values are measured (Lal, 1976; Lal, 1983; Sidiras and Pavan, 1985; Crovetto, 1996).

5.2 Influence of NT on physical soil properties

Under no-tillage higher infiltration rates have been measured compared to conventional tillage (Roth, 1985), and this results in a drastic reduction of erosion. In no-tillage a higher soil moisture content and lower soil temperatures as well as higher aggregate stability have been measured (Kemper and Derpsch, 1981; Sidiras and Pavan 1986; Derpsch et al., 1988). At the same time a higher soil density occurs under no-tillage (Lal, 1983; Derpsch, et al., 1991), which is considered negative by many scientists. Despite this fact, higher yields of crops are obtained in Paraguay, Brazil and Argentine with this system, as compared to conventional tillage.

5.3 Influence of NT on biological soil properties

Due to the fact that no mechanical implements are used that destroy the "nests" and channels built by micro-organisms, higher biological activity occurs under the no-tillage system. Also, micro-organisms do not die because of famine under this system (as is the case under bare soils in conventional tillage) because they will always find organic substances at the surface to supply them with food. Finally, the more favourable soil moisture and temperature conditions under no-tillage also have a positive effect on micro-organisms of the soil. For these reasons more earthworms, arthropods, (acarina, collembola, insects), more micro-organisms (rhyzobia, bacteria, actinomicetes), and also more fungi and micorrhyza are found under no-tillage as under conventional tillage (Kemper and Derpsch,1981; Kronen, 1984; Voss and Sidiras, 1985). Despite the fact that chemicals are used to kill weeds, higher biological activity occurs under no-tillage, an indicator of a healthier soil.

5.4 Water quality

Water quality is improved in no-tillage. While drainage water from conventional tillage watersheds are brown in colour and carry a lot of sediments, watersheds in Brazil that have changed to no-tillage have been found to drain clear water even in times of heavy rainfalls.

5.5 Sanitary aspects

Some diseases of crops increase under no-tillage (Igarashi, 1981; Homechin, 1984; Reis, 1985; Reis et al., 1988). For this reason no-tillage should not be practised in monoculture. In general, a well balanced crop rotation with the use of green manure crops is sufficient to neutralise this negative aspect of no-tillage. In relation to pests, no-tillage can have positive or negative effects, and this depends on the specific pest and also on prevailing climatic conditions. In general, the diversity of insects, spiders, etc., increases under the mulch covered soil, where they find more favourable conditions for reproduction. As a result, many useful insects (predators) develop, and this leads to a better biological equilibrium, where pests may be controlled by predators, thus reducing the necessity for chemical pest control.

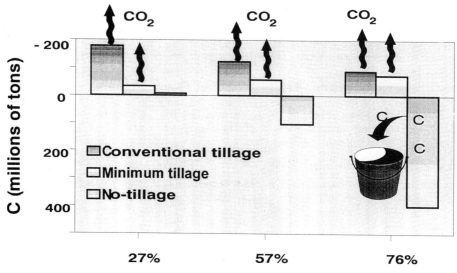

Figure 2: Impact of tillage systems on fate of carbon by year 2020

5.6 Environmental aspects

Intensive soil tillage accelerates organic matter mineralisation and converts plant residues in carbon dioxide, which is liberated into the atmosphere contributing to the green house effect and to global warming. Recent research performed in the USA by USDA/ARS shows that soil carbon is lost very fast -as carbon dioxide- within minutes after the ground is intensively tilled, and the amount is directly related to the intensity of tillage. After 19 days, total losses of carbon from ploughed wheat fields were up to five times higher than for unploughed fields. In fact, the loss of carbon from the soil equalled the amount that had been added by the crop residue left on the field the previous season (Reicosky, 1997). While fossil fuels are the main producer of carbon dioxide, estimates are that the widespread adoption of conservation tillage could offset as much as 16% of world-wide fossil fuel emissions (CTIC, 1996).

Figure 2 (prepared by Reeves, 1995) illustrates the fate of soil carbon considering three hypothesis of adoption of conservation tillage in the USA until the year 2020. In the first hypothesis, in which conservation tillage adoption rates of 1993 (27%) are maintained, and where conventional tillage prevails, almost 200 million tons of carbon are lost to the atmosphere. In the second hypothesis, in which conservation tillage adoption would increase to 57%, some improvement can be observed in relation to the first. In the third hypothesis, when conservation tillage adoption rates would reach 75%, in conventional tillage almost half the carbon is lost in relation to hypothesis one, while no-tillage would contribute to increase carbon deposits into the soil by almost 400 million tons, where it contributes to increase soil fertility (Kern and Johnson, 1993a). Minimum tillage apparently is not able to retain additional carbon in the soil, but it does avoid a net loss.

Widespread conversion of agricultural production from conventional tillage to conservation tillage, would change the whole soil manipulation system from a source of atmospheric carbon to a sink of organic carbon into the soil (Kern and Johnson, 1993b)

6 Conclusions

1. Under tropical conditions tillage will generally mineralise organic matter at rates greater than possibilities of reposition, resulting in decreasing organic matter content of the soil and diminishing yields over time.
2. High rainfall intensities prevailing in the tropics are generally associated (even on gentle slopes) with soil loss rates greater than natural soil regeneration, resulting in physical, chemical and biological soil degradation, that leads to diminishing yields over time.
3. Because organic matter degradation and/or soil erosion cannot be avoided when the soil is tilled, sustainability of agricultural production cannot be achieved with this system in the tropics. Ploughing and soil tillage are in opposition to sustainable land use.
4. No-tillage into reasonable amounts of crop residues in general improves chemical, physical and biological soil properties, making sustainable agriculture possible.
5. Soil tillage results in unacceptable CO_2 emissions into the atmosphere, and instead of carbon being deposited in the soil improving its fertility, tillage contributes to the greenhouse effect and to global warming.
6. The utilisation of the no-tillage system with permanent soil cover not only improves soil quality for the farmer, but it also improves the environment for all.

References

Cannel, R.Q. and Hawes, J.D. (994): Trends in tillage practices in relation to sustainable crop production with special reference to temperate climates. Soil & Tillage Res. **30**, 245-282

Crovetto, C. (1996): Stubble over the soil. The vital role of plant residue in soil management to improve soil quality. American Society of Agronomy, 238 pp.

CTIC (1996): Cons. Tech. Inf. Centre, CTIC Partners, April/May 1996, Vol. 14 N° 3.

Derpsch, R., Sidiras, N. and Roth, C.H. (1988): Results of studies made from 1977 to 1984 to control erosion by cover crops and no-tillage techniques in Paraná, Brazil. Soil & Tillage Research **8**, 253-263.

Derpsch, R., Roth, C.H., Sidiras, N. and Köpke, U. (1991): Controle da erosão no Paraná, Brasil: Sistemas de cobertura do solo, plantio direto e preparo conservacionista do solo. GTZ, Eschborn, SP 245.

Hailu, Z. and Runge-Metzger A. (1993): Sustainability of land use systems: the potential of indigenous measures for the maintenance of soil productivity in Sub-Sahara African agriculture; a review of methodologies and research results. J. Margraf, Weikersheim, Germany, 168 p.

Hoffmann, U., Kippenberger, C., Krauß, U., Kruszona, M., Schmidt, H., Thormann, A., Hoffmeyer, M. and Schrader, J. V. (1983): Untersuchungen über Angebot und Nachfrage mineralischer Rohstoffe, BGR, XVIII Phosphat. In Kommission: E. Schweizerbart'sche Verlagsbuchhandlung (Nägele und Obermiller).

Homechin, M. (1984): Influencia do plantio direto na incidencia de doenças. Plantio Direto, Ponta Grossa **2** (6), 2.

Igarashi, S. (1981): Ocorrência e controle de doenças. Cultura do trigo: doenças foliares. IAPAR (Ed.), 1981.

Kemper, B. and Derpsch, R. (1981): Results of studies made in 1978 and 1979 to control erosion by cover crops and no-tillage techniques in Paraná, Brazil. Soil and Tillage Research **1**, 253 - 267.

Kelly, H. W. (1983): Keeping the land alive. Soil erosion its causes and cures. FAO Soils Bulletin N° 50, FAO, Rome.

Kern, J. S. and Johnson, M. G. (1993a): Conversion to conservation-till will help reduce atmospheric carbon levels. Fluid Journal, Vol. 1, N° 3, 11-13.

Kern, J. S. and Johnson, M. G. (1993b): Conservation tillage impacts on National soil and atmospheric carbon levels; Soil Sci. Soc. Am. J. **57**: 200-210

Kochhann, R. A. (1996): Alterações das Características Físicas, Químicas e Biológicas do Solo sob sistema de Plantio Direto. Resumos, I Conferencia Anual de Plantio Direto, p 17-25, 4 - 6. 9. 1996, Aldeia Norte Editora, Passo Fundo, RS., Brazil

Kronen, M. (1984): Der Einfluß von Bearbeitungsmethoden und Fruchtfolgen auf die Aggregatstabilität eines Oxisols. Z. f. Kulturtechnik und Flurbereinigung **25**, 172- 180.

Lal, R. (1976): No-tillage effects on soil properties under different crops in Western Nigeria; Soil. Sci. Soc. Am. J. **40**, 762-768.

Lal, R. (1983): No-till Farming. Soil and Water Conservation and Management in the Humid and Subhumid Tropics. Monograph N° 2, IITA, Ibadan, Nigeria.

Mannering, J. V., and Meyer, L. D. (1963): The effects of various rates of surface mulch on infiltration and erosion; Soil Sci.. Soc. Am. Proc. **27**, 84-86.

Rasmussen, P.E. and Smiley, R.W. (1989): Long-term management effects on soil productivity and crop yield in semiarid regions of eastern Oregon. Station Bull. 675. USDA-ARS, Agriculture Experiment Station, Oregon State University, 58 p.

Reeves, D. W. (1995): Soil management under no-tillage: soil physical aspects. In Resumos "I Seminario Internacional sobre o Sistema de Plantio Direto", August 7 - 10, 1995, EMBRAPA - Aldeia Norte Editora, Passo Fundo, RS, Brasil, 127-130

Reicosky, D.C. (1997): Tillage-induced CO_2 emissions from soil. Fertilizer Research, Kluwer Academic Publishers (In print)

Reis, E. M. (1985): Doenças em plantio direto: ocorrência e seu controle. In: III Encontro Nacional de Plantio Direto, Ponta Grossa, Anais, Castro, Fundaçao ABC, 104-117.

Reis, E. M., Fernandes, J.M.C. and Picinini, E. C. (1988): Estrategias para o controle de doenças do trigo. EMBRAPA, CNPT, Passo Fundo, RS. 50 p.

Roth, C. H. (1985): Infiltrabilität von Latossolo-Roxo-Böden in Nordparaná, Brasilien, in Feldversuchen zur Erosionskontrolle mit verschiedenen Bodenbearbeitungssystenen und Rotationen. Göttinger Bodenkundliche Berichte **83**, 1 -104.

Sanchez, P. A. (1976): Properties and Management of Soils in the Tropics, Wiley, New York.

Sidiras, N. and Pavan, M.A. (1985): Influencia do sistema de manejo do solo no nivel de fertilidade. R. bras. Ci. Solo **9**, 249 - 254.

Sidiras, N. and Pavan, M.A. (1986): Influencia do sistema de manejo na temperatura do solo. R. bras. Ci. Solo **10**, 181 -184.

Steiner, K. (1996): Causes of Soil Degradation and Development approaches to Sustainable Soil Management, Pilot Project Sustainable Soil Management, GTZ Eschborn, Margraf Verlag, 50 p.

Stocking, M. (1986): The cost of soil erosion in Zimbabwe in terms of loss of three major nutrients. FAO Consultants' Working Paper, Rome.

Stocking, M. (1988): Quantifying the on-site impact of erosion. In: Rimwanich, S. (de.) Land Conservation for Future Generations, Proceedings of the Fifth International Soil Conservation Conference. Dep. of Land Development, Bangkok, 137-161

Thomas, G.T. (1990): Labranza Cero, resultados en EEUU y Observaciones en campos Argentinos, AAPRESID, Rosario, Argentina, 16 p.

Venialgo, N. (1996): Paper presented at: "II Encuentro Latinoamericano de Siembra Directa en Pequeñas Propiedades" Edelira, Itapua, Paraguay 11. - 14. 3. 1996. Proyecto "Conservación de Suelos" MAG - GTZ, Asunción. (Proceedings, in print).

Voss, M. and Sidiras, N. (1985): Nodulação da soja em plantio direto em comparação com plantio convencional. Pesq. agropec. bras. **20**, 775 - 782.

Addresses of authors:
R. Derpsch
Deutsche Gesellschaft für Technische Zusammenarbeit (GTZ) GmbH
Soil Conservation Project
Casilla de Correo 1859
Asuncion, Paraguay
K. Moriya
Dirección de Extensión Agraria (DEAG)
Ministerio de Agricultura y Ganadería
Ruta Mariscal Estigarribia Km 11
San Lorenzo, Paraguay

Development of Conservation Tillage Techniques Through Combined On-Station and Participatory On-Farm Research

E. Chuma & J. Hagmann

Summary

The project 'Conservation Tillage for Sustainable Crop Production Systems' (ConTill) has been testing and developing conservation tillage systems in the semi-arid areas of Masvingo Province, Zimbabwe, since 1988 based on a dual approach of on-station and adaptive on-farm research.

Out of five tillage techniques (conventional tillage, mulch ripping, clean ripping, hand hoe and tied ridges), only mulch ripping could be considered as being ecologically sustainable. No-till tied ridging followed closely due to its sound performance in soil and water conservation. Two other minimum tillage systems, hand-hoeing and clean ripping, were able to reduce soil loss as compared to conventional mouldboard ploughing, but their level of soil loss was still above tolerable levels.

Performance of the different tillage techniques under farmers' conditions proved to be highly variable depending on soil, site and farmer specific conditions. To address the problem of high variability of conditions, it was concluded that different techniques and systems should be promoted as options rather than as standardised blanket recommendations. Farmers capacity to chose, develop and to adapt techniques should be enhanced through encouraging farmer experimentation on technical options in order to adapt technologies to their specific needs and conditions.

The results of on-station and on-farm research were often contradictory but the integration of the two enabled us to understand more factors for performance than one component alone could have done. This understanding revealed that conservation tillage alone without considering further aspects of crop husbandry and soil fertility can only result in minor yield increases and does not necessarily increase the sustainability of the crop production system. Extension and further research should focus on an integrated approach for land husbandry.

Keywords: Conservation tillage, tied-ridging, participatory research, Zimbabwe, semi-arid areas

1 Introduction

About seventy five percent of Zimbabwean resource poor farmers live and farm in the semi-arid parts of the country (Madondo, 1992). These areas are characterised by unreliable rainfall with an annual average of about 600 mm. Drought-induced crop failure on average occur in one out of every four years. The soils are generally sandy and inherently infertile. High water losses from arable lands through surface runoff and rampant soil loss through sheet erosion further diminish the low agricultural potential of these areas. Soil and water conservation is of paramount importance under these conditions and tillage offers the most promising tool to improve soil productivity in the fragile semi-arid environment. Addressing this opportunity, the Department of Agricultural, Technical and Extension Services (AGRITEX) sited one of its two sites of the GTZ-supported Conservation Tillage Project (Contill) in the semi-arid area of Masvingo, based on the Makoholi Research Station.

The Contill Project has been testing and developing conservation tillage systems applying a combined on-station and on-farm research approach. This paper summarises technical results which emanated from six years of on-station and three years of on-farm research.

2 Methodology

2.1. On-station research

On-station, five tillage treatments were tested and developed by assessing their merits with regard to soil and water conservation, and crop yield. The five tillage treatments evaluated were: (1) conventional inversion ploughing using a single-furrow ox-drawn mouldboard plough, (2) no-till tied-ridging (semi-permanent ridges tied at regular intervals), (3) clean ripping into bare ground, (4) mulch ripping into stover mulch on the surface, (5) hand hoeing into bare ground before the onset of the rains. The tillage treatments were arranged in a randomised block design with three replicates.

Runoff and soil loss were collected and measured on a daily basis from 300m^2 plots laid out at a 4.5 % slope (Wendelaar and Purkins, 1979). At the beginning and five years after the installation of the filed plots, clay, organic carbon, aggregate stability and penetration resistance were assessed on the different tillage treatments in order to monitor changes. Clay and organic carbon samples were collected to a depth of 25 cm. Clay was determined by the hydrometer method after dispersion with monosodium phosphate. Organic carbon content was determined using the dichromate method. Aggregate stability was determined with the mechanised wet sieving method (Wells, 1978) and penetration resistance was measured with a hand held Bush penetrometer (Anderson et al., 1980)

2.2 On-farm research

2.2.1 Research approach

The on-farm programme was based on farmer participatory research using adaptive trials. The no-till tied-ridging system (nttr) was evaluated for socio-economic, technical and environmental feasibility in farmer-managed and farmer-implemented trials. Testing and development of the nttr system served as an entry point to participatory technology development based on farmer experimentation. The implementation of participatory research also served the purpose of developing an approach for innovation development and spreading which was named "Kuturaya" (let's try) by farmers (Hagmann et al., 1997a). Methods to initiate the process of innovation development and to catalyse farmer participation have been based on principles of 'Training for Transformation' (Hope & Timmel, 1984), an approach to encourage rural people to self organisation. To understand factors influencing farmers' decisions in the adoption and adaptation of innovations, hence participation in technology development, a methodology combining formal surveys, informal interviews and observations, farmer evaluation tours and technical measurements was applied. This comprehensive methodology is described in Chuma (1996) and Hagmann (1993).

2.2.2 Technical measurements

Technical evaluation of the nttr system was based on a completely randomised paired plot design where the conventional tillage was put side by side with the no-till tied-ridging.
  Crop establishment, soil profile moisture, penetration resistance and yield were assessed. Crop yield was assessed from five replicates of check plots (5x5 m) which were marked out in treatment pairs two weeks after establishment. Plant population, grain yield, stover yield, prolificacy and grain per cob were assessed. Grain yield was reported at 12.5 % moisture.

3 Results and discussion

3.1 On-station research

3.1.1 Soil loss and runoff

Soil loss due to sheet erosion was generally low for all treatments during the first four years with annual totals being less than 6 tons/ha for all treatments. The low soil loss was attributed to the fact that all plots were opened from dense woodland in 1988 and also because most erosive, high intensity storms generally occurred later in the season after the development of vegetal cover (Vogel, 1994). In the fifth season (1992/93) the deterioration of soil structure under continuous tillage and cultivation started to become evident, notably under conventional inversion tillage. Up to 11 tons/ha soil loss was recorded on conventional inversion tillage. It was however only in the sixth year (1993/94) that soil losses increased drastically; up to 40 tons/ha were recorded on conventional tillage (Fig. 1). This was due to increased soil erodibility (Chuma, 1993) and high intensity storms that fell early in the season. Highest soil losses occurred under conventional mouldboard ploughing because this treatment leaves a fine unprotected surface susceptible to splash and sheet erosion. No-till tied-ridging and mulch ripping showed lowest soil and water losses throughout all the years. Sheet erosion was reduced under no-till tied ridging because runoff was minimised by microcatchments formed by cross-ties in the furrows. The protective effect of stover left on the surface greatly reduced erosion under mulch ripping. Clean ripping and hand hoeing showed slightly lower soil losses than conventional tillage because soil disturbance was minimised. However, their surfaces are also exposed and susceptible to raindrop splash.
  Surface runoff showed similar trends than soil loss. For all treatments surface runoff was very low in the first year (1988/89) due to high surface roughness and good soil structure after the land was opened up. The hand-hoe treatment yielded highest runoff losses in 1989/90, in 1991/92 and in 1992/93. Conventional tillage had the second highest in these two years. It had highest water losses due to runoff in 1990/91 and 1993/94. These two tillage treatments provide a smooth surface which encourages surface runoff. Clean ripping was third in all years because the rip lines provide a rough surface but the lack of any protection on the soil surface leads to splash and crust formation. Mulch ripping and tied ridging showed the lowest runoff in all years. Under mulch ripping, the rip line and mulch cover provide a rough protected surface which reduces flow velocity and splash induced crust formation. Runoff was reduced under no-till tied ridging because water was retained in microcatchments formed by the ties.

3.1.2 Changes in erosion-related soil properties after five years of cultivation

Clay contents in the upper root zone (0-25cm) showed no significant difference (at $p<0.05$) between treatments. Hence tillage treatments and related soil erosion did not have a measurable effect on the topsoil clay content. Differences in structural stability due to different clay levels are not to be expected.

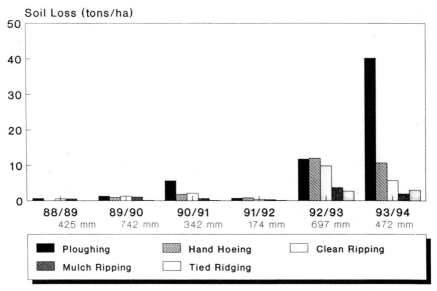

Fig. 1: Soil loss from five different tillage treatments at Makoholi

Organic carbon decrease after opening the bushland for cultivation was 9% on mulch ripping and 41% on conventional mouldboard ploughing after 5 years of cultivation. Reductions of organic carbon were highest on conventional tillage as organic matter which is bound to soil particles was eroded and washed down with particles (the highest soil loss occurred in conventional tillage) and as soil disturbance in inversion tillage causes high mineralisation of organic carbon. This confirms the concern that with present tillage practices large amounts of organic carbon (of up to 2.5 million tons annually country wide) are being lost leading to increased soil erodibility and reduced productivity (Elwell & Stocking, 1988). Reductions in organic carbon were lowest on mulch ripping because of low soil loss and minimum soil disturbance which led to slow oxidation rates. Moreover, stover mulch on the surface added organic matter to the soil and generally caused lower soil temperatures which resulted in lower mineralisation rates. Mulch ripping appears to be the most feasible way of maintaining and improving organic carbon levels in the sandy soils.

The structural stability indicated by the mean weight diameter (MWD) and aggregates greater than 2 mm (AGG2) was highest under mulch ripping. This is attributed to the higher organic carbon content maintained on this treatment. A difference of 0.22% organic carbon between mulch ripping and conventional tillage showed a difference in aggregate stability of almost 100%, an evident sign of a lower erodibility.

Soil penetration resistance increased under all treatments with depth and reached a maximum at a distinct layer (hard pan) below the plough zone (Figure 2).

In the hand hoe treatment the position of the compacted layer was less than 30 cm below the surface. However, there was no significant difference between all treatments (p<0.05%). Absolute soil strength values, however, were significantly higher (exceeding 2000 kpa) on hand hoe treatment than on all the other treatments. Root penetration inhibition under the hand hoe treatment at low moisture is expected since soil strength values exceeding approximately 2000 to 3000 kpa have been reported to limit crop emergence and root penetration (Smith, 1988; Vogel, 1994).

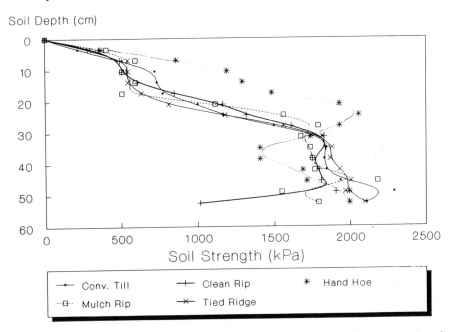

Fig. 2: Soil strength profiles for five different tillage treatments at Makoholi (gravimetric soil moisture content 5-7%)

3.1.3. Crop yield

For all treatments, grain yields were highly variable from season to season. Conventional mouldboard ploughing produced the highest grain yield in the first two years (1988/89 and 1989/90) (Fig. 3). In these years this treatment performed well because the soils were still in good structural condition after having been opened from virgin land in 1988. In 1990/91 and 1992/93, mulch ripping treatment produced highest yields. By then, the amount of mulch on the surface had accumulated enough to have an effect on soil temperature reduction and moisture conservation. No-till tied-ridging produced the highest yields in 1993/94. In this particular year there were highly erosive storms early in the season after fertilizer application. It was apparent that the applied fertilizer was only conserved on the tied ridges, but was lost through sheet erosion and leaching on the other treatments. In 1991/92, due to the devastating drought, all treatments yielded zero.

Tied ridging produced the least grain yield in 1988/89, 1989/90, 1990/91 and 1992/93. However, this was only significantly different from the best treatment in 1989/90 and 1992/93. The weaker performance of the ridged treatment could be due to a bias in the plot layout and due to moisture stress of the crop on tied ridges. The raised ridges generally tended to be slightly drier than the surfaces of the other treatments despite the water harvested between the ties. This lack of increased moisture availability is due to the low water holding capacity of the sandy soils. The hand hoe treatment performed worst in 1993/94 and in all the other years it was second to the lowest in yields. The poor performance of the hand hoe treatment could be due to the combined disadvantages of high fertility losses through erosion and of a limited rooting volume. The clean ripping treatment occupied the middle position in the ranking because it does not have the moisture stress experienced with tied ridging but, however, high erosion and runoff reduced its potential.

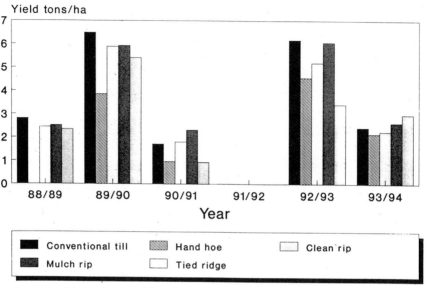

Fig. 3: Grain yield on five different tillage treatments at Makoholi

On-station results showed the advantages and disadvantages of the different tillage techniques and explained the underlying biophysical processes well. In terms of soil and water conservation, mulch ripping and tied ridging have shown high potential. In a high risk environment however, it is particularly important to further develop technologies with farmers in a systems context. As stover mulch is normally fed to cattle, mulch ripping was seen as a difficult option to promote and tied ridges were preferred as an entry point for further development together with farmers.

3.2 On-farm research

3.2.1 Approach to farmer participatory research

A major result of the on-farm work was the development of an approach for participatory technology development which enables farmers and researchers to evaluate technologies together in a collegial relationship. Besides an evaluation of nttr, the approach facilitated the development of several other innovations in soil and water conservation and tillage implements. It further showed high potential for extension (Hagmann et al., 1997b).

3.2.2 Technical evaluation of no-till tied ridges

The height of ridges implemented by farmers hardly reached 250mm as initially recommended by Elwell and Norton (1988) and as implemented on-station. In most cases farmers' ridges were approximately 150 mm high. This in itself was a farmer adaptation and innovation which proved to question many of the assumptions which emanated from on-station research. This is to be considered when evaluating results from on-farm trials. The conservation effect of ridges of 150 mm height proved to be high provided the slope layout was correct.

Crop establishment, contrary to the measurements on-station, was generally not worse on tied ridges than on conventional tillage. Results of four years (1991/92 to 1994/95) have not shown any significant difference (P<0.05) in establishment between tied ridges and conventional tillage in all communal areas evaluated. There were high variations between farmers and seasons. Close monitoring of planting by farmers and evaluation of crop establishment showed that seed emergence, particularly on tied ridges, depends on the planting procedure e.g. the depth of the hole, the method of covering and the time elapsed between hole/furrow opening and closing (evaporation).

To have good establishment, it is very important to open a hole/furrow deep enough and cover properly soon after opening. The recommended method of planting on tied ridges by digging planting holes resulted in the high emergence on tied ridges. On conventional tillage most of the farmers opened furrows with the plough, left the furrow open for a period allowing high evaporation, and often did not cover properly. Therefore they obtained emergence rates below optimum. However, hand planting tends to be slow and is the traditional planting method so is looked down upon as being backwards. Therefore, some farmers were innovative and planted by opening a rip-line on top of the ridge and planted during re-ridging. This has also resulted in establishment problems due to delayed and incomplete covering. It became apparent that appropriate implements for planting on tied ridges were required to overcome the above constraints.

Soil moisture measurements showed that moisture content on tied ridges was not higher than on conventional tillage on the sandy sites (clay less than 10%). Tied ridges revealed to be drier in the surface layers. In 1992/93 the average moisture in the depth range 0-20cm was about 10% lower on tied ridges than on conventional tillage and in 1993/94 the average moisture in depth range 0-5cm was 14% lower on tied ridges but in the layer of 5-15 cm the difference was very minor with 4.2% only. The differences in moisture between tied ridges and on conventional tillage decrease with depth for all seasons analysed. At a depth of more than 20 cm there was no longer any significant difference ($p<0.05$).

These results show that due to increased evaporation the top of the ridges is generally drier than on conventional tillage. The water harvested in the cross ties does not result in an increased moisture content in the sandy soils. Due to the poor water holding capacity of the coarse textured sands it is lost to rapid and deep drainage. However, the water harvesting effect is beneficial to groundwater recharge. Similar results have been reported on-station (Vogel, 1994) and in trials in sub-humid areas north of the country (Nyagumbo, 1993).

On the heavier textured soil, moisture in the top 15 cm was lower on tied ridges than on conventional tillage but the differences narrowed down with depth. A higher soil water storage on tied ridging could only be measured directly after storms. In some cases however it was observed that plants on conventional tillage were wilting despite higher soil moisture under this treatment, whereas no wilting occurred on tied ridging. It appeared as if soil water in the ridges would be more easily available resulting in a higher plant water uptake than on conventional tillage which could be linked to the hard pan in the subsoil and the larger rooting volume of ridges. However, further observations and analyses have to be carried out to understand these processes.

3.2.4 Crop yield

Yields on farmers' fields were highly variable between seasons, between blocks and within farms. Crop performance on tied ridges was different on periodically waterlogged wetlands and in well drained uplands sandy soils. On the wetlands, tied ridges performed significantly better than conventional tillage and up 66% higher grain yields were recorded. This was due to the drainage effect of the raised ridges that resulted in better aeration when waterlogging conditions prevailed on conventional tillage. Upland tied ridges still performed better than conventional tillage, but only marginal increases were recorded on the sandy soils. The better crop on tied ridges could be attributed to the advantage in rooting volume as well as to a more effective protection of nutrients and soil fertility in the ridge from leaching and erosion. However, on well-drained sands, the lower moisture in the upper layer on the tied ridges possibly suppressed yields. It appears that the advantages of protected fertility and better rooting volume were more important than the disadvantage of reduced moisture resulting in better yields on tied ridges. Farmers provided explanations for the observed yield differences. The rooting volume advantage was confirmed by measurements on-farm (Chuma 1993). Soil fertility protection under tied ridges was confirmed through lysimeter measurements (Hagmann, 1994).

The high variation in the performance of tied ridging between farmers (in all years the farmer as a factor influencing yield was statistically significant) makes it rather difficult to make definitive blanket statements on the performance of the no-till tied-ridging system. The benefit of the system highly depends on farmer management which is a function of resources and skills.

4 Conclusion

With regard to ecological sustainability it was confirmed that conventional mouldboard ploughing is not a sustainable tillage system. High sheet erosion, rapid breakdown of organic matter and structure are the major factors which contribute to rapid soil depletion which cannot be sustainably compensated for with inorganic fertilisers. To maintain a minimum level of productivity, organic matter has to be added continuously. Soil degradation on reduced tillage treatments like clean ripping and hand-hoeing is less severe. However, due to high surface runoff and erosion rates which are still above tolerable levels, these tillage systems cannot be considered sustainable. No-till tied ridging has proven to be sustainable in terms of soil erosion and protection of nutrients from leaching, but the organic matter status of the soil could not be maintained at the desired level. The only truly sustainable tillage technique is mulch ripping. This technique was able to maintain high organic matter levels, reduce soil erosion to a minimum and to achieve the highest water use efficieny.

From an agronomic point of view, it has been confirmed that the massively structured sandy soils tend to develop a hard pan which limits the rooting volume. This pan must be loosened by either tillage or biological measures. Annual ploughing can not alleviate this problem, nor can hand-hoeing. Only deep ripping can break this pan and tied ridging is able to reduce its effect through an increased rooting volume and improved soil structure. Crop establishment has shown to be highly dependent on the planting procedure in all tillage treatments. Establishment problems on ridges can occur, in particular with small seeds like cotton and smallgrains, but also depend on the planting procedure and experience of the particular farmer. Crop yields indicated that mulch ripping has the highest yield potential. Reduced tillage in form of ripping proved to have no major disadvantage as compared to ploughing, whereas hand-hoeing appeared to reduce yields.

Contradictory yield results from on-station and on-farm trials for tied ridging (slightly reduced yields on-station, significant yield benefits on-farm) as well as a high variability from farmer to farmer, show that treatment effects in general are extremely site, soil and farmer specific and therefore it is impossible to develop standardised blanket recommendations. The most relevant technical and socio-economic information could only be obtained through interactive farmer participation in on-farm trials. This process linked research and extension and produced valid quantitative and qualitative results which allowed us to understand the factors for performance of the techniques. It revealed that technical aspects in land management are less important for development and spreading of soil and water conservation techniques than socio-cultural and socio-economic constraints (Nyagumbo, 1995).

On light textured soils with low inherent fertility, conservation tillage alone without considering other aspects of crop husbandry and soil fertility management can only provide minor increases in yield and will not necessarily result in sustainable crop production systems. An integrated approach for land husbandry, which considers site specific improvements in crop husbandry, soil fertility and the conservation system in the farming systems context is required to achieve that goal.

The main conclusion of these insights is that unless farmers are permitted to chose their best options, develop them and experiment with them in order to understand processes and factors themselves, the output in terms of soil and water conservation and production will be low. Success in soil and water conservation depends more on farmers' capacities than on the promotion of one standardised technique found to perform well by researchers.

Acknowledgements

We would like to thank Adelaide Moyo for her contribution to this paper with regard to some on-station data. The work of the field and laboratory technicians O. Gundani, M. Diza and K. Masunda is particularly acknowledged, as is the work of the field foremen J. Jumah, P. Nheya and the assistants B. Matingo, O. Dziwore, S. Vandira, L. Wafawarova and J. Chuma. Without their dedication and commitment we would not have been able to carry out the field work.

References

Anderson, G., Pigeon, G.I., Spencer, H. B and Parks, R. (1980): A new hand held recording penetrometer for soil studies. J. Soil Sci. **31**, 279-296.
Chuma, E. (1993): Effects of tillage on erosion-related soil properties of a sandy soil in semi-arid Zimbabwe. In: Kronen, M. (ed.), Proceeding of the Fourth Annual Scientific Conference,. SADC Land & Water Man. Res. Prog., SACCAR, Gaborone, Botswana, 319-330
Chuma, E. (1996): The contribution of evaluation methods to understanding adoption of innovations in Zimbabwe. In: Budelman, A. (ed.): Agricultural R&D at the Crossroads: Merging systems research and social actor approaches. KIT Publications, Amsterdam.
Elwell, H.A. & Norton, A. J. (1988): No-till tied ridging: a recommended sustained crop production system. Agritex Handbook, Harare
Elwell, H.A & Stocking. M.A (1988): Loss of nutrients by sheet erosion is a major hidden farming cost. The Zimbabwe Science News **22**, 7/8): 79-82.
Hagmann, J. (1993): Farmer participatory research in conservation tillage: approach, methods and experiences from adaptive on-farm trial programme in Zimbabwe. In: Kronen, M. (ed.), Proc. of the Third Annual Scientific Conference, October 5-7 1992. SADC Land and Water Management Research Programme, SACCAR, Gaborone, Botswana. 217-236
Hagmann, J. (1994): Lysimeter measurements of nutrient losses from a sandy soil under conventional-till and ridge-till in semi-arid Zimbabwe. In: Jensen, H.E. et. al. (eds.), Soil Tillage for Crop Production and Protection of the Environment. Proc. of the 13th Int. Conf. of the Int. Soil Tillage Res. Org., Aalborg, Denmark, July 24 to 29 1994, 305-310.
Hagmann, J., Chuma, E., Murwira, K. (1997a): Kuturaya; Participatory Research, Innovation and Extension. In: van Veldhuizen, L., Waters-Bayer, A., Ramirez, R., Johnson, D. & Thompson, J.: Farmers' Research in Practice: Lessons From the Field. IT publications, London
Hagmann, J., Chuma, E., Murwira, K. (1997b): Strengthening peoples capacities in soil and water conservation in southern Zimbabwe. Paper publ. in this proceedings).

Hope & Timmel (1984): Training for Transformation; a handbook for community workers. Mambo Press, Gweru, Zimbabwe.

Madondo, B.S. (1992): The impact of the structural adjustment programme on existing crop technologies in the semi-arid zones of the communal farming sector of Zimbabwe. Paper presented to the Third Symposium on Science and Technology 6-8 October 1992, Harare.

Nyagumbo, I. (1993): Farmer participatory research in Conservation tillage: Part ii: Practical experiences with no-till tied-ridging in communal areas lying in sub-humid North of Zimbwe. Proc. of the Third Annual Scientific Conference, October 5-7 1992. SADC Land and Water Management Research Programme, SACCAR, Gaborone, Botswana, 319-330

Nyagumbo, I. (1995): Socio-cultural constraints to development projects in communal areas of Zimbabwe; a review of experiences from farmer participatory research in conservation tillage. Research Report 14, Conservation Tillage Project, Inst. of Agric. Eng., Harare

Smith, R.D., (1988): Tillage trials in Zimbabwe: 1957 to 1988. Report commissioned for the project 'Conservation Tillage for Sustainable Crop Production Systems' (Contill), IAE, Harare, 220pp.

Vogel, H. (1994): Conservation Tillage in Zimbabwe; Evaluation of several techniques for the development of sustainable crop production systems in smallholder farming. African Studies Series, A 11, Berne.

Wells, J. D. (1978): A review of methods used to determine soil structural stability. Chemistry and soil Research Institute Report, Department of Research and Specialist Services, Harare.

Wendelaar, F.E and Purkins, A.N. (1979). Recording soil loss and run off from 300m^2 erosion field plots. Reserch Bulletin No. 24, Department of Conservation and Extension (now Agritex) Harare, Zimbabwe.

Addresses of authors:
Edward Chuma
University of Zimbabwe
IES
MP 167 Mount Pleasant
Zimbabwe
Jürgen Hagmann
Natural Resource Management Consultant
Talstraße 129
D-79194 Gundelfingen, Germany

The Contribution of Banana Farming Systems to Sustainable Land Use in Burundi

Th. Rishirumuhirwa & E. Roose

Summary

Two field surveys were conducted in the Kirimiro region in the centre of Burundi. The goal was to study the banana-based cropping systems in the area especially traditional practices of soil fertility management. The region is densely populated and average farm size is 0.88 ha. The number of cattle is declining due to fodder shortage resulting in a decline of manure production. Soil fertility and consequently production is low.

Farmers use all sources of organic manure in order to maintain soil fertility such as farmyard manure, compost and crop residues. The limiting factor for manure and compost production is the availability of organic matter. Mulching is used for banana groves and coffee plantations. Under these crops there is hardly any soil erosion.

Soil fertility is declining with distance from the homestead. The study revealed four distinct rings of soil fertility status which are again managed in different ways. The area around the homestead is the most fertile, receiving the highest amount of organic manure. The next ring receives already mulch lower rates while distant fields are not fertilised at all, except for some incorporated crop residues or weed. The fourth ring is constituted by pasture and forest, which are sources of organic matter. There is thus a net transfer of fertility from distant fields towards the homestead. Crops are chosen according to the fertility status, i.e. the most demanding crops and especially banana are grown close to the homestead. The authors propose ways of increasing productivity, such as agroforestery, mineral fertilisers, more efficient organic matter recycling and erosion control.

Keywords: Traditional farming systems, tropical highlands, banana, soil fertility management, organic manure

1 Introduction

Most parts of Burundi are densely populated and land is permanently cultivated. The average population density is 232 people/km^2 and the average farm size 0.88 ha (in 1995). Burundi belongs to the tropical highlands of central Africa. Large areas of Burundi have a high erosion hazard. Erosion and leaching of nutrients reduce soil fertility and productivity of the land.

Facing these limitations, how do the farmers perform in controlling their negative effects? What are their skills and know-how in land husbandry for sustainable food production? To try to answer these questions, traditional agricultural practices have been studied with a particular emphasis on land fertility restoration techniques.

This paper presents the results of these studies. It gives the main characteristics of the Burundian Central Plateau farming systems, in which banana is the basic component, and their impact on sustainable land use. In addition, the efficiency of the residues of this crop in soil conservation and in soil properties improvement, as pointed out by previous studies, are highlighted.

2 Materials and methods

The study was conducted in Kirimiro, one of the main ecological districts of the Burundi Central Plateau.

This plateau is characterised by a tropical climate modified by altitude (1600 - 2000 m above sea level). According to Rishirumuhirwa et al. (1989), the mean annual temperature is 18.7° C and mean annual rainfall is 1157 mm. The rains are bimodal with the maxima in November and in April, respectively. There is a dry season from June to August. More than 80% of the soils are acid and leached with pH < 5.0 and C.E.C. < 5 meq/100 g. They belong to the Hygro-xero-ferralitic soils and to Ultic Haplustox according to the INEAC (Institut National pour l'Etude Agronomique du Congo) soil classification system and to the Soil Taxonomy, respectively. Slopes are gentle (less than 8%) to moderate (8 - 20%).

Two fields surveys were carried out. The first was designed to characterise and to assess farm size, family composition, labour availability, cropping systems and livestock (kind, number, management). Forty farms, with household heads, grouped according to years in charge (0 - 5 years; 6 - 10 years; 11 - 20 years and more than 20 years), were selected for this survey.

The second survey was conducted in 16 farms selected out of the 40 farms (4 farms per age group). This survey allowed us to study traditional agricultural practices including recycling of organic matter, production and use of ashes, manure and mineral fertilisers and their flow within the farm.

For the two surveys, field and farm surfaces were determined with a tape-measure of 20 m +/- 0.01 long. The ash production was observed for one month, the ash was collected and weighed once per week. The annual production was computed on the basis of these data. The manure production was calculated by its weigh determined from the stall and the compost pit when emptied (twice per year, October and February). Crop residues used as fodder and litter contribute to manure production. Some of these residues are also used as raw material for compost production or as mulch primary on coffee plantation. The quantities of residues removed from the farm and used as mulch were weighed. The residues remaining in the fields were not taken into account.

3 Results and discussion

3.1 Farm characteristics

3.1.1 Size

The agricultural extension services classify the farm land in the following categories (AFRENA, 1987): (1) total area, physical area of the farm; (2) cropped area; (3) the annual exploited area, the total area exploited during the two seasons. Table 1 presents the characteristics of the farms in Burundi, in the Kirimiro region and Kiremera site.

The average farm size in Burundi and Kiremera, in particular, is very small due to the high population density. Land is intensively cropped with a land intensification rate varying from 147 to 159%. The farms have been grouped into 4 classes according to their sizes. The results are presented in Table 2.

	Burundi (1)	Kirimiro (1)	Kiremera (2)
Total area (TA)	1.74	1.35	0.88
Cropped area (CA)	0.88	0.61	0.66
Annual exploited area (AEA.)	1.29	0.97	1.01
Land use rate (CA/TA) in %	51	45	75
Land intensification rate (AEA/CA) in %	147	159	153
Total population per farm	4.6	4.5	6.1
Active population per farm	2.8	2.9	4.0

Source: (1): République du Burundi (1981) and (2) survey data.

Table 1: Farm size and land use in Burundi, Kirimiro - Kiremera (ha).

Farm group limits (ha)	Total farms per group	% per group	Mean farm size per group	Total farm size per group	% per group
< 0.5	11	28	0.30	3.33	9
0.5 - 1	14	35	0.73	10.26	29
1 - 1.5	8	20	1.21	9.70	28
> 2	7	17	1.72	12.01	34

Table 2: Farm size in Kiremera (in ha).

The farm size recorded at Kiremera ranges between 0.18 to 2.57 ha with an average of 0.88 ha per farm. The smallest (less than 1 ha) represent 63% but cover only 39% of the investigated areas. The biggest (more than 1.5 ha) represent 17% of all the farms and 34% of the total area.

3.1.2. Livestock husbandry

The livestock density in Kiremera and Kirimiro as compared to Burundi is presented with the cattle situation in Burundi (Table 3).

	Burundi (1)	Kirimiro (1)	Kiremera (2)
Cattle density/ha	0.23	0.29	0.79
Cattle density/habitant	0.14	0.11	0.13
Small ruminants density/ha	0.48	0.80	1.93
Small ruminants density/ habitant	0.36	0.30	0.32

Source: (1): PNUD (1986) and (2): Survey data of Kiremera.

Table 3: Cattle and small ruminants density in Burundi, Kirimiro and Kiremera.

According to PNUD (1986), average cattle density decreases while small ruminants density increases as the farm size declines due to the increasing population. At Kiremera this reduction is confirmed since 30% of the investigated farms do not dispose of cattle. Small ruminants are prominent in farms smaller than 0.80 ha. In this case, cattle husbandry is possible only with intensification practices including permanent stalling and fodder production. The declining number of cattle and the shift to small ruminants cause a reduction of the production of farmyard manure with negative impact on soil fertility maintenance.

3.1.3 Land use and cropping systems

When studying the land use and cropping sytems, 6 sub-units or rings could be identified. The first ring, around homestead, is highly fertilised by manure, ashes, plant residues and domestic wastes provided by the other sub-units and by communal lands. It is cropped with densely planted bananas sometimes associated with taro or cassava. The second ring is characterised by a multi-cropping system, comprising banana, beans, maize, cassava, sorghum. This ring receives manure but much less as the first one. The third ring is not fertilised at all and soils are unfertile. It is planted with less demanding crops such as cassava and sweet potato. Intercropping is no longer practised here. The forth ring is occupied by pasture, fallow and wood and the fifth comprises valley bottoms. Coffee plots constitute a separate unit, which is always mulched. Table 4 describes ring size and cropping systems within Kiremera farms.

Ring	Ring size in ha			Main crops	Associated crops
	Min.	Mean	Max.		
Banana grove	0.03	0.11	0.25	Banana	taro, beans, cassava
Multi-cropping system	0.07	0.34	1.11	Banana	bean, maize, cassava, groundnuts
Cassava/sweet potato	0.00	0.15	0.42	Cassava / sweet potato	-
Pasture and wood-lands	0.00	0.22	1.15	pasture, fallow and wood	-
Valley bottom	0.00	0.07	0.36	Bean	Maize, gourd

Table 4: Land use systems of farms in Kiremera (1st season 1995).

The first 3 rings are present in all farms, while the others may be absent. The relative size of each ring depends on the total farm area, the presence of cattle, the soil fertility and the investment capabilities of the farmer.

The associated crops vary with the season. In Kiremera and Kirimiro, maize is the prominent component of crop associations during the first season while bean predominates during the second season.

3.2 Farms operations and fertility transfers

The ring farming is closely related to organic matter recycling and to fertilising practices, which again is related to the choice of crops. The following methods have been identified within the survey areas.
a) Farmyard manure production: organic matter is used as fodder or litter. The litter, the dung and the cattle urine are mixed and decompose together for at least six months. This manure can be of good quality if the shed is deep enough or associated with the compost method.
b) Compost production: organic matter is dumped into pits. Three pits are recommended by extension services. Organic residues are put in the first pit to decompose and are removed from one pit to another every 2 months. They must be covered to avoid nitrogen losses by ammoniac evaporation and sprinkled during dry season to enhance biological processes. Compost production is labour demanding and farmers rarely meet its requirements. In average, one compost-pit was recorded for 2 farms in the survey areas. Fifty five % of the farms, particularly the smallest, did not have any pit. The compost was never sprinkled and rarely covered. Thus, compost production is rather limited and its quality is low.
c) Restitution of crop residues: crops residues, particularly banana leaves and pseudo - stems, maize, sorghum, cassava and sweet potatoes stalks are left in the field after harvest.

d) Incorporation of weeds and crop residues: this is a particular case of the previous method and consists in incorporating weeds and crop residues when ploughing. According to Simonart (1992), this method contributes to neutralise aluminium in the soil and to increase phosphorus availability.
e) Burning: dry weeds and crop residues are gathered and burnt before harrowing for planting.

All these methods generally are applied simultaneously in the same farm with manuring being the most productive one. Quantities applied could not be exactly determined during the survey. Production of manure, compost, mulch and ashes at Kiremera is presented below (Table5) and correlated with farm size and livestock.

Farm size (ha)	Cattle in UBT[1]	Manure (kg)	Compost (kg)	Mulch (kg)	Total O.M.[2]	Ashes (kg)	NPK fertilisers (kg)
< 0.5	0.5	33	154	362	549	294	35
0.5 - 1	0.8	204	423	290	917	257	24
1 - 1.5	1.3	343	157	274	774	261	42
> 2	1.8	415	381	440	1236	286	56

(1) UBT = Unité de bétail tropical (tropical cattle unit, 250 kg)
(2) O.M. = organic matter

Table 5: Relation between farm size, number of cattle and manure, compost, and ashes production at Kiremera (kg of dry matter/farm/year).

There is a clear correlation between farm size and number of cattle. As stated by several authors, manure production varies with husbandry practices. In the traditional system it is estimated at 500-600 kg/cattle/year (Roose, 1994; Pozy, 1989) while it is 5-8 tons/cattle/year and 8-12 tons/cattle/year in semi-intensive and intensive systems, respectively (Simonart, 1992). In Kiremera manure production is low since traditional system still prevails even though semi-intensive and intensive systems are already applied at a small scale.

No effect of farm size and number of cattle have been noted on compost production and on use of crop residues in Kiremera. These practices seem to be related to the farmer's skills. Ash production does not vary between farms. Organic matter used for manure or compost production is provided by the different units (rings) of the farm. It is transported to the homestead where stall and compost pits are located. The mulch is primarily provided by bananas of ring 1 and 2, but also by ring 4 (pasture and woodlands) as well as the communal areas. It is generally applied to coffee plantations.

Ashes can be considered as a traditional mineral fertiliser. They are provided by wood combustion for cooking purposes from ring 4 or from communal areas. Their production comes up to 275 kg/year/family. It seems to depend neither on farm size nor on cattle numbers but on the number of meals per day and the kind of food to cook. The homestead is the main manure and ash production area. According to traditional agricultural practices, these products, as well as the domestic wastes, are used as fertilisers on bananas in the first ring and, at a lesser scale, on banana and associated crops of ring 2. Manure is generally placed around the banana stems or spread in the seed-bed when planting. In the other rings, the recycling of organic matter by burning or by incorporating grasses and residues prevails.

Traditional practices thus create a gradient of soil fertility by transfer of organic matter from the other rings to the most fertile area situated around homestead (ring 1). Communal lands can contribute to this transfer. Crops are located in this gradient according to their requirements. The functioning of the system is presented in figure 1.

Mineral NPK fertilisers are used in Kiremera on maize and bean in very limited quantities (< 50 kg/farm/year) only.

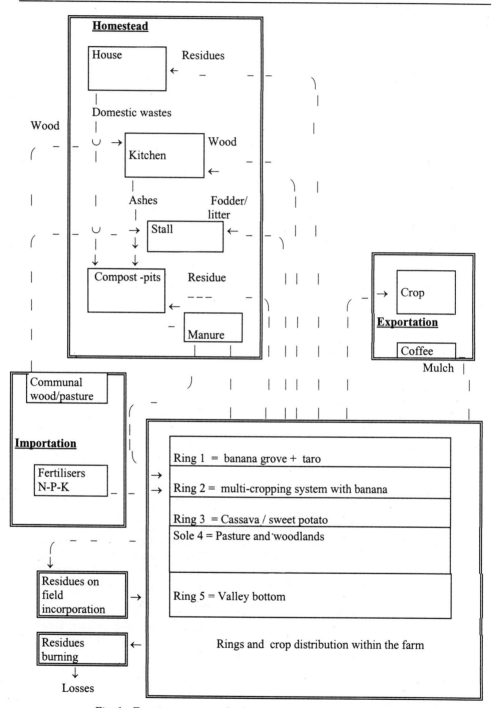

Fig. 1 : Farming system in the densely populated areas of Burundi.

3.3 Role of banana in farming systems

Banana is the main crop in the traditional farming systems of Burundi in all areas below 1900 m altitude. It plays a key role from an economic and ecological point of view. According to Nkurunziza (1991), it occupies 17% of agricultural land and represents about 40% of the total food production in the country.

There are three main types of bananas grown in Burundi: cooking bananas (+/- 15%), beer bananas (more than 80%), and dessert bananas (less than 5%). As a consequence of its prominent position in the farming systems, banana is given particular care. It is planted around the homestead and receives important quantities of manure and other fertilisers. These practice leads to a high production of fruits and crop residues. These residues (leaves and pseudo-stems) are used as mulch for coffee, to cover roofs, as fodder or litter for cattle and for some handcraft. Banana is then the main provider of organic matter in the farm and plays a key role in land fertility restoration. Banana residues used as mulch control runoff and erosion and improve soil chemical and physical properties (Rishirumuhirwa, 1993a). Banana plants protect the soil when densely spaced and mulched with banana leaves, the erosion crop index (C index as defined by Wischmeier and Smith, 1978) ranges from 0.3-0.6 with row spacing of 2 to 5 m. Soil losses increase with a wider spacing, while run-off is little affected. The efficiency of mulch is significantly increased when disposed in rows across the slope. Annual mean runoff is reduced to +/- 2% and banana erosion C index from 0.2-0.6 to 0.03-0.06 (Rishirumuhirwa, 1994).

4 Conclusions

The productivity of the traditional system is very limited. The quantity of dry matter residues produced in Kiremera does not exceed 1.5 tons/ha/farm even with cattle and in the largest farms. Substantial improvements can be achieved by several practices among which the following are recommended:

a) Agroforestery: This technique consists in introducing trees in the farming systems. Trees can be planted at the borders between farms or associated with crop production. They can provide fodder, litter, mulch, firewood and for other purposes. Roose et al. (1993), have recorded a production of 1 to 4 tons/ha/year of leaves with an average of 200 trees per farm and 3 to 9 tons/ha/year with *Calliandra* and *Leucaena* hedges (4 m spacing). In addition some leguminous trees used in agroforestery can provide over 160 kg of N/ha/year by biological fixation (Kang et al., 1984).

b) Mineral fertilisers: The use of mineral fertilisers is very limited in Burundi and in Kiremera even though FAO trials and demonstration tests (FAO,1982) have pointed out its efficiency. Banana never receives mineral fertilisers, even though fruit and biomass (crop residues) production can be significantly increased by fertiliser application, from 1-2.7 tons/ha/year residues (Rishirumuhirwa, 1993b)to 17 tons/ha/10 months (Martin-Prevel et al., 1968).

c) Efficient organic matter recycling: Cattle keeping increases organic matter recycling, the manure production depending on the intensity of the production system. Declining farm size in Kiremera are linked with a shift to stall keeping. In this system, cattle remains permanently in stall and can be kept with 0.32 ha of forage plants. Farms without cattle (30% recorded at Kiremera) can produce compost in pits.

d) Erosion and runoff control: In intensive production systems biomass yields of about 17 tons of dry matter/ha/year can be obtained by combining agroforestery and mineral fertilisers. This is sufficient for fodder and mulching, with a soil cover rate of more than 25%. According to Rishirumuhirwa (1993b) this soil cover rate is achieved with 2.7 tons of mulch 7 cm thick. When mulch is disposed in rows across the slope the annual runoff is reduced to 2% and soil loss +/- 30 times.

The adoption of these methods could significantly improve the sustainability of traditional farming systems which are affected by the increasing population and the declining number of cattle, both leading to a decline of soil fertility and yields.

References

AFRENA (1987): Sytèmes d'utilisation des sols au Burundi: Description, analyse et stratégies agroforestières.- Pub. ISABU-ICRAF, 164 p.

FAO (1980): Programme engrais et des intrants connexes (phases I et II) au Burundi: Conclusions et recommandations du Projet. Rome.

Kang, B.T., Wilson, G.F. and Lawson, T.L. (1984): Alley cropping - A stable alternative to shifting cultivation. International Institute of Tropical Agriculture, Balding & Mansell Ltd, 22 p.

Martin-Prevel, P., Lacoeuilhe, J.J. and Marchal, J. (1968): Les éléments minéraux dans le bananier "Gros Michel" au Cameroun. Fruits **23 (5)**: 259-269

Nkurunziza, F. (1991): Population et espace agraire au Burundi - Les limites de l'ajustement. Cahiers du CIPED n° 9, 69 p.

PNUD (Programme des Nations Unies pour le Développement) (1986): Le secteur vivrier au Burundi. Tome I, II et III, PNUD, New-York.

Pozy, P. (1989): L'atelier du Bututsi. ISABU, Bujumbura.

République du Burundi (1981): Ivème Plan Quinquennal de Développement Economique et Social du Burundi 1983-1987. Annexe I, Vol. 1, Secteur Agricole, 173 p.

Rishirumuhirwa, T. (1993a): Contribution des résidus du bananier en conservation de l'eau et du sol.- in Bulletin Réseau Erosion, n° 13, Montpellier: 63-70.

Rishirumuhirwa, T. (1993b): Potentiel du bananier dans la gestion et la conservation des sols ferrallitiques du Burundi. Cah. ORSTOM, sér. Pédol., vol. XXVIII, n° 2, 1993: 367-383.

Rishirumuhirwa, T. (1994): -Bananeraie, modes de paillis et restauration des sols acides.- in Bulletin Réseau Erosion, n° 14, Montpellier: 115-122.

Rishirumuhirwa, T., Birasa, E.C., Bigura, C., Lunze, L. and Kurayum, M. (1989): Etude pédologique des 8 sites repères pour les essais engrais au sein de la CEPGL (Moso, Mashitsi, Rubona, Karama, Yangambi, Mulungu, Gandajika et M'Vuazi).

Roose, E. (1994): -Introduction à la gestion conservatoire de l'eau, de la biomasse et de la fertilité des sols (GCES)- Bulletin Pédologique de la FAO n°70, Rome.

Roose, E., Ndayizigiye, F. and Sekayange, L. (1993): Agroforesterie et GCES pour restaurer la fertilité des sols acides sùr les collines très peuplées du Rwanda. Cah. ORSTOM Pédo. **28, 2**: 327-350.

Simonart, T. (1992): La conservation des sols en milieu paysan burundais. Etude et hiérarchisation des stratégies antiérosives. Mémoire Univ. Catholique Louvain, Fac. Agro., 141 p. + Annexes.

Wischmeier, W.H. and Smith, D.D. (1978): Predicting rainfall erosion losses - A guide to conservation planning U.S. Department of Agriculture, Agriculture Handbook n° 282, 58 p.

Addresses of authors:
Théodomir Rishirumuhirwa
EPFL-DGR
IATE-Pédologie, Ecublens
CH-1015 Lausanne, Switzerland
Eric Roose
Réseau Erosion
Centre ORSTOM
BP 5045
34032 Montpellier, France

Towards Sustainable Agriculture with a No-tillage System

A. Calegari, M.R. Darolt & M. Ferro

Summary
The majority of soils in Paraná State have a high clay content. The high rainfall intensities contribute to erosion hazard when the soil is tillled and unprotected by plant residues. Mulch-producing crops and adequate rotation programs constitute one of the most efficient production systems that have been developed in tropical and subtropical conditions.

It is estimated that Paraná State has more than 2.5 million hectares under a no-tillage system. The main factors for adopting no-tillage (animal traction) are less labour and better soil conservation. The no-tillage system is viable when its basic principles are well understood by farmers.

No-tillage systems contribute to a better labour distribution during the year, practically eliminating ploughing, harrowing, and mechanical weed control. This leads to more available time to arrange, plan and manage different activities for better farm diversification. The system substantially reduces soil losses, improves soil fertility, increases crop yields, and enhances production stability making permanent land use possible, thus leading to sustainable land management.

Keywords: Sustainable agriculture, no-tillage system, cover crops, crop rotation, soil characteristics.

1 Introduction

The State of Paraná is situated in Southern Brazil and has approximately 6.5 million hectares cultivated with summer crops, such as soybean, maize, beans, cotton, upland rice, wheat, sugarcane, cassava, etc. Approximately half of this area remains under fallow during the Winter season and it is subject to serious erosion problems under agricultural production. Average soil losses of 10 to 40 t of fertile soil ha^{-1} were observed for traditional soil tillage systems (Sorrenson and Montoya, 1984).

Immediate and indiscriminate land use was decisive in the different agricultural regions of the Paraná since the beginning of colonization. Most often, the natural vocation of areas was not respected, thus speeding a degradation process by water erosion and a decrease in organic matter and nutrients in the soil. This situation began to change, mainly due to relevant research, together with rural extension work to help farmers adopt new practices to manage and conserve the soil. Farmers have become aware of natural resource conservation. The understanding and use of new technologies of soil management and conservation has prompted an increase in land use under no-till systems, green manure species, and crop rotations. This change has contributed to a higher production efficiency and increased farm net income.

The objective of this paper is to present some no-tillage results and discuss the efficiency of the no-till system compared to the ploughed system.

2 Material and methods

A field trial was carried out at Santo Antonio farm, district of Floresta, Paraná, in a eutrophic Red Latosol, with gently rolling topography. Climate according to the Koeppen classification is "Cfa". Latitude and longitude are 23°36'32' S and 52°04'54' W, respectively. The altitude is 360m.

A complete randomised block design with 3 replications and a 15 × 20 m plot was adopted. Maize and soybean were planted as Summer crops with wheat, black oat, white lupines, and lupine-oat as Winter crops. Two tillage systems were compared: conventional tillage (CT) and no-tillage (NT) under the following management systems: wheat-soybean (NT); wheat-soybean (CT); wheat-maize-oat-soybean (NT); white lupines - maize - oat - soybeans - wheat - soybeans (NT); wheat - maize - oat mixed with lupines - soybeans (NT).

Green and dry matter of the aerial parts of the cover crops were evaluated as well as pests and diseases, soil chemical analysis, and soybeans, wheat and maize grain yields. In addition, in year 7 of the experiment, the physical changes in soil aggregation and water infiltration were also evaluated. To evaluate soil aggregation, soil samples were collected in small pits in all plots at two depths: 0-10 and 10-20 cm. After this the soil samples were evaluated by the Water Sieving Method. Water infiltration was measured using a rainfall simulator (Göttingen Simulator), following the method described by Roth et al. (1986).

At the onset of the experiment the soil had the following chemical characteristics: pH, 4.4; Al^{3+}, 0.27; Ca^{2+}, 5.3; Mg^{2+}, 1.2; K^+, 0.3; C (%), 3.3; P(ppm), 4.6. Texture was 75% clay, 9% silt, and 16% sand.

The continuous use of no-tillage associated with increasing amounts of crop residues over the surface can considerably decrease soil erosion (Lal, 1975; Triplett et al., 1968; Wilson, 1978). High aggregate stability indexes in Red Latosols may reduce soil clay dispersion, thus reducing runoff and decreasing the erosion risk (Roth et al., 1986). In addition, the use of no-tillage plus cover crops and crop rotations will significantly improve the physical (water infiltration, aggregation, etc.) and chemical characteristics over time as well as yields of Summer crops.

3 Adoption of the no-till system by Small Farmers:

Until 1990 only a few examples of no-tillage with small farmers in Latin America were found in the literature. Since then a number of projects have been carried out and many were presented in 1996 at the II Latin American Meeting On No-Tillage On Small Farms. According to FAO, approximately 4 million farms in Brazil use animal traction combined with human traction. These farms are characterised by family labour, animal traction, limited natural resources, unsuitability for intensive agriculture, high slope steepness, shallow soils and low natural fertility. Since 1985, the IAPAR developed no-till technologies with animal traction such as the direct sowing-fertilising machine (Gralha-Azul), in addition to technologies for the management of low fertility soils and for weed control in small farms. In 1991 on farm experiments were established with the aim of: a) verifying technical and economical factors that restrict the adoption of no-till system; b) monitoring the generation and technological adaptation process; c) evaluating ecological, technical, economic, and social sustainability of no-tillage systems in small farms. In 1993, the small farm no-tillage project started with the participation of farmers, extension, research, and agricultural industry. The project resulted in an improvement of animal-drawn no-till implements for the management of the system, better labour distribution during the year, superior maize and beans yields, varying 78-106%, respectively, compared to conventional systems and positive net income and system adoption by 75% of the farmers involved in the project. The results show that the no-tillage system is technically, economically, and socially viable for small farmers, and an important tool towards small farm sustainability.

4 Results with no-tillage system and crop rotation system in Paraná State:

The use of green manure such as black oat, radish, lupine, vetch, rye, mucuna, pigeon pea, cowpea, crotalarias, millet, etc. and of crop rotation is increasing in different regions of Paraná State (Calegari et al.,1993). These practices are important to complement the no-tillage system. Crop rotation is the basis for sustainable agriculture and for the no-tillage systems in the tropics, mainly in respect to pest and desease control, but also the production of crop residues to cover and protect the soil surface. These aspects are to be considered part of an efficient no-till system.

According to different regions having specific soil and climatic conditions, several crop rotations systems have been developed in Paraná State.

For example, maize can rotate with oat-beans/beans- oat and vicia-maize-radish- beans/beans, and onion goes with maize residues/oat- onion- oat and vicia-onion/black mucuna- mucuna residues-onion/maize.

A study conducted in a 50 ha experimental site in Northern Paraná showed that an adequate no-till system with soybean in crop rotation can generate an important net income compared to conventional systems (Table 1).

This result shows that no-till combined with a crop rotation including soybean gives significantly higher return than conventional tillage and mono-cropping of soybean. This results were based on a soybean price of U$166.00 per ton of grains.

No-till and Crop Rotation System Advantages	Values of yield Increments (U$)	Reduction of production costs (U$)
Crop yield improvement	3960	
Cheaper machine maintenance		1145
Less fuel use		731
Less labor use		2880
Less fertiliser use		186
Benefits	3960	4942
TOTAL BENEFIT	**8902**	

Table 1: Economical evaluation of soybean crop (50 ha) in no-till crop rotation system compared with conventional tillage in northern Paraná.

Results obtained in an experiment in Northern Paraná (Calegari et al., 1995) showed an increase in soil water infiltration from 20 mm/h under conventional tillage to 45 mm/h under no-tillage (soybean-wheat system). The increase in infiltration was due to an improvement in the physical properties, through leaving mulch on the soil surface, higher density of plant root systems and an increase in biological activity which yielded more soil pores.

In addition, a higher percentage of aggregates > 2 mm was observed under crop rotation. The aggregate stability index (AS%) and the mean weight diameter (MWD) were significantly different in no-tillage when compared to conventional tillage (soybean-wheat), which explained the higher water infiltration and lower soil erosion (Table 2). The effects of organic residues (roots, leaves, etc.) on soil micro-organisms activity, including root exudates and fungal hyphae, are important for maintaining and increasing aggregate stability (Roth et al., 1986).

Crop rotations, which included white lupine, contributed with nitrogen input to the systems and thus increased the maize yield. The areas planted with black oat plus white lupine accumulated more mulch at the soil surface, compared to the conventional system. Crop growth was positively affected because of improvements in the physical characteristics of the soil which contributed to greater water availability, less erosion, and a significant decrease in weed infestation.

Management	Tillage Systems	A.S.% Depth (cm)		MWD Depth (cm)	
		0-10	10-20	0-10	10-20
Wheat-maize-oats-soybeans	No-till	41.46A	40.29B	1.74B	1.76A
Lupins-maize-oats-soybeans-Wheat-soybeans	No-till	41.06A	37.42B	1.83B	1.69A
Wheat-soybeans	No-till	41.66A	38.80B	1.75B	1.66A
Wheat-maize-oats + lupins - Soybeans	No-till	40.20A	43.24B	1.66AB	1.61AB
Wheat-soybeans	Conv. Tillage	26.02B	34.26B	1.63B	1.29B

Tukey test at 0.05 - Capital letters comparisons columns.

Table 2: Effect of different tillage systems and crop rotations on aggregate stability (AS%) and mean weight diameter (MWD). Average of 3 replications after 7 years. Floresta-Paraná State 1991.

Comparing both systems, Summer under no-tillage systems yielded 34.4% and 13.7% more soybean and wheat, respectively, compared to conventional tillage systems. In addition, the crop rotation increased yields of soybean and wheat by another 19.2% and 5.8% in comparison to monoculture (Table 3).

	Crops			
	Soybeans	Yield %	Wheat	Yield %
No-tillage	2816	134.4	2121	113.7
Convent. tillage	2094	100.0	1864	100.0
* Crop rotations	3040	119.2	2200	105.8
* Monoculture	2550	100.0	2078	100.0

* Average.

Table 3: Average grains yield (kg/ha) under different tillage and cropping systems. St. Antonia Farm Floresta, Paraná State (Average of 1985/86-91/92).

These results show that after 7 years with these systems under local conditions it is possible to conclude that the no-tillage system and crop rotations are more efficient than conventional (ploughed system) and monoculture for soybean and wheat crops.

5 Conclusions

- No-tillage combined with crop rotation had on average a significantly higher percentage of aggregates > 2mm, AS% and MWD.
- Soil water infiltration in no-tillage had higher values when compared to conventional tillage in monoculture.
- No-tillage and crop rotation including white lupines and black oats had significantly higher yields of soybeans and wheat as compared to conventional tillage and monoculture.
- A production system which includes green manure, crop rotation and no-tillage, is economically viable and can be regionally adapted and therefore can contribute to the sustainability of soil management in this region.

References

Calegari, A., Mondardo, A., Bulisani, E.A., Wildner, L. do P., Costa, M.B.B., Alcântara, P.B., Miyasaka, S. and Amado, T.J.C. (1993): Adubação verde no sul do Brasil. 2a. ed.- Rio de Janeiro, AS-PTA.

Calegari, A., Ferro, M., Grzesiuk, F., Jacinto Junior, L. (1995): Plantio direto e rotação de Culturas. Cocamar, Maringá, Paraná. 64p.

Lal, R. (1975): Role of mulching techniques in tropical soil and water management. IITA Tech. Bull. no. 138p.

Roth, C.H., Pavan, M.A., Chaves, J.C.D., Meyer, B., Frede, H.G. (1986): Efeito das aplicações de calcário e gesso sobre a estabilidade de agregados e infiltração de água em um Latossolo Roxo cultivado com cafeeiros. R. bras. ci. Solo, Campinas, **10**:163-166.

Sorrenson, W. J. and Montoya, L. J. (1984): Economic implications of soil erosion and soil conservation practices in Paraná, Brazil. Report on a consultancy, Iapar, Londrina GTZ, Eschborn.

Tripplett, G.B. Jr., Van Doren, D.M. Jr. & Schmidt, B.L. (1968): Effect of corn (*Zea mays* L.) stover mulch on no tillage corn yield and water infiltration. Agron. J. **60**: 236-239.

Wilson, G.F. (1978): Potential for no-tillage in vegetable production in the tropics. Acta. Hort. 84: 33-51.

Addresses of authors:
A. Calegari
M.R. Darolt
Agronomic Institute
P.O. Box 481
86001-970 Londrina-Paraná, Brazil
M. Ferro
Cocamar
87020-010 Maringá-Paraná, Brazil

Biological Recuperation of Degraded Ultisols in the Province of Misiones, Argentina

H. Morrás & G. Piccolo

Summary

Three ecosystems are compared in a Typic Kandihumult of the Province of Misiones: a) a 50-year old "yerba mate" plantation (*Ilex paraguariensis*) in which mechanical weed control has produced a compacted subsurface layer, b) a 50-year old "yerba mate" plantation in which elephant grass (*Pennisetum purpureum*) was introduced as green manure between the rows 6 years ago and c) virgin soil under subtropical forest. The results show a progressive decrease in clay and organic matter content, soil aggregate size and A-horizon depth between the virgin forest soil, the maté plantation with elephant grass, and the traditional maté plantation. In both plots with "yerba mate" the densification of the B1 horizon is conspicuous; this phenomenon is more intense and deeper in the B1 without cover. In the plot with elephant grass, bulk density is somewhat lower and decreases rapidly with depth. The data suggest that elephant grass increases porosity due to root development and incorporation of organic matter, thus diminishing the compaction of B1 layer. Moreover, the green cover most probably also exerts its influence diminishing the impact of the rain drops and surface run-off, therefore reducing erosion processes.

Keywords: Compaction, elephant grass, biological recuperation, Argentina

1 Introduction

Misiones Province has deep clayey red soils derived from basaltic rocks in a subtropical forest environnment. An important area of this province is cultivated with "yerba mate" or maté (*Ilex paraguariensis*); this is a perennial shrub whose leaves are used to make an infusion typical for the countries of the region. High precipitation values, a predominant hilly landscape, and regional cultivation practices, make the land highly susceptible to water erosion. Erosion processes in Misiones have been evaluated by several studies (e.g. Quevedo and Bellón, 1954; Grüner, 1955; Casas et al., 1983). These studies have shown that the agricultural techniques employed produce a reduction on the organic matter content and a deterioration of the structure of the surface horizons.

Although surface degradation of those red soils has been considered in some detail, except for some rare references practically no attention was given to subsoil degradation caused by tillage (i.e. Lasserre and Ríos, 1983). Therefore the aim of this paper is to present some characteristics of compacted horizons resulting from the "yerba mate" monoculture, as well as to determine possibilities for reversing this physical condition by means of adequate agricultural practices.

2 Materials and methods

Three situations corresponding to an Ultisol (Typic Kandihumult) located in Campo Ramón, province of Misiones (Argentina) were studied: a) a "yerba mate" plantation with 50 years of monoculture in which weed control between the rows performed with repeated annual passages with the disk-harrow has generated a compacted subsurface layer, b) the same 50 years old "yerba mate" plantation in which elephant grass (*Pennisetum purpureum*) was introduced between the rows as green manure 6 years ago, and c) virgin soil under the characteristic subtropical forest of the area. The three sites are located along a transect of 150 m, at the same topographical position.

Soil profiles were described in the field and standard laboratory analysis were made for organic matter, pH, cation exchange capacity, equivalent humidity, texture, bulk density and total porosity. Dry and wet sieving were performed to obtain the macro- and micro-aggregate fractions.

3 Results

The macrostructure of the A horizons shows differences for the three situations (Table 1). In the virgin soil, the A horizon has well-developed medium subangular blocks. The aggregates are quite rounded, being by sectors transitional to very coarse crumbs. In the soil with "yerba mate" associated with elephant grass, the structure is also medium subangular blocky; however, there is a fine granular substructure. In the soil with maté under mechanical weed control, the structure of the A horizon is mainly coarse granular. In all three B1 horizons, including the compacted layer of soils with "yerba mate", the structure is coarse subangular blocky. In native forest soil the transition between A1 and B1 horizons is less evident. In contrast, in the plots cultivated with maté the contact between the Ap and the compact layer is notorious due to the change in structure, the abrupt and weavy boundary, and the horizontal development of part of the roots.

Horizons	Ecosystems					
	Native Forest		Maté + Eleph. Grass		Maté	
	1° Order	2° Order	1° Order	2° Order	1° Order	2° Order
A	Strong medium subangular blocky	Strong very coarse crumbs	Strong medium subangular blocky	Strong fine granular	Strong coarse to medium granular	....
B_1	Strong medium to coarse subangular blocky	Very fine subangular blocky	Strong coarse to very coarse subangular blocky	....	Strong coarse subangular blocky	Medium subangular blocky
B_{21}	Moderate subangular blocky	....	Coarse angular to subangular blocky	....	Coarse subangular blocky	....

Table 1: Macrostructure of main soil horizons

Organic matter content in the surface horizon is characterized by a marked and progressive reduction from the virgin soil (67 g.kg^{-1}), "yerba mate" with elephant grass (42 g.kg^{-1}) and unconsociated "yerba mate" (38 g.kg^{-1}) (Table 2). The rest of the horizons have equivalent quantities among the studied situations. The clay content in A also diminishes progressively from the forest to the uncovered maté plantation, while the absolute values of clay in B1 and B21 horizons are similar in all the three cases. The progressive reduction in organic matter and clay content among the A horizons of three situations is reflected in a parallel decrease of equivalent humidity and cation exchange capacity.

The difference in base saturation of A and B1 horizons of the three situations is remarkable. Correlation with the pH can be drawn as higher values are observed in the virgin soil, medium values in the uncovered maté plantation and the lower ones in the remaining case.

All bulk density and porosity values from B2 and B3 horizons in the three profiles are similar. In contrast, the first two horizons have differences among the three cases. In the forest soil bulk densities of the A and B1 horizons are lower than in both "yerba mate" plots. In the soil with elephant grass the bulk density is intermediate and decreases in the lower part of the B1 horizon. In the uncovered maté plantation, the B1 has the largest bulk density value of all the samples without differences between the upper and lower parts of the horizon: measured values (B.D.=1.40 Mg.m^{-3}) are close to the penetration limit of roots in clay soils (Veihmeyer and Hendrickson, 1948).

The aggregate stability analysis in dry conditions shows that in the A horizon (0-0.05 m. layer) the proportion of macroaggregates is significantly higher in the virgin soil and in the one under elephant grass than in the more degraded soil under monocultivation with maté (Table 3). In the top layer of B horizons (0.05-0.15 m) the aggregate distribution is similar in the three situations, an increase of aggregates of greater size (2000-8000 (m) being recognized with respect to the surface layer. Wet sieving reveals a marked increment of the proportion of macroaggregates in the elephant grass treatment soil respect to the other two situations, in the surface as well in the subsurface layers.

4 Discussion

Soil compaction implies a compression which increases the density of the soil body with a simultaneous reduction in porosity. In other words, the effect of the compaction in the soil is a change in its structure (Gupta et al., 1989). Compacted layers are influential in several aspects. First of all because they restrict plant root grow (Trouse, 1971). Secondly they modify the surface water dynamics because of reduced infiltration and increased runoff. Larger water erosion processes become a consecuence (McCormack, 1987; Voorhees, 1987).

For the studied soil profiles, the gradual diminution of horizon thickness, aggregate size and stability, clay and humus content in the A horizon, from the position of the virgin forest soil to the one with maté without inter-row cover, reflect the empoverishment in colloidal particles produced by the erosion that affects the cultivated soils.

In the B1 horizons, the macrostructure is relatively similar among the three profiles (coarse subangular blocky). However in the forest soil these blocks have a very fine substructure that it is not observed in the maté profiles. On the other hand, in these last situations bulk density data clearly shows the subsurface compaction which is larger and deeper in the B1 horizon of the uncovered "yerba mate" plot.

Since both "yerba mate" situations were initially similar, the intermediate condition of Ap and B1 horizons of the soil with elephant grass should be considered the result of a reaggregation process, which could be due to the elimination of tillage and more importantly to the influence of elephant grass. This grass may act through root development, an increase of organic matter and a

Table 2: Physical and chemical data of studied profiles

Ecosystems	Native Forest				"Yerba mate"+Elephant Grass						"Yerba Mate"				
Horizons	A_1	B_1	B_{21}	B_{22}	A_p	B_{ld}	B_1	B_{21}	B_{22}	B_3	A_p	B_{ld}	B_1	B_{21}	B_{22}
Depth (cm)	0-10	10-34	34-105	105-130+	0-10	10-17	17-45	45-105	105-155	155-205+	0-6	6-15	15-35	35-107	107-127+
Bulk Density (Mg m^{-3})	1.03	1.30	1.25	1.27	1.15	1.37	1.28	1.26	1.30	1.29	1.10	1.40	1.39	1.25	1.27
Porosity (m^3 m^{-3})	0.61	0.51	0.53	0.52	0.57	0.49	0.52	0.53	0.51	0.51	0.59	0.47	0.47	0.53	0.52
Humidity Equivalent (g kg^{-1})	308	304	333	326	257	271	272	321	334	343	240	258	278	335	342
Organic Matter (g kg^{-1})	67	30	12	11	42	34	25	13	9	7	37	34	22	12	8
Clay <2 μm (g kg^{-1})	627	632	839	839	579	571	650	827	870	838	554	590	639	834	834
Silt 2-50 μm (g kg^{-1})	242	234	100	96	276	300	230	100	61	96	270	284	236	110	82
Sand 50-2000 μm (g kg^{-1})	131	134	71	65	145	129	120	73	69	66	176	126	125	56	75
C.E.C. (cmolc kg^{-1})	23.9	16.8	16.3	17.3	16.5	14.8	14.5	15.1	14.7	13.7	15.8	15.4	14.3	15.2	14.6
V (%)	76.5	66.1	50.9	30.6	41.8	35.8	51.0	27.8	30.6	24.1	58.2	50.6	52.4	26.3	25.3
pH in water (1:2.5)	6.0	5.3	5.3	5.5	5.0	5.0	5.0	5.1	5.0	5.0	5.2	5.2	5.3	5.0	4.9

Table 3: Distribution of aggregates in soil samples (% of sample dry weight)

Treatment	Dry Sieving								Wet Sieving							
Depth	0-0.05 m				0.05-0.15 m				0-0.05 m				0.05-0.15 m			
Aggregate fractionsμ	>8000	8000-2000	2000-250	250-100	<100	>8000	8000-2000	2000-250	250-100	<100	>8000	8000-2000	2000-250	250-100	<100	
Native Forest	15 a†	40 a	39 b	4 b	2 c	12 a	44 a	37 a	5 b	2 b	17 ab	45 a	31 ab	1 b	6 a	15 a	41 a	36 ab	1 a	6 a
Maté +Elephant Grass	14 a	36 a	40 b	6 b	3 b	14 a	42 a	36 a	6 ab	3 b	28 a	40 a	26 b	2 b	4 a	21 a	53 a	20 b	2 a	4 a
Maté	6 b	27 b	52 a	10 a	5 a	14 a	42 a	31 a	9 a	4 a	9 b	30 a	52 a	4 a	6 a	13 a	38 a	41 a	2 a	7 a

† Values followed by the same letter within each aggregate fraction are not significantly different at $P > 0.05$.

higher biological activity (Píccolo, 1995; Píccolo et al., 1996) thus increasing soil porosity and particle aggregation. In the soil with elephant grass a decrease in the base saturation and in pH was also observed. This may be due to an increase in the microbiological activity and a strong nutrient extraction by the green manure.

5 Conclusions

The subsoil compaction in red soils of Misiones cultivated with "yerba mate" is the result of weed control between rows, performed with repeated annual passages with the disk-harrow. A further consequence has been an increase in soil loss by water erosion. Therefore, it is considered that planting of elephant grass in degraded soils under "yerba mate" culture has several beneficial aspects. First, elephant grass is an aggressive and competitive grass that controls weed and reduces the need for mechanical tillage. Second, this plant increases soil exchange capacity and soil aggregation by incorporation of organic matter. Finally, elephant grass appears to improve the physical conditions of these soils in two more ways: through root development producing a greater porosity and reversing the subsoil compaction and through a vegetation cover, which diminishes raindrop impact, increases the infiltration/run-off ratio and as a result reduces erosion processes.

References

Casas, R., Michelena, R. and Lacorte, S. (1983): Relevamiento de propiedades físicas y químicas de suelos sometidos a distintos usos y manejos en el sur de Misiones y NE de Corrientes. INTA-EEA Misiones, Nota Técnica N° 35, 18 p.
Grüner, G. (1955): La erosión en Misiones. Ministerio de Agricultura y Ganadería, Buenos Aires, Publicación Miscelánea N° 141, 70 p.
Gupta, S., Sharma, P. and De Frandi, S. (1989): Compaction effects on soil structure. Advances in Agronomy **42**: 311-337.
Laserre, S. and Ríos, M. (1983): Evaluación de los recursos del suelo del área de frontera de la Provincia de Misiones. INTA-EEA Corrientes, 69 p.
McCormack, D. (1987): Land evaluations that consider soil compaction. Soil Tillage Research. **10**: 21-27.
Piccolo, G. (1995): Efecto de diferentes cultivos utilizados como abonos verdes sobre características ed ficas de un Rodudalf típico (Misiones, Argentina). Ciencia del Suelo, **13** (2): 101-103.
Piccolo, G., Rosell, R. and Galantini, J. (1996): Transformaciones de la materia orgánica en un suelo laterítico (Misiones, Argentina). I-Distribución del carbono orgánico en fracciones de agregados. Resúmenes, XV Congreso Argentino de la Ciencia del Suelo, La Pampa, 147-148.
Quevedo, C. and Bellon, C. (1954): Conservación del suelo en Misiones, con especial referencia al cultivo de Yerba Mate. Ministerio de Agricultura y Ganadería, Instituto de Suelos y Agrotecnia, 21 p.
Trouse, A. (1971): Soil conditions as they affect plant establishment, root development and yield. In: Compaction of agricultural soils. American Society of Agricultural Engineers, Michigan, 225-276.
Veihmeyer, F. and Hendrickson, A. (1948): Soil density and root penetration. Soil Science **65**: 487-493.
Voorhees, W. (1987): Assessment of soil susceptibility to compaction using soil and climatic data bases. Soil Tillage Research **10**: 29-38.

Addresses of authors:
H. Morrás
INTA-CIRN, Instituto de Suelos
(1712) Castelar, Buenos Aires, Argentina
G. Piccolo
INTA-EEA Cerro Azul
(3313) Cerro Azul, Misiones, Argentina

The Effect of Grazing, Surface Cover and Tillage on Erosion and Nutrient Depletion

D.K. Malinda, R.G. Fawcett, D.Little, K. Bligh & W. Darling

Summary

Rainfall simulation studies were conducted on red-brown earths in cropping regions of southern and western Australia. The simulations were carried out in summer after the harvest of crops. Two adjacent target areas (each 0.5 x 1.0 m) were tested at each simulation with the existing crop or pasture residues being retained on one target area, and removed on the other.

Wheat stubble sites: Heavy grazing and trampling of the stubble at 2 sites (5-7% slope) pulverised the dry soil to depths of 30-40 mm and resulted in high runoff of 57-81% of the 30 mm of simulated rain applied in 18 minutes. This is in contrast to only 4-42% at the 2 other sites (1-2% and 4-5% slope) with ungrazed stubble. Values for soil loss were 1.0-8.3 t/ha (grazed) and 0.1-0.7 t/ha (ungrazed). A second rainfall simulation at Saddleworth increased soil loss from 0.3-3.4 t/ha to 4.6-8.3 t/ha with 3 and 0 t/ha of stubble cover respectively. The range of soil and nutrient losses in the runoff sediment were: soil (0.04-2.1 t/ha), OC (2.5-172 kg/ha), Total N (0.1-11.6 kg/ha), and total P (0.03-1.9 kg/ha). However, nutrient concentrations e.g. OC were 7.9-10.6% at grazed sites compared with 2.3-3.6% for non grazed sites. Surface cover at simulation and No till reduced runoff and soil losses compared with no cover, Reduced Tillage and Tillage.

Pasture sites: Heavy grazing of pasture in summer by sheep at 1 site (5% slope) resulted in medium runoff values (45-48%), relatively low soil losses (0.4-0.6 t/ha) but nutrient concentrations in the sediment were similar to those from the stubble grazed plots at the same site. Results from the other pasture site (6% slope) showed that the removal of grasses in two successive growing seasons compared with grass retention before simulation, increased runoff from 15% to 30-60%, and soil loss from 0.2-0.3 t/ha to 1.2-2.2 t/ha respectively; the effect on nutrient concentrations in the runoff sediment was small. The removal of existing cover at simulation only increased runoff and soil loss at this site where grasses had been removed for 2 years prior to simulation (30 to 60% and 1.2 to 2.2 t/ha respectively). Implications of the results are discussed in relation to on-site and off-site effects.

Keywords: Sheep, wheat and pasture cover, graze, Mediterranean climate, soil loss, nutrient loss, soil structure, simulated rainfall

1 Introduction

Sheep and cattle have been an integral component of most dryland farming systems in the cropping regions of southern Australia since the existing vegetation on these lands began to be cleared for agriculture in the 19th and early 20th Centuries by European settlers. The region has a Mediterranean type climate with cool wet winters and hot dry summers. The growing season for pastures and crops (including weed control before sowing crops such as wheat, barley, lupins, and peas) extends from

April-March to October-November with average annual rainfall values ranging from about 300 to 800 mm. Potentially erosive storm rains (e.g. 25 mm in 15 minutes) can fall during the growing season and during the summer months. The traffic by sheep and cattle can result in both the compaction of soils such as red-brown earths during wet spells in winter, and the pulverising of surface soil to depths of 20-40 mm in the dry summer period.

This traffic (trampling) of surface soil by sheep can considerably reduce the infiltration rate (Proffit et al., 1993). Excessive grazing of pastures, residues and crop stubbles, especially by sheep, also increases the risk of water erosion because of inadequate surface cover and damaged top soil particularly rich in nutrients. Top soil structure under controlled grazing practice was not only superior to that found under the traditional stocking practice, but similar to that found in the ungrazed treatment (Proffit et al., 1995). Our rainfall simulation studies have shown conclusively that the infiltration capacity of red-brown earths and similar soils is positively related to surface cover particularly within a 0-80% cover, e.g. 0-3 t/ha of wheat stubble (Malinda, 1995, Malinda et al., 1994). The beneficial effects of No Till systems have been more apparent at low levels of surface cover.

This paper presents results from our studies using the Northfield rainfall simulator (Malinda et al., 1992) in summer following the wheat harvest in order to assess the effects of grazing sheep, tillage systems, and surface cover on the rainfall infiltration capacity of the soil and the potential water ero-sion. The studies were made at two sites in Western Australia (WA) and three in South Australia (SA).

2 Methodology

2.1 Details of the field experiments

The sites were on red-brown earths (fine, mixed, thermic Calcic, Haploxeralf) at tillage and rotation experiments. The existing surface cover at the time of simulation comprised pasture and wheat stubble residues. Details of the sites are contained in tables 1 and 2.

At Turretfield grass removal treatments were imposed in August 1990 and September 1991. The treatments used for simulation were No herbicide (grass retained), and grass removed by applying herbicide in both 1990 and 1991.

At Saddleworth and West Dale, heavy grazing of the stubble by sheep and the associated trampling of the soil prior to simulation had resulted in the soil surface being pulverised to a depth of 20-40 mm. At Saddleworth, treatments for the simulation experiment included the amount of stubble cover (adjusted values placed on the soil surface ranged from 0-6 t/ha; values for 0 and 3 t/ha are reported here), and 2 successive simulations 6 days apart Site 4(a) and 4(b) results refer to the first and second simulations, respectively, on the same target areas. The second simulation was made to check general observations in the field that the rainfall infiltration capacity of red brown earths can be relatively high during an initial rain spell after cultivation, but can be markedly reduced during subsequent rain spells because of surface sealing and compaction of the soil aggregates.

Site	Lat. (S)	Long. (E)	Rainfall (mm)	Slope (%)	Clay (<2 μm)	Silt (2-20μm)	Sand (20μm-5mm)
1	28° 30'	114° 49'	464	4-5	16	14	70
2(a,b)	32° 18'	116° 50'	580	5	17	13	70
3	34° 04'	138° 38'	450	1-2	22	17	61
4(a,b)	34° 06'	138° 47'	495	5-7	21	18	61
5(a,b)	34° 30'	138° 55'	500	6	22	18	60

Sites 1: Chapman, WA; 2(a,b): West Dale, WA; 3: Halbury SA; 4(a,b) Saddleworth SA and 5(a,b) Turretfield SA

Table 1: Experimental sites details

Effect of grazing on erosion and nutrient depletion

Site	Period of experiment	Rotation	Tillage	Grazing	Date of simulation	Herbicide
1	8	WBGL	NT, RT	No	Feb 1995	Yes
2(a)	8	W-GL	NT, T	Yes	Feb 1995	Yes
2(b)	8	Pasture	No	Yes	Feb 1995	No
3	15	WBGL	NT, T	No	March 1993	Yes
4(a,b)	>15	W/BGL	T	Yes	Feb 1990	No
5(a)	13	Pasture	No	Yes	April 1992	No
5(b)	13	Pasture	No	Yes	April 1992	Yes

Herbicides: Herbicides at sites 1, 2 and 3 were used only for weed control during a short fallow before NT or RT seeding
NT: Plots were seeded with a drill fitted with narrow points (< 30 mm wide)
RT: Plots were seeded with a drill fitted with wide shares (100-150 mm wide)
T: Plots were cultivated approximately 3 times and seeded with wide shares (100-150 mm wide)

Table 2: Treatment details

2.2 Rainfall simulation

The Northfield rainfall simulator was used to assess the effect of tillage and rotation systems, grazing and surface cover, on the infiltration capacity of, and associated soil and nutrient loss from, red-brown earths. Simulations were replicated four times for each treatment being tested at a given site.

The simulator is based on a microprocessor-controlled slotted disc, programmed to apply rainfall with an energy of 29 $J/m^2/mm^{-1}$ and an intensity of 100 mm/hr^{-1} for 18 minutes (30 mm in total). The target areas for simulation comprised two adjacent subplots, each 50 x 100 cm in area. The existing surface cover at all sites except site 4 (a,b) was removed from one target area and weighed, leaving this target area bare during simulation; the existing cover was retained on the other target area during simulation. Runoff water and detached sediment were collected from each target area, the volume of runoff water being measured at 3 minute intervals and associated aliquots of sediment plus water being taken to determine soil loss (all sites) and nutrient loss (% and weight, except Site 4) - Organic Carbon (OC), total Nitrogen (N) and total Phosphorus (P).

3 Results

3.1 Runoff and soil loss

Cropping land - The results in Fig. 1a illustrate the substantial increases in % runoff with a simulated rainstorm associated with heavy grazing by sheep of wheat stubbles on red-brown earths, compared to ungrazed sites. The effect of surface cover was most pronounced with successive simulations at Site 4; the results at this site also illustrate the adverse effect of a further rain storm on the infiltration capacity of a red-brown earth that has been subject to structural degradation caused by heavy sheep traffic. In the absence of grazing (Sites 3 and 1), runoff was higher with the T or RT treatments, but with grazing (Site 2(a), runoff was substantial for both tillage systems and surface cover treatments. Comparable soil losses (Fig. 1b) followed the overall trends in % runoff with losses of up to 1-2 t/ha at Sites 1, 2(a) and 3. However the losses at Site 4 were particularly high with values up to 5-8 t/ha.

Pasture land - The percentage runoff values from the grazed pastures ranged from 15-60% (Fig. 2a). The low values at Site 5(a) (grass retained during the previous 2 years) were comparable to those for Sites 1 and 3, with NT and ungrazed wheat stubble (Fig. 1a). Where grasses had been

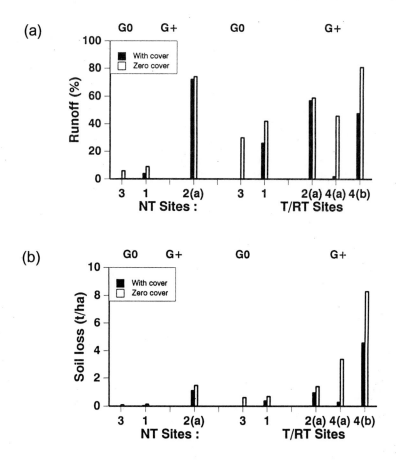

Fig. 1: The effect of grazing wheat stubble by sheep (G0, G+: zero or plus grazing), surface cover at the time of simulation, and tillage system (NT or RT, T) on (a) runoff during simulated rain (% of 30 mm of rain applied); and (b) the associated soil loss from red-brown earths used for cropping. The surface cover at Site 1 was 3.2 and 3.5 t/ha for NT and RT, respectively; Site 2(a): 2.5 and 0.8 t/ha for NT and T; only the 'zero cover' treat-ment tested at Site 3.

removed for 2 years prior to simulation (Site 5(b)), runoff increased from 15 to 30 % for 'with cover' and 'zero cover' respectively. Runoff values for Sites 2(b) and 5(b) were comparable to values at Sites 1 and 3 (ungrazed wheat stubble) and T or RT (Fig. 1a) and Site 4(1) with grazed wheat stubble and 3 t/ha of surface cover during simulation. The range of values for weight of soil loss (Fig. 2b) were comparable to those at the wheat stubble sites with the exception of Site 4 (Fig. 1b).

3.2 Nutrient loss

Cropping land - The concentration of OC and total N in the sediment lost at the wheat stubble sites ranged from 2-5% to 7-11% and were related closely to comparable values of soil loss (Fig. 3a). The

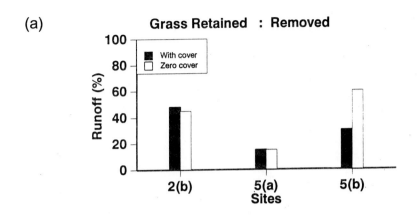

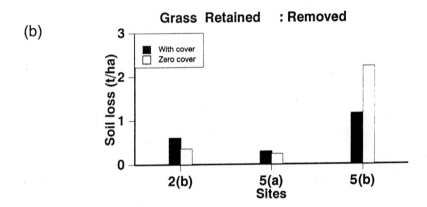

Fig. 2: The effect of grazing of pasture by sheep, grass removal, and surface cover at the time of simulation on (a) runoff during simulated rain (% of 30 mm of rain applied); and (b) the associa-ted soil loss from red brown earths used for permanent pasture. Grass removal: grass removed during the previous 2 years (Site 5(b) only, or retained (Sites 2 and 5(a)). The surface cover at Site 2(b) was 1.0 t/ha, at Site 5(a): 1.5 t/ha and Site 5(b): 0.4 t/ha.

lower set of values were associated with the ungrazed stubble at Sites 1 and 3, and the higher set with values from Site 4 where the stubble had been heavily grazed. In contrast, the concentration of Total P was relatively constant (0.08-0.13 %) for all values of soil loss (Fig. 3a).

Pasture land - The concentrations of total P were about 0.5-1% irrespective of site and soil loss values (Fig. 4a). Values for OC and total N at Site 5 were about 4-6% and 0.4% respectively, with displayed little or no response to increasing soil loss. Values for OC and total N at Site 2(b) were relatively high (8-11% and 0.5-0.7%, respectively) with soil losses of less than 0.6 t/ha. There was however, a general increase in the weight of nutrient losses with increased soil loss (Fig. 4b) particularly for OC and total N (about 10-80 and 2-10 kg/ha, respectively).

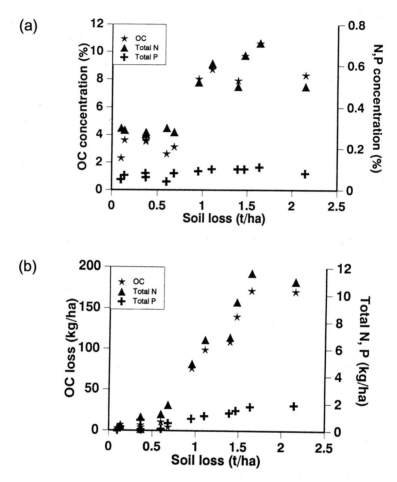

Fig. 3: The relation between soil loss during simulation on wheat stubbles and (a) associated concentrations of OC, Total N and Total P in the detached sediment; and (b) associated weights of OC, Total N and Total P in the detached sediment; (data from Sites 1, 2(a) and 3).

4 Implications

Our research clearly shows that conservation land management practices such as the retention of adequate surface cover (of pasture and crop plants including residues), avoidance of heavy grazing by sheep of crop and pasture residues, and the use of No Tillage systems will increase or maintain the water infiltration capacity of red-brown earths and similar soils. This will decrease the risk of runoff and associated soil erosion and nutrient loss from sloping land. There will be less water-logging of cropping land, so that weed control and seeding operations can be carried out more timely, with subsequent increases in crop yields. It should be noted that the adverse effects of trampling of surface soils by sheep with regard to infiltration capacity of soils should be considered when the grazing of pastures or stubble residues are included in the management of tillage and rotation experiments. Otherwise any potential beneficial treatment effects regarding infiltration and

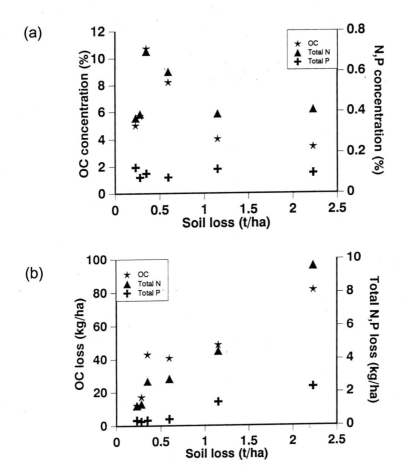

Fig. 4: The relationships between soil loss during simulation on permanent pastures and (a) associated concentrations of OC, Total N and Total P in the detached sediment; and (b) associated weights of OC, Total N and Total P in the detached sediment; (data from Sites 2(2), and 5 included).

erosion could be nullified. Furthermore, while NT systems have clear advantages with regard to improving soil structure, the systems rely heavily on the use of herbicides which must be used with care if problems with herbicide resistant weeds, and residues affecting subsequent crops, are to be avoided.

Our results regarding the soil and nutrients removed in runoff during storm rains suggest that the adoption of conservation farming systems can reduce both sedimentation and nutrient pollution of water ways, dams and reservoirs.

References

Malinda, D.K. (1995): Factors in conservation farming that reduce erosion. *Australian Journal of Experimental Agriculture* volume **35**, No 7, 969-78.

Malinda, D.K., Schultz, J.E. and Dyson, C.B (1994): The effect of rainfall intensity, tillage and stubble cover on crust thickness, runoff and soil loss in South Australia. Proceedings, Second International Symposium on "Sealing, Crusting, Hardsetting Soils: Productivity and Conservation" 107-112, University of Queensland, Brisbane, Australia, February 7-11, 1994.

Malinda, D.K., Fawcett, R.G., Dubois, B.M. and Darling W.M. (1992): Improved equipment and methods for the study of rainfall infiltration, run-off, soil and nutrient loss. South Australian Department of Agriculture Technical Paper No. 33. 27pp.

Proffit, A.P.B, Bendotti, S., Howell, M.R. and Eastham, J. (1993): The effect of sheep trampling and grazing on soil physical properties and pasture growth for a red-brown earth. Australian Journal of Agricultural Research volume **44**(2), 317-331.

Proffit, A.P.B., Bendotti, S. and Riethmuller, G.P. (1995): A comparison between continuous and controlled grazing on a red duplex soil, 2: subsequent effects on seedbed conditions, crop establishment and growth. Soil and Tillage Research **10**, 35(4), 211-225.

Addresses of authors:
D.K. Malinda
R.G. Fawcett
D. Little
W. Darling
South Australian Research and Development Institute and
Cooperative Research Centre for Soil and Land Management
PMB 2 Bld 2N
Glen Osmond, South Australia 5062
K. Bligh
Western Australian Department of Agriculture
Baron Court
Perth 6001, Australia

Preliminary Results of the Grass-Tree System for Rehabilitation of Severely Eroded Red Soils

Lin Kai Wang, Xiao Tian, Li Yili, Su Shuijin & Xie Fuguang

Summary

Masson pine (*Pinus massoniana Lamb.*) on shallow soils grow slowly and soil erosion occurs frequently in Longan orchards. An experiment on grass establishment under trees (both Masson pine and Longan fruit) to rehabilitate severely eroded sloping land was conducted from 1984 to 1994. It was found that grass planted under pine and fruit trees can promote the growth of these trees. Annual height growth of pine trees increased from 5.46 cm/year on original severely eroded land (control plots) to 62.24 cm for pine and grass. The biological mass increment of pine roots and shoots on the grassland were 9.7 and 5.92 times of that on control plots (CK), respectively. Total quantity of bacteria in soil of grassland increased 5.5 times over that of CK. The quantity of soil loss was reduced from 5444 t/(km^2.yr) on CK to 466 t/(km^2.yr) on grassland. There was no negative impact of grass on fruit tree growing. The grasses did not compete with fruit trees for soil nutrients and moisture but increased contents of nutrients and moisture. Soil moisture of grassland in orchard was increased by 8.8-20.7% compared with topsoil of bare land of orchards in different seasons because of the organic matter increase and water evaporation decrease by the grass cover.

The soil surface temperature of grass and Longan orchard was 13.3°C lower than that of bare soil surface of orchard reaching 50°C at 13:^{00} in August 1993. The soil surface temperature at 13.^{00} in grass and pine tree land was 42.0°C which was 4.8°C lower than that of CK on average.

The grass and Longan system obtained a higher economic benefit than pure fruit, grass, crops, and grass and pine. The grass and pine plantation had a lower cash return from forage 3-5 years after planting the grass. There was a benefit only after harvesting the pines. The experimental results showed that grass interplanting with fruit and timber trees is an attractive measure to rehabilite eroded red soils and allows for sustainable land use.

Keywords: Grass, Masson pine, Longan, red soil, rehabilitation

1 Introduction

The province of Fujian is located at the southeastern coast of China. It belongs to the sub-tropical monsoon climate zone and has a dense population. Because of agricultural use in mountainous areas, severe soil erosion has become the main factor restricting local agricultural development. In 1993, soil conservationists started to plant fuel grass on the bare hillsides in Fujian. By doing this they were able to succeed in controlling erosion in some of the affected areas. However, due to the low economic value, these technologies were not able to be applied in the whole area. On the other hand, some experiences had been obtained from the development of pasture on sloping land (Zhenju Huan, 1984), but because of the exuberant vitality of native shrubs, the problems of managing and stabilizing pasture has not yet been solved. The problems impeded the extension of grass planting that would have affected normal development of the livestock. In addition, parts of the mountainous areas suffer under severe soil erosion so that even the pilot Masson pine, the best kind of drought-enduring and barren-resistance tree, could hardly grow there.

Accompanying the increase in population, the amount of arable land has gradually decreased. In the last decade, soil erosion has been very serious in the new orchards. At the same time, farmers retained the custom of weeding in the old orchards, also causing soil erosion of the old orchards. Since the 1960s, there were some successful technologies for interplanting green manure in the orchards. However, the low value of green manure and the lack of suitable perennial species of green manure which forced farmers to replant each year, resulted in an unwillingness by farmers to accept the technologies. In 1985, we had interplanted perennial forage grass in the sparse Masson pine forests and the new Longan orchards on sloping land, hoping to solve the above-mentioned problems. The result of the experiment showed that planting of forage grass can promote the growing of Masson pine, control soil erosion and improve the conditions for the micro-ecological environment. Interplanting forage grass in the orchards would not affect the growing of Longan fruit trees, and can improve soil fertility and land productivity per unit area. It has been used as a successful example for developing eroded sloping land and has been used in large rural areas.

2 Materials and methods

The experimental site is located in south Fujian at N 24°30', and E 117°39'. Annual precipitation is about 1600 mm, 52.2% of which is distributed during April and June. Annual non-frost period is about 329 days. The highest recorded temperature is 40.9°C and the lowest is -2°C. Soils are severely eroded red soils from granite rock in the sloping land area..

Pine trees (*Pinus massoniana Lamb*) originally restricted in growing, have on average an age of 6.2 years, a height of 50 cm and a width of crown of 42 cm. Tree density was about 320 plants per ha.

Several varieties of grasses including annual grass (*Digitaria sanguinalis*), perennial grass (*Paspalum orbiculare*), legume (*Stylosanthes humilis, Lespedeza bicolor, Indegofera andlyantha*), *Paspalum notatum , Paspalum wetterieinii, Digitaria sanguinalis, Indigofera* were used. The grass and legume mixture contained 60% annual grasses, 30% perennial grasses and 10 % legumes. Thirty kg per ha were seeded.

All soil was phoughed by hoe in the depth of 15-30 cm, 5.25 t/ha manure was spread on the loosened soil. The manure contains 7.2 % organic matter, available N, P, K was 325 ppm, 1636 ppm and 365 ppm, respectively. Sweet potatoes and soybeans were planted in the orchard as a control comparing grass planting in the orchard. The Longan fruit tree was planted at 300 plants per ha in a level terrace.

The experiment was divided into two groups of grass-Masson pine timber tree and grass-Longan fruit tree. Each group was divided into several treatments of replicates of planting grass or not. The sediments from the runoff test plots were sampled and determined each time after rainfall. Runoff test plots for determining soil erosion were 20 m long, 5 m wide, 15 % of slope, and were subjected to three treatments including CK, grass and pine land, and grassland cultivated by hoe without fertilization.

Soil was sampled for bulk density in 0-5 cm and 5-10 cm depth by using corers of 2.21 cm diameter. Soil samples for chemical analyses and micro-biological determinations were taken in 12 replicates at various locations of each treatment.

3 Results

3.1. Effect of grass on erosion and fertility of red soil

3.1.1 Erosion control during pasture establishment period

During the pasture establishment on growth-restricted Masson pine land, the seeding combination of annual grass, perennial grass as well as legumes was applied in order to reach a fast cover during the rainy season in Spring and Summer. The crab grass (*Digitaria sanguinalis*) grows rapidly and covers loose eroded bare red soil. By Fall perennial grass such as *Paspalum orbiculare* and *Paspalum wettrieinii* replace the annual crab grass. Legume shrubs such as *Indigofera emkigantha* and *Lespedeza bicolor* grow vigorously after the second and third year. It was shown that the stable grass and pine tree system had been formed containing perennial grass, legume shrubs and young Masson pine trees.

Treatment	Period (year. month)				Soil loss	
	1985.03-04	1985.05-08	1985.09-1986.03	1986.04-1987.02	Total	(T/km² Yr)
CK	65.8	3746.2	3198.6	3876.0	10887.6	5444.00
Grassland	405.4	367.6	27.0	75.6	875.6	465.83

Note: Grassland was cultivated by hoe before seeding, soil loss was sampled from March 1985 to February 1987

Table 1: Soil erosion [t/cm²] during pasture establishment period

The soil on cultivated eroded bare land is loose. Soil erosion by rain occurs easily in March and April before the bare land is covered by grass during the pasture establishment period. The quantity of eroded soil on cultivated grassland was 5.2 times more than that on CK, but on grassland it was only 10 % of that on CK during May to August (Table 1). It was shown that soil erosion on newly established pasture was substantially reduced.

3.1.2 The effects of grass on soil fertility

Table 2 shows that cultivation and fertilizing, but also grass growth, affect the soil. The soil fertility had been improved apparently by the grass growth especially in organic matter, P_2O_5, K_2O and soil bulk density.

Treatment	Organic Matter (%)	pH	Total nutrients (%)			Available Nutrients (ppm)			Soil bulk Density (g/cm³)
			N	P	K	NH_3	P_2O_5	K_2O	
Grassland	1.63	5.02	0.046	0.036	3.4	208.34	9.5	75.6	1.42
CK_1	0.43	5.08	0.026	0.019	3.1	216.36	2.8	53.0	1.53
CK	0.41	4.65	0.015	0.0015	1.2	93.56	2.8	52.2	1.60

Note: CK_1-Bare land cultivated & fertilized without grass seeding; 0-10 cm depth was sampled.

Table 2. Soil improvement by grass on severely eroded Masson pine forested land

The soil nutrients had been increased through the grass growing on bare land of the Longan orchard (Table 3). The dead leaves and roots of grass had decomposed so that the organic matter content was increased. Furthermore the moisture content of the orchard with grassland was 8.8%-20.7% more than that in CK_2 orchard land in different seasons. This is because evaporation was reduced and grass covers also intercepted the runoff so that the soil moisture increased.

Treatment	Year	pH	Organic matter(%)	Total Nutrient (%)			Available Nutrient (PPM)		
				N	P	K	NH_3 (ppm)	P_2O_5	K_2O
Grass cover	1992	4.9	1.71	0.0971	0.103	0.683	77.21	39.20	21.10
	1993	5.2	1.76	0.0993	0.112	0.699	81.21	41.45	25.34
CK_2	1992	4.8	1.47	0.0816	0.091	0.598	86.23	23.83	13.23
	1993	5.2	1.43	0.0767	0.092	0.687	60.63	18.70	25.30

Note: CK_2. bare land of the orchard, 0-20 cm depth layer was sampled

Table 3: Nutrient content increase in soil of orchard

3.2 Influence of pasture establishment on tree growth

3.2.1 Pasture effect on the Masson pine growth

The height of original masson pine increased annually only 5.46 cm on severely eroded land. On the grassland with planting grasses for three years, the pine grew 62.24 cm annually, which is 2.9 times that of bare pine land cultivated by hoe and fertilizing (CK_1), as shown in Table 4. The pine roots had been evidently developed. Their capability to uptake soil nutrients on eroded sloping land had been greatly improved. The biomass of roots, timbers and branches were 6.93 times, 10.7 times and 3.46 times of that on CK_1 respectively. The crown of pines was greener after planting the grasses than that of pines on originally eroded sloping land. Thus the system turns the red bare slopes to green forest.

Treatment	Tree height growth(cm)	Root weight (kg/tree)	Timber volume (dm^3 /tree)	Dry weight of tree branch (kg/tree)
CK_1	21.36	0.12	0.48	0.65
Grassland	62.64	0.83	4.88	2.25
Increasing times	2.93	6.93	10.17	3.46

Table 4: The effect of grass on Masson pine

3.2.2 Effect of grass on Longan fruit tree in orchard

The results in Table 5 showed that there was no negative impact of grass on the growing of fruit trees in the orchard. The trees were sustainably promoted by grass. Because grass root grew within 30 cm depth and most of fruit tree roots were below that layer, the grasses do not compete for soil fertility and moisture with fruit trees, but also increase contents of soil nutrients and moisture due to an increase of organic matter and decrease of water evaporation by grass cover. Soil moisture of grass land in orchard was 8.8 %-20.7 % higher than that of bare land in the orchard in different seasons (see 3.1.2).

Treatment/Item	Jul. 1993			Nov. 1994			Increase					
	H.	D.	C.	H.	D.	C.	H.	%	D.	%	C.	%
Fruit+grass	134.7	19.0	41.9	181.9	3.77	115.5	47.2	40%	1.86	10%	73.5	173%
Fruit	136.1	19.7	44.8	181.9	3.61	107.9	45.8	33%	1.64	8%	63.1	141%

Note: H.-tree height; D- tree diameter at ground line; C.- tree crown diameter.

Table 5: Effect of grass on Longan fruit tree growth (cm) in orchard

3.2.3 Effect of grass on the micro-climate of the eroded red soil

3.2.3.1 Microorganism activities

After grassland was established the soil physical conditions improved and bulk density was reduced because of annual grasses and their roots dying in the soil, as shown in Table 2. Furthermore the temperature at the soil surface was also lower than eroded bare land, which provides good conditions for micro-biological survival and development. Based on the determination of micro-biological activity in February 1987, the total quantity of bacteria was 5.5 times than that in the originally bare land; the quantity of ammonia releasing bacteria was 17.1 times higher compared to CK_1, as shown in Table 6. It promotes available NH_3 release in the soil. The number of soil fungi also increased by 9.3 times.

Treatment	total bacteria	fungi in soil	Ammonia releasing bacteria
Grassland	99.5×10^6	65.1×10^4	202.0×10^5
CK_1	18.0×10^6	7.0×10^4	11.8×10^5
Increasing times	5.5	9.3	17.1

Note: total bacteria includes ammonia releasing bacteria; determined in Feb.1987

Table 6: Effect of grass growth on micro-organism activities [Nos/kg.soil]

3.3.2 The temperature changes by grass

Temperature measurements conducted at 13.00 in August 1993 show that the soil surface temperature of grassland was lower than that of bare land in orchard, as shown in Fig. 1. The soil surface temperature in orchard was 50°C, which is 13.3°C higher than with grassland in orchard. The grass cover had a great effect on soil temperature within the 30 cm soil layer. The temperature above 37°C in soil has negative impact on root development of fruit trees (Li Lairong 1983).

Due to the growth of grass on the severely eroded red soil, the temperature at 13.00 in grass-pine tree land is 42.0°C which is 4.8°C lower than average on the original bare pine land, according to statistic data for July from 1987 to 1988.

3.4 Grass-tree economic benefit

Table 7 shows that output per ha and economic benefit of fruit tree with grass was higher than others. The output values for soybeans and sweet potatoes were much lower than for the fruit tree and all combinations.

Treatment	forage output (Kg/ha.yr)	fruit/grain output (Kg/ha.yr)	timber output (M^3/ha.yr)	output value (RMB/ha.yr)
Soybean	4050.0	600.0		3225.0
Sweet potato	4500.0	13500.0		6300.0
Pine tree	7520.0		1.5	5620.0
Longan +grass	9687.6	2205.0		18984.1
Longan		2230.0		13234.3
Grass	9148.5	0		5489.1

Note: Forage includes green residues of soybeans or sweet potatoes, the data are determined from 1989 to 1994.

Table 7: Output comparison among fruit tree-grass, timber tree-grass and crops on sloping land

Grass-timber tree plantation has a lower benefit from of forage harvest during 3-5 years and obtains cost recovery from tree after timber harvest due to grass improving the fertility of the eroded slope. This helps the Masson pine grow vigorously and later produce timber (Kaiwang Lin, 1988).

4 Discussion and conclusion

Grass in the grass-tree system plays an important role in the rehabilitation of red soils on severely eroded sloping land. It has a positive effect on erosion control and soil improvement and is superior to pure stands of trees. Grass-tree system establishment was an effective measure for improving severely eroded sloping land and for bringing it under sustainable land use.

The experimental results show that grass interplanting with fruit trees is a successful model, but the following management practices should be adopted:
(1) The grass should be cut both for high quality fodder and for keeping the soil surface clean under the tree crown in order to protect the fruit tree from grass competition;
(2) The grass type with short roots and medium height should be selected;
(3) It is better to plant grass with perennial legume for good cover and high quality pasture;
(4) Grass growing up within the area under the tree crown along the fruit trees, should be cut to prevent grass competition with fruit trees.

The bahiagrass (*Paspalum notatum*), *Paspalum wettrieinii* and style legume are the best varieties of forage for pasture establishment with fruit tree in the orchards of the southeast coastal area of China.

Grasses have a better effect on soil improvement and on rehabilitation of severely eroded land based on vegetation establishment per unit area of land because of their good root systems and fast catabolism of biomass (Li Yili, 1994).

The selection of grass or legume varieties and its combination as well as grassland management techniques are also important factors that influence the establishment of successful grass-tree system and efficient production.

The pasture establishment must provide economic benefits from animal products or production of edible fungus by use of grass (Li Yili, 1994). Therefore the utilization of grass is an important factor that affects the existence of the grass-tree system and its economic benefit. The economic benefit is the key in motivating farmers' interest in participating in soil conservation activities in which sustainable land use will be carried out in areas of degraded red soils.

References

Li Lairong (1983): Longan Fruit Cultivation, Agriculture Press House, Beijing;
Zhenju Huan (1984a): Study on Natural Hilly Grassland Rehabilitation in Fujian. Journal of Fujian Agricultural College **13(1):** 47-51.
Kaiwang Lin (1988): A preliminary Study on Establishing Grassland for the Seriously Eroded areas in Fujian. Acta Conservationis Soli et Aquae Sinica, 49-54.
Li Yili (1995): Discussion on Technique and Economy of Grassland Establishment for Soil Conservation Fujian Soil Conservation, No 4.

Addresses of authors:
Lin Kai Wang
Xiao Tian
Li Yili
Su Shuijin
Xie Fuguang
Fujian Soil and Water Conservation Committee
6 Tongpan Road
Fuzhou 350003
Fujian, P.R. China

Reducing Soil and Nutrient Losses from Furrow Irrigated Fields with Polymer Applications

R.D. Lentz, R.E. Sojka & C.W. Robbins

Summary

Irrigation furrow runoff contains sediment, associated organics, and nutrients that enter surface waters as non-point source contributions. We compared the effects of anionic polyacrylamide (PAM) applications on furrow runoff losses of sediment, nitrate, ortho-phosphorus (ortho-P), total-phosphorus (total-P), and chemical oxygen demand (COD). Dry bean was planted on Portneuf silt loam (Durixerollic Calciorthids) after conventional tillage. Initial irrigation inflows of 23 L/min were cut back to 15 L/min after runoff began. Control furrow streams contained no PAM. PAM was applied continuously at 1 mg/L during the PAM **C1** treatment. In the PAM **I10** treatment, 10 mg/L PAM was applied to inflows early in the irrigation, then stopped once runoff began. Runoff from PAM-treated furrows was 37% less than for controls. Runoff, ortho-P, and total-P concentrations in control furrows were 5 to 7 times that of the pulsed-PAM treatment, and control COD levels were 4 times those of the PAM **I10** treatment. The **C1** concentration values for all components except nitrate were about twice as large as those of the **I10** treatment. Total seasonal soil loss was 3.14 Mg ha^{-1} for control furrows, 0.35 for PAM-**C1**, and 0.25 for PAM-**I10** treatments. Relative to controls, PAM markedly reduced total furrow losses of sediment, ortho-P, total-P, and COD (60 to 92%), but had little influence on runoff nitrate.

1 Introduction

Irrigation-induced erosion threatens agricultural sustainability and surface water quality worldwide. Soil losses resulting from 50 to 80 years of continuous furrow irrigation have reduced the productivity potential of some northwest U.S. farms by 25% (Carter, 1993). Furrow runoff contains sediment and associated organics and nutrients that enter surface waters as non-point source contributions and damage downstream users and environments. Alternative, more efficient sprinkler irrigation systems are usually less erosive, but are more costly. Moreover, in southern Idaho, current inexpensive water prices provide little motivation to irrigate more efficiently. Nonetheless, gradual conversion to sprinkler systems is occurring, primarily because of associated labor savings (Dennis Kincaid, personal communication, 1996). Still, 50 to 60% of farm managers in south-west and south-central Idaho use surface irrigation. An effective, economical, and favorably received erosion control practice is needed to conserve soil in this and other irrigated areas worldwide.

Lentz et al. (1992) showed that applications of a water-soluble anionic polyacrylamide (PAM) of 12-15 Mg mol^{-1} and 18% charge density reduced furrow irrigation-induced sediment losses by up to 97%. Over three years, an average soil-loss reduction of 94% was achieved by applying an initial 10 mg L^{-1} PAM pulse in irrigation water mainly during furrow advance, i.e. while water first traversed the length of the dry furrow. When we applied 0.5 mg L^{-1} PAM to irrigation water continuously for the entire irrigation, we used 75 to 86% less PAM, and achieved a 70 to 80% soil-loss reduction, relative to controls (Lentz and Sojka, 1996).

Both these application techniques were enthusiastically accepted by irrigators in the western US. PAM first became generally available in 1995, and was used on about 20,000 ha that year. In 1996 PAM was used on nearly 200,000 ha!

Seybold (1994), Barvenik (1994), and Barvenik et al. (1996) reviewed safety, toxicity and environmental regulation aspects of PAM-use. When PAM is employed at concentrations recommended for agriculture and other industries (1-10 mg L^{-1}), the polymer is non-toxic and no adverse environmental affects have been observed. PAM is authorized for use in treatment of potable water supplies, in food processing, and in paper products used to store food. PAM formulations used for treating irrigation water are food-grade quality and compositions are strictly regulated.

A preliminary study by Lentz and Sojka (1994) indicated that PAM-use not only reduced field soil losses, but also appeared to diminish nutrient and organic losses in furrow runoff. In their study, a continuous < 0.5 mg L^{-1} PAM treatment was applied to irrigation inflows. This less-than-optimal PAM treatment produced slight to moderate reductions of total-P, ortho-P, nitrate, and BOD (biochemical oxygen demand) in runoff.

Our objective was to compare the effects of zero, continuous, and pulsed-PAM applications on furrow runoff losses of sediment, nitrate, ortho-P, total-P, and chemical oxygen demand. The COD provides a measure of organic matter losses occurring in runoff.

2 Methods

The field study was conducted at the USDA-ARS Northwest Irrigation and Soils Research Laboratory at Kimberly, Idaho, USA. Dry bean (*Phaseolus vulgaris* L .'Viva Pink') was planted on Portneuf silt loam, coarse-silty, mixed, mesic Durixerollic Calciorthids. The seedbed was prepared with disk and roller harrow. Surface soil texture was silt loam (10% clay, 70% silt), organic matter was 10-13 g kg^{-1}, cation exchange capacity was 190 mmol$_c$ kg^{-1}, saturated-paste-extract electrical conductivity (EC) was 0.7 dS m^{-1}, ESP was 1.5, pH was 7.7, and calcium carbonate equivalent was 5%. Furrows were 175 m long, with a 1.6% slope. Furrows were shaped with a weighted wedge-shaped forming tool. To avoid infiltration differences between wheel-tracked and nonwheel-tracked furrows, only trafficked furrows were irrigated and monitored.

Irrigation water was from the Snake River. Its electrical conductivity was 0.5 dS m^{-1} and its SAR was 0.5. A gated pipe conveyed water to the each furrow, and adjustable spigots controlled inflow rates. A cutback irrigation strategy was employed. Initial irrigation inflows were relatively high at 23 L min^{-1} to move water across the field quickly, then flows were cut back to 15 L min^{-1} to reduce runoff. Irrigations were 8 to 24 hours in duration. The field was irrigated five times during the season.

Furrow inflows and outflows were monitored, and runoff sediment concentrations were measured throughout each irrigation. Measurements were made at one-half hour intervals early in the irrigation, and every hour or every several hours in the later half of the irrigation, after outflows and sediment loads had stabilized. Inflow was measured by filling a known volume, and outflows were measured with v-notch flumes (Trout and Mackey, 1988). Sediment was measured using Imhoff cones (Sojka et al., 1992). Details of the flow and sediment monitoring procedure were given by Lentz et al. (1992).

The study compared three treatments. Control furrow-streams contained no PAM. PAM was applied continuously at 1 mg L^{-1} in a continuous PAM treatment (**C1**), and was applied at 10 mg L^{-1} during the advance phase only in a pulsed-PAM treatment (**I10**). PAM injection in the pulsed treatment was curtailed at an average 111 min after the irrigation began, ie. shortly after the end of the advance phase, and untreated water was used for the remainder of the irrigation set. PAM stock solutions, prepared one to two days prior to the irrigation (Lentz and Sojka, 1996), were metered into the head of each furrow with positive displacement pumps. Turbulence created by the incoming water stream mixed and dispersed the aqueous PAM concentrate into the flow.

Three runoff samples were collected from each furrow during an irrigation. Samples were taken from outflow monitoring flumes. Runoff nutrient content was not determined for the last irrigation. Since 97% of the field's sediment losses occurred in the first four irrigations, we believed nutrient losses produced by the final irrigation were also very small. One runoff sample was collected at 1 to 2 h into irrigation, a second at 5 to 6 h, and the third at 8 to 10 h into the irrigation. Samples were analyzed for total-P (Greenberg et al., 1992), ortho-P (Watanabe and Olsen, 1965), chemical oxygen demand, COD (American Public Health Association, et al., 1971), and NO_3-N (2.0 mM potassium benzoate eluent and liquid ion chromatography). Runoff samples were stored in a refrigerator for <8 days before being analyzed.

The experimental design was a complete randomized block, with three replications. Furrow infiltration and field loss calculations were made with the computer program, WASHOUT (Lentz and Sojka, 1995). The program integrated runoff and pollutant losses over the duration of the irrigation. Net infiltration was calculated as the difference between total inflow and total outflow. Total nutrient and COD losses were computed assuming that runoff constituent concentrations were constant between sampling intervals. Treatment means were compared using the Duncan multiple range test (Snedecor and Cochran, 1980).

3 Results and discussion

The **C1** and **I10** PAM treatments influenced both material concentrations and hydraulic characteristics of the furrow streams. PAM's significant impact on total field-loss values reflected the combined effect of these factors.

Net infiltration for **C1** was 1.5 times that of control furrows and net infiltration for **I10** was 1.3 times that of control values (Table 1). Therefore, runoff from PAM-treated furrows was smaller than that of the control furrows. On average, the outflow rate of PAM-treated furrows was 37% less than that for control furrows (Table 1). By reducing runoff, PAM applications decrease field soil and nutrient losses, assuming furrow-stream material concentrations remained unchanged, or decline.

	Control	I10	C1
Mean Outflow (L/min)	8.9 a[†]	5.9 b	5.3 b
Total Outflow (mm)	44 a	30 b	22 c
Net Infiltration (mm)	49 c	63 b	71 a

[†] similar letters across rows indicate nonsignificant differences ($P < 0.05$).

Table 1: Outflow and infiltration values, averaged over all irrigations.

PAM applications reduced concentrations of sediment, ortho-P, total-P, and COD in furrow runoff (Table 2). Runoff nitrate concentrations did not differ among treatments. Runoff ortho-P and total-P concentrations in control furrows were 5 to 7 times that of the PAM-**I10** treatment, and control COD levels were 4 times those of the PAM-**I10** treatment (Table 2). Material concentrations in PAM-**I10** furrows were about one-half those of PAM-**C1**.

Thus, PAM treatments both decreased furrow outflows and, with the exception of NO_3-N, reduced furrow-stream pollutant concentrations. The combined effect decidedly reduced material field losses.

Total soil loss for the five irrigations was 3.14 Mg ha^{-1} for control furrows, 0.35 for PAM-**C1**, and 0.25 for the PAM-**I10** treatments. That is, PAM-**I10** reduced total soil loss 92%, and PAM-**C1** 89%, as compared to control furrows. When computed on a per irrigation basis, soil-loss reduction for the PAM-**I10** treatment was 91% vs. 85% for PAM-**C1** (Fig. 1D). The per-irrigation C1 value was smaller than the total-season value because PAM-**C1**'s control of soil erosion was less consistent among irrigations than that of PAM-**I10**.

Runoff component		Control	I10	C1
Ortho-P	(mg/L)	0.43 a[†]	0.09 c	0.20 b
Total-P	(mg/L)	0.88 a	0.12 b	0.24 b
NO3-N	(mg/L)	0.05 a	0.07 a	0.06 a
COD	(mg/L)	119.7a	31.5 b	88.5 a
Sediment	(mg/L)	1800 a	300 b	500 b

[†] similar letters across rows indicate nonsignificant differences ($P < 0.05$).

Table 2: Mean material concentration in furrow runoff

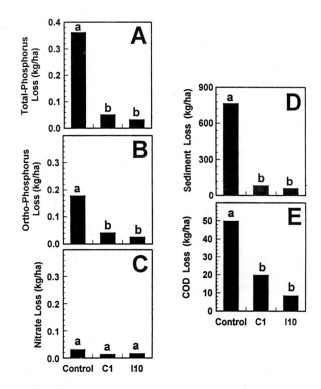

*Fig. 1: Total-P (A), ortho-P (B), NO3-N (C), sediment (D), and COD (E) losses per irrigation from nontreated and continuous 1 mg L-1 (**C1**) and pulsed 10 mg L-1 (**I10**) PAM-treated furrows. In each graph, columns with common letters do not differ (P=0.05)*

Total-P and COD losses from the first four irrigations were 60 to 91% lower from PAM-treated furrows than from control furrows (Figs. 1A, 1B, 1E). Total NO_3-N losses from all furrows was uniformly low. Total NO_3-N losses from PAM-treated furrows were one-half that of controls, although the difference was significant at P = 0.24 (Fig. 1C). Again, PAM-I10 most effectively reduced total nutrient losses. Relative to controls, the PAM-I10 application reduced total-P losses by 89% vs. 83% for PAM-C1, and reduced total COD losses by 83% vs. 60% for PAM-C1 (Fig. 1).

4 Conclusions

1. PAM additions to furrow inflows substantially reduced furrow-irrigation field-losses of sediment, total-P, ortho-P, and organic matter, compared to untreated furrows.
2. The most effective treatment for reducing sediment and nutrient losses was PAM-I10, where 10 mg L^{-1} PAM was metered into furrow irrigation inflows during the furrow advance (during water's initial advance down furrow). PAM-I10 reduced total field losses of sediment by 92%, total-P by 91%, ortho-P by 86%, and organic matter (COD) by 83%, compared to untreated furrows.
3. The PAM treatment reduced field losses by decreasing material concentrations in runoff and by reducing runoff volume. PAM accomplished the latter by maintaining higher net infiltration rates in treated furrows than was present in nontreated furrows.

Acknowledgements

This work was supported in part by a grant from CYTEC Industries, through a Cooperative Research and Development Agreement (contract 58-3K95-4-216). We thank Mr. Larry Freeborn, who ran most of the analyses, Mr. Jim Foerster and Mr. Ron Peckenpaugh for their technical support, and Ms. Emily Aston, Ms. Elizabeth Whitchurch, and Mr. Paul Miller for their lab and field assistance.

References

American Public Health Association, American Water Works Association, Water Pollution Control Federation (1971): Oxygen demand (chemical). In: Standard Methods for the Examination of Water and Wastewater (ed.), 13 American Public Health Association, New York, 495-499.
Barvenik, F.W. (1994): Polyacrylamide characteristics related to soil applications. Soil. Sci. **158**: 235-243.
Barvenik, F.W., Sojka, R.E., Lentz, R.D., Andrawes, F.F. and Messner, L.S. (1996): Fate of acrylamide monomer following application of poly-acrylamide to cropland. In: R.E. Sojka and R.D. Lentz (eds.), Managing Irrigation-Induced Erosion and Infiltration with Polyacrylamide. Proc., College of Southern Idaho, Twin Falls, ID. 6-8 May 1996. Univ. of Idaho Misc. Publ. No. 101-96, 103-110.
Carter, D.L. (1993): Furrow irrigation erosion lowers soil productivity. J. Irr. Drain. Eng. **119**: 964-974.
Greenberg, A.E., Clesceri, L.S. and Eaton, A.D. (eds.) (1992): Standard methods for the Examination of Water and Wastewater. 18th ed. Am. Public Health Assoc., Washinton, DC. P4-112
Lentz, R.D., Shainberg, I., Sojka, R.E. and Carter, D.L. (1992): Preventing irrigation furrow erosion with small applications of polymers. Soil Sci. Soc. Am. J. **56**: 1926-1932.
Lentz, R.D. and Sojka, R.E. (1994): Field results using polyacrylamide to manage furrow erosion and infiltration. Soil. Sci. **158**: 274-282.
Lentz, R.D. and Sojka, R.E. (1995): Monitoring software for pollutant components in furrow irrigation runoff. In: L. Ahuja, J. Leppert, K. Rojas, and E. Seely (eds.), Proc. Workshop on Computer Applica-

tions in Water Management, 23-25 May, 1995, Colorado State University Water Resources Research Institute Info. Series No. 79, 123-127.

Lentz, R.D. and Sojka, R.D. (1996): Five-year research summary using PAM in furrow irrigation. In: R.E. Sojka and R.D. Lentz (ed.) Managing Irrigation-Induced Erosion and Infiltration with Polyacrylamide. Proc., College of Southern Idaho,Twin Falls, ID. 6-8 May 1996. Univ. of Idaho Misc. Publ. No. 101-96, 20-27.

Seybold, C.A. (1994): Polyacrylamide review: soil conditioning and environmental fate. Commun. Soil Sci. Plant Anal. **25**: 2171-2185.

Snedecor, G.W. and Cochran, W.G. (1980): Statistical Methods. Iowa State University Press, Ames, IA.

Sojka, R.E., Carter, D.L. and Brown, M.J. (1992): Imhoff cone determination of sediment in irrigation runoff. Soil Sci. Soc. Am. J. **56**: 884-890.

Trout, T.J. and Mackey, B.E. (1988): Furrow flow measurement accuracy. J. Irr. Drain. Eng. **114**: 244-255.

Watanabe, F.S. and Olsen, S.R. (1965): Test of an ascorbic acid method for determining phosphorus in water and $NaHCO_3$ extracts from soils. Soil Sci. Soc. Am. Proc. **29**: 677-678.

Addresses of authors:
R.D. Lentz
R.E. Sojka
C.W. Robbins
USDA-ARS
Northwest Irrigation and Soils Research Laboratory
3793 N. 3600 E.
Kimberly, ID 83341, USA

The Effects of Tillage and Cover Crops on Some Chemical Properties of an Oxisol and Summer Crop Yields in Southwestern Paraná, Brazil

A. Calegari & I. Alexander

Summary

Evidence shows that when nutrients removed by crop harvest are not replaced by mineral weathering and organic inputs, systems became unsustainable. Historically crop residues have played an important role as mulch for soil and water conservation and as an input for maintaining soil organic matter and returning nutrients to soil.

Data show that crop residues and tillage regime caused significant alteration and redistribution of nutrients within the soil profile. There are also effects in nutrient cycling and changes in soil physical and biological properties. In spite of the fact that no-till system caused a nutrient concentration in the upper soil layer, this was important and not a disadvantage for maize development. Thus the no-tillage system promoted better soil conditions for N and P that became more available and retrievable for maize crop.

The better physical and biological soil conditions promoted by the application of organic residues, N and P probably contributed to the greater maize grain yield in the no-tillage system in this experiment.

Keywords: Cover crop, no-till, crop residue, soil organic matter

1 Introduction

In Brazil, since colonisation began, a lack of adequate global planning in natural resource use in all different regions, as well as the use of land without taking into consideration its agricultural aptitude, have led to misuse and exploitation of land. Consequently, in many areas the soil has not received any efficient management and conservation, resulting in a decline of natural fertility and organic matter levels.

This also resulted in changes in the physical, chemical, and biological characteristics, such as lower infiltration rates and aggregate stability, a decrease in nutrient concentration, and in microbiological activity, respectively, mainly due to excessive mechanization of soil tillage and monocultural cropping, which in turn, increased soil erosion by water.

The results obtained with a no-tillage system in the field experiment showed that maize as well as soybean grain yield increased compared to conventional tillage.

The objective of this study was to assess the effect of winter cover crops on soil chemical properties under conventional and no-tillage systems followed by soybean and maize, the latter with and without applications of nitrogen.

2 Material and methods

The field experiment was established in 1986 at the Iapar (Agronomic Institute) experimental station at Pato Branco, Southwestern Paraná State, Brazil (52° 41' W, 26° 07' S and 700 m altitude). The experimental area had been covered by subtropical forest that was cleared and cultivated with maize for 10 years before the experiment started. Climatologically the area belongs to the zone of subhumid tropical climate or Koeppen's Cfb (climate without dry season, with fresh summer and the average of hottest month < 22°C). Annual rainfall averages 1200-1500 mm. The soil of the experimental site is an Oxisol, very acid with a high clay content (75% clay, 13 % silt, and 12 % sand) dominated by Kaolinitic minerals (clay 1:1). The chemical analysis in the begining of the experiment was: pH 4.7, Al-0.12, Ca-3.67, Mg-2.10, K-0.43, EC-11.94, C-2.54 and P-3.8. The soil site was not fertilised and limed at the beginning of the trial.

Experimental treatments combined winter cover crops, two levels of nitrogen and two tillage systems. The winter cover crops used were: blue lupine (*Lupinus angustifolius* L.), hairy vetch (*Vicia villosa* Roth), black oat (*Avena strigosa* Scherb), corn spurrey (*Spergula arvensis* L.), oilseed radish (*Raphanus sativus* L.), winter wheat (*Triticum aestivam* L.), and fallow. At flowering stage cover crops were controlled with a knife roller (lupine, hairy vetch, black oats and oilseed radish) or by the application of herbicides (spergula and fallow). Wheat grain was harvested and the residues were left on top of the soil as mulch or incorporated before planting the summer crop. The tillage treatments were conventional tillage (one disc plough and 2 disc harrowings) every year before crop planting and no-tillage.

Summer crops of maize and soybean followed by a winter cover crop were planted during nine years. Phosphate and potassium fertilisers were broadcasted before planting the summer crops. Nitrogen (0 and 90 Kg N ha^{-1}) was applied to maize plots only, applied as urea in two split dosages: 1/3 broadcast at planting time and 2/3 banded six weeks after planting.

The treatments were laid out using a split-plot design in three blocks. The winter cover crops were the main plots (240 m²) and the tillage treatments and N-rates were subplots (120 m²).

Samples of shoot plant tissue of the different cover crops and fallow vegetation was collected during flowering in 1994; after harvest wheat straw was collected for dry matter determinations. The samples were dried at 60°C, ground, and sieved (1 mm), digested by nitric perchloric acids, and analysed for P, K, Ca, and Mg. Nitrogen was determined using a micro-kjeldahl block digestion apparatus.

The soil samples were collected in small pits opened in all sub-sub plots. A total of 4 points in each sub-sub plot was sampled, sieved (2 mm) and then ground by mortar & pestle to pass a 0.2mm mesh.

Organic carbon content of the soil was determined by wet oxidation with potassium dichromate in concentrated sulphuric acid, and inorganic phosphorus by spectrophotometry at 880 nm (Nelson and Somners, 1982).

Soil samples were digested by wet oxidation method using concentrated H_2SO_4, H_2O_2 as an additional oxidising agent and $LiSO_4$ to raise the boiling point (Anderson and Ingram, 1993).

Total N was determined by colorimetry (Auto-Analyzer), total P by flow injection analysis (Stannous Chloride Method), K by flame photometry, and Ca and Mg by atomic absorption spectrophotometry at 422.7 and 285.2 nm.

Treatment	P inorg. (µg/g)		P total (µg/g)		O. Carbon (%)		Nitrogen (µg/g)		Calcium (µg/g)		Magnesium (µg/g)		Potassium (µg/g)	
	No-T	Conv	No-T	Conv	No-T	Conv	No-T	Conv	No-T	Conv	No-T	Conv	No-T	Conv
Fallow	7.71	3.82	664	497	3.01	3.14	3.00	2.77	0.56	0.43	0.34	0.42	0.89	0.90
Lupin	7.17	8.13	658	642	3.57	3.32	3.26	3.01	0.46	0.47	0.46	0.42	0.95	1.00
Vetch	7.10	2.34	687	536	3.75	3.68	3.52	3.32	0.68	0.48	0.44	0.45	1.28	0.96
Oat	7.86	5.65	668	557	3.57	3.47	2.94	2.85	0.71	0.50	0.45	0.42	1.36	0.88
Spergula	7.78	4.58	686	645	3.66	3.44	3.12	2.83	0.57	0.43	0.40	0.39	0.94	1.00
Radish	6.97	4.66	621	541	3.46	3.47	2.90	2.96	0.53	0.61	0.33	0.45	0.91	0.99
Wheat	6.13	4.47	725	731	3.73	3.38	2.92	2.74	0.59	0.58	0.40	0.33	0.85	0.86

Table 1: Values of some soil nutrients (0-5 cm) after 9 years under no-till and conventional system

Treatment	P Inorg. (µg/g)		P total (µg/g)		O. Carbon (%)		Nitrogen (µg/g)		Ca (µg/g)		Mg++ (µg/g)		Potassium (µg/g)		pH		Al³⁺+H⁺ (meq/100 g)	
	No-T	Conv	No-T	Conv	No-T	Conv	No-T	Conv	No-T	Conv	No-T	Conv	No-T	Conv	No-T	Conv	No-T	Conv
Fallow	1.87	2.40	483	679	2.82	3.02	0.34	0.42	0.34	0.50	0.40	0.33	0.92	1.01	4.92	5.22	3.73	1.30
Lupin	4.66	6.20	721	619	3.40	3.61	0.46	0.42	0.41	0.51	0.39	0.40	1.06	1.07	4.73	4.88	5.54	3.69
Vetch	1.47	1.60	529	631	3.96	3.41	0.44	0.45	0.34	0.47	0.44	0.37	0.94	0.91	4.75	5.04	4.20	2.08
Oat	2.52	4.70	626	634	3.08	3.17	0.42	0.40	0.47	0.51	0.39	0.44	1.08	0.92	4.79	5.04	3.61	0.97
Spergula	1.93	4.54	741	725	2.87	3.01	0.40	0.39	0.34	0.41	0.36	0.26	0.86	0.75	4.80	4.90	4.56	3.03
Radish	2.62	2.05	521	538	3.55	3.59	0.33	0.45	0.41	0.43	0.39	0.42	1.06	1.20	5.01	4.80	4.03	2.01
Wheat	1.47	3.06	639	578	3.47	3.34	0.40	0.33	0.35	0.54	0.28	0.36	0.87	0.91	4.80	5.12	3.31	1.56

Table 2: Values of soil nutrients (5-15 cm) after 9 years under no-till and conventional system

3 Results and discussion

3.1 Soil properties

The results show that at the depth of 0-5 cm an accumulation of organic material at the soil surface and the unploughed regime in no-tillage system play an important role in maintaining higher levels of inorganic P (Pi) and total P (Pt) compared to conventional system (Table 1).

The no-till treatments had higher levels of nutrients and organic carbon than conventional tillage system. Apparently under the no-till system more organic residues were accumulated, thus increasing nutrients in the soil profile and improving soil fertility.

After 9 years of field experiment the plots with cover crops had a higher organic carbon content in both layers than those left under fallow.

The higher phosphorus content (Pi and Pt) of the surface layer (0-5 cm) indicate that the different cover crops have an important P recycling capacity and this was improved when the residues were concentrated on surface annually.

At the depth of 5-15 cm, due to the decrease of organic matter content and pH values (acidifying process), an increase in exchangeable acidity (Al^{3+} and H^+), lower Pi, Pt, Ca and Mg in the soil profile was observed (Table 2).

The different cover crops showed significant effects on the organic carbon level for both depths (0-5 and 5-15 cm). The means of all winter crops presented greater values than fallow in both depths. This reflects the important effect of winter cover crops on the increase of the organic carbon content. The results also showed that there was no difference in calcium contents for tillage and cover crop regime (0-5 cm), but at 5-15 cm significant differences were observed and greater values were found in conventional tillage except for spergula. This means that after liming in the no-till system calcium was either leached (5-15 cm) or remained above this layer which contributed to an increase of Al^{3+} and H^+ levels, and a decline of pH values, Ca and Mg and Pi contents.

The magnesium values presented no difference in 0-5 and 5-15 cm layers between tillage and cover crop regimes, although a slight decrease in the second layer was observed. Probably the release of calcium and magnesium by organic inputs in the upper layer (0-5 cm) can be related to the reduction of Al^{3+} and H^+. Calcium added by lime also could have contributed to this.

The potassium concentration presented significant difference for cover crops regime (0-5 cm) and tillage × cover crop in 5-15 cm. Some cover crops, mainly radish and lupin, contributed to increased K content in the soil. In general there was a trend for higher values in conventional tillage compared to no-tillage suggesting K leaching for this system. Liming increased Ca and Mg concentration displacing K into the soil solution and promoting leaching of this nutrient.

3.2 Winter plant tissue elements

According to environmental conditions, tillage regime, and specific characteristics of soil behaviour under different cover crops varied, resulting in different nutrient absorption capacity (Derpsch and Calegari, 1985; Calegari et al., 1993).

Results presented in Table 3 showed differences among cover crops for total nitrogen, P, K, Ca and Mg content in dry matter. The different chemical composition of legume and non-legume residues leads to different decomposition and N mineralization rates.

N levels were highest in lupin and vetch and lowest in oat, spergula and wheat. The ability of cover crop to absorb P from the soil solution as well as the tillage regime influenced phosphorus plant tissue concentrations (Table 3). P was highest in no-till plots. Overall concentrations were highest in fallow, vetch, and oat, and lowest in wheat.- The different cover crops showed significant differences in the K values accumulated in plant tissue. Calcium and magnesium values were

Treatment	N (µg/g)		P (µg/g)		K⁺ (µg/g)		Ca⁺⁺ (µg/g)		Mg⁺⁺ (µg/g)		C (%)		C:N ratio	
	No-T	Conv	No-T	Conv	No-T	Conv	No-T	Conv	No-T	Conv	No-T	Conv	No-T	Conv
Fallow	2.57	2.56	0.34	0.25	2.59	3.10	1.18	0.86	0.36	0.30	38.24	38.93	15.03	15.23
Lupin	3.19	3.10	0.19	0.19	2.29	1.98	1.20	1.08	0.49	0.26	37.65	37.63	11.86	12.14
Vetch	3.82	3.49	0.30	0.28	2.03	1.98	0.78	0.81	0.27	0.26	37.87	38.97	10.05	11.19
Oat	1.93	2.08	0.28	0.25	2.15	2.21	0.43	0.44	0.21	0.20	39.69	41.63	20.76	20.20
Spergula	2.13	1.71	0.22	0.16	3.45	2.76	0.52	0.51	0.77	0.70	41.92	43.07	19.78	25.62
Radish	2.68	2.76	0.17	0.14	2.80	2.66	1.54	1.46	0.76	0.66	38.58	36.91	14.45	13.40
Wheat	0.77	0.87	0.06	0.06	1.15	0.87	0.22	0.23	0.10	0.09	40.38	43.58	52.71	49.99

Table 3: Nutrients concentration (%) and C:N ratio in cover crops dry matter

altered not only by different cover crops but also by the tillage system. Fallow, lupin, and radish had highest Ca tissue concentrations and reflected a positive effect of no-till. Radish, lupin, and vetch presented highest Mg tissue concentrations. Therefore, the concentration of this nutrient in radish was higher under conventional tillage than in no-till.

3.3 Maize yield

Results of maize grain yield showed significant differences for tillage systems and nitrogen rates. No-tillage system without nitrogen resulted in higher maize yields than conventional tillage in all treatments except for wheat. All treatments compared maize yields were significantly ($p=0.05$) higher in no-tillage systems than in conventional tillage. These differences in maize yields were even greater when nitrogen was applied. Undoubtedly the maize yield was directly related not only to the previous crop, but also to crop residues accumulated in previous years (Table 5). As can be seen from Table 5 the treatments with lupin, spergula, oat, radish and vetch presented higher dry matter input than fallow. The different nutrient inputs could be directly related to maize performance in this last year.

Treat.	Maize yield (kg ha^{-1})			
	No-T 0 N	No-T 90 N	Conv 0 N	Conv 90 N
Fallow	2829	3506	2420	3069
Lupin	3988	4163	3160	3404
Vetch	3958	4049	2834	3047
Oat	3496	4605	3086	3997
Spergula	3856	3966	3269	3515
Radish	3473	3835	2723	3543
Wheat	2894	3993	3240	3475

Table 4: Maize grain yield (kg ha^{-1})

Greater nitrogen absorption in the no-tillage system compared to conventional tillage led to different nitrogen concentrations in maize residues (Table 4). Also total P values were influenced by tillage systems; no-till contributed to higher accumulation in maize residues than conventional regimes.

The no-tillage legume treatments (lupin, vetch) and spergula without nitrogen showed significantly higher maize yields as compared to fallow. The maize yields after wheat, radish, and oat presented no difference with others treatments. The results showed that cover crops (legumes and spergula) can accumulate nitrogen quantities comparable to 90 Kg N ha^{1} as urea fertiliser, compared to conventional tillage. Thus with no-tillage systems significant amounts of fertiliser can be saved. This maize yield increase was correlated to soil phosphorus availability and phosphorus content in maize plant leaves. The greater phosphorus uptake is attributed to higher moisture content below the mulch and consequently higher phosphorus diffusion rate by plants.

The data showed that crop residues and also the tillage regime caused significant alteration and redistribution of nutrients within the soil profile. Also, there were effects in nutrient cycling and certainly soil physical and biological properties which were not evaluated in this study. In spite of the fact that no-till systems caused a nutrient concentration on the upper soil layer, this was important and not to the disadvantage for maize development. Thus, the no-tillage system promoted soil conditions for which improved the availability and the uptake by maize of N and P.

Treat.	N (µg/g) No-T	N (µg/g) Conv	P (µg/g) No-T	P (µg/g) Conv	K⁺ (µg/g) No-T	K⁺ (µg/g) Conv	Ca⁺⁺ (µg/g) No-T	Ca⁺⁺ (µg/g) Conv	Mg⁺⁺ (µg/g) No-T	Mg⁺⁺ (µg/g) Conv	C (µg/g) No-T	C (µg/g) Conv	C:N ratio No-T	C:N ratio Conv
Fallow	0.53	0.43	0.42	0.27	0.850	0.800	0.124	0.122	0.145	0.146	54.40	54.39	107.76	126.04
Lupin	0.63	0.55	0.47	0.42	0.933	0.867	0.141	0.139	0.139	0.146	54.26	54.75	87.44	101.58
Vetch	0.52	0.46	0.38	0.32	0.750	0.867	0.146	0.131	0.166	0.147	54.58	54.26	107.30	120.50
Oat	0.48	0.48	0.43	0.38	0.883	0.867	0.120	0.125	0.145	0.148	54.30	54.39	114.99	114.65
Spergula	0.60	0.49	0.43	0.36	0.883	0.850	0.144	0.140	0.152	0.147	54.34	54.49	93.67	112.95
Radish	0.51	0.51	0.44	0.32	0.917	0.900	0.141	0.123	0.151	0.138	54.74	54.50	109.96	109.48
Wheat	0.51	0.48	0.43	0.37	0.883	0.850	0.137	0.130	0.146	0.141	54.56	54.01	110.66	117.53

Table 5: Nutrient concentrations (%) and C:N ratio in maize shoot dry matter after harvest

The improved physical and biological soil conditions promoted by the accumulation of organic residues (Lal, 1990; Ehlers, 1975; Killham, 1994), and the better availability of N and P probably contributed to the greater maize grain yield in the no-tillage system in this experiment.

Treatment	Winter cover crops		Maize		Soybean		TOTAL INPUT	
	not-till	conv.	not-till	conv.	no-till	conv.	No-till	conv.
Fallow	20015	14971	29305	27930	11805	11209	61125	54110
B. lupin	35195	28198	34900	32341	12685	11581	82780	72120
h. vetch	27157	23038	32054	27691	11879	11044	71090	61773
B. oat	36991	32989	27940	28944	12824	12068	77755	74001
C. spurrey	38815	35023	31539	28655	12179	11139	82533	74817
O. radish	27833	19767	30957	30663	12399	11643	71189	62073
W. wheat	25418	24453	27357	25539	11644	11561	64419	61553

Table 6: Total input of shoot dry matter of winter crop (total flowering), maize and soybean residues (after harvest)(Kg.ha^{-1}) since 1986 until 1995.

4 Conclusions

- No-tillage when compared to conventional tillage caused alteration in the distribution of some soil nutrient distribution within the soil profile.
- The winter cover crops produced different amount of dry matter and had different tissue nutrient concentrations;
- Residues accumulation, including summer crop residues, increased the levels of some soil nutrients and soil organic carbon.
- Maize residues after harvest showed higher P and N concentration in no-tillage system.
- The better soil conditions brought about by the no-tillage system promoted higher maize and soybean yields.
- Using some winter legumes (blue lupin and hairy vetch) in no-tillage, an input equivalent to 90 Kg N ha^{-1} of fertiliser when compared with fallow in conventional system is possible.

References

Anderson, J.M. and Ingram, J.S.I. (1993): Tropical Soil Biology and Fertility. A Handbook of Methods. Second edition. C.A.B. International, Wallingford, UK.
Calegari, A., Mondardo, A., Bulisani, E.A., Wildner, L. do P., Costa, M.B.B.; Alcantara, P.B., Miyasaka, S. and Amado, T.J.C. (1993). Adubação verde no sul do Brasil. 2a. edição- Rio de Janeiro, AS-PTA.
Derpsch, R. and Calegari, A. (1985): Guia de plantas pare adubação verde de inverno. Documentos Iapar no.9, Londrina.
Ehlers, W. (1975): Observations on earthworm channels and infiltration on tilled and untitled loess soil. Soil Science. **119**(3): 2429.
Killham, K. (1994): Soil Ecology. Cambridge University Press.
Lal, R. (1990): Soil erosion in the tropics: principles and management. McGraw-Hill, New York.
Nelson, D.W. and Somners, L.E. (1982): Total carbon, organic carbon, and organic matter. In: A.L.Page, R.H. Miller & D.R. Keeney (eds), Methods of Soil analysis) Part 2. ASA/SSSA, Madison, USA, 539-579.

Addresses of authors:
Ademir Caligari
Agronomic Institute of Parana, P.O. Box 481, 86001-970 Londrina, Brazil
Ian Alexander
Plant and Soil Science Department, University of Aberdeen, Aberdeen, UK

Turnover of Green Manure and Effects on Bean Yield in Northern Zambia

T.N. Mwambazi, B. Mwakalombe, J.B. Aune & T.A. Breland

Summary

During a three-year field experiment conducted at two locations in northern Zambia, the effects of incorporating *Crotalaria zanzibarica*, *Crotalaria juncea*, *Sesbania sesban* and *Hyperrhenia* grass spp. (Control) on bean yield was studied under the traditional mounding system. Decomposition and N mineralisation dynamics of the residues were determined in a litter bag field study and a laboratory incubation, respectively. N concentration in the green manures ranged from 0.4 % to 1.91 %, it was lowest for *Hyperrhenia* spp. and highest for *C. zanzibarica*. Bean yields were highest when *C. zanzibarica* was used as green manure and lowest in the control. Daily first order decomposition rates in 1995 were 0.043, 0.030 and 0.024 for *C. zanzibarica*, *C. juncea* and *Hyperrhenia* spp., respectively. This ranking corresponded with decreasing initial N concentration of the plant materials. Initial N immobilisation was observed with all residues. Following immobilisation a net release of N occurred after 7 days for *C. zanzibarica* and *C. juncea*, and after 21 days for *Hyperrhenia* spp. Whereas only 25 % of the initial N content of *Hyperrhenia* spp. was mineralised in five weeks, 60 and 65 %, respectively, was mineralised from *C. juncea* and *C. zanzibarica*.

Keywords: Green manure, decomposition, N mineralisation, *Crotalaria* spp., bean yield

1 Introduction

In Northern Zambia, grass-mounding (*fundikila*) is one of the most important traditional farming systems. Unlike slash and burn cultivation (*chitemene*), which is woodland based, *fundikila* is grassland based. Its carrying capacity is estimated at 20-40 persons km^{-2} (Mansfield et al., 1975). Under *fundikila* cultivation, fallow grass is incorporated in mounds as manure towards the end of the rains. At the onset of the next rains, the mounds are levelled and planted with finger millet. Following harvest, mounds are not made until the beginning of the next rains. The mounds are then planted with beans. Upon harvesting the crop, the mounds are not levelled until prior to sowing finger millet at the start of the next rains. This rotation between legumes on mounds and cereals on the flat continues until the garden is abandoned for fallow regeneration (Stromgaard, 1991).

Unless supplemented with reasonable amounts of mineral fertiliser, the productivity of the *fundikila* system is often very marginal. This is explained by the low nutritive value of *Hyperrhenia* grass, the main plant material buried in the mounds. The inability to purchase adequate inputs, such as mineral fertiliser, is one of the major constraints facing most farmers. Introduction of high quality and fast-growing legume green manures may reduce this problem.

The purpose of this study was two-fold: 1) to determine the potential of the legumes, *Crotalaria zanzibarica*, *Crotalaria juncea*, and *Sesbania sesban* to improve soil fertility and crop yield relative to *Hyperrhenia* grass spp. and 2) to determine the rates of decomposition and N mineralisation of the plant materials. To address these objectives field experiments, litter field studies and a laboratory incubation were carried out.

2 Materials and methods

2.1 Site description

The field experiments were conducted at two locations in the northern province of Zambia; at Misamfu Regional Research Centre in Kasama (10° 10' S 31° 12' E, 1384 m altitude) and at Kaka in Mbala (8 ° 51' S 31° 20' E, 1673 m altitude). The rainfall pattern is unimodal with a mean annual rainfall of 1360 mm at Misamfu and 1500 mm at Kaka falling between November and April. Mean monthly temperatures during the rainy season range from 19 to 27 °C (FAO 1993).

Site	Depth (cm)	$PHCaCl_2$	OC (%)	Tot N (%)	Av. P (mg kg^{-1})	Ca^{2+}	Mg^{2+}	K^+	Na^+	Al^{3+}	Al sat. (%)
						(Cmolc kg^{-1})					
Misamfu	0-15	4.4	1.5	0.06	18.2	0.9	0.67	0.20	0.1	0.3	13.8
	15-30	3.9	0.7	0.03	4.1	0.5	0.38	0.10	0.1	0.8	42.6
Kaka	0-15	4.7	0.9	0.05	10.2	0.7	0.49	0.35	0.1	0	0
	15-30	4.5	0.6	0.04	4.3	0.4	0.36	0.30	0.1	0	0

Table 1. Chemical properties of soil sampled at the beginning of the growing season

The soils are deep, well-drained and largely derived from non-basic parent material. These are strongly leached, acidic, and low in inherent fertility and classified as *Ferralsols* (FAO/UNESCO, 1994) or *Haplorthox Paleustult* (USDA 1975). Selected chemical properties are listed in Table 1. above. Tree vegetation is dominated by *Brachystegia* and *Julbernardia* genera attaining heights of 12-15 m. The forest floor is dominated by *Hyperrhenia* and *Digitaria* grass species (Lawton, 1982; Stromgaard, 1990).

2.2 Design and management of the field experiments

Selected trial sites were long secondary fallow or primary forest. Sites were cleared, hand-hoe prepared, and the legume green manures were sown at he onset of the rains in November. The green manures were cut at flowering stage and incorporated into mounds in February. Bean (*Phaseolus vulgaris*) was planed as a test crop shortly after incorporation. The study was carried out over a period of three growing seasons from 1992/93 to 1994/95. The design was a randomised complete block with three replications. The green manure treatments were as follows: *C. zanzibarica*, *C. juncea*, *S. sesban*, and Control (natural grass fallow of *Hyperrhenia* spp.) Phosphorus treatments were 0 and 11.9 kg P ha^{-1} (not applied to control treatment). Phosphorus was added as single super phosphate (SPP) in two equal splits at sowing of the green manures and he rest at bean planting. Total and harvest plot sizes were, respectively, 4m × 5m and 3m × 3m.

In February, at the end of the vegetative stage, green manure biomass was harvested, weighed and buried in mounds. The bean crop was planted (35 kg ha^{-1}) on the mounds immediately after biomass incorporation. Following bean plant emergence, stand counts were recorded and hand weeding was carried out. During the last two growing seasons, the experiment was repeated on adjacent sites. Due to the poor performance of *S. sesban* during the first two seasons, *S. macrantha* was used in the third season.

Owing to early withdrawal of rains at Misamfu, bean grain did not reach full maturity by harvest time in two out of three seasons. When no grain yield was obtained, treatments performance were based on whole above-ground dry matter biomass of the bean crop.

2.3 Plant biomass and soil samples

Samples of above-ground green manure biomass were collected from each plot. Dry matter yields were determined after oven drying to constant weight at 65°C. To determine chemical composition of the plant materials, samples were collected, washed in distilled water, air-dried, finely ground to pass through 0.15 mm mesh, and analysed for content of lignin, polyphenols, C, N, P, K, Ca, and Mg. Total nitrogen was determined by the micro-kjeldahl method, lignin by the acid detergent fibre method (Van Soest and Wine, 1968) and soluble polyphenols by a revised Folin-Denis method (King and Heath, 1967).

Soil samples were collected 0-15 and 15-30 cm) for chemical analyses at the beginning (composite), during and at the end of the growing season. The samples were air-dried and sieved through a 2 mm mesh and analysed for pH (0.01 M CaCl$_2$), available P (Bray-1), organic carbon (Walkey-Black method), total N (Kjeldahl method), exchangeable bases and Al (0.1 M ammonium acetate method). Following extraction with ammonium acetate, Ca and Mg were determined by atomic absorption and K and Na by flame photometry.

2.4 Decomposition studies

Decomposition studies were conducted with *C. zanzibarica* and *C. juncea* in 1993, 1994 and 1995. *Hyperrhenia* spp. was included in 1995. For each plant material, a representative sample about 40 g of the whole plant (leaves, twigs and stem) was placed in the standard TSBF litter bag of 7 mm mesh in a flat configuration (Anderson and Ingram, 1993). The litter bags were buried in the top soil to simulate field conditions.

Each year, five litter bags of each plant material were randomly placed in each of the three replicates following harvesting of green manures and five samplings were conducted per year. At each sampling, the plant material was carefully removed from the litter bag and soil particles removed by hand. The remaining plant material was then washed with distilled water and oven-dried to constant weight at 65 °C to determine dry matter content. The material was then ground and analysed for concentrations of lignin, soluble polyphenols N, P, K, Ca and Mg.

2.5 N mineralisation

Samples of fresh plant materials was collected as described above. The material were cut into smaller pieces, dried and finely ground. For each treatment, 1 g of ground leaves was thoroughly mixed with fresh soil equivalent to 500 g dry weight to approximate a mulch rate of 4 t ha^{-1}. Each plant material was replicated three times. A control (soil only) was also included. Deionised water was added to bring the moisture content to about 70 % water holding capacity. Subsamples of 10 g were immediately extracted in 100 ml 2M KCl for 30 minutes. The extracts were analysed for ammonium (Gentry and Willis, 1988) and nitrite plus nitrate (Technicon, 1977). The remaining was incubated in the dark at room temperature (approximately 25 °C). Subsamples were taken at 2, 3, 4 and 5 weeks after addition of plant material and analysed for exchangeable NH_4^+-N and NO_3^-. Net mineralisation from plant residues was calculated as the difference in exchangeable NH_4^+ and NO_3^- between amended and unamended soil.

2.6 Statistical analyses

Yield data were analysed using the MSTAT statistical package (Nissen et al. 1994). A first-order equation, $y_t = Y_o e^{-kt}$ where y_t is fraction of initial plant material ($Y_o = 1$) remaining at time t (days) and k is the rate constant, was fitted to decomposition data by non-linear regression (JMP 1989).

Treatment	Misamfu				Kaka			
	1993	1994	1995	Mean	1993	1994	1995	Mean
C. zanzibarica	3925	1300	4225	3150	6650	630	4382	3887
C. zanz+SSP	5600	2415	6625	4880	7650	515	9295	5820
C. juncea	250	1050	1050	783	2400	285	1538	1408
C. juncea+SSP	570	1325	1300	1065	7950	610	2688	3749
S. sesban	20	0	425[b]	10[c]	2650	250	115[b]	1450[c]
S. sesban+SSP	0	0	626[b]	0[c]	2800	575	184[b]	1688[c]
Control (Hyp.)	2815[a]	2103[a]	3475	3068	6497[a]	1958[a]	4538	4331
LSD (5%)	818	651	4286		1630	201	3443	

SSP - Single Super Phosphate
[a] Estimated from 1995 results and not included in LSD calculation
[b] *Crotalaria macrantha*
[c] Mean for 1993 and 1994

Table 2. Mean dry matter production (kg ha^{-1}) of the various organic inputs at the two locations

3 Results

3.1 Biomass and composition of green manure

Irrespective of phosphorus treatment, *C. zanzibarica* invariably produced more dry matter biomass than *C. juncea* which, in turn, performed better than *S. sesban* (Table 2). Continued poor emergence and low stand counts recorded for *S. sesban* were largely responsible for the low biomass yield. Over the entire trial period, there tended to be a positive and, in some cases, a substantial effect of SSP on green manure biomass production at both sites (Table 3). Phosphorus additions increased average yield of *C. zanzibarica* by 54 and 50 % at Misamfu and Kaka, respectively.

Initial chemical characteristics of the plant materials are given in Table 3. The total N accumulated in the above-ground dry matter at incorporation varied from 177 to less than 20 kg N ha^{-1}. For *C. zanzibarica*, the average N accumulation in above-ground matter when no SSP was applied was 60 and 74 kg N ha^{-1} at Misamfu and Kaka, respectively.

	Concentration (%)							Ratio			
Species	N	P	K	Ca	Mg	Polyp	Lign	Lign/N	Polyp/N	(Polyp+Lign)/N	Strd portion
C. zanz	1.91	0.27	0.20	0.11	0.37	1.21	8.21	4.32	0.63	4.95	0.23
C. juncea	1.25	0.26	0.19	0.40	0.51	1.96	11.52	9.2	1.50	10.78	0.32
S. sesban	1.30	0.27	0.17	0.21	0.33	nd	nd	nd	nd	nd	nd

nd Not determined   Lign - Lignin   Strd - Structured

Table 3. Chemical composition of the green manures

3.2 Decomposition

In all three years. *C. zanzibarica* decomposed faster than *C. Juncea* (Fig. 1 a, b, and c). *Hyperrhenia* was included in the third year. Its decomposition rate was slower than that of the other species. In 1995, the time taken for the materials to lose 50 % of their initial mass was 16, 23, and 29 days for *C. zanzibarica*, *C. juncea*, and *Hyperrhenia* spp., respectively.

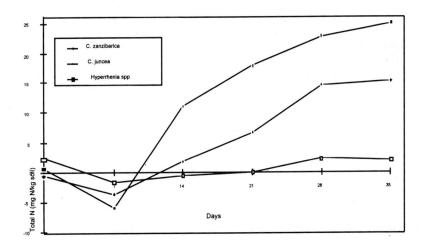

Fig. 1: Net total N mineralised

3.3 N-mineralisation pattern

There was N immobilisation with all three species during the first seven days of incubation (Fig. 2). With *Hyperrhenia spp..* Immobilisation corresponded to 21 % of initial N content. Corresponding values for *C. zanzibarica* and *C. juncea*, net release of N started between 7 and 14 days of incubation and for *Hyperrhenia* spp. after 21 days.

Cumulative total N mineralised at the end of the incubation was 65, 60, and 25 % of initial N for *C. zanzibarica*, *C. juncea* and *Hyperrhenia spp.*, respectively.

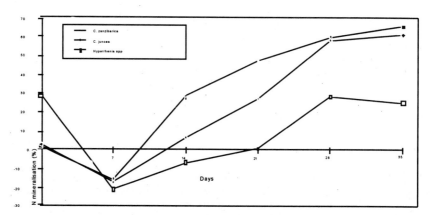

Fig. 2: Cumulative % N mineralised

3.4 Crop yields

In the first year, there was significant ($p < 0.05$) bean yield response to *C. zanzibarica* at both sites (Table 4). The effect of *C. juncea* was only statistically significant when phosphorus was added. Relative bean yield response with green manuring was in the order *C. zanzibarica* > *C. juncea* > *Hyperrhenia spp*. Relative performance on bean yield of the various organic inputs, when fertiliser was not added was 100, 56, 43, 29 % at Misamfu (whole stover) and 100, 67, 44, 18 % at Kaka (grain).

Treatment	Misamfu				Kaka			
	1993 biomass	1994 grain	1995 biomass	mean biomass (2 years)	1993 Grain	1994 grain	1995 grain	mean grain
C. zanz.	1684	177	681	1182	2072	322	454	949
C. zanz + SSP	3285	320	764	2024	2234	167	410	937
C. juncea	947	237	791	869	1384	173	505	893
C. jun + SSP	1285	260	780	1033	1999	195	486	688
S. sesban	717	124	488[d]	603	905	233	319[d]	569[e]
S. ses + SSP	867	125	453[d]	660	1336	177	551[d]	756[e]
Control	485	90	402	444	370	103	333	269
LSD (5%)	763.4	NS	NS	NS	1457.2	NS	NS	NS

[d] *Sesbania macrantha* treatment   [e] Mean for 1993 and 1994

Table 4: Response of bean ($kg\ ha^{-1}$) to green manuring with and without SSP

In the last two seasons no treatment effects on bean yield were recorded. Positive effects appeared to be masked by severe bean stem maggot attack in the second season and by severe flower damage due to heavy rain storms in the third season. Bean stem maggot which accounted for over 60 % loss in stand count within one month of planting, was due to use of untreated bean seed which is the usual traditional farmer practice. There was no significant effect of treatment across the years.

4 Discussion

Initial N concentration of the green manures was the factor which correlated best with decomposition rate. Mineralisation of nitrogen from green manures followed the same pattern. The results are in agreement with those of Tian et al. (1992), who reported nitrogen mineralisation was directly related to rate of decomposition. Palm and Sanchez (1991) and Melillo et al. (1982) have demonstrated that lignins and polyphenols can modify the N supply of plant residues. Even though *C. juncea* had higher polyphenol and lignin content than *Hyperrhenia spp.*, a greater portion of its initial N content was mineralised.

A net immobilisation was observed during the initial decomposition of material with 1.9 % N. This is in agreement with results from other studies suggesting that critical N concentration for N mineralisation may be up to 1.7-2.0 % N (Harmsen and Van Schreven 1955, Constantinides and Fownes 1994, Aber and Melillo, 1980).

As in our study, Palm and Sanchez (1991) also observed rapid initial immobilisation. They attributed this to nitrosation, a reaction in which soluble polyphenolics that leach out from the leaves into the soil would bind to NO_2 formed during nitrification and produce a chemical N immobilisation. This phenomenon has also been observed in acid soils (Nelson and Bremner, 1969; Azhar et al., 1986). However, it is equally likely that the initial immobilisation may also be microbial. A study on fitting a mechanistic model to net mineralisation from legume materials indicated that N concentration Of substances readily available to micro-organisms may be higher than normally assumed (T.A. Breland, Agric. Univ. Norway, pers. Comm.).

The highest bean yield was produced when *C. zanzibarica* was used as a green manure. This can be attributed to a greater biomass production, a higher nitrogen concentration, rapid decomposition and N mineralisation of *C. zanzibarica* than of the other species. Green manure and bean yield was increased as a result of application of SPP. The chemical analyses showed a relatively low level of P (Table 1) and it is therefore expected that there is a response of green manures and bean to P application. In another study in Kasama, finger millet yield doubled compared to a control when *C. zanzibarica* was used in combination with SSP (Mwambazi and Gauslaa 1994).

Even though *Hyperrhenia spp.* are not able to release much nitrogen, it may have a positive effect by contributing to maintaining soil organic matter level. This can be attributed to its higher level of structural material as defined by Parton et al. (1987) in *Hyperrhenia spp.* (56 %), than that of *C. zanzibarica* (23 %). It is the structural components of the organic material which specially contributes to the build up of soil organic matter. This has practical implications for soil fertility management options. While mounding with *Hyperrhenia* grass, with a structural fraction of more than twice that of *zanzibarica*, may not provide enough N for plant growth in the short term, it may be important in the maintenance of soil physical properties and build up of soil organic matter in the long run.

Acknowledgements

The authors wish to thank Mr. Alfred Mkonda for his contribution towards field experimentation and data handling. Thanks also go to the management of Misamfu Regional Research Centre for its role in making this study possible. We also express gratitude to the Ecology and Development programme at the Agricultural University of Norway and the Soil Productivity Research Programme, Kasama, Zambia, for the necessary financial support.

References

Aber, J.D. and Melillo, J.M. (1982): Nitrogen immobilisation in decaying hard wood leaf litter as a function of initial nitrogen content. Can. J. Bot. **60**, 2263-2269.

Anderson, J.M. and Ingram, J.S.I. (eds) (1993): Tropical soil biology and fertility. A handbook of methods, second edition. CAB international, 36-40.

Azhar, E.L., Verhe, R., Proot, M., Sandra, P. and Verstraete, W. (1986): Binding of Nitrate-N on

polyphenols during nitrification. Plant and Soil **94**, 369-386.
Constantinides, M. and Fownes, H.J. (1994): Nitrogen mineralisation from leaves and litter of tropical plants: Relationship to nitrogen, lignin and soluble polyphenol concentration. Soil Biol. and Biochem. **26**: 49-55.
FAO (1993): A climatic database for irrigation planning and management (CLIMAT for CROPWAT). FAO Irrigation and drainage paper **49**, 63.
FAO/UNESCO (1974): Soil map of the world. Vol 6: Africa. United Nations Educational, Scientific, and Cultural Organisation, Paris.
Gentry, C.E. and Willis, R.B. (1988): Improved method for automated determination of ammonium in soil extracts. Comm. in Soil and Plant Analyses **19**: 721-737.
Harmsen, G.W. and Van Schreven D.A., (1955): Mineralisation of organic nitrogen in the soil. Advances In Agronomy **7**, 299-398.
JMP (1989): JMP® User guide. Version 2 of JMP. SAS Institute Inc., Cary, NC, USA.
King, J.G.C. and Heath, G.W. (1967): The chemical analysis of small samples and of leaf material and the relationship between the disappearance and composition of leaves. Pedobiologia **7**: 192-197.
Lawton, R.M. (1982): Natural Resources of Miombo woodland and recent changes in agricultural land-use practices. Forest Ecology and Management **4**, 287-297.
Mansfield, J.E., Bennet, J.G., King, R.B., Lang, D.M. & Lawton, R.M. (1975): Land resources of Northern and Luapula province, Zambia. Land resource study 19, Land Resources Division, Ministry of Overseas Development, England.
Melillo, J.M., Aber, J.D. & Muratore, J.F. (1982) Nitrogen and lignin control of hardwood leaf litter decomposition dynamics. Ecology **63**: 621-626.
Mwambazi, T.N. and Gauslaa, I. (1994): Improvement of the grass-mound system using a legume and inorganic inputs. Zambian J. Agric Sci. **4**, 27-31.
Nelson D.W. and Bremner, D.W. (1969): Factors affecting chemical transformation of nitrite in soils. Soil Biol. Biochem. **1**, 229-239.
Nissen, Ø., Hove, K. and Krogdahl, S. (1994): Statistical programme ENM. New version of MSTAT written in turbo pascal. Department of Plant Sciences, Agricultural University of Norway, ÅS, Norway.
Palm, C.A. and Sanchez, P.A. (1991): Nitrogen release from some tropical legumes as affected by lignin and polyphenol contents. Soil Biol. and Biochem. **23**, 83-88
Parton, W.J., Schimel, D.S., Cole, C.V. and Ojima, D.S. (1987): Analysis of factors controlling soil organic matter levels n great plains grasslands. Soil Sci. Soc. Am. J. **51**, 1173-1179
Stromgaard, P. (1990): Effect of mound cultivation on concentration of nutrients in Zambian miombo woodland soil. Agriculture, Ecosystems and Environment **32**, 295-313.
Stromgaard, P. (1991): Soil nutrient accumulation under traditional African agriculture in the miombo woodland of Zambia. Trop. Agric. **68**, 74-80.
Technicon (1977): Nitrate and nitrite in water and sea water. Industrial method 158- 71W/A, Technicon Industrial Systems, Tarrytown, NY, USA.
Tian, G., Kang, B.T. & Brussaard, L. (1992): Biological effects of plant residues with contrasting chemical composition under humid tropical conditions-Decomposition and nutrient release. Soil Biol. and Biochem. **24**:1051-1060.
USAID (1975): Soil Taxonomy: A basic system of soil classification for making and interpreting soil surveys. Soil Conservation Service, US Dept. of Agriculture, Washington DC Agricultural Handbook 436.
Van Soest, J. & Wine, R.H. (1968): Determination of lignin and cellulose in acid detergent fibre with permanganate. Journal of the Association of Official Analytical Chemists **51**: 780-785.

Addresses of authors:
T.N. Mwambazi
B. Mwakalombe
Soil Productivity Research Programme, Misamfu Regional Res.Centre, P.O. Box 410055,Kasama, Zambia
J.B. Aune
Centre for International Environment and Development (CIED), Noragric,
Agricultural University of Norway, Box 5001, 1432 Aas, Norway
T.A. Breland
Department of Biotechnology, Agricultural University of Norway, Box 5001, 1432 Aas, Norway

Influence of Runoff Irrigation on Nitrogen Dynamics of *Sorghum bicolor* (L.) in Northern Kenya

J. Lehmann, F. v. Willert, S. Wulf & W. Zech

Summary

In tropical drylands, low and variable rainfall severely limits crop production. Runoff irrigation is reported to increase and stabilize crop production in arid and semiarid regions. At the same time, the runoff water may affect the nutrient status of the soil. This study was carried out in a runoff irrigation system with *Sorghum bicolor* (L.) in semiarid Northern Kenya. In this paper, we examine the effect of runoff water on soil chemical properties and on the mineral nutrition of *Sorghum*. The nutrient contents of the soil before and after irrigation were compared under runoff irrigation and rainfed agriculture. Plots with good and poor growing *Sorghum* were selected and analysed for their foliar nutrient contents. The net nitrification was measured with and without irrigation using resin cores. In order to test for N deficiency, a fertilization experiment was conducted with the application of 100 kg N ha^{-1}. Foliar N contents of sorghum are lower under runoff irrigation than under rainfed agriculture. This can be explained by the lower soil NO_3 content and the lower net nitrification in the irrigated plots. The total N content decreased more than 10% within one year after establishing the runoff irrigation system. N fertilization increased the thousand grain weight by 33% and the total grain yield by 350%. Leaching of nutrients and especially N is the main problem limiting crop yields in the studied irrigation system. Runoff irrigation can only be successful if the available nutrients are kept in the system or additional nutrients can be provided.

Keywords: Nitrogen, Northern Kenya, nutrient leaching, runoff irrigation, *Sorghum bicolor*.

1 Introduction

In semiarid tropical areas, precipitation may be extremely variable. Without irrigation, crop yields are generally low and very unreliable. Water harvesting has shown to be an inexpensive and effective way of increasing biomass production in many arid and semiarid regions (Klemm, 1989; Tabor, 1995; Lehmann et al., in press).

In runoff irrigation, the surface runoff water after a heavy storm is guided into levelled basins. The standing water is allowed to percolate deep into the soil profile, increasing the plant available soil moisture reserves. Crops, which are sown after the flood, are then able to utilize this additional soil water and, in this way, produce higher yields (Evenari et al., 1968).

The nutrient contents of soils in arid and semiarid tropical environments are usually low especially of N (Aggarwal and Praveen-Kumar, 1994). The influence of water harvesting on soil fertility has not been the subject of specific research to date. Runoff irrigation may have various effects on soil chemical properties: (1) the high water percolation rates may remove excess salts from the topsoil and therefore improve the soil chemical properties; on the other hand, essential nutrients may be leached from the rooting zone of the crop. (2) The irrigation water can serve as a source of undesirable salts or desirable plant nutrients (Nabhan, 1984). (3) Another effect of irrigation could be a plant nutrient deficiency induced by accelerated biomass production when soil water is not limiting plant growth (Sanchez, 1976).

The objectives of the present study were: first, to identify the limiting nutrients for crop growth in the runoff irrigation system; second, examine the reason for reduced plant growth using relationships between foliar nutrient contents and total soil nutrients, analyses of soil extracts and a fertilizer experiment; and to discuss third, the effects of runoff irrigation with respect to dynamics of nutrients and salts, and to evaluate strategies to improve the nutrient supply for the crops.

Depth [cm]	Horizon	pH H_2O	EC* [mS cm^{-1}]	SAR*	Corg [g kg^{-1}]	N [g kg^{-1}]	K* [mg l^{-1}]
0-7	Ah	8.6	0.54	3.8	5.3	0.42	5.4
7-14	2A	8.9	0.35	3.1	2.5	0.21	7.6
14-30	3Ah	8.6	0.43	3.0	6.4	0.62	5.0
30-60	3Bt	8.9	0.56	6.2	5.1	0.43	2.9
60-107	3Btn	9.2	1.85	20.1	8.0	0.56	6.3
107-170	4Btz1	8.7	5.81	27.1	5.3	0.32	10.8
170+	4Btz2	8.2	13.62	26.6	2.3	0.24	11.7

Depth [cm]	water content [%]				particle size distribution [%]					
	pF 0	pF 1	pF 1.5	pF 2	c-sand	m-sand	f-sand	c-silt	f-silt	clay
0-7	25	25	24	18	0	3	35	33	17	12
7-14	26	25	22	11	1	11	65	12	4	6
14-30	30	29	27	24	5	2	18	40	21	18
30-60	27	27	25	22	0	0	11	47	27	16
60-107	25	24	24	24	0	0	4	41	27	29
107-170	26	26	26	25	2	7	24	31	15	22
170+	27	26	26	25	1	2	16	41	23	17

Depth [cm]	Ca* [mg l^{-1}]	Mg* [mg l^{-1}]	Na* [mg l^{-1}]	Cl* [mg l^{-1}]	SO$_4$* [mg l^{-1}]	B* [mg l^{-1}]
0-7	18.1	4.0	68.1	11.0	22.0	0.072
7-14	17.2	1.3	49.5	13.9	15.5	0.017
14-30	25.2	5.8	63.4	8.0	15.4	0.021
30-60	15.1	5.6	110.0	11.9	37.6	0.033
60-107	17.9	3.8	357.8	78.1	444.8	0.190
107-170	141.5	29.9	1361.0	559.4	1185.0	0.177
170+	555.0	111.7	2622.0	2080.8	1210.0	0.105

* measured in the saturation extract.

Table 1: Soil chemical properties at the experimental site.

2 Study site and methods

2.1 Location

The study was carried out near Kakuma in Northern Kenya (34° 51' East and 3° 43' North, altitude 620 m a.s.l.). The natural vegetation is a thornbush savanna with *Acacia tortilis*, *Acacia reficiens* and *Ziziphus mauritiana*. The rainfall distribution is generally bimodal with an annual precipitation of 318 mm (from 14 years; W.I. Powell, and Rutkana Drought Control Unit, unpubl. data). The soils are classified as endosodi-calcaric Fluvisols; they are deep and usually loamy, sometimes sandy, with high pH and EC, low organic carbon and nitrogen contents (Table 1). The sodium adsorption ratio (SAR) of the saturation extract is very high especially in the subsoil. Low K concentrations in the saturation extract may indicate nutritional imbalances for plants due to the high content of other cations.

2.2 Experimental design and treatments

During March to June 1994, the runoff irrigation system was built using a design of levelled basins. The level of the irrigated plots was lowered by 40 cm compared to the original soil surface. Dams surrounding the plots were constructed with the soil material from the excavation (Lehmann et al., in press).

After the flood had infiltrated, *Sorghum* was sown in rows 50 cm apart and 25 cm apart in the row on 19 May 1995. Microplots were chosen for good and poor growing *Sorghum* under runoff irrigation and under rainfed conditions in five replicates. The *Sorghum* was harvested on 4 September, and grain yield and thousand grain weight (TGW) were measured.

For the fertilizer experiment, microplots of 4 m^2 each were chosen before the flood and 100 kg N ha^{-1} was applied as $(NH_4)_2SO_4$ in 4 replicates. After the flood, *Sorghum* was sown on 1 to 6 May 1996, harvested on 5 August, and total grain yield and TGW were determined.

2.3 Leaf and soil analyses

At *Sorghum* heading, the second blade from apex was sampled for plant analyses, washed with deionised water, dried at 60°C for 48 hours and ground. C, N and S were analysed with an automatic C/N/S Analyser.

Soil samples were taken from 0-15 cm depth before the first flood in June 1994 and after 3 floods in June 1995, mixing at least 6 subsamples from one plot. Up to that time, no crops had been grown in these fields. C and N were analysed with an automatic C/N Analyser.

At the end of the dry season in March 1995 and after two floods in May, soil samples were taken from 0-15 cm depth from soils under runoff irrigation and rainfed agriculture; the samples were kept in a cool box and deep frozen within two hours. Fifty g of the moist soil were mixed with 100 ml 2 N KCl and shaken well. The soil extracts were treated with H_2SO_4, frozen and transpor-ted to Germany. NO_3^- was determined photometrically with a Rapid Flow Analyser.

For the determination of the amount of NO_3^- in the soils without plant uptake (net nitrification), resin cores with mounted tubes were inserted inside and outside the irrigation system. The unit consists of a 10 cm PVC tube filled with soil mounted on a 5 cm resin core (both 7.8 cm diameter); the resin cores were filled with a mixture of acid washed sand and an anion and cation exchange resin Amberlite MB20 (Wulf, 1996). The resin units were installed before the flood on 12 April and removed on 9 June. Eight replicates were used under runoff irrigation and three under rainfed conditions. After the removal, the resin-sand columns were cut into three segments to verify that the analysed NO_3^- derives from the soil core, and no NO_3^- had left with the percolating water. Upon removal, soil samples were taken, extracted with 2 N KCl and analysed for NO_3^- as described above. The results of the resin core and the soil sample analyses were combined and given as net nitrification.

Soil N and EC were compared by analysis of variance using a randomized complete block design (Little and Hills, 1978). A multiple comparison of means was included with the Duncan-test. TGW, NO_3^- and net nitrification were compared by t-test.

3 Results and Discussion

3.1 Relationship between foliar nutrient content and irrigation

The microplots with good and poor growing *Sorghum* were chosen in order to assess the growth limiting nutrients in the runoff irrigation system. The biomass production, grain yield and TGW were significantly lower in the poor growing *Sorghum* than in the good growing *Sorghum* (data not shown; $p<0.05$). The N content was lower in the leaves of the poor growing *Sorghum* (Fig.1); N seems to be an important limiting nutrient for plant growth in the semiarid tropical environment. Soils in arid regions are generally low in N, and N deficiency is likely to occur (Aggarwal and Praveen-Kumar, 1994). The foliar N contents of the *Sorghum* under runoff irrigation is significantly lower than under rainfed agriculture for the poor growing stands ($p<0.01$; Fig.1).

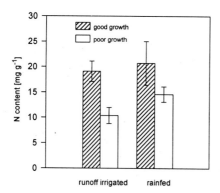
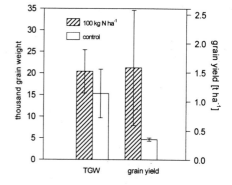

Fig.1: Foliar N content of Sorghum at heading under runoff irrigation and rain-fed agriculture in good and poor growing stands (n=5); means and standard errors.

Fig.2: Thousand grain weight (TGW) and total grain yield with and without application of 100 kg N ha^{-1} (n=3); means and standard errors.

3.2 Influence of N fertilization on crop yields

The application of 100 kg N ha^{-1} increases the grain yield by 350 % and the TGW by 33 % at an optimal water supply in 1996 (Fig.2). With 1.6 t ha^{-1} in our study, the grain yield from fertilized plots is comparable to results from runoff irrigation in Mali with 1.8 to 2.9 t ha^{-1} (Klemm, 1989) or Niger with 1.9 t ha^{-1} (Tabor, 1995). Without N fertilization, the grain yield of 1996 (control: 0.35 t ha^{-1}; Fig.2) is in the range of the results from 1995 when crops severely suffered from water stress (Lehmann et al., in press). A better water supply for *Sorghum* is not able to increase the yields without the application of N fertilizer. A sufficient N supply seems to be the prerequisite for high yields also with an ample water supply under runoff irrigation.

3.3 Influence of runoff irrigation on soil NO_3^-, N and EC

The soil NO_3^- content decreases after the flood in the runoff irrigated plots (not significant), but increases under rainfed agriculture (p<0.001; Fig. 3). As there is no vegetation at this point and thus no plant uptake (the *Sorghum* was just emerging), the lower NO_3^- content can be entirely explained by leaching and a lower nitrification. Seyfried and Rao (1991) found high N leaching losses of 57 kg N ha^{-1} 9 months^{-1} under a monocropped maize in Costa Rica. In our study, reduced nitrification seems to be an important factor for decreasing soil NO_3^- content as well, because the net nitrification is considerably lower under runoff irrigation in the first weeks after the flood (p<0.05; Fig. 3). A reduction of O_2 availability during the flood may have reduced the nitrification in the soil (Wild, 1988), although most of the particles are in the fine sand and coarse silt fraction indicating good soil aeration. Also, a lower total N content may have been the reason for the reduced mineralisation, since irrigation has been shown to reduce the N content (Fig. 4).

Runoff irrigation significantly decreased the total N content by 10 % in 0-15 cm depth during the first year after establishing the irrigation system (p<0.05; Fig. 4); the irrigation system was filled three times during this period with a total amount of 1400 mm. At the same time, the electric conductivity (EC) decreased significantly (p<0.05). Nabhan (1984) could not find a decline of soil fertility in fields irrigated with runoff water in the Sonoran desert. Our results do not confirm the findings that the flood water maintains levels of macronutrients.

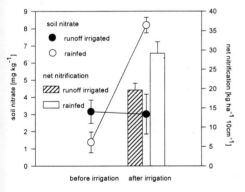

Fig. 3: Soil nitrate in 0-15 cm depth before and one month after the flood under runoff irrigation and rainfed agriculture (n=4); and net nitrification in 0-10 cm depth under runoff irrigation (n=8) and rainfed agriculture one month after the flood (n=3); means and standard errors.

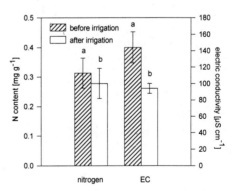

Fig. 4: Total soil N and electric conductivity (EC) before and after the flood (n=4); bars with different letters are significantly different at p<0.05; means and standard errors.

3.4 Consequences for nutrient management under runoff irrigation

Leaching of nutrients and especially of N is the biggest problem limiting crop yields in the studied irrigation system. Runoff irrigation can only be successful, if the available nutrients are kept in the system or additional nutrients can be provided. Apart from the successful use of expensive chemical fertilizers, cheap organic fertilizers such as leaf mulch or animal manure should be tested. Agroforestry may provide useful strategies to reduce nutrient leaching and increase soil N content by biological N_2 fixation.

Acknowledgements

We are indebted to the German Ministry of Economic Cooperation (BMZ) for funding this work within the German Israel Agricultural Research Agreement (GIARA), the Deutsche Gesellschaft für Technische Zusammenarbeit (GTZ) and the International Centre for Research in Agroforestry (ICRAF) for their cooperation and invaluable logistic support.

References

Aggarwal, R.K. and Praveen-Kumar (1994): Availability and management of nitrogen in soils of arid ecosystems. Annals of Arid Zone **33**: 1-18.
Evenari, M., Shanan, L. and Tadmore, H. (1968): Runoff farming in the desert. I. Experimental layout. Agron. J. **60**: 29-32.
Klemm, W. (1989): Bewässerung mit Niederschlagswasser ohne Zwischenspeicherung im Sahel. PhD Dissertation. University of Karlsruhe, Karlsruhe, Germany, 117 p.
Nabhan, G. (1984): Soil fertility renewal and water harvesting in Sonoran desert agriculture. Arid Lands Newsletter **20**: 21-38.
Lehmann, J., Droppelmann, K. and Zech, W. (in press): Runoff irrigation of crops with contrasting root and shoot development in the semi-arid North of Kenya: water depletion and above- and below-ground biomass production. J. Arid Environm.
Little, T.M. and Hills, F.J. (1978): Agricultural experimentation. John Wiley; New York, 350p.
Sanchez, P. (1976): Properties and Management of soils in the tropics. Wiley, New York, 618p.
Seyfried, M.S. and Rao, P.S.C. (1991): Nutrient leaching losses from two contrasting cropping systems in the humid tropics. Trop. Agric. (Trinidad) **68**: 9-18.
Tabor, J.A. (1995): Improving crop yields in the Sahel by means of water-harvesting. J. Arid Environm. **30**: 83-106.
Wild, A. (1988): Russell's soil conditions and plant growth. Longman, New York, 991p.
Wulf, S. (1996): Stickstoffdynamik in *runoff*-bewässerten Jungpflanzungen von *Acacia saligna* im semi-ariden Kenia. MSc Thesis, University of Bayreuth, unpublished, 59p.

Addresses of authors:
Johannes Lehmann
Frank v. Willert
Sebastian Wulf
Wolfgang Zech
Institute of Soil Science and Soil Geography
University of Bayreuth
D-95440 Bayreuth, Germany

Nitrogen Management, Soil Nitrogen Supply and Farmers' Yields in Sahelian Rice Based Irrigation Systems

M.C.S. Wopereis, C. Donovan, B. Nébié, D. Guindo, M.K. Ndiaye & S. Häfele

Summary

Studies on nitrogen management, soil nitrogen supply and corresponding farmer rice yields were conducted in two Sahelian irrigation schemes: the Kou Valley in Burkina Faso (40 farmers) and the Office du Niger in Mali (34 farmers). In Burkina Faso, rice yields ranged from 3.4 to 6.9 t ha^{-1}. N uptake from non-fertilized plots was considered a proxy for soil N supplying capacity and ranged from 32 to 83 kg N ha^{-1}. In Mali, yields ranged from 2.7 to 8.7 t ha^{-1}, whereas N uptake from non-fertilized plots varied from 23 to 91 kg N ha^{-1}. Farmer fertilizer N strategies were very diverse, and did not follow blanket fertilizer recommendations. Malian farmers applied on average twice as much N as farmers in Burkina Faso (126 kg ha^{-1} as compared to 76 kg ha^{-1}). No relationship was found between rice yield and N applied or soil N supplying capacity. Farmer and field specific N management is needed to achieve a better match between soil N supply, N applied, and crop N demand.

Keywords: Soil nitrogen, nitrogen uptake efficiency, sustainability

1 Introduction

The demand for rice in West Africa is increasing rapidly, outgrowing domestic supply, and placing great pressure on existing land under irrigated rice (520,000 ha in West Africa according to Matlon et al., 1996). Nutrient management and fertilizer use (especially N, P, and K) are far from optimal, due to a range of agronomic and socio-economic constraints (WARDA, 1996). Recent long-term monitoring studies of intensified irrigated rice production systems in Asia indicate stagnant yields and mining of soil fertility (Cassman et al., 1995; Cassman and Pingali, 1995a,b). It is likely that similar sustainability problems are occurring (or will occur in the near future) in irrigated rice-based cropping systems in West Africa. We conducted a soil fertility management survey in two key Sahelian irrigation schemes to assess yields, fertilizer strategies, and soil fertility issues, focusing on N fertilizer use.

2 Materials and methods

The surveys were implemented in irrigation schemes in the Kou Valley in Burkina Faso and in the Office du Niger in Mali during the 1995 rainy season. In Burkina Faso, 20 farmers were selected with fields located on a clay loam soil (CL); 20 other farmers were selected with fields on a sandy clay loam soil (SCL). These soil types represent, respectively, 38% and 24% of the total area under rice cropping in the Kou Valley. All 40 farmers transplanted and double-cropped (two rice crops per year on the same plot), which is the local norm. Most farmers used rice variety ITA 123. The current blanket fertilizer recommendation for the wet season in the Kou Valley is 300 kg 14-23-14 and 100 kg urea per ha. In Mali, farmers practice both single-cropping (SC) and double-cropping (DC), so we selected 18 single-cropping farmers and 16 double-cropping farmers. All sample farmers work on a 'Danga' clay loam soil, which occupies 43% of the total area under rice cropping in the Office du Niger. They transplant, using mostly the improved rice variety BG 90-2. The blanket fertilizer recommendation for Office du Niger for the wet season is 100 kg 18-46-0 and 260 kg N as urea per ha.

In both sites, we requested farmers to manage their irrigated rice plots normally, with their own choice of fertilizer applications (TP), with one exception. Each farmer established a 10 m × 10 m plot (T0) at the irrigation canal side, separated from the main field by bunds. The T0 plots had no fertilizer applications, but were otherwise managed by the farmers using the same practices as the TP plot. All cultural practices were surveyed, including input use. N uptake (standard Kjeldahl procedure), and grain and straw yields were measured at maturity for T0 and TP plots. Soil N supply was estimated from plant N uptake in T0 plots and recovery rates for applied N based upon the difference in N uptake between T0 and TP. Physiological efficiency of N was calculated as yield increase over increase in N uptake with applied inputs. Partial factor productivity of N was determined as yield in TP plots over applied N (Cassman and Pingali, 1995a, Cassman et al., 1997).

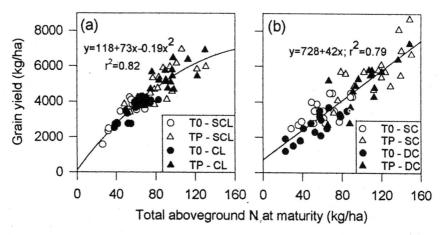

Figure 1: Relation between grain yield and total crop N in T0 and TP plots in 40 farmer fields in the Kou Valley, Burkina Faso (a) and in 34 farmer fields in the Office du Niger, Mali (b).

3 Results and discussion

The data show high variability in grain yield and total N uptake among the farmer-managed TP plots and the T0 plots, indicated in Figures 1a-b. There were strong relationships between yield and total N uptake in both Burkina Faso and Mali, indicating the importance of N as a yield limiting factor. Similar relationships were reported by Cassman et al. (1993) and Cassman et al. (1997) for irrigated rice fields in the Philippines. Yield outcomes with the farmer-managed TP plots varied greatly: from 2.7 to 8.7 t ha^{-1} in Mali and from 3.4 to 6.9 t ha^{-1} in Burkina Faso.

Site	Type	No. of cases	T0 yield (t/ha)	TP yield (t/ha)	N uptake: T0 (kg/ha)	N uptake: TP (kg/ha)	N fertilizer (kg/ha)	N recovery efficiency (%)	N physio-logical efficiency (%)	N partial factor productivity (kg/kg)
Burkina Faso	Sandy clay loam soil	20[1]	3.4 (0.73)	4.9 (1.0)	51 (12.5)	83.4 (21.1)	75.4 (25.9)	46 (25)	50 (29)	71 (22)
	Clay loam soil	20[1]	3.6 (0.59)	5.1 (0.92)	60 (12.0)	88.4 (19.4)	77.4 (29.3)	47 (32)	59 (37)	78 (35)
	Difference between types		n.s.	n.s	**	n.s.	n.s.	n.s.	n.s.	n.s.
Mali	Single cropping	18	3.5 (0.70)	5.8 (1.27)	62 (17.9)	113.9 (25.3)	142.7 (32.9)	37 (17)	51 (28)	42 (12)
	Double cropping	16	2.6 (0.77)	5.0 (1.22)	54 (19.1)	103.5 (22.0)	110.0 (18.1)	46 (18)	50 (19)	46 (12)
	Difference between types		***	**	n.s.	n.s.	***	n.s.	n.s	n.s.

Note: Source is authors' unpublished data. Figures in parentheses are standard deviations. "Differences between types" indicates the results of two-tailed t-tests of the means using the assumption of equal variances. "n.s." indicates not significant; ** indicates significant at the 5% level; *** indicates significant at the 1% level. N recovery efficiency is the increase in plant N as a percentage of applied fertilizer N. N physiological efficiency is (yield TP - yield T0)/(N uptake TP- N uptake T0). Partial factor productivity is (yield TP / applied N).

Table 1: *Descriptive statistics for yield, N uptake, N fertilizer use, N recovery efficiency, N physiological efficiency and N partial factor productivity in the Kou Valley, Burkina Faso and Office du Niger, Mali, wet season, 1995.*

[1] Recovery rate, applied N, physical efficiency, and partial factor productivity for two extreme cases were excluded.

In Burkina Faso, there were no significant differences between the two soil types for the average T0 yields nor for the average TP yields, although there was greater variability in TP yields compared to T0 yields (Table 1). The average N uptake for the T0 plots in CL soil was significantly higher than the N uptake from T0 plots in SCL soil.

In Mali, the average yield in double-cropped T0 plots was significantly lower than the average yield in single-cropped T0 plots (Table 1), although the N uptake in double-cropped T0 plots was only slightly lower than the N uptake in single-cropped T0 plots. This indicates that supply of nutrients other than N was affecting yield as observations around flowering revealed no significant weed, insect pest, or disease pressure in both surveys. Intensified rice cropping (growing two rice crops per year on the same field), therefore, negatively affected soil quality. Some farmers may have experienced yield losses due to late transplanting and subsequent cold-stress-induced spikelet sterility around flowering.

Farmer fertilizer N strategies were very diverse, with farmers in Mali on average applying twice as much as farmers in Burkina Faso (126 kg ha^{-1} compared to 76 kg ha^{-1}). If farmers wish to use applied fertilizer to substitute for a perceived lack of soil N supplying capacity, a negative relationship would be expected between applied N and soil N. Figures 2a-b show no such relationship. Cassman et al. (1993) reported similar results for irrigated rice farming in the Philippines. Either farmers did not take into account the soil N in their fertilizer N strategies, or they were unaware of the true soil N supplying capacity. The variety of fertilizer N strategies pursued indicates that farmers were not just following blanket recommendations, so that cannot be used to explain the lack of consistency between plant needs and N fertilizer applications.

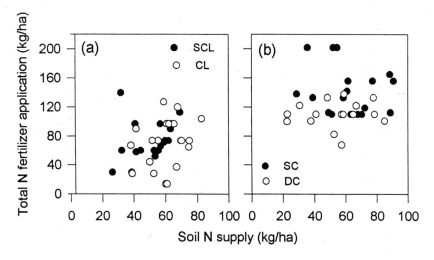

Figure 2: N fertilizer rates applied as related to soil N supply, proxied by T0 plant N uptake for 40 farmers in the Kou Valley, Burkina Faso (a), and for 34 farmers in the Office du Niger, Mali (b).

There was no significant relationship between total N application and yield (graph not shown). We found a large variability in N fertilizer recovery rates (Table 1 and Figures 3a-b). Some farmers were able to achieve over 50% recovery rates for applied N while other farmers recovered less than 5% of the applied N. There were no significant differences in average recovery rates between sandy clay loam soil and clay loam soil in Burkina Faso, nor between single-cropped and double-cropped plots in Mali, but the recovery rates were highly variable.

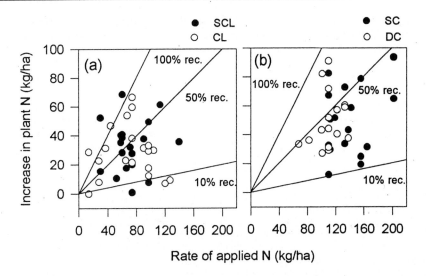

Figure 3: Recovery of applied N in 40 farmer fields in the Kou Valley, Burkina Faso (a) and in 34 farmer fields in the Office du Niger, Mali (b). Lines with 100%, 50% and 10% recovery are indicated.

The N physiological efficiency ranged from 50 to 59 kg grain / kg N, which is typical for modern rice varieties grown in the tropics (Yoshida, 1981; Kropff et al., 1994; Cassman et al., 1997). In some cases, however, very low physiological efficiencies were found, indicating that other factors than N were limiting yield. For a few farmers, extreme physiological efficiencies were found (higher than 80 kg grain / kg N). Such values seem unrealistic and are most probably due to an underestimation of straw weight at the maturity sampling. Partial factor productivity of applied N was high for farmers in Burkina Faso (around 75 kg grain kg^{-1} applied N) and rather low for farmers in Mali (around 44 kg grain kg^{-1} applied N). Burkinabé farmers get, in other words, more value out of a kg of N fertilizer than farmers in Mali. Cassman et al. (1997) reported average partial factor productivities of applied N of 61, 59 and 46 kg grain kg^{-1} N for irrigated rice farmers in the Philippines.

4 Conclusions

Grain yields were closely related to total crop N at maturity, however no relation was found between yield and total amount of N fertilizer input. The strategies of the farmers with regard to N fertilizer input varied without any apparent relationship to soil N supplying capacity. As a Malian researcher put it, 'there appears to be chaos in fertilizer application strategies'. Plant recovery of fertilizer N was highly variable, with some plots recovering less than 5% of the applied N. Thus, there is great importance to determine the causes for these differences in recovery rates.

This work provides evidence that in Mali there is a decline in soil quality under double-cropping, with lower average plant N and significantly lower average yields (both T0 and TP). In spite of this, the average fertilizer application is lower for the double-cropped plots than for the single-cropped plots, suggesting that the difference may increase over time, with important implications for intensified agriculture in the region.

The results indicate that current blanket recommendations for regions and entire countries do not respond to the variability within soil types and cultivation practices. Farmers could achieve substantial gains in efficiency of input use along with the maintenance of soil fertility if field-specific soil fertility management packages were developed. Such packages should focus on (*i*) adjusting the quantity of applied N to soil N supply and (*ii*) optimizing timing and quantity of applied N.

Acknowledgements
We thank the farmers of the Kou Valley in Burkina Faso and the Office du Niger in Mali. This study was supported by a grant of the Federal Ministry of Cooperation (BMZ) and the German Agency for Technical Cooperation (GTZ) in Germany.

References
Cassman, K.G., , M.J., Gaunt, J. and Peng, S. (1993): Nitrogen use efficiency of rice reconsidered: What are the key constraints? Plant and Soil **155/156**: 359-362.

Cassman, K.G., De Datta, S.K., Olk, D.C., Alcantara, J.M., Samson, M.I., Descalsota, J.M. and Dizon, M.A. (1995): Yield decline and the nitrogen economy of long-term experiments on continuous, irrigated rice systems in the tropics. In: R. Lal and B.A. Stewart (eds), Soil Management: Experimental Basis for Sustainability and Environmental Quality. Lewis / CRC Publishers, Boca Raton, Florida, 181-222.

Cassman, K.G. and Pingali, P.L. (1995a): Extrapolating trends from long-term experiments to farmers' fields: the case of irrigated rice systems in Asia. In: V. Barnett, R. Payne and R. Steiner (eds), 63-84. Agricultural Sustainability: Economic , Environmental and Statistical Considerations. John Wiley & Sons Ltd., London, UK.

Cassman, K.G. and Pingali, P.L. (1995b): Intensification of irrigated rice systems: learning from the past to meet future challenges. GeoJournal **35** (3): 299-305.

Cassman, K.G., Gines, G.C., Dizon, M.A., Samson, M.I., Alcantara, J.M. (1997): Nitrogen-use efficiency in tropical lowland rice systems: contributions from indigenous and applied nitrogen. Field Crops Research **47**, 1-12.

Kropff, M.J., Cassman, K.G. and van Laar, H.H. (1994): Quantitative understanding of the irrigated rice ecosystems and yield potential. In: Virmani, S.S (ed). Hybrid rice technology: new developments and future prospects. Selected papers from the International Rice Research Conference. International Rice Research Institute, P.O. Box 933, 1099 Manila, Philippines.

Matlon, P., Randolph, T. and Guei, R. (1996): Impact of rice research in West Africa. Paper prepared for the International Conference on the Impact of Rice Research, held in Bangkok, Thailand, 3-5 June 1996, 28 pp.

Yoshida, S. (1981): Fundamentals of rice crop science. International Rice Research Institute, Los Baños, Philippines, 269 pp.

WARDA (West Africa Rice Development Association) (1996): Annual Report 1995 Bouaké, Côte d'Ivoire, 134 pp.

Addresses of authors:
M.C.S. Wopereis
C. Donovan
West Africa Rice Development Association, BP 96
St. Louis, Senegal
B. Nébié
Institut d'Etudes et de Recherches Agricoles, Station de Farako-Bâ, BP 910
Bobo-Dioulasso, Burkina Faso
D. Guindo
M.K. Ndiaye
Institut d'Economie Rural, Programme Riz Irrigué, BP 07
Niono, Mali
Stephan Häfele
Institut für Bodenkunde, Universität Hamburg, Allende-Platz 2, D-20146 Hamburg, Germany

Soil Conservation on Rainfed Bench Terraces in Upland West Java, Indonesia: Towards a New Paradigm

Edi Purwanto & L.A. Bruijnzeel

Summary

Artificially bounded plots tend to underestimate runoff and sediment yield from backsloping bench terraces. An alternative approach is the so-called Natural Boundary Erosion Plot (NBEP) which comprises a single backsloping bed plus its adjacent upslope riser, with the measurements being made at the outflow point of the drain running at the foot of the riser.

Amounts of runoff and sediment from four NBEPs were determined on an event basis during two consecutive 5-month rainy seasons in rainfed upland terrain near Malangbong, West Java, Indonesia. Depending on original slope gradient, net sediment outputs ranged from 83-137 t/ha (10° slope) to 146-179 t/ha (20° slope). The erosion intensity on the terrace risers was very high (225-360 t/ha riser surface/5.5 months). Contributions from the risers made up 61-68% of the total soil loss from the two NBEPS on the gentler slope. Corresponding values for the steeper slope were all above 84%. It is probable, therefore, that the terrace risers (rather than the terrace beds) were the main producers of sediment. If stream sediment loads in the study area are to be reduced, conservation measures should aim primarily to reduce riser erosion. Proposed measures include (i) planting fodder grasses on the risers, (ii) leaving coarse harvesting residues (corn and cassava stalks) as mulch in the drainage gutter, possibly aided by (iii) the establishment of silt pits at the downstream end of the gutter.

Keywords: Bench terracing, Java, riser erosion, volcanic upland, watershed management

1 Introduction

In response to widely observed erosion and sedimentation problems, the Indonesian Government launched a major programme for the reforestation of State Forest land and the regreening of privately owned land (particularly on slopes above 50%) in 1976 (Pickering, 1979). As the programme developed, an increasingly large proportion of the budget was spent on the construction, rehabilitation and maintenance of check dams and bench terraces (Anonymous, 1989). The usefulness of the programme in reducing stream sediment loads has been questioned by some who consider that the role of geological erosion (e.g. landslides, bank erosion) has remained undervalued (e.g. Diemont et al., 1991). Similarly, attention has been drawn to the allegedly large volumes of sediment generated by rural roads and villages (Rijsdijk & Bruijnzeel, 1990, 1991). Partly as a result of such observations, watershed management in Java is currently undergoing considerable revision in terms of assumptions and approaches. However, there can be little doubt that, without quantifying the respective sediment sources in specific catchment areas, the debate will continue. Moreover, without such information it will also be impossible to evaluate the effectiveness of expensive soil conservation programmes (Bruijnzeel & Critchley, 1996).

A start was made in October 1994 to address some of these questions under the umbrella of a collaborative project of the Ministry of Forestry of the Republic of Indonesia and the Vrije Universiteit Amsterdam, in the 105-ha upland agricultural Cikumutuk catchment near Malangbong, West Java. The main objective of the research is to collect baseline data on catchment hydrological response, on-site erosion and sediment delivery for the prevailing land use and soil conservation setting (Edi Purwanto, 1996). This paper discusses results on runoff and sediment production on two bench terraced slopes of contrasting gradients during two five-month periods covering the bulk of the 1994/95 and 1995/96 rainy seasons, using an alternative technique known as the 'natural boundary erosion plot' (NBEP; Bruijnzeel & Critchley, 1996). The implications of the present findings for watershed management and soil conservation programmes in humid tropical volcanic upland terrain are discussed in detail.

2 Study area

The study catchment is situated at an elevation of 575-750 m in the headwater area of the Cimanuk River, close to the town of Malangbong, Garut District, some 55 km ESE of the city of Bandung. The area receives an average rainfall of ca. 2370 mm/yr (Malangbong, 1971-1987) distributed over about 125 rain days. Rainfall is generally <100 mm/month in June, July and August. Various types of oxisols (reddish brown latosols) have developed in the Holocene andesitic volcanic ashes and Plio/Pleistocene volcanic breccias underlying the area. Almost 80% of the catchment is covered by rainfed bench terraced fields (*tegal*) that are mostly used for the mixed cultivation of hill rice, maize and cassava (rainiest months: November-February), followed by groundnuts, maize and the remaining cassava (using residual moisture: March-June). Irrigated rice fields (*sawah*) occupy about 15%, mostly in the valley along the main stream while the remaining 6% are made up by residential areas (5%) and a single mulberry plantation (1%). Most of the homegardens and *sawah* are privately owned but for historical reasons the ownership status of the majority of the rainfed fields is unclear (Edi Purwanto 1996).

3 Plot instrumentation

Use was made of the 'natural boundary erosion plot' (NBEP) technique developed by Bruijnzeel & Critchley (1996) for the measurement of runoff and sediment yield from backsloping bench terraces. Basically, an NBEP comprises a single backsloping terrace bed plus its adjacent upslope riser, with the measurements being made at the outflow point of the toe drain running along the foot of the riser. Bruijnzeel & Critchley (1996) provide a detailed discussion of the pros and cons of the NBEP and of the traditional artificially bounded runoff plot.

Two complexes of backsloping bench terraces were selected for the construction of NBEPs in October/November 1994. Two NBEPs (nos. A and B) were built on a relatively gentle slope (original gradient 8-10°) and an additional two (nos. C and D) on a steeper slope (original gradient 20-28°). The basic characteristics of the NBEPs are listed in Table 1.

Terrace risers in the study area make up ca. 20-40% of the total area (depending on original slope gradient) on a projectional basis. Expressed as riser face area, these figures may increase to more than 60% on the steeper slopes (cf. Table 1). Adding the (projected) areas of the toe drains would raise the respective fractions by another 6-12%.

Terrace	Bed area (m²)+	Riser area (m²)#	Total plot area (m²)	Fraction occupied by riser*
A	136	46 / 34	183	0.25/0.19
B	89	32 / 26	128	0.25/0.20
C	20	27 / 19	46	0.59/0.42
D	28	44 / 29	69	0.63/0.42

+excluding toe drain.
#expressed as riser face area and projectional area, respectively.
*expressed as the ratio of actual and projected riser surface area, respectively, to total projected plot area.

Table 1: Basic characteristics of four 'natural boundary erosion plots' in the Cikumutuk catchment, West Java.

Rainfall at the two terrace pairs was measured daily using a standard raingauge placed at 1.5 m to avoid rainshadow and surface splash effects. The four terrace units were equipped with a runoff and sediment collecting device situated at the lowest point of the drainage gutter on each terrace. The NBEPs drained via a concrete 90° V-shaped inlet into a 210-litre capacity stilling basin in which coarse material could settle. The basins had a divider system made of 102 mm diameter pipes, allowing one-fifth of the excess runoff to be drained into a first collecting drum of 180 litre capacity. The latter, in turn, was equipped with a divider system made of 38 mm diameter pipes which discharged one seventh of the excess runoff into a second collecting drum of 200 litre capacity. All collecting devices were covered against rainfall.

Measurements of *runoff* volumes were made in the morning after each event using a graduated yardstick and converting waterlevels to volumes by pre-determined conversion equations. Depth-integrated 0.5 litre or 1.5 litre samples were taken from the stilling basin and, where necessary, from the collection drum(s) for the determination of the *concentration of sediment in suspension*. The water in the drums (but not in the stilling basin) was stirred thoroughly before sampling. The volume of any *coarse material* that had settled in the stilling basin was determined separately after gradually draining off the supernatant water after which a 100 cm³ core sample was taken for the volume to weight conversion. All collectors were cleaned after sampling was completed. Concentrations of sediment in suspension were measured by filtering through pre-weighed 'Melita' coffee filters, oven-drying the residues at 105 °C for 24 h and weighing to the nearest 0.01 g (0.001 g in 1995/ 96). A comparison of sediment concentrations in streamflow obtained with the paper filter method and 0.45 μm Millipore filters suggested an underestimation of 3-6%. The present data were adjusted accordingly. It is possible, however, that this represents an underestimate because of the presence of finer particles in the collected runoff compared to the streamwater.

Riser erosion was estimated using hundreds of 50 cm steel erosion pins of 5 mm diameter inserted horizontally into the riser. During the 1994/95 season, two pins were used per metre riser length, one in the topsoil section (fine sandy loam, slope 31-37°) and one in the subsoil section (fine sandy clay, slope 50-63°). All pins stuck out by 50 mm at the start of the observations (18 November). They were remeasured on 4 May 1995, i.e. after 167 days. Detailed measurements of riser dimensions (one cross section per metre riser length) were made as well on these two days, enabling the estimation of the volume of soil lost between the two dates. To improve the estimates, four erosion pins were used per metre of riser length during the 1995/96 season, which were read five times between 10 November 1995 and 24 April 1996. On three occasions, the dimensions of the risers were measured as well. In addition, a number of bench mark pins with a concrete base were installed vertically at the top of the risers, connected by horizontal wiring to facilitate the measurement of terrace dimensions. The lowermost pins stuck out 100 mm to improve the quantification of the deposition of loose material at the foot of the riser. Finally, the rate of gutter fill-up by eroded material was monitored separately in 1995/96: by means of three vertical pins per drain.

Because the precision with which the pins could be read was ±1 mm at best, this implies an uncertainty equivalent to 8-10 t/ha per reading for the prevailing bulk density of 0.8-1.0 g/cm^3.

3 Results

Runoff and sediment yield data for the four NBEPs have been processed for two consecutive five-month periods covering the main part (November-March) of the 1994/95 and 1995/96 rainy seasons. Seasonal rainfall, runoff and sediment yield totals are listed in Table 2. The amounts of rainfall recorded during the two seasons differed by about 10%, with the 1994/95 season being slightly wetter. Runoff response varied considerably between plots, even for one and the same terrace pair (A and B, with runoff coefficients of ca. 0.18 and ca. 0.27, respectively), although daily amounts of runoff were highly correlated for the two terrace pairs (values of R-squared typically around 0.80). Runoff coefficients for the two NBEPs situated on the steeper slope (C and D) were consistently higher (0.33-0.37 in 1994/95 and 0.31-0.33 in 1995/96) than for terraces A and B on the gentler slope.

Terrace	1994/95			1995/96		
	Rainfall (mm)	Runoff (mm)	Erosion (kg/m^2)	Rainfall (mm)	Runoff (mm)	Erosion (kg/m^2)
A	2196	415	8.3	2019	359	9.2
B	2196	539	13.7	2019	589	11.4
C	2213	807	14.6	1935	633	16.1
D	2213	686	17.9	1935	593	17.5

Table 2: Rainfall (mm), runoff (mm) and sediment yield (kg/m^2 projected area) totals for four 'natural boundary erosion plots' during two consecutive five-month (November - March) rainy seasons in the Cikumutuk catchment, West Java.

To examine to what extent such differences might be related to potential differences in runoff contributions by the riser and gutter areas of the respective terraces, the runoff totals were expressed as a fraction of the latter areas (on a projectional basis). If the resulting 'riser cum gutter runoff coefficient' (RGRC) would be less than unity, then this could be taken as evidence that all the runoff could be provided (at least in theory) by the riser cum gutter area. If, on the other hand, the value of the RGRC would exceed unity, then contributions from other parts would be required. On a seasonal basis, the RGRC remained below unity in all cases, except for terrace B in 1995/96, which lends some support to the contention that the majority of the runoff is indeed generated in the riser cum gutter area. Nevertheless, in about 20 and 10% of the rainfall events, runoff contributions from outside the riser cum gutter area (presumably the terrace bed) were inferred for storms larger than 40-60 mm on terraces A and B, and C and D, respectively (Edi Purwanto & Bruijnzeel, unpublished).

The sediment yield data in Table 2 illustrate the fact that surface erosion on backsloping bench terraces in humid tropical volcanic steeplands can still be unacceptably high. To obtain an idea of the source of the eroded material, a preliminary comparison has been made in Table 3 of the overall seasonal sediment losses from the respective NBEPs and the corresponding amounts of sediment generated on the risers. Although the periods of the respective observations do not match entirely (per season), such a comparison is still valid because the amounts and size distribution of rainfall during the weeks for which the NBEP output data have not yet been processed (April 1995 and 1996) were similar to those received during the 18 and 10 days in November 1994 and 1995, respectively, which were not included in the riser retreat observations.

Plot	1994/95			1995/96		
	Riser erosion (kg/m^2)	Input from riser (kg)	(%)*	Riser erosion (kg/m^2)	Input from riser (kg)	(%)*
A	22.6	1033	68	28.8	1148	67
B	32.6	1053	61	35.8	925	62
C	40.9#	1096#	165#	28.7	735	95
D	36.0	1584	128	32.1	1002	84

#believed to be overestimated
*computed as input from riser (kg) divided by sediment loss from plot (kg)

Table 3: Riser erosion intensity (kg/m^2 projected), gross sediment input from the riser in kg and as a percentage of the net overall sediment loss from four 'natural boundary erosion plots' during two consecutive rainy seasons (November-March) in the Cikumutuk catchment, West Java. Riser erosion data pertain to the periods 18 November 1994 - 4 May 1995 and 10 November 1995 - 24 April 1996.

The findings listed in Table 3 are of particular interest. Firstly, the erosion intensity on the unprotected terrace risers of the study plots was very high. Also, the total amounts of sediment produced by the risers of the two NBEPs on the steeper slope exceeded (1994/95) or approached (1995/96) overall sediment losses from the plots. Even on the gentler slope (where the risers constituted a smaller fraction of the total area (Table 1), did riser erosion still make up 61-68% of the overall sediment loss (Table 3). The quoted percentages become even more remarkable if the volumes of freshly eroded riser material that went into storage in the gutter (8-13% at A,B; 2-6% at C,D) are added (1995/96 only). Although the pin-based estimates of riser erosion are admittedly crude (±8-10 t/ha), they are nevertheless of the same order of magnitude between terraces and years (Table 3). Work is currently in progress to refine the above estimates using more sophisticated techniques to determine the volumes of runoff and sediment produced by the riser on an event basis (cf. Critchley & Bruijnzeel 1995).

5 Implications for soil conservation strategy

Although in the absence of actual observations of erosion on the terrace beds the data presented thus far cannot be taken as direct evidence that the terrace risers are the chief producers of sediment in the steeper parts of the study area and in similar rainfed bench-terraced volcanic steeplands, it is nevertheless highly probable (cf. Bruijnzeel & Critchley 1996). The implications for soil conservation strategies are potentially profound. For example, it has been stated that the costs of soil erosion in Java mainly manifest themselves as on-site losses of plant productivity (estimated by Doolette & Magrath (1990) at about 300 million US$/yr) rather than as off-site losses (such as reduced efficiency of irrigation systems and reservoirs due to siltation, estimated at 25-90 million US$/yr). At first sight, the large contrast in on-site and off-site costs associated with soil erosion would seem to justify the high investments involved in on-site mechanical conservation works like bench terracing. However, these frequently quoted cost figures may well need to be revised if the bulk of the eroded sediment derives from the terrace risers rather than the beds. After all, most terrace risers in Java are largely unproductive!

The consequence is that, if stream sediment loads in these volcanic uplands are to be reduced via soil conservation, practices should aim primarily at reducing riser erosion and trapping eroded material deposited at the foot of the riser. In view of the absence of readily available rock material to be used as riser support it is proposed here to apply a combination of (i) planting fodder grass on the risers, (ii) leaving coarse harvesting residues (corn and cassava stalks) as mulch in the drainage gutter (called *'vertical' mulching* in the Javanese literature; Edi Purwanto 1996), possibly aided by (iii) the establishment of silt pits at the end of the gutter.

Technically speaking, the *planting of grass* is probably the best way to protect the vulnerable risers of the study area against erosion. Also, there would be other benefits such as increased plant productivity per terrace unit and additional income. However, the farmers in the Cikumutuk area are generally reluctant to plant fodder grass on their land. The most frequently cited reasons include: the promotion of pests and diseases, fear of theft, fear of shortage of fodder during the dry season forcing the selling of livestock, and the general feeling that grassed risers are 'untidy' (Edi Purwanto 1996). Clearly, some major incentives are needed if the planting of fodder grasses is to become widely accepted as a standard conservation practice in the area.

Similarly, whilst the use of *surface mulch* has been demonstrated to reduce surface erosion as well as boost plant productivity in many tropical areas (Young, 1989; Hudson, 1995), the use of crop residues as a mulch on terrace beds is hardly practised in the study area. This is mainly because most of the leafy crop residues are already used as fodder or composting materials while in addition a thick layer of mulch presents problems during the planting and sowing of the mixed crops (Edi Purwanto 1996). However, coarse crop residues such as corn and cassava stalks (i.e. the ones not used for the next rotation of cassava) often remain on-site, usually in stacks along field boundaries. When such coarse crop residues were arranged along the contour (at a rate of 3 t/ha and at 11 m intervals) on a 15% slope in the lowlands of West Java, plot-based sediment losses were reduced considerably, even more so than in the case of ridge terracing at equally large intervals (Kamir, 1995).

The concept of 'vertical mulching'

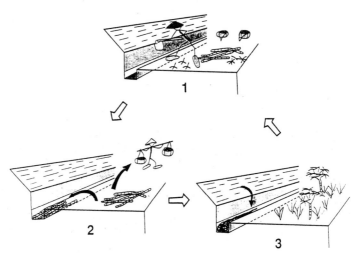

Figure 1: The 'vertical mulching' technique on backsloping bench terraces (modified from Edi Purwanto 1996). Stage 1: Crop harvest and deepening of the drain; Stage 2: Distributing the dug up soil over the terrace bed while maintaining the reverse slope; filling the gutter with coarse harvesting residues; Stage 3: Decomposition of harvesting residues while new crops are growing.

A variant of the 'vertical' mulching technique of Kamir (1995) for use on backsloping bench terraces has been proposed by Edi Purwanto (1996) and is illustrated in Figure 1. The method is both simple and cheap. In addition, levels of organic matter on the terrace beds are maintained while at the same time runoff and sediment coming from both the terrace riser and bed are trapped. However, if the 'vertical' mulching technique is to be effective, the corn stalks and other coarse material harvested during the second half of the rainy season should especially be used to trap material during the more critical initial part of the rainy season. Finally, the *traditional silt pits at the outlet of the drainage gutters* (typically ca. 0.5 m deep with a surface area of about 1 m^2) could be added to the above two measures as an extra precaution. Again, the technique is simple and only requires the regular cleaning of the pits. Although the majority of the local farmers appreciated the function of the pits, only 20% of the people had actually constructed them on their lands (Edi Purwanto 1996).

Now that baseline conditions of hillslope runoff and erosion in the Cikumutuk area have been determined with sufficient accuracy, a sound basis is available for the evaluation of the effects of specific soil conservation methods on on-site erosion or, depending on the scale at which such measures would be applied, off-site sediment yield. An experiment involving the planting of *Paspalum conjugatum* on two NBEPs was initiated at the start of the 1996/97 rainy season.

Acknowledgements

We thank Ir Soetino Wibowo (Head of the Regional Centre for Land Rehabilitation and Soil Conservation, Bandung, and Ir Yuliarto (Head, Sub-Centre for LR & SC, Garut) for their continued support to the project. The data that form the backbone of this article were painstakingly collected under often trying conditions. Our sincere thanks go to the project's field assistants - Bapak[2] Aceng, Cucu, Kasdi, Sony H. and Tata - and the students from VUA that participated, notably Peter Bakker, Bart van Eijk, Paul van 't Hoff, Hero Holwerda, Olaf van der Kolk, and Caspar Schoorl. Special thanks are due to Ir Beben Chandra, Albert van Dijk and Edo Jubaedah. Mr Henry Sion provided the diagrams.

References

Anonymous (1989): Cisadane-Cimanuk Integrated Water Resources Development (BTA-155). Volume IX-A (Erosion), Pusat LitBang Pengairan, Bandung, and Delft Hydraulics, Delft.
Bruijnzeel, L.A. & Critchley, W.R.S. (1996): A new approach towards the quantification of runoff and eroded sediment from bench terraces in humid tropical steeplands and its application in Sout-Central Java, Indonesia. In: M.G. Anderson & S.M. Brooks (eds.). Advances in Hillslope Processes. J. Wiley, New York, 921-937.
Critchley, W.R.S. and Bruijnzeel, L.A. (1995): Terrace risers: erosion control or sediment source? In: J.B. Singh & M.J. Haigh. (Eds.). Headwater Control. Balkema, Rotterdam, 529-541.
Diemont, W.H., Smiet, A.C. and Nurdin (1991): Rethinking erosion on Java. Netherlands Journal of Agricultural Science **39**: 213-224.
Doolette, J.B. and Magrath, W.B. (1990): Watershed Development in Asia: Strategies and Technologies. World Bank Technical Paper No. 127.
Edi Purwanto (1996): Upland Agriculture in Perspective. Research Methodology and Soil Conservation Strategy. Cikumutuk Project Working Paper No. 1, Malangbong.
Hudson, N.W. (1995): Soil Conservation, 3rd ed., Iowa State University Press, Ankeny, Iowa.
Kamir R. Brata (1995): The effectiveness of vertical mulching as a soil and water conservation measure in upland agriculture on the latosol of Darmaga. Indonesian Journal of Agricultural Science **5**: 13-19 (in Indonesian, with an abstract in English).
Pickering, A.K. (1979): Soil conservation and rural institutions in Java. IDS Bulletin **10**: 60-65.

Rijsdijk, A. & Bruijnzeel, L.A. (1990, 1991): Erosion, Sediment Yield and Land-Use Patterns in the Upper Konto Watershed, East Java, Indonesia. Volumes I-III. Konto River Project Communication no. 18. Konto River Project, Malang, Indonesia.

Young, A. (1989): Agroforestry for Soil Conservation. CAB International, Wallingford, U.K.

Addresses of authors:
Edi Purwanto
School of Environmental Conservation Management
Bogor, Indonesia
and
Faculty of Earth Sciences
Vrike Universiteit
De Boelelaan 1085
1081 HV Amsterdam, The Netherlands
L.A. Bruijnzeel (for correspondence)
Faculty of Earth Sciences
Vrije Universiteit
De Boelelaan 1085
1081 HV Amsterdam, The Netherlands

Impact of Soil and Water Conservation Practices on Stream Flows in Citere Catchment, West Java, Indonesia

N. Sinukaban, H. Pawitan, S. Arsyad & J. Armstrong

Summary
Established agricultural practice in upland vegetable growing areas in Indonesia includes cultivation up and down the slope and burning of crop residues. Combined with high rainfall intensities, these practices have contributed to high runoff and severe erosion. Collaborative research between Indonesia and Australia has been carried out to study the impact of applying soil and water conservation practices (SWCP) in vegetable growing areas on stream flows in a small upland catchment (10.2 ha) in West Java. The type of SWCP employed were planting on the ridges across the slope or planting on ridges constructed at 15 to 30 degrees with respect to the contour.

Hydrograph analysis from three years of observation appear to support the premise that the adoption of SWCP results in a significant reduction in quick flow, while substantially increasing and prolonging low flows. Corresponding to the introduction of SWCP, on 31% of the catchment the magnitude of quick flow was reduced from 72 % to 39 % while low flows increased from 28 % 61 % of total water yield. The runoff coefficient decreased from 35 % to 24 % of annual rainfall.

The trend indicated by these data may have positive benefits for upland farmers, as a result of improved moisture status during the early part of the dry season, as well as improving the water supply for downstream areas during the dry season.

Keywords: Rainfall, peak flow, baseflow, quick flow, low flow, and water yield.

1 Intoduction

Stream flow in catchments is a variable that can range widely. It is highly correlated with rainfall and antecedent soil moisture. In tropical areas flow in small agricultural catchments often ceases in the dry season, while in the wet season large flows are recorded following heavy rainfalls. Other factors affecting stream flow are terrain configuration, soil properties (texture, structure, depth, infiltration rate), the level of surface storage, and vegetative cover (Kostadinov and Mitrovic, 1994; Viessman et al., 1977).

In West Java, Indonesia, there is some anecdotal evidence, based on observations of stream levels, to suggest that peak flows have increased, while dry season flows have decreased. It has been speculated that this is a result of the very intensive agricultural development, and consequent cultivation of steep land, that has occurred in these areas.

Land use and land management can significantly affect catchment water yield and peak flows from a given catchment, principally by the modification of the infiltration rate and surface storage (Raudkivi, 1979; Viessman et al., 1977).

High levels of vegetative cover in agricultural areas protect the soil from raindrop impact and create conditions favourable for infiltration. At the same time vigorous growth increases evapotranspiration rates and depletes soil moisture, and therefore may reduce the volume of groundwater recharge, and hence catchment baseflow levels (Raudkivi, 1979). From a study conducted in Queensland, Australia, Wockner and Freebairn (1991) and Barker (1989) reported a reduction in mean annual runoff of up to 30 % resulting from optimal management of residues. High levels of surface cover also increase surface roughness and reduce flow velocities. Titmarsh et al. (1991) in a Queensland study indicated that peak discharge could be reduced by up to 60% if strip cropping was adopted in the catchment

Changes in management practices which modify surface detention storage have been shown to significantly affect runoff. Practices such as cross-slope cultivation and tied ridging (basin tillage), described by McFarlane et al. (1991), can trap significant amounts of runoff on slopes. Sinukaban et al. (1994) demonstrated that surface runoff from steeply sloping plots, located in the catchment described in this paper, could be reduced by up to 50 % if the orientation of ridges used for growing vegetables was changed from up and down the slope to across the slope.

This paper presents three years of stream flow data from a small upland catchment used for vegetable production in West Java where SWCP have been partly (only on 31 %) introduced. The apparent trend in the data is discussed.

2 Materials and methods

This research was carried out in three consecutive growing seasons (years), from 1992 to 1995, in a small catchment located at Citere (latitude 7°29'S, longitude 107°35'E), Pangalengan subdistrict, West Java, Indonesia.

The area of the catchment is 10.2 ha. Annual rainfall at the site ranges from 1700 mm to 2400 mm, with approximately 90 % falling in the wet season (main growing season). Elevation of the catchment ranges from 1430 to 1480 m.a.s.l. The soils in the catchment are deep well-structured clay loams of volcanic origin (Andosols). Infiltration rates are high and drainage is very good. The topography is hilly with slope ranging from 3 to 50 %.

The main landuse in the catchment is agriculture, principally for vegetable crops; about 10 % of the catchment is used for housing. The agricultural portion of the catchment is subdivided into blocks which are owned and cultivated by different farmers or groups of farmers. The size of the blocks ranges from 0.01 to 2.5 ha. Each block can be planted with different crops in any given season. Two crops are usually grown in the wet season while a third, opportunity crop, is sometimes grown in the dry season.

SWCP were implemented in the catchment as follows: first year (Oct 92 to Sep 93), no SWCP employed; second year (Oct 93 to Sep 94), SWCP in 2.15 ha (21 per cent of catchment), third year (Oct 94 to May 95) SWCP in 3.1 ha (31 % of catchment) i.e. 69 % of the catchment did not receive SWCP that year.

The type of SWCP employed on the 31 % of catchment were planting on ridges across the slope (cross-slope working) or planting on ridges constructed at 15 to 30 degrees with respect to the contour. The latter was a compromise introduced as some farmers are very resistant to tilling across the slope. All crop residues were used as mulch or incorporated into the soils during cultivation. The height of ridges ranged from 11 to 19 cm.

The major cropping pattern employed was potatoes or cabbages, followed by a second crop of cabbages or potatoes, respectively. In a small portion of the catchment other vegetable crops were grown such as tomatoes, chilli, carrots, onions, and corn. Corn or tomatoes were mostly grown as a third crop. Cultivation was carried out by manual hoeing

Rainfall was recorded using a tipping bucket raingauge with a Wesdata^R datalogger and a manual raingauge. Runoff or stream flow from the catchment was measured using an automatic AUS1[R] chart water level recorder which was installed on a small cut-throat flume (40 cm width and 80 cm height). The level of flowing water was also manually measured each day. Flow rates were calculated from water levels using the following rating equation (checked by laboratory calibration): $Q = 0.8733 \, H^*1.72$; where Q = flow rate (m^3/s), and H = flow level (m).

The recorded water levels were used to determine daily flow discharge and annual water yield. The method described by Hewlett and Hibbert (1967) was used to separate the delayed flow from the total flow.

3 Results and discussion

3.1 Rainfall

Annual rainfall for the 1992/93 season was 2173 mm, 2020 mm for 1993/94 and for 1994/95 (up to May) 1910 mm. Monthly rainfall (Fig. 1) ranged from 1 mm to 360 mm in 1992/1993, from 0 to 356 mm in 1993/1994, and from 0 to 310 mm in 1994/1995. The wet season, characterised by total rainfall more than 100 mm per month occurred between November to April in 1992/1993 and from November to May, respectively, in 1993/1994 and 1994/1995. The dry season with less than 60 mm rainfall per month occured from June to August 1992/1993 and from June to October in both 1993/1994 and 1994/1995.

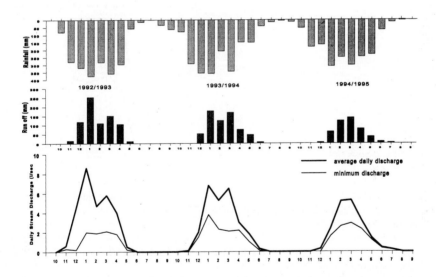

Fig. 1: Total monthly rainfall and runoff, and average and minimum (base flow) discharge. October 1992 - September 1995

Approximately 88 - 92 % of the annual rainfall fell during the wet seasons; 1916 mm in 1992/1993, 1860 mm in 1993/1994 and 1701 mm in 1994/1995. The decrease in total rainfall during the wet season was 3 per cent in 1993/94 and 11 per cent in 1994/95. Rainfall in 1994/1995 was therefore slightly below average.

3.2 Stream flow characteristics

Monthly runoff for the three seasons is shown in Fig. 1. There is a gradual decline in average discharge which could simply reflect differences in rainfall amount and intensity between the three seasons. However, there also seems to be a slight upward trend in baseflow levels which suggests that the introduction of SWCP has resulted in improved infiltration (and less runoff).

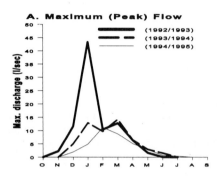
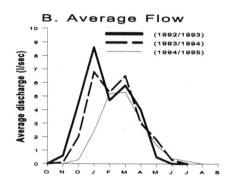
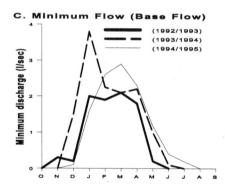

Fig. 2: Comparison of maximum average and minimum flows for three 'wet' seasons.

Maximum, average and minimum daily flows in each month for the three wet seasons are compared in Fig. 2. The peak flow recorded in the 1992/93 season was 43.3 l/sec, while the peak flows recorded following the introduction of SWCP were 14.3 l/sec and 11.2 l/sec in 1993/1994 and 1994/1995, respectively (Fig. 2a). The highest average monthly flow also decreased progressively from 8.6 l/sec before SWCP were applied, to 6.8 l/sec and 5.3 l/sec in 1993/1994 and 1994/1995, respectively (Fig. 2b).

The decrease in high flows during the wet seasons was partly due to the decrease in rainfall in 1993/1994 and 1994/1995. However during the wettest months (Dec, Jan, and Feb), where there was little difference in total rainfall, the maximum and average flows were reduced following the introduction of SWCP. On the other hand, the minimum flow (base flow) increased after applying SWCP even only on 31 % of the catchment and the flow was prolonged (Fig. 2c). The minimum flow was prolonged from June in 1993 (before SWCP was applied) to July in 1994 and 1995 (after SWCP was applied). The highest minimum flow was increased from 2.1 l/sec in 1992/1993 (before SWCP was applied) to 3.8 l/sec and 2.9 l/sec in 1993/94 and 1994/1995 respectively (after SWCP was applied) (Fig. 2c)

3.3 Annual water yield

The annual water yield (runoff) was calculated from daily stream flow. Annual water yield from the catchment decreased progressively through the monitoring period; from 766.9 mm in 1992/1993 to 657.5 mm in 1993/1994 and 467.0 mm in 1994/1995 (Table 1).

Annual runoff coefficients (proportion of rainfall leaving the catchment as runoff) also decreased progressively as the use of SWCP increased. Runoff decreased from 35 % before SWCP were applied, to 33 and 25 % after SWCP were applied on only 21 and 31 % of the catchment, respectively (Table 1).

Table 1 also shows that quick flow decreased dramatically from 72 % of annual water yield before SWCP to 49 % and 39 % as the use of SWCP increased progressively. There was a corresponding increase in delayed flow. The magnitude of delayed flow increased from 217 mm (28 % of annual yield) before SWCP to 334 mm (51 % of annual yield) and 286 mm (61 % of annual yield) as the use of SWCP increased to 21 and 31 % of the catchment, respectively.

Water Yield Characteristics		Without SWCP	With SWCP	
		1992/1993	1993/1994	1994/1995
Rainfall	(mm)	2173	2018	1901
Annual water yield	(mm)	767	658	467
Runoff Coefficient	(%)	35	33	25
Quick flow	(mm)	550	324	181
Percent of water yield	(%)	72	49	39
Delayed flow	(mm)	217	334	286
Percent water yield	(%)	28	51	61
Annual infiltration and evapotranspiration	(mm)	1406	1361	1434
Percent of rainfall	(%)	65	67	75

Table 1: Effect of SWCP on annual water yield in Citere catchment

Although it is recognised that further monitoring and analyses are required to confirm the above trends, the differences observed following the introduction of SWCP into the catchment, particularly the reduction in quickflow and the corresponding rise in delayed flow, suggest that the SWCP's employed may be partly responsible for these differences. The amount of annual infiltration plus evapotranspiration (the difference between annual rainfall and water yield) was slightly increased from 1406 mm in 1992/1993 to 1434 mm in 1994/1995 even though total rainfall decreased (Table 1). Since there was no significant change in land use pattern, cropping pattern and weather during these three years, it seem that there was constant evapotranspiration. Therefore, the increase in annual infiltration plus evapotranspiration in 1994/1995 is likely due to the increase of annual infiltration. The ability of the applied SWCP even on only 31 % of the catchment could well account for part of the reduction in quick flow for this well-structured, well-drained soil. Improved infiltration could be attributed to the ability of ridges and furrows created by SWCP to retain overland flow. The ridges and furrows act as micro-dams, retaining flow and allowing more time for water to infiltrate into the soil. This may also partly account for the prolonged base flow following the introduction of SWCP.

The trends discussed have important implications for agricultural catchment management planning in the region, as well as having benefits for local farmers. Reductions in quick flow will reduce the soil erosion rate and, in turn, save nutrients and maintain productivity. The benefit of these practices as erosion control measures and to sustain productivity has already been established

(Sinukaban et al., 1994). Secondly, by increasing soil moisture and increasing and distributing low flow further into the dry season, the SWCP will improve the chances of a succesful third crop, as well as improving downstream water supply (for drinking water and irrigation) during the early part of the dry season.

However, since established methods of cultivation are well-entrenched, long-term on-going extension activities are required to convince farmers of the benefits of applying SWCP.

4 Conclusions

Data from three years of observations indicate that cross-slope cultivation and residue incorporation and mulching can reduce total runoff (reduced runoff coefficients) and quickflow, while increasing and prolonging delayed flow in this catchment. Potentially this has important implications for farmers and catchment managers in the region. Further monitoring is required to confirm the trends discussed.

Acknowledgments

Financial assistance for this study was provided by the Australian Centre for International Agricultural Research (ACIAR). Grateful acknowledgement is extended to the following for their assistance in the field and laboratory: Adi Jaya, Irwan Sukri Banuwa, David Guluda, Jemmy Rompas, Yayat Hidayat, Ali Yasmin Wiralaga, and Wawan.

References

Barker, P. (1989): Runoff and soil loss from conventionally tilled and direct drilled wheat in southern New South Wales. Aust. J. Soil and Water Cons. 2 (1), 28-31.

Hewlett, J.D. and Hibbert, A.R. (1967): Factors affecting the response of small watersheds to precipitation in humid areas. In International Symposium on Forest Hydrology. W.E. Sopper and H.W. Hull (Eds.). Pergamon Press. 275-290.

Kostadinov, S.C. and Mitrovic, S.S. (1994): Effect of forest cover on the stream flow from small watersheds. J. Soil Water Cons. **49(4):** 382-386.

McFarlane, D., Delroy, N. and McKissock, I. (1991): Water erosion of potato land in Western Australia; II. Methods of reducing soil loss. Aust. J. Soil and Water Cons. 4 (2), 26-32.

Raudkivi, A.J. (1979): Hydrology - An Advanced Introduction to Hydrological Processes and Modelling. Pergamon Press. 341-343.

Sinukaban, N., Pawitan, H., Arsyad, S., Armstrong, J. and Nethery, M. (1994): Effect of soil conservation practices and slope lengths on runoff, soil loss, and yield of vegetables in West Java, Indonesia. Aust. J.Soil Water Cons. 7 (3), 25-29.

Titmarsh, G., Connolly, R. and McLatchey, J. (1991): Predicting the influence of cropping patterns on discharges from a distributed stream in a brigalow landscape. Aust. J.Soil Water Cons. 4 (4),30-37.

Viessman, W., Knapp, J.W., Lewis, G.L. and Harbaugh, T.E. (1977): Introduction to Hydrology. Harper and Row. 564-565.

Wockner, G.W. and Freebairn, D.M. (1991): Water balance and erosion study on the eastern Darling Downs - An update. Aust. J. Soil & Water Conserv. 4: 41-47.

Addresses of authors:
Naik Sinukaban
Hidayat Pawitan
Sitanala Arayad
Bogor Agricultural University, Jurusan Tanah - IPB JI.
16144 Bogor, Indonesia
Jim Armstrong
Department of Land and Water Conservation, Goulburn, Australia

Response to Conservation Measures in a Red Soil Watershed in a Semi Arid Region of South India

R.N. Adhikari, M.S. Rama Mohan Rao, S. Chittaranjan, A.K. Srivastava, M. Padmalah, A. Raizada & B.S. Thippannavar

Summary

A watershed management approach to resource conservation and economic productivity was initiated in 1976 and monitored as of 1979 for its impact on resource conservation and productivity increase in an alfisol watershed located in a semi-arid region of south India. Conservation measures viz., graded *bunds* in arable lands, contour trenches in non-arable lands combined with gully treatments, reduced annual runoff from 17 to 2.3 per cent, peak rate of runoff from 15 to 4.4.m^3/sec and soil loss to 1.1 t/ha/year. Conservation of rain water, coupled with improved practices, increased food production from 42 to 78 t/year over 103 ha under dryland conditions. As a result of conservation and gully control measures, a surface water storage of 6110 m^3 was created across the watershed which resulted in 14% of the annual rainfall becoming ground water recharge. This is reflected in the increase in irrigated area under the wells located in the watershed from 19 ha in 1985-86 to 56 ha in 1993-94. The average cost of such development worked out to rupees 752/- per ha at the 1979 price level. An economic evaluation of the overall project gave a cost-benefit ratio of 1.53.

Keywords: Watershed management, soil and water conservation, runoff, soil loss.

1 Introduction

Red soils in India (mainly alfisols) and their associates occupy an area of 117.2 m.ha (Anonymous, 1988). A major portion of these soils occur in south India under semi-arid conditions. These soils have a poor water holding capacity and enfeebled fertility due to the absence of soil conservation measures. These areas suffer continuously from problems of runoff and soil erosion aggravated by undulating topography, making crop production very risky. Annual soil losses from these areas are estimated at 2.4 to 3.6 t/ha/year (Dhruvanarayana and Rambabu, 1983) and 5 to 9.5 t/ha (Krishnamurthy, 1971), while runoff from these soils varied from 9 to 33 per cent (Katyal et al., 1994). Though the problem is serious, often no attention is paid to control soil erosion as the amounts lost with each rain is almost imperceptible, about 15 t of soil from a hectare of land removes only 1.0 mm of soil from the surface.

The average reduction in grain yield in red soils of the semi-arid region due to the loss of 1.0 cm of soil was observed to be 140, 53 and 50 kg/ha with respect to sorghum, pearl millet and castor, respectively, with production losses being much higher in years of unfavourable rainfall (Vittal et al., 1990). Such decline in yields undermines our ability to increase production and alleviate poverty. Hence the paradigm for agriculture in the mid 1960's of meeting the challenge of increasing food and fibre production must now change to one of sustainability and respect for natural resources. To achieve this, development of new agricultural technologies and a better appreciation of the existing but under-utilized knowledge of resource management will be crucial in meeting the ecological needs and in meeting the anticipated food demands of the growing population.

The research programmes carried out by the Central Soil and Water Conservation Research and Training Institute, (CSWCRTI) and its regional research centres have shown that integrated development through scientific land use planning adopting a watershed approach is one sure way to conserve resources and improve productivity. This approach was tested at the G.R.Halli watershed which represents semi-arid conditions of the Deccan plateau to demonstrate the economic benefits, apart from serving as a model. The results are presented in this paper.

2 Study area and methodology

The G.R.Halli watershed has an area of 314 ha (lat. 14° 17' 30" N, Long. 76° 23' 55" E, at an altitude of 724 m above msl) and is located 6 km north of Chitradurga, Karnataka State. It lies in a semi-arid region, where the average annual rainfall in the watershed area is 612 mm. The slope angle varies from 3 to 5% in agricultural and marginal lands, and between 10 and 20% in hill areas. Soils are red loam, varying in depth from 7.5 to 45 cm. Out of 314 ha, only 158 ha were arable and the remaining were non-arable and totally denuded (Anonymous, 1981).

With the coordination of the Drought Prone Area Programme authorities, a master plan was prepared envisaging resource development and restoring the ecological balance after matching the peoples' needs in relation to resource potential determined through necessary physical and biological surveys. The conservation treatments were implemented by the different line departments of the Karnataka State government during the period 1977 to 1979 under the technical supervision and guidance of CSWCRTI, Research Centre, Bellary.

The programme consisted of providing a diversion drain at the foot hills to drain excess rain water at a non-erosive velocity into the natural drains, and providing graded *bunds* of 0.75 m^2 cross section at 1 m vertical interval in arable lands over an area of 116 ha, connected to main gullies through waterways. Land smoothening, contour cultivation, levelling wherever feasible, and dead furrows formed inter-terrace measures in arable lands for *in situ* conservation of rain water. Checkdams, *nala bund*, drop structures, in addition to vegetative checks at appropriate places, were provided for conserving excess runoff to enhance ground water recharge and to stabilize gullies along the drainage course. In the case of non-arable lands, staggered contour trenches of $4.0 \times 0.5 \times 0.5$ m with 0.25 m staggered equilizers at 4 m apart at 10 m horizontal interval and pits of $0.5 \times 0.5 \times 0.5$ m inbetween contour trenches at 4 m distance were formed for rain water collection. Planting was done on the trench mounds and in pits with suitable tree species. Impact of conservation measures on hydrological, ecological and economic growth was studied regularly by following standard procedures.

3 Results and discussion

The resource conservation programme, as envisaged in the master plan, was implemented for achieving sustainability during the period 1976-79 by investing Rs.2,36,087, consisting of Rs. 16,550 for *bund*ing in the agricultural sector and Rs. 2,19,537 for gully control and afforestation, in forest sector. The benefits of such investments have been monitored since 1979 after adopting recommended soil and water conservation measures along with crop improvement technology and pasture development.

3.1 Rainfall, runoff and water balance

Rainfall in the watershed during the study period (1976-94) varied from 353 mm in 1976 to 832 mm in 1977, with an annual average of 570 mm. The month of October recorded the highest average rainfall of 110 mm followed by 93 mm in September, 71 mm in May, with the lowest rainfall of 0.9 mm occurring in February. The period between May to September recorded 351 mm which was 62 per cent of the annual rainfall, while October and November contributed 165 mm which was 29 per cent of the total.

The distribution pattern suggests the possibility of increasing cropping intensity from 100 to 200%. The crop growing period could be extended from May to December with appropriate conservation measures, as compared to the existing system of mono-cropping. The effectiveness of the conservation measures in reducing runoff and erosion was monitored through the installation of stage level recorders in non-arable areas having a catchment of 120 ha, and compared with the runoff recorded from another micro-watershed having a 7.0 ha catchment receiving no conservation treatment.

The runoff from the treated area was 2.3 per cent of the annual rainfall while that of the untreated area was 17 per cent. The soil loss from the treated catchment was 1.1 t/ha while the soil loss from the untreated area could not be recorded as the soils were gravelly. The low runoff from the treated watershed is due to staggered contour trenches which would have trapped runoff water up to 132 m^3 during each event depending on the intensity and frequency of rainfall (Rama Mohan Rao and Chowdary, 1994). Maximum peak rate of discharge was observed to be 15 m^3/sec for the untreated watershed while it was only 4.4 m^3/sec for the treated area.

3.2 Land use pattern

Data on changes in land use following conservation measures are presented in Table 1. Provision of a diversion drain at the foot hills helped to bring an area of 17 ha under productive utilization. As a result of a significant increase in *in situ* moisture and ground water availability, the number of bores and in-well bores increased from 9 in 1985-86 to 19 in 1993-94, increasing the gross irrigated area under wells from 19 ha to 56 ha. Soil and water conservation measures based on scientific principles have been reported to improve profile moisture storage and ground water resources (Rama Mohan Rao & Chowdary, 1994; Sarkar, 1994; Das, 1996).

Land use	Pre-project (1976-77)	Post-project (Average of 15 years)
Rainfed	113	122
Irrigated	17	24
Cultivable waste	22	6
Village/School building	4	4
	156	156

Table 1: Land use changes (ha)

3.3 Production changes

The bio-mass production from non-arable lands increased from almost nothing to 6000 kg/ha/year initially, but subsequently dropped to 855 kg/ha/year due to afforestation and their shade effect, and fuel production to 9.22 t/ha/10 year rotation (Pathak and Roy, 1994). In case of arable lands the yields of major crops.such as sorghum, pearl millet, ragi (*Eluerine coracana*) and setaria (*Setaria italica*) within the watershed increased over the pre-project period to 78, 50, 37 and 47%, respectively, while the yields of the above crops from inside the watershed registered a surplus of 21, 15, 31 and 17%, respectively, to yields outside the watershed.

The evaluation period was classified into normal and below normal years and crop yields were compared with those outside the watershed. The yields of sorghum, ragi, pearl millet and setaria during below normal years increased by 21, 27, 32 and 19% respectively, compared to 21, 5, 30 and 15% in the outside watershed in normal years. This clearly reflects the impact of soil and water conservation measures. The yield increase due to resource conservation measures as influenced by rainfall is presented in Table 2. Across the seasons, total production from rainfed and irrigated agriculture increased by more than double when compared to production during the pre-project period. This suggests the success of the watershed programme in improving the economic well-being of the people (Table 3). Due to higher production, the use of nutrients increased from 140 kg N, 129 kg P_2O_5 and 73 kg K_2O during the pre-project period, to 2770 kg N, 4940 kg P_2O_5 and 180 kg K_2O during 1993-94.

	Normal season			Below normal season			Average (1980 - 1994)		
	With graded bund	Without graded bund	% increase with bund	With graded bund	Without graded bund	% increase with bund	With graded bund	Without graded bund	% increase with bund
Jowar CSH5	874	722	21	865	717	21	869	719	21
Jowar local	533	456	17	460	400	16	496	428	16
Ragi	646	499	30	470	355	32	558	427	31
Bajra	465	441	5	462	364	27	463	402	15
Setaria	472	411	15	422	356	19	448	383	17
Ground nut	822	625	32	521	458	14	671	542	24
Sunflower	326	286	14	257	234	10	289	258	12

Table 2: *Impact of conservation measures on grain yield (kg/ha)*

	Cropped area		Food production	
Land use	Pre-project	As of 1994	Pre-project	As of 1994
Rainfed	113	103	42	87
Irrigated	17	55	13	117
Total	130	158	55	204

Table 3: Cropped area (ha) and food production (tonnes)

3.4 Socio economic impact

The income levels and distribution of income presented in Table 4 indicate the economic well-being of the people as a result of resource conservation. The per capita income increased by 350% over the pre-project period value of Rs.252/-.

	1994-95			Pre-project period (1976-77)		
Size group	Rainfed	Irrigated	Overall	Rainfed	Irrigated	Overall
<1.0ha	2311	20440	8079	671	6561	2545
1-2 ha	7281	29316	13577	1142	8209	3161
2-4 ha	8303	34737	15646	1984	10491	4347
> 4.0 ha	14860	137201	88265	7089	11907	9980
Overall	6359	41865	17546	1457	8686	3734

(Values reported are at 94-95 price level)

Table 4: Income (Rs) level and distribution

3.5 Economic evaluation

An economic evaluation of the watershed was carried out by comparing the actual costs and accomplishments in relation to the original plan. The present cost benefit ratio worked out to be 1.68 and 1.44, respectively, for agriculture and forest sectors for a 20 year project life and a 15% discount rate. The over all cost benefit ratio of the project was calculated to be 1.53. This clearly indicates that the rehabilitation of semi-arid regions with soil and water conservation technologies at a watershed scale can be highly economical.

References

Anonymous (1981): Watershed planning and development for red soils G.R.Halli, Chitradurga. Jt.pub. of CS&WCR&TI.R.C., Bellary and DPAP, Govt.of Karnataka, Chitradurga.

Anonymous (1988): Indian Agric. in Brief. Pub. by Department of Economics and Statistics. Min. of Agri. Government of India, New Delhi, 22nd edition.

Das, D.C. (1996): Critical appraisal of soil and water conservation measures undertaken to combat land degradation. In: Soil Management in relation to land degradation and environment. Bull no.17 Indian. Soc. Soil Sci. New Delhi, 27-31.

Dhruvanarayana, V.V. and Rambabu (1983): Estimation of soil erosion in India. J. Irrg. and Drainage Engineering, ASCE: **109**: 419-433.

Katyal, J.C., Shrinivas Sharma., S.K. Das and Mishra, P.K. (1994): Moisture conservation and rainwater management in red soil regions. Indian. J. Soil Conserv. **22**: 15-25.

Krishnamurthy, K. (1971): Dryland farming problems in Mysore. UAS Research series, No.12, Hebbal, Bangalore.

Pathak, P.S. and Roy, M.M. (1994): Silvipastoral system of production. Research bull. IGFRI, Jhansi. p.53

Rama Mohan Rao, M.S. and Narayana Chowdary, P. (1994): Watershed management for sustainable development (Chinnatekur watershed - a case study). Jt .Pub. of CS&WCR&TI.,Research Centre, Bellary and DRDA, Kurnool, Govt.of A.P. pp.69.

Sarkar, A.N. (1994): Rainfed farming and watershed development; a holistic approach for sustainability. Indian Fmg. **44**: 30-33.

Vittal, K.P.R., Vijayalakshmi, K. and Bhaskar Rao, M. (1990): The effect of cumulative erosion and rainfall on sorghum, pearlmillet and castor bean yields under dry farming conditions in Andhra Pradesh. India. Experimental Agric. **26**: 429-439.

Foot Note:
Nala	= Drain
Bund	= Small earthern dam (less than a meter height) on agricultural land.
Nalabund	= Small earthern dam (around 3 m height) across the drain.
One Dollar	= Rs.8/- during 1979-80.

Address of authors:
R.N. Adhikari
M.S. Rama Mohan Rao
S. Chittaranjan
A.K. Srivastava
M. Padmalah
A. Raizada
B.S. Thippannavar
Central Soil and Water Conservation Research and Training Institute
Research Centre
Bellary-583 104, Karnataka, India

Rehabilitation of Degraded Land in the Sahel: An Example from Niger

K. Michels, C. Bielders, B. Mühlig-Versen & F. Mahler

Summary

Efficient strategies for the rehabilitation of degraded soils in the Sahel region can play an important role in rebuilding land productivity. We tested the following strategies on experimental plots with different degrees of degradation for their effect on soil properties and millet production: (1) millet stover mulch, (2) overnight kraaling of small ruminants, and (3) manual soil tillage. Soil tillage was supposed to improve infiltration by breaking surface crusts and through the incorporation of soil amendments. For most soil and yield parameters, however, tillage had no significant effect. Compared to the remaining treatments the application of small ruminants manure and urine resulted in significantly lower soil penetration resistance and bulk density, improved soil nutrient status, enhanced root growth up to 2 m depth and higher millet yields. These effects occurred on both formerly non-degraded and severely-degraded plots. The yield increase using this strategy was similar on both degradation levels, when compared to plots without any improvement strategy. The application of crop residue mulch more than doubled the yield on the degraded soils when compared to the soils with no amendments. The yield level with crop residue, however, was less than 150 kg ha^{-1} and thus insufficient to compensate for the high amount of stover applied. Farmers can rehabilitate degraded soils using organic material. Manure shows clear, immediate effects whereas millet stover is less effective.

Keywords: Sahel, soil degradation, land rehabilitation, crop residue, kraaling

1 Introduction

Farmers in Niger are often aware of the declining soil productivity on their sandy soils and some are using adequate methods to ensure their survival within a context of very limited resources (Lamers and Feil, 1995). With declining fallow periods, however, farmers are facing a decline in land productivity. Understanding the causes of soil degradation and further developing adapted and efficient measures to counteract this process is a challenge to agricultural research. The economic viability of mineral fertilizer use is still unclear for many subsistence-oriented farming systems. The sustainability of intensified cropping systems needs further research not only in economic terms. The technical options for the rehabilitation of degraded land are also important. The interaction between livestock herders and crop producers plays a crucial role in maintaining soil fertility (Breman and De Wit, 1983). The yield-increasing effects of manure applications is well documented (Powell, 1986). A series of field and pot experiments were therefore conducted at the ICRISAT experimental station at Sadoré in Southwest Niger with the objective to test appropriate soil productivity regeneration strategies and to understand their mechanisms. A unique opportunity was provided by the fact that another completed experiment had resulted in plots with three well-defined degrees of degradation arranged in a randomized complete block design.

2 Materials and methods

In a previous experiment on the field station of the ICRISAT Sahelian Center in Sadoré, Niger, a field susceptible to wind and water erosion was continuously cropped with pearl millet during the rainy seasons 1992 through 1994. Two treatments to halt erosion were tested during these three years: a mulch of 2 t ha^{-1} crop residue or a similar mulch with plastic tubes (Buerkert, 1995). Plots without mulch served as a control. After the three years, the treatments resulted in three different levels of physical and chemical soil degradation: the bare plots had more erosion crusts, higher penetration resistance and lower fertility levels when compared to mulched soil (Figure 1). Several centimeters of topsoil were lost due to erosion (Buerkert, 1995). These trends were also reflected in crop yield differences. On the continuously bare soil, millet grain yield in the last season was much less than half the yield of the plots with crop residue application (Figure 2). This occurred despite the yearly application of nitrogen and phosphorus doses that are recommended for the millet systems on the sandy soils in Niger.

In a subsequent study three rehabilitation strategies were implemented in May 1995: an application of 2 t ha^{-1} millet stover, or of 10 t ha^{-1} manure of small ruminants plus 1250 l ha^{-1} of urine. Nutrient content of the manure was 18 kg N, 4 kg P, and 1.5 kg K per ton. Both amendments were applied with or without manual soil tillage on the day of millet sowing. Manual tillage was done using a hand-hoe immediately before sowing on 0.5-m wide strips 0.5 m apart from each other. The layout was a split-split-plot design. Millet variety ICMV 89305 was sown manually in planting holes spaced 1×1 m. No mineral fertilizers were used for the regeneration in 1995. The amount of the applied stover corresponded to a quantity that can be found on farmers field after a very good harvest. The quantity of manure and urine is representative for fields that have been *kraaled* over-night with cattle or small ruminants as it is done by farmers in the Niamey region. A survey among farmers revealed that goats, sheep and cattle were kept in kraals overnight between one and three weeks. In order to obtain 10 t of manure per hectare, an average of 2.5 nights is needed when one goat occupies a square meter.

In order to investigate the mechanisms of the manure effects, two pot trials were done. In an aluminum toxicity experiment root growth of millet seedlings was measured 72 hours after sowing in small pots using the method described by Ahlrichs et al. (1991). In a subsequent pot trial using degraded and undegraded soil, the effects of the six following factors on millet growth were tested in a factorial design: the application of N, P, K, the nematicide carbofuran (methyl-carbamate), lime, and micronutrients.

3 Results and Discussion

The application of small ruminant manure in quantities representative of overnight *kraaling* resulted in an enhanced millet growth which was visible already in the juvenile stage. Plant height on the kraaling treatments was the same on undegraded and degraded plots whereas millet height on bare soil and crop residue mulch was less on degraded soil compared to undegraded soil. (Figure 3). Tillage had no significant effect on millet growth. These trends continued through the growing season and were also reflected in the negligible grain and stover yields of the mulched and the bare plots (Figure 4). Rainfall during 1995 was 485 mm and thus sufficient to produce an average yield.

On undegraded plots kraaling improved root growth in the top 1.2 m of the soil compared to the other treatments. On degraded plots root growth was enhanced by *kraaling* up to a depth of 2.0 m (Figure 5).

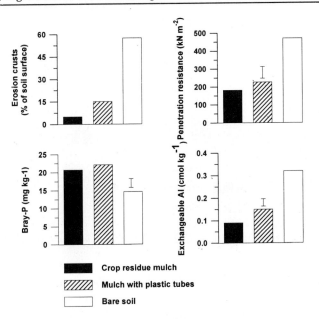

Fig. 1: Soil physical and chemical properties after three years of mulching and millet cultivation.

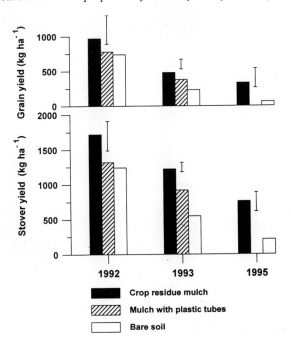

Fig. 2: Millet yields 1992 - 1995 as affected by mulch type (Source: Buerkert, 1995; own data).

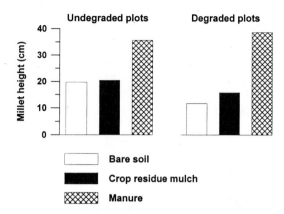

Figure 3: *Millet height at 25 days after emergence; first year of land rehabilitation.*

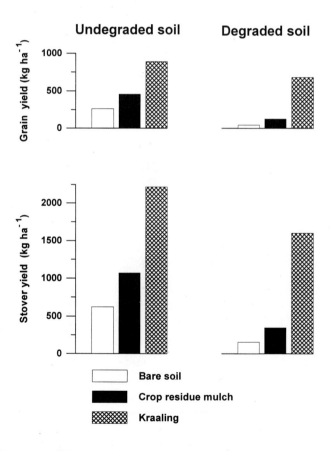

Figure 4: *Millet grain and stover yields; first year of land rehabilitation.*

Rehabilitation of degraded land in the Sahel, Niger

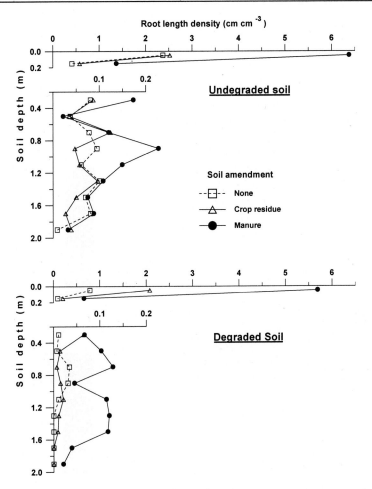

Figure 5: Millet root growth at flowering; first year of land rehabilitation.

Manure and urine thus had a clear growth enhancing effect, but the responsible mechanisms were not yet obvious. It was possible that manure reduced the amount of exchangeable aluminum in the soil below a tolerable level, or that soil microorganisms benefited from the high organic matter application and improved the soil physical properties. However, the soil tillage and its crust-breaking effects had no impact on millet growth. This indicated that the soil physical characteristics alone could not explain the growth and yield differences. Simply, the high nutrient input provided the plants with otherwise limited elements. Two pot experiments were designed to answer the remaining questions.

The aluminum toxicity test resulted in no difference in millet root growth among different degradation levels despite different exchangeable aluminum contents. Rooting depth was between 9.0 and 9.5 cm after 72 hours with the variety ICMV 89305. When the local millet variety was used, rooting depth on the degraded soil attained 11.5 cm.

Among the nutrients tested in the second pot trial, phosphorus and nitrogen fertilizer showed the most significant effects (Figure 6). The unfertilized control showed poor growth even on undegraded soil. On degraded unfertilized soil, plant growth stopped almost completely a few days after emergence when only one of the two minerals, nitrogen or phosphorus, was provided. On both undegraded and degraded soil, phosphorus was the most limiting element until a few days before harvest. Because of the limited size of the pots, nitrogen may have become limiting when plants were larger in size. Only the combination of nitrogen with phosphorus improved growth on degraded soil to a level comparable to undegraded soil (Figure 6).

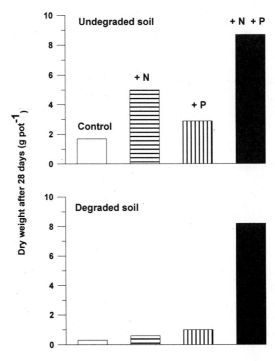

Figure 6: Millet dry weight in pots after 28 days; pot trial.

These results led us to conclude that the high amounts of phosphorus and nitrogen present in the manure and urine may have been the main factors responsible for the rapid growth enhancement on degraded soil. Although there is no clear indication of aluminum toxicity per se, this conclusion does not preclude that P-deficiency in the severely degraded soil may be linked to the high levels of exchangeable Al. The results of this pot experiment is no proof that millet growth on degraded land is limited by the lack of plant-available soil nutrients only, but that nutrients are the main limiting factors. Manure also improves soil physical conditions and decreases crusting.

As mentioned by Carter and Murwira (1995) for another region in Africa, it is unrealistic to expect quick rises in soil fertility over large areas given the extensive nature of the present farming systems and the number of livestock. The optimum amounts of manure applied is not yet known for the sandy Sahelian soils. In order to get a better insight into resource allocation by farmers, an understanding and scientific analysis of local resource use strategies are urgently needed (Carter and Murwira, 1995).

4 Conclusions

Soil tillage can be ineffective to rehabilitate land, even on compacted soil, when growth is restricted mainly by lack of nutrients. In contrast, manure is a very efficient measure to regenerate soil chemical and physical fertility in the short term and it procures high yield levels. The application of millet stover as mulch may improve long-term fertility but was unable, in our case, to show significant short-term effects. High exchangeable aluminum levels and low pH values seem not to be toxic to pearl millet plants but may reduce phosphorus availability. The optimum management of manure deserves further research for the development of sustainable farming systems in the Sahel.

References

Ahlrichs, J.L., Duncan, R.R., Ejeta, G., Hill, P.R., Baligar, V.C., Wright, R.J. and Hanna, W.W. (1991): Pearl millet and sorghum tolerance to aluminium in acid soil. *In* Wright, R.J., Baligare, V.C., and Murrman, R.P.: Plant - soil interactions at low pH. Advances in Plant and Soil Sciences **45**. Kluwer Academic Publishers, Dordrecht, The Netherlands.
Breman, H. and De Wit, C.T. (1983): Rangeland productivity and exploitation in the Sahel. Science **221**: 1341-1347.
Buerkert A. (1995): Effects of Crop Residues, Phosphorus, and Spatial Soil Variability on Yield and Nutrient Uptake of Pearl Millet (*Pennistum glaucum* L.) in Southwest Niger. PhD thesis Hohenheim University. Verlag Grauer, Stuttgart, Germany.
Carter, S.E. and Murwira H.K. (1995): Spatial variability in soil fertility management and crop response in Mutoko communal area, Zimbabwe. Ambio **24(2):** 77-84
Lamers, J.P.A. and Feil, P.R. (1995): The many uses of millet crop residues. ILEIA Newsletter **9**:15.
Powell, J.M. (1986): Manure for cropping: a case study from central Nigeria. Expl. Agric. **22**: 15-24.

Addresses of authors:
Karlheinz Michels
International Crops Reseach Institute for the Semi-Arid Tropics (ICRISAT)
Sahelian Center
BP 12404
Niamey, Niger
Present address:
Institute of Plant Production and Agroecology in the Tropics and Subtropics (380)
University of Hohenheim
D-70593 Stuttgart, Germany
Charles Bielders
International Crops Research Institute for the Semi-Arid Tropics (ICRISAT)
Sahelian Center
BP 12404
Niamey, Niger
Bernhard Mühlig-Versen
Institute for Plant Nutrition (330)
University of Hohenheim
D-70593 Stuttgart, Germany
Fritz Mahler
Research Program 'Adapted farming in West Africa' (793)
University of Hohenheim
D-70593 Stuttgart, Germany

Evaluation of the Effectiveness of Soil and Water Conservation Measures in a Closed Sylvo-Pastoral Area in Burkina Faso

W.P. Spaan & K.J. van Dijk

Summary

A degraded sylvo-pastoral area near the village of Zana Mogho, Bam Province, Burkina Faso, was safeguarded from land use activities for a number of years in order to rehabilitate the land by stimulating regeneration of its vegetation and improving the physical and chemical properties of the soil. In this exclosure a number of soil and water conservation measures were adopted to accelerate these processes. The effectiveness of these measures as to the extent to which they had improved the soil water, soil nutrient and biomass balances, was evaluated in an exploratory study from May to September 1993.

The measures consisted of (1) lines of tree trunks and stones serving as low barriers to run-off, and of (2) mulching of surfaces with branches, herbs and grasses.

Stone and trunk lines were most effective in conserving water. Soil moisture content on plots where these measures had been taken was always higher than on mulched plots. The measures had no effect on soil fertility. The enrichment ratio of the sediment was almost 1.0. All measures were observed to increase vegetation cover. Mulching with dry grass was very effective in capturing soil particles. Mulching with branches had no measurable effect. Of all measured variables, thickness of accumulated sediment and soil moisture content were the most useful ones for evaluating the impact of the measures.

Keywords: Degradation; exclosure; regeneration; conservation measures; effectiveness

1 Introduction

The aim of this exploratory study was to evaluate the effectiveness of soil and water conservation measures for the rehabilitation of degraded sylvo-pastoral land. The study was carried out in Zana Mogho village, 20 kms north-west of the town of Bourzanga, Bam Province, Burkina Faso, from May to September 1993 (Van Dijk, 1994). Bam Province is located on the Central Plateau, one of the flattest parts of Burkina Faso. It is situated in the northern fringe of the northern Sudanian zone, with an annual rainfall ranging between 600 - 900 mm. The rainy season lasts from May to September and is characterized by showers with high rainfall intensities and large spatial variation (Vlaar, 1992). The population density of Bam province is high by local standards (41\km^2). Population growth is 2.8 %. The present farming system is no longer capable of feeding the growing population of people and cattle. The resulting over-exploitation of the natural resources has led to the degradation of soil and vegetation (Claude et al., 1991). In turn, this degradation threatens the viability of agriculture and livestock raising practiced in the area. Therefore, for the survival of its people, it is necessary to protect the soil (Hien et al., 1992).

In this paper, degradation is defined as the reduction of the capacity of the soil to conserve a certain mode of exploitation (Blaikie and Brookfield, 1987). Rehabilitation is defined as the restoration of the productivity of degraded soils (Critchley et al., 1992). Rehabilitation can be monitored by looking at changes in soil water, soil nutrient and biomass balances. The effectiveness of a soil and water conservation measure can be defined as the extent to which it improves these balances. To assess the influence of the measures on these balances, the following variables were measured: (1) infiltration capacity and soil moisture content (water balance); (2) sedimentation, bulk density, and soil nutrient content (soil and soil nutrient balance), and (3) quantity, structure and composition of biomass (biomass balance).

In 1989 an area of about 25 ha was demarcated and henceforward safeguarded from any kind of land use (*mise en défens*, exclosure) in order to stimulate natural regeneration of the vegetation and the restoration of the physical and chemical properties of the soil. The exclosure was established in agreement with the local people (Kempen, 1992). Prohibited activities included cutting wood and grasses, burning the vegetation and grazing. The area had been used as a cotton field and was abandoned as such about 15 years earlier. After that it was used as sylvo-pastoral land. When it was closed for use, it was seriously degraded. Exclosure as a single measure for land rehabilitation is only effective in slightly to moderate degraded areas. In extremely degraded areas it requires additional measures. Therefore, in 1990, supporting soil and water conservation measures were undertaken on a trial basis.

2 Materials and methods

The following soil and water conservation measures are considered in this paper: line measures consisting of low barriers of trunks or stones, and area measures involving the application of woody or grass mulch.

The trunk and stone lines were laid out along the contour. The height of the barriers varied between 0.20 m and 0.40 m. The trunks and stones were gathered from the surrounding area. The barriers were meant to:
* slow down and spread run-off;
* increase infiltration and soil moisture;
* increase sedimentation of soil particles and litter;
* capture seeds in order to stimulate their germination,
* to protect fields situated down-slope.

The woody mulch consisted of a cover of dead branches and small trunks. The grass mulch was made up of a cover of dry grass and herbs collected at the end of the dry season. For an adequate protection of the soil about 8 tons/ha was needed. Branches and grass were gathered in the vicinity of the exclosure. Mulching was meant to:
* attract termites, which make the soil porous and thereby increase the infiltration rate;
* reduce soil temperature and evaporation;
* slow down the velocity of run-off and to spread it over the area;
* capture sediment, litter and seeds to stimulate regeneration of the vegetation;
* reduce rainfall energy (Roorda and Lamers, 1991).

The measures were evaluated on test plots of 10 x 10 m. The test plots were left open to runoff from up-slope. The effect of line measures was established by measuring the above mentioned variables 0.5 m up-slope of each measure and compared with conditions 8 m up-slope. On the mulched plots measurements were conducted on spots with vegetation and compared with conditions on spots without vegetation.

At the time of the evaluation the western part of the exclosure (Figure 1) was found to be heavily degraded with 80 % bare soil, sparse herbs, nearly all of which were annual species. The eastern part of the area consisted of savanna vegetation with shrubs, widely spaced herbs and annual grasses. Here, 50 % of the soil was bare. All measures were replicated four times, with two replications in the most degraded part of the area and two replications in the less degraded part. The measures with woody mulch were replicated five times.

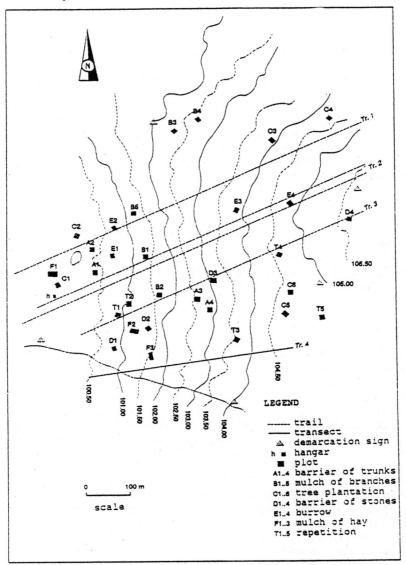

Fig. 1: Outline of exclosure with experimental plots

2.1 Measurements

The following measurements were made:
1) Infiltration measurements were carried out with a disk infiltration-meter. They were restricted to three plots.
2) Soil moisture content measurements were made after big rain storms on the next day. They were carried out six times all during the wet season. Soil samples were taken and weighed before and after drying. On the plots with line measures, samples were taken 0.5 m down-slope of the barrier, and 0.5 m, 3.0 m and 8.0 m up-slope of the barrier. On the mulched plots, samples were taken from the bare soil and from soil with a vegetation cover. The samples were taken at a depth of 0.15 - 0.20 m and of 0.35 - 0.40 m.
3) In order to measure bulk density and establish water retention curves (pF), undisturbed samples were taken 0.5 m and somewhere between 8 and 18 m up-slope of the measure at various depths in four plots. At many places it was impossible to take samples because of the hardness, stoniness or loose structure of the soil.
4) Sedimentation was measured with the aid of erosion pins placed at regular distances up-slope and down-slope of the line measures and cross-wise in the mulched plots. The length of the pins above the soil surface was measured once a month with a precision of 1 mm.
5) Soil fertility was analyzed on samples taken at different depths in four plots at the beginning of the wet season. The content of N (%), P (ppm) and organic matter (%) was analyzed with a Hach-field-kit.
6) Biomass quantity was determined using the 'Comparative Yield' method whereby biomass is collected from sites with similar vegetation located outside the exclosure so as not to disturb the vegetation in the test plots.
7) Photographs were taken in all plots at the beginning and the end of the wet season to record changes in vegetation growth during the study period. Other qualitative visual observations were recorded as well.

3 Results

The results of the study can be broken down as follows:
1) Most of the test plots showed moderate to high infiltration rates (20-40 mm/h). However, rates over time at the same spot differed considerably. This variation appeared to be larger than the variation between plots. The only conclusion which could be drawn with certainty is that infiltration rates on soil covered with vegetation was two to four times higher than on bare soil.
2) Showers which occurred at 29 June (5 mm), 9 July (34 mm), 22 July (45 mm), 29 July (20 mm), 26 August (17 mm) and 15 September (26 mm) had a similar effect on soil moisture content. Soil moisture content at different locations on the plots and at different depths did not show much variation over time. Moisture content values on plots with trunks lines and woody mulch are presented in Figure 2.

On the plots with line measures, soil moisture content diminished beyond 8 m up-slope of the barriers. Soil moisture content on mulched plots was about 50% less than on plots with line measures.
3) Bulk density was very high (> 1.60 g/cm^3) at almost all locations. The available soil moisture varied between 8.5 and 13.3 %.
4) Sedimentation on plots with trunk lines is presented in Figure 3.

Differences in sedimentation are mainly caused by topography of the terrain. Sedimentation on the plots with grass mulch in the most degraded part of the area was higher than on the other plots, probably as a result of the vast bare area up-slope. Dry grass proved very effective in capturing sediment. Because of termite activity, dry grass had to be renewed each year.

SWC measures, closed sylvo-pastoral area, Burkina Faso 1299

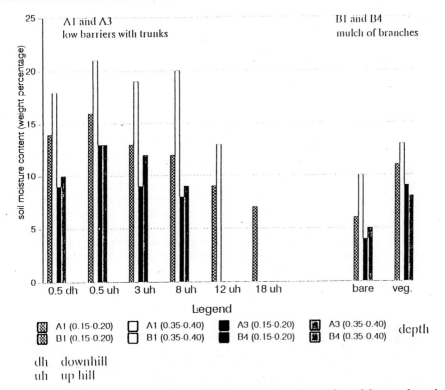

Fig. 2: Values of soil moisture content (in weight percentages) on plots with trunk lines and woody mulch in the wet season of 1993.

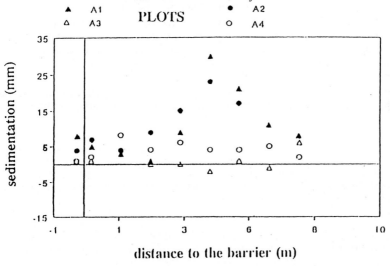

Fig. 3: Thickness of accumulated sediment on plots with trunk lines (A1 and A2 in the severely degraded part of the area; A3 and A4 in the less degraded part) in relation to the distance from the barrier

5) As for soil fertility, nitrogen content (0.03 %), available phosphorus content (4 ppm) and organic matter content (0.5 %) were low for all locations. No effect of the measures was observed.
6) Because of the difficulty in finding a comparable site outside the exclosure, biomass was analyzed at only four plots. The differences in biomass between plots were large. Since the amounts of biomass present on the plots before the measures were taken was not known, it was not clear how much the measures had actually contributed to the available quantities of biomass at the time of the evaluation study.
7) According to visual observations the change in vegetation between the dry and the wet season was impressive. At the end of the dry season the plots looked bare. At the end of the wet season there was a strip of rather dense vegetation, with a height of about 1.50 m along the line measures and with a width ranging from several meters up-slope to half a meter down-slope of the barriers. There was a large variety of plant species. On mulched plots the vegetation cover was comparable to that of the surrounding area. It was not very extensive. It seems that the presence or absence of vegetation is not so much caused by the soil and conservation measures but by the variable terrain conditions, such as presence of a crust, slope or presence of trees. Characteristics of the terrain around the place where the soil conservation measure had been taken seemed to influence its impact on the rehabilitation of the land to a large degree. For example, a soil conservation measure applied at a place with a large bare up-slope area appeared to conserve more soil and water because it received more run-off water.

4 Conclusions

Infiltration, bulk density, water retention, soil fertility and biomass measurements did not show clear effects caused by the soil and water conservation measures. For example, soil fertility was very low at all locations, close to the soil and water conservation measure as well as at its reference point. On the other hand, there were great differences due to the spatial variability of the terrain. In this exploratory study, the above-mentioned variables were not very suitable to assess the effectiveness of soil and water conservation measures because of the location of the measures. Photos were useful to record changes in vegetation cover in the area over time. They were useful as qualitative indicators of the effectiveness of soil and water conservation measures. Soil moisture content and sedimentation were easy to measure. Thickness of sediments clearly showed the effect of measures on soil conservation, whereas soil moisture content showed the effects on water conservation. Therefore, these two variables can be deemed suitable for evaluating conservation measures in an exploratory study such as here. It can also be concluded that line measures were effective soil and water conservation measures. There was no difference in effectiveness between the two types of line measures. In comparison to line measures, mulching was less effective, although grass mulch was more effective than woody mulch.

References

Blaikie, P.M. and Brookfield, H.C. (1987): Land Degradation and Society. Methuen & Co., Ltd., London, England, 296 p.

Claude, J., Grouzis, M. and Milleville, P. (1991): Un espace Sahélien, la mare d'Oursi, Burkina Faso. ORSTOM, Paris, 241 p.

Critchley, W.R.S., Reij, C.P. and Turner, S.D. (1992): Soil and water conservation in Sub-Saharan Africa; Towards sustainable production by the rural poor. International Fund for Agricultural Development, Rome, 110 p.

Hien, F., Stroosnijder, L. and Zoungrana, I. (1992): Les mesures de conservation des eaux et des sols pour la régénération des espaces sylvo-pastoraux: diversité et effectivité dans le plateau central du Burkina Faso. Université Agronomique Wageningen, Pays-Bas, 16 p.

Kempen, M. (1992): Contribution au cadre conceptuel des forêts villageoises et des mises en défens au Centre Nord, Burkina Faso. MET, SPET, Kongoussi, SNV, PAFV, Ouagadougou, Burkina Faso, 64 p.

Roorda, T. and Lamers, B. (1991): Projet mise en défens 1990 (MET, SNV, AFVP): La mise en défens de Zanamogo: Une étude pour la régénération des sols et végétations sylvo-pastoraux sahéliens au Burkina Faso. CIEH, Ouagadougou, Burkina Faso, UAW, Wageningen, Pays Bas (rapport de fin d'études), 80 p.

Van Dijk, K.J. (1994): Evaluation des mesures de régénération des terres dégradées à Zana Mogho, Burkina Faso. Antenne Sahélienne de l'Université Agronomique Wageningen Pays-Bas et de l'Université de Ouagadougou Burkina Faso, 40 p.

Vlaar, J.C.J. (ed.) (1992): Les techniques de conservation des eaux et des sols dans les pays du Sahel. Rapport d'une étude effectuée dans le cadre de la collaboration entre le Comite Interafricain d'Etudes Hydraulique (CIEH), Ouagadougou Burkina Faso, et l'Université Agronomique Wageningen, 99 p.

Addresses of authors:
W.P. Spaan
Wageningen Agricultural University
Department of Irrigation and Soil & Water Conservation
Nieuwe Kanaal 11
6709 PA Wageningen, The Netherlands
K.J. van Dijk
Ministério para a Coordenção de Aceção Ambiental
(Ministry for Coordination of Environmental Affairs)
Maputo, Mozambique

Gully Reclamation in Mafeteng District, Lesotho

F.A. Mayer & E. Stelz

Summary

This paper describes the concept, strategies and techniques applied by the Natural Resources Management component of the Mafeteng Development Project in Lesotho.

The Mafeteng Development Project (MDP) is financed jointly by the German and Lesotho Governments and implemented by the German Agency for Technical Co-operation (GTZ). MDP was developed in response to the increasing challenges confronting the people of Mafeteng, namely declining household incomes and continued environmental degradation. Since 1986 MDP has worked alongside the Ministry of Agriculture and the Ministry of Local Government through the District Agricultural Office and the District Secretary in the southern district of Mafeteng. After an orientation phase in one of six wards the project expanded to cover the entire district. It offers a multiple advisory service covering district planning, extension, community development, small scale business, agronomy, livestock and natural resources management.

Keywords: Gully reclamation, Lesotho, land tenure, public works programme

1 Introduction

The Kingdom of Lesotho is a small mountainous country covering 30,355 km^2 (app. size of Belgium) with about 2.1 Mio inhabitants. The country is entirely surrounded by the Republic of South Africa.

Lesotho has one of the world's lowest income economies (World Bank, 1995). The country has few natural resources of which water is most important. At present, Phase I of the Lesotho Highlands Development Project is being implemented. The aim is to export water to the Johannesburg area of South Africa and thereby generate electricity for domestic consumption.

Lesotho's economy is heavily dependent on the Republic of South Africa with 95 % of its imports coming from there. About 40 % of Lesotho's labour force is employed in South African gold and coal mines. The remittances account for one third of its GNP (World Bank, 1995).

Only 9% of Lesotho's total area is considered to be arable land. The minor role of agriculture in the country's economy is explained by poor degraded soils, frequent droughts and severe winters (13.7 % to the GDP; Economist Intelligence Unit, 1996). Even in years with normal rainfall, Lesotho is only able to produce 40 % of its food requirements. Maize and other food therefore must be imported. Nevertheless most Basotho depend strongly on agriculture in terms of employment and livelihood (Mhlanga, 1994).

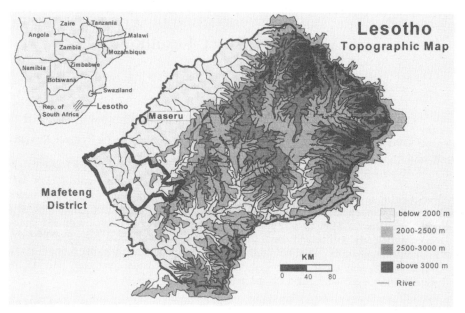

Figure 1: Topographic and location map

2 The Mafeteng District

The Mafeteng district, one of ten districts in Lesotho, is situated in the South West of Lesotho and covers an area of 2119 km² (29°30'-30°05' S 27°00'-27°45' E). About 256,000 people live in the district; the average population density is 121 inhabitants per km².

The population in Mafeteng is heavily depending on remittances of migrant miners working in South Africa and on subsistence farming as major sources of income. The industrial base is limited and can provide employment for only a few. The economic situation is becoming worse due to declining migrant income, deteriorating land resources and frequent droughts. This has resulted in a downward spiral of poverty and environmental degradation.

2.1 Climate, geology and soils

Three agro-ecological zones can be distinguished in the district. The lowlands 1500 m to 1800 m asl comprise 82 % of the area, the foothills (1800 m - 2000 m asl) 15 % and 3% mountains (2000 m - 2900 m asl).

The climate is temperate, semiarid to subhumid with a unimodal summer rainfall pattern. Annual precipitation ranges from 600 mm in the lowland areas to 800 mm in the mountains. The mean annual temperature ranges from 15° C in the lowlands to 11° C in the mountains. The duration of the frost period is 127 days in the lowlands and only 90 frost-free nights in the mountains.

The parent materials of the lowlands are predominantly Karoo sandstone and shale, on which mainly solodic Planosols, gleyic and haplic Solonetz and Lithosols (Albaqualfs and Natraqualf) developed. These soils are usually of low fertility and highly susceptible to water erosion.

Cambisols or Mollisols of basaltic origin are dominant in the foothills and mountains. These soils are less erodible and have a higher fertility status.

2.2 Land use and land tenure

About one quarter of the district area is regarded as arable land. The main crops are maize and sorghum in summer and wheat in winter. Yields are low, usually not exceeding 0.7 t/ha. The non-arable land and cultivated fields after harvest are used as communal rangeland. In 1988 only 0.6 % of the lowland areas were covered by forest (Conservation Division, 1988).

Although the Land Act of 1979 and several amendments provide for security of tenure by enabling allocations of 99 year leaseholds, these legal provisions have never been fully implemented. Currently the Land Act and customary laws are applied, the later being predominant in the rural areas.

Population growth and declining arable land has left one-third of the rural population in the Mafeteng district without access to land. A "Form C", the application for a lease, is the standard land tenure document and a prerequisite for a land title deed providing secure usufructuary rights for agricultural purposes. These "Form Cs" are usually allocated to individuals for degraded land like gullies.

2.3 Environmental degradation

The lowlands of Lesotho, including the Mafeteng district are heavily affected by environmental degradation. The natural resource base of fragile soils and scarce vegetation cover is constantly deteriorating due to:
- erratic high intensity rainfalls on steep slopes
- overgrazing
- relatively high population density
- inappropriate cultivation of marginal areas

An estimated 20,000 gullies, most of them still expanding, dissect the landscape and have occupied about 5 % of the lowland areas (Conservation Division, 1988). Sheet and rill erosion are even more severe. Estimates of average annual soil losses of up to 100 t/ha have been reported (Schmitz & Rooyani, 1987). Rills, gullies, badlands and bare rock have become a common feature especially in the lowlands.

Crop residues, cow dung and even weeds are used as fuel inducing a steady decline of organic matter in the soil. Free-roaming cattle, goats and sheep overgraze the communally used rangeland as well as harvested and fallow fields. This depletes the vegetation cover thus contributing to the environmental problems. The National Rangeland Inventory (Range Management Division, 1988) estimates on overstocking vary from 42 to 300 %.

3 Gully reclamation

MDP is supporting the District Agricultural Office (DAO) Department of Conservation, Forestry and Land Use Planning by implementing soil and water conservation measures. Since 1989 DAO/MDP has emphasised gully rehabilitation. This programme is complemented by other conservation measures such as terracing, conservation farming and afforestation. During the orientation phase, several techniques of gully reclamation were tested. These tests have shown that a gully can be stabilised by structural and biological means and may be turned into productive land again through tree and fodder planting.

3.1 Approach

MDP supports communities, groups and individuals implementing conservation measures. This is done via the DAO's extension service and the Department of Conservation, Forestry and Land Use Planning. A guiding principle of the project is to actively involve the land users during the planning and implementing phase. The project provides the planting material and where needed gabions and cement. Tools are available on a loan basis. The gully owners carry out the reclamation works under supervision of the extension and conservation staff. An important activity of the Department is the training of farmers and staff.

3.2 Land tenure

One of the most striking features of the project's intervention is the emphasis on ownership of the gully. The project made a significant contribution to the practice of allocating land titles for gullies to individuals and groups (Turner, 1995).

A land title, or lease application form (Form C) for the respective land is a prerequisite to participate in the gully rehabilitation programme. This precondition encouraged many villagers to apply for a gully. The Government of Lesotho and principal chiefs promote and support the allocation of land titles in gullies. Secure land tenure is an incentive, especially for landless people, to participate.

3.3 Lesotho Highlands Revenue Development Fund

Since 1995 a public works programme financed through the Lesotho Highlands Revenue Development Fund (LHRDF) has been implemented. This had a major impact on the approach, inducing a shift from supporting mainly individual initiatives to a more communal emphasis. The LHRDF was established to contribute to poverty alleviation through labour intensive infrastructure and conservation projects. Participants in this programme are paid on a daily basis for building dams, gully rehabilitation structures and rural roads.

Projects implemented through this fund are identified by the communities. The technical departments of the District Agricultural Office assist the democratically elected Village and Ward Development Councils in designing, costing and implementing projects. The project proposal including a quotation are forwarded to the LHRDF-Board for approval. The organisational set-up of the projects is part of the local governments (VDCs) role. The technical supervision and guidance is under the responsibility of the Department of Conservation, Forestry and Land Use Planning.

The establishment of this programme provides a basis for more comprehensive conservation measures in catchment areas on communal land. The number of applications from individual farmers however has declined. Many farmers would rather seek additional cash income in the public works programme. If they build their own gully structures material benefits for the individual can only be expected in the medium or long term. Rock (1994) estimates the period to achieve cost recovery to 10-18 years depending on the size of gully, labour and material used.

3.4 Gully rehabilitation measures

The process of gully reclamation has two phases. In the dry winter months the check dams or silt traps are built. The biological rehabilitation phase takes place in the rainy season. Where enough silt has accumulated in gullies, grasses, legumes and trees are planted.

3.4.1 Physical rehabilitation

Five types of structures are promoted as follows:
- Sandbags and small loose stone structures across rills or small gullies in the fields

Erosion rills are usually located in the fields and often caused by damaged contour bunds or terraces. It is essential to prevent these rills from growing as they will turn into deep gullies further eroding upstream.

The placing of black plastic-sandbags across the rill is promoted to stabilise the terrace. The sand bags have a volume of about 50 l and are filled with earth. Up to two layers are laid out across the rill so that a centre spillway is formed. The sandbags are linked to one another by strong twine. The lifetime of the sandbags is three years. After this period the sandbag structures have usually filled up with sediment. If necessary another layer of sandbags may be added. Small loose stone structures are used where stones are easily available.
- Cemented walls or archweirs in horseshoe shape for larger gullies

The construction of stone masonry check dams/arch weirs in large gullies was very popular during the initial phase of the project. The farmer had to provide the locally available materials like sand, water and stones. The project provided tools and cement and a trained "donga foreman" who assisted the farmer in constructing the wall. The cost for this type of structure is half the cost of a gabion structure (see Figure 2). This technique is presently only applied on the more stable soils of basaltic origin and where the availability of stones is limited. The aim being to reduce costs and to enhance sustainability.

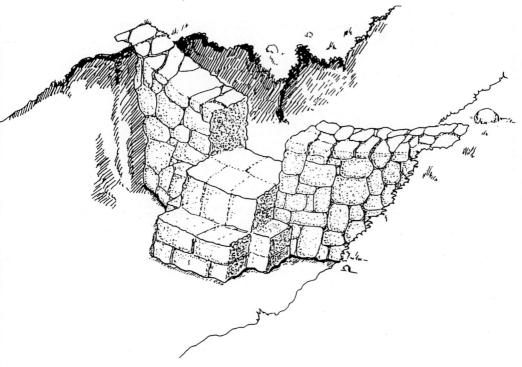

Figure 2: Cement archweir (from Rock 1994)
Dimensions: height 1.5 m; width 0.5 m; length across gully 5 m.

- Gabions in large gullies with very high water flow

Gabion structures are long-lasting but expensive. Considerable know-how is required for proper design and construction. The most critical issues is the anchoring of the gabions in the side walls. Gabion structures tend to be bypassed therefore a minimum anchoring depth of 1.5 m into the walls is recommended. A central spillway of half gabion cages is built in the central part of the structure. This type of structure is only used in gullies with catchments of more than 10 ha and where high runoff is expected. This is due to the high costs and the risk of by-passing.

- Loose stone structures in large gullies

The most popular structures are made of loose stones. This technique is very simple, cheap and easily adopted by the land users. Strong emphasis is put on proper anchorage of at least 1 m into the gully walls and 0.3 m into the foundation. A central spillway 1.5 m wide and 50 cm deep is built to reduce water pressure on the sides and to dispose of runoff safely. Steps and an apron are built underneath the spillway to avoid undercutting of the structure. The average size of loose stone structures built in the Mafeteng district is 15 m^3. This technique is the most appropriate especially under lowland conditions with fragile soils. Even if damaged this type of structure is easier to repair than gabions or cement walls.

Figure 3: Loose stone check dam.
Dimensions: height 1.5 m; width 2 m; length across gully 6 m.

3.4.2 Biological rehabilitation

After 1-2 years the structures in the rills and gullies have usually filled up with sediment and are ready to be planted.

In the deeper gullies a variety of trees are planted. In moist gully beds willows *(Salix babylonica, S. fragilis, S. caprea)* and poplars *(Populus deltoides, P. x canescens, P. deltoides)* are used. Closer to the gully walls, *Robinia pseudoacacia, Gleditsia triacanthos, Acacia dealbata, A. decurrens* are planted. On the side walls and surrounding areas outside of the gully, *Eucalyptus* div. spec., *Pinus radiata* and *P. halepensis* are planted.

To ensure a dense multi-storey vegetation cover, fodder grasses and legumes are planted. Kikuyu grass (*Pennisetum clandestinum*) is a hardy, soil-covering species well-suited for nearly all soils in the Mafeteng district. Legumes like hairy vetch (*Vicia dascycarpa x villosa*) and arrowleaf clover *(Trifolium vesiculosum)* are valuable fodder crops and show good growth in the stabilised gullies. Vegetative contour barriers of Vetiver grass (*Vetivera zizanioides*) are planted in the gully to collect more sediment. Vetiver grass, *Eragrostis curvula* and various legumes are also planted on the rehabilitated contour terraces.

Biological rehabilitation is not only of major importance to ensure long-term stabilisation but can also contribute to the income of the gully owners through fodder and wood production. The main problem with biological rehabilitation is the poor survival rate of plants due to untimely grazing. Despite the fact that cut and carry harvesting of fodder is recommended the gully beds are an attractive grazing ground particularly during the dry season.

Under Lesotho's conditions, with a fodder shortage in 8 months even Vetiver grass is grazed down to a height of a few centimetres. The grass survives the cold winters but does not develop its full potential like in tropical countries.

The gully rehabilitation programme is the most popular component of MDP's Natural Resources Management Section, in terms of the number of applications. Since 1989 about 1000 check dams have been built and planted with trees and grasses by individuals and groups. The gully rehabilitation programme seems to be well-suited as a starting point for awareness creation. However it has to be kept in mind that it does not offer a solution for the pressing problems induced by overgrazing and inappropriate cultivation. This is a function of the ever-increasing number of people trying to make a living from a finite resource base.

4 Additional conservation activities

MDP's Natural Resources Management Section considers erosion preventive measures as a primary task and supports several additional conservation techniques.

4.1 Establishment and rehabilitation of terraces and contour bunds

Terraces or grass strips are found in the fields all over Lesotho. Improper ploughing and the trampling of cattle have damaged many of these terraces. MDP is supporting farmers in re-establishing and maintaining the terraces in their fields.

4.2 Small earthen dams and ponds

The construction of small dams and ponds is a water conservation measure supported by MDP. These are used for irrigation of vegetable gardens and as livestock drinking reservoirs. DAO/MDP is providing technical advice, ox-drawn scoops and other tools like wheelbarrows on a loan basis.

4.3 Intercropping (Machobane Farming System)

The main focus of the Machobane Farming System is to reduce the risk of entire crop failure through diversification. It involves crop rotation and intercropping using maize, sorghum, pumpkins, beans, watermelons and fruit trees with potatoes as the leading crop. The Machobane

Farming System is labour intensive but requires only low cost inputs. The system also involves the use of ash and cow dung instead of chemical fertiliser.

5 Concluding comments

The Mafeteng Development Project (MDP), implemented jointly by the Ministry of Agriculture and GTZ, is operating in the Mafeteng District in the SW of Lesotho.

The combination of fragile ecosystems, high human as well as livestock population leads to severe soil degradation in the form of sheet, rill and gully erosion. A characteristic feature of the foothills and the lowlands are numerous gullies dissecting the landscape.

Since 1989 the Natural Resources Management Section of the Mafeteng Development Project has emphasised gully rehabilitation complemented by conservation measures like bund stabilisation, contour ploughing, intercropping and afforestation.

The quick and impressive results of structural measures which can successfully stop the gullies from growing and cause them to fill up with sediment makes gully rehabilitation a good starting point to create awareness on soil erosion.

MDP supports communities, groups and individuals implementing gully rehabilitation measures. The active involvement of the land users during the planning and implementing phase is a guiding principle of the project. A land title for the respective land is a prerequisite to participate in the gully rehabilitation programme. This precondition encouraged many villagers to apply for a gully. Secure land tenure is the main incentive especially for landless people (30 % in rural areas) to participate in gully rehabilitation.

The technical measures applied in the gully rehabilitation programme comprise stone arch weirs, gabions, dry stone structures and sandbags.

After the structures have filled up with sediment a variety of trees, fodder grasses and legumes are planted.

References

Conservation Division, Ministry of Agriculture (1988): National Resource Inventory of Lesotho. Maseru, 15 pp
Economist Intelligence Unit (1996): Country Report Botswana Lesotho 2nd quarter 1996. 31 pp
Mhlanga, M. L. (1994): The Mafeteng District: A Baseline Survey, 56 pp
Range Management Division, Ministry of Agriculture (1988): Lesotho National Rangeland Inventory: Methodology, Results and History from 1981 through 1988. 27 pp
Rock, F. (1994): The Donga Rehabilitation and Bund Stabilization Programme in Matelile, unpublished, 23 pp
Schmitz, G. and Rooyani, F. (1987): Lesotho Geology, Geomorphology, Soils, National University of Lesotho, 204 pp
Turner, S. (1995): Gully reclamation in the lowlands and foothills of Lesotho: the Matelile Rural Development Project and the Mafeteng Development Project. In: Centre for Development Cooperation Services, Vrije Universiteit Amsterdam (1995): Successful Natural Resource Management in Southern Africa, Windhoek, 204 pp
World Bank (1995): Lesotho Poverty Assessment, Report No. 13171-LSO, 246 pp

Addresses of authors:
Frank Axel Mayer
Elke Stelz
Mafeteng Development Project
P.O. Box 988, Maseru, Lesotho

Linking the Production and Use of Dry-Season Fodder to Improved Soil Conservation Practices in El Salvador

R.G. Barber

Summary

There has been a long history of soil conservation projects in El Salvador with overriding emphasis on physical structures and financial incentives, but the adoption rates of physical structures have been low. The FAO-CENTA "Sustainable Agriculture on Hillsides Project" is therefore emphasizing agronomic practices, e.g. crop residues and live barriers, with physical structures assuming a complementary role. Investigations showed that a minimum 75% surface cover of crop residues, equivalent to about 3.5 t/ha, is needed to ensure low erosion risks. However, farmers use crop residues as dry-season fodder which creates a conflict in their use. From a knowledge of the quantity of residues produced (1.8-10 t/ha depending on cropping system and productivity), the quantity required for soil protection (3.5 t/ha), and the quantity required as fodder in 6 months of dry season (1.7 t/l.u. for production), the deficit in residue production can be calculated. The choice as to which options for producing additional fodder are most appropriate will depend on farm size, crop productivity and number of livestock. For farms of 1.4 ha and larger, additional dry-season fodder can be produced by promoting silage or hay production from improved pastures, silage crops with a cover crop, or forage trees. However, for properties of only 0.7 ha it is necessary to resort to crop residues, harvested from grain crops with an intercalated cover crop, for fodder production.

Keywords: Soil conservation, crop residues, fodder production, farming systems, El Salvador

1 Introduction

This article presents the technical strategy of an agricultural development project, "Sustainable Agriculture on Hillsides," that is being executed by the Centro Nacional de Tecnología Agropecuaria y Forestal (CENTA) of the Ministry of Agriculture and Livestock, El Salvador, with FAO technical assistance and Dutch Government funding. The area of the project covers the Departments of Cabañas, Morazán and northern Usulután where small-scale farmers practise a mixed farming system of grains and livestock on steeply sloping lands. Crop residues play a critical role in that they are used as livestock feed but are also needed for soil conservation. Farming systems are being promoted that produce more biomass for livestock consumption in the dry season so that a greater proportion of crop residues can be left on the soil surface as protection against erosion.

2 Characteristics of El Salvador

El Salvador is a small hilly country in Central America of only 21,000 km^2 situated between latitudes 14° 27′ N and 13° 09′ N. Average annual rainfall is 1,800 mm which falls mainly from May to October, followed by six months of virtual drought. At least 75% of the country is hilly, 58% of the grain crops are grown on 12 to 50% slopes, and estimated erosion rates are 130-140 t ha^{-1} y^{-1}. The soils are fine-textured, well-drained alfisols and inceptisols of moderate to high chemical fertility derived from volcanic ashes and basic lavas. Very stony and shallow soils occupy almost 20% of the country.

The total population of El Salvador is 5,600,000, of which about 50% live in rural areas and 60% exist at extreme poverty levels. Within the project area 90% of farmers cultivate less than 1.4 ha, and only 53% are land owners; the others rent land for periods of 1-3 years. Most tenancy agreements require the tenant farmer to leave crop residues for grazing by the land owner's cattle. The typical farming system comprises maize and beans which constitute the basic diet, sorghum which is used mainly as cattle feed, and up to 6 head of cattle. Twelve annual cropping systems have been identified which vary in the number of associated crops, the form of association and time of planting. No-tillage is practised using a planting stick for seeding, and crop residues are either burned, grazed, or cut-and-carried for livestock depending on how many livestock the farmer possesses, the scarcity of dry-season fodder, and whether or not the land is rented.

3 Previous soil conservation strategies in El Salvador

There has been a long history of soil and water conservation in El Salvador since the 1960's, but the overriding emphasis has been on the construction of physical structures and the use of financial incentives. A recent study of 21 land-owning farmers of an FAO agroforestry and soil conservation project showed the percentage of farmers continuing to maintain soil conservation structures three years after the conclusion of the project was 100%, 67% and 20% for individual terraces, rockwall barriers, and hillside ditches, respectively (Vides, 1996). The adoption rates coincide reasonably well (75% agreement) with whether or not farmers observed yield benefits from the conservation structure. Some 75% of farmers claimed they installed the structures because of the provision of subsidised credit. In another study of an ongoing USAID project that includes a soil conservation component where no significant financial incentives are given (Streed, 1995), only 8% of farmers trained in the construction of physical conservation structures, tested them on their properties. Long-term adoption rates would be expected to be even lower. Thus the adoption rates of physical structures are low without financial incentives, and even with the incentive of subsidised credit, only physical structures which require little maintenance such as rockwall barriers, or those associated with high value crops, e.g. individual terraces for citrus, are likely to be maintained. Hillside ditches which require considerable maintenance and which are associated with low value grain crops are unlikely to be maintained. The main reasons for farmer reluctance to construct physical soil conservation structures are believed to be the high labour requirements and opportunity costs, and the long period for investment recuperation. It may also be that farmers are unconvinced of the benefits of physical structures, even in the long term !

Despite the many failures of soil conservation projects in El Salvador there has been a notable success at Guaymango-Metalío in the SW of the country. By a combination of intense "no burning" campaigns, and making credit availability conditional upon the adoption of a recommended technology package that included zero tillage, improved varieties and fertilizers, at least 2,300 ha of rolling to hilly lands are now being cultivated to maize relay cropped with Criollo sorghum using conservation tillage (Calderón et al., 1991). Yields have increased by a factor of 2-3, and erosion has been effectively controlled on slopes up to 45% without the use of physical conservation measures. The maize-sorghum cropping system yields large quantities of residues in this area, and controlled grazing leaves 6-7 tons ha^{-1} of residues on the soil surface which is sufficient to reduce soil losses to acceptable levels (Choto and Sain, 1993).

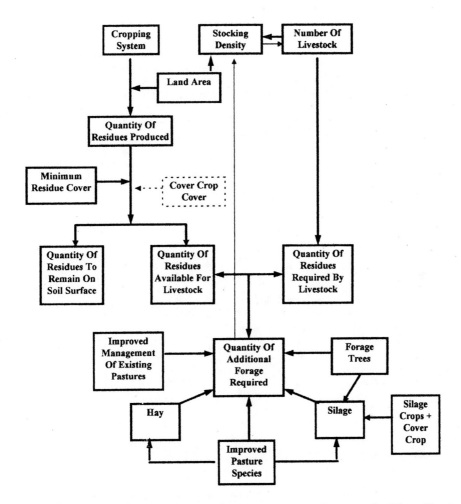

Fig. 1: Relationship between the factors influencing residue production and residue utilization

4 The project's strategies for promoting soil conservation

Given the reluctance of farmers to implement physical soil conservation structures and the success of agronomic conservation measures on slopes of up to 45% in Guaymango-Metalío, the project is emphasizing the promotion of agronomic conservation practices. The implementation of physical structures is only recommended where agronomic practices are unable to adequately control erosion, and are considered as complementary to agronomic practices.

There is considerable evidence throughout the world that leaving crop residues on the soil surface promotes effective soil and water conservation (Shaxson et al., 1989). However, for the majority of mixed farmers in El Salvador, the priority use for crop residues is as fodder for the dry season when there is a serious fodder deficit. Hence a conflict arises between leaving residues on the soil surface as protection against erosion, and feeding the residues to livestock. Thus the problem changes from a cropping system problem to a farming system problem.

The relationships between factors that determine the quantity of residues produced, the quantity needed to protect soils against erosion, and the quantity required as fodder are presented in Fig. 1. The quantity of residues produced depends on the cropping system, crop varieties, productivity and cropping area, and can vary from 1.8 t/ha for low yields of maize-beans to 10 t/ha for moderate yields of maize-Criollo sorghum (Choto and Sain, 1993).

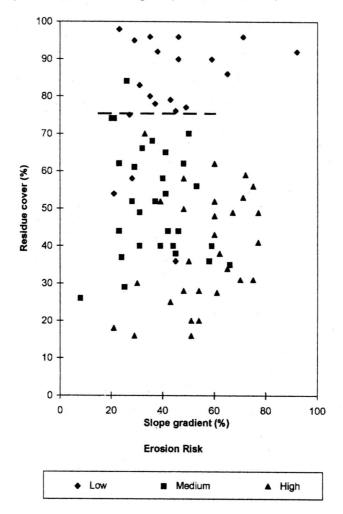

Fig. 2: Relationship between soil erosion risk, percentage residue cover and slope gradient.
(Argueta 1996)

According to Shaxson et al. (1989) a 40% surface cover is required to reduce soil losses to values less than 10% of those expected from bare soil. However, the percentage residue cover that will ensure low erosion risks would be expected to vary with slope gradient, soil type, and rainfall erosivity. According to Lal (1995, Pers. Comm.) a 20% slope requires 70% residue cover to control soil erosion. In a recent study in the department of Morazán, El Salvador, soil erosion risks were evaluated semi-quantitatively on the basis of soil surface morphological features, using a method similar to that of Bergsma (1992). The erosion risks were rated, and then related to slope gradient and percentage surface cover (Argueta, 1996). Results showed that a minimum 75 % residue cover, equivalent to about 3.5t/ha, is necessary to ensure low erosion risks (Fig. 2). No clear evidence was found for increasing percentage residue covers at higher slope gradients. Cover crops and weeds can also contribute to surface cover, and the Project is promoting farmer participatory investigations into the insertion of leguminous cover crops in the grain cropping systems.

During the 6 months of rainy season there is an abundance of forage, but a serious fodder shortage occurs in the subsequent 6 months dry season. The forage requirement per livestock unit of 390 kg is estimated at 2.5% of live weight/day for a production level of 2 to 5 liters milk/day, and 1.5% of live weight/day for maintenance feeding (Alfaro, Personal Communication). Thus, the forage requirement for 180 days of dry season is 1.7 t/l.u. for production and 1 t/l.u. for maintenance.

From a knowledge of the quantity of crop residues produced and the quantity required for soil protection, the amount available for livestock fodder in the dry season can be calculated. A comparison of this value with the quantity of fodder required by the livestock will indicate how much additional fodder needs to be produced to satisfy the requirements of soil conservation and livestock productivity in the dry season.

5 The production of additional fodder for the dry season

There is an abundance of fodder in the rainy season which can be used to produce additional fodder for the dry season through silage or hay production from improved pastures, silage crops or forage trees. To ensure an adequate protection of the soil surface after harvesting a silage crop, a cover crop that is not harvested needs to be intercalated within the silage crop. Forage trees, established as live fences along farm boundaries and field divisions, can be used for hay production, or for silage if mixed with 75% of a graminaceous crop. Extending the period of grazing beyond the end of the rains may be achieved to a limited extent by using more drought-resistant pasture species and by better management of existing pastures. Unfortunately no acid-tolerant forage tree species have been identified that continue producing forage well into the dry season. In situations where land is very limiting, it may be unavoidable to utilize part of the crop residues for silage. The upper part of maize plants harvested just above the cob at physiological maturity can be used for silage production by a rustic method known as "hornos forrajeros" (forage ovens) (Mejía, 1996). This practice leaves some 60% of the vegetative biomass on the soil surface. To ensure adequate soil protection against erosion an intercalated cover crop within the grain crops will be necessary.

6 The choice of options for producing additional fodder

The choice as to which options for producing additional fodder are most suitable for farmers depends mainly on farm size, crop productivity and number of livestock. An average family of 7.3 persons will consume 0.9 t maize per year in the form of tortillas. Assuming average maize yields with low fertilizer inputs of 1-1.3 t/ha, an area of 0.7-0.9 ha is therefore required for maize production.

For farmers with 0.7 ha, the area is barely sufficient to produce the family's annual maize requirement. Assuming typical yields of 1.3 t/ha for maize and 1 t/ha for sorghum, about 3.45 t/ha of crop residues are produced from maize relay-cropped with Criollo sorghum. This quantity of residues is only sufficient for soil protection. However, many farmers with 0.7 ha possess 2 l.u. of cattle, but without any land for pastures or silage crops. For these farmers, forage trees can supply a part of the dry season fodder, but it will also be necessary to resort to crop residues for fodder. Some 400 m of *Gliricidia sepium* live fences would provide 600 kg/yr of tree forage (Alfaro et al., 1996), and harvesting the tops of maize plants and all of the sorghum foliage would provide 2,500 kg of dry matter. Thus 3,100 kg of dry matter would be available for silage production, sufficient for the production of 1.8 livestock units (of 390 kg live weight) during 180 days of dry season. To avoid soil erosion problems due to the utilization of most of the crop residues for silage, a leguminous cover crop would need to be incorporated into the grain cropping system.

For farmers with 1.4 ha, about 0.5 ha are available for silage production or pastures. Silage production from 0.5 ha of sorghum intercalated with a cover crop for soil protection, and using minimal fertilizer application, would yield 2.5 t of dry matter, sufficient for the maintenance feeding of 2.4 l.u.or the production of 1.4 l.u. for 180 days of dry season without needing to use crop residues as fodder.

For farmers with 2 ha or more, there is much greater flexibility in the choice of options shown in Fig. 1. About 1.1 ha are available for forage production, and a combination of 0.7 ha of pastures for grazing in the rainy season and 0.4 ha of silage crops for dry-season fodder production would support 2 l.u. Alternatively, silage production from 1.1 ha of sorghum would permit the production feeding of 3 l.u. for 180 days of dry season.

7 Conclusions

A quantification of crop residue production in comparison to the quantity of crop residues needed for erosion control and livestock fodder, will enable the deficit in dry-season fodder production to be estimated. For properties of 1.4 ha and larger, promoting additional fodder production from silage or hay made from improved pastures, forage trees, or silage crops, will result in higher livestock productivity in the dry season as well as improved soil conservation. However, for properties of only 0.7 ha, it will be necessary to resort to crop residues for fodder, and to sow an intercalated cover crop in the grain crops to provide soil protection.

Acknowledgements

The author expresses his gratitude to Ing. Manuel Alfaro Ticas, Coordinator of the Livestock and Pastures Programme of CENTA for his help and advice.

References

Alfaro, M.A., Araujo, G.A. and Mejía, N. (1996): Arboles forrajeros: Una alternativa para la alimentación del ganado. Programa de Producción Animal, Centro Nacional de Tecnología Agropecuaria y Forestal, Ministerio de Agricultura y Ganadería, Izalco, Sonsonate, El Salvador.

Argueta, M.T. (1966): Análisis de la producción y utilización de rastrojos y su efecto sobre el riesgo de erosión en el departamento de Morazán. Tésis de Ingeniero Agrónomo, Facultad de Ingeniería, Universidad Centroamericana "José Simeon Cañas," San Salvador, El Salvador.

Bergsma, E. (1992): Features of soil surface microtopography for erosion hazard estimation. In: H. Hurni and Kebedi Tato (eds.), Erosion, Conservation and Small-Scale Farming. Selection of papers presented at the 6th. International Soil Conservation Conference of the International Soil Conservation Organization (ISCO), Ethiopia and Kenya, Nov.1989.

Calderón, F., Sosa, H., Mendoza, V., Sain, G. and Barreto, H. (1991): Adopción y difusión de la labranza de conservación en Guaymango, El Salvador: Aspectos institucionales y reflexiones técnicas. In: Agricultura sostenible en las laderas centroamericanas: Oportunidades de colaboración interinstitucio-nal, San José, Costa Rica: Instituto Interamericano de Cooperación para la Agricultura (IICA). Pp 189-210.

Choto de Cerna, C. and Sain, G. (1993): Análisis del mercado de rastrojo y sus implicaciones para la adopción de la labranza de conservación en El Salvador. In: Programa Regional de Maíz: Síntesis de resultados experimentales 1992. Guatemala City, Guatemala: CIMMYT Regional maize program for Central America and the Caribbean.

Mejía N. (1996): Construyendo hornos forrajeros con puntas de maíz. Carta Informativa, Vol. 1. No. 2. Centro Nacional de Tecnología Agropecuaria y Forestal, Ministerio de Agricultura y Ganadería, San Salvador, El Salvador.

Shaxson, T.F., Hudson, N.W., Sanders, D.W., Roose, E. and Moldenhauer, W.C. (1989): Land Husbandry: A Framework for Soil and Water Conservation. World Association of Soil and Water Conservation and Soil and Water Conservation Society, Ankeny, Iowa, USA.

Streed, E. (1995): Land tenure and agroforestry - soil and water conservation in El Salvador. Unpublished report, PROMESA, San Salvador, El Salvador. USAID.

Vides, M.A. (1966): Evaluación participativa de los sistemas de conservación de suelos y agua en zonas de influencia de proyectos agroforestales. Tésis de Ingeniero Agrónomo, Facultad de Ingenieria, Universidad Centroamericana "José Simeon Cañas," San Salvador, El Salvador.

Address of author:
Richard G. Barber
25, Elsley Road
Tilehurst, Reading RG31 6RP, UK

Influence of Demographic, Socio-Economic and Cultural Factors on Sustainable Land Use

B. Rerkasem & K. Rerkasem

Summary

Land degradation is not always inevitable even within an unfavourable socio-economic context. This paper illustrates the varied effects of demographic and socio-economic conditions on the land use by a review of the situation in the mountainous areas of mainland Southeast Asia.

The problems of land degradation in the Montane Mainland of Southeast Asia (MMSEA), which extends from south-western China, through Myanmar, Thailand, Laos, Vietnam and a small part of Cambodia, have received wide publicity since the 1980s. Evidence presented here are drawn from case studies and field research carried out in all of the six countries by local researchers and in regional collaborative activities. Although land degradation is common under adverse conditions of population increase, land tenure uncertainty, increasing pressure of market forces and unfavourable government policies, it does not always happen as a matter of course.

In all the countries, regardless of political regimes or diverse ethnic cultures, local farmers and communities have been able to develop coping strategies to increase productivity in a sustainable manner. From the analysis of case studies we propose a hypothesis that three key elements play a significant role in sustainable land use management, (i) the presence of appropriate and cost effective technology, (ii) social organization for communal resource management, especially as a moderating influence against market forces and (iii) the ability of local people to participate in making crucial decisions related to land management.

Keywords: Erosion, land degradation, social and economic factors, sustainable land use

1 Introduction

The fact that the topic of "Influence of demographic, socio-economic and cultural factors on sustainable land use" was chosen as a major theme for this ISCO conference on soil conservation, clearly illustrates the acceptance by soil scientists of the roles played by human and social factors in land use management. In this paper we attempt to show that the manifestation of these effects are by no means straight forward. The influence of social and economic factors cannot be assumed. Increased population density does not always lead to land degradation; farmers who are without a legal title to land ownership do not always mistreat the land and despite claims to the contrary by some anthropologists and social scientists, conservative or exploitative cultivation practices are not determined by people's ethnicity or culture. In this analysis we will draw substantially on our experience and data from the Montane Mainland of Southeast Asia (MMSEA). Apart from our own familiarity with local conditions, the situation in the MMSEA offers an ideal case

study where land use has come under increasing pressure caused by social and economic processes. A lesson emerges that should be applicable to land use management elsewhere, although the combinations of the ecological and human factors may vary:
- land degradation is but a symptom of a complexity of natural and human processes, highly dynamic local institution and political situations;
- solutions in soil conservation are effective only when developed on the basis of understanding this complexity.

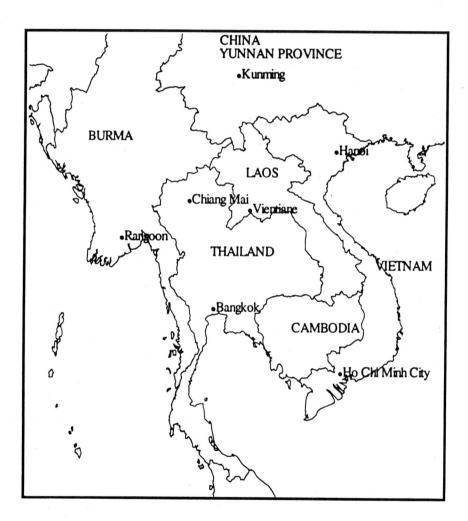

Figure 1: Mainland Southeast Asia

2 Demographic and socio-economic pressure on the land: the MMSEA example

The MMSEA is roughly defined as the adjoining mountainous regions of south-western China (Yunnan Province), Myanmar, Thailand (Northern), Laos and Vietnam (Fig. 1). Compared with in other parts in the respective countries, the area has been relatively inaccessible and isolated until recently. The indigenous population is made up of people who belong to ethnic groups that are distinct from the majority population in the lowlands. Land use is characterised by cultivation of steep slopes. The most widespread system of traditional land use is shifting agriculture, with small highland valleys developed for wetland rice where possible. Many detailed studies have shown that shifting agriculture to be highly diverse in their use of resources: some forms are exploitative (Keen, 1978) while others are conservative (Kunstadter, 1978). Pressure on the land has been escalating in the last 30 years, both from population increase and the rapid pace of economic development.

2.1 Demographic pressure

The annual growth rate of the national population of the MMSEA, Laos being the exception, has declined significantly in the last 15 years (Table 1). The picture in the mountains is, however, very different. In Thailand, for example, population in the mountains grew at the rate of 7% annually between 1986 and 1991 compared with 1.4% in the lowlands (TDRI, 1994). This was brought about by high natural growth rate of about 3%, combined with migration from the lowlands and across the borders. Some areas, possibly because of their strategic location in relation to economic development, have received more migrants than others. Migration into the mountainous area of the Chiang Mai Province (Thailand) has raised the population growth to an incredible annual rate of 12% (TDRI, 1994).

Similarly in south-western China, the population in the mountain area has tended to increase faster than in the lowlands (Guo, 1993). Xishuangbanna Prefecture saw its population almost quadrupled between 1949 and 1990 (Wu and Ou, 1995). This can be explained by major waves of southward migrations between 1960 and 1974 coupled with preferential policy towards family planning for ethnic minority groups. In Vietnam, with encouragement and support from the government, 5 million people from the lowlands have moved into the Central Highlands in the last 30 years, half of them between 1980 and 1990 (Sam, 1994).

Much lower population densities are found in Myanmar and Laos (Table 1), but land use has been affected by wars. The presence of live bombs that still remain after the US/Vietnam war has severely restricted access to agricultural land in Laos (SUAN, 1991). Despite its relatively low population density, rotation cycle for shifting cultivation in Laos is now half of that 30 years ago. Another result of local wars is cross-border migration of people seeking refuge from the fighting. Waves of migration from Myanmar have added to the population pressure in the border region in both Thailand (Rerkasem and Rerkasem, 1994) and China (Xu et al., 1996).

	Average Annual Growth of the National Population(%)		National Population Density (1993), per km^2
	1980-85	1990-95	
China	1.44	1.42	129.2
Laos	2.29	3.00	20.0
Myanmar	2.09	2.14	67.8
Thailand	1.83	1.27	111.3
Vietnam	2.18	2.03	217.8

Source: Adapted from WRI 1994

Table 1: Population growth in MMSEA countries

2.2 Social and economic development

In addition to population increase, pressure on the land was further accentuated by recent economic and social development. For a long time, the mountains were of little interest. In recent years, countries become more conscious of the need for conservation, and the national governments enacted laws and legislations to set aside areas for national parks, wildlife sanctuaries and other types of reserves. The land available for agriculture in the mountains has suddenly been much reduced. In Northern Thailand, since 1972 some 2 million hectares out of a total of 17 million hectares have been designated national parks, wildlife sanctuaries and other types of reserves (Rerkasem and Rerkasem, 1994). Further land use restriction has been imposed by the zoning of forest areas into various classes of watersheds (IUCN, 1984; Woolridge et al., 1985), with conservation aims reflecting increasing demand for water by the lowland and urban population, for irrigation, domestic water supply and electricity generation. For Vietnam, the Party's Central Executive Meeting on January 21, 1994 declared that "In forestry the strategic task is to resolutely protect the existing the 9.3 million hectares of forests" (Sam, 1996).

Problems related to land use in the mountain is further complicated by ambiguities in the institution of land ownership. The MMSEA countries differ somewhat in the official laws, policies and actual implementation regarding ownership and use rights of mountain land. In Thailand, by the provision of the Land Code of 1954, all mountain land and land within 40 m of the foot of a mountain has been declared by the Ministry of Interior to be out of bounds for any one to take possession (Ratanakhon, 1978). In China, since the early 1980s, rights to mountain land has been allocated and auctioned off (Zuo, 1996). In order to attract external investment for afforestation and other types of "hoped for" sustainable development of mountain land, individuals and enterprises are allowed to buy long term use rights, 40-100 years, without restriction based on residency or nationality, which inevitably means that less land will be available to the local population.

Market forces bring new opportunities but also another set of pressures on the mountain land. Until recently mountain communities of the MMSEA were hardly ever touched by the outside world. Development of the transportation network and mass communication during the past 30 years in Thailand and 10-15 years in the other countries have brought access to both public services and markets in local and national urban centres. Cabbages and other vegetables grown in the mountains in Northern Thailand are now trucked to Bangkok and beyond. In China, increasing demand for tropical and subtropical crops from the rest of the country has resulted in expansion of production areas in the mountains of Yunnan (Wu and Ou, 1995). The decade from 1981 to 1991 saw sugarcane production rose 5 times and tropical and subtropical fruits 3.5 times in Xishuangbanna, China's most tropical prefecture. Encouraged by a domestic price that is about twice the world price, provided by government support under a policy to increase domestic production of rubber, area planted to rubber in Xishuangbanna has increased to almost 100,000 ha in 1990s, from almost nothing 30 years ago. The emergence of cross-border trade adds another dimension to the development of mountain agriculture in the MMSEA.

Also the construction of dams and reservoirs and the creation of forest plantation and other afforestation schemes add to the pressure on mountain land. For example, in land scarce Vietnam, the building of the Hoabinh Dam, Vietnam's largest hydroelectric facility, forced the relocation of some 52,000 people, and many of them have settled on the steep slopes above the reservoir (Rambo, 1996). Similar effects have been observed elsewhere in the region. In Thailand, efforts to regain the national forest cover lost in recent years have led to allocation of land classified as "degraded forests" to various afforestation schemes. In reality, these lands are rarely unoccupied, and the result is further reduction of agricultural land (Rerkasem and Rerkasem, 1994).

2.3 Cultural factors

With the great diversity of land use types and ethnicity of the farmers in the MMSEA, it is tempting to associate a certain land use pattern with a specific ethnic group. Indeed many ethnographic studies in the area from the 1950s to 1970s tended to do so. However, there are now ample evidence from all of the countries in the region to clearly indicate that there is no such simple correlation between culture and land management. In Thailand, the Akha, Hmong, Lahu and Lisu had for a long time been known as opium growing, "pioneer" shifting cultivators who were said to be continuously migratory and practice a type of shifting agriculture in which a field is cropped for several years and then abandoned when fertility has declined and/or weeds become too dense (Kunstadter and Chapman, 1978). By 1960s, however, many villages of people belonging to these groups had begun to acquire paddy land and settle down to permanent, often irrigated, agriculture (TDRI, 1994). Pioneer shifting cultivating groups of Vietnam such as the Hmong and Dzao have become sedentary since the 1950s and 1960s (Cuc, 1996). Throughout the region, the close association between types of shifting agriculture and certain groups of people described in 1970s has under gone marked changes. Migratory villages common in the region until the 1970s are now hard to find.

3 Impact on land use sustainability

In many instances the link between land degradation and its social and economic causes has been assumed rather than proved through rigorous analysis.

3.1 What degradation?

Losses of forest cover throughout the region have been documented by satellite imageries and aerial photographs (Table 2). Almost always automatically, this is said to be followed by land degradation, especially soil erosion. Although this assumed "chain of events" is commonly found as a major justification for development projects, direct evidence is rare. With the aid of satellite images the extent of forest cover losses at the macro level is now relatively easy to assess. But field level studies have shown that picture on the ground can be very complicated.

Region/country	Period	Forest loss ha	%[a]	Reference
Xishuangbanna, China	1950-1983	600,000	49	Calculated from
	1983-1993	80,000	12	Wu and Ou (1995)
Northern, Thailand	1982-1989	741,000	8	Rerkasem and Rerkasem (1994)
Laos	1981-1989	457,000	2	Souvanthong (1995)
North-West, Vietnam	1965-1985	472,358	49	Calculated from Sam (1994)

[a] % of forest area at beginning of period

Table 2: Forest cover changes in the MMSEA.

Extensive deforestation in the Yunnan Province in south-western China is universally accepted as the result of mismanagement that had prevailed during the period 1950-1979. Field studies of six villages in Lijiang County in 1980s have revealed extremely complicated forest land use and management dynamics that precludes broad generalisation (Ives and He, 1996). It was shown that although some deforestation has taken place relatively recently, in many places tree cover has not changed in the last 50 to 60 years, in other areas the forest cover had increased, and mature stands of fir that are more than 700 years old can still be found at high altitudes. Land use dynamics and their associated problems of degradation can sometimes be hidden in macro level forest cover data. Natural forests that have been replaced by tree crops, e.g. rubber plantations in Yunnan, although vastly different in their ecological functions, are still counted as forests (Xu et al., 1996).

3.2 Land use sustainability in the MMSEA

When some areas in the MMSEA were studied in detail in 1960s, shifting cultivation in the forms prevailing then were at least as effective as irrigated agriculture in terms of labour productivity, e.g. 40-50 kg of rough rice/manday (Kunstadter, 1978). Also, at least with some forms of shifting agriculture, communities were found to have existed in the same location for over one hundred years, with well preserved natural and managed forests, and a well developed communal institution and a wealth of indigenous knowledge that were effectively employed in systems of mountain land management that were conservative towards natural resources (e.g. Kunstadter, 1978; Nakano, 1980). Land productivity, however, was low, e.g. these systems of shifting agriculture required twenty times or more land than irrigated agriculture to feed the same population (Rerkasem and Rerkasem, 1994).

The various pressures exerted on mountain land discussed above, supported by wide publicity on land degradation in the mountains of Southeast Asia of 1980s (Allen, 1993), would seem to suggest an inevitability of environmental disaster throughout the region by the 1990s. Papers published in the 1970s and 1980s indeed reported of mountain villages on the brink of starvation, in the midst of denuded and eroded mountains (e.g. Hinton, 1978; Keen, 1978; Cooper, 1984). In 1990s, however, instead of a continuing decline, a very different picture is emerging, from many observations and detail studies of mountain villages in the region (Table 3).

Country	References
China (Yunnan)	Guan et al., 1995; Guo and Padoch 1995; Menzies 1996; Xu et al., 1996
Laos	SUAN 1990; SUAN 1991; Rerkasem 1992; Souvanthong 1995
Thailand	TDRI 1994; Rerkasem and Rerkasem 1994; Thong-Ngam et al., 1996; Turkelboom et al., 1996; Ekasingh et al., 1996
Vietnam	Cuc et al., 1990; Sam 1996; Cuc 1996; Rambo 1996

Table 3: Example of mountain village studies in the MMSEA in the last 10 years.

In spite of a great diversity in conditions, land use in the different parts of montane Southeast Asia all now appear to be sharing some common outstanding features. The first is their highly dynamic nature. As reported above, through out all the countries, former migratory communities have become sedentary. Former "pioneer shifting cultivators" are now practising some form of rotational cropping as well as cultivating wetland rice. Opium cultivation has declined sharply, in Thailand by the second half of 1980s, and is currently declining in Vietnam. Almost all villages have entered into the market economy to varying degrees. Another common feature is diversity of land use systems within villages, within farming systems, and even within single fields. Villages living on shifting agriculture alone have become exceptions. Common land use types and other "livelihood activities", some old and some new, that are practised in various combinations by individual farming systems are listed in Table 4. Households typically engage in a combination of 6-10 of these activities.

These "new" land use types can be seen as moves by mountain communities and farmers to respond to the various pressures and opportunities discussed above. Improvement in the performance of montane agricultural systems, including effective control of land degradation, has been the results of interactions between these diverse activities as well as their combined effects. The improvement itself, on the other hand, has been possible because of (i) a set technological innovations that have either been imported from the outside, transferred from different locations within the region, or indigenously developed (ii) the presence of communal institution for resource management (iii) participation by local communities in resource management decisions. Soil and water conservation, which contributes to sustainable land use management, is very much integrated into these production technologies.

Land use type/ Livelihood activities	Examples of products/services
Swiddening/rotation	Upland rice (food)
	Maize, cassava, buckwheat (feed, food)
	Associated domesticated, semi-domesticated, wild species during cropping/fallow phase (food, feed, firewood, cash)
Wetland paddy	Rice (food, cash)
	Other crops in rice-based cropping systems (e.g. wheat, soybean, garlic, cabbages - cash, food)
Livestock	Cattle, goats, pigs, poultry, (draught, food, ceremonial, investment)
Gardens[a]	Vegetables, fruits, bamboo (food, cash)
Orchards/plantations[a]	Fruits, nuts, timber, rubber, tea, coffee
Agroforestry[a]	Bamboo (shoots, wood), tea, timber, *Amomum spp. Cinnamomum cassia*
Forest - gathering[a]	Bamboo shoots (food, cash), bamboo (home use, cash), broom grass (cash), mushrooms (food, cash), various greens (food, cash), rattan (food, cash), various wild species (e.g. medicinal, dye - own use, cash)
Handicrafts	Bamboo, rattan and wood works, metal (silver, iron), embroidery (home use, cash), broom making etc.
Trade	Assembling and transportation of various local products
Wage employment	Labour for intensive, high value crops

[a] some products may come out of any of these types of land use, e.g. tea is found both in plantations and as agroforestry system integrated into natural forests, bamboo (for edible shoots as well as the wood) can be found in home gardens, plantation, agroforestry or from the wild).

Table 4. Common land use types/livelihood activities of montane farming systems of mainland Southeast Asia.

3.3 Production technology

The set of agricultural production technology that have contributed to recent transformation of mountain agriculture in the MMSEA include wetland rice, irrigation, alternative high value cash crops, rotation and multiple cropping, agroforestry, gardening, orchards and plantation, and integration of livestock.

The adoption of high yielding hybrid rice in China (Rerkasem and Guo, 1995), high value vegetables and flowers in Thailand (TDRI, 1994) reflect the introduction of modern technologies. On the whole, however, development is often the result of adaptation of traditional practices. The transfer of technology has largely been local, i.e. between different ethnic groups or different localities. These points are illustrated below with the cases of wetland rice and irrigation development.

Some former migratory pioneer shifting cultivator villages in Thailand took up wetland rice simply by buying the paddy land, presumably with the silver and gold accumulated during previous good opium seasons, and picking up accompanying management skills along with it (Rerkasem and Rerkasem, 1994). Similarly, the H'mong in Vietnam (who not long ago were said to have an innate aversion to wetland rice cultivation) have developed paddies (Cuc, 1996), by learning from wetland rice growing tribes, e.g. the Dzao, who share the same speaking language. There are many similar stories of spontaneous local transfer of irrigation technology (Zhang, 1994; Gao, 1992; Sektheera and Thodey, 1975). Since virtually all of Yunnan is mountainous, irrigation along with terracing had become an important component of its agriculture even from these earlier times. In 1950s, when many former dryland farmers were introduced to wetland rice cultivation they were exposed to irrigation at the same time. Further south, e.g. in Thailand, development in mountain irrigation was more recent, but has also involved transfer of management skills from those with longer experience. The potential of wetland rice (in highland valleys and terraces) as a sustainable land use alternative for the mountains has been recognised by government agencies and development projects and farmers themselves (Sam, 1994; Rerkasem and Rerkasem, 1994 and Souvanthong, 1995). But the need for effective transfer of the skills involved is frequently not so well appreciated, as witnessed by the numerous paddies developed by various assistance schemes and then abandoned.

Gravity fed sprinkler irrigation is a "modern" development that has spontaneously swept the mountains of Northern Thailand in the last 15 years. It has enabled mountain farmers to grow high value crops such as vegetables throughout the dry season and take advantage of the cooler temperatures at higher altitudes especially during the height of the tropical summer (March - May) to meet growing demands in increasingly rich cities. The potential to help transform mountain agriculture in other countries in the region, i.e. a cross-border transfer, should not be overlooked. The system makes use of plastic water pipes, which are cheap, easy to transport and assemble. For individual farmers, management is a simple affair of moving a hose and turning on the tap. Development, which involves bringing water from the water source in a main pipe, and maintenance and regulation however, requires a community level of organisation (see below).

Many of the current practices that are now contributing to increased productivity and land use sustainability in the MMSEA have come from the farmers themselves. This is especially true in Thailand where the agricultural systems that contribute directly to the national economy are all in the lowlands, mountain agriculture is considered too marginal to warrant major research support. Highland development projects (generally with substantial funding from foreign governments) on the other hand, were given a mandate that strictly prohibited research.

In addition to filling the gap left by the absence of support from publicly funded research, farmers' understanding of their own set of ecological, economic and social environment can come up with technical solutions to problems of sustainable agriculture much more effectively than research scientists, who sometimes still have yet to come to grip with the ecological environment, and being mostly from the natural science disciplines such as agronomy or soil science, have little idea of the social and economic context of the farm. Following examples of indigenous technologies encountered in Thailand illustrate these points.

3.4 Improving upland rice production

Farmers' yield of upland rice has generally been low, less than 1,000 kg/ha (Rerkasem and Rerkasem, 1994). Improving this productivity has been widely recognized to be central to sustainable mountain land use (Sabhasri, 1978), but getting mountain farmers to "improve" their upland rice production is probably among the most difficult of agricultural extension efforts. The commonly held belief that fertilisers are not used because they are not cost effective has been proved with economic analyses (TDRI, 1994; Renaud, 1995). But even when they are given fertiliser by development project, farmers still refuse to apply it to the upland rice. This has been interpreted as due to the lack of response of traditional rice varieties to management, especially fertilisers (Aneksamphant and Tejajai, 1996). The evidence that upland rice yield can actually be increased came from farmers. In 1990 we began to be shown upland rice fields that were "high yielding", by the farmer's own estimate as well as our agronomic calculations (Rerkasem et al., 1992).

Crop cutting surveys later verified that the farm yield has indeed increased, to 2,000-3,000 kg/ha.

The upland rice, of traditional Karen varieties, was grown in rotation with cabbages that received heavy doses of chemical fertilisers (mainly N and P). For the cause of this improvement, residual fertilisers would be the first guess but clean weeding under cabbages probably also led to less weeds, another pernicious problem in the upland rice. Since these first observations, upland rice rotations are now widely seen, with other "high" input crops, with varying levels of input including soybean and other grain legumes, in other areas of Northern Thailand.

3.5 Use of soil improving species

That mountain farmers traditionally make use of a great variety of plant species is well documented (Kunstadter et al., 1978; Sutthi, 1985). Some of these have made contribution towards soil improvement, either as a by-product of cash cropping, e.g. the grain legumes, or purposely so. Grain legumes have always been part of shifting cultivation. Seeds of cowpeas (5-8 varieties of *Vigna unguiculata*), rice bean (*Vigna umbellata*), *Lablab purpureus* are commonly found among the "genebank" kept, under care of the women, above the fireplace of every mountain farm household. Traditionally these were planted in the "swiddens", fields of upland rice, maize or opium that occupy the productive phase of shifting agriculture, to provide harvests for the kitchen through the year. Commercialisation has seen many of these, along with introduced grain legumes (e.g. soybean, Azuki bean or *Vigna angularis*), incorporated on a larger scale into cropping systems, i.e. mixed or in sequence with maize or upland rice. Cash is no doubt the primary incentive for this transformation. However, detail studies have shown that contributions of these to soil fertility maintenance and improvement can be considerable. For example, rice bean intercropped with maize not only provides substantial additional harvest of the beans with no detrimental effect on the maize yield, but also adds significant amounts of nitrogen from the atmosphere (Rerkasem and Rerkasem, 1988).

The use of plants specifically for soil improvement has been found in agricultural systems that are still largely subsistence as well as those that have become cash oriented. As an example of the former, *Mallotus* spp. are well known amongst rotational shifting cultivators in Northern Thailand, from the H'tin villages on the border with Laos to Karen and Lua on the Myanmar border, as a fallow-enriching species. *Mimosa invisa*, a recently arrived exotic leguminous "weed", capable of contributing large amounts of nitrogen by biological fixation, has been incorporated into high valued cash cropping systems based on vegetables (Rerkasem and Rerkasem, 1994).

3.6 Soil conservation technology

Mountain farmers in the region understand the idea of land degradation, especially soil erosion, when it has obvious and major effects on productivity, e.g. wetland paddies at the base of slopes buried by sand and silt from higher up (Rerkasem et al., 1992). Some sedentary groups have developed wetland rice paddies from the soil deposition fan at the base of slopes after several seasons of shifting cultivation on the upper slopes (e.g. see Zinke et al., 1978). Hillside ditches are employed by others to redirect and channel water run-off. However, most extension efforts to get farmers to adopt various soil and water conservation practices such as terracing, alley cropping, contour strips have, at best, been met with mixed results. Terraces, when directed and paid for by development projects, are too often poorly constructed. Many of these are abandoned soon after completion, when soil fertility is annihilated by the subsoil that is brought up. The region also has its share of soil and water conservation practices adoption that was sustainable only as long as the project was paying for it by direct and indirect incentives. Preventing soil erosion has never been found to be among the farmer's primary land use objectives (Ashadi, 1992; Rerkasem et al., 1992).

Where they have been most effective and sustainable, soil and water conservation practices were perceived to serve some obvious and practical functions. In Laos an allocation of land with legal title is the primary reason for the popularity of an agroforestry system involving planting of teak, although the financial return from it is extremely long-term (Souvanthong, 1995). Because ownership of mountain land in Thailand is tenuous, soil and water conservation is generally practised in the belief, even though largely unsubstantiated, that it will eventually lead to at least long term use rights (Rerkasem et al., 1992). The recognition of the value of irrigation in raising land productivity, especially in the dry season, and the association made between water yield and forest protection has led to conservation of head water forests in many areas (TDRI, 1994). The fact that land productivity has been significantly raised by the new production technologies has made it easier for farmers to absorb the investment cost of soil and water conservation. Effective reduction of soil erosion came about largely because farmers have dropped the upland rice, at least in the way it has been traditionally grown, for more productive alternatives. Once the primary component of shifting cultivation, area under upland rice has been declining throughout the region, and in many villages it is no longer grown (Cai, 1996; Sam, 1995; Thong-Ngam et al., 1996). Installation of contour strips and other erosion control measures are more readily accepted into highly productive fields such as irrigated cabbages and other vegetables (Rerkasem et al., 1992; TDRI, 1994).

3.7 Communal institution for natural resource management

A decision to adopt any of the above technology depends only partly on individual farmers. Influences of incentives (e.g. financial support for developing wetland paddies, promises of land ownership titles) and disincentives (e.g. restriction on land clearing and burning) provided by the implementation of government's development and conservation policies are acknowledged. However, the role played by communal institution has also been crucial. Because of their different political structure and social history, "communal institution" in the mountains differ somewhat among the countries and among groups of different ethnicity. In the socialist countries, there are functioning local governments down to the village level. In Thailand, the official governing structure is reaching into mountain villages only very slowly (TDRI, 1994). In general, however, village's traditional institution deals with all civic matter as well as management of common resources. The notable exception is Vietnam, where the traditional village institution has been superseded by the socialist state structure of co-operatives.

In general, village level management deals with development and management of land, irrigation and forest resources, which are shared within the village, and sometimes between neighbouring communities. A communal level of organisation is essential when a farmer's private economic goals come into conflict with his/her neighbours or the community at large. In most places the production of high value crops such as vegetables near the village is possible only where the population of free-ranging livestock is very small or there is an effective communal regulation concerning crop damages by roaming animals. For example, we found in a Karen village south of Chiang Mai that the rule is that during the day it is the responsibility of the owner of the crop to make sure that the fences are cattle-proof. At night, when the animals should be tied or locked up in stalls, the responsibility for damage is shifted to the cattle owner (Rerkasem and Rerkasem, 1994).

Development of paddy land is generally a private activity, and wetland rice fields are privately owned. Development and management of an irrigation system, however, is a communal affair. Communal development and management of irrigation system of the Dais in Yunnan (Gao, 1992) and Thais in Northern Thailand (Sektheera and Thodey, 1975) have been studied and documented more than others, partly because of their long history.

Recent economic liberalisation in socialist countries, especially China and Vietnam, is having some impacts on land use in the mountains. Local villages sometimes revert back to their old institutions when faced with the numerous, and ever changing, policy directives (e.g. for China see Zuo, 1996; Zhao, 1996) from the various government agencies on the one hand, and local pressures on the land, on the other.

3.8 Local participation in resource management decisions

Until now, villagers and communities in the MMSEA have been able to participate in major land use management decisions largely only by default on the part of the government. Indigenous technologies and innovations are in use when no real improvements were forthcoming from modern alternatives. Communities must make their own rules or revise traditional rules to cope with the management of common resource. There is some good in this, as cultural belief and values are sometimes more effective towards sustainable resource management, especially when economic development is over-emphasised by government policies (for example from China see Xu et al., 1996, and Thailand see Ganjanapan, 1996). Unfortunately a communal institution without legal or official basis is sometimes limited in its effect.

4 Conclusion

These examples from the mountains of mainland Southeast Asia have shown that land degradation is but a symptom of complex human and ecological processes. From the recent MMSEA experience, it is concluded that farmers, villagers and local communities have been successful in reducing the risk of land degradation because they understand the interacting human and natural processes involved far better than scientists, development workers and erosion control specialists. In this paper we have purposely tried to concentrate on efforts that have been effective under the various pressures of population, socio-economic and cultural factors, in order to reach for some understanding on how land degradation could be slowed down. On the whole, however, land degradation remains a serious problem as well as a challenge to "soil conservationists". But technical solutions risk being irrelevant unless they are based on an understanding of the highly dynamic nature of local institutional and political processes.

References

Allen, B.J. (1993): The problems of upland management. In: Brookfield, H. and Byron, Y. (eds.), South-East Asia's Environmental Future: The Search for Sustainability, Tokyo: United Nations University Press and Kuala Lumpur: Oxford University Press, 225-237.

Aneksamphant, C. and Tejajai, U. (1995): Management of sloping lands for sustainable agriculture in Northern Thailand, Thai Jour. Soils and Fertilisers **17**, 168-179.

Ashadi (1992): Socio-economic Evaluation of Integrated Soil-Water Conservation and Cropping Systems, M.S. Thesis. Graduate School, Chiang Mai University, Chiang Mai.

Cai, K. (1996): Changes in Land Use in Response to Socio-economic Change in Xishuangbanna China, M.S. Thesis. Graduate School, Chiang Mai University, Chiang Mai.

Cooper, R.G. (1984): Resource scarcity and the Hmong response: a study of resettlement and economy in northern Thailand. Singapore University Press, Singapore.

Cuc, L.T. (1996): Swidden agriculture in Vietnam (in case of Hmong, Dao and Tay ethnic groups. In: Proceedings of the Symposium on Montane Mainland Southeast Asia in Transition. Chiang Mai: Chiang Mai University, 104-119.

Cuc, L.T., Gillogly, K. and Rambo, A.T. (1990): Agroecosystems of Midlands of North Vietnam. East-West Environment and Policy Institute, Occasional Paper No. 12. East-West Center, Honolulu.

Ekasingh, M. Shinawatra, B., Onpraphai, T., Promburom, P. and Sangchayosawat, C. (1996): Role of spatial information in the characterisation of highland watersheds and communities. In: Proceedings of the Symposium on Montane Mainland Southeast Asia in Transition. Chiang Mai: Chiang Mai University, 402-425.

Ganjanapan, A. (1996): Will community forest law strengthen community forestry in Thailand? In: Proceedings of the Symposium on Montane Mainland Southeast Asia in Transition. Chiang Mai: Chiang Mai University, 349-365.

Gao, L. (1992): Preliminary exploration of the Dai's traditional water-conservancy and its social meaning in Xishuangbanna. In: Indigenous Land Use Management Systems in Yunnan. Proceedings of the Seminar on Traditional Land Use Management. October 8-14, 1992. Kunming, China, 129-146.

Guan, Y., Dao, Z. and Cui, J. (1995): Evaluation of the cultivation of Amomum villosum under tropical forest in southern Yunnan China, PLEC News and View March, 22-28.

Guo H. (1993): Land Transformation: farmers adaptations in Southern Yunnan, China, Paper presented at PLEC: Thailand/Yunnan 2nd cluster meeting, 4-6 November 1993. Kunming, China.

Guo, H. and Padoch, C. (1995): Patterns and management of agroforestry systems in Yunnan, Global Environmental Change **5**, 273-279.

Hinton, P. (1978): Declining production among sedentary swidden cultivators: the case of the Pwo Karen. In: Kunstadter, P., Chapman, E.C., and Sabhasri, S. (eds.), Farmers in the Forest, Honolulu: The University Press of Hawaii, 185-198.

IUCN (1984): Thailand Watershed Classification, IUCN Bulletin Supplement **1**, 4.

Ives, J.D. and He, Y. (1996): Environmental and cultural change in the Yulong Xue Shan, Lijiang District, NW Yunnan, China. In: Proceedings of the Symposium on Montane Mainland Southeast Asia in Transition. Chiang Mai: Chiang Mai University, 1-16.

Keen, F.G.B. (1978): Ecological relationships in a Hmong (Meo) economy. In: Kunstadter, P., Chapman, E.C. and Sabhasri S. (eds.), Farmers in the Forest, Honolulu: The University Press of Hawaii, 210-221.

Kunstadter, P. (1978): Subsistence agricultural economics of Lua' and Karen Hill Farmers, Mae Sariang District, Northwestern Thailand. In: Kunstadter, P., Chapman, E.C. and Sabhasri S. (eds.), Farmers in the Forest, Honolulu: The University Press of Hawaii, 71-133.

Kunstadter, P. and Chapman, E.C. (1978): Problems of shifting cultivation and economic development in Northern Thailand. In: Kunstadter, P., Chapman, E.C. and Sabhasri, S. (eds.), Farmers in the Forest, Honolulu: The University Press of Hawaii, 3-23.

Menzies, N.K. (1991): 'The role and consequences of livestock in swidden agroecosystems in Sepone: implications for development interventions. In: Swidden Agroecosystems in Sepone District, Savannakhet Province, Lao PDR, Southeast Asian Universities Agroecosystems Network, Khon Kaen, 122-124.

Menzies, N.K. (1996): The changing dynamics of shifting cultivation practises in upland southwest China. In: Proceedings of the Symposium on Montane Mainland Southeast Asia in Transition, Chiang Mai: Chiang Mai University, 51-68.

Nakano, K. (1980): An ecological view of a subsistence economy based mainly on the production of rice swiddens and in irrigated fields in a hilly region of northern Thailand. Tonan Ajia Kenkyu (South East Asian Studies) **18**, 40-67.

Rambo, A.T. (1996): 'The composite swiddening agroecosystem of the Tay ethnic minority of the north-western mountains of Vietnam. In: Proceedings of the Symposium on Montane Mainland Southeast Asia in Transition. Chiang Mai: Chiang Mai University, 69-89.

Ratanakhon, S. (1978): Legal aspects of land occupation and development. In: Kunstadter, P., Chapman, E.C. and Sabhasri S. (eds.), Farmers in the Forest, Honolulu: The University Press of Hawaii, 45-53.

Renaud, F. (1995): Financial cost-benefit analysis using data from the Doi Tung soil-conservation experimental site in Northern Thailand, Presented at the Seventh Annual Meeting of the Management of

Sloping Lands for Sustainable Agriculture in Asia Network. Organized by Thai Department of Land Development and International Board on Soil Research and Management. 16-20 October 1995, Chiang Mai, Thailand.

Rerkasem, K. (1992): Agroecology Report for Luang Prabang, Multiple Cropping Centre, Chiang Mai University, Chiang Mai.

Rerkasem, K. and Guo, H. (1995): Report on a workshop on agroecosystems and biodiversity in Montane Mainland Southeast Asia, PLEC News and Views September, 5-10.

Rerkasem, K. and Rerkasem, B. (1988): Yields and nitrogen nutrition of intercropped maize and ricebean (Vigna umbellata [Thunb.] Ohwi and Ohashi), Plant Soil **108**, 151-162.

Rerkasem, B. and Rerkasem, K. (1990): Legumes of the Highlands 1989, A Report to the Thai-Australia Highland Agricultural and Social Development Project, Chiang Mai University, Chiang Mai.

Rerkasem, K. and Rerkasem, B. (1994): Shifting Cultivation in Thailand: its current situation and dynamics in the context of highland development, IIED Forestry and Land Use Series **No**. 4. 140 p.

Rerkasem, B., Rerkasem, K. and Shinawatra, B. (1992): Watershed Development Demonstration Program: A Monitoring and Evaluation Report, Agricultural Systems Programme, Chiang Mai University, Chiang Mai.

Sam, D.D. (1994): Shifting Cultivation in Vietnam, Country Report, IIED Forestry and Land Use Series No. 3. 65 p.

Sam, D.D. (1996): Shifting agriculture practices today - in Vietnam. In: Proceedings of the Symposium on Montane Mainland Southeast Asia in Transition. Chiang Mai: Chiang Mai University, 31-50.

Sabhasri, S. (1978): Opium culture in Northern Thailand: social and ecological dilemmas. In: Kunstadter, P. Chapman, E.C. and Sabhasri, S. (eds.), Farmers in the Forest, Honolulu: The University Press of Hawaii, 160-184.

Sektheera, R. and Thodey, A.R. (1975): Irrigation systems in the Chiang Mai Valley: Organization and Management. In: Agricultural Economics Report No. 6, Chiang Mai: Faculty of Agriculture, Chiang Mai University, 81-98.

SUAN (1990): Two Upland Agroecosystems in Luang Prabang Province, Lao PDR: a Preliminary Analysis, Southeast Asian Universities Agroecosystems Network, Khon Kaen.

SUAN (1991): Swidden Agroecosystems in Sepone District, Savannakhet Province, Lao PDR, Southeast Asian Universities Agroecosystems Network, Khon Kaen.

Sutthi, C. (1985): Highland Agriculture: from better to worse. In: MacKinnon, J. and Vienne, B. (eds.), Hill Tribes Today. White Lotus, Bangkok, 107-142.

Souvanthong, P. (1995): Shifting cultivation in Lao PDR, Country Report, IIED Forestry and Land Use Series No. 5. 38 p.

TDRI (1994): Assessment of Sustainable Highland Agricultural Systems, Natural Resources and Environment Program, Thailand Development Research Institute, Bangkok.

Thong-Ngam, C., Shinawatra, B., Healy, S. and Trebuil, G. (1996): Farmer's resource management and decision-making in the context of changes in the highlands. In: Proceedings of the Symposium on Montane Mainland Southeast Asia in Transition. Chiang Mai: Chiang Mai University, 462-487.

Turkelboom, F., Van keer, K., Ongprasert, S., Suthigullbud, P. and Pelletier, J. (1996): The changing landscape of the Northern Thai Hills: adaptive strategies to increasing land pressure. In: Proceedings of the Symposium on Montane Mainland Southeast Asia in Transition. Chiang Mai: Chiang Mai University, 436-461.

Woolridge, D.D., Chakao, K. and Thangtham, M. (1985): A method for watershed classification in Thailand. Mimeograph, Royal Forestry Department, Ministry of Forestry and Cooperatives of Thailand.

WRI (1994): World Resources 1994-95. Oxford University Press, London.

Wu, Z. and Ou, X. (1995): The Xishuangbanna Biosphere Reserve: A Tropical Land of Natural and Cultural Diversity, China, UNESCO South-South Cooperation Programme on Environmentally Sound Socio-economic Development in the Humid Tropics. Working Paper No. 2. UNESCO, Paris.

Xu, J., Lu X,, Fox, J., Podger, N. and Ai, X. (1996): Impacts of population, policy, and property rights on community forest management in Xishuangbanna, China. In: Proceedings of the Symposium on Montane Mainland Southeast Asia in Transition. Chiang Mai: Chiang Mai University, 150-180.

Zhang, N. (1994): Dali, (Thai translation by Ratanapon, P.), Muang Boran, Bangkok.

Zhao, J. (1996): Farmers' hill-land use rights change in 1980s in Yunnan Province, China. In: Proceedings of the Symposium on Montane Mainland Southeast Asia in Transition. Chiang Mai: Chiang Mai University, 90-95.

Zinke P, Sabhasri S. and Kunstadter P. (1978): Soil fertility aspects of the Lua forest fallow system of shifting cultivation. In: Kunstadter P., Chapman E.C. and Sabhasri S. (eds.), Farmers in the Forest, Honolulu: University Press of Hawaii, 134-159.

Zuo, T. (1996): Forestry Liangshan Daohu and its impacts on land resource management in 1980s, Yunnan, China. In: Proceedings of the Symposium on Montane Mainland Southeast Asia in Transition. Chiang Mai: Chiang Mai University, 96-103.

Addresses of authors:
Benjavan Rerkasem
Kanok Rerkasem
Multiple Cropping Centre
Chiang Mai University
50200 Chiang Mai, Thailand

Demographic Growth and Sustainable Land Use

M. Tiffen

Summary

The paper discusses the relationship between population growth, accompanying social changes, soil fertility and methods of environmental management. As land becomes more scarce, people generally seek new technologies and make investments to improve land quality. Different levels of population density lead to differences in appropriate management strategies. These propositions are discussed in relation to:
- a case study of a particular area that has experienced a fivefold increase in population between 1930 and 1990 (Machakos District, Kenya)
- a sample of cases, world-wide, where population increase or decline has been thought to affect soil erosion
- a case study of attitudes to soil conservation in Kenya in 1996, in areas of different levels of population density

It is argued that there is not an inevitable correlation between population increase, poverty and land degradation. Rather, there is a natural tendency for farmers to invest in new technologies and soil improvement strategies as land becomes scarce and more valuable. Their greatest difficulties occur not when densities first start growing from a very low base, nor when densities have reached a relatively high level. Erosion is most likely to increase visibly at intermediate but growing densities. In response, farmers will try to find remedies, or to develop new land. Often, they will first seek to do the latter. If this option is not available, they invest in land improvement, but their efforts can be either supported, or thwarted, by government policies. It is useless to promote soil conservation either where low population densities mean land is not valued, or where government policies remove the incentive and means to invest. It is also essential that farmers and those who seek to advise them share the same view of what is meant by degradation or improvement of land and water resources, or they will be acting at cross-purposes.

Keywords: Population density, soil fertility, farming systems, climate, markets

1 Demographics and the diagnosis of the status of soil management

1.1 Vicious circles or trend lines?

It is often assumed that there is a vicious circle or downward spiral which makes degradation of the resource base the inevitable result of population growth. The possibility of soil improvement in these circumstances is only gradually being recognised in current scientific literature. The IBSRAM Paper *Soil Water and Nutrient Management Research* (Greenland et al., 1994) declared that it was possible to arrive at a "win-win" situation in which sustainable farming systems provide both increased productivity and wealth, and effective soil and water conservation (p.4), through

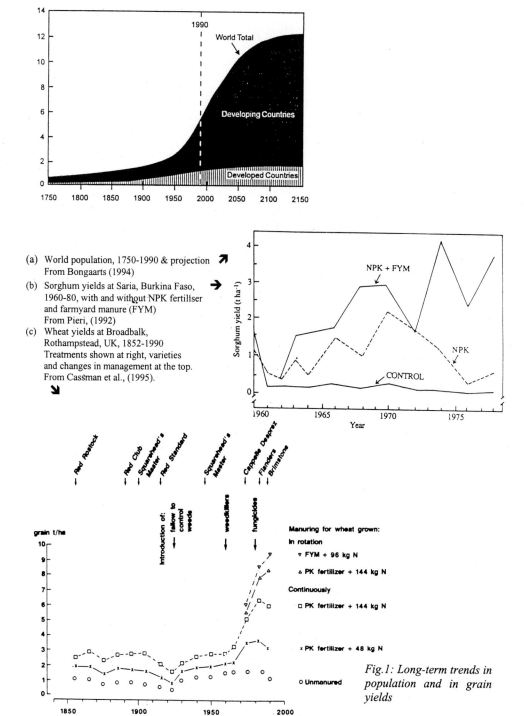

Fig.1: Long-term trends in population and in grain yields

technologies that are economically viable and socially acceptable (p.9). It nevertheless reproduced two diagrams of the downward spiral to the poverty trap. The title of the study, *Reversing the Spiral* at least indicates that degradation is not inevitable, and its cross-country time series analysis showed that yields are higher as the availability of cropped land per capita declines (Cleaver and Schreiber 1994, 68).

Historically, changes are more likely to occur along a trend-line, rather than to be either cyclical or static. Changes also interact with each other. To understand trends we need firstly to look at time measured in decades, and secondly at the interactions with other forces. The shape of the trend is important. Those who draw downward spirals tend to think of population growth as a straight upward trend, and of crop yields under their scenario on a straight downward trend to zero. In fact, neither is the case. The shape of these curves is illustrated in Figure 1.

1.1.1 Population growth

Figure 1a shows an s-shaped curve. Population grew slowly from the low densities prevalent before the nineteenth century, so countries spent centuries in slow growth. Growth in Europe accelerated from about 1800. In Europe, the acceleration took the form of an increase in average growth rates to around 1% per annum during a relatively long period from 1800 to about 1950. In Asia and Africa the annual growth rates have been higher, and the period of acceleration has been shorter. In Africa the acceleration has been confined to this century, and is already in some countries showing signs of the flattening out that has already occurred in many Asian countries. As density increases and societies become more complex, children become more expensive to rear. Their education costs more and their entry into paying occupations is delayed. People therefore desire smaller families (Tiffen, 1995 discusses this in relation to Kenya). The definitive shape of the global curve is as yet uncertain; it may start levelling out sooner or later than as depicted in Figure 1a, according to the speed with which people opt for smaller families.

1.1.2 Soil productivity

The trend line for soil productivity under constant cropping without artificial replacement of nutrients is not a straight line to zero. In many ways it has the opposite shape to the demographic curve: a fairly rapid fall initially, followed by a bottoming out at a low level of productivity. "All farm systems which are not sufficiently maintained tend to drift into a low-output steady state" (Ruthenberg, 1980: p.10). This curve is shown clearly in two long-term experiments, for sorghum in Burkina Faso (Figure 1b) and for winter wheat in the United Kingdom (Figure 1c). The initial drop is not shown in the latter, perhaps because it had occurred before the graph begins at 1850.

1. 2 Interactions: Technological change and social institutions

Fairly obviously, a rising population trend is likely to have a disastrous collision with a static yield trend. However, interactions occur between trends. Human beings normally take action to avoid disaster. They adapt and invent. Population growth leads to fundamental changes in the relationship between man and natural resources, in society, and in technology The yield results depicted occurred on research stations, not farms. The normal response to inadequate yields is the invention and dissemination of new farming technologies, as Boserup showed (Boserup, 1965). Technological change is the component that is missing from the famous analysis of demographic disaster made by Thomas Malthus - and from some of the vicious circle diagrams. Figures 1b and 1c illustrate the impact on yields of some simple changes in technology. At higher population densities it is easier for people to learn from each other and to spark new ideas (Simon, 1986). It is also cheaper to provide institutions for the spread of learning, such as schools and agricultural extension (Tiffen et al., 1994: 267).

Globally, the pace of technological change, and accompanying social change, is now in an accelerating phase. The technological change curve may be similar in shape to the population curve, but we do not yet know if there are circumstances under which it will flatten out. So far, on a global basis, the acceleration of technical inventions and improvements has enabled us to outpace population change. Globally, we are meeting total food requirements and incomes are rising, although there are inequities of distribution and national or local difficulties due to poor policies, wars and climatic variations.

1.3 Change in farming systems and in land management as population grows

Demographic growth is the inevitable result of the present population structure of many developing countries. Even though more countries are reaching the point at which people start to have smaller families, the large numbers of girls and young women already born who are reaching the child-bearing years ensures total world population will continue to grow for some decades. This is so even though they are likely to choose to have fewer children than their mothers had.

This growth necessarily leads to change in farming systems, social institutions and aspirations. The type of change we witness will depend upon the point which the society we are observing has currently reached. Population growth creates different problems for land management according to whether present densities are less than $30/km^2$ or more than $300/km^2$. It is therefore not sufficient to blame land degradation on "over-population" and "over-stocking" (domestic animals normally increase with the number of domestic hearths). We have rather to understand where society is now, and where it is likely to want to go, if we are to assist people to diagnose their present problems, and to find acceptable solutions.

Table 1 shows a rough approximation of the changes affecting agricultural systems and land management at different stages of population density. The density levels are very approximate, and will differ according to the main agro-climatic zones. Density will always be low in the arid zones, where pastoralism or ranching is are the only feasible farming systems. There is now a better appreciation both of pastoralists' rationality, and the resilience over the decades of grazing areas that may appear devastated by a few years of drought (Beinart, 1996, Behnke et al., 1993), but these will not be discussed here. This paper concerns farming areas where cropping is important.

Table 1 shows how population density, farming systems, and social and economic practices are all interrelated. Boserup (1965) showed not merely how labour inputs increase, and fallows diminish, as population density increases, but also how, simultaneously, tenurial customs evolve towards firmer and more extensive private rights. There are also increases in specialisation. This means that there are specialist teachers, traders, manufacturers, etc., who create an increasing market for farm products.

Boserup pointed out that shifting cultivation gives the best returns to labour when land is plentiful. It is now generally agreed that shifting cultivation does no long term damage where population densities are low (Okigbo, 1977). However, as population rises, fallows become shorter, and fertility falls. People adapt either by moving out to new land, or, by dividing existing land amongst their heirs, and adopting new husbandry techniques. It is logical to expect that as farms become smaller, and land becomes scarce and more valuable, people will want to improve, not devastate, their crucial resource. They are more able to invest labour per hectare as they manage a smaller rather than a larger land area. At the same time, the improved access to information and to markets, which comes from the increase in specialisation which accompanies higher population density, provides them with the stimulus and the means to do so. It must be stressed that land improvements cost the farmer money or labour time; they have to be worthwhile in terms of their benefit : cost ratio.

Additional pressure for change may come from changes in markets and prices, or from climatic variation. To quote Ruthenberg again: "... in farming there is usually a sequence of important changes (disturbances) over time, and most productive farm systems are not very stable. Most farm systems are therefore in a moving state" (Ruthenberg, 1980: p.10). He classified three types of change:

- inputs, activities and outputs change, but in a way that soil fertility is always maintained
- investments to improve soil and output are greater than required for maintenance, so productivity improves
- inputs are insufficient for maintenance, leading to the low-level steady state, or in extreme cases, man-made deserts.

Table 1 assumes that farmers will prefer to maintain or improve fertility, particularly as land becomes a more scarce and valuable commodity. However, whether farmers have both the means and the incentive to invest in adequate inputs and land improvement depends very much on the policy and infrastructural environment, as will be discussed later. It also seems that the period when population density is growing from a low to a medium level is in many ways the most difficult for farmers, both in terms of being able to recognise that the situation necessitates change, and in collecting the information and investment resources needed to make effective changes.

1.4 Definitions - ours and theirs

This paper follows Ruthenberg (1980) in taking soil fertility as the chief indicator of sustainable management. The question is whether, over a time period, the soil's productive capacity is being raised, maintained, or damaged. Time is the test of sustainability, but we have to think in terms of decades, not years. Productivity will vary from year to year, according to climatic conditions, as Figures 1b and 1c show. These variable yields may make it difficult for farmers to recognise falling fertility over a short period such as five years. When they do recognise it, their tactic may be to put on a large dose of manure in one year, and to let fertility run down for two or three years before putting on another dose. Many fallowing systems are based on cultivation for 3-7 years, with declining yields, before allowing the land to rest and recover. In both cases a visiting observer may witness the deterioration, but not the recovery. Older farmers are generally good at recognising falling trends that have developed over ten or more years, and can discuss their management tactics and responses, and the extent to which they have been able to halt, alleviate or reverse it.

Contrary to general belief, farmers' time horizons are generally much longer than ours. They may have in mind a twenty year period over which they plan gradually to improve their farm, as they obtain resources. The authors of vicious circle diagrams often incorporate poverty and inability to invest, because they would like to see a whole farm revamped in a year or two. Few farmers can manage this, but many poor farmers have a strong desire to pass the land to their children in good shape. They succeed by small incremental investments and improvements over twenty years.

However, most farmers define as development activities that some observers classify as degradation. An example is the clearing of land for new farms, especially during the medium density period when farming is expanding on to inferior or difficult land, such as that having steeper slopes, or swampy bottom land. The first necessary step is to cut down existing vegetation, and to plant an annual crop as quickly as possible to get food and money for the settler's immediate needs. In the humid zones of west Africa and Asia permanent tree crops, such as cocoa, coffee, oil

Zone	Arid, semi-arid, humid	Semi-arid, humid	Semi-arid, humid
Population per km^2	Low densities < 30/km^2.	Medium densities - 30-100/km^2	High densities: 100 - 600/km^2
Agricultural system	Arid areas: pastoralism; Semi-arid: agro-pastoralism and shifting cultivation; Humid: shifting cultivation, no animals if disease present.	Fallows shorten; annual cultivation at 50-70/km^2. Crops and animals becoming more integrated.	Double or treble cropping. Varying role for livestock.
Fertility management	No manuring, animals freely graze, long fallows or opening of new land.	Manure used; animals herded and kraaled. Mixed cropping or rotations including legumes.	Manure scarce in relation to demand; supplements purchased.
Land availability	On a district basis, less than 10% of potentially cultivated land farmed.	Percentage of cultivated land in older-settled areas steadily increasing; clearance of new land (of lesser quality).	No unoccupied, unclaimed land. Land has a high value.
Typical problems	Not serious: opportunistic use of pastures leads to "over-grazing" till herds move or die, when grasses recover; forest clearance may leave soil temporarily unprotected, but regrowth soon occurs.	Falling fertility as fallows shorten may lead to loss of vegetative cover and erosion problems; conflicts between herders and farmers as latter expand into new semi-arid areas, valley-bottom soils etc. Erosion if farmers move up slope.	Urban pollutants and sediments on agricultural land; roadside erosion; pollution from agro-chemicals; competition for water.
Investments	No land-related investments.	Investments in new land clearance; soil and water conservation begins after a delay. Tree protection and planting.	Land improvement and water conservation continues; new investment in irrigation and tree crops.
Tenure	Clearer has temporary use rights; community may exclude others from its territory; other rights unimportant as land plentiful.	Family head establishes firm rights to cleared and cultivated land, then to stubble, then to private grazing land, first by inheritance, later by other means.	Land can only be acquired by inheritance, purchase or lease. Active land market.
Market access	Marketing difficult; transport costs high; self-sufficiency necessary.	Market access slowly improves. As imported food is still costly, farmers combine family food crops and sales.	Market access good. Farmers begin to risk buying some of their food needs, and specialising.
Division of labour	Few specialists; distance makes access to them difficult.	More specialists, as large villages and small towns develop (tool-makers, teachers, builders, artisans, traders etc.)	Large non-farm sector providing services, goods and an expanding market; urbanisation.
Technology & information	Low technology level.	Technology improved by experimentation, learning, market access.	Access to information good; technology rapidly improving.
Education	Education by parents.	Primary schools cheaper and nearer.	Secondary education and training common.
Family size preference	Large families; high child death rate.	Large labour force needed; more survivors	Costs of children increase; small family.

Table 1: Population density and change in agricultural and socio-economic systems

palms, citrus, etc., will later be planted over an expanding area. These will give protection to the soil from the heavy rains, comparable to that offered by the original trees. In semi-arid areas farmers migrating up steeper slopes will often, out of the necessity of immediately feeding their families, cultivate carelessly. Manuring, terracing and other improvements will take place gradually, first in the fields nearest the homestead, and, later, in the more distant ones.

The period when forest is cut down has its obvious dangers for soil productivity, but some observers will view both the intermediate period of shifting arable cultivation, and the later replacement of native trees by cropped trees, as "degradation". The farmer will call the latter "development". The same situation occurs where farmers drain swamps, replacing the natural vegetation with rice, sugar, vegetables, or dairy farms. Personally I would agree with the farmer that this is development, provided that the productive capacity of the soil is being maintained and that down-stream damage is not being incurred.

Clearly, advisers should consult with farmers on their long-term objectives, and avoid as far as possible conflict between farmer aims and project aims. It is likely that the farmers who are the permanent inhabitants will ultimately win, rather than the short-term project.

2 The Machakos case study

Machakos is a District in Kenya, where changes in physical, biological and socio-economic systems between 1930 and 1990 were as far as possible measured and analysed by a team of researchers from the Overseas Development Institute, London and the University of Nairobi. Their reports have been synthesised in Tiffen et al., 1994. At the time the study began in 1990 the District boundaries included some 10% of semi-humid highland, the area of first settlement, some 40% of better semi-arid lower land, where settlement increased in the first fifty years of the century, and 50% of poorer semi-arid land, mainly settled since 1960. There are two growing seasons per year, but the rain is so variable that in most of the district there is only a 60% chance of obtaining 250mm in a season. We were fortunate that the sites of photographs taken by a soil scientist in 1937 had been sufficiently well described to enable us to photograph the same areas in 1991. There was also a series of air photographs, 1948-78, as well as very good archival records and other reports. Elderly community leaders in a sample of five villages provided further insights, as did farmers interviewed during transects.

The results showed that population had increased from 240,000 in 1930 to nearly 1,500,000 in 1990. Density in the three main agro-ecological zones had increased as shown in Figure 2a. Much more land was used for cultivation. Most of this cultivated land had been terraced, by 1978 in the older areas, (Figure 2b) and by 1991, when our observations were made, in the areas settled since 1960. Terracing, and other methods of concentrating and conserving water, has enabled farmers to diversify, adding small areas of coffee, fruit and vegetables to the main staples of maize and legumes. Livestock numbers are considerably higher than in 1930, despite the reduction of the grazing area, since they are also fed crop residues, and from cut grass planted on the terrace lips. In the semi-humid zone they are invariably stall-fed or tethered. However, there has been a reduction in the number of livestock per farmer, and there is a manure shortage now in relation to demand. Some is bought in from neighboring ranches, and farmers also buy fertiliser for coffee and vegetables. The wood-fuel crisis often predicted has not materialised. Households have every incentive to economise on their use of fuel, but it remains available from planted or protected trees in fields and hedges, and, in the semi-arid areas, from the trees on the grazing land. The latter, now all privately owned, is managed to produce fodder, fuel, timber, charcoal and honey; the last three as a source of revenue. Figure 2c shows the increase in the value of production per head. This excludes the fuel and timber, which we could not value, and also disregards the 50% of rural income that now comes from the expanding non-farm sector of services, trade and small businesses.

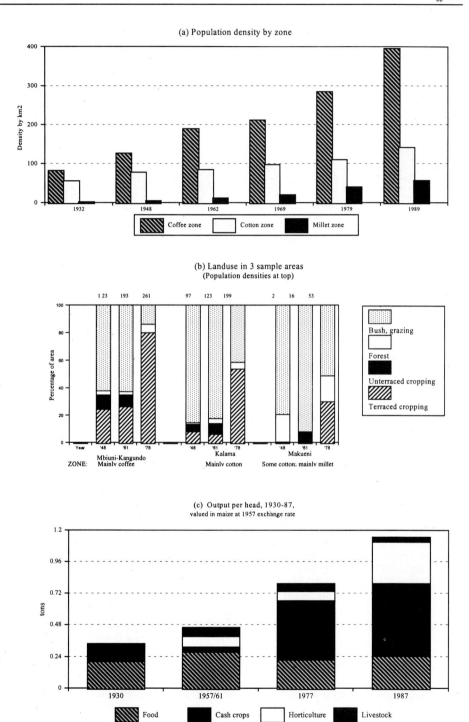

Fig. 2: Population density, land use and agricultural output, Machakos District, Kenya, 1930-90

Average agricultural output increased despite the fall in the average quality of the settled land, as farming spread into the dryer zones of the district.

Machakos certainly suffered horrendous erosion problems in the 1930s and 1940s in the two better zones where the population was then concentrated, and where densities averaged around 75 and 50 per km^2 respectively in 1948 (Figure 2a). The government launched a compulsory programme of work. Improvement began during the late 1940s, when the northern part of the district got better roads to the growing town of Nairobi, and farmers saw that money could be made from vegetables. In the 1950s they were allowed to grow coffee, previously a "European" crop. In the conditions of Machakos, both coffee and vegetables require terraces to conserve water. Farmers selected by preference a labour-intensive terrace constructed by throwing soil from the ditch uphill, which starts the formation of a bench. The compulsory programme had recognised the shortage of labour, teaching the labour-saving method of throwing the soil downhill from ditch dug along the contour. The ditch led water to a point of safe discharge, such as a grass strip or natural water way. The ditch was difficult to maintain, as it often broke in heavy rains. It was also less effective than a bench in directing scarce water to the crop roots. During the 1950s there was a substantial advisory input from agricultural officers during a special government programme, which also provided tools to work groups. This programme became more successful when the work groups were encouraged to appoint their own leaders and choose their preferred technology. This was usually the bench.

For a few years, leading up to independence in 1963, farmers abandoned terrace construction, as a sign of their opposition to colonialism, but by the mid 1960s' individuals and groups resumed it, because it was effective. The special programme came to an end in 1962, and there was little official advice on conservation until 1978, when a new government programme began. Nevertheless, air photographs show how much was accomplished between 1961 and 1978 (Figure 2b).

Following independence, people also surged into the dryer areas. Much of this area had been Crown Land, which they had only been permitted to graze on licence during the colonial period. The process of settlement horrified observers in the early 1970s. Settlers cut down trees and burned the bush. They were also accused of shifting cultivation, since they cleared new areas each year. The farmers, (who came from hills where they had practised permanent annual cultivation) explained to us in 1990 that they had never practised shifting cultivation. They cropped a new area each year, because this gave them a permanent land right according to local custom, even if they subsequently let it revert to grazing land.

The government had begun a process of land title registration in the 1950s. This only reached Machakos in 1968. The process is very slow and was still incomplete in 1990. People had their boundaries unofficially demarcated by elders, which is sufficient to enable them to buy, sell and lease land. In some cases, they had cleared more than they could manage, and they gave or sold land to friends and former neighbours, who helped them in the battle against marauding animals.

By 1990 the average new farm was about 10 hectares (compared to 1-3 hectares in the highlands). Part was under crops, usually terraced, and part was used for combined grazing and timber. Many could not completely control bush regeneration on the grazing area and woody species were expanding at the expense of grass. In the better semi-arid zone, longer settled and with higher population densities, grazing areas were smaller, and bush management more intensive. In general, the smaller the farm, the more trees per hectare, since they produce valuable products and services.

In the older areas, that were the worst eroded in the 1930s, erosion has largely been controlled on arable land (which is now a higher proportion of total land). In the remaining areas of grazing land, management is improving, although erosion has not yet been totally halted. Farmers use their limited investment resources first to protect the land giving the highest returns (the arable). They only improve grazing areas at a later stage. However, fencing and hedging means that the land is only grazed by the farmer's own animals, and he or she can limit and control fodder off-take. The

Date	Country	Comment in summary	R
	Asia		
1989	China, Loess Plateau (LP)	Population control the key to land protection. Circa 113/km²	+
1993	China, LP	Population stated; no comment.	?
1986	China, LP	50% increase in erosion due to population increase.	+
1988	China, Yellow River	Geological and accelerated erosion present. Worst erosion not where population highest.	-
1991	Sri Lanka	People moving on to steep slopes above tea plantations.	+
1985	Indonesia, Java	1 million ha critical due to population pressure [project proposal].	+
1989	North Yemen	Destruction of terraces, out-migration.	-
1989	Syria	Population pressure > overgrazing.	+
1944	India, Punjab	Expanding population trying to survive by shifting cultivation; overgrazing.	+
1990	"	Siwalik hills eroded after opening for cultivation in mid 1800s.	+
1991	India, Kerala	Only small fraction of land adequately treated.	+
1991	Thailand, N.	Expanding population moving up hills; shifting cultivation.	+
1991	Taiwan	Expanding population cultivating steep slopes.	+
1985	Himalayas	Expanding population moving upslope. Roads create landslides. Overcropping, overgrazing, overcutting.	+
	Europe		
1990	England	Increases in population and erosion 3,000BC, 500BC, Roman era, 1200, 18C. Drop now, because of less fallow.	±
1987	Europe	Massive erosion in 18C with rising population; serious also in 14C with falling population and economic decline. Now less, new techniques.	±
1990	France, Italy	Land abandonment in Mediterranean areas > fires, deforestation.	-
1987	France	Population increase and increased cereals for subsistence > erosion on cultivated mountain land, floods on plain, due to deforestation. After 1871, mountain emigration.	+
1987	Greece	Erosion most frequent near high density areas.	+
1987	Spain	Population increase in 18 and 19C led to woodcutting and clearing > erosion. 20C: abandonment of marginal land >accelerated erosion.	±
	Africa		
1977	Africa	No erosion if fallows long. Serious as they shorten.	+
1988	Africa	Huge problems in SSA, but not due to population density.	-
1990	Kenya, Machakos	Population 1.1 million, carrying capacity for subsistence farming 530,000.	+
1993	Kenya, Machakos	Tiffen et al. See text, Section 3 of paper.	-
1984	Kenya, Machakos and Nyeri	Machakos holding its own with population increase; Nyeri alleged to be dropping. [Nyeri would have much higher population density than Machakos].	±
1977	Kenya, Machakos	Population increase in first half of 20C led to erosion.	+
1983	Kenya, Baringo	Said to be worst-eroded district in 1930s; tragic in 1963. Farms now expanding on steep slopes. Semi-arid.	+
1991	Kenya	85% of country experiencing desertification due to population pressure.	+

Table 2: Population growth, positively (+) or negatively (-) related to soil erosion (R)

1993	Kenya	70% of arable land terraced; agricultural output has outpaced population growth.	-
1983	Kenya, Samburu	Overpopulation and overgrazing [pastoral].	
1989	Sudan, dryland	Short fallows, overgrazing.	+
1953	"	"The creeping desert".	+
1972	Nigeria, northwest	Land husbandry better when density high. Danger point 77-96/km²	±
1951	Nigeria, Eastern	Gullies may be geologic, rather than due to population. Fertility loss.	±
1965	"	Due to over-farming.	+
1964	"	Population pressure, gullies, fertility loss.	+
1972	"	Shortened fallows.	+
	Nigeria, Tiv	Fallows shortened to 2 years instead of 10 in heavily populated area. Fertility loss.	+
1972	Tanzania, Uluguru Mts	Deforestation due to expansion of peasant farming. High density(?).	+
1982	Tanzania, Mbula	Fallows shortening.	+
1937	Malawi	Population spreading uphill.	+
1960	Morocco	Rapid population increase> clearing new land > deforestation for fuel and goats. Also deforestation in 7C near cities.	+
1981	Morocco, Rif	50 years of mostly unsuccessful conservation; population increase from 30 to 50/km², 1900-60.	+
1972	Algeria	Population expansion> wood-cutting, began in mountains in mid 18C.	+
1990	North Africa	Population pressure > deforestation.	+
1961	Burkina Faso	Especially round city.	+
1942	Zimbabwe	Little with shifting cultivation. Increased with use of plough.	+
1975	Lesotho	Shorter fallows, over-grazing, falling yields.	+
1982	"	Mounting population pressure.	+
1956	South Africa, near Durban	Erosion in dam catchment area due to absence of men, poor farming.	-
1972	Uganda, Karamoja	Pastoralists with shifting cultivation, shortening fallow. High population density (?)	+
1954	Uganda, West Nile District	No erosion with shifting cultivation. In south dense (?) population. Shorter fallows and tobacco-growing > erosion.	+
1972	Ghana, North East	Shortening fallow, sheet erosion. Uneven population density, average 91/km², 386/km² in parts. 38% unoccupied due to river blindness.	+
1984	Ethiopia	Intact natural cover, no erosion. Erosion with intensively cultivated grains. Only small part of potential rainfed area cultivated.	+
1983	"	Expansion up steep slopes; deforestation.	+
	Americas		
1987	Haiti	Shifting cultivation, reduced fallows, uncertain tenure.	+
1987	Ecuador	Erosion worst on 15% of land with highest population density (?).	+
1978	Guatemala	Accelerated deforestation in last 3-4 decades. Case study of Totonicapan [170/km²?]; high density with less deforestation.	±

Source: derived from CIESIN (1995).

Tab. 2: continued

worst eroded grazing land is often that which is the subject of a dispute about ownership, and where, in consequence, an investment in fencing has not been worthwhile.

In the newer settled areas erosion probably increased initially, but it is now being brought under control on the cultivated portions. Farmers have become aware of their problems, and they know that there is no new land left. The time between initial settlement and the beginning of protective measures is now shortening. This has probably been assisted by advisory soil conservation programmes in the 1980s, started by the government, or the churches and NGOs. However, about half the work done between 1981-5 seems to have been outside these programmes.

3 A global sample

CIESIN (1995) provides a bibliography on erosion of some 600 items. I did a search, extracting all those items where the word "population" occurred in the abstract. These 58 cases have been grouped under continent and country in Table 2. They do not purport to be a full search of all the items where increase or decrease of population was believed to have affected erosion rates, since sometimes the process might be referred to as increase in cropping intensity or in grazing pressure. Nevertheless, they represent a fair sample of items in the bibliography where population growth or decline was suspected to be a cause of change in soil erosion rates.

At first sight Table 2 appears to confirm a relation between population growth and increase in erosion. The positive association is shown by a + in the final column; 43 out of the 58 cases. In only 7 cases is a negative association noted - either that a smaller population cannot cope, or that a larger population has resulted in better care, or less erosion. In six cases both positive and negative associations are noted, either in different parts of the area, or at different periods. Closer inspection shows that erosion is most often mentioned in cases where population is low but increasing, rather than when densities are high.

Where the summary gives figures from which population density can be derived, this is noted. In most cases there is only reference to expansion, or, in a minority of cases, to decline through emigration. In a few cases the term high density is applied to areas which, according to the classification used for Table 1, come into the medium density zone - or even the low density zone, (for example in the two Ugandan cases). There are 5 references to shifting cultivation and pastoralism, which implies low densities. There are 8 references to shorter fallows, which implies the low end of the medium densities, with the stage of permanent cultivation not yet reached. Nine cases apparently refer to the initial stages of settlement - this overlaps with six references to steep slopes (population expanding up-slope). Erosion associated with population increase in Europe usually refers to the eighteenth century, when densities would have been at most in the medium range. In India, the description of the Punjab in 1944 with an expanding population trying to survive by shifting cultivation seems odd for the province taken as a whole. A search of the entire bibliography for other descriptions of the Punjab revealed only six different references to the Siwalik Hills. These hills are obviously a special case, and in one instance were definitely described as sparsely populated.

In general, Table 2 supports the hypothesis that the stage of transition between long fallows and permanent cultivation is a difficult time, and that erosion may be especially marked in the first stages of new settlement. It does not tell us what happened later, because in most cases experts visit the areas where bad erosion is occurring, and do not revisit them when erosion has been mitigated by better husbandry. It is interesting that Prothero (1972) noted a danger zone in north-western Nigeria at around 77-96/km^2, with better husbandry when densities exceeded this level. The area around Kano city in northern Nigeria has a population density now of some 300/km^2. The land has been under permanent cultivation since at least the beginning of this century, with a high standard of husbandry, and no sign of a falling yield trend (Mortimore, 1993a, 1993b).

It is worth noting that special cases occupy soil erosion experts greatly - the Loess Plateau in China and the Awka area of Eastern Nigeria obtain a similar tally to the Siwalik Hills. In these cases there is probably a geological contributing factor, due to the nature of the soils, and it can be quite difficult to separate man-made erosion from natural processes. A couple of the more alarmist reports may have been affected by a desire to secure funding - in the Java case the reference document was a project proposal.

4 Kenya

Visits to 3 to 4 sites in each of 6 different districts in Kenya, where the Government's Soil Conservation and Water Branch had undertaken concentrated work in defined catchments, led to the observation (in January 1996) that erosion was most prevalent, and most difficult to cure, in areas where population density was low. One example was an inland area of Coastal Province, where land was still plentiful, land rights uncertain, and livestock wandered freely over their owner's and neighbours' farms. Cut-off drains and trash lines put in four years earlier had almost disappeared. Another area was the notorious lower slopes of Baringo, in the Rift Valley, where a small and scattered population could barely make a dent on huge gullying that originated in the steep upper slopes. In the high population density areas farmers had generally taken several measures to prevent erosion and to improve fertility. The standard of land husbandry was especially good in the densely populated parts of Eastern and Central Kenya, where there was good access to markets. It was markedly less so in the equally densely populated area in the far west, where markets were more distant, and cash was derived mainly from the remittances of absent menfolk (and some daughters). In Eastern Kenya, in a semi-arid zone where farmers were establishing new farms, and where population density fell into the middle bracket, fallows were diminishing, and work on conservation measures was being initiated. This area appeared set to follow the Machakos path (Tiffen et al, 1996).

5 Other types of soil decline

This paper has concentrated on soil erosion. There are, of, course, other forms of soil deterioration. Some of these occur in the high density areas, for example, the salinisation of irrigated areas. However, often this is not because of high population density, but rather, to poor management and maintenance of the canals and drains. This may be due to the failure either, to set an adequate irrigation fee, or, to ensure that the irrigation agency has the necessary budgetary control and incentives to give adequate resources to maintenance.

What is important in many high density areas is a fall in fertility due to inadequate replacement of nutrients, or to a deterioration in soil structure due to falling organic content. These again are often not due simply to population pressure per se, but are often associated with market distortions, or to difficulty in obtaining chemical fertiliser, or adequate information about its use. In Western Kenya, for example, the manure from the cow most families kept was not utilised as carefully as in central Kenya. This seemed to be due to greater difficulties in marketing profitably, and therefore, less incentive to undertake the associated hard work. Everywhere in Kenya, in 1995 and 1996, farmers were in difficulty because the government Milk Marketing Board had been failing to pay them for the milk they had delivered, disrupting their ability to buy fertiliser.

6 Necessary policy conditions for sustainable intensification

This paper argues that there is a natural tendency for populations to invest in land improvements as the population grows. Management becomes more intensive as farms become smaller. However, as already stated, land improvement always requires investment of hard effort and often, also money. It is only undertaken willingly when it is profitable, and when it is known that the improved land will benefit the improver and his or her children. Necessary conditions for a growing population to invest in land improvements include: security of tenure and private control both of land and livestock; moderate taxation of agricultural output to make investment possible and attractive; promotion of trade and the spread of information (for which peace and internal security are essential); and investment in, and maintenance of, transport infrastructure to improve marketing. Where these conditions are lacking, as in Ethiopia and countries formerly under Soviet influence, population may increase without inducement to invest in land improvement, and the result is often land deterioration or land pollution. The customary evolution of secure land rights has sometimes been impeded by botched or unwise land reforms or villagisation programmes, although land reform has succeeded in some Asian countries. It is also helpful to have an extension policy which leads to frank discussion with farmers of their aims, knowledge and resources, and constraints, and which then offers them a range of technical options to suit individual circumstances. Top-down imposition of one technology for all has a record of failure, particularly when it is tried in advance of need, when land is still plentiful and the labour to manage it is limited.

7 Conclusion

The relationship of population to land conservation is not straightforward. Land improvement does not occur smoothly. Even under good policies, periods of deterioration normally precede and spur periods of improvement. This is particularly so when population density is low but growing; at around 30-50/km^2 fallows have to be reduced and farmers may need time to realise that new methods of restoring fertility are necessary. At these levels, marketing conditions are still difficult, and the means for investment are lacking. Conservation programmes can only succeed in such areas when they run in conjunction with programmes to improve market access and profitability. As population levels build up, access to market and information tends to improve, and the increased shortage of land means there is an incentive to improve the smaller areas that will be inherited. Tenure systems take time to develop from common access to fallows and grazing land, to controlled grazing on private land. Most soil conservation measures are incompatible with free grazing, as Kenya has found despite substantial efforts in such areas with skilled staff.

At higher population densities the incentive and the means to conserve and improve soils are present. Steep slopes can be well managed, as in Nepal and Java. In Nepal there are landslips due to natural processes, but farmers can and do repair these in a few years. Improvement is still taking place in Java, as evidenced by a comparative study of villages in 1974 and 1988, which were still experiencing population increase (Prabowo and McConnell, 1993). Management may deteriorate if labour is drawn off to more profitable occupations elsewhere. The deterioration of terracing in the Yemen due to much labour out-migration to Saudi Arabia is well known. At very high densities new problems occur, perhaps in the management of soil fertility, or the management and maintenance of the irrigation systems which are necessary for double cropping and/or higher yields. However, in such cases, policies may be more critical than population density. Sustainability is a characteristic that can only be judged over decades, not years.

References

Behnke, R., Scoones, I. and Kerven, C. (1993): Range Ecology at Disequilibrium. New Models of Natural Variability and Pastoral Adaptation in African Savannas,: ODI, IIED and Commonwealth Secretariat, London.

Beinart, W. (1996): Soil Erosion, Animals and Pasture over the Longer Term. In: M. Leach and R. Mearns (eds.), The Lie of the Land: Challenging Received Wisdom on the African Environment, London: Heinemann and Currey.

Bongaarts, J. (1994): Global and Regional Population Projections to 2025. Paper presented at the Round Table on Population and Food in the early 21st century. IFPRI, Washington D.C.

Boserup, E. (1965): The conditions of agricultural growth: The economics of agrarian change under population pressure, Allen and Unwin, London.

Cassman, K.G., Steiner R. and Johnston, A.E. (1995): Long-term Experiments and Productivity Indexes to Evaluate the Sustainability of Cropping Systems. In: V. Barnett, R. Payne, and R. Steiner, Agri-cultural Sustainability: Economic, Environmental and Statistical Considerations. Chichester, UK, John Wiley.

CIESIN (Consortium for International Earth Science Information Network) (1995):.Land Degradation and Desertification Bibliography. Version used was downloaded from desert-l@wcmc.org.uk via hdgec@ciesin.org. [Compiler H. Dregne (personal communication)].

Cleaver, K. M. and Schreiber, G.A. (1994): Reversing the Spiral: The population, agriculture and environment nexus in sub-Saharan Africa. The World Bank, Washington D.C.

Greenland, D.J., Bowen, G., Eswaran, H., Rhoades, R. and Valentin, C. (1994): Soil, Water and Nutrient Research - A new Agenda, IBSRAM Position Paper, Bangkok.

Mortimore, M. (1993a):The intensification of peri-urban agriculture: the Kano-Close-Settled Zone, 1964-1986. In: B.L. Turner, R.W. Kates and D.L. Skole (eds.), Population Growth and Agricultural Change in Africa, Gainesville: University of Florida Press, 358-400.

Mortimore, M. (1993b): Northern Nigeria: land transformation under agricultural intensification. In: C. Jolly, and Torrey, B.B. (Eds), Population and land use in developing countries, Washington, D.C: National Academy Press.

Okigbo, B. N. (1977): Farming systems and soil erosion in Africa. In: D. J. Greenland and R. Lal (eds.), Soil Conservation and Management in the Humid Tropics, New York, John Wiley & Sons, 151-163.

Pieri, C.J.M.G. (1992): Fertility of Soils. A future for farming in the West African Savanna, Berlin: Springer Verlag.

Prabowo D. and McConnell, D.J. (1993): Changes and development in Solo Valley farming systems, Indonesia, FAO, Rome.

Prothero, R. M. (1972): Some observations on desiccation in north-western Nigeria. In: R. Mansell Prothero (ed.), People and Land in Africa South of the Sahara, London, Oxford University Press, 38-48.

Ruthenberg, H. (1980): Farming systems in the tropics, Clarendon Press, Oxford.

Simon, J. L. (1986): Theory of population and economic growth, Basil Blackwell, Oxford.

Tiffen, M., Purcell, R., Gichuki, F., Gachene, C. and Gatheru, J. (1996): External Evaluation of the National Soil and Water Conservation programme, Kenya. Overseas Development Institute, London.

Tiffen, M. (1995): Population Density, Economic Growth and Societies in Transition: Boserup Reconsidered in a Kenyan Case-study. Development and Change **26,1**: 31-66.

Tiffen M., Mortimore, M. and Gichuki, F. (1994): More people, less erosion: Environmental recovery in Kenya, Wiley, Chichester, UK.

Address of author:
Mary Tiffen
Drylands Research
17 Market Square
Crewkerne, Somerset TA18 7LG, UK

Culture and Local Knowledge - Their Roles in Soil and Water Conservation

W. Östberg & Ch. Reij

Summary

Opening one's eyes to the cultural dimensions of land management makes one aware of the diversity of solutions developed around the world by individual farmers or groups of farmers. Blue-print approaches to soil and water conservation stand meagre chances of succeeding. After a brief discussion of different interpretations of 'culture', examples are given of local conceptualisations of natural resources in different settings. One conclusion drawn is that things are seldom what they seem. Another is that the perspective has changed fundamentally from a „we" studying „them" to different voices making themselves heard. Research findings are but one version among several. A different „we" emerges, encompassing both „us" and „the other" which is much better adapted to practical work for improved management of natural resources.

The paper then proceeds to outline local technologies in soil and water conservation (SWC) in Africa. Indigenous SWC techniques commonly continue to be in use for long periods of time and are in many cases also improved upon, whereas modern techniques more often than not are inadequately maintained and rarely modified by the farmers who are supposed to benefit from the improvements. Since the late 1980's there is a growing consensus that one should build on local techniques whenever possible. In a limited number of cases this has been done, and the results are promising. Improved traditional planting pits in Burkina Faso and Niger exemplify this process.

The way towards marrying traditional and modern knowledge goes via direct cooperation between farmers and researchers. While this may be a rather obvious point to make it is little adhered to in practical conservation work. The paper thus concludes on the consoling note that farmers, agricultural engineers, physical and social scientists all have contributions to offer. If only they come together. This means proceeding slowly and cautiously trying solutions as we move ahead.

Keywords: Culture, indigenous soil and water conservation, Africa

1 Introduction

The concept of culture has taken on new meanings in anthropology and related fields of the social sciences during the last ten years or so. While it may appear an academic question of little concern to practitioners of soil and water conservation, this changed understanding of culture may in fact provide new insights into SWC and help us to become better aware of what goes on.

A common-sense understanding of 'cultural aspects of soil conservation' is very likely to be that it deals with how traditions, values and life styles determine the way land is managed: the way

we do things around here; the way we care, or do not care, for our land. However, to such a reasonable interpretation a number of questions can be asked:
- Do all members of a local community know the same things ? Is it not rather the case that they know quite different things depending on age, class, ethnicity, gender, professional specialization, etc.
- Is there such a thing as 'the local community'? Most villages are linked to a wider society influencing much of their way of life. And what about The Ubiquitous Market? Does it not by and large determine what local communities are to grow and sell? Can we really talk about local cultures when national and international forces strongly influence life in any village we can think of?
- Everywhere young people leave their home villages hunting new opportunites in urban environments. Returning home for visits they display new life styles, such as the acquisition of spouses originating from other parts of the country, adherence to alien religions and occasionally even demonstrations of conspicious affluence - all of which influence life in their home villages.

What we somewhat paradoxically encounter when we look for the cultural dimensions of land management, for 'traditions', thus turn out to be changes, differences, variations, mixes. Cultures are in the plural (Hannerz, 1993: 97) and they are changing. All cultures have their locally anchored histories, of course, but they are also created by distant powers and markets, by educational institutions, media, migrations and generally by the flux of time. And by individuals acting, innovating.

Critics of the received understanding of culture (i.e. that which in a given society is being passed down from one generation to the next) argue that the preoccupation of anthropologists with describing and explaining cultural differences have resulted in constructing and maintaining such differences. Describing a society's culture almost inevitably is to make it appear more static, more homogeneous than what people's lived realities are. Culture therefore becomes „a tool for making other", in the words of Abu-Lughod (1991), who goes on to say that cultural studies are dehumanizing, to the evident surprise of generations of anthropologists who themselves have liked to think that their work gives justice and even respect to peoples otherwise little considered.

More than 30 years ago Leach (1965: 27) pointed out that the blunder may be „the misleading idea that culture exists independently of those who inherit it". This easily leads one to describe „culture" as a localized assemblage of various items, different for each culture. This is „a thesis which conveniently reinforces the author's feeling that he is intellectually superior to the others whom he studies ... „we" act in a rational manner; it is only the „others" who are ignorant slaves to irrational custom" (*loc. cit.*).

So far we have said that in the current debate within anthropology „culture" may mean something different from what we perhaps expect it to mean. What, then, does „culture" stand for?

2 Hedging in the „cultural"

Instead of trying to decipher more or less enduring values and customs near and far, anthropologists have increasingly come to understand culture as „meanings which people create, and which create people, as members of societies" (Hannerz, 1992: 3). Not least the last four words are important: it is in a social context we consider culture.

Many factors other than logical thought are behind how people understand themselves and create meaning, including for instance feelings, taste, smell and sound. If so, it must be simplistic to base accounts of people's ways of life merely on interviews and even more so on information gathered only via questionnaires. „Indeed, we should treat all explicit knowledge as problematic, as a type of knowledge probably remote from that employed in practical activities under normal circumstances" (Bloch, 1991: 194). Research into social processes becomes far more interesting and realistic, but also demanding, as we approach all that which is beyond standard answers to standard questions.

If cultural studies currently overflow with accounts of continuously changing situations, relations, multiple meanings, there is of course in any society also something permanent, all that which goes on day by day, year by year, as people live their lives and look after their land, or do not do so.

Sometimes strongly expressed feelings of local belonging, of „our tradition", develop in opposition to the surrounding world, for instance against a soil and water conservation project, to defend local interests against what people fear to be take-over attempts by outside forces (cf. for instance Östberg, 1995: 82ff.). This may result in strongly argued versions of the past where variations and conflicting elements are minimised. Culture is often about political rights (Weiner, 1995: 18). In such situations it is obvious that culture is not about how things used to be as much as about what is coming into being.

Two processes thus occur simultaneously. People build collective identities with the help of common experiences, views, rules and symbols at the same time as they keep changing their views depending on context and try to steer clear of collective representations. To study cultural aspects of soil conservation is therefore not about documenting a number of supposedly typical and eternal „traditional practices" but to understand both permanence and change, both homogeneity and differentiation. While this may seem plain enough when expressed as a general argument it remains an insight little utilized in studies of local natural resource management.

3 Reflexive studies

As social scientists now increasingly find that they are trying to capture floating relations and influences between acting individuals, rather than documenting cultural rules, it automatically follows that people's subjective experiences come into focus in a much more decisive way than before. Typified categories like 'villagers', 'farmers', or 'land managers' of earlier studies give way to real people who make their voices heard and who demand to participate in both research and development work. Those whom we used to think of as 'informants' have become 'actors', or 'stake holders', with whom researchers interact rather than merely study their way of life.

Social scientists are well aware that absolute objective truth is unattainable. Certainly research work should be as unbiased as possible. The documentation should be open to inspection by others and the reporting as truthful as possible. Still, it is unavoidable that information is gathered in particular situations involving specific relations to people in a community under study. The findings we report are not the same as The Truth about a certain place, or a certain land use practise, even if they represent knowledge about them.

Researchers construct realities as do other actors on the local scene. One may compare different versions, agree on what are reasonable interpretations, but scientists cannot claim the monopoly of understanding the situation. Neither for that matter can planners and politicians (cf. Blaikie, 1995a). Instead of being authoritative guides to other people's lives, social scientists increasingly find themselves involved in new kinds of relationships where mutual exchange, negotiation and continuous learning are more important than lecturing.

The reorientation towards a more dynamic concept of culture, as well as the current preoccupation with reflexivity (i.e. problematizing the role of the researcher and of science), and an increasing focus on local variations, could inspire soil and water conservationists to examine further their own historical, cultural, and political roles.

Three brief examples of what outside observers hold to be the gist and implications of local traditions could serve to bring home the point that (the study of) local cultures are not always what we expect them to be.

4 Exemplifying culture

Case 1. Among the *Marakwet* of north-western Kenya a son inherits land from his father via his mother. If a man has more than one wife his land will be divided equally between the "houses" (or „wombs"). This means that a son with few full brothers will inherit a larger share than those of his half-brothers who happen to have several full brothers. A son born out of wedlock has the same right to inherit as legitimate sons. His mother will be regarded as a separate house. This means that a single extra-marital son may get a larger share of his father's estate than the legitimate sons. Each wife cultivates "her" fields with the help of her children while they also "help" the husband/father with his fields. The wife/wives are responsible for feeding the family. The grains kept in the husband's granary are for his private use, for feeding guests, for ceremonies and as an extra security against famine. For the husband it makes sense to have many children supporting him with farm work and if he has more than one wife his security increases, particularly if the different households can be based in different ecological zones. Likewise a woman gains from having several children helping her. Everything functions better if there are many arms to do the work. And when it comes to settling political and judicial matters it may be a considerable advantage for all to belong to a populous lineage. However, currently increasing pressure on land makes small family size an obvious choice - at least in the eyes of planners and experts. But, as we have seen, other factors suggest that large families make sense. There is furthermore the contentment of being part of a large family. Both men and women can tell of the proudness they feel when they set off for a celebration as a sizeable group. Or the comfort one enjoys from having other family members around or the sense of satisfaction at belonging to a growing descent group. Many things count when it comes to deciding whether large families make sense or not.

To summarize: both the mounting land scarcity and the inheritance rules point to monogamous and small families. Everyone in Marakwet is well aware of the conflicts that may reign between co-wives and their respective sons, for obvious material reasons. Yet, many Marakwet men and women appreciate having a large family. The family represents stability, cooperation and unity, not only strife over resources. Looking at the family from the point of descent, increasing population is not a problem but on the contrary demonstrates successful relations, people living decent lives. Offspring are seen as a confirmation that people respect God and one another - that things are the way they should be. It cannot be determined on the basis of principles whether the descent ideology or regarding the family as an arena for competition over natural resources will empirically provide security and good living conditions. Things are not quite the way family planners expect them to be. And this is not because of dated „customs".

Case 2. *The Zafimaniry* are a group of shifting cultivators in eastern Madagascar. They are currently running out of forest as a result of overswiddening caused by both growth in population and the creation of nature reserves. People have had to turn to irrigated rice cultivation. Their closed, forested landscape is changing into cleared, wide, levelled valleys, terraced for irrigated rice fields. Maurice Bloch, on whose account (1995) this short summary is based, writes that he tried as hard as he could to get villagers to tell him how much they deplored the change in their environment. „I failed to get the slightest response, though people occasionally, and without much interest, noted a few minor inconveniences which deforestation causes" (p. 64). Once an old woman appeared almost to have swallowed his bait when she agreed that she liked the forest. Why? „Because you can cut it down." (p. 65). As Bloch's analysis unfolds we learn that the reasoning behind is not quite what outsiders may expect, i.e. that the Zafimaniry apprehend the landscape in a utilitarian manner. No, they „are as enthusiastic as are the *Guide Michelin* and municipal authorities about good views" (p. 65). Cleared mountain summits, where villages are built, is one such much appreciated view, since it represents the achievements of ancestors „who have inscribed themselves on to the unchanging land, especially on those points which rise out of the chaos of the forest" (p. 74). Similarly, the new terraced rice valleys tell of people who have successfully made their mark on the land.

So, to these swidden culivators deforestation in fact tells of their successes and not of a development running out of hand, as their visiting ethnographer would prefer them to think.

Case 3. *The Rangi* of central Tanzania live in a very severely degraded area (Christiansson, 1988; Christiansson et al., 1993; Payton et al., 1992) and have for more than fifty years been exposed to determined efforts of land rehabilitation by the authorities. Whole villages have been forcefully removed from denuded land (Fosbrooke, 1950/51) and all livestock expelled from an area of 125 square kilometres to allow the land to rest and revive. Through the decades experts have agreed that the so-called Kondoa Eroded Area cannot support its population. Yet many Rangi have been anything but cooperative and grateful „beneficiaries" of this concern for their plight. Instead of putting their efforts into land rehabilitation work, as demanded by the authorities, many Rangi have preferred to combine resources from the degraded core area with extensive cultivations on the plains below the Kondoa Eroded Area (Mung'ong'o, 1995; Östberg, 1986). Some illegally keep herds of livestock within the enclosed area, thereby obstructing offical policy. Thus, many Rangi disagree with the experts on what makes for stable and prosperous life in the Kondoa Eroded Area. Many farmers of this badly degraded area seem to prefer to make their living through means other than soil conservation. Adapting to changing circumstances is what through time has characterized Rangi farming. As in many other parts of Africa farmers find it advantageous to move back and forth between different modes of production, depending on changing natural as well as political and economic conditions, whereas planners would prefer them to stay put and invest in soil conservation.

The three examples we have now rushed through, Marakwet, Zafimaniry, Rangi, have one common theme, that things often are not quite the way we outsiders expected them to be. People may see the world differently from what our reasoning makes us believe. Nevertheless aid agencies invest substantial sums to make their understanding of the situation come true, paying the bills for family planning in Marakwet, organizing fund raisings to save the rain forests, supporting a major soil conservation project in Central Tanzania. But development projects often disappoint their instigators, as we all know. The point here is not that family planning or soil conservation are in themselves malicious inventions by oppressive authorities but that it makes sense to open one's eyes to how people themselves try to make ends meet and to reach an understanding together with them (incorporating the conflicting interests of men and women, rich and poor, young and old, etc.) of both what the problem is and what its remedies may be. The rest of this paper reports on such experiences.

5 Indigenous soil and water conservation

During the last decades there has been a steadily growing interest for indigenous soil and water conservation, to no small extent caused by the poor performance of modern SWC techniques imposed on farmers. Chambers (1979: 1) concluded almost twenty years ago that „to neglect the stock of indigenous technical knowledge, and the processs whereby rural people can assimilate, adapt, communicate and create knowledge, is both inefficient and wrong". A benchmark study like

Paul Richards' „Indigenous Agricultural Revolution" (1985) helped to open the eyes of many researchers and we now have a number of overviews and analyses of a wide range of indigenous SWC practices in the South (cf. for instance Brokensha et al., 1980; Critchley et al., 1994; Rajasekaran et al., 1991; Reij et al., 1996). A narrow definition of indigenous SWC would be techniques which have been developed locally, as distinct from techniques imposed from outside. But Scoones et al. (1996: 10) point out that many local SWC techniques, today regarded as local, have been derived from elsewhere. The water harvesting techniques used in the rice fields of the Shinyanga Region in Tanzania were developed through interactions between Asian migrants working in the area and local farmers (Shaka et al., 1996.). The *trus* water harvesting technique used in the Kassala region of Sudan may have its origin in West Africa or in the Nile region (van Dijk, 1995: 76f.). The improved traditional planting pits, since 1990 rapidly being adopted by Haussa farmers of the Illéla District of Niger, have their origin in the Yatenga region of Burkina Faso where Mossi farmers developed the technique in the early 1980s. The Mossi farmers may themselves have picked up the idea from Dogon who came to the Yatenga region as drought refugees. Cases such as these reveal a considerable external influence in the development of what are taken to be local SWC techniques.

In some countries farmers have adopted, maintained and expanded SWC techniques introduced by experts to the extent that these practices are now considered indigenous. Grass strips were introduced in Swaziland during the 1940s by the colonial agricultural service. By the end of the 1950s almost 114,000 km grass strips had been installed, protecting virtually all cultivated land against erosion. The King supported the idea and in 1954 issued an Order-of-the King obliging all Swazi farmers to have grass strips in their fields. Without the King's personal commitment adoption would certainly have been slower. Most grass strips are still in place despite unsubstantiated claims by conservation experts in 1968 that grass strips are not efficient and subsequent proposals to replace them by broad-based terraces (Osunade and Reij, 1996: 153).

Indigenous SWC techniques are often equated with physical structures, like stone bunds and terraces, but they also include for instance agroforestry and tillage practices. When preparing a study on indigenous SWC practices in the lakes depression zone of Northern Zambia, Sikana and Mwambazi (1996) failed to identify any physical structures for harvesting water or controlling erosion in Kaputa District. They had expected local skills and innovations to flourish as the low and erratic rainfall severely limits agricultural production in the District. However, giving up their ambition to find *the* clever technology they were subsequently able to produce a fascinating analysis of continuously changing agricultural practices and dynamic interactions between fishing and farming in the region.

6 Differences between indigenous and modern SWC practices

Although a growing number of specialists recommends that soil and water conservation interventions should „build on traditions" rather few cases can be found where this is being done. The reason may be that starting from local practices requires a resolute change in the way most conservation projects have been designed and implemented. This may not be acceptable to donor agencies, and it would also require a change of mind among many conservation experts. Projects rarely share power over resources with local interests. Land managers are still, by and large, without influence over both how projects work and how research is conducted.

Projects promoting modern SWC techniques tend to be target-oriented and conservation works are often implemented over large contiguous areas. Having a specified number of hectares that should be treated each year makes it easy to calculate the inputs required, to estimate funding needs and to organize disbursements. Development Banks tend to take the process one step further and use the annual targets as a basis for the calculation of costs and benefits. This is to a considerable degree an excercise in futility, because practice shows over and over again that targets are seldom achieved (cf. Blaikie, 1989: 34).

Traditional SWC techniques are often visually unimpressive. They are usually limited to individual fields rather than realized on contiguous blocks of land. Some conservation experts label conservation efforts in individual fields a „post-stamp treatment"; they prefer to work on entire slopes or watersheds. However, from the point of view of individual farmers such „stamps" can be perfectly rational. From Armaniya in Ethiopia, Krüger et al. (1996: 178) describe how stone bunds are constructed: „The stone content of soil tends to increase in areas with a high rate of erosion. Above a certain threshold, the accumulation of stones can seriously impede cultivation and reduce crop production. Farmers then start collecting stones and create a pile of a maximum of four stones. The second stage in the creation of stone bunds is the transfer of stones from these piles to semi-permanent heaps ... and after some years, farmers convert these heaps into stone bunds".

Many other examples can be found where individual farmers each year treat fractions of their fields with conservation measures. How much they do depends on their own motivation and means. Such gradual investments in the land is something that donor agencies and many experts find difficult to handle, and even to accept.

Farmers who invest in SWC measures on their land rely on family labour, on traditional work parties or on hired labour. Work parties imply an element of reciprocity. If you come and work on my fields, I will also help you. Where improved traditional SWC measures are rapidly adopted by farmers, one often also observes a revival of traditional work parties. In other cases, men, women and youth organize themselves in separate groups and undertake SWC on the fields of individual farmers. They are paid by the owner at an agreed rate for every metre of conservation work undertaken. This happens for instance on the Central Plateau in Burkina Faso where groups divide part of the proceeds amongst their members and reserve another part in a collective fund. Again all decisions are made by individual farmers.

Farmers who have invested their labour in traditional SWC works (terraces, bunds or trash lines) may well decide one day to destroy or to move them. The reason behind this is usually that they want to distribute the fertile sediments accumulated behind the conservation works. This kind of flexibility may be hard to accept by experts.

7 How transferable are indigenous SWC practices?

Indigenous SWC practices are usually site-specific, which means that they have emerged in a particular physical, socio-economic and demographic context. Does that exclude their transfer to other regions? The answer is an ambiguous yes and no.

Traditional planting pits in the Yatenga region of Burkina Faso had a reputation of being inefficient until they were improved in the early 1980s by one or more Mossi farmers. They were subsequently adopted by Haussa farmers from the Illéla district in Niger, who visited the Yatenga in 1989 (Hassan, 1996: 57), but planting pits are rarely used by other Mossi farmers elsewhere on the Central Plateau of Burkina Faso (Ouedraogo and Kaboré, 1996: 83). There may be several reasons behind this, which we cannot discuss in detail here. However, thousands of farmers in the Yatenga and the Illéla District now adopt planting pits as the most simple and cost-effective technique currently known in the Sahel for the rehabilitation of strongly degraded land. It produces considerable gains in plant production (grain and fodder) from the first year and in that sense it performs better than any conventional soil and water conservation or water harvesting technique introduced by experts.

Many local SWC techniques such as bench and step terraces in mountain areas represent the cumulative efforts of several generations and it is highly unlikely that such techniques will readily be adopted by farmers elsewhere, because they represent an investment which is too high in relation to the benefits that can be derived from them. They will at best be maintained in the regions where they are found at present, but even that is not certain due to the growing importance of non-farm income and the migration to cities. It may be more cost-effective for farmers to find employment than to invest in land of low productivity (Blaikie, 1989: 27). Where, on the other hand, the land produces a surplus, it has become common all over Africa that income from agriculture is invested elsewhere, for instance in trade or social networks (cf. Berry, 1993).

8 Does it make sense for farmers to invest in improved traditional SWC techniques?

A primary objective of many farmers in drylands is to improve food production not only in years of good rainfall but also in drought years. It is estimated that from 1990 to 1995 about 6000 hectares of strongly degraded land were rehabilitated with improved traditional planting pits in Niger's Illéla district (Hassan, 1996: 57-60). Even in a drought year (1993, rainfall less than 300 mm) an average yield of almost 400 kg/ha of millet was obtained against 144 kg/ha on untreated land. The costs of hired labour to dig pits on 1 hectare are in the order of 12,500 CFA (=25 $US); the cash value of the millet in a drought year is 64 $US and in a year of good rainfall 160 $US. This explains why a land market has emerged in this region and why heavily degraded land is sold at increasingly higher prices. When farmers in an area north of Niger's capital Niamey, where a number of different SWC techniques have been introduced by a project, were asked which technique they preferred, they did not need to think long: planting pits (*tassa*), and it is now clear why.

9 Conclusions

Studying innovative farmers in Malawi, Segerros et al. (1996: 110) concluded that „it has been proven that there are farmers who can outdo the best soil conservationists and irrigation engineers". Niemeijer (1996: 90f.) makes the same point from a historical perspective. African small-scale farmers often proved themselves more successful than both settler farmers and large-scale mechanized schemes. They were more flexible, more knowledgeable about diverse micro-environments and better able to adapt to dramatically changing weather conditions.

Much can be gained from linking farmers and researchers in an effort to improve the efficency of the wide range of indigenous techniques. Researchers involved in development initiatives discover, somewhat paradoxically, that there is a growing demand for their contributions because their general answers often do not work. To a much greater extent than before, information is now in demand on *specific* realities and *diverse* experiences, for instance of men and of women, and of farmers commanding different resources. This is at the same time one more reason why farmers should be involved in research. There are not enough scientists around to record the diversity of situations and experiences.

As farmers become more active partners in the research process we should get results that are more relevant to people's needs and also sustainable for the future. The diversity and complexities of every-day farming will enter the research process much more directly than in conventional on-farm research. Through intensive interaction in the field, researchers are finding new roles for themselves and thereby also gaining new perspectives on their own specialities. Academia starts to become influenced by the villages, just as the South's music, literature and art for decades have been enriching our lives in the West. If these are gains, there may also be a few concessions to be made, for instance the recognition that the complexities of processes which cannot be directly observed may be under-estimated.

Our discussion of indigenous soil and water techniques concluded that there are many different truths, and interests, and that these need to be negotiated. These points could equally well be linked to the critiques of the concept of culture we briefly set out in the first half of this paper.

References

Abu-Lughod, Lila (1991): Writing against culture. In: Richard G. Fox (ed.), Recapturing Anthropology. Santa Fe: School of American Research Press.
Berry, S. (1993): No Condition is Permanent: The Social Dynamics of Agrarian Change in Sub-Saharan Africa. Madison: The University of Wisconsin Press.
Blaikie, P. (1989): Explanation and policy in land degradation and rehabilitation for developing countries. Land Degradation and Rehabilitation **1**: 23-37.
Blaikie, P. (1995): Understanding environmental issues. In: Stephen Morse and Michael Stocking (eds.), People and Environment. London: UCL Press, 1-30.
Bloch, M. (1991): Language, anthropology and cognitive science. Man (N.S.) **26(2)**:183-198.
Bloch, M. (1995): Peole into Places: Zafimaniry Concepts of Clarity. In: Eric Hirsch and Michael O'Hanlon (eds.), The Anthropology of Landscape. Perspectives on Place and Space. Oxford University Press, 63-77.
Brokensha, D., Warren, D.M. and Werner, O. (1980): Indigenous knowledge systems and development. Madison: University Press of America.
Chambers, R. (1979): Editorial. IDS Bulletin **10(2)**: 1-3. (Special issue on „Rural Development: Whose Knowledge Counts?", R. Chambers, (ed.))
Christiansson, C. (1988): Degradation and Rehabilitation of Agropastoral Land - Perspectives on Environmental Change in Semi-arid Tanzania. Ambio **17(2)**: 144-152.
Christiansson, C., Mbegu, A.C. and Yrgard, A. (1993): The Hand of Man. Soil Conservation in Kondoa Eroded Area. Nairobi: SIDA's Regional Soil Conservation Unit, Report No 12.
Critchley, W.R.S., Reij, Ch. and Willcocks, T.J. (1994): Indigenous soil and water conservation: a review of the state of the knowledge and prospects for building on traditions. Journal of Land Degradation & Land Rehabilitation **5**: 293-314 .
Dijk, J.van (1995): Taking the waters. Soil and water conservation among the settled Beja nomads in Eastern Sudan. Leiden: Africa Studies Centre Series.
Fosbrooke, H.A. (1950/51): The Fight to Rescue a District. East African Annual, 168-170.
Hannerz, U. (1992): Cultural Complexity. Studies in the Social Organization of Meaning. New York: Columbia University Press.
Hannerz, U. (1993): When Culture is Everywhere. Reflections on a Favourite Concept. Ethnos **58(1-2)**: 95-111.
Hassan, A. (1996): Improved traditional planting pits in the Tahoua Department, Niger. In: Chris Reij et al. (eds.), Sustaining the soil: traditional soil and water conservation in Africa. London, Earthscan, 56-61
Krüger, H.J., Fantaw, B., Michael, Y.G. and Kajela, K. (1996): Creating an inventory of indigenous soil and water conservation measures in Ethiopia. In: Chris Reij et al. (eds.), Sustaining the soil: traditional soil and water conservation in Africa. London, Earthscan, 170 - 180.
Leach, E.R. (1965): Culture and Social Cohesion: An Anthropologist's View. Daedalus **94**: 24-38.
Mung'ong'o, C.G. (1995): Social processes and ecology in the Kondoa Irangi Hills, Central Tanzania. Stockholm University, Department of Human Geography, Meddelanden series B 93.
Niemeijer, D. (1996): The Dynamics of African Agricultural History: Is it Time for a New Development Paradigm? Development and Change **27**: 87-110.
Östberg, W. (1986): The Kondoa Transformation. Coming to grips with soil erosion in Central Tanzania. Uppsala: Scandinavian Institute of African Studies, Research Report no. 76.
Östberg, W. (1995): Land is Coming Up. The Burunge of Central Tanzania and Their Environments. Stockholm Studies in Social Anthropology, 34.
Osunade, M. and Reij, Ch. (1996): „Back to the grass strips": a history of soil conservation policies in Swaziland. In Chris Reij et al. (eds.), Sustaining the soil: traditional soil and water conservation in Africa. London, Earthscan, 151 -155.

Ouedraogo, M. and Kaboré, V. (1996): The Zaï: a traditional technique for the rehabilitation of degraded land in the Yatenga, Burkina Faso. In: Reij et al. (eds.), Sustaining the soil: traditional soil and water conservation in Africa. London, Earthscan: 80 -84.

Payton, R.W., Christiansson, C., Shishira, E.K., Yanda, P. and Eriksson, M. (1992): Landform, soils and erosion in the North-eastern Irangi Hills, Kondoa, Tanzania. Geografiska Annaler **74 A**(2-3): 65-79.

Rajasekaran, B.D., M. Warren and S.C. Babn. 1991. Indigenous natural resource management systems for sustainable agriculture: a global perspective. Journal of International Development **3**: 1-15.

Reij, Ch., Scoones, I. and Toulmin, C. (1996): Sustaining the soil: traditional soil and water conservation in Africa. London: Earthscan.

Richards, P. (1985): Indigenous Agricultural Revolution. Ecology and food production in West Africa. London: Hutchinson.

Scoones, I., Reij, Ch. and Toulmin, C. (1996): Sustaining the soil: indigenous soil and water conservation in Africa. In: Chris Reij et al. (eds.), Sustaining the soil: traditional soil and water conservation in Africa. London, Earthscan, 1 - 27.

Segerros, M., Prasad, G. and Marake, M. (1996): Let the farmer speak: Innovative Rural Action Learning Areas. Pp. 91-116 in Successful natural resource management in Southern Africa. Amsterdam: Vrije Universiteit, Centre for Development Cooperation Services (CDCS); Windhoek: Gamsberg MacMillan.

Shaka, J.M., Ngailo, J.A. and Wickama, J.M. (1996): How rice cultivation became an „indigenous" farming practice. In: Chris Reij et al. (eds.), Sustaining the soil: traditional soil and water conservation in Africa. London, Earthscan, 126-133.

Sikana, P. and Mwambazi, T. (1996): Environmental changes and livelihood responses: shifting agricultural practices in the lakes depression zone of Northern Zambia. In: Chris Reij et al. (eds.), Sustaining the soil: traditional soil and water conservation in Africa. London, Earthscan, 107-116.

Weiner, A.B. (1995): Culture and Our Discontents. American Antropologist **97(1)**: 14-21.

Addresses of authors:
Wilhelm Östberg
Environment and Development Studies Unit
School of Geography
Stockholm University
S-10691 Stockholm, Sweden
Chris Reij
Centre for Development Cooperation Services
Vrije Universiteit Amsterdam
De Boelelaan 1115
1081 HV Amsterdam, The Netherlands

Indigenous Soil and Water Conservation Practices in Ethiopia: New Avenues for Sustainable Land Use

Yohannes Gebre Michael

Summary

In Ethiopia, land degradation problems caused mainly by soil erosion have been chronic. To alleviate these problems, a great number of soil and water conservation programmes have been carried out by the government and by non-governmental organisations (NGOs) over the last two decades. There has been, however, little adoption of these conservation techniques by Ethiopian farmers. At the same time, indigenous soil and water conservation (ISWC) techniques are being widely practised.

An inventory of ISWC techniques was conducted in different parts of Ethiopia, selected on the basis of agro-climatic conditions, farming systems and prevailing traditional practices. More than twenty techniques were identified, including stone and soil bunds, traditional ditches, micro basins, trash-lines, mulching, mixed cropping, contour ploughing and agroforestry. The reasons why farmers apply these techniques include the colonisation of steeper slopes for cultivation, the need to drain off excess water and/or the desire to harvest run-off water. The techniques are site-specific, and secure land-use rights appear to favour their application.

The different ISWC techniques are not practised to an equal extent within or across the communities studied, on account of local ecological and socio-economic variations. To ensure sustainable land use, it is essential that site-appropriate ISWC practices are taken into consideration by development planners.

Keywords: Indigenous soil and water conservation, Ethiopia

1 Introduction

In Ethiopia, about 1.5 billion metric tons of soil are eroded annually. About 45% of it comes from arable land. The average soil loss from cropland is 42 tons/ha/year. At test-plot level, soil erosion rates as high as 200 tons/ha/annum or more have been recorded in the northern part of Ethiopia (see Table 1). However, such figures are probably exaggerated, as they overlook the process of soil deposition.

Crop yields per year are expected to decline by one to three percent while the population is expected to double in less than 25 years. On account of the cumulative effect of land degradation, the incidence of famine and drought is also expected to increase in the future. This scenario presents a major challenge in feeding the present and future population while ensuring sustainable land management (Hurni, 1986; Wood, 1989).

Administrative region	Station	Mean annual rainfall (mm)	Mean annual erosivity (J/mh)	Cultivated mean annual runoff (mm)	Test-plot mean annual soil loss (t/ha)
Shewa	Andit-Tid	1379	506	251-582	86-212
Gojam	Anjeni	1690	633	533-828	131-170
Illubabor	Dizi	1512	646	13-74	1-4
Sidamo	Gununo	1300	599	155-171	48-80
Hararge	Hunde Lafto	935.1	346	23-49	22-25
Wello	Maybar	1211	420	103-191	32-36

Source: Soil Conservation Research Program, 1996.

Table 1: Magnitude of erosion on cultivated plots in Ethiopia (1981 - 1992 average)

To respond to this challenge, a number of major soil and water conservation (SWC) programmes have been undertaken both by the government and by many NGOs over the last twenty years. However, the conservation techniques they promoted have been neither adopted nor adequately maintained by the intended beneficiaries (farmers). The major reason why the newly introduced SWC technologies were not being accepted was the prevailing 'top-down' approach, which marginalised the participation of the land-users and ignored their local practices. A review of traditional conservation practices in Ethiopia indicates that many farmers in Ethiopia have developed a wide range of ISWC practices (Table 2) such as stone terraces, traditional ditches, diversion ditches, micro basins, trash-lines, mulching, mixed cropping, agroforestry and contour ploughing (Tahal, 1988; Yohannes, 1992; Kruger et al, 1996; Reij et al, 1996).

2 Description of ISWC technologies: case studies from high- and low-rainfall areas

2.1 Armaniya, Northern Shoa (high-rainfall area)

Armaniya is situated about 250 km north of Addis Ababa. The annual rainfall varies from 1000 to 1300 mm, and is bimodal. The first cropping season, locally known as *'Belg'* (in the Amharic language), lasts from February to May, and second cropping season (*'Kremt'*) from mid-June to October. The major crops grown are barely, wheat, teff, beans, peas, linseed, lentils and maize. Sugarcane, onion and chat are grown as cash crops. The region has an ox-plough culture. The major ISWC techniques practised in Armaniya are the following:

Traditional ditches, locally known as *'fessase'* or *'boye'*, are deep-rooted in the culture and widely practised. To indicate the importance of these ditches in everyday farming, the farmers have a proverb '*Arso yele fesese, temaguto yale wase ayehoneme',* **w**hich means 'Farming without a drainage ditch is like a lawsuit without bail'. Usually, the ditches are made by the household head using a pair of oxen drawing the *'maresha'* plough, and aligned towards the natural waterways. There are local institutions like the so-called *'yeweha abate'* (father of the water) or group of elders who supervise the appropriate use of ditches and resolve related conflicts between

Technology	Predominant Agroclim. zone	Level of permanency	Specifications of the technology (range)						Functions					
			Slope (%)	Spacing (m)	Length (m)	Width (m)	Area occupied per plot (%)	Water harvesting	Run-off management	Soil trapping	Fertility improvement	Slope modification	Others	
Stone/soil Bund	Wurch	fixed/not f.	0-5	1-30	1-30	0.40-1.5m	2-15	no	yes	yes	yes	yes	Boundary	
	Dega	fixed/not f.	"	"	"	"	"	yes/no	"	"	"	"	Fence	
	Weyna Dega	fixed/not f.	"	"	"	"	"	yes/no	"	"	"	"	More space	
	Kolla	fixed/not f.	"	"	"	"	"	yes	no	"	"	"		
Traditional Ditches	Wurch	not fixed	5-20	6-20	9-93	0.20-0.30	1-4	no	yes	"	no	no	Demarcation line in seed broadcasting and protect seed from washing down slope	
	Dega	not fixed	"	"	"	"	"	"	"	"	"	"		
	W.Dega	not fixed	"	"	"	"	"	"	"	"	"	"		
Diversion Ditches	Wurch	fixed	2-15	-	-	> 0.50	<1	no	yes	no	"	"		
	Dega	fixed	"	-	-	"	"	"	"	"	"	"		
	W.Dega	fixed	"	-	-	"	"	"	"	"	"	"		
Micro-basin	Kolla	not fixed	-	-	-	0.30-0.50	30-60	yes	"	yes	"	yes	Protect soil surface	
Trash-lines	W. Dega	not fixed	0-3	2-15	1-20	0.20-0.50	10-40	"	"	"	yes	no	Food, cash crop, fodder, construction, fire wood, traditional medicine	
	Kolla	not fixed	"					"	"	"	"	"		
Agroforestry	W.Dega	fixed/not f.	Trees and bushes					"	"	"	"	yes		
	Kolla	fixed/not f.						"	"	"	"	"		
Contour Ploughing	Wurch	not fixed	Only the last ploughing (seeding) is along the contour					"	"	"	no	no		
	Dega	not fixed						"	"	"	"	"		
	W. Dega	not fixed						"	"	"	"	"		
	Kolla	not fixed						"	"	"	"	"		
Mixed Cropping	W. Dega	not fixed	Two or more crops combined together					"	"	"	yes	"		
	Kolla	not fixed												
Mulching	W. Dega	not fixed	After harvesting, the residue remains in the soil (control grazing)					"	"	"	"	"	Protect soil surface Diversify production Risk minimisation	
	Kolla	not fixed												

Table 2: *Major indigenous soil and water conservation technologies on arable land in Ethiopia: technical and functional classifcations/summary*

land-users. These traditional institutions lost much of their importance with the establishment of Peasant Associations in the 1970s. The purposes of such ditches are to demarcate boundaries within the field, to protect newly-sown seeds from being washed downslope, and to drain excess water from the field. Generally, the number of ditches per plot varies with crop type, slope, soil type and shape of the plot.

Diversion ditches, locally known as *'golenta'*, are made in plots located adjacent to different land uses such as grassland, forests, footpaths etc., in order to protect from run-on to the field. They are constructed by ploughing very deeply or using hand-digging tools. Usually, farmers with adjacent farms cooperate in constructing the diversion ditches, which are aligned towards the natural waterways.

Stone/soil bunds, the former known locally as *'kabe'* while the soil bunds covered with grass are known as *'weber'* or *'dibi'*, are associated in high-rainfall areas with traditional ditches. Different types of such bunds can be found, which include the following:

- **Stone piles and heaps**, locally known as *'gwazh'*, are the first and second stages in the process of bund construction. Their primary objective is usually to minimise the amount of stones on the surface of a field, to the extent that these stones reduce the yield of crops per unit area. Moreover, the stones accumulated in piles or heaps serve as an emergency source of materials for repairing destroyed bunds or for controlling rills and gullies which might develop in fields. Unlike the case with standard bund construction, not all stones lying on the soil surface are removed; some are left purposely to serve as a stone mulch and to minimise the splash effect of rainfall. The Amharic proverb *'Konjo liji kagule tiru mirt keguri yegenyale'* means 'The beautiful are borne from the ugly and good yield is obtained from stony plots'.

- **Re-vegetated bunds** are a common practice; they are permanently located at various spots in the field such as at depressions, near gullies and on field boundaries. Some are left uncultivated to allow indigenous grass to grow (in the Sidamo area); some are planted with cash crops such as chat (in Hararge) and enset ('false banana'), sugarcane and banana (in Sidamo), and cops and acacia trees (in Shoa).

- **Moving bunds** within permanent terraces in order to improve soil fertility is a practice that is widely applied in many parts of Ethiopia. After deposition of sediments behind a temporary bund has reached a desired level, the farmer systematically dismantles it and spreads the sediment throughout the plot. For each dismantled bund, an alternative one is constructed immediately. Dismantling the bunds starts three to ten years after they were constructed, depending on the speed of sedimentation, the land pressure (the poor dismantle their bunds very frequently) and the frequency of land distribution. Many conservation professionals strongly criticise this strategy of moving bunds regularly. However, one should also appreciate that the bunds are being moved within permanent conservation structures, each dismantled bund is immediately replaced, and the deep soil which had been confined to a narrow strip above the bund is spread over a larger area of land to allow better use of built-up soil fertility. In other words, this strategy optimises both the production and the protection component of land management.

- **Bench terraces (back-slope terraces)**, locally known as *'weber'* or *'dibi'*, are usually constructed along depressions and near the homestead. They can be up to 3 m high. They are not commonly found, possibly because of a lack of land-tenure security.

2.2 Konso, Southern Gamgofa (low-rainfall area)

The rainfall in the Konso area is usually less than 500 mm per annum and is very erratic. The major crops grown in this area are sorghum, maize, millet, barley, wheat and haricot beans. It is a hoe-farming culture. The major ISWC techniques practised in Konso are:

Stone (back-slope) terraces. Because hand-hoeing is widely practised, the spacing between the stone terraces is narrower than in areas where ox-ploughing is practised. In Konso, the technique of back-slope terracing has been applied far from the homestead for many generations. In this mountainous region with considerable population pressure, terrace construction is a prerequisite for bringing more land into cultivation. The technique is not self-standing: terracing is integrated with the use of micro-basins, trash-lines, mixed cropping and agroforestry.

Micro-basins, locally known as *'kaha'* (in the Konso language), are constructed within the stone terraces during land preparation, in order to harvest and concentrate water.

Trash-lines are prepared at the ridges of the bunds and micro-basins, using the straw from maize and sorghum. They serve as mulch (to reduce splash effect and minimise evaporation) and to improve soil fertility along the bunds and micro-basins (see Figure 1).

Mixed cropping involves the simultaneous growing of different types of crops such as maize, sorghum, millet, wheat, barley, beans and sunflower, as a component of land-use intensification. The seeding rate depends on the moisture balance. During periods of moisture stress, thinning is done; the removed plants serve as fodder for livestock.

Agroforestry. Perennial plants such as coffee, chat and multi-purpose trees such as moringa are planted at the foot of the bunds.

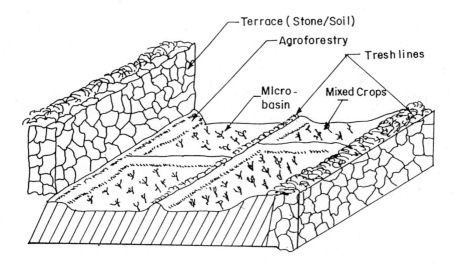

Figure 1: Integrated indigenous soil and water conservation techniques in Konso

3 Major characteristics of the ISWC techniques

The ISWC techniques investigated in the two case-study areas had the following characteristics in common:

- **Site-specific:** The farm plots are perceived by the farmers to be heterogeneous; as a consequence, different parts of a plot are treated differently. This means that many ISWC techniques can be found in a single plot. Care for the adjacent downslope plots is also considered under the traditional rules and regulations.

- **Complex:** On problematic areas such as depressions, run-on sites and shallow soils with many stones, trash-lines and deeper ditches can be applied systematically. Furthermore, a plot cultivated twice a year requires the use of water-harvesting techniques during the small rainy season (*Belg*) and drainage techniques during the main rainy season (*Kremt*). This requires the seasonal use of different techniques or a combination of techniques. None of the ISWC techniques have any standard slope, spacing and length within and between plots.

- **Gradually developed components of the farm system:** Constructing or establishing ISWC techniques such as stone piles, bunds, trash-lines and mixed cropping is carried out as a component of farming activities from ploughing to harvesting. Any one structure is not made at once but rather developed gradually over time, which allows the introduction of adjustments.

- **Multi-functional:** In all the ISWC techniques investigated, their functions were not confined only to soil and water conservation; they also served other purposes, for instance, they functioned as fencing or to manage run-off water.

- **Both short-term and long-term benefits:** Some ISWC techniques have already short-term impacts, while others have relatively medium- or long-term impacts. For example, trash-lines have a short-term impact on water harvesting and a medium- to long-term impact in soil fertility improvement. Such extended benefits of ISWC techniques provide a reason for their wide application.

- **Integration of several techniques:** As a rule, no single ISWC technique stands on its own within a small plot. Usually, two or more techniques are combined to give a synergistic effect. For example, in the high-rainfall area the traditional bunds are combined with drainage and diversion ditches, while in the low-rainfall area the back-slope terraces are combined with micro-basins, trash-lines, agroforestry and mixed cropping.

- **Reduced labour inputs and risks:** Traditional conservation practices are realised as part of regular farming operations, using local tools and materials, the ox-plough and human labour over an extended period of time. This helps to avoid peaks in labour inputs and to minimise risk.

- **Involvement of local institutions:** Local institutions such as *'mekenajo'* (reciprocity) and *'debo'* (invitation of workgroups) and application of community rules and regulations through the *'shemagles'* (elders) are still viable and widely used in conservation and other farming operations.

- **Heritage of ancestral tradition:** Technologies which have been handed down from generation to generation have gained considerable confidence within the community.

- **Specific prerequisites:** Most of the ISWC techniques are practised where land tenure is secure. Certain techniques such as mulching, trash-lines and agroforestry are possible only if grazing is controlled. The availability of materials like stones and crop residues (maize and sorghum), but also food habits, can influence the choice of techniques.

4 Conclusions

To ensure the sustainability of SWC in the smallholder farming households in Ethiopia, the following conflicts between modern and indigenous technologies should be resolved:

- **Standardisation versus site-specificity:** Modern SWC techniques are introduced in a standardised form with constant vertical intervals and similar techniques throughout a catchment, whereas farmers under the same conditions would apply a wide range of indigenous SWC techniques which are complementary and take into account micro-variability in soil conditions and in topography.

- **Concentrated versus gradual work:** The construction of modern physical SWC structures is done all at once by removing the stones available at plot level, whereas indigenous SWC starts with making piles and heaps of stones, retaining some on the field for mulching, and gradual construction at specific sites where the problems within the plot are regarded as serious and in need of priority treatment. This approach is cost-effective and avoids concentration of labour inputs and risks.

- **Protection versus production:** The main emphasis in modern SWC technologies is to conserve soil (protection) and to develop bench terraces, while in indigenous SWC the techniques, such as moving bunds, are meant for both production and protection.

- **Mass versus household/small-group labour:** Under modern SWC, the work is achieved through mass mobilisation, whereas the source of labour in the construction of traditional SWC structures is the household head, other household members and small neighbouring groups.

- **Institutions for conflict management:** Conflicts which arise with regard to the misuse of indigenous SWC technologies are resolved through local institutions such as the *'yeweha abate'* (father of water), whereas under modern SWC regimes such mechanisms do not exist or are not effective.

For these reasons, among others, the intermarriage of indigenous and modern SWC techniques is essential to ensure better land husbandry.

References

Chambers, R. and Jiggins, J. (1986): Agricultural research for resource-poor farmers: a parsimonious paradigm. Discussion Paper 220. Institute for Development Studies (IDS), University of Sussex, Brighton.

Hurni, H. (1986): Degradation and conservation of soil resources in Ethiopia highlands. Paper presented at the 1st International Workshop on African Mountains and Highlands, October 18-27, Addis Ababa.

Kruger, H.J., Yohannes G.M., Berhanu F., and Kefini K. (1996): Inventory of indigenous soil and water conservation in Ethiopia. In: C. Reij et al. (eds.), Sustaining the soil: indigenous soil and water conservation in Africa, Earthscan Publications Ltd., London, pp. 170-180.

MOA (1989): A preliminary report on land reallocation. Ministry of Agriculture, Addis Ababa (Amharic edition).

Reij, C. (1991): Indigenous soil and water conservation in Africa. Gatekeeper Series No. 27. International Institute for Environment and Development (IIED), London.

Richards, P. (1985): Indigenous agricultural revolution: ecology and food production in Africa. Hutchinson, London.

Soil Conservation Research Program (1996): Data base report of SCRP Research Units. SCRP, Addis Ababa.

Tahal Consulting Engineers Ltd (1988): Study of traditional conservation practices in Ethiopia. Consultancy report, Addis Ababa.

Wood, A. and Stahl, M. (1989): Ethiopia: national conservation strategy. Consultancy report, Addis Ababa.

Yohannes G.M. (1992): Conservation aspects of indigenous farming systems in Wolayita, southern Ethiopia. Paper presented at the International Workshop on Soil and Water Management for Sustainable Smallholder Development, 2-11 June 1991, Arusha (Tanzania) and Nairobi (Kenya).

Address of author:
Yohannes Gebre Michael
Soil Conservation Research Program (SCRP)
P.O. Box 33569
Addis Abeba, Ethiopia

Traditional Views of Soils and Soil Fertility in Zimbabwe

H.K. Murwira & B.B. Mukamuri

Summary

The aim of this study was to appraise traditional views of soil fertility as constructed within a larger socio-economic and cultural context. Perceptions of soil, and the management of soil fertility can be characterised as a response to the complexity of the farming systems and the diverse and risk-prone environments, albeit fashioned by socio-cultural and political beliefs of the people. Though the study is not representative of all communal areas in Zimbabwe, the outline of the few cases studied is comprehensive enough to show the complexity of soil fertility management perceptions in communal areas. Communal areas are diverse in terms of their historical and cultural backgrounds and this is manifested by the differences in perceptions of soil fertility across the three areas that were studied. The paper also provides a detailed analysis of influences such as beliefs which underlie local people's practices as regards soil fertility management.

Keywords: Perceptions, beliefs, soil fertility, Zimbabwe

1 Introduction

Mainstream soil fertility management prescriptions often contrast with the socio-economic, cultural and technological horizons of the people to whom the packages are designed to serve. This paper is an attempt to describe the traditional and cultural perspectives of how local people in the study areas construct soil fertility within a typical rural context. The authors also attempt to describe how farmers deconstruct "official and scientific" soil fertility management prescriptions. For example the use of inorganic fertilizer and total removal of trees from fields has always been resisted by rural farmers despite official instructions to do so (Wilson, 1989; Mukamuri, 1995).

Soil fertility management is traditional, historical and evolutionary. By traditional we mean that the practices are handed down within generations. The historical dimension relates to the different experiences the communities have undergone since colonialism, latifundialisation which followed it and now the barrage of extension messages which have been instituted to and after independence. The practices are evolutionary in the sense that they have changed over time, particularly with regards to the different 'modern' influences which they have inevitably fused with.

Modernization has had considerable impact on people's beliefs and practices. For example, western religions and education have heavily influenced the way people behave and interpret life. Often, traditions are interpreted as backward and basically modifiable practices. As the paper will show there is no way in which local beliefs and practices though cult based can be equated to anti-civil society as many social and political commentators have tended to say. Instead, we see them as alternative actions based on specific contexts.

A local cultural perspective is essential in understanding soil fertility management in rural agricultural systems. Soil fertility management practices are fashioned to meet the site-specific requirements of particular soil types and environments. Because they are site-specific, practices tend to be intertwined with local beliefs arising from the heuristic experience of the local populace.

2 Study methods

This study was undertaken in three areas, Chivi, Mangwende and Mutoko, from the period June 1995 to September 1995. The work involved carrying out informal and formal interviews with key informants in the study areas. Key informants included traditional chiefs, headmen, witch doctors (n'angas), spirit mediums and some elderly farmers. The key informants were culturally knowledgeable people who are considered within society as the custodians of local culture.

The key issues of interest were: farmers' perceptions of soils and soil fertility; cultural and spiritual beliefs and how they impact on soil fertility management; historical changes in local people's perceptions of soil fertility.

3 Results and discussion

3.1 Definition of soil fertility

Soil fertility is described in scientific terms as the capacity of a soil to provide nutrients and water to plants. Biological, chemical and physical properties of the soil are the major determinants of soil fertility. Local concepts view this definition of soil fertility as rather limited, as it does not recognise the role of the supernatural in the process of plant growth.

There is no specific local definition of soil fertility, but there are many terms used to describe it. It is the symbolism behind the terms that brings about the definition. Words such as, gombo (virgin land), kwegura (old), sakara (wasted), munda mutsva (new field), ivhu rakakora ('fat' or rich soil), ivhu rakasimba ('strong' rich soil with well developed structure) are used to describe soil fertility. Soil fertility is also defined indirectly in terms of the capacity of a soil to give good yields. Such concepts are not new and have been reported in the early development of soil fertility in western science (Tisdale and Nelson, 1975). We argue that the local concepts of soil fertility provide a basis from which researchers can develop a greater understanding of soil fertility management by rural communities.

3.2 Soil fertility in the wider context

Local concepts of soil fertility are mired with the supernatural. Many of the interviewees believed that the strength of soil comes from the creator. It follows according to the lore that people have to pay their respects to God, they have to abide by his laws and have good neighbourly relationships. The paying of respects is by annual post-harvest ceremonies (beer festivals) at which the spirits are thanked for providing adequate food and also requested to continue doing so in future. It was recognised that a good harvest depends on rain and the soil. Hence there is a relationship between soil fertility and the creator (Fig. 1). In recognition of the creator people often left large tracts of land called 'marambotemwa' or conservation areas. These were considered sacred and were left to conserve the soil and trees especially fruit trees. Fig. 2 summarises the various factors (biophysical and cultural) which affect the social construction of soil fertility by rural people. Specific factors are analysed in the sections that follow.

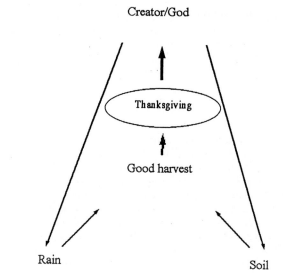

Fig. 1: Rural peoples perception of the relationship between soil fertility and the Creator. A combination of good rains and soil fertility gives good harvest. post-harvest ceremonies have to be carried out to pacify the Creator, so He can give good rains and keep soils fertile in future.

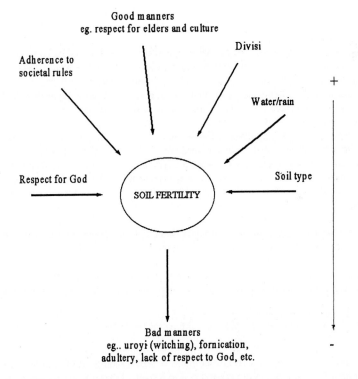

Fig. 2: Rural people's perceptions of factors affecting soil fertility. They are a combination of biophysical and cultural factors. Respect for God is central (see Fig. 1). +/- are perceived factors that enhance/diminish soil fetility.

3.3 Cultural perspectives and practices

The use of divisi or tsvera[1] is a wide spread belief and practice among local people. Divisi is supposedly a magical practice that enhances crop yield. During the interviews in Chivi, Mangwende and Mutoko four types of divisi were highlighted by a majority of people.

The first one is to enable the farmer to get large harvests. The second type works by protecting the farmer's crops from being taken by others who have divisi or tsvera. The third type is one which "takes" other people's crops to enhance one's harvest. Fourth, is the divisi which gives a lot more than what one would normally get from the field but it makes the food 'weak'; it (the food) does not satisfy the family irrespective of how much and how the food is prepared. This divisi was classified as a bad one. The good type is one which satisfies the family despite a small harvest. People in the study areas still believe that there is a lot of makwenzi (bushes from which medicine is obtained) which can be used to protect or take crops from the fields. Interviewees seemed to point towards a continuity between protection of the home-stead, individuals, family and the field.

It was also customary in the study areas that farmers received divisi and medicine to 'cure' fields which have been 'bewitched' every year from their territorial cult centres and herbalists.

Interviews with spirit mediums and herbalists led to very interesting discussions. At first they denied that people still come to them to look for divisi. At least none of them agreed having seen anybody who consulted them for these medicines. However, they all mentioned that when they are called to people's homes (to treat the sick and the bewitched) it is then that they notice their patients' fields have also been bewitched. In most cases they advise their clients to get their 'fields cleansed of these bad things'. The informants mentioned having been involved in digging up mativisi[2] and burning them.

A faith healer from Mutoko alleged that people often think that yields from their fields are low because of the poor soils, yet it is because their fields have been bewitched (kushinhira). Many faith healers use a mixture of salt, water and milk when dealing with divisi, witchcraft or any disturbing spirits. Salt is believed by the healers to be a neutralizing agent. Readers must bear in mind that salt is not wanted at all traditional places and ceremonies. Salt is therefore associated with passivity and death. Water is believed to have cooling effects which may help in minimizing the wrath likely to be caused by divisi. Milk is used because of its natural qualities of giving life.

From the discussions with ordinary farmers, it became clear that divisi is declining in importance, however, the traditional healers, n'angas and faith healers reinforce the practice. Our view is that they want to be in charge of a larger spectrum of people's lives as much as possible. This is also a way of guaranteeing their continuity and functional roles in society.

Views about traditional medicines and their effectiveness are varied. This is in part a reflection that people's beliefs are not static but rather dynamic, dialogical and dialectic. Though some still believe in and practice divisi, some informants reiterated that divisi was now a thing of the past. Another farmer explained that what now matters are 'rain, good soils, and manure'. It was interesting to note that most of the farmers interviewed thought that yields were declining, and that they rate their fore-fathers as better farmers. It was put forward to the farmers that the reason why they are getting lower yields now was because they were not using divisi. Farmers disagreed with

---

[1] Note that the concepts of divisi and tsvera are different in the different parts of the country. The former is largely used in the south while the latter is used in north eastern Zimbabwe. To the Budya of north eastern Zimbabwe, tsvera means the malicious use of magic or medicines to hurt/destroy other farmers and their fields. To the Karanga of southern Zimbabwe divisi is not necessarily harmful to other people but could actually improve or harm the farmer's welfare. The Karanga have a distinction between good and bad divisi. To the Budya tsvera has something to do with hurting other people. However, they have another distinction of magic which can be used to actually improve yields called chikwambo. This, too, targets other farmers and in particular other family members.

[2] Mativisi is the plural form of divisi.

this notion. They attributed the higher yields obtained in the past to larger family sizes which provided labour, and to the growing of appropriate crops especially small grains which were drought resistant.

3.4 Rain and soil fertility

Local people in the study areas perceive rainfall as an important component of soil fertility. They argue that no matter how much manure or fertilizers is applied to the field, if there is no rainfall no meaningful yield can be obtained. Local people try by all means to maintain good social and cultural behaviour to ensure that Mwari/God or the ancestors will bring rains. Lack of rains and therefore lack of adequate harvests is a punishment from God and the ancestral spirits[3]. Rain-making ceremonies are therefore central to Shona people's lives and politics. In the three areas studied the practices were standard and done almost every year. The ceremonies, though directed to local territorial cults (Matonjeni and Mutimuchena) are done in praise of local autochthonous and apical lineage members.

Despite the real importance of rainfall and the doubtful ability of the ceremonies to actually provide it, the rainmaking ceremonies have been largely used to maintain hegemony over local people by the traditional leadership. They have also used them to legitimate their existence and gain monopoly to scarce resources like land and wetlands. Use of rain making ceremonies and claims of the ability to actually provide are understandable considering the dry climates of the study areas (Mukamuri, 1987, 1988, 1995 and also Mukamuri and Wilson, 1995).

3.5 Local and official perceptions of soils

A former conservation officer who used to work in the extension service in Mutoko from 1959 to 1962 was interviewed to get past official views of soil fertility management. His main task in extension was to design and implement mechanical conservation and contour bunds focussing more on the arable fields in order to prevent soil erosion. The informant noted that people's attitudes towards contour bunds was negative because they (people) felt being forced to do what they did not want. Although they involved chiefs and headmen in the implementation of the soil conservation works, all the planning was done by the Conservation Officers without consulting with chiefs nor locals.

The informant recalled that he worked at a time when people in Mutoko were forced to destump all trees from arable lands particularly those who wanted to get master farmer certificates. The official argument was that destumping made ploughing easy. Destumped fields were believed to reduce damage to machines and ploughs. The other argument was that people could also use smaller animals and the pace of ploughing could be increased. Plant spacing was considered neater in destumped fields. Destumping was also thought by extension to enable rural farmers to make good crop estimates. Another official belief was that trees interfered with crops particularly by competing with them for nutrients. However, people saw things differently. For example, people were not keen to destump trees because of the labour involved. Locals also saw advantages in keeping certain tree species in their fields. The informant also now agrees with some of the farmers' views. For example, he noticed that trees like muhacha (*P. Curatelifolia*) actually improved crop yields under their canopy. He added that fruit trees are not cut from fields because of their valuable fruits. Unfortunately, extension workers ignored all the benefits locals get from indigenous trees.

---

[3] An informant from Mangwende thought that rainfall is getting less because of too many churches, adultery and other negative things.

Some informants pointed out that farmers know little about soil compositions, however, this lack of knowledge is compensated for by for example, carefully looking at the grass which grows in specific areas. For example, soils on which zangibarium (soso) grows are associated with infertility. Rural farmers also tell the fertility of the soil by its coarseness. Coarse soils according to locals are not fertile. Coarse soils are often sandy, and are called 'musheche' in the local classification systems. Most fertile soils are believed to be fine textured. They can also tell the quality of the soils by examining the weight[4]. They look at soil characteristics such as, weight, colour, coarseness and hydrological qualities, and crop performance. Farmers also look at indicator tree species to determine if a soil is fertile or not (FSRU, 1993).

Another informant, a woman farmer in Mutoko, Zimbabwe, gave her own views on soils and soil fertility. She recognises soils in her field in terms of good and bad. Good and bad soils are judged by their moisture retention abilities. During heavy rains, good soils drain up well, while bad soils get waterlogged resulting in crops turning reddish. Sandy soils can sometimes be good when there is little rain, but with higher rainfall, they need more fertilizer compared to heavier soils. She thought that sandy soils which give good harvests contain 'lime'. She noticed the 'lime' in the soil because the ground-water is difficult to wash with. It does not easily form a lather. However, she blames most of the sandy soils in her area for being infertile and lacking natural lime. In her sandy portions of the field she grows groundnuts. People in her area fail to adequately deal with problems of soil fertility because of lack of advice from extension.

Farmers do not want sandy jechajecha soils as they have low inherent fertility. They prefer black and red soils because they are more fertile hence they give better yields. The most preferred soils are the "fat soils". For example, termitaria (churu) soils, black and reddish soils are classified as fat. Fat soils make crops grow faster and bigger even without nutrient inputs. They are "strong" when digging. However, some fat soils can be light when handling them. Another method farmers use to determine fertile from less fertile soils is by dropping the soils from a height. "Those without food get blown by the wind easily".

However, they distinguish between "new" and "old" sandy soils. New sandy soils range from virgin lands in generally sandy areas. These have a potential for giving food without fertilizers for a few years. The old sands range from old fields which have turned into "makura", characterised by high presence of zangibarium (soso) which is an indicator of low fertility, to sandy fields which have been cultivated over a long period of time. It is these sandy soils that informants allege are a "headache". This is because of the ever present problem of thinking about improving soil quality in such fields.

4 Conclusions

The interviews showed that traditional perceptions are increasingly under pressure from forces of change particularly from the young who now feel that their system alone cannot address all their soil fertility and general livelihood needs. The management of soils and soil fertility in communal areas can be characterised as a response to the complexity of the farming systems and the diverse and risk-prone environments, albeit fashioned by socio-cultural and political beliefs of the people. We contend that for researchers to provide a meaningful contribution to improved soil fertility management, the development process should start with the knowledge systems of farmers. A definition of problems, analysis and priorities of farmers should be an integral of the development process. This study may not have dwelt with all the aspects, nevertheless it is an attempt to look at

---

[4] This claim has prompted us to start thinking about participatory soil fertility research workshops dealing amongst other issues with local soil classification systems and their relationship to soil fertility management.

the knowledge base in some of the communal area farming systems and in particular how that knowledge is influenced by cultural and spiritual beliefs.

Understanding rural people's knowledge is not sufficient on its own unless one applies that new-found knowledge. Full cognizance has to be taken not to over-emphasise revalidating past practices at the expense of its limitations and the implications for the future in the face of a changing environment. On this note, (Bebbington, 1994) recognises that pressures on agriculturally based livelihoods are intensifying hence undermining the relevance of some earlier agricultural practices.

How we interpret farmers' strategies strongly influences the direction that research takes (Carter and Murwira, 1995). The challenge is not to incorporate farmers' knowledge into our science but to interpret and translate what the farmers know into the actual developmental process. Often the application of farmers' knowledge and practices takes on subtle forms. For example, farmers solemnise particular days during the week for religious or spiritual reasons 'chisi', therefore researchers who want to work on-farm have to gain the confidence of the farmers by observing these days. We would like to argue that applying the farmers' knowledge involves appreciating the farmers' environment, concepts and practices. Often it is easier to incorporate new technologies within the research design if there is appreciation that some local practices could be potentially more beneficial.

Acknowledgements

The authors are grateful for financial support from the African Development Foundation through grant 813-ZIM, and the European Union. Support from the DSE (Germany) for the senior author to attend the ISCO96 conference is gratefully acknowledged.

References

Bebbington, A.J, (1994):Composing rural livelihoods: from farming systems to food systems. In: I.Scoones and J.Thompson (eds). Beyond Farmer First, 88-93.
Carter, S.E. and Murwira, H.K. (1995): Spatial variability in soil fertility management and crop response in Mutoko communal area, Zimbabwe. AMBIO 22 (2):77-84.
F.S.R.U. (Farming Systems Research Unit) (1993): Soil Fertility management by smallholder farmers: A participatory Rapid Appraisal in Chivi and Mangwende Communal areas. Department of Research and Specialist Services, Zimbabwe.
Mukamuri, B.B. (1987): Karanga religion and environmental protection. Unpublished B.A. thesis, University of Zimbabwe.
Mukamuri, B.B. (1988): Rural conservation strategies in south central Zimbabwe: an attempt to describe Karanga thought patterns, perceptions ands environmental control. In: R.Grove (ed), Conservation and rural people in Zimbabwe. Cambridge African Monographs and Baobab Press.
Mukamuri, B.B. (1995): Making sense of social forestry: a political and contextual study of forestry practices in southern Zimbabwe. Acta Universatis. Ser A vol 438.
Mukamuri, B.B. and Wilson, K.B. (1995): Dessication and environmental religion in south central Zimbabwe. (James Curey, 1995, in press).
Tisdale, S.L. and Nelson, W.L. (1975): Soil fertility and fertilisers. 3rd Edition, (Macmillan Publishing Co. New York).
Wilson, K.B. (1989): Trees in fields in southern Zimbabwe. Journal of Southern African Studies **12** (2).

Addresses of authors:
Herbert K. Murwira
Chemistry and Soil Research Institute, Box CY 550, Causeway, Harare, Zimbabwe
Billy B. Mukamuri
Institute of Environmental Studies, University of Zimbabwe, Box MP167, Mt. Pleasant, Harare, Zimbabwe

Socio-Economic Constraints to the Use of Organic Manure for Soil Fertility Improvement in the Guinea Savanna Zone of Ghana

A.S. Langyintou & N. Karbo

Summary

The high population pressure on land in the Guinea savanna zone of Ghana has resulted in low land productivity because farmers can no longer practise the fallow system of farming which ensures soil fertility restoration. Consequently, land productivity can only be increased with the use of external inputs such as inorganic and organic fertilisers. The use of inorganic fertilisers received a high adoption rate in the 1960s when agricultural inputs were subsidised. Removal of input subsidies in the 1980s made fertiliser use uncompetitive. The use of organic fertilisers, especially animal droppings, remains the option for the resource-poor farmer. Its adoption at the farm level is a concern for most development agents.

The results of a study conducted in northern Ghana in 1993/94, involving 420 farmers to identify constraints to the use of animal droppings as manure, indicate that the practise is known among farmers but inadequate manure, inappropriate seed-bed type for ease of spread of dung, the joint ownership of livestock, the high labour demand to manage the dung and the bulky nature of the manure remain bottlenecks for adoption of this approach. Effective integration of crops and livestock would facilitate the management of dung. A change in the method of seedbed preparation to ease fertiliser incorporation, as well as a reform in the traditional system of ownership of livestock emphasising individual ownership may facilitate the effective use of animal droppings.

Keywords: Socio economics, organic manure, crop-livestock, soil fertility, Guinea savanna

1 Introduction

The Guinea savanna zone of Ghana occupies over 40% of the total area of Ghana of 239,000 km^2 and stretches southwards from the northern border (PPMED, 1991). The Northern, Upper West and Upper East administrative regions make up the zone. Nearly 30% of the country's population of about 16 million inhabits the area. An estimated 85% of this population lives in rural areas where the production of crops and livestock is their main source of livelihood. Population density varies from an estimated 23 and 30 persons/km^2 in the Northern and Upper West regions respectively. In the Upper East region, it varies from 113 to 200 persons/km^2.

The zone experiences mono-modal rainfall from April to September with an annual precipitation ranging from 700 to 1200 mm. It has a comparative advantage for the production of cereals and legumes. Northern Ghana, the main "cereal basket" of the country, produces all the national supply of sorghum and millet as well as groundnuts and cowpea (Langyintuo et al., 1995; PPMED, 1994). About 56%, 45% and 23% of yam, rice and maize are also produced, respectively. Livestock are predominant in the area with about 79% and 41% of large and small ruminants respectively (VSD, 1993).

Low population pressure on land in former times made it possible for farmers to practise a fallow farming system whereby a piece of land was cultivated for three to four years and fallowed for up to 15 years to ensure regeneration of soil fertility. Current demand on land for farming and infrastructural development exert pressure on land making the fallow system of farming almost impossible. Consequently land is intensively used evidenced from the estimated R-values[1] of 0.9 and 0.7, implying that 90% and 70% of the arable lands are cultivated annually in the Upper East and Upper West regions, respectively. In the northern region, however, nearly 50% of cultivable land is fallowed annually (Albert, 1995; Langyintuo, 1989). A direct consequence of intensive cultivation of land, if not accompanied by proper soil management is land degradation, especially declining soil fertility leading to reduced and variable crop yields (Ruthenberg, 1980). Therefore, to ensure increased productivity of the soils the use of external inputs, organic and inorganic fertilisers, coupled with good soil management are imperative.

As a measure to restore soil fertility the use of inorganic fertilisers was promoted by the government in the 1960s. Adoption rates were high due mainly to the high subsidisation of agricultural inputs including mineral fertilisers. Removal of input subsidies and the depreciation of the Ghanaian "cedi" in the early 1980s resulted in high input prices relative to output prices thus making fertiliser unaffordable by farmers. For example, between 1979 and 1982 the maize compound fertiliser (NPK) price ratio was 1 : 7.2 but between 1983 and 1986, it was 1 : 3.5, representing 106% relative increase in input price. It was not surprising, therefore, that in his survey of over 4000 fields in northern Ghana in 1993, Albert (1995) observed that only 16% received some mineral fertilisers. This poses a threat to the national supply of cereals if soils are not appropriately ameliorated hence the effort by most development agents to promote the use of animal droppings for soil fertility restoration.

While the use of cover crops to ameliorate soil fertility is still at an experimental level, that of animal droppings is well known among farmers. Nevertheless its practice is marginal among farmers in the Guinea savanna zone of Ghana. This study identifies the main socio-economic constraints militating against the use of animal droppings as manure in the Guinea savanna zone. To achieve this objective, 420 farmers were interviewed in the districts of Bole (n = 146), Tamale (n = 143) and East Mamprusi (n = 131) of the northern region as part of the Savanna Agricultural Research Institute (SARI) Northern Region Farming Systems Research Team's socio-economic program for 1993/94.

2 Results and discussions

2.1 Crop-livestock integration within the farming systems of northern Ghana

There are benefits that could be derived from effectively integrating crops and livestock. Animal traction is thought to improve quality and timeliness of farming operations thus raising crop yields and incomes. Droppings from the animals as manure could improve soil fertility while crop residue may serve as feed for animals. In the study area an examination of crop-livestock integration linkage points of (1) spatial distribution of crops and livestock, (2) perceived roles of crops and livestock in the farming systems, (3) the use of crop residue to feed livestock, (4) the use of animal power in crop production, and (5) the use of animal droppings to improve soil fertility, indicated a limited extent of integration.

All farmers interviewed grow crops on pieces of land ranging from 1.3 to 5.8 ha. Main crops grown are cereals (68%), legumes (18%), fibre crops (2%) and root and tuber crops (12%). Except in some parts of the Bole district, all cereals are planted on either ridges or flat land. In most parts of the

---

[1] R-value = crop years/(crop years + fallow years)

Bole district, cereals are planted on mounds as in the case of yam and cassava. In addition to crop production, farmers keep livestock such as cattle, goats, sheep and poultry. Average tropical livestock units per farm family ranged from 2.8 to 12.9 (Langyintuo, 1995).

Ownership by type of animal indicated that household heads owned about 90% of all cattle, and 80% each of sheep and goats. In most cases household heads declared sole ownership of cattle whether they were jointly owned or belonged to their children. Joint ownership may result through inheritance, that is, a father bequeathing his cattle to his children. While small ruminants and poultry are taken care of by members of the household, large ruminants are sometimes left in the custody of Fulani herdsmen who feed and water them. These Fulani herdsmen may keep herds belonging to a number of households.

Although farmers agreed that livestock played a very important role in the family risk management they considered the enterprise secondary to crop production. Livestock are important during festivals such as the annual yam festival, funeral celebrations, donation to in-laws and dowrying of wives, etc. (See Table 1). Large ruminants were considered as symbols of social superiority of the family. Even though bullocks are owned, farmers do not consider the use of cattle as farm power very important. Bullock ownership is very popular in the East Mamprusi district (77%) as against the other two districts where the use of mounds as seed-beds, limited cash for initial investment in bullock traction and joint ownership of cattle constrain the adoption of the technology.

Activity	District			Total (n=420)
	Bole (n=146)	Tamale (n=143)	East Mamprusi (n=131)	
3 important roles of cattle				
1. Social obligations*	79	100	-	60
2. Sales	74	100	50	75
3. Savings	63	100	53	72
4. Power	-	-	50	17
Bullocks ownership	10	5	77	30
Use of dung	10	40	52	34
Use of crop residue	10	8	78	32

Source: Field survey, 1993/94
Note: Social obligations include dowrying of wives, funeral celebrations, sacrifices, etc.

Table 1: *Farmers' response to crop livestock integration (% of farmers)*

The use of crop residue to feed livestock is important in the East Mamprusi district where about 78% of farmers practice the technology. This is often linked to the presence of bullocks as farmers make efforts to provide supplementary feed during the dry season to ensure maximum output from the animals at the beginning of the season. Reasons such as lack of adequate labour, available range land, and reptiles attack during management of crop residue accounted for the low adoption rate of the technology (Langyintuo, 1995).

While all farmers admitted using wet cowdung to plaster their houses, only 38% of them indicated using animal droppings on their farms with the view to fertilising them. At the district level, the proportion of farmers not using animal droppings as manure varied from 86% in the sparsely populated district of Bole to 48% in the relatively densely populated East Mamprusi district (Table 2). Some socioeconomic and technical reasons, as discussed below, were thought to militate against the adoption of the technology.

Source of manure	District			Total (n=420)
	Bole (n=146)	Tamale (n=143)	East Mamprusi (n=131)	
Cattle	10	40	52	34
Goats	4	8	12	8
Sheep	5	8	14	9
Pigs	1	0	2	1
Poultry	2	3	4	3
None	86	52	48	62

Source: Field survey, 1993/94

Table 2: Use of animal droppings as manure in northern Ghana (%)

2.2 Constraints to the use of animal droppings to improve soil fertility

As indicated by Fernanadez-Rivera et al. (1993), the amount of manure available for crop production in a given area depends on the number of animals, the proportion of animals used for manuring, the amount of faeces excreted per animal and the ability of the farmers to collect it. The spatial location of livestock during the manuring season is thus an important factor that influences the amount of manure available for cropping. The type of animal enterprise and degree of integration into the farming system also influence the type and quantity of manure available (Stangel, 1993). In the study area, cattle are the main manuring animals supported by goats and sheep (Table 2). Of the 38% of farmers who use animal manure, 34% used cowdung while only 1% used droppings from pigs. In the East Mamprusi district all the 52% adopters used cowdung.

The use of cowdung as manure was thought to be associated with a number of problems, such as the introduction of foreign weeds into the farm by 50%, 32%, and 20% of farmers in East Mamprusi, Bole and Tamale districts, respectively (Table 3).

Activity	District			Total (n=420)
	Bole (n=146)	Tamale (n=143)	East Mamprusi (n=131)	
Introduction of foreign weeds	33	20	50	34
High labour demand	74	80	80	78
Joint ownership of cattle	45	6	3	18
Soils fertile	58	2	0	20

Source: Field survey, 1993/94

Table 3: Problems with the use of animal droppings (cowdung) as manure (% of farmers)

Besides introducing foreign weeds into the farm, the use of manure was thought to demand high opportunity cost of labour. Collecting and preserving the dung as well as manually transporting the bulky product to the fields demanded considerable labour time as was indicated by 78% of farmers.

The absence of any form of mechanisation to assist in the incorporation of the manure into the field further accentuated the labour problem. Use of mounds as seedbeds also inhibits the ease of incorporating manure into the field.

The joint ownership of cattle tended to present some problems in manure management. Where cattle were jointly owned as in the Bole district, sharing of manure from the kraal was considered a problem. The practice of free ranging during the dry season does not permit the deposition of adequate dung for use on the farms, further exacerbating the distribution problems.

With Fulani herded cattle in the Bole district, farmers expressed similar distributional problems, as cattle in the kraal may belong to a number of farmers. In areas of the Tamale district where non-permanent kraals are used, as well as where Fulani herdsman do not cultivate cereals, cattle are kraaled in a dynamic fashion on farmers' fields in turns and the system of rotation agreed upon by all stakeholders. Where Fulani herdsmen cultivate cereals, the benefit of dynamic kraaling is enjoyed by they (Fulani) themselves in addition to the milk in exchange for their labour.

About 58% of farmers in Bole district were of the opinion that their soils were relatively fertile enough to support crop growth and therefore did not need to apply any manure. Furthermore, poor and degraded soils could be fallowed until their fertility was restored. The abundance of farm land to afford fallowing was demonstrated by their not cultivating compound farms in most parts of the district.

	Units	District						Northern region ('000)	
		Bole (n=146)		Tamale (n=143)		East Mamprusi (n=131)			
		Cattle	Goat/sheep	Cattle	Goat/sheep	Cattle	Goat/sheep	Cattle	Goats/sheep
Average number of heads	singles	4	11	2.5	12	2	9.4	490,800	903,800
Manure production/head/year (fresh wt)	mt	6	0.5/0.6	6	0.5/0.6	6	0.5/0.6	6.0	0.5/0.6
Moisture content	%	75	45	75	45	75	45	75	45
Total manure production/year (dry wt)	mt	6	3.3	3.8	3.6	3	2.8	736,200	273,400
Potentially available manure to farmers*	mt	3.6	2.0	2.3	2.2	1.8	1.7	441,700	164,00
Nitrogen content of manure dry weight	%	1	1.8	1	1.8	1	1.8	1.0	1.8
Potential manure nitrogen produced per year	kg	36	36	22.5	39	18	31	4,417,300	2,952,800
Total cropped area on cereals	ha	2.7	2.7	3.1	3.1	1.5	1.5	293,500	293,500
Available manure (dry wt) for cereals	mt/ha	1.3	0.7	0.74	0.7	1.2	1.2	1.5	0.56
Manure nitrogen available for cereals**	kg/ha	12.8	12.8	7.3	12.6	12	20.7	15	10

Sources: PPMED (1993, 1994); Langyintuo (1995), Karbo et al (1994)
Note: * 40% lost due to open grazing of livestock (Schleich, 1986).
 ** Cowdung is stored loosely by farmers. To account for N losses during storage, reduce values by 29% if used soon after collection and 59% after three months of storage (Kwakye, 1980).

Table 4: Manure production at the regional and household levels

One other important technical problem observed was the unavailability of sufficient manure for use. Apart from a limited number of cases where some farmers practice dynamic kraaling by rotating the position of the kraal on the farm during the dry season, the system of keeping livestock is free ranging with overnight kraaling. Consequently the amount of dung produced is less by about 40% (Schleich, 1986). An estimate of the quantity of dung applied by farmers in the Guinea savanna zone was in the range of 0.1 to 1 t/ha on the average. While this figure may appear small, an estimate of the potential manure production with available livestock figures does not appear any encouraging.

At the regional level if all potentially available manure was applied to the cereal plot of 293,500 ha, the rate would have been 1.5 t/ha against the recommended rate of 4 t/ha (Anane-Sekyi et al., 1995). Ignoring losses in nitrogen during storage, the amount of nitrogen available would have been 15 kg/ha, about 25% of the recommended rate of 60 kg/ha N for cereals in the zone. An attempt to achieve the set target would mean increasing the cattle herd size by as much as 75%. When the contribution of goats and sheep are added, 40% of recommended nitrogen is met.

At the household level, farmers can fertilise their cereal fields at a rate ranging from 0.74 t/ha in the Tamale district to 1.3 t/ha manure in the Bole district (Table 4) a rate much lower than the 4 t/ha recommendation. Available manure nitrogen from cattle could fertilise cereal plots at a rate ranging from 7.3 kg/ha in the Tamale district to about 12.8 kg/ha in the Bole district. If goats and sheep are considered, the rate could be doubled. This level is still inadequate to meet the requirements of 60 kg/ha N or 4 t manure.

3 Conclusion

While the available manure from cattle, goats and sheep are inadequate for manuring due to low livestock numbers and open grazing, high labour demand for its management compels some farmers not to adopt the technology. Where mounds are used as seed beds, manure use is further constrained. In some traditional settings, the joint ownership of cattle, introduces distributional problems of the dung.

To facilitate the adoption of the technology, an effective crop-livestock integration is imperative. Feeding livestock in the kraal with crop residue could provide more and better manure with high nitrogen content from the crop waste, animal faeces and urine. Appropriate storage methods for the manure to minimise nitrogen losses should be focused on during extension work. The drudgery of manually managing the dung may be eased by the use of bullock carts. Ownership of bullocks could be promoted by credit through the public sector mainly. The private sector could also be encouraged for support. Emphasis on individual ownership of livestock or a workable methodology for the distribution of dung may be of relevance in promoting the use of animal manure.

References

Albert, H. (1995): Farm household systems in Northern Ghana and the problem of *Striga*. Unpublished Manuscript. Savanna Agricultural Research Institute, Ghana.

Anane-Sekye, C., Dennis, E. A., Afful-Pungu, G. (1995): Soil fertility problems in Upper East region of Ghana. The role of organic and inorganic fertilisers. Unpublished Manuscript. Soil Research Institute, Ghana.

Fernandez-Rivera, S. Williams, T.O., Hiernaux P. and Powell J.M. (1993): Faecal excreation by ruminants and manure availability for crop production in Semi-Arid West Africa. Proceedings of an international conference on Livestock and sustainable nutrient cycling in mixed farming systems of sub-saharan Africa. Vol. II: Technical Papers. Edited by J.M. Powell, S. Fernandez-Rivera, T.O. Williams and C. Renard. Addis Ababa, Ethiopia, 22-26 November, 1993, 149-169.

Karbo, N., Alhassan, W. S. and Adongo, S. A. (1994): Small holder lamb fattening based on crop residues and agro-industrial by-products in northern Ghana. Animal Research Institute, Ghana.

Kwakye, P.K. (1980): The effect of method of dung storage and its nutrients (NPK) content and crop yield in the northeast savannah zone of Ghana. FAO Soils Bulletin **43**, 282-288.

Langyintuo, A.S. (1995): Constraints to the effective integration of crops and livestock in the northern region of Ghana. Savanna Agricultural Research Institute, Ghana. 1995 Annual Report.

Langyintuo, A.S., Abatania, L., Asare, E. and Albert, H. (1995): The economic of alternative methods of soil fertility improvement in the guinea savanna zone of Ghana. Proceedings of seminar on Organic and Sedentary Agriculture, organised in Accra, Ghana, 1-3 November 1995, 295-315

PPMED (Policy Planning Monitoring and Evaluation Division). (1991, 1994) Statistical information from the Ministry of Food and Agriculture, Ghana.

Ruthenberg, H. (1980): Farming Systems in the Tropics. Clarendon Press, Oxford.

Schleich, K. (1986). The use of cattle dung in in agriculture. Natural Resources and Resource Development **23**: 53-87

Stangel, P.J. (1993): Nutrient cycling and its importance in sustaining crop-livestock systems in sub-saharan Africa: An overview. Proceedings of an international conference on Livestock and sustainable nutrient cycling in mixed farming systems of sub-saharan Africa. Vol. II: Technical Papers. Edited by J.M. Powell, S. Fernandez-Rivera, T.O. Williams and C. Renard. Addis Ababa, Ethiopia, 22-26 November, 1993, 37-59.

VSD (Veterinary Services Department), Ghana (1993): Livestock census figures for 1993.

Addresses of authors:
A.S. Langyintuo
Savanna Agricultural Research Institute
P.O. Box 52
Tanale, Ghana
N. Karbo
Animal Research Institute
P.O. Box 52
Tamale, Ghana

Ecological and Socio-Economic Reasons for Adoption and Adaptation of Live Barriers in Güinope, Honduras

J. Hellin & S. Larrea

Summary

In the 1980s, the non-governmental organisation World Neighbors promoted live barriers of *Pennisetum purpureum* (Napier grass) and *P. purpureum* × *P. typhoides* (King grass) in the Güinope region in Honduras. Since the end of the World Neighbors programme in 1989, farmers have adapted the technology and now use more species than those originally promoted. The authors conducted a Rapid Rural Appraisal (RRA), interviewed 68 farmers, held semi-structured interviews with 10 additional farmers, and organised a workshop in order to identify the species being used in live barriers and the criteria used by farmers when selecting these species. Farmers recognise that the live barriers promoted by World Neigbors are effective in retaining soil, but also point out that the species are invasive and there is little demand for the amount of fodder produced. Farmers are increasingly using *Saccharum officinarum* (sugar cane) and fruit trees in live barriers. These species are not as effective in retaining soil but contribute to the farm household in terms of domestic consumption and/or the sale of the products of the live barriers. The study highlights the importance in soil and water conservation programmes of working with farmers in identifying their needs and selecting appropriate technologies.

Keywords: Land husbandry, live barriers, small-holder hillside farmers, Honduras, farmer decision-making.

1 Introduction

The Güinope region in southern Honduras lies between 500 and 1800 metres above sea level and covers an area of 204 km². The population is approximately 5,500 of which 80% is engaged in agriculture. Annual rainfall is 1100-1300 mm and the main rainy season is from May to October. An impenetrable subsoil underlies the 15-50 cm deep top soil and farming takes place on slopes which are often 15-30% (Bunch and López, 1994). These slopes are subject to severe erosion. Between 1981 and 1989, the non-governmental organisation World Neighbors promoted a number of soil and water conservation (SWC) technologies in 41 communities in the Güinope region. The SWC technologies included live barriers of *Pennisetum purpureum* (Napier Grass) and *P. purpureum* × *P. typhoides* (King grass), contour drainage ditches below the live barriers, and the use of chicken manure as a natural fertiliser. Prior to 1981, farmers in the Güinope region practiced very few SWC technologies and live barriers were unknown (Roland Bunch pers. comm.).

Studies by Bunch and López (1994) and López et al. (1995) have documented that five years after World Neighbor's programme finished in the Güinope region, farmers continue to use the SWC technologies that had been promoted. They also noted that in several communities farmers are adapting the live barriers and are including different species to those originally promoted. Neither study detailed the reasons for the adoption and adaptation of the live barrier technology.

2 Methodology

The authors carried out a Rapid Rural Appraisal (RRA) in the Güinope region towards the end of 1995 in order to identify some of the species being used in live barriers. At the beginning of 1996, the authors began a more detailed study to explore the rationale behind species selection for use in live barriers. The study focused on the 15 principal communities in the municipality of Güinope as opposed to the larger Güinope region.

The focus of the study in the Güinope municipality was a structured interview. The first step was to determine the number of farmers in the 15 communities who had adopted SWC technologies, such as live barriers, since 1981 irrespective of whether they had rejected, continue to use or have adapted the SWC technologies. The authors did not have access to a list of farmers who had adopted one or more of the SWC technologies during the World Neighbors programme. Three farmers were therefore interviewed in each of the 15 communities in order to identify these farmers. A total of 299 farmers were identified in the 15 communities. The limitation of this approach was that the farmers tended to remember those who continue with the SWC technologies rather than those who have abandoned the technologies.

The sample of farmers selected for interview in each community was based on a maximum error of 30% and a probability of 0.2. The formula used was:

$$n = (t^2 * p * q)/E^2 \quad \text{(Cochran, 1976)}$$

where:

n = the size of the sample
t = the value with a probability of 0.20 found in the Student Tables
E = the maximum error permitted (30%)
p = the proportion of adopters and adapters of SWC technologies
q = the proportion of those who have rejected the SWC technologies

N.B. $p + q = 1$

In the above formula $p*q$ is at a maximum when $p = q = 0.5$ and this also gives the maximum sample. Although it is far from certain that $p = q = 0.5$ in each community i.e. that the proportion of adopters/adapters of live barriers is the same as those who have rejected the technology, it was decided to use these figures because they gave the maximum sample size per community. Sixty-eight farmers in the 15 communities were subsequently interviewed.

The authors also consulted secondary data on the Güinope region, carried out semi-structured interviews with 10 farmers and held a workshop for a further 10 farmers who had received extension training from World Neighbors and who are still involved in SWC extension work. The workshop and semi-structured interviews were designed to shed more light on the motivation of farmers using live barriers and the criteria that they used/use when selecting species for the live barriers.

3 Results and discussion

3.1 Species being used in live barriers in the Güinope region

The farmers interviewed are using a total of 12 species in live barriers however, through the RRA the authors documented 19 species that are being used in live barriers in the Güinope region (see Table 1).

The authors have defined "adoption" as the use of the live barriers promoted by World Neighbors i.e. *P. purpureum* (Napier Grass) and *P. purpureum* × *P. typhoides* (King grass) and "adaptation" as the use of species other than the two grass species. Farmers who have adapted the live barriers have not been separated into first-time users of other species live barriers and those who started with live barriers of *P. purpureum* (Napier Grass) and *P. purpureum* × *P. typhoides* (King grass) and whom subsequently selected other species.

Scientific name	Common name (in Spanish)
Grass species	
Vetiveria zizanioides	Zacate valeriana
Cymbopogon citratus	Zacate limón
Pennisetum purpureum	**Zacate napier**
Pennisetum purpureum x Pennisetum typhoides	**Zacate king grass**
Setaria geniculata	Setaria
Panicum maximum	Zacate guinea
Saccharum officinarum	Caña de azúcar
Brachiaria mutica	Pasto pará
Paspalum notatum	**Pasto bahía**
Trees, shrubs and other plants	
Cajanus cajan	Gandul
Manihot esculenta	**Yuca**
Coffea arabica	Café
Citrus limetta	Lima
Citrus sinensis	Naranja
Prunus persica	**Durazno**
Gliricidia sepium	Madreado
Colocasia esculenta	**Quíscamo**
Musa acuminata	**Guineo**
Ananas comosus	**Piña**

[a] The species being used by the farmers who were interviewed in the Güinope municipality are in bold. Many of the farmers are using a mixture of several species in the same live barrier.

Table 1: Species being used in live barriers in the Güinope region [a]

Of the 68 farmers in the municipality of Güinope who were interviewed, 63 had established live barriers between 1980 and 1996[1] and only three had subsequently rejected them. In 1996, there were 29 farmers who had established live barriers of *P. purpureum* (Napier Grass) or *P. purpureum* × *P. typhoides* (King grass), 20 farmers had adapted the technology and had established live barriers of other species, and 11 farmers had live barriers of the two grass species and other species (Figure 1).

---

[1] Farmers who worked with World Neighbors at the beginning of the programme were sometimes unable to remember the exact year when they established a live barrier. Several cited 1980 even though World Neighbors started work in the Güinope region in 1981. The authors have decided to use the responses of the farmers.

From 1980-1996, the majority of farmers established live barriers of *P. purpureum* (Napier Grass) or *P. purpureum* × *P. typhoides* (King grass). However by 1988, farmers were establishing more live barriers of other species (Figure 2). This trend has continued to the extent that in the period 1992-1996, farmers established five live barriers of *P. purpureum* (Napier Grass) or *P. purpureum* × *P. typhoides* (King grass) compared to 18 live barriers of other species.

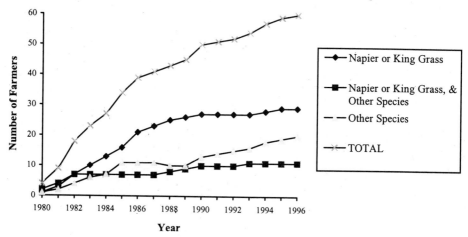

Fig. 1: The cumulative number of farmers using different species in live barriers 1980-1996 (sample size 68 farmers)

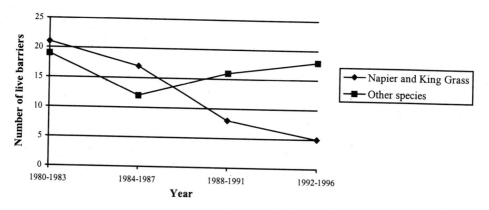

Fig. 2: The number and types of live barriers established by 63 farmers 1980-1996

3.2 Criteria used by farmers for selecting different species for use in live barriers

Farmers in Güinope recognise that live barriers of *P. purpureum* (Napier grass) and *P. purpureum* × *P. typhoides* (King grass) retain soil and that over a period of 5-10 years natural terraces are formed. Farmers, however, have also pointed out that there are several disadvantages with using these two grass species in live barriers. These disadvantages can be divided into three categories.

- The most cited disadvantage is that both species are invasive if not regularly managed. This is especially the case with *P. purpureum* (Napier grass) and precludes the use of both species as a green manure.
- The two grass species provide excellent fodder and therefore the maintenance of the live barriers need not be a problem from the farmers' point of view if the farmers are regularly cutting the live barriers to feed to their cattle. However, few farmers in Guinope have cattle and therefore there is little demand for the amount of fodder produced.
- The two grass species have an extensive root system and therefore compete with agricultural crops. There are several examples in the Güinope region where the maize plants on either side of *P. purpureum* (Napier grass) and *P. purpureum* × *P. typhoides* (King grass) live barriers are stunted compared to the maize between the live barriers. This deleterious effect can be seen up to 2 metres on either side of the live barrier.

The species being selected by farmers in the Güinope region for use in live barriers are those that are not invasive and, more importantly, those that contribute to the farm household in terms of domestic consumption and/or the sale of the products of the live barriers. Farmers are increasingly using *S. officinarum* (sugar cane) in live barriers and a number of fruit trees including lemon, orange and pear trees (Table 1). Many of the species being used in live barriers are not as effective as *P. purpureum* (Napier grass) and *P. purpureum* × *P. typhoides* (King grass) in controlling soil erosion but they make far greater contributions to the farming system than the two grass species.

The results of the study point to the fact that species selection for use in live barriers in the Güinope region is based on socio-economic and ecological criteria. Farmers are clearly interested in the ability of the live barriers to retain soil and contribute to soil fertility, otherwise they would not be using the species in live barriers, but they show a preference for species that confer other benefits. This conclusion fits into the philosophy of the *Land Husbandry* approach to soil conservation which recognises that farmers are primarily concerned with stable and economic production and not with the conservation of soil and water *per se* (Shaxson, 1996).

The study in the Güinope region also highlights the importance in SWC programmes of collaborative work with farmers in analysing individual farming systems, identifying farmers' needs and selecting appropriate technologies.

Acknowledgements

The study in the Güinope region was partly funded by the British Government's Department for International Development (DFID) through the Forestry Research Programme. The authors are very grateful for DFID's financial support.

The authors would like to thank all the farmers in the Güinope region who participated in the structured and semi-structured interviews and the workshop. In particular they are very grateful to Mario Zavala from the community of Casitas.

The authors are also very grateful to Roland Bunch and Gabino López (Asociación de Consejeros para una Agricultura Sostenible, Ecológica y Humana, Honduras) for the background information on the Güinope region and the following at the Escuela Agrícola Panamericana, El Zamorano for their contributions to this study: Dennys de Moreno, Armando Medina, Miguel Avedillo, Nelson Gamero, Marco Granadino and the students of "Social Research Methods".

The authors received invaluable assistance in the design of the structured interview from Karen Dvorak (International Centre for Tropical Agriculture) and Steve Sherwood (Cornell University).

References

Bunch, R. and López, G. (1994): Soil recuperation in Central America: measuring the impact four to forty years after intervention. Paper presented at New Horizons: The economic, social and environmental impacts of participatory watershed development. International Workshop organised by the International Institute for Environment and Development (IIED), Bangalore, India 28 November--2 December 1994.

Bunch, R. (pers. comm.): Asociación de Consejeros para una Agricultura Sostenible, Ecológica y Humana (COSECHA).

Cochran, W. (1976): Técnicas de muestreo, Translated by Eduardo Casas Díaz. 6 edition. Mexico. Contintental. 507 pp.

López, G., Garcia, J. and Bunch, R. (1995): Adoption of soil and water conservation technologies in the Güinope district, Honduras. Study funded by Silsoe Research Institute, UK. 23 pp.

Shaxson, T.F. (1996): Principles of good Land Husbandry: achieving the conservation of the land's productive potentials. Newsletter of the Association for Better Land Husbandry. No. 5, June 1996, 4-14.

Addresses of authors:
Jonathan Hellin
Natural Resources Institute
Chatham Maritime, Kent ME4 4TB, U.K.
S. Larrea
Escuela Agricola Panamericana
Apartado Postal 93
Tegucigalpa, Honduras

Farmers and Their Perception of Soil Conservation Methods

J. Currle

Summary

The paper focuses on improving the quality of agricultural extension on environmental issues. An effective extension approach should start with a sound understanding of farmers' perception. A case study is presented from the Kraichgau region in Southwest Germany that describes how farmers perceive both soil erosion and the recommendations of researchers on soil conservation. In this region, as in many others, increased soil loss through water erosion has taken place over the last three decades. This has corresponded with the modernisation and mechanisation of agriculture. For the farmers, the menace is not very visible as erosion is perceived as isolated catastrophic events rather than as a continuous process. This is one reason for the great reluctance in accepting erosion control systems of mulching and minimum tillage that have been developed by agricultural researchers. The preconditions for implementing such systems are at variance with the organisational pattern of medium-sized farms and also contradict, in various ways, the social norms of the farming communities. The paper concludes with a plea for extension to focus on the long-term effects of soil erosion and for environmentally-sound erosion control techniques to be developed, together with farmers, in order to fit local conditions.

Keywords: Environmental perception, agricultural extension, farmer knowledge systems, soil erosion, soil conservation, participatory technology development.

1 Introduction

The Kraichgau, a region in the south-western part of Germany situated roughly in the triangle contained by Heidelberg, Karlsruhe and Heilbronn, is known as a very fertile farming area. The fertile loess-layers led to a long history of settlement in and cultivation of the rolling landscape. The region also has remains of centuries old water-erosion and evidence of sedimentation (Bleich, 1978). The history of erosion in this area shows that the amount of eroded soil fluctuated dramatically according to the prevailing land-use patterns (Quist, 1984). The very small lots, the man-made barriers like hedges and ridges between the lots, and the high levels of grass production, clover and alfalfa for fodder, limited the danger until the 1960s. Thus the traditional agricultural production pattern itself kept the risk of erosion low in this potentially vulnerable area. Three decades ago, however, the production pattern began to change dramatically allowing for a steady growth in erosion. This was due to:

a) The "Flurbereinigung", i.e. the consolidation of farmland and its re-allotment, was an administrative attempt to increase the profit of family-farms. The small lots of individual farmers were combined resulting in an average increase of parcel size from about 0.5 ha to about 2 ha.

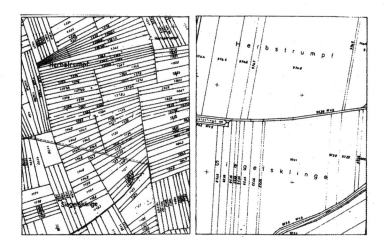

Figure 1: Part of the village district before (left) and after (right) reallotment.

These bigger farm parcels also changed the ploughing direction. Formerly they were ploughed at right angles to the slope and ploughing was done with the help of animals which were not strong enough to pull a plough uphill. The situation changed with the introduction of tractors. Working up the slope was no longer a problem, but contour ploughing with heavy machinery caused severe problems with downhill side-drift.

	hilltop, 2° slope	upper slope, 4-5° slope, 50m length	lower slope, 6-7° slope, 150m length
Before reallotment	1.7mm	2.4mm	1.7mm
After reallotment	1.6mm	4.3mm	7.7mm

Table 1: Average erosion losses in the Kraichgau before and after reallotment (Quist, 1984)

b) With the enlargement of fields, most of the traditionally grown hedge and ridge field boundaries were removed.
c) Maize and sugar beet, two crops that leave the soil extremely vulnerable, are gaining more and more importance in agricultural production.

As a result of these developments, erosion was no longer controlled as part of normal agricultural practices. Erosion control now required more conscious consideration within the agricultural production process.

The acceleration in erosion was first noticed around 1970 on one of the big estates with large fields (Eichler, 1974). Beginning in 1980, researchers developed measures to stop or at least diminish soil erosion mainly on these big farms. These measures are almost exclusively cultivation methods, in particular mulching and reduction in soil tillage. They were subsequently promoted by the public agricultural extension system, yet their adoption was not overwhelming (about 10-15% of the farmers practised these mulching methods in 1990). This did not change much even after the introduction of a state government programme to provide support to farmers for buying machinery necessary for the desired cultivation methods.

Hence, there are two major causes preventing farmers from using recommended erosion control methods; the farmers' views of soil erosion as a problem, and, the farmers' attitude towards the recommended erosion control methods. The purpose of the investigation reported upon here was to examine both these issues and to elaborate improvements for extension on soil erosion control.

2 Methodology

Three neighbouring villages no more than six kilometres apart in the central Kraichgau were chosen for study (Currle, 1994). Despite their proximity, the villages differed considerably in the implementation of the officially recommended erosion control measures. The farmers' perception of soil erosion and decision-making was examined using an open qualitative approach. Semi-structured interviews provided a flexible tool for covering central topics while providing an opportunity for the farmers to express new ideas. All forty-seven full-time farmers of the villages were interviewed; the interviews were tape-recorded, transcribed and evaluated by hermeneutic text analysis (Lamnek, 1989).

3 Farmers' view of soil erosion

3.1 Soil erosion as a natural phenomenon

It was surprising to find that most farmers considered soil erosion as normal, as something that was and will always be part of the region. As stated earlier, erosion is indeed a phenomenon that has always existed in the region and, even in spite of traditional structures, catastrophes could never be totally eliminated. Since farmers viewed soil erosion as a natural phenomenon which their grandfathers also had to live with, they concluded that they too will have to learn to live with it because it is a phenomenon which they felt unable to influence. This conclusion is strengthened by the way they assessed their risk of being hit by soil erosion: they perceive the loss of fertile topsoil as a result of extraordinary catastrophic rainstorms which, by bad fortune, might occur and hit their own fields once or twice a year.

3.2 Perception of changes in farming practices

Farm sizes and production structures have changed tremendously during the last thirty years. All farms investigated were affected by these changes. They had expanded in area, produced considerably more maize and sugar beet and now ploughed parcels that were two to three times the size of those in 1969.

Interestingly, most of the farmers interviewed did not mention these changes. They did not perceive these as dramatic events but as a slow, yet normal development with practically no influence of its own. That farmers do not perceive the changes in their own farming methods and in the landscape can explain why they did not correlate it with an increase in soil erosion. On a day-to-day basis the details of these changes are not apparent. Erosion processes as well as gradual changes in farming only become apparent if we evaluate the consequences over a relatively long period of time and make a direct comparison with the past.

	cultivated land per farm (ha)		average parcel size (ha)		crops that further soil erosion[b] (% of cult. area)		crops that stop soil erosion[c] (% of cult. area)		farms with dairy cattle		total of farms investigated
	1969	1990	1969	1990	1969	1990	1969	1990	1969	1990	
A-village[d]	16.8	29	0.63	0.73	17.6	38.5	30.8	17.1	15	11	15
B-village	17.6	43	0.61	1.9	16.7	30.3	21.6	12.9	12	8	12
C-village	21.6	39	0.98	1.8	14.5	35.8	32.6	12.8	16	8	16

a according to the farmers investigated
b potatoes, sugar-beet, maize
c fodder (without maize), meadows, pasture land
d A, B and C-village refer to the three communities under study (s. p. 4)

Table 2: Changes in farming 1969 - 1990[a]

3.3 Erosion as one problem among many others

This attitude explains why fighting erosion is not a priority for the farmers. Yet another reason for paying so little attention to this problem is their ever-present fear of bankruptcy and farm closure. As one of the farmers interviewed put it:

"Soil erosion is something that occurs maybe once in a few years, but falling prices for our products, that's something that hits all of us every year".

This insecurity about the future for most of the medium-sized family farmers causes them to pay little attention to soil erosion.

4 Farmers' attitude towards recommended erosion control methods

Widespread implementation of erosion control measures requires a change in the farmers' lack of interest in the problem. Yet even given this change, broad acceptance of the promoted erosion control measures would be doubtful. For the majority of farmers the effort needed to implement the minimum tillage and mulching techniques is too high.

4.1 Comparison of the traditional and recommended new tillage system

Instead of just tilling the land after harvest and leaving the weeds to sprout and be ploughed down in late autumn, the erosion control techniques recommend the sowing of green manure after harvest. This green manure is not re-ploughed into the soil in late autumn, but is either left to freeze, mulched, or worked into the soil during winter time.

Before drilling, the summer crop weeds and unfrozen green manure plants must be chemically destroyed. Drilling is done on a prepared seedbed, or as direct drilling. The farmers generally considered three criteria in evaluating this new procedure: first, the risk that accompanies the introduction of a new technique, second, the question of how the new technique fits into the existing farm and labour organisation and, third, though usually little considered during the development of new techniques, is the important question of how it fits into the farmers' thinking and acting patterns and whether it challenges social norms.

4.2 Farmers' risk assessment with regard to erosion control techniques

The life of a farmer is extremely dependent on nature. Nobody experiences the whims of nature and man's limited influence on this process more directly than the farmer. Government policies and the whims of the Department of Agriculture are also regarded as being unpredictable and a source of risk over which farmers have little influence (Illien & Jeggle, 1978 pp. 40; Kölsch, 1990 p. 125). It is these two sources of risk which farmers aim to minimise. The traditional tillage procedure provides clear evidence for this. It starts with shallow tillage after harvest, allowing weeds to sprout and providing a long period for mechanical weeding in Autumn before the ploughing is done from the beginning of November until mid-December. After ploughing, the field is ready for a problem-free drilling in Spring. Each of these traditional steps is well defined and carried out within a period of time long enough to guarantee adequate weather conditions of several days. In this way dependency on the climate is reduced to a minimum.

This is different from the mulching system recommended by researchers to control erosion. Here low soil moisture after harvesting produces poor conditions for green manure to sprout. Poorly developed green manure in turn cannot prevent germinating weeds from growing. Many weeds as well as lush green manure need tillage during Winter and a herbicide surface treatment before drilling. Tillage can only be done as long as the soil is frozen enough to support the tractor, whereas the topsoil should not be too frozen in order to allow at least mulching or shallow tillage. Only a few days each year, if at all, provide these ideal climatic conditions. When it comes to drilling in Spring, the plant residue needs more time to dry enough so as not to clog the machinery.

For the farmer, these considerations require constant observation and the flexibility to be ready to perform the necessary steps that the day-to-day combination of climatic and crop conditions demand. Thus, the recommended techniques contrast greatly with the traditional system of having a regular sequence of cultivation steps that reduce the dependency on climatic factors to a minimum.

The natural risks are worsened by an increased dependency on the Department of Agriculture. Applying mulching techniques almost always requires the application of herbicides before seedbed preparation. The use of these chemicals is strongly controlled by the government and permission for use of a specific substance is given or withdrawn in a way the farmers find unpredictable. One of them put it in these words:

> "Erosion control, very well...but I doubt that everything works so smoothly as they say. A colleague of mine who's doing it sprayed his sugar beet with "Roundup" (herbicide) already in spring time. That's something I've never done. It's OK, but what if they withdraw permission for "Roundup"? We used to fight grass weeds with "Galant" (herbicide) and the other year it was prohibited just like that".

In this respect, it is the traditional tillage system that provides more independence from external forces, whereas the erosion control system, in the way the farmers experience it, just adds to insecurity and dependence.

4.3 Compatibility with existing machinery and labour organisation

It proved necessary for many farmers to invest in specialised tillage mechanisation in order to implement the recommendations. As most of the farms had little specialised production, buying specialised machinery would not usually be profitable. For the farmers with cattle another difficulty arose. In order to guarantee good germination of green manure, the sowing must be done shortly after the grain harvest in order to have residual soil moisture and a well-prepared seedbed for the plants. However, the grain harvest for the cattle-holding farmers is just before the time when slurry must be spread and maize for silage has to be harvested. These two points highlight the difficulties inherent in the recommended erosion control methods. They are difficult to implement on traditional family farms which practice a wide range of production activities. The minimum tillage and mulching methods are techniques for specialised farms that focus on crop production and whose labour profile gives additional time after harvest.

4.4 Mulching techniques and social norms

The compatibility of new techniques with routines of farmers is a third criteria in their appraisal. These routines unconsciously direct the "do's" and "dont's" of everyday life. Customarily this way of thinking is sanctioned socially and thus a new technique faces difficulties in being accepted. Having a clean field is a strong social indicator of being a good farmer. Hence, the mulching and minimum tillage techniques collide with the traditional norm and this generates strong resistance from the farmers.

Until recently, one of the most important problems of crop cultivation was how to effectively fight weeds. Apart from using crop rotation, weeds could only be fought mechanically, mostly by hand labour. The risk of low yields due to weeds was high and so it was advisable to fight even low levels of weed infestation to prevent it from getting out of control. This was important for one's own parcels of land and those of neighbours, because high levels of weed infestation easily crossed the limits of individual fields. Thus keeping ones fields clear of weeds produced an ideal image of the "clean field". This ideal is the farmers' standard of measurement for their own and as well as for the fields of their colleagues. A young farmer interviewed stated:

"We've learnt that good ploughing is the core of crop cultivation. By looking at the way a farmer has ploughed you can judge whether he is a good farmer or not. A good farmer should be able to plough straight, should have his field even and the field should be well frozen in spring. And even more, to have a clean field means that the sugar beets have to look like a garden in spring time. You should be able to see each row of the seedlings, each track of the tractor".

These words show the rigidity of the ideal image. The "clean field" is clearly defined and has to meet several well-established criteria. For the farmer to deviate from this image means endangering his reputation in the village. When asked whether his colleagues would talk bad about him if his fields were full of weeds, one of the farmers interviewed expressed the influence of village opinion this way:

"No, it's not that I pay any attention to the talk of the village. By the way it not only depends on the mulching. You see, we have other careless fellows here in the village, whose fields look awful even without mulching".

With his first sentence this farmer denies the impact of social pressure on himself. But with his second and third sentence he makes clear what he thinks about colleagues who don't respect the social norm of having a clean field. In denying the impact of social pressure on himself, this farmer highlights the way this pressure works. Nobody has to say anything to someone who has a field full of weeds because that person knows that the community considers him a bad farmer. And even he himself shares this valuation because the "clean field" is a commonly-shared and individually internalised perception pattern (Berger and Luckmann, 1966).

When one looks at the recommended mulching methods with this background in mind, the communication difficulties between technically-oriented agricultural advisors and farmers become evident. Under present technical conditions, the rigidity of the norm for reinforcing effective weed control is not necessary because the availability of herbicides allows for higher weed tolerance without running the risk of non-manageable competition. This is usually a point for extension and is a technical change minimum tillage methods take advantage of. For the farmer on the other hand the implementation of these methods means giving up the image of the "clean field"; it means to both incorporate the technical requirements of herbicide application and live with the consequences of ignoring a social norm. Thus, mulching methods can help protect the soil and yet are difficult to implement in the social reality of the farmer.

5 Consequences for extension and development of soil conservation techniques

Soil erosion is both a world wide and a local problem, and one which is receiving increasing attention in an age that strives for sustainable agriculture. It is also a problem of society as a whole and its control has, therefore, significant public good characteristics that can require public sector intervention through mechanisms such as extension (Kidd et al., 1997). Solutions, however, can only be implemented by the farmers. Little can be done against soil erosion without the co-operation of the farmers. Taking this and the above findings into consideration, two points become clear:
- We have to start with the farmers' perception and look for ways to enable them see soil erosion as their problem as well.
- In addition to being technically effective, erosion control measures must fit the economic, social and psychological framework of farmers.

5.1 Strategies to change farmers' perception of the problem

Gradual soil erosion processes are obviously difficult to perceive as they take place. Except for very big disasters, soil losses in the field are hardly visible and the deterioration of soil fertility is seen only after a long period of time. For the next generation of farmers, lower soil fertility will be an accepted environmental fact rather than a consequence of the actions of their fathers and grandfathers. Thus, the distribution of soil and fertility losses over a long period of time makes them much less visible to normal human perception. This effect is strengthened by the distribution of often considerable soil losses over large areas. Fifteen tonnes of fertile topsoil lost on one hectare is equivalent to about a one millimetre layer. Both time and space perception problems can be addressed by extension. The actual measurement of soil losses in one of the villages investigated is an example of how this could be done. This measurement was done with simple, understandable methods on the fields of some of the farmers. It was these farmers who, on realising that their own fields were losing up to 150 tonnes of fertile top soil a year, began to understand erosion as their own problem.

5.2 Farmers' participation in technology development as a precondition for erosion control

A big estate in the region was the first collaborator for the development and testing of the new mulching and minimum tillage methods. The manager of this estate, a competent crop farmer, could act independently and without the pressure of social village norms and other restrictions that small farmers face. It is striking that it is exactly these types of farm, i.e. medium-sized to big farms, in the study region that are successfully applying these mulching and minimum tillage methods. It is obvious that the technical solution offered, which was in fact not aimed at a certain target group but at controlling erosion in crop cultivation in general, fitted the farms on which the experiments were carried out as if they were developed specifically for them. Given the conditions of these farms (i.e. more specialised in crop production) and given the thinking and perception patterns of their farmers, it was easy to implement the new techniques. Where these conditions do not exist for economic, social or personal reasons the implementation becomes difficult or near impossible.

A general conclusion would be that an innovation fits conditions of the situation where it was developed. In addition, for successful and sustainable further implementation, it requires these same conditions. Extension is not in a position to reorganise existing farms and patterns of perception and thinking to such an extent for ready-made technical innovations to be spread. If we want to tackle environmental problems that are not possible to solve without the introduction of new techniques and working procedures we should develop and adapt these techniques in the way described by Hagmann (1993).

- A single technical solution to a problem does not help in different conditions. However, if we want it to work, it is important to assess its feasibility together with the farmers concerned.
- The development and adaptation of new techniques must be performed in close co-operation with the target groups. It is important in this case to have these groups relatively homogeneous according to their social and economic conditions.
- Different social, economic and environmental conditions, even within the same area, may lead to different solutions with variation in the effectiveness of soil conservation. This is the case in Germany as with many other parts of the world (e.g. Lamers, 1996). This should be accepted even if some of the solutions do not have the greatest technically possible impact on soil conservation. A lower level of soil erosion control may well have a more optimal fit with the multitude of objectives small farmers must manage.

Acknowledgements

The research described in this paper was undertaken while at the Department of Agricultural Communication and Extension, Institute for Social Sciences of the Agricultural Sector, University of Hohenheim (430A), 70593 Stuttgart, Germany. The author would like to thank the German Research Foundation (DFG) for funding this research.

References

Berger, P.L. and Luckmann, Th. (1966): The social construction of reality. Doubleday Inc., N.Y.
Bleich, K. E. (1978): Erosion von Böden infolge Bodennutzung. In: Daten Dokumente Umweltschutz, Umwelttagung, Heft 22, Universität Hohenheim.
Currle, J. (1994): Landwirte und Bodenabtrag. Empirische Analyse der bäuerlichen Wahrnehmung von Bodenerosion und Erosionsschutzverfahren in drei Gemeinden des Kraichgaus. Weikers-heim, Margraf Verlag.
Eichler, H. (1974): Bodenerosion im Kraichgauer Löß. In: *Kraichgau,* Folge 4, S.174-189.
Hagmann, J. (1993): Farmer participatory research in conservation tillage; approach, methods and experiences from an adaptive on-farm trial programme in Zimbabwe. In: Kronen (ed.), Proceedings of the fourth Annual Scientific Conference of the SADC-Land and Water Management Research Programme, held in Windhoek, Namibia, October 11-14, 1993. Gabarone, Botswana
Ilien, A. and Jeggle, U. (1978): Leben auf dem Dorf. Zur Sozialgeschichte des Dorfes und Sozialpsychologie seiner Bewohner. Westdeutscher Verlag, Opladen.
Kidd, A.D., Lamers, J.P.A. and Hoffmann, V. (1997): Towards pluralism in agricultural extension: A growing challenge to the public and private sectors, Entwicklung + Ländlicher Raum, 2/1997.
Kölsch, O. (1990): Die Lebensform Landwirtschaft in der Modernisierung. Grundlagentheoretische Betrachtungen und empirische Deutungen zur Agrarkrise aus der Lebnswirklichkeit von konventionell und ökologisch wirtschaftenden Landwirten aus Niedersachsen. Frankfurt a.M.
Lamers, J.P.A. (1995): An assessment of wind erosion control techniques in Niger: Financial and economic analysis of windbreaks and millet crop residues. Verlag Ullrich E. Grauer, Stuttgart, 208 pp..

Lamnek, S. (1988): Qualitative Sozialforschung. Vol. 2: Methoden, Techniken. München, Weinheim.
Quist, D. (1984): Zur Bodenerosion im Zuckerrübenanbau des Kraichgaus. Diss. Fakultät für Biologie, Universität Heidelberg. Unveröffentlicht.

Address of author:
Jochen Currle
PACTeam GbR
Nufringer Str. 4
D-70563 Stuttgart, Germany

Farmer Perception of Soil Protection Issues in England and Wales

R.J. Unwin

Summary

In 1993 the Ministry of Agriculture Fisheries and Food (MAFF) and the Welsh Office Agriculture Department published a Code of Good Agricultural Protection for the Protection of Soil in England and Wales. The Code set out general principles of soil management to give guidance to farmers and other managers of agricultural land. In 1995 a survey of farmers was conducted to determine their awareness of the Code and their attitudes to it. Whilst 28% claimed to know of the Code only 5% could prove ownership despite it being available free of charge. The style and content of the Code was generally well received. During the winter of 1995/6 a pilot study of farmers attitudes to soil protection was undertaken in four areas. The aim was to assess the occurrence and severity of any soil protection issues and farmer appreciation and reaction to them. The results confirm that on sandy soils water erosion is locally an important factor. Farmers in these areas are showing signs of becoming increasingly concerned about the situation; some off-site effects are of more concern than the loss of the soil resource. On clay soils the management practices that have been developed in recent years appear to be maintaining soils in a satisfactory condition.

Keywords: Soil protection; soil erosion; Codes of Good Practice; attitude survey; extension.

1 Introduction

In the United Kingdom there is little specific legislation to protect soils. Good soil management practice has traditionally been achieved by voluntary compliance with advice provided by extension services targeted to local needs and emphasising the self-interest of farmers to protect their soils. In 1970 a report was published (MAFF, 1970) which expressed concern about organic matter levels in soils and of structural conditions due to the use of heavy machinery under unsuitable conditions. This report made little mention of water erosion but in the years that followed it is generally agreed that in certain areas this problem increased (e.g. Boardman 1995). In 1996 the Royal Commission on Environmental Pollution issued a report on the sustainable use of soil (RCEP, 1996) and reviewed all aspects of soil protection in the United Kingdom. The Commission accepted that whilst soil erosion was important locally there was no evidence of widespread problems.

Government advice on management practices to protect the environment are contained in Codes of Good Agricultural Practice for the Protection of Water (MAFF, 1991), Air (MAFF, 1992) and Soil (MAFF, 1993). The Soil Code has sections on Soil Fertility, Physical Degradation,

Contamination and Restoring Disturbed Soils. They provide the general principles of soil management which should be followed and adapted to individual circumstances.

In 1995 a survey was undertaken to determine farmer awareness of and reaction to all three Codes. Subsequently during the winter of 1995/96 a pilot study was undertaken in four areas of England to consider the occurrence of soil protection problems and farmer perception and reaction to them. Results from both of these studies are presented and discussed in this paper.

2 Survey of the Codes of Good Agricultural Practice

2.1 Method and results

The objectives of the survey were to assess ownership and use of the Codes, to assess reactions to their content and style and to gauge interest in acquisition amongst non-owners. A random sample of farmers stratified according to farm size was selected and 1155 received a screening interview by telephone. More detailed telephone interviews were held with 337 of these farmers whilst 264 personal interviews were held on farms. Just over half of those visited (145) owned one or more of the Codes. As well as being asked about ownership, contacts were asked for their opinion on the style and content of the Codes, and if they had acted on the advice. An assessment was made to determine if farmers who had copies of the Codes were farming more closely to recommended practice than those that did not.

Over a quarter of all farmers contacted (28%) claimed to be aware of the Soil Code and 12% to have a copy. Only 5% could prove they actually owned a copy. Of those farmers owning the Soil Code virtually all claimed to have read it (94%) but only 13% had passed it to any of their farm staff to read. Nearly half (47%) claimed to have taken action since reading the Code.

The 60 farmers who assessed the Soil Code were overwhelmingly supportive of its style and content. Over 90% considered it comprehensive and clearly presented. Around 80% found it easy to read, understand and find the section of interest. Over 70% described it as not too technical.

Of those farmers already judged to be following Good Practice, and who had not previously seen a Soil Code, nearly two thirds expressed an interest in acquiring a copy when shown one during the interview. Across all Codes there was a clear indication that farmers who did not appear to be following Good Practice were less likely to see the need for advice or to read the Codes.

2.2 Conclusions

The style and content of all of the Codes are effective in communicating their message once they are in the hands of farmers. However despite being readily available free of charge the Soil Code has not become widely distributed among its target audience. Of more concern is the clear indication that farmers who do not appear to be following Good Practice are less likely to recognise the need for advice in general or the Codes in particular.

3 Pilot project on soil protection issues in four areas of England

2.1 Method

The objectives of this study were to assess soil conditions in four pilot areas, to consider any changes over the last ten years, to assess the attitudes and knowledge of farmers with respect to soil protection and to consider the need for further guidance to alleviate any problems found.

The criteria for selecting the areas to be studied were that they should be representative of soils and farming systems, not all should be known to have soil problems and the areas should not have been extensively studied by other workers. It was decided to concentrate on four areas. Two with similar sandy soils in areas with contrasting relief, rainfall and farming systems. These were the lower Otter Valley in East Devon with grass and arable systems on rolling relief with 900 mm annual rainfall and an area in the West Midlands with more subdued relief, and predominantly arable systems in 700 mm annual rainfall.

The third area on shallow chalk soils in north Hertfordshire with continuous arable farming on rolling relief and a low rainfall (570 mm). This contrasts with similar soils on the South Downs (Boardman, 1995) where soil erosion is common on steep slopes and there is 790 mm annual rainfall. The fourth area was in west Lincolnshire to represent the clay soils of central and eastern England with subdued relief, predominantly cereal cropping and rainfall less than 700 mm.

Ten farm visits were made in each area between November 1995 and February 1996. A questionnaire was completed by a soil scientist in discussion with the farmer. The consultants were well received by the farmers who welcomed the opportunity to discuss soil matters on a one-to-one basis. The farmers' views of soil management and soil protection issues on their farms were recorded. For each possible issue that they recognised on their farm they were asked to rate the problem as serious, significant or minor. They were also asked if it was increasing, decreasing or staying the same. The soil scientists then walked the land and recorded their assessment of the situation. An overall assessment of the farmers knowledge, interest and soil management abilities was made and ownership of soil maps and the Soil Code was recorded. No information was provided by farmers concerning the financial consequences of any problems and no soil measurements were made to determine any deterioration in soil condition

3.2 Results

3.2.1 Clay soils of Lincolnshire

Soils were predominantly non-calcareous clays, sandy clays or clay loams (Hodgson, 1974) and slopes commonly less than 4^0. Combinable crops (winter cereals and oilseed rape) accounted for 86% of the area, set-aside 7% and grass 5%. Virtually all of the soils were under-drained with regular programs of subsoiling and three farms were also mole-drained regularly.

Problems of soil management were only reported in some fields in some years. Straw incorporation has become more common following the ban on straw burning in 1993, and several farmers felt the return to ploughing or discing from tined systems for the primary cultivation had improved soil conditions. Three out of the ten farmers believed low organic matter was affecting workability but several commented that more powerful machinery and earlier sowing meant that autumn cultivations were more likely to be completed in good conditions. This increase in the capability of machinery and the swing to autumn sown crops has improved soil conditions since the concerns of the 1960s (MAFF, 1970).

3.2.2 Calcareous silty clay loams of Hertfordshire

Soils were predominantly (50%) calcareous shallow silty clay loams over chalk with associated clay loams and silty clay loams. All farms had at least 10% of their land with slopes of more than 11^0. Winter cereals (63%) dominated the cropping with around 10% each of spring cereals and oilseed rape with 6% set-aside and only 3.5% grass. Break crops of grain legumes accounted for most of the remainder.

Any heavier and deeper soils are usually under-drained with regular programmes of secondary treatments. Straw incorporation was generally believed to have improved workability although one farmer considered it had resulted in unacceptable "puffiness" (lack of consolidation) of seedbeds. Three farmers felt that capping due to low organic matter had decreased since the increase in straw incorporation.

Water erosion was recognised as a potential problem but one that only occurred under extreme conditions or where run-off from roads and tracks concentrated water onto soils. No significant erosion was observed on these farms but rainfall during the period of the study was below average and it so was not possible to check the farmers, appreciation of what might be occurring. Fine flat seedbeds which are known to encourage run-off and erosion were avoided for cereals but were considered essential for oilseed rape. The area is not generally recognised as suffering from water erosion and therefore contrasts with similar soils in other areas (Boardman, 1995). This difference between regions is not always acknowledged when attempts are made to estimate the areas of soils at risk from erosion on the basis of soil type without adequately considering rainfall, slope and cropping pattern.

3.2.3 Sandy soils of the West Midlands

Soils were predominantly loamy sands or sandy loams with some clay loams. Seven of the ten farms had at least 25% of the land on slopes of 5^0 or greater. Winter cereals only accounted for 35% of the land with another 7% spring sown, 5% uncropped set-aside, 29% potatoes or sugar-beet and small areas of oilseeds and other break crops. Grass was on 10% of the area and one farmer had put his steepest land into permanent set-aside.

Returns of organic manures were greater than the previous two areas with eight of the ten farms having significant quantities to recycle. Applications were commonly made before root crops and ploughed under in the autumn or in the spring.

Wind erosion was a problem in parts of some fields in some of the years when sugarbeet was grown. Fine, firm seedbeds were regarded as essential. Most affected farmers had already attempted to provide more shelter to affected areas by planting hedges or leaving existing ones to grow up. Capping or slaking of seedbeds was reported as a problem on eight of the ten farms. However few related this to low organic matter in the soil.

Water erosion was recognised as a problem on all farms with its significance varying between serious, significant and minor. On at least two farms the soil scientist believed that the problem was being understated. When water erosion occurred it was generally associated with cereal crops and steep slopes.

3.2.4 Sandy soils in East Devon

The average field size at 5 ha was considerably smaller than in the other areas despite a program of hedge removal that has continued up to the last five years. Sandy loams are the predominant soil with small areas of loamier variants. Seven farms had at least 25% of the area on slopes of 5^0 or more and five had more than 25% on more than 8^0.

Nearly half of the area was in grass (31% short leys, 7% long leys and 8% permanent pasture) with another 14% in other forage crops, chiefly maize. A further 25% is used for winter cereals with smaller areas of roots, oilseeds and spring cereals. Outdoor pigs are kept on four farms. Nine farms had significant quantities of their own manure whilst the other regularly received sewage sludge. Seven of the nine farms spreading their own manure did so throughout the year with the resulting risk of soil compaction and potentially polluting run-off.

Eight farmers recognised water erosion as a problem. Maize, outdoor pigs, vegetables harvested in winter, wheelings and heavy machinery were variously mentioned as the cause of problems. Low organic matter was not. One farmer who did not believe there was a problem was ignoring apparently serious erosion. This was probably because any runoff from his land flowed straight into the river. Farmers were making more attempts to intercept or trap runoff than to prevent it. Only one farm had introduced grass buffer strips to control flow. Half of those who recognised a problem felt that it was getting worse. Two who felt the situation was improving said that they had received less complaints of sediment on roads. One of the outdoor pig keepers was having to limit the fields that he used to those with lesser slopes.

3.3 Discussion and conclusions

Over 90% of the farmers visited displayed a satisfactory knowledge of the principles of soil management. The assessment by the soil scientist of conditions in the field usually confirmed the farmers' opinion. As with the survey on the Soil Codes those with the biggest unrecognised problems were the least likely to see the need for change. Only four farmers (10%) had copies of the Soil Code. Farmers were well aware of the need to maintain chemical fertility and recognised the particular problems of their own soils.

The results present a reassuring picture in Lincolnshire and Hertfordshire. These pilot areas are believed to be representative of large areas of arable farming in lowland England. In the two sandy areas water erosion is causing farmers concern. They have not introduced adequate measures to control it. The importance of declining soil organic matter is not generally recognised by the farmers. Off-site effects are commonly of greater concern to them than loss of topsoil. Their concern is usually limited to redeposition of silt on the roads rather than the environmental consequences of sediment, pesticides or phosphorus being discharged to surface waters. The increasing willingness of local authorities to recover the cost of cleaning roads is responsible for this attitude.

4 Future initiatives

At the outset of the pilot study a self-help booklet to help farmers prepare a Soil Management Plan was thought to be a possible development. The farmers perception of the problems of water erosion was generally poor, particularly of the relevance and need to control off-site effects. It was concluded that self-help plans were not appropriate until farmers had a better appreciation of the seriousness of the problem on their farm.

Further work has been commissioned by MAFF to identify more precisely areas where erosion may be threatening the long-term fertility of soil. Decisions can then be made on the need for increased information activities. There is an on-going publicity campaign to increase awareness of the Codes and the need for Good Practice to be followed. The Soil Code is being revised and there are plans to relaunch it in 1998 together with the Water and Air Codes. The relaunch will take into account the results of the surveys reported in this paper.

References

Boardman, J. (1995): Damage to property by runoff from agricultural land, South Downs, Southern England, 1976-1993. The Geographical Journal **161**, 177-191.

Hodgson, J. M. (1974): Soil Survey field handbook. Technical Monograph No. 5, Soil Survey, Harpenden.

MAFF (1970): Modern farming and the soil. Report of the Agricultural Advisory Council on Soil Structure and Soil Fertility. HMSO London.
MAFF (1991, 1992, 1993): Codes of good agricultural practice for the protection of water (1991), air (1992), soil (1993). MAFF Environment Matters. London.
RCEP (1996): Royal Commission on Environmental Pollution, Nineteenth Report, Sustainable use of soil. Cm 3165, HMSO, London.

Addresses of authors:
R.J. Unwin
FRCA
Ministry of Agriculture, Fisheries and Food
Nobel House
17 Smith Square
London SW1P 3JR, England

Using Incentives and Subsidies for Sustainable Management of Agricultural Soils - A Challenge for Projects and Policy-Makers

M. Giger

Summary
The paper presents the results of a working group consisting of Swiss development organisations and universities involved in rural development activities in the South. Incentives are used very frequently in projects aimed at improving sustainable management of soils. The paper explores the problems that are involved with this practice. Several strong reservations are expressed from economic, sociological and development-oriented perspectives. Improving price incentives and the institutional framework, and developing technologies that are economically attractive and environmentally friendly, as well as improving access to credits, markets and eco-labels, are presented as possible alternatives for improving incentives without creating new subsidy systems. Several arguments are presented to show why subsidies may be a useful instrument despite the problems involved. One conclusion is that subsidies could be applied if the new technologies that they are meant to promote are economically more attractive than the ones they are intended to replace.

Keywords: Incentives, subsidies, sustainable soil management, soil conservation, projects

1 Introduction

Many projects use "incentives" with the objective of assisting farmers in adopting new soil conservation technologies. Inputs like seeds, seedlings, fertilisers, and equipment are offered free of charge or are heavily subsidised. Conservation work is very often financed through cash payments or food-for-work programs.

From an economic point of view, these incentives are created and financed through the use of subsidies. Although projects frequently use such subsidies, in practice very few projects systematically analyse either the effects or the side-effects of subsidies. Do these subsidies really bring about long-lasting improvements in sustainable management of agricultural soils? And if they do, are they cost-effective instruments whose approaches can be replicated? Finally, are there side-effects which must be taken into account in an appropriate manner?

The observations made in this paper should not be interpreted as an attempt to deny the need to support farmers in using their soils in a more sustainable way. The primary aim is to offer some reflections on how to find more efficient ways to manage soils sustainably while avoiding some of the negative experiences of the past. The author takes sole responsibility for any errors of fact or interpretation. The views expressed here do not represent the official policy of any of the concerned organisations.

ISBN 3-923381-42-5
© 1998 by CATENA VERLAG, 35447 Reiskirchen

2 Scope and background of the paper

2.1 Scope

The scope of this paper is restricted to economic incentives. Other incentives based on cultural factors, beliefs, etc. are certainly important, but will not be discussed here. Incentives that influence soil management practices are not only used by projects, but are also created by powerful factors at other levels (i.e. price fluctuations and price policies, land tenure, etc.). This paper concentrates on incentives used by projects and programs at the local level, but also suggests other opportunities to improve incentives for farmers.

2.2 Background

Within a network consisting of Swiss NGOs involved in rural development activities and the Swiss Agency for Development and Cooperation (SDC), a working group was set up to look into the question of incentives. The issue had been identified as relevant to organisations working in the field of sustainable soil management, and the need to explore and analyse problems related to this topic had been clearly expressed. The objectives are to raise the awareness of project planners and counterpart organisations with regard to the problems of using incentives, and to formulate recommendations or practical tools for dealing with these problems.

This paper will review some of the findings of the working group and discuss the implications for project planning and implementation. The working group based its findings on a review of the literature and on the experiences its members and their organisations have had with different projects.

3 Definition of terms

Very often, project reports and related literature use the terms "incentives" and "subsidies" almost as synonyms. In actual practice the incentives mentioned in reports very often consist of subsidies allocated for seeds, fertilisers, equipment or labour by the projects or programs. Sometimes incentives related to soil conservation also consist of small gifts or awards given to farmers in order to promote specific activities introduced by extension services. For our purposes, the relevant terms will be defined as follows:

A **subsidy** is an instrument used by the state or by private actors to reduce the cost of a product or increase the returns from a particular activity (Kerr, 1994). It may be provided in cash or in kind and usually serves a specific purpose.

Incentive is a much broader term which encompasses everything that motivates or stimulates people to act.

Therefore, subsidies are the **instrument** commonly used by projects to create incentives for soil conservation activities.

Compensations are payments by the state to a private entity for delivering a service to society (producing something for the public good). For some authors this represents a specific form of subsidies, whereas for others it represents a category by itself. There is some consensus, however, that such compensation payments are indeed necessary in many instances, although side effects may be similar to those discussed here for subsidies in general.

4 Creating incentives through subsidies

4.1 What are the problems?

Projects that use subsidies should systematically analyse the problems and potentials of these instruments. Do such subsidies help to promote sustainable soil management? And do they effect changes that are sustainable? What are the side effects of the use of such subsidies? Unfortunately, these issues are seldom explored systematically in the literature dealing with sustainable management of soils (some exceptions include GTZ, 1995; Kerr, 1994; Smith, 1994; Camino Velozo, 1987).

Many reports on projects related to soil conservation do take note of the issue of incentives, although usually only in a summary analysis. In recent years these rather cursory treatments have quite frequently indicated general disappointment in experiences with incentive systems in the past (Douglas, 1994; Hudson, 1991). On the other hand, the effectiveness and efficiency of subsidy systems have been regarded very critically by many economists (for a discussion see Ellis, 1992).

Reservations with regard to the use of subsidies are summarised below.

4.2 Reservations from an economic perspective

Allocating subsidies for soil conservation activities may mean a **departure from the polluter-pays** principle, which is central to environmental policies based on economic principles. The polluter-pays principle (OECD, 1995) means that those who cause environmental damage ought to pay for it. A departure from this principle may in fact reduce incentives for soil conservation.

Allocating such subsidies may mean that society is **implicitly** giving agriculture a right to degrade the environment, a privilege which has far-reaching consequences. Subsidies for soil conservation could imply that society is giving farmers a right to degrade soils and to cause on-site and off-site damage (Baur, 1995)

In the long run, subsidies have a tendency to become **very costly systems**. In 1985, for example, pesticide subsidies alone amounted to about US$ 150 million in Indonesia (Kenmore, 1991). Even when they have outlived the original purpose for which they were created, subsidies tend to be very difficult to discontinue. Subsidy systems tend to be inflated and unnecessarily prolonged because of the interests of all the actors in a society who benefit directly or indirectly from them and who have more influence than those who pay for them - usually tax payers or society as a whole.

Subsidising inputs for agriculture may create **distortions at the level of the individual farm** and influence management decisions in many ways: the allocation of production factors may become inefficient. Scarce (expensive) resources may be used where less expensive ones could have been used (subsidising inorganic fertilisers may result in less use of organic fertilisers like compost). Farmers may even be persuaded to do things which they do not really see as viable solutions, just in order to be able to qualify for the subsidies which they receive in the form of food-for-work or cash-for-work.

Subsidies may reduce incentives in still another way (Kerr, 1994): those who see that others receive subsidies may delay their own efforts in conservation work in the hope of receiving subsidies themselves at a later stage.

Given the precarious livelihood that many farmers earn from subsistence agriculture, considerations about subsidies may not seem highly relevant to some observers. However, the consequences of subsidy systems may prove very costly to the state in the long run and thus to farmers as well. Issues such as unequal access to land or unequal distribution of wealth, for instance, should probably be addressed through other, more suitable instruments. In fact, subsidies for soil conservation may not benefit the poorest: they will usually benefit only those who have access to land. But let us look at still another category of arguments.

4.3 Reservations from a sociological or development-oriented perspective

Subsidies may make farmers and communities dependent on outsiders. The capacity of communities to help themselves may be weakened (Bunch, 1982). Communities may start to wait for outside support, and not undertake necessary conservation efforts on their own.

Subsidies may signify a paternalistic attitude characterised by outsiders telling people what to do (Bunch, 1982; Kerr, 1994). In this sense, they are implicitly or explicitly grounded in the assumption that people cannot solve problems on their own.

Subsidies can also be associated with cliental networks. Politicians, planners and populations may be caught in such relationships, and subsidies can in fact make it easier to lose sight of real development objectives (Kerr, 1994). Subsidies may therefore be allocated on political grounds rather than on the basis of real needs and priorities.

Some observers have noted that subsidies can also become a tool of competition between different development organisations that are under obligation to their donors to show quick results (for instance, as reported in working papers by Kai Schrader from PASOLAC, Nicaragua).

Based on these considerations, one may conclude that there are many problems associated with subsidies related to soil conservation, most of which are very basic problems linked with development co-operation and government intervention in general. These problems are worth considering when planning conservation activities.

Category	Example	Possible problems
Price-incentives: (Prices of fertilisers, land, agricultural products): These prices could be influenced through respective price policies, taking in account external costs of production and consumption (Pearce et all, 1988).	Prices for agricultural products could be influenced while taking account of soil degradation associated with their production (tree crops could be favoured, thus influencing production structure and soil use; von Maydell, 1994).	Prices have to influence supply and demand in order to clear markets. At the same time, prices cannot always reflect ecological objectives.
Institutional Framework: (Land rights, property rights, laws): Creation or strengthening of property rights systems, in order to create incentives for sustainable use of soils (Wachter, 1996).	*De facto* open access situations due to lack of government control over publicly owned land should be corrected, and new common or private property rights created.	Changing institutional frameworks is often a slow and very difficult process. Those who stand to lose from such changes will resist any attempts at reform.
Resource-conserving technologies: Development of technologies that are environmentally friendly and economical at the same time.	Systems such as no tillage cultivation with cover crops.	For many systems, solutions so far exist only in theory. Development will be long and costly.
Direct Incentives: (Credits, access to means of production or inputs, marketing, voluntary agreements): The objective is to enable sustainable production by farmers.	Credit schemes, eco-labels, improving market systems.	Very diverse.

Table 1: Overview of starting points for creating better incentives for sustainable soil management

5 Are there alternatives?

In the light of the problems associated with subsidies, it is sensible to start looking for alternatives to subsidies even in the initial phase of programs and projects. Alternatives may not often be found when concentrating on individual plots or even watersheds. But they may be found when looking for other approaches and other strategies, perhaps strategies to *indirectly* influence soil erosion or degradation. The objective should be to create incentives without creating new subsidy systems.

Possible starting points for creating better incentives for sustainable management can be usefully categorised as in table 1.

6 Implications

6.1 Is a compromise possible?

In the light of reservations based on the theoretical and practical considerations presented above, various positions may be taken regarding possible implications.
- Should subsidies for soil conservation be discontinued altogether? This recommendation was made recently in a broad study on Participatory Watershed Development (Hinchcliff et al., 1995; Kerr, 1994). Similar, but less categorical recommendations can be found in the subsector policy on Sustainable Land Use formulated by the Swiss Agency for Development and Cooperation (SDC, 1995).
- Are there conditions under which some subsidies are justified on economic grounds or based on other considerations? Arguments have recently been proposed in relation to subsidising rock phosphate fertilisers in Africa (Lele, 1996). Other institutions such as GTZ have also published papers which advise caution but cite certain instances under which subsidies might be used or be advisable (Current et al., 1995; GTZ, 1995). But what precisely are these conditions?

6.2 Under what conditions are subsidies justified?

To answer this question we might turn back to the arguments used in the long debate about the rationale and efficiency of input subsidies, especially fertiliser subsidies. For a summary description of these arguments, see, for example, Ellis (1992).
- One argument in favour of such subsidies has been termed **"dynamic disequilibrium"**. This means that *temporary* incentives are needed in order to overcome risk-aversion by farmers and enable them to take advantage of a new technology. Once the subsidy is phased out, farmers are expected to continue the new technology, if the particular new technology is in fact more productive and hence more attractive to them. Under these conditions, the use of the subsidy could be considered as economically justified.
This argument, if it is accepted, carries a very important message: **the new technology must be economically more attractive than the one it is replacing**. The "dynamic disequilibrium" argument applies only under this condition.
Herein lies the problem of many soil conservation techniques which have been advocated in the past: many of them are not economically profitable from the point of view of the individual land owner (for an example from Ethiopia, see Kappel and Ludi 1996).
Such an evaluation from the point of view of the individual farmer (or group of farmers) could be taken as one minimum condition which must be fulfilled before starting a subsidy system.

The subsidy would then just be used to enable the adoption of a new technique and would be terminated once farmers have had their first experiences in using this technique.
- Another argument in favour of subsidies can be found in benefits society stands to gain from farming activities which are not compensated by market prices (protection of watersheds, for instance, or conservation of biodiversity of landscapes). The farming practices generate external costs which are paid by others, i.e. society at large. Diminishing these **external costs** could therefore justify intervention by the state (von Maydell 1994), although the economic literature does not clearly indicate concrete cases involving such external costs or their relevance in developing countries. The difference between soil functions which should be produced and preserved by farmers as a by-product of their work and functions which are an extra burden inflicted on farmers by society is in practice very diffuse. For example, the on-farm costs of soil erosion are probably not examples of external costs, although the loss of biodiversity may be such a cost. In practice, the two may be closely linked.

  Nevertheless, this argument can be of value in certain instances. If possible, the costs of these subsidies should then be borne by those who actually benefit from them. This could be a local city council (in the case of a watershed), or the international community (in the case of nature reserves of global importance). In addition, the question whether such subsidies should be used for inputs and equipment for farmers or to finance extension activities, training and research, needs to be answered in order to avoid some of the negative implications of subsidies.
- A broad category of arguments in favour of protecting agriculture in developing countries addresses the issue of the **unsustainable economic environment** in which agriculture is practised today. Depressed international commodity prices are due to export and production subsidies in the developed world and also to unsustainable agricultural practices in the developed and the developing world. The economic environment for agriculture itself is probably unsustainable. Based on these considerations, government intervention in favour of agriculture is needed, especially in the developing world. Arguments against subsidies for soil conservation are not intended to deny this urgent need. However, government intervention in favour of agriculture may need to take other forms than direct subsidies by being more concerned with creating positive incentives using other forms of policy intervention and reforms.

7 Conclusions

From the above discussion we can conclude the following:
1. There are a number of arguments which illustrate the need to exert caution when using subsidies in order to improve incentives for farmers to adopt soil conservation technologies. These reservations stem from both economic and sociological considerations.
2. It is possible to envisage approaches other than subsidy systems which aim to improve incentives for farmers. Such options should always be considered a priority.
3. If social objectives such as assisting the poor are the real reasons for these subsidies, projects other than soil conservation projects (for instance rural credit, education, infrastructure) may present a better chance of success in meeting these objectives. Such alternatives would need to be examined.
4. There are also some arguments which call for the use of subsidies as a legitimate instrument in certain instances.Possible negative side-effects should, however, be anticipated and minimised.
5. It is necessary for project planners and conservation specialists working in development projects or planning institutions to be very conscious of the challenge of finding new incentives for soil conservation.
6. Co-ordination between different development organisations present in an area is greatly needed.

7. Considering the unfavourable economic environment in which agriculture is practised, reforms and policy interventions are needed to remove negative incentives for agriculture which impede agricultural production. There is also a great need for attempts to promote sustainable agricultural practices.

References

Baur, P. (1995): Ökologische Direktzahlungen - ein Diskussionsbeitrag aus ökonomischer Sicht. In: Agrarwirtschaft und Agrarsoziologie No. 2/1995, Schweizerische Gesellschaft für Agrarwirtschaft und Agrarsoziologie, Zürich.

Bunch, R. (1982): Two Ears of Corn. Oklahoma, World Neighbours, USA.

Camino Velozo, R. de (1987): Incentives for community involvement in conservation programs. FAO Conservation Guide No. 12, FAO, Rome.

Current, D., Lutz, E. and Scherr, S. J. (1995): The Costs and Benefits of Agroforestry to Farmers. The World Bank Research Observer, Vol. 10, No. 2. August 1995. Reprint IFPRI Washington DC.

Douglas, M. (1994): Sustainable Use of Agricultural Soils. A Review of Prerequisites for Success or Failure. Development and Environment Reports No 11, Group for Development and Environment, Berne.

Ellis, F. (1992): Agricultural Policies in Developing Countries. Cambridge University Press.

GTZ (Hrsg) (1995): Die Rolle von Anreizen bei der Anwendung von Ressourcenmanagment über Selbsthilfeansätze als Vorgehensweise. GTZ, Abteilung 402, Umwelt und Ressourcen-schutz, Verbreitung angepasste Technologien. Eschborn, Germany. 17 S. plus Annexe.

Hinchcliff, F., Guijt I., Pretty, J.N. and Shah, P. (1995): New Horizons: The Economic, Social and Environmental Impacts of Participatory Watershed Development. IIED Gatekeepers Series No. 50. London: London Environmental Economics Centre.

Hudson, N. W. (1991): A study of the reasons for success or failure of soil conservation projects. FAO Soils Bulletin 64, FAO, Rome.

Kappel, R. and Ludi, E. (1996): Economic Analysis of Soil Conservation in Ethiopiy. ISCO Conference Proceedings, Bonn.

Kenmore, P. E. (1991): How Rice Farmers clean up the environment, conserve biodiversity, raise more food, make higher profits. Indonesia's IPM - A Model for Asia. Workshop Paper. FAO Rice IPC Progamme. Manila.

Kerr, J. M. (1994): How Subsidies Distort Incentives and Undermine Watershed Development Projects in India. Draft paper, IIED New Horizons Conference on Participatory Watershed Management, Bangalore, India

Lele, U. (1996): Cited by Heidhues: Nahrungsicherung - eine Herausforderung an Wissenschaft und Technik. In: entwicklung+ländlicher raum 3/96, Frankfurt, Germany.

OECD (1975): The Polluter Pays Principle. OECD, Paris.

Pearce, D., Barbier, E. and Markandaya, A. (1988): Environmental Economics and Decision Making in Sub-Saharan Africa. LEEC Paper 88-01. London: London Environmental Economics Centre.

Perich, I. (1993): Umweltökonomie in der entwicklungspolitischen Diskussion. Berichte zu Entwicklung und Umwelt Nr.8. Berne: Geographisches Institut der Universität Bern.

SDC - Swiss Agency for Development and Cooperation (1994): Sustainable Management of Agricultural Soils. Sectoral Policy Paper. SDC, Bern, Switzerland.

Smith, A. (1994): Incentives in community forestry projects: a help or a hindrance? Rural Development Forestry Network Papers 17c, Overseas Development Institute, London.

Wachter, D. and North, N. (1996): Land Tenure and Sustainable Management of Agricultural Soils. Development and Environment Reports No.15. Centre for Development and Environment. Berne, Switzerland.

von Maydell, O. (1994): Agrarpolitische Ansätze zur Erhaltung von Bodenresourcen in Entwicklungsländern. Landwirtschaft und Umwelt. Schriften zur Umweltökonomik. Kiel, Wissenschaftsverlag Vauk, Germany

Address of author:
Markus Giger
Centre for Development and Environment, University of Berne, Hallerstraße 12, CH-3012 Bern, Switzerland

How to Increase the Adoption of Improved Land Management Practices by Farmers?

Ch. Reij

Summary

Projects with a target-oriented approach to soil and water conservation (SWC) do not have much impact on the problems of land degradation. It is argued that the major challenge is to induce large numbers of farmers to invest voluntarily in improved land management practices. One way to achieve this would be to identify triggering mechanisms or single measures, which are considered so attractive by farmers that they decide to invest their labour and their scarce financial resources in improved land management. For example, credit and farmer-to-farmer visits can function as triggering mechanisms and these also offer opportunities for addressing gender and equity issues in SWC, aspects which are commonly ignored by SWC projects. Building on traditional SWC practices may also lead to higher rates of adoption and more farmer-innovators who spontaneously improve their SWC practices and who could be an important source of inspiration.

Keywords: Articipatory approaches in soil and water conservation, farmer-innovators, triggering mechanisms, credit, gender, equity issues

1 Introduction

Many soil and water conservation projects (SWC) have failed because land users were not willing to maintain soil conservation works, which were sometimes constructed at a heavy cost per ha. Most SWC activities usually come to a grinding halt as soon as external funding is no longer available. Non-adoption of improved land management practices is the rule and rates of voluntary adoption are usually negligible. The major reasons for failure have been analysed extensively (Reij, et al. 1986; Hudson, 1991). It is remarkable, however, that no systematic analysis of reasons of failure seems to have been made prior to the mid-1980's.

A major reason for failure emerging as a common thread in all analyses is the lack of land user participation in all phases of the project cycle. Most SWC projects now claim to promote participatory SWC, but despite the right discourse, they often continue to impose conservation techniques on farmers and their participation is essentially limited to providing labour for the construction of conservation works designed by outsiders. Labour is often not provided on a voluntary basis, but land users are either coerced to provide their labour, or they are paid in the form of „Food-for- Work" or „Cash-for-Work". Experience shows that although such incentives can be appropriate in specific conditions, they often have counter-productive effects in the sense that without them land users will neither maintain nor expand SWC works.

Projects have managed to treat a few thousand ha or sometimes tens of thousands of ha in situations in which millions of ha require improved land management practices. Usually a few thousand farmers are 'adopting' conservation practices, mainly because of the incentives attached to it, but they represent only a tiny minority of the land users, often less than 5 % of the total rural population. In most cases projects have such low rates of implementation that it would take a century or more before all land requiring investment in SWC would be treated. The conclusion seems inescapable that it does not make any sense for most SWC projects to continue in the same old ways.

The challenge is to create conditions which induce large numbers of resource-poor farmers, to invest their own scarce labour and financial resources in improved land management practices. Unless a situation can be created in which the majority of stakeholders will voluntarily invest in improved land management practices, the battle against land degradation cannot possibly be won except of course in particular villages, or even in districts, but not at the national level. If large-scale voluntary adoption of improved land management practices is accepted as the major challenge then SWC projects will have to radically change their current practices and design new strategies which should lead to the mobilisation of hundreds of thousands of land users in each country.

2 Elements of success

An analysis of successful SWC and watershed management projects does offer some insight into what can be done to increase the attractiveness of SWC to farmers (Rochette, 1989; Critchley et al., 1992; Adolph, 1994; Hinchcliffe et al., 1995; CDCS, 1996; Reij et al., 1996). It would, for instance, help to:
- introduce SWC techniques which are attractive to farmers, which usually means that they should produce perceptible short-term increases in plant production;
- build as much as possible on traditional SWC practices;
- provide adequate support and appropriate incentives to farmers;
- put more emphasis on farmer training (men and women);
- move away from a top-down approach and promote more participatory approaches to SWC, which means involving farmers in all phases of the project design cycle;
- move from a target-oriented to a process approach;
- improve land tenure security;
- create an enabling macro-policy framework (for example, create a legal basis for the empowerment of local institutions for natural resource management).

The discussion in this paper will be limited to five aspects: (i) a more participatory approach to SWC; (ii) the choice of SWC techniques (iii) the role of farmer-innovators (iv) farmer-to-farmer visits and (v) the identification of triggering mechanisms.

2.1 A more participatory approach to SWC

In particular since the early 1980's some SWC and watershed management projects started selecting SWC techniques together with farmers rather than by imposing techniques. An interesting example is found in Zimbabwe where SWC techniques (graded earth bunds) were imposed on farmers since the early 1930's. In 1991 the ITDG Food Security Project and the Agritex/GTZ Conservation Tillage Project introduced participatory research approaches in their projects in the Chivi and Zaka Districts (Hagmann and Murwira, 1996: 102). The projects emphasized the importance of participation and co-operation in organizational development in order to build institutions which enable people to become self-reliant. At the same time they promoted a process of innovation by farmers and technical options were developed either by farmers or at research stations and then offered to farmers for testing. The following table indicates the farmer's perceptions of the old and the new approach and the most important aspects of the participatory approach. A comparison of the old and the new approach reveals such vast differences that one can speak of a paradigm shift. Rates of adoption of conservation techniques have increased with the new approach: out of a total of 1136 households in Ward 21 of Chivi District, 80 % are now practising SWC in one form or another (Hagmann and Murwira, 1996: 104-105).

Characteristics of approaches

Old approach
Forceful methods were used.
Only few people could benefit (eg.literate).
Intercropping was forbidden.
Failed to address SWC convincingly.
We were told to do things without questioning.
Usefulness of conservation works never explained.
No dialogue between farmers and extensionists.
Little co-operation among farmers.
Extension agents treated our fields as theirs.

New/participatory approach
Everyone to benefit as all are free to attend meetings now.
There is dialogue.
Process is well explained (teaching by example).
Farmers are the drivers now.
Intercropping is encouraged to boost yields.
Farmers are being treated as partners and equals.
No discrimination against poor or rich, educated and uneducated.
We are given a choice of options.
They pay attention to us and take time to find solutions to farmers' problems.
We are being encouraged to try out new things.
It helps farmers to work co-operatively.
Farmers practise SWC with enough knowledge of why they should do it.
Learning from others through exchange visits/learning through sharing.
It helped farmers to develop the ability to encourage each other in farm activities.
Encouragement to practise SWC techniques through participation.
Farmers are free to ask for advice.
Yields have increased through SWC techniques.
The dedication of modern extension agents/researchers.
It has brought development in the area.
It is very effective in the conservation of trees, soil and water.

Source: Hagmann and Murwira (1996: 104)

Since the end of the 1980's, SWC projects in the West African Sahel are increasingly evolving into Community-based land use management projects („Gestion de Terroir"). The reason behind this trend is the growing awareness that governments neither have the means nor the technical capacity to deal with land degradation at village level and village communities should be mobilized, trained and empowered to decide when, where and how they want to manage the natural resources within their village boundaries. The village organisations created for Community-based land management do not only deal with SWC on already cultivated land, but they also address the management of non-cultivated land as well as socio-economic priorities expressed by villagers, such as education, drinking water and health. Experience in the Sahel shows that whereas technical successes in SWC on cultivated land can be achieved relatively easily, the process of building and empowering local institutions for Community-based land management takes more time, because specific skills (planning, conflict management, financial management, organisational skills) have to be developed[1].

2.2 SWC technologies

One important lesson drawn from successful SWC projects is that the techniques promoted should, whenever possible, be simple, low-cost and effective, which means that they should control runoff and erosion as well as produce perceptible yield increases (30 % or more) within one or two years. Several papers to this IXth ISCO Conference make it abundantly clear that this is often not the case (see Clark et al., for Sri Lanka) and sometimes SWC even leads to lower incomes (Huszar et al., for Java)

It is impossible to pretend that cost-effective SWC techniques can be identified for each and every situation, but the question should always be asked who selected the techniques promoted by the project and are they the ones preferred by the local farmers ? The current move away from mechanical SWC to an emphasis on biological conservation (vegetative barriers; agroforestry) and agronomic measures (mulching, minimum tillage, intercropping and multiple cropping systems) is promising because it has the potential to produce „win-win" situations in the sense that better conservation leads to more production.

The possibilities to produce substantial short-term yield increases are considerable in semi-arid areas (300 - 700 mm), which, curiously enough, are often regarded by donor agencies as regions of low potential. The trick is that water harvesting techniques relieve the water constraint for crops, which generates substantial yield increases from year 1. The challenge then becomes to improve soil fertility management in order to prevent yields from falling in subsequent years. In parts of the West African Sahel two water harvesting techniques have been largely adopted by farmers during the last decade: they are contour stone bunds and improved traditional planting pits. In particular the latter are becoming increasingly popular, because they allow farmers to rehabilitate strongly degraded land on which they obtain yields that vary from 0 kg per ha in the without situation to 300 - 400 kg/ha of millet or sorghum in dry years and 1.000 - 2.000 kg/ha in years of good rainfall[2]. There is a growing consensus amongst SWC specialists that it is important to build as much as possible on traditional SWC practices rather than introducing modern techniques, and in the few cases that this was done, the rates of adoption by farmers were very promising (Hassan, 1996; Hagmann and Murwira, 1996).

2.3 The role of farmer-innovators

The most neglected source of inspiration for the selection of conservation techniques are **farmer-innovators**, who on their own initiative improve conservation practices. Very few conservation specialists deliberately try to identify these farmer-innovators in order to find out what they have achieved or whether their experiments could be built upon. In Southern Africa initiatives in this direction were made by the IFAD-funded SWC-AGF project in Lesotho (Critchley and Mosenene, 1996) and by the Lesotho-based Environment and Land Management Sector Coordination Unit of SADC (Segerros et al., 1996).

---

[1] See also recent interesting papers on organising for local-level watershed management in Colombia (Ravnborg and Ashby, 1996) and on local institutions and farming systems development in India (Mosse, 1996).

[2] The case of improved traditional planting pits is analysed for the Yatenga region of Burkina Faso by Ouedraogo and Kaboré (1996), for the Djenné region in Mali by Wedum et al. (1996) and for the Tahoua region in Niger by Hassan (1996).

The impact of an outstanding farmer-innovator can be demonstrated by the case of Mr. Sawadogo Yacouba who lives in the Yatenga region of Burkina Faso (see also: Ouedraogo and Kaboré, 1996).

Around 1980 Mr. Yacouba started experimenting with the improvement of traditional planting pits or *zay*. He increased the diameter of the pits from about 15 cm to 25 - 30 cm and their depth from 5 to 15 - 20 cm. Their spacing was about 80 cm. These pits were dug during the dry season on barren degraded land and some manure was put into the pits. This manure attracted termites, whose holes increased the water holding capacity of the soil. The concentration of water and nutrients in one spot allowed some yield in dry years and a very good yield in years with 'normal' rainfall. The OXFAM-funded Agro-Forestry Project decided to promote the *zay* and they were adopted quickly by farmers in the Yatenga region. The impact of Mr. Sawadogo's experiments with *zay* has been substantial. He has probably contributed more to progress in SWC in the Sahel than any SWC researcher or SWC project.

2.4 Farmer-to-farmer visits

There is a growing consensus that farmer-to-farmer visits can be powerful tools to spread conservation messages. Farmers visits to other farmers who operate under more or less similar conditions are more efficient than any speeches made by techniciens or experts.

In 1989 a group of 10 farmers from the Illéla District in Niger visited the fields of Mr. Sawadogo in Burkina Faso and upon return some of them decided to improve their traditional planting pits or *tassa*. In 1989 4 ha were treated with *tassa*, in 1990 70 ha, in 1991 400 - 500 ha, in 1992 1000 ha. It is estimated that by mid-1995 about 6000 ha of strongly degraded land in Illéla District had been rehabilitated with the help of *tassa* (Hassan, 1996: 57).

Although not all farmer-to-farmer visits will be as successful as the above-mentioned example, it is clear that it does make sense for SWC projects to invest more resources in farmer-to-farmer visits, while at the same time avoiding that successful farmers and successful projects will be overrun and suffer from too much attention.

2.5 The identification of triggering mechanisms

Even if SWC technologies are relatively low-cost and efficient, there may be reasons why they are not rapidly adopted, such as labour constraints and transport. In such cases projects could try to identify **triggering mechanisms** or single measures, which farmers consider so attractive that they decide to invest their labour and their scarce financial resources in improved land management. The following two cases which are situated in very different socio-economic and agro-ecological contexts look at credit as a potential triggering mechanism.

Case 1 Farmers and donkey carts on the Central Plateau of Burkina Faso

Donkey carts are an essential tool for the intensification of agriculture on the Central Plateau, but their costs are high (about 400 US $). Most farmers can't afford buying one, but they are all keen to have a donkey cart, which can be used for the transport of stones, manure, water, firewood, etc. Virtually all SWC projects on the Central Plateau now use lorries to transport stones (free of cost) to farmers fields for the construction of stone bunds. This practice, however jeopardizes post-project continuation of bund construction by farmers. Some SWC projects now test, or are about to test, variations of the following model: small groups of farmers (4 - 5 farmers, ideally belonging to the same extended family) are equipped with a donkey cart on the following conditions:

1. They get access to a highly subsidized donkey cart (part gift, part credit at below market interest rates).
2. They pay a minimum financial contribution of their own (10 - 15 % of the value of the donkey cart, in order to ensure proper use and maintenance.
3. In return they commit themselves to build stone bunds on x hectares, to construct compost pits, to plant or protect trees, etc. during 4 or 5 years. What they do in return can be defined jointly by the farmers and the project and should depend on the specific agro-ecological and socio-economic situation in each village (no blanket prescriptions).
4. Progress in implementation has to be monitored (a sample of x % of farmers each year) and if progress is well-below expectation the donkey cart will be withdrawn; if progress is as agreed then the donkey cart is theirs after 4 years, if not it will be taken away by the programme.

Case 2 Terrace construction in the Uplands of Java

The standard approach to watershed management on Java is based on capital subsidies for terrace construction and on subsidies for farm inputs. Three to four years after project support has stopped, the incomes of model farmers have fallen below those of non-project farmers (Huszar et al., 1996). However, despite major terracing programmes on Java, soil erosion continues to be substantial, because many terrace risers remain bare. Soil loss measured from bare terrace risers was 6 - 8 kg per m^2 (Critchley and Bruijnzeel, 1995: 10). Terrace construction and terrace rehabilitation is expensive and it does not substantially reduce siltation of reservoirs in downstream areas. The question is what can be done to stimulate farmers to improve themselves both traditional and modern terraces ? The single most important technical improvement would be to plant grasses on terrace risers and terrace lips, but this is hardly attractive to resource-poor farmers who are hardly affected themselves by soil erosion, because the soils are deep and fertile. Stall-feeding of small livestock, however, is an important source of income for many small farmers and in parts of Java resource- poor farmers have little or no livestock. The trick could be to make credit available to small farmers for livestock raising, but on condition that they plant and protect terrace risers and terrace lips (Critchley and Reij, 1994).

In both cases credit to farmers (subsidized or not) is suggested as the main triggering mechanism. In a previous example from Niger, a farmer-to-farmer visit triggered large-scale adoption of improved traditional planting pits. It is clear that what farmers perceive as triggering mechanisms may vary from region to region.

3 Triggering mechanisms, gender and equity aspects

SWC projects generally continue to ignore gender and equity aspects and by doing so they benefit the resource-rich more than the resource-poor farmers and they tend to increase the workload of women without a matching increase in their benefits. Projects should make more efforts to ensure that also resource-poor farmers and women benefit from their activities. **A major advantage of credit is that it can be targeted to resource-poor farmers and to women and thus promote equity.** Tools-for-work can be used to equip those who need it most and care should be taken during the selection of participants for farmer-to-farmer visits that also resource poor-farmers and women are represented.

4 Constraints to increased stakeholder participation in SWC

The culture of donor agencies and of governments poses a major constraint to increased stakeholder participation in SWC as well as to a more rapid adoption of improved land management practices. Both donors and governments want quick and tangible results and for that reason **they too easily fall into the trap of a target-oriented approach to SWC**, which means that they try to define how many ha of land will be treated in each year with which particular techniques and what level of inputs is required to produce this level of land treatment. These data and the estimated benefits of the various SWC techniques form the basis for the calculation of costs and benefits of the proposed project. Practice has shown over and over again that this does not work; it is nothing less than an unproductive ritual as targets are usually not achieved. It is not only an unproductive ritual, the large number missions fielded to design a project are often also a waste of resources (Critchley et al., 1992: 75, 76). **The interests of the land users would be served better by a shift from a target-oriented to a process-oriented approach**, which means that the first step would be to test and evaluate SWC techniques with farmers and based on those results activities would be scaled-up in the following years. The challenge is to create sufficient flexibility to be able to respond to perceived and changing priorities of land users. In the appraisal report for the IFAD-funded SWC project in the Illéla District of Niger, improved traditional planting pits are not mentioned at all. The farmers were supposed to construct stone bunds and half moons, but as they were free to chose, the farmers in most cases preferred to invest in improved traditional planting pits.

5 Final remarks

If SWC projects start to support farmer's priorities and build on existing knowledge and local dynamics rather than by imposing their own techniques and modalities on farmers, and if they start providing appropriate forms of support to resource-poor farmers (men and women) the rates of (voluntary) adoption of improved land management practices may increase beyond expectation. This will improve both farmer's incomes and the environment.

References

Adolph, B. (1994): What does it take to make a successful project ? Some findings from 13 watershed management projects in South India. Paper presented to the Workshop on New Horizons: the economic, social and environmental impacts of participatory watershed development. Bangalore, November 28 - December 2, 1994.

CDCS (1996): Successful natural resource management in southern Africa. Windhoek, Namibia, Gamsberg Macmillan Publishers.

Critchley, W., Reij, C. and Turner, S.D. (1992): Soil and water conservation in sub-Saharan Africa: towards sustainable production by the rural poor. IFAD Rome, CDCS Amsterdam.

Critchley, W. and Reij, C. (1994): Watershed management on Java (Indonesia) and Cebu (Philippines): observations, issues and opportunities. Report of a study tour.

Critchley, W. and Bruijnzeel, L.A. (1995): Natural boundary erosion plots: a novel methodology for measuring erosion from bench terraces. Contour **7**: 8 - 11.

Critchley, W. and Mosenene, L. (1996): Individuals with initiative: network farmers in Lesotho. In: CDCS, Successful Natural Resource Management in Southern Africa, Gamsberg MacMillan Publishers, Windhoek, 71 -81.

Clark, R., Manthrithalike, H., White, R.J. and Stocking, M.A. (1996): Economic Valuation of Soil Erosion and Conservation: a case study of Perawella, Sri Lanka. Chapter 11, this book.

Hagmann, J. and Murwira, K. (1996): Indigenous SWC in Southern Zimbabwe: a case study of techniques, historical changes and recent developments under participatory research and extension. In: Reij, C., Scoones, I. and Toulmin, C., Sustaining the Soil: Indigenous Soil and Water Conservation in Africa. Earthscan Publications, London, 97 - 106.

Hassan, A. (1996): Improved traditional planting pits in the Tahoua Department (Niger): an example of rapid adoption by farmers. In: Reij, C., Scoones, I. and Toulmin, C., Sustaining the Soil: Indigenous Soil and Water Conservation in Africa. Earthscan Publications, London, 62 - 68.

Hinchcliffe, F., Guijt, I., Pretty, J.N. and Shah, P. (1995): New Horizons: the economic, social and environmental impacts of participatory watershed development. IIED Gatekeeper Series No.50.

Hudson, N.W. (1991): A study for the reasons for success and failure of soil conservation projects. FAO Soils Bulletin 64. FAO, Rome.

Huszar, P.C., Ginting, S.P. and Paribu, H.S. (1996): Including Economics in the Sustainability Equation: Soil Conservation in Indonesia. Chapter 11, this book.

Ouedraogo, M. and Kaboré, V. (1996): The zaï: a traditional technique for the rehabilitation of degraded land in the Yatenga, Burkina Faso. In: Reij, C., Scoones, I. and Toulmin, C., Sustaining the Soil: Indigenous Soil and Water Conservation in Africa. Earthscan Publications, London, 80 - 84.

Mosse, D. (1996): Local institutions and farming systems development: thoughts from a project in tribal western India. ODI Agricultural Research & Extension Network Paper No. 64.

Ravnborg, H.M. and Ashby, J.A. (1996): Organising for local-level watershed management: lessons from Rio Cabuyal watershed, Colombia. ODI Agricultural Research & Extension Network Paper No. 65.

Reij, C., Turner, S.D. and Kuhlman, T. (1986): Soil and water conservation in sub-Saharan Africa: issues and options. IFAD Rome and CDCS Amsterdam.

Reij, C., Scoones, I. and Toulmin, C. (1996): Sustaining the Soil: Indigenous Soil and Water Conservation in Africa. London, Earthscan Publications.

Rochette, R.M. (1989): Le Sahel en Lutte contre la Désertification. Josef Margraf Verlag.

Segerros, M., Prasad, G. and Marake, M. (1996): Let the farmer speak: Innovative Rural Action Learning Areas. In: CDCS, Successful Natural Resource Management in Southern Africa. Gamsberg MacMillan, Windhoek, 91 - 116.

Wedum, J., Doumbia,Y.,.Sanogo, B.,.Dicko, G. and Cissé, O. (1996): Rehabilitating degraded land: zaï in the Djenné Circle of Mali. In: Reij, C., Scoones, I. and Toulmin, C., Sustaining the Soil: Indigenous Soil and Water Conservation in Africa. Earthscan Publications, London, 62 - 68.

Addresses of authors:
Chris Reij
Centre for Development Cooperation Services
Vrije Universiteit
De Boelelaan 1115
1081 HV Amsterdam, The Netherlands

Soil Conservation Extension: From Concepts to Adoption

D. Sanders, S. Theerawong & S. Sombatpanit

Summary
Whether a soil conservation project succeeds or fails depends largely on its extension programme. This paper discusses the findings of a workshop held in Chiangmai, Thailand, June, 1995 which examined this subject. The workshop found that conservation projects will only succeed if they provide some immediate and obvious benefit to farmers. To do this, programmes must promote practices which not only conserve soil but which also increase production, reduce labour or costs or provide some other immediate benefit. Farmer participation is essential throughout the whole process of identifying the problems, finding solutions and implementing the programmes. Short term projects seldom succeed and there is need for long-term involvement. Gender issues are important and extension programmes must provide for the needs and capabilities of women. Conservation practices are unlikely to be accepted unless they can be easily absorbed into the existing farming system. For this reason, it is best to build on traditional conservation systems which farmers already understand. There are a number of social and economic factors, such as land tenure, which are usually the real reasons for land degradation. These issues must be tackled if conservation is to work.

Keywords: Extension, concepts, strategies, implementation, adoption.

1 Introduction

Extension invariably plays an important part in soil conservation programmes. In fact, whether a programme succeeds or fails frequently depends on extension. With this in mind, the Soil and Water Conservation Society of Thailand organized a workshop in Chiangmai, Thailand, from 4-11 June, 1995, entitled "International Workshop on Soil and Water Conservation Extension: Concepts, Strategies, Implementation and Adoption".

The objectives of the workshop were to review and evaluate the past and present modes of extension in soil conservation projects, relate these to the success of projects and develop guidelines for effective soil conservation extension in the future. This paper summarizes the findings of the Workshop.

2 The old approach to soil conservation extension

In the past, soil conservation tended to be implemented in a very "top-down" manner. Usually the problems of soil erosion were seen as physical problems that could best be overcome by the application of engineering works. The solution for erosion was therefore largely seen in the design and implementation of works like contour banks, terrace systems and gully control structures. It was believed that the answers lay in technology which had to be developed by research workers. Once perfected on the research stations, the task of extension was then simply to pass the techniques on to the farmers. The farmers were usually seen as the recipients in the process and were not expected to contribute other than to help with the installation and maintenance of the advocated works.

This approach did not work because farmers are primarily concerned with striving to earn a reasonable income, feeding their families, paying off their debts and educating their children. Few see erosion as an immediate problem. Unfortunately, most conventional soil conservation measures did little if anything to increase yields or contribute to solving the more immediate needs of the farmers. As a result, farmers were usually not interested in the soil conservation programmes that were offered to them. They therefore tended to be slow and expensive to implement. Even worse, once the erosion control measures had been installed, the farmers usually showed little interest in preserving or maintaining them. So critical did this problem become that the governments of some countries tried to force farmers to instal and maintain conservation works. This did not work either and in some countries farmers deliberately destroyed conservation works in open displays of defiance. Clearly, a new approach to the implementation of soil conservation extension was needed.

3 A better approach

It is now realized that how a country's land is managed ultimately depends upon the perceptions and actions of its many thousands of land users. It is only these people who can bring about the desired changes quickly and over large areas. It is also now accepted that soil erosion is the foreseeable result of poor land management and use. As it is in the farmers' interest to maintain the productivity of the land from which they have to make their living, they do not degrade it intentionally. Bad land use and management are either due to ignorance or, more likely, to economic, social and political pressures.

It is also realized now that soil conservation programmes can only work effectively if there is large-scale participation of rural people and that this will only be achieved if the measures advocated are able to provide some obvious and tangible benefit.

For soil conservation programmes to work, the role of the extension worker must change. It must no longer be that of one who just passes on information to the farmer but that of a facilitator who helps the land users to identify the real problems and then to develop their own solutions. The farming communities must be able to put these solutions into practice themselves with a minimum of external support.

FAO's scheme for land conservation and rehabilitation provides guidelines for countries wishing to put this new approach into practice (Dent, 1997). It outlines a "framework for action" based on three underlying principles: improving land use, obtaining the participation of the land users and developing the necessary institutional support.

Nowhere is this approach to soil conservation working better than in Australia, where the Landcare programme was introduced by the Government in the mid-1980s. Landcare is based on a "bottom-up" extension approach under which communities are actively encouraged and empowered to develop and implement their own land conservation programmes. So successful is this programme proving to be that a quarter of the country's farming community is already involved. About 2,000 Landcare groups have been formed and the number is increasing at an exponential rate. Landcare is supported by a national partnership between government, farmers and the conservation movement in a large-scale, non-coercive approach to improving land management. A remarkable feature of Landcare is that it is succeeding in a period when Australian agriculture has been devastated by a combination of harsh climatic events and unfavourable economic conditions. Under this programme, extension workers provide advice and support to the groups but their function is only that of facilitators - helping to bring people together, stimulating discussion and offering advice when asked, but leaving the initiative with the land user groups themselves. The Landcare movement has provided a model which can be adapted and applied in other countries, both developed and developing (Knowles-Jackson and Truong, 1997). A feature of Landcare is that it aims at group extension activities and moves away from the one-to-one extension which is time consuming and expensive. This is important as, arguably, the commonest problem facing extension services worldwide is a shortage of resources, particularly staff.

Another example of this new approach comes from the Philippines where a new programme called Sloping Agricultural Land Technology (SALT) has been developed and is being promoted. Here the technology is based on a system of agroforestry, contour cultivation and a number of practices which not only control erosion, but also lead to increased production and farm incomes. Extension workers provide training for cooperating farmers who, in turn, introduce the new technology to their districts. The secret of success appears to be the fact that the technology quickly brings obvious benefits to the farmers in the way of better production, as well as overcoming soil erosion (Cruz, 1997; Medina et al., 1997). SALT has a potential use in other parts of the tropics and it is now being tested in several other countries.

The way in which the SALT technology is being promoted in the Philippines raises an interesting issue: the use of the leading farmer in extension programmes. It seems from the Philippines experience that the SALT technology can be effectively disseminated by farmer leaders who are used by extension to "observe, teach and organize other farmers". On the other hand, the Workshop heard criticism of the use of "master farmers" in Zimbabwe (Hagmann et al., 1997). Perhaps the answer lies in the way in which the lead or master farmer is used. This is an important subject which requires investigation.

4 Issues to be considered

From successful programmes, a number of factors were identified in the course of the Workshop that should be taken into consideration by extension workers in soil conservation programmes.

4.1 Making soil conservation attractive to farmers

Conservation practices are only likely to be attractive to farmers if they are productive or meet some immediate and obvious need. To help develop such practices, extension workers must look at the whole farming system and not just confine themselves to the problems of erosion. This means looking at the integration of crops and livestock and paying particular attention to organic matter and soil fertility maintenance, as well as keeping the soil in place. Management practices such as minimum tillage in Brazil, pasture improvement in Australia and fodder production with stall feeding in Rwanda have all been used, under very different conditions, as alternatives to the intensive use of physical erosion control structures. All have proved effective ways of increasing farm incomes and at the same time controlling erosion.

Shah et al. (1997) listed some of the productive measures that are being promoted in Malaysia by the extension service in cocoa, rubber and fruit growing areas. They include hoeing to increase infiltration, mulching and the use of grasses in and around trees. All these practices can increase yield while controlling erosion.

In practice, taking the broader perspective to conservation is seldom easy for the extension worker. For one thing, it means that the extensionist cannot work in isolation but must cooperate closely with specialists working in other fields. It also requires a much closer coordination of government and non-government programmes than those existing in most countries at present.

4.2 The importance of farmer participation

It was generally agreed that soil conservation programmes can only succeed if there is wide-spread and genuine participation in the programmes by the land users in the whole planning and implementation process. For this to happen the role of the extension worker must become that of a facilitator, constantly encouraging the land users to analyze their problems, seek solutions and develop self reliance. In most cases they will need some assistance in putting these into practice but the resulting programmes must remain their programmes, run and implemented by the farmers themselves to the greatest extent possible.

While this is most promising and provides hope for much more success in programmes in the future, there is need for caution because, with the possible exception of the Landcare programme in Australia, nowhere yet has the participatory approach been applied on a large scale and over a large area. Hagmann et al. (1997) pointed out that setting up a participatory extension programme can be time-consuming, difficult and requires skilled and dedicated staff. At present few extension workers have the experience or skills to develop a participatory programme, while some new programmes which claim to be participatory appear to be that in name only. Much more work will be needed in developing the necessary methods, knowledge and skills before we will see a true participatory approach applied successfully over large areas.

4.3 The need for continuity

Continuity also appears to be an important element of successful conservation extension. Change is not an event but a process. Extension workers must be patient and their programmes must be long term. This has been demonstrated in Kenya where the success of the national soil conservation programme can be largely attributed to the strong support which the Swedish International Development Agency (Sida) has provided for over more than 20 years (Thomas, 1997). Only with time can different methods and practices be tested and adapted, staff adequately trained and the confidence and trust of the farming community gained.

4.4 The importance of women in soil conservation extension programmes

An integral part of the farm is the farm family and the extension worker must consider the structure and capabilities of those running the farm. Gender issues are extremely important. In many farming communities it is the women who are doing most of the farm work. This fact is usually overlooked and most of the extension workers are men who tend to work with the male members of the farmer families. As a result, schemes are often developed which cannot work because they do not take into consideration the capabilities of the women. For example, in many parts of Africa women spend many hours each day carrying water as part of their duties. It is therefore unrealistic to expect them to take on other labour intensive work like stall feeding cattle. The first step may be to provide a better water supply so that they have time to take part in other, more productive activities. More female extension workers are needed if problems such as these are to be recognized and dealt with. This is particularly so in those places where the culture does not allow male extension staff to work directly with women.

4.5 The promotion of traditional conservation measures

Many traditional farming systems contain very effective conservation practices. These can be seen in the old terracing systems of Yemen, in nomadic grazing systems in the Middle-East and shifting cultivation in tropical Africa. Unfortunately, many traditional systems are going out of use because of population pressures, economic conditions and other reasons.

Farming systems, particularly small-scale and subsistence systems, tend to be complex and it is often very difficult to integrate completely new practices into them. However, there are usually possibilities of adapting and improving the traditional systems so that they can once more be used effectively by farmers. This is because the underlying principles are known and understood and farmers already have the skills to put them into practice. This was demonstrated in the 1960s when efforts were made to introduce soil conservation practices into the badly eroded dryland areas of Jordan (Sanders, 1988). A number of different practices were tried. None of those based on technology developed in other countries were accepted by the farmers. However, one based on the traditional practice of stone wall terracing was readily taken up. Since then thousands of hectares have been treated in this way on hundreds of individual farms. The lesson here is that extension workers should try to improve on locally known practices before attempting to introduce something completely new.

4.6 The need to record what has been done

An associated need is to document what is already known and what has been done. Some conservation-effective traditional practices are going out of use in many places and will be lost forever unless recorded. Even more serious is the fact that the results of many present-day projects are not well recorded or widely made available. Not only the success stories but also the failures should be documented as it is common to find new projects replicating work that has already been carried out and failed only a few years before in the same country.

4.7 Does the "watershed" concept work?

With changes in the way in which extension is being implemented, the limitations of the "watershed management" approach are being exposed. This concept was developed in the days when the "top-down" approach to extension predominated. Although there is a sound technical rationale for the watershed approach, it usually makes little sense to farmers to whom administrative, village and farm boundaries are far more important than hydrological boundaries (Thomas, 1997). Nevertheless, many countries still base their soil conservation programmes on the watershed approach. India is an example and Grewal(1997) described a number of cases where soil conservation projects have been very successfully implemented on a watershed basis.

The lesson to be learned from this is that for overall soil conservation planning it may be necessary to think in terms of watersheds or catchments, but for community mobilisation the extension worker is far better employed working to administrative and farm boundaries which are known and understood by the local populations. In this way, whole watersheds can eventually be covered as adjacent farms and administrative units are dealt with.

4.8 The importance of land tenure

The reason for erosion, and other forms of land degradation, are often complex. It is now recognized that the land tenure system plays an important part in how well land is managed and conserved. Farmers with no long-term rights of access to land are very unlikely to invest in improvements or labour intensive work like terracing or tree planting. If the land users are granted long-term rights of use, the position can change dramatically.

Vietnam provides a good example of this and Phien and Siem (1997) reported on how farmers have changed their systems of management once they have been officially allocated land. With long-term land rights, farmers are growing more perennial crops, there is better ground cover, erosion is being reduced and the land is generally being used in a more sustainable way.

4.9 Other underlying causes of erosion

There are a number of other social and economic factors which may be hidden causes of land degradation besides the land tenure system. These include a pricing structure that does not encourage increased production, lack of access to markets and a poor infrastructure. Frequently, the extension worker can do little until these underlying problems are solved but programmes can fail if these problems are not at least recognized.

4.10 The use of mass media

The mass media could be more effectively used in soil conservation extension. The media include newspapers, magazines, radio, television and interpersonal communications such as field days. Research in the USA and other places shows that farmers obtain their information from a variety of sources. This indicates that extension programmes should use a combination of different forms of the media in their work. This is now possible everywhere as radios are now present even in the poorest rural communities, television is reaching an ever increasing audience and increased literacy rates in most countries make the use of printed material increasingly important.

4.11 Farmer-extension worker-researcher links

With the changing approach to extension in soil conservation programmes, the relationship between farmer, extensionist and researcher must also change. No longer can the relationship be simply that of new technology being passed down from the researcher through the extensionist to the farmer. Soil conservation still needs research stations and some of the traditional types of research. But, the demand now is for a new relationship under which the problems are first defined by consultations between the three parties who then work together to develop solutions. In practice, this means both extension workers and farmers taking an active part in research with more on-farm trials and the development of a close dialogue between all the parties concerned. Here a vital role has to be played by the extension workers who have to have the skills and abilities to bring the other two parties together and to express and explain things in terms that all can understand.

The IBSRAM has accepted this challenging role and under their Sloping Lands Network national agricultural research centres in Asia are being assisted to test and develop different technologies, not only for their technical soundness, but also for their acceptability to farmers - something that has been missing in the past (Sajjapongse, 1997).

4.12 The need for sound policy, co-ordination and organization in soil conservation extension services

One of the most important issues to come out of the Chiangmai Workshop was the need for a well structured and well organized extension service. This implies a clear policy and adequate provision for the training of staff. Without this, little can be achieved. Vietnam has realized the importance of this and is concentrating on developing a strong extension structure as one of its first priorities in tackling land degradation. Gamage (1997) pointed out that some of the problems of Sri Lanka were caused through lack of sound policy and co-ordination in extension and stresses the need for a high level committee to exercise inter-ministry and national co-ordinating functions. FAO's "Framework for Action" provides guidelines on what is required and an indication of how the necessary policies can be developed and put into practice (Dent, 1997).

4.13 The need for incentives and subsidies

There is growing concern about the use of incentives and subsidies in soil conservation programmes. Most programmes depend heavily on these in one form or another. The extremes vary from merely providing free advice to fully paying farmers to instal whatever conservation works may be considered necessary. In recent years food aid has been extensively used in the poorer developing countries to pay farmers to participate in conservation schemes.

The most common problem is that soil conservation programmes start out with the use of incentives and subsidies. Frequently this attracts farmers into the programme for the wrong reason - just for the sake of obtaining the subsidy or incentive. When the subsidies and incentives are withdrawn - as nearly always happens - farmers lose interest and leave the scheme.

For this reason, there is a growing feeling that all incentives and subsidies are bad and that programmes should be designed without them. On the other hand, a quick review of successful soil conservation programmes reveals that all of them provide incentives and/or subsidies in one form or another and it is difficult to envisage a successful programme without their use in some form. For example, the highly praised Landcare programme in Australia depends largely for its success on the provision of grants by the government. Perhaps all that can be safely said at this stage is that great care should be exercised in introducing incentives and subsidies into any soil conservation programme. In particular, thought must be given to any counter-productive effects and whether the funds are likely to be available in the long term.

4.14 Monitoring and evaluation

Finally, the workshop recognized a growing need for the monitoring and evaluation of soil conservation extension programmes. Procedures for doing this have not been given sufficient attention. This is partly due to the problem of defining clearly what is meant by soil conservation and also the great diversity of conservation practices. Nevertheless, a number of methods do exist, ranging from the use of aerial photos to baseline surveys of what precisely is happening on sample farms. Certainly the land users themselves should be more closely involved in the monitoring and evaluation process if it is to be effective. Until there is systematic monitoring and evaluation, we will continue to experience the present difficulties: programmes continuing to repeat the same mistakes which are only discovered when it is too late to bring in changes.

References

The following references are to be found in the book,"Soil Conservation Extension:From Concepts to Adoption". S. Sombatpanit, M. Zöbisch, D.W. Sanders and M.Cook (eds.), Oxford & IBH Publishing Co. Pvt. Ltd. New Delhi, 488 pp.

Cruz, E.B. (1997): The Adoption of Hedgerow as Soil Conservation Measure in the Phillipines.
Dent, F.J. (1997): Land Conservation and Rehabilitation Scheme of FAO.
Gamge, H.(1997): Land Degradation - Problems and Prospects of Soil Conservation Extension in Sri Lanka.
Grewal, S.S. (1997): Land and Water Management in North India: an Assessment of Success and Failure in Implementation of Programs.
Hagmann, J., Chuma, E. and Murwira, K. (1997): Indigenous Soil and Water Conservation and Extension in Southern Zimbabwe.
Knowles-Jackson, C.H.D. and Truong, P.N.V. (1997): The Australian Landcare Movement: Its Formation and Operation in a Soil Conservation District.
Medina, S.M., Anase, M. and Narioka, H. (1997): Promotion of Soil Conservation through Participatory On-farm Research in the Philippines.
Phien, T. and Siem, N.T. (1997): Soil Conservation in Agricultural Extension in Vietnam.
Sajjapongse, A. (1997): The Sloping Lands Network and the Opportunity for Technology Transfer.
Shah, R.B., Eusof, Z. and Zamnis, M.M. (1997): Soil Conservation Extension in Malaysia.
Thomas, D.B.(1997): Soil Conservation Extension: Constraints to Progress and Lessons Learned in Eastern Africa.

Other references:

Sanders, D.W. (1988): Food and Agriculture Organization Activities in Soil Conservation. In: Moldenhauer and Hudson (eds.), Conservation Farming on Steep Lands, Soil and Water Conservation Society. Ankeny, Iowa, 54-61.

Addresses of authors:
David William Sanders
Queen Quay
Flat No. 1
Bristol BS1 4SL, U.K.
Sompong Theerawong
Samran Sombatpanit
Department of Land Development
Phaholyithin Road
Chatuchak
10900 Bangkok, Thailand

Contributions of Research On Soil and Water Conservation in Developing Countries

K. Herweg

Summary

Scientists alone will not solve the problem of land degradation, but they can play an important part in the process of developing efficient soil and water conservation measures. The model presented in this paper is a potential tool for assisting researchers from different disciplines in self-evaluation. Such a model can be used for both developing research and implementation strategies and a critical review of the role of research.

Three major points for change in the role of research were identified. First, the scientific perception of soil erosion must be modified to become more holistic, supplemented by other stakeholders' views. This is possible only through close co-operation between different disciplines. Second, researchers' communications must move away from one-way transfers to an exchange of information, ideally between all stakeholders, but at least with land users. Third, research must modify its outputs. Presently, dissemination of research takes place predominantly in the form of complicated research papers. Different potential users of research findings, such as policy-makers, planners, extension workers, land users, etc., need to be addressed through media appropriate to their level.

Keywords: Soil and water conservation, interdisciplinary research, conceptual model, dissemination of research results, appropriate outputs

1 Introduction

Despite some examples of successful and efficient soil and water conservation (SWC), it can often be observed that measures which are newly introduced to an area are not adopted, or are rejected or neglected, as soon as the implementing project is phased out (Hudson, 1991). In cases when implementation of SWC is accompanied by research, a question must be raised about what impact the research may have on implementation and what its future role will be. This paper presents a tool for designing research in soil erosion and SWC.

Soil erosion research focuses on land which is utilised for production of crops, livestock, or wood. The common term "accelerated soil erosion" describes a basically natural process compounded by human land use activities. Thus, accelerated erosion is a phenomenon of a both biophysical and socio-economic character, and problems of soil resource management need to be addressed through interdisciplinary work.

Although this is a widely-accepted concept, in practice the social and natural sciences speak different languages. For example, hydrologists base their research on plots, sub-watersheds, and watersheds, while social scientists recognise individuals, households, or communities as relevant levels of decision-making. The question is how to overcome discrepancies and how to find a common basis. Many essential quantitative parts of the systems approach, such as water, nutrient or energy flows and balances, touch both the biophysical and socio-economic subsystems, but a number of socio-cultural aspects remain to be considered. For example, soil and nutrient losses as well as labour inputs, etc. can be expressed in terms of costs, but they largely exclude indigenous norms and values, taboos, etc. These, however, may influence decisions on SWC at least as much as economic factors.

2 A conceptual model as a tool for systems analysis

According to Ahnert (1981: 6), geomorphological research cannot involve quantitative research methods alone; they must be supported by semi-quantitative and qualitative methods and models. Extending this principle to soil erosion and conservation research, an interdisciplinary group committed to a common goal - managing soil erosion - requires a common view on a qualitative level, before taking a specialised, detailed approach. This first step is often not given sufficient attention. Examples of how to integrate biophysical and socio-economic sciences as well as implementing agencies on a conceptual level can be found in Messerli and Messerli (1979) or Wiesmann (1995) and Wiesmann (forthcoming). The following simplified "conceptual model" could be a tool for developing such common views if it is discussed and developed by an interdisciplinary team.

Figure 1 encompasses a biophysical environment, a socio-economic framework, and a land use subsystem. The land use subsystem originates through the overlap of the two main components. The scheme reflects the present structure of disciplinary research at university level and is therefore taken as an entry point. Since SWC neither deals with a pure biophysical nor with a pure socio-economic subsystem, the land use subsystem is virtually the only existing component. Figure 1 (Model 1) describes a situation at an initial stage of discussion, including only the most important elements, based on discussions in the Soil Conservation Research Programme (SCRP) in Ethiopia. It is used as an example to explain how a common understanding of a specific situation, a qualitative systems analysis, could be achieved by an interdisciplinary group.

It is important to note that researchers do not appear as independent observers but as an integral part of the system. In general, the elements can be seen as "black boxes", the details of which become important later on. The arrows mark some of the relations or influences which are relevant in the context of SWC. At this stage it is more crucial that the team agrees on which components and relations to include in the discussion. At a later stage, a higher degree of detail may be developed depending on the needs and desired outputs of the study. A methodology to further assess and evaluate potential resources is presented by Chambers et al. (1989).

SWC is an integral part of sustainable soil or land management. Ultimately, it is not the researchers but the land users who make a final decision on how to utilise the land. Therefore, the latter group appears as the key element of this model. Land users - individuals, households, associations or communities - have specific perceptions and experience. They have goals and needs which can explain the logic and rationale behind their strategy of action.

On the one hand, land users are influenced by biophysical factors such as climate, topography, soils, etc. which provide the conditions for potential land use and production. This is the natural setting which at the same time permits and restricts the growth of a specific selection of crops. On the other hand, the socio-economic framework, that is, the political, economic, social and cultural setting, may even have a stronger influence on land users' decisions. In combination with the land users' knowledge, experience, etc., both biophysical and socio-economic aspects imply potentials and limitations for land management. Wiesmann (forthcoming) describes the conceptual approach to peasants' range of options.

Together these potentials and limitations define the range of options, in other words, the possibilities that a land user actually has to use the land and to prevent degradation. The actual land use and production reflect the choice the land user made or was forced to make. Generally, land use implies acceleration of soil erosion, due to temporary removal of the protecting vegetation cover. In the long run, erosion changes the biophysical potential for the land users. If production is reduced as a result of erosion, the socio-economic framework will be affected as well. The impact of erosion on production may become apparent very late, and land users may not initially perceive erosion as a problem. Perception is therefore a key issue in decision making on SWC.

Researchers tend to perceive erosion as a technical issue, and on the basis of soil loss measurements, they assess it as a problem. Such information is then disseminated to planning and decision-making bodies, commonly published as scientific reports and papers, which may not be the optimal way to address planners and decision-makers. It is definitely an inappropriate means to address farmers. The scientific perception of soil erosion may lead to top-down decision-making which often causes additional limitations in the range of options of the land users. Researchers mostly develop on-station solutions and bring them to people via technology transfer, a one-sided way of communication with restricted feed-back mechanisms.

This compact analysis of a real situation may omit some factors, but it indicates important keys to improvement. The model is at no stage considered "final", it is rather a means to visualise and reflect on the state of discussion.

3 Entry points for a change in the role of research

The next stage, Model 2 (Figure 2) shows basically the same elements as Model 1; changes are highlighted (bold text and symbols). They indicate entry points for an improved research approach to SWC. A major change is the differentiation of the stakeholder groups. The researchers themselves are not a homogeneous group; they represent disciplines with different research emphases. Similarly, other "stakeholders" have to be distinguished. For instance, farmers with higher income have a wider range of options to select SWC technologies than do poor farmers. In this model, the team agreed to give land users special attention in the research set-up.

The model can now be developed in many ways: to agree on a common research basis; to define the research focus or topics of emphasis; to jointly discuss individual research results, or to interpret their impact on the whole system. In this paper, the model is used to discuss the role of research and possible improvements in view of a better integration of SWC research and implementation.

First, the scientific perception of soil erosion needs modification; it has to be replaced by a more holistic perception. In this context, holistic implies that neither the technical analysis of the erosion process and the concentration on technical solutions on the one hand, nor the socio-economic analysis on the other hand, can independently solve the problem. The term "holistic" suggests that the research approach involves both sciences, and each side must understand the philosophy, language, terminology and methodology of the other. It also implies that SWC is not seen as an isolated activity but as an integral part of the overall land use, land management or land husbandry activities.

Second, arrows going both ways represent a process of mutual learning, understanding, planning and decision-making. Ideally, a forum is created to integrate experience and awareness of all stakeholders, and most importantly, the land users.

Third, research has to reflect upon the way it disseminates information. Research reports are good for researchers, but other target groups such as policy-makers, planners, extension staff, land users or the public, must be addressed through other means. Generally, researchers consider reearch reports the end of their task. It is dangerous, however, to leave the transfer and interpretation

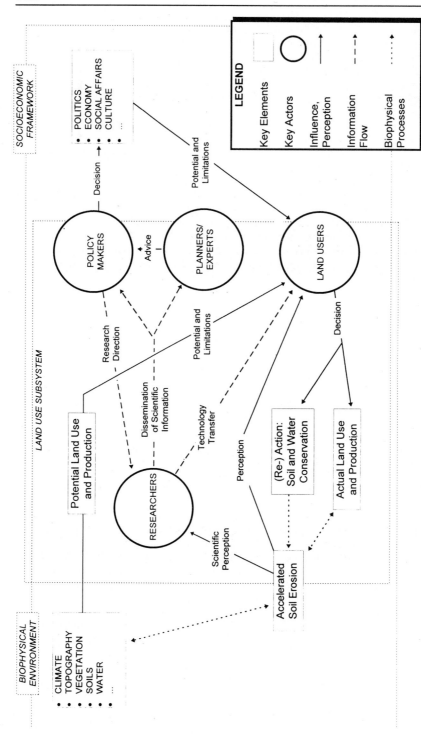

Figure 1: A concept of soil erosion/soil and water conservation (Model 1)

Soil and water conservation in developing countries

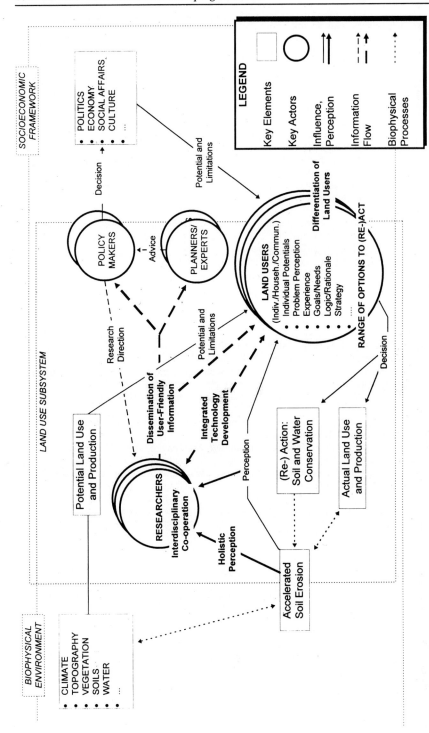

Figure 2: The role of research in soil and water conservation (Model 2)

of research results to others who do not know the accuracy of the information, or the limits of extrapolation and validity of the forecast (Herweg & Ostrowski, 1997). Research must collaborate with other stakeholders in order to bring its messages into a transferable form, e.g. as shown in Table 1.

Target groups	Outputs	Contents
Land users	Commonly understandable articles, exchange of experience over technology transfer	Technical, economic
Extension services, technicians	Manuals, training courses	Methods, checklists, technical details
Department/ministry staff, experts, researchers	Decision support systems, data base, GIS, selected analysis	Criteria for assessment, assessment schemes, setting priorities
Policy makers	Decision support systems, political recommendations	Criteria for assessment, assessment schemes, priority setting
Public	Videos, simplified information	Overview

Table 1: Dissemination of research results

References

Ahnert, F. (1981): Über die Beziehung zwischen quantitativen, semi-quantitativen und qualitativen Methoden in der Geomorghologie. Z. Geomorph. N.F., Suppl. Bd. 39: 1 - 28; Berlin - Stuttgart.

Chambers, R., et al. (1989): Farmer first. Farmer innovation and agricultural research. Intermediate Technology Publications; London.

Hudson, N. W. (1991): A study of the reasons for success or failure of soil conservation projects. FAO Soils Bulletin, 64; Rome.

Herweg, K. & Ostrowski, M. (1997): The influence of errors on erosion process analysis. Soil Conservation Research Programme; Research Report 33; Bern.

Messerli, B. & Messerli, P. (1979): Wirtschaftliche Entwicklung und ökologische Belastbarkeit im Berggebiet. Fachbeitrag zum Schweiz. MAB Programm, Nr. 27, Bundesamt für Umweltschutz; Bern.

Wiesmann, U. (1995): Nachhaltige Ressourcennutzung im regionalen Entwicklungskontext. Konzeptionelle Grundlage zu deren Definition und Erfassung. Berichte zu Entwicklung und Umwelt, Nr. 13; Bern.

Wiesmann, U. (forthcoming): Sustainable regional development in rural Africa: conceptual considerations and case studies from rural Kenya. Geographica Bernensia, African Studies; Bern.

Address of author:
Karl Herweg
Centre for Development and Environment
Institute of Geography
University of Bern
Hallerstraße 12
CH-3012 Bern, Switzerland

Training for Better Land Husbandry

M. Douglas

Summary

Peoples' participation and better land husbandry are the key elements in the new approach to sustainable land use. Training in the participatory approach to better land husbandry should be targeted at:
- farm households;
- development workers;
- policy makers and senior officials; and
- college and university students.

At the farm household level the promotion of better land husbandry can draw upon a growing body of participatory learning techniques, notably the farmer field school approach. Such training should enable farmers to learn for themselves the costs and benefits of better land husbandry. In-service training should enable development workers to learn, through 'hands on' guided practical field exercises, what is involved in 'working with farmers for better land husbandry'. Familiarising policy makers and senior officials with the better land husbandry approach calls for awareness creation or sensitization rather than training as such. At the college/university level the need is for students to learn how to understand and work with the land users as much as how to classify the properties of the land itself. This in turn requires a recognition in academic teaching circles that rapid rural appraisal (RRA) and participatory rural appraisal (PRA) methods are 'scientifically' valid.

Keywords: Land husbandry, training, participatory learning

1 Better land husbandry - A new approach to sustainable land use

It is possible to recognise two key elements underlying the new approach to sustainable land use that has emerged from the lessons learnt from past failures and successes - namely peoples' participation and better land husbandry (Douglas, 1993, 1994 & 1995, Shaxson, 1996).

The first element of the new approach - **peoples participation** - derives from a realisation of the failings inherent in the past 'top down' methods used to plan and implement soil conservation and agricultural production projects (Chambers et al., 1989). Sustainable rural development requires a 'bottom up' approach, one in which the project beneficiaries actively participate in the process. For too long farmers have been the passive recipients of externally derived research and extension recommendations. The new approach aims to enhance farmers' inherent skills and capability to develop and disseminate their own technologies (Chambers, 1993a).

The second element - **better land husbandry** - represents a shift in emphasis away from soil conservation per se to a more holistic approach. The concept of husbandry is widely understood when applied to crops and animals. As a concept signifying understanding, management and improvement, it is equally applicable to land. Thus (ABLH, 1994):

Land husbandry is the care and management of the land for productive purposes - only through sound land husbandry can the land's productive potential be sustained and enhanced.

2 Concepts and principles of better land husbandry

The concept of better land husbandry derives from the belief that it is possible for farmers to manage and improve (husband) their land resources in ways that can ensure their use for productive purposes on a sustainable basis. What is best described as the participatory approach to better land husbandry is based on two key principles (Douglas, 1995):
- that it is possible to combat land degradation through the adoption of management practices which yield production benefits while being conservation-effective;
- that rural people, educated or not, have a greater ability than previously assumed by outside experts to analyze, plan, implement and evaluate their own research and development activities.

The focus of this paper is on the training requirements for better land husbandry, for further details on the underlying concepts and principles see Douglas (1995) and Shaxson (1996).

3 Targets for training

For the above 'new thinking' to have any real impact it needs to get beyond a few practitioners and out of the conference halls and into the field. Critical to the effective promotion of better land husbandry is the development and implementation of innovative training programmes for the following priority groups:
- farm households and other rural based natural resource managers - i.e. those directly involved in managing the land for productive purposes;
- development workers (notably government and NGO technicians) - i.e. those expected to work with the land managers in an advisory and facilitating capacity;
- college and university students - i.e. potential future development workers.

Given that the approach has yet to be widely adopted within government, NGO and educational circles there is a further group that need to be educated about the principles, practice and benefits of better land husbandry. These are the policy makers and senior officials responsible for policy formulation, budget allocation, approval of work plans, and technical direction for natural resource related development programmes. There is little point in training the field level development workers in a new approach if those who take the decisions, on what their development strategies and work priorities should be, have little if any understanding of what is involved in the new thinking on sustainable land use.

4 Types of training

To assist in the widespread adoption of the better land husbandry approach at the field level it is believed that individual countries will need to develop their own specific package of training activities. Such a comprehensive programme can be expected to group individual training activities according to the following four broad categories:
- farmer training;
- in-service training of government and NGO development workers;

- awareness creation/sensitization of policy makers and senior officials;
- pre-service training in colleges and universities.

4.1 Farmer training

The underlying principle of much of the past farmer training conducted by soil and water conservation specialists has been to demonstrate to farmers the error of their past ways and to show them how they could do it better in the future. It is now clear that there is little point in designing educational programmes intended to explain to farmers the problems of soil degradation and the need for them to do something about it. They usually already know but have more urgent and pressing problems requiring all their energies (Hudson, 1991).

Farmers have largely been passive recipients in externally conceived training programmes delivered by 'outsiders' (i.e. government or NGO development workers). Typically there has been a one way flow of information from the technician to the farmer with the implicit assumption that the technical knowledge resides with the technician. There is a need for new innovative teaching methods which respect and build on the wealth of local knowledge residing within the farming community. The aim should be to create a learning environment in which farmers have the opportunity to learn for themselves through observation, discussion and participation in practical learning-by-doing field exercises (e.g. conducting their own comparative technology trials).

When developing farmer training programmes it needs to be remembered that in most countries the numbers of extension staff available to work with farmers at the village level are very limited. Given that the revenue budget resources available for such work are also limited this situation is unlikely to change in the near future. In addition to the failure of past top down teaching methods there is thus a practical reason underlying the need to look for alternative extension approaches to disseminate the better land husbandry message. It is believed that future farmer training programmes should investigate and make greater use of farmer-to-farmer training methods.

There are several informal and formal methods that have been used in a number of different countries to enable farmers, rather than extension workers, to take the lead in disseminating information and directly training other farmers, notably (after Reijntjes et al., 1992):

- **informal individual peer teaching** - farmers who have developed, or adopted an improved land husbandry practice, could pass on their ideas and knowledge about the practice (through discussion, demonstration and practical teaching) to farmers in their immediate neighbourhood.
- **informal group training** - farmers who have developed, or adopted an improved land husbandry practice, could act as trainers for visiting groups of farmers from other villages, or they could participate as key resource persons in farmer workshops held in other communities.
- **formal group training** - experienced and innovative farmers could act as trainers in formal farmer training courses. This may require that they receive training on how to train other farmers and that they are involved in joint planning of the training courses (content, methodology, etc.).
- **study tours** - small groups of farmers, selected by their community, could visit local research stations, neighbouring agricultural/forestry projects, or farmers and farmer groups in other areas who have developed, or already adopted one or more improved land husbandry practices. The group would observe and discuss trials and experiences with adopting and adapting particular practices, in order to pick up ideas (about both the technology and experimental methods) they could try out in their own area. On return to their village they would share what they have learnt with others in the community.

- **field days/exchange visits** - field days could be arranged whereby a group of farmers visit one or more farmers fields to observe and discuss the progress and results of on-farm trials or experience with the adoption of an improved practice. Exchange visits would involve farmers from two or more different areas visiting each others farms to observe, discuss and exchange information on what each group is doing.
- **farmer workshops** - farmers could be brought together within adhoc workshops to systematically assess their situation, prioritise their problems and to seek solutions. Such participatory appraisal and planning could be an important learning process for rural communities. Individual workshops may be restricted to groups of farmers selected on the basis of common problems and interests, or they may seek to be representative of the village as a whole. The latter being needed when the issues to be addressed require community, rather than individual household action (eg. the better management of communal grazing and woodland resources).
- **regular group meetings (farmer field schools)** - farmers involved in investigating one or more potential solutions to a certain problem would meet together regularly (e.g. every 2-3 weeks during the growing season for crop experiments, or critical periods for fodder availability for pasture improvement experiments). Preferably in the fields or grazing lands where an experiment is being conducted. During each meeting farmers would observe and discuss progress with the experiment and look at the possibilities for adapting and improving individual treatments. Such meetings would provide a forum for farmers to review problems with the trials and to seek possible solutions. The experience farmers gain from observing, measuring and discussing the trials could be expected to improve farmers' understanding of how and why to conduct and monitor experiments. (For further details on the farmer field school approach see, below).

4.1.1 The Farmer Field School approach

The Farmer Field School approach (FFS) is an innovative approach to participatory farmer centred learning. It was developed by an FAO Project in South East Asia as a way for small-scale rice farmers to investigate for themselves the benefits of adopting integrated pest management (IPM) practices in their paddy fields. It is believed that this specific approach, although originally developed for IPM purposes, could be used as a way to enable farmers to learn about, and investigate for themselves, the costs and benefits of alternative better land husbandry practices. The characteristics of the approach are as follows (after FAO-IPM 1993):
- **Farmers as Experts.** Farmers `learn-by-doing' ie. they carry out for themselves the various activities related to the particular farming/forestry practice they want to study and learn about. This could be related to annual crops, livestock/fodder production, orchards or forest management. The key thing is that farmers conduct their own field studies. Their training is based on comparison studies (of different treatments) and field studies that they, not the extension/research staff conduct. In so doing they become experts on the particular practice they are investigating.
- **The Field is the Primary Learning Material.** All learning is based in the field. The rice paddy, yam plot, maize field, banana plantation, coffee/fruit orchard, woodlot or grazing area is where farmers learn. Working in small sub-groups they collect data in the field, analyze the data, make action decisions based on their analyses of the data, and present their decisions to the other farmers in the field school for discussion, questioning and refinement.
- **Extension Workers as Facilitators Not Teachers.** The role of the extension worker is very much that of a facilitator rather than a conventional teacher. Once the farmers know what it is they have to do, and what it is that they can observe in the field, the extension worker takes a

back seat role, only offering help and guidance when asked to do so. Presentations during group meetings are the work of the farmers not the extension worker, with the members of each working group assuming responsibility for presenting their findings in turn to their fellow farmers. The extension worker may take part in the subsequent discussion sessions but as a contributor, rather than leader, in arriving at an agreed consensus on what action needs to be taken at that time.

- **The Curriculum is Integrated.** Crop husbandry, animal husbandry, horticulture, silviculture, land husbandry are considered together with ecology, economics, sociology and education to form a holistic approach. Problems confronted in the field are the integrating principle.
- **Training Follows the Seasonal Cycle.** Training is related to the seasonal cycle of the practice being investigated. For annual crops this would extend from land preparation to harvesting. For fodder production would include the dry season to evaluate the quantity and quality at a time of year when livestock feeds are commonly in short supply. For tree production, and conservation measures such as hedgerows and grass strips, training would need to continue over several years for farmers to be able to see for themselves the full range of costs and benefits.
- **Regular Group Meetings.** Farmers meet at agreed regular intervals. For annual crops such meetings may be every 1 or 2 weeks during the cropping season. For other farm/forestry management practices the time between each meeting would depend on what specific activities need to be done, or be related to critical periods of the year when there are key issues to observe and discuss in the field.
- **Learning Materials are Learner Generated.** Farmers generate their own learning materials, from drawings of what they observe, to the field trials themselves. These materials are always consistent with local conditions, are less expensive to develop, are controlled by the learners and can thus be discussed by the learners with others. Learners know the meaning of the materials because they have created the materials. Even illiterate farmers can prepare and use simple diagrams to illustrate the points they want to make.
- **Group Dynamics/Team Building.** Training includes communication skills building, problem solving, leadership and discussion methods. Farmers require these skills. Successful activities at the community level require that farmers can apply effective leadership skills and have the ability to communicate their findings to others.

Farmer Field Schools are conducted for the purpose of creating a learning environment in which farmers can master and apply specific land management skills. The emphasis is on empowering farmers to implement their own decisions in their own fields. Within this form of training problems are seen as challenges, not constraints. Farmer groups are taught numerous analytical methods. Problems are posed to groups in a graduated manner such that trainees can build confidence in their ability to identify and tackle any problem they might encounter in the field. Putting the classroom in the field allows the field to be the learning material and the farmer to be able to learn from real live examples. Putting the classroom in the field means that the educator (extension worker) must come to terms with the farmer in the farmer's domain.

4.1.2 Principles not packages

The need is to move away from package type extension training programmes such as the conventional T&V (teach and visit) system, which rely on weekly atomised messages disseminated through contact farmers. Training for better land husbandry should take a broad integrated approach to working with farmers based on the principle that farmers have a desire to learn how to be better farmers in order to best optimise the returns to their factors of production (land, labour, capital and knowledge).

The strategy should be to teach principles, any activity encompasses several principles, principles bring out cause and effect relationships, principles help farmers discover and learn, principles help farmers to learn so that they can continue to learn. Standardised extension packages have nothing to do with learning and do not encourage learning. In the long run they are neither cost effective nor effective at improving the quality of farmers management skills. Skilled farmers can optimise yields and maintain soil productivity independently of others. Furthermore packaged approaches increase the dependence of farmers on central planners (FAO-IPM 1993).

4.2 In-service training

Promotion of a participatory approach to better land husbandry generates a need to substantially modify current training approaches and curricula so as to create new attitudes, skills and awareness within extension staff and other development workers. Changing from a top-down to bottom-up approach creates retraining needs at all levels (Segerros & Cheatle, 1993).

A recent FAO technical cooperation programme in Jamaica concluded that before the better land husbandry approach could be adopted, the existing soil conservation officers would need retraining (FAO, 1996). This was because they required a major reorientation in their thinking and way of working, away from the conventional top down physical planning approach to soil conservation, to a bottom-up holistic better land husbandry approach. One in which the conservation dimension is integrated into productive farming systems.

In addition to the orientation training targeted specifically at the Jamaican soil conservation specialists a need was identified for a broader in-service training programme that should additionally be targeted at the parish based Extension Officers and Agricultural Assistants. This need was for practical training covering the following (FAO, 1996):

- methodological training on how to follow the process of participatory planning for better land husbandry (this to include the use of Rapid and Participatory Rural Appraisal (RRA/PRA) tools and the Farming Systems Development (FSD) approach);
- simple field survey methods for land evaluation including qualitative techniques for the field assessment of land degradation; and
- practical field based technical training on the use of simple tools (eg. the A'frame) and better land husbandry technologies (eg. inter-cropping, alley cropping, green manuring, cover crops, compost and other organic matter management practices, as well as other appropriate soil and water conservation measures) for improving the conservation effectiveness of hillside farming systems.

In the short term it is believed that a major effort must be put into the development of short practical in-service training along the lines of the above. The aim should be to start with the currently serving development workers in both government and NGO agencies and by means of appropriate training to develop the necessary cadre of subject matter specialists with the inter-disciplinary planning skills for working with farmers for better land husbandry.

4.2.1 Training in the participatory approach to better land husbandry

The participatory approach to better land husbandry involves small inter-disciplinary teams meeting with farmers in a specific geographic area. The task of such a team is to appraise and understand the natural and socio-economic circumstances in which a particular farming community operates, and then to use this information to assist the constituent farm households to develop improved farming systems, that are both productive and conservation effective. The approach is problem driven and the wide range of subject matter that is likely to be encountered means that the team members must work together, each contributing to the analysis and development of recommendations from their different disciplinary backgrounds.

In Africa and S.E. Asia the Commonwealth Secretariat, SADC and FAO have to date conducted several 3-5 week in-service training workshops to introduce development workers to the concepts and procedures involved in the `participatory approach to better land husbandry' (Douglas, 1988, Douglas et al., 1991, FAO, 1994, Douglas 1996). The aim of each training exercise has been to reproduce, for the participants, as close to a `real life' situation as possible, hence the teaching emphasis throughout was on `learning by doing'.

At the start of each workshop the participants were divided into working groups with each group taking on the role of an inter-disciplinary participatory appraisal team. Each group was told at the outset that its members were expected to work together, throughout the workshop, and by the end, to have identified potential technical and policy options that would improve the productivity and conservation-effectiveness of the existing farming systems. The group's final conclusions were to be based on an analysis of the actual problems and circumstances facing farmers within a specific area (each working group was assigned its own area of 1-2 villages to study). The training thus provided the participants with the opportunity to learn most of the basic concepts and procedures involved in adopting a participatory approach to better land husbandry through a `hands on', guided practical learning experience. Whereas most of the guidance came from the resource persons much of the learning in practice came from their interaction with the farmers and the other members of their working group.

The training included some semi-formal classroom sessions designed to introduce the participants to the key concepts and procedures. However the main means of instruction was through topic focused group discussions and plenary sessions, and practical group work - both in the classroom and the field. The role of the lectures and handouts was to introduce, at key points in each workshop, the issues involved in the various steps of the better land husbandry approach. The aim was to get the participants thinking about the issues involved and to give them enough background to undertake the next stage in their practical exercise. No more than 20% of the time was taken up with formal teaching. Some 40% of the time was spent in the field in direct contact with farmers. The remaining 40% was taken up with group and plenary discussions, classroom based group work and report writing.

Whereas these training workshops have been used to introduce the participants to a new approach to soil and water conservation they have also been used to introduce a new participatory learning approach. Conventional in-service training of development workers typically relies heavily on a classroom based approach, where most of the teaching is by way of formal lectures with the lecturer seen as the major source of information to be accepted without any serious challenge. The methodology of the Commonwealth Secretariat and FAO workshops has been based on a more participatory style of training which invites challenge - indeed depends on it, accepting that there is no monopoly of wisdom on the side of the resource persons (FAO, 1987). Instead such training encourages and expects all the participants to contribute ideas and opinions from their own field experience. Hence the emphasis on topic focused group discussions and plenary sessions, and practical group work - both in the classroom and the field. The role of the resource persons in such training is one of drawing out and giving due weight to the experiences and insights of the participants.

Experience from these training workshops has shown the benefits of starting them off with an initial group and plenary discussion session rather than the more usual keynote paper. The approach was to divide the participants into small groups and ask them to discuss and then report back on one or more specific topics such as:
- what is meant by farmer participation?
- what is land husbandry?

There are two key reasons for starting a training workshop in this way. Firstly it immediately breaks the participants expectations that they have come to sit in rows listening to a teacher lecturing at them. Secondly it gives the resource persons an opportunity to assess the participants current awareness on the issues to be covered during the workshop. Furthermore starting a workshop by having a resource person present his/her ideas on the theme undervalues the experience and ideas of the participants, and encourages them to become passive recipients of the teacher's `words of wisdom'. In addition having the initial discussion in smaller groups rather than a plenary session encourages better individual participation from the start.

One of the characteristics of such in-service training is that all of the participants will be serving officers with often many years practical field experience. A participatory style of teaching can usefully draw on this experience for considering specific topics through focused discussion rather than the traditional classroom lecture. This approach can be illustrated with reference to a recent FAO sponsored training exercise in Jamaica (FAO, 1996) where the workshop participants were divided into four groups with each group assigned one of the following topics to consider:

- farm household bio-physical circumstances;
- farm household economic and financial circumstances;
- social/cultural and gender (sex and age) considerations within the farm household; and
- elements of the farming system.

Each group was asked to draw upon its members field experience to discuss and determine for its specific topic what were the key issues and why they were important for the promotion of better land husbandry at the small-scale farm level. Rather than have them sit through a lecture on the topic of farm household circumstances it was considered more effective from a learning point of view to have the participants consider the issues for themselves. Each group was required to make a presentation of their conclusions to the rest of the participants. During the discussion following each presentation it was possible for the resource persons to draw the groups attention to any issues that might have been overlooked or misunderstood, enabling them to 'learn from their mistakes'. In addition the resource persons were able to use various points raised during the discussions to show the variety of different factors that need to be taken into consideration when seeking to increase farm production on a sustainable basis.

4.2.2 RRA/PRA training

A variety of organisations have trained development workers in the use of rapid rural appraisal and participatory rural appraisal methods. Some of which has been directed at natural resource management issues.

From a review of a number of published reports of such training (examples being IIED, 1989, 1990 & 1991; CFSCDD, 1993) it would appear that the duration of such training exercises has typically been between 5 - 14 days. The reports emanating from the short, 5-7 day, PRA exercises appear to be very superficial in terms of their appraisal of farmers' circumstances and analysis of the problems. The longer RRA/PRA exercises, 12 - 14 days, tend to have more information on the farming systems and socio-economic circumstances within the study area. However the best bet solutions to the identified problems are generally broadscale development options that require a lot more design work before any field level activities could be initiated. A feature of most of the published reports is that whereas participatory techniques have been used to elicit information from farmers the analysis has still largely been done by the `outside' team. Greater involvement of the farmers in the problem analysis and development of solutions would require more time and flexibility than is possible within the confines of a 1 - 2 week training programme.

The limitations of much of the PRA training on offer can be illustrated by reference to training conducted by an NGO for an FAO watershed management project in Nepal (reviewed in FAO

1995). The training was conducted in December 1993 and lasted 7 days. It followed the following programme:
Day 1 - Introduction to PRA
Day 2 - Mapping and Transect Walk
Day 3 - Semi-structured Interview, Wealth Ranking and Seasonality Analysis
Day 4 - Time Line, Time Trend and Flow Diagram
Day 5 - Preference Ranking
Day 6 - Mobility Map and Venn Diagramming
Day 7 - Report Writing

The emphasis was on the teaching of PRA tools rather than a participatory approach to watershed management. The value of such short training is questionable particularly when it involves the participants using farmers in order to learn how to use a limited number of PRA techniques. The farmers used in this way would have received very little benefit from the training exercise. Furthermore PRA would have been taught as an end in itself rather than as one of a range of development 'tools' that can be used to work with farmers for the promotion of upland conservation and development. Furthermore 7 days is too short a time to gain practical experience with the overall 'process' of participatory watershed management planning.

Many of the advocates of PRA teaching suggest a minimalist approach in which the participants are given a very basic orientation on behaviour and attitudes and a sketchy idea of methods and sent out into the field and told to get on with it. The emphasis is on short training involving an experiential approach in which the participants learn by doing, make mistakes, embrace error and so improve (Chambers, 1993b). Such an open approach may work for well educated and highly motivated NGO staff. However experience suggests that it is less successful when dealing with field level government extension workers who are more used to being told what to do. Left to their own devices they may well flounder uncertain as to what to do next. Hence when training government staff there is often a need for a more organised and formal structure in which the resource persons can provide a degree of guidance and direction while encouraging the trainees to be innovative and to reappraise their past ways of working with farmers.

4.3 Awareness creation

There is little scope for incorporating the better land husbandry approach into government natural resource management strategies where senior officials and policy makers are unfamiliar with the concepts, principles and practices involved. Such officials therefore need to learn about them. However because of their seniority any programme to educate them about the approach will need to be developed in a diplomatic and subtle way as they can be expected to react negatively to the idea that they need to be trained. Hence the emphasis at this level is on awareness creation or sensitization rather than training as such.

Senior officials can be invited to participate in high level half to one day 'consultations' to consider the better land husbandry approach. The invitation should preferably come from someone of equal or preferably higher rank and stress that their participation is sought because the organisers would value their expert opinion on how the new approach might be adopted. The success of such meetings will depend on the ability of the key speaker(s) being able to put over the material briefly, without using technical jargon, and in a way that stresses the social, economic and political benefits of the approach. The highlights of the presentations should be readily available in the form of 1-2 page briefing papers.

Where a pilot government project or innovative NGO has been successful in working with farmers to promote the use of better land husbandry practices at the field level, senior officials can be invited to participate in one or more field days in order to 'learn' about the approach. As the guest(s) of honour such officials can be entertained by the farmer experts who can demonstrate the practices and talk about the production with conservation benefits realised.

There is scope for linking the in-service training of field staff with awareness creation amongst senior officials. This was done successfully in Ethiopia as part of an FAO programme that was piloting the participatory approach to better land husbandry (FAO, 1994). Two regional level in-service training workshops of 3 weeks duration were conducted in different parts of the country. The awareness creation strategy was to invite the regional senior decision makers and heads of the technical bureaux to attend what was effectively a one day review workshop, held on the last day of each training exercise. This final day began with the trainees, in their working groups, making semi-formal presentations to the invited senior officials of the conclusions and recommendations of their field study exercises. The presentations then led into a general discussion on the policy, institutional and technical implications of the recommendations for further action. These one day review workshops also provided a forum in which the invited officials could review the better land husbandry approach with regard to its relevance to the development programmes and institutional structure of the various technical agencies within the region.

Senior officials at the national level will rarely have the time to participate in more than a one day consultation or seminar. However it is often possible to conduct 2-3 day seminars/workshops for what can be described as middle level officials. That is for project managers or the technical heads of regional/district/parish offices with direct responsibility for the work of the field level development workers. Such meetings can be used as a forum for introducing the concepts, principles and practices of better land husbandry with the participants expected to review and discuss them in the context of their appropriateness to the work of their staff and geographic area of responsibility.

4.4 Pre-service training

There is a critical need to improve the pre-service training of future development workers. In many countries this will require a major review of the present syllabuses and curricula of the agricultural schools, colleges and universities from which such workers are recruited. This is to ensure that any future graduates recruited by government or NGO development agencies will already have the requisite knowledge and mental attitude for promoting a participatory approach to better land husbandry prior to entering service, rather than as at present having to be retrained on entry.

Changing the material and way subjects are taught within the pre-service arena can be difficult as many academic institutions are jealous of their independence and resent interference from so called non-academics. Resolving this issue may require an official government backed national task force which can address the different concerns of the educational establishments and the relevant government ministries and NGO agencies.

Many of the procedures and practices associated with better land husbandry are based on practical field experience rather than the results of conventional scientific research. This can be a hindrance in getting the approach accepted in academic circles because of the lack of scientific proof. In this regard it is worth noting that as a result of the work of people like Robert Chambers the use of rapid rural appraisal (RRA) and participatory rural appraisal (PRA) methods have gained a degree of validity within social science institutions. However such methods have yet to be widely accepted amongst agricultural and natural science institutions. There is a need for closer cooperation between the field practitioners and the academics if the experiences of the former are to be validated and taught by the latter.

The following are some of the key issues that need to be considered when reviewing the current pre-service training on offer:

- need for recognition in academic teaching circles that rapid rural appraisal (RRA) and participatory rural appraisal (PRA) methods are 'scientifically' valid;
- need for students to learn how to understand and work with the land users as much as how to scientifically study and classify the properties of the land itself;
- need for students to learn more than just the agronomic requirements of different crops but also about the constraints and opportunities faced by farmers when making crop and land management decisions;
- need for students to be put into practical field based learning situations where they have to talk with and learn from farmers about the rationale behind rural household decision making and its effect on influencing the nature of individual small-scale farming systems; and
- need for more inter-disciplinary linkages between college/university departments so that students can understand both the bio-physical and socio-economic dimensions to sustainable land use.

5 Conclusion

The concept of better land husbandry represents a relatively new approach to soil and water conservation. Before the approach can be widely adopted there is a need to introduce the principles and practices to farmers, extension workers and others. This requires the development of new training programmes for farmers, development workers and college students as well as awareness creation programmes intended to 'educate' senior officials. There is a growing body of experience within development circles with the use of a variety of new, innovative and participatory learning methods that can be used for training rural household members and development personnel. The promotion of sustainable land use requires not only a shift in emphasis from soil conservation to better land husbandry but the adoption of new learning methods within future training programmes.

References

ABLH (1994): Association for Better Land Husbandry. Publicity flyer for the Association for Better Land Husbandry UK.
Chambers, R. (1993a): Sustainable Small Farm Development: Frontiers in Participation. In: Hudson N.W. & Cheatle R.J. (eds.), Working with Farmers for Better Land Husbandry. Intermediate Technology Publications, London.
Chambers, R. (1993b): Challenging the Professions - Frontiers for Rural Development. Intermediate Technology Publications London.
Chambers, R., Pacey A. & Thrupp L.A. (eds) (1989): Farmer First Farmer Innovation and Agricultural Research. Intermediate Technology Publications, London, England.
CFSCDD (1993): Report of Training Workshop on Participatory Rural Appraisal Techniques April 21-29 1993, Nazareth Ethiopia. Community Forestry and Soil Conservation Development Department, Ministry of Natural Resources Development and Environmental Protection, Addis Ababa, Ethiopia.
Douglas, M.G. (1988): Integrating Conservation into the Farming System, Land Use Planning for Smallholder Farmers. An Outline for a Training Workshop. Commonwealth Secretariat London.
Douglas, M.G. (1993): Making Conservation Farmer Friendly. In: Hudson N.W. & Cheatle R.J. (eds.), Working with Farmers for Better Land Husbandry. Intermediate Technology Publications, London.
Douglas, M.G. 1994. Sustainable Use of Agricultural Soils. A Review of the Prerequisites for Success or Failure. Development and Environment Reports No. 11. Group for Development and Environment, Institute for Geography, University of Berne, Switzerland.
Douglas, M.G. (1995): A Participatory Approach to Better Land Husbandry. Paper presented at the International Workshop on Soil Conservation Extension: Concepts, Strategies, Implementation and Adoption, Chiang Mai, Thailand 4-9 June 1995.

Douglas, M.G. (1996): Integrating Conservation into the Farming System - Report of a Consultancy Mission to Zomba, Malawi. Commonwealth Secretariat London.

Douglas, M.G., Carandang, D., Felizardo, B. and Ranola, R. (1991): Development of Conservation Farming Systems, An Outline for a Training Course. Training materials for the ASOCON Regional Action Learning Programme on the Development of Conservation Farming Systems. Jakarta Indonesia.

FAO (1987): Training Workshop in Farming Systems Development - Tutors Manual. AGSP FAO Rome.

FAO (1994): Consultancy Reports of the First, Second and Third Missions of the Participatory Technology Development Consultant for Project ETH/85/016. AGLS FAO Rome.

FAO (1995): Consultancy Report of Mission to Nepal, 29 October - 11 November 1995, of the Integrated Watershed Management and Land Husbandry Consultant. FAO/Government of Italy Cooperative Programme Inter-regional Project for Participatory Upland Conservation and Development, GCP/INT/542/ITA. Forestry FAO Rome.

FAO (1996): Formulation of a Soil Erosion Control Programme - Jamaica. Technical Report TCP/JAM/4451(A) Technical Report. FAO Rome.

FAO-IPM (1993): IPM Farmer Training: The Indonesian Case. Indonesian National Integrated Pest Management Program. FAO-IPM Secretariate Yogyakarta Indonesia.

Hudson, N.W. (1991): A Study of the Reasons for Success or Failure of Soil Conservation Projects. FAO Soils Bulletin No.64 FAO Rome.

IIED (1989): Participatory Rapid Rural Appraisal in Wollo, Ethiopia Peasant Association Planning for Natural Resource Management. IIED Sustainable Agriculture Programme & Ethiopian Red Cross Society UMCC - DPP.

IIED (1990): Rapid Rural Appraisal for Local Level Planning Wollo Province, Ethiopia. IIED Sustainable Agriculture Programme & Ethiopian Red Cross Society.

IIED (1991): Farmer Participatory Research in North Omo, Ethiopia. Report of training course in Rapid Rural Appraisal. Girara PA and Abela PA. IIED Sustainable Agriculture Programme & Farm Africa.

Reijntjes, C., Haverkoort, B. & Waters-Bayer, A. (1992): Farming for the Future - An Introduction to Low-external Input and Sustainable Agriculture. ILEIA Macmillan Press.

Segerros, K.H.M & Cheatle, R.J. (1993): Improved Training Approaches for Sustainable Land Husbandry. In: Hudson, N.W. & Cheatle, R.J. (eds.), Working with Farmers for Better Land Husbandry. Intermediate Technology Publications, London.

Shaxson, T.F. (1996): Principles of Good Land Husbandry. ENABLE No. 5 June 1996. Newsletter of the Association for Better Land Husbandry.

Address of author:
Malcolm Douglas
Langenfeld
Easingword Road
Huby, York YO6 1HN, UK

Strengthening Peoples Capacities in Soil and Water Conservation in Southern Zimbabwe

J. Hagmann, E. Chuma & K. Murwira

Summary

The paper highlights constraints of the top-down extension of soil and water conservation in semi-arid Zimbabwe. As a response, a participatory approach to innovation development and extension was developed and tested in an iterative process together with farmers.

The approach aims at developing farmers' capacities in land husbandry through a learning process based on farmers' own experimentation. Learning by trying out (experiential learning) instead of being taught 'foreign knowledge' was the concept which led to confidence building and to the revival of indigenous knowledge. Innovations for soil and water conservation were developed as a synthesis of traditional and modern ideas jointly by farmers, researchers and extensionists. Spreading of innovations was facilitated by strengthening the capacities of local institutions and improving communication structures within communities.

The activities were successful in generating appropriate conservation technologies and to spread them. In some areas, the approach managed to involve up to 80% of the farmers in soil and water conservation related activities. The adoption increased drastically and self-organising capabilities in the communities were strengthened.

Keywords: Participatory Technology Development, participatory extension, soil and water conservation, local institutions, learning process approaches

1 Introduction

1.1 Background

Soil and water conservation played an important role in the development of smallholder agriculture in Zimbabwe. The introduction of the plough early this century triggered a drastic change in the farming system and in practices. This change, equal to an agricultural revolution, virtually eradicated indigenous technical knowledge in soil and water conservation and in other farming related fields. As a consequence, degradation through plough-based agriculture became so severe, that the colonial administration forced smallholders to implement externally-developed conservation works. This often went along with the transition from shifting cultivation to permanent agriculture and settlement.

At present, a growing concern about continuing degradation despite the existence of mechanical conservation systems with contour ridges manifests itself in various studies which indicate that production cannot be sustained with plough-based agriculture (Elwell, 1991) and alternative systems for soil and water conservation are being studied (Norton, 1987; Elwell, 1993). Innovations are required not only in terms of the technologies, but also with regard to extension approaches.

Since 1991, two projects have developed and tested new approaches to innovation development and extension in the semi-arid areas of southern Zimbabwe (Chivi and Zaka Districts in Masvingo Province) where subsistence farming is the predominant livelihood of smallholder farmers. The area is marginal for cropping with poor sandy soils and erratic rainfall. Both projects work together in close collaboration with farmers and now use a similar approach. One project, the ITDG Food Security Project orientates more towards extension of soil and water conservation techniques, whereas the other, the AGRITEX/GTZ Conservation Tillage Project is more research orientated. The paper gives a brief overview of the history of conservation extension and the rationale behind the change which now takes place. The new approach to development and extension of soil and water conservation technologies is presented and the impact is analysed.

1.2 History of conservation and extension in Zimbabwe

Already in 1926, the colonial government devised a centralisation policy which sought to consolidate all arable land in the 'native reserves' into large permanent blocks and to protect the soil with conservation works (Norton, 1995). Contour ridges and storm drains which were designed for high rainfall areas and large-scale agriculture were transferred to smallholder farmers even in semi-arid areas and became the synonym for soil conservation. Even nowadays, hardly any conservation technologies other than contour ridges are being practised.

Conservation and agricultural extension as a whole started in the 1920's by an American missionary with promotion of contour ridges and storm drains. The paradigm of the superiority of western technology over African knowledge was used to 'civilise and develop natives along European lines by entering the cash economy through a system of intensive crop production in native reserves. ...' (Alvord, 1926). Eradication and replacement of indigenous farming practices by commoditisation and commercialisation of African agriculture was the goal. The strategy to extension was top-down teaching by demonstrations based on a 'master farmer programme', where elitist demonstration farmers were taught ready-made packages of 'modern' farming practices and would then demonstrate the efficacy of European civilisation (Madondo, 1995).

In addition to the extension efforts, conservation was enforced by law as from the 1940's. During the liberation struggle (1976-1980) conservation suffered a major drawback as it was seen as a colonial measure to oppress African farmers. Since independence (1980) legislation still obliges farmers to maintain contour ridges but de-facto the control does not exist any more. The major focus of agricultural extension has shifted from conservation to production and commercialisation of smallholder farming. Madondo (1995) claims that the colonial heritage is still prevalent in the present extension system in the sense that knowledge is believed to flow from one source to the user in a hierarchical order and the mode of knowledge generation is based on teaching not on educating the farmer. The master farmer model has been the most legitimate symbol for emulation.

1.3 Rationale for a change of the extension approach

From a quantitative point of view conservation extension in Zimbabwe appears to be very successful. More than 90 % of arable lands in 'Communal Areas' are conserved by contour ridges. The effective output of these adopted measures in terms of soil and water conservation however, is rather poor. In a recent study in a semi-arid area in southern Zimbabwe, it was found that the mechanical conservation system of 66% of the surveyed fields could not guarantee protection from rill erosion (Hagmann, 1996). Deep rills and depressions were abundant and contour ridges spilled over regularly which concentrated water and caused rill erosion. This process made contour ridges the major cause for rill erosion instead of being the cure. In terms of technology the failure can be

attributed to the following factors:
- contour ridges and drains were inappropriate in the semi-arid area as they are to drain water out of the fields instead of infiltrating and retaining the water. This can be attributed to the simplified generalisation of recommendations disregarding ecological zones and the different farming systems
- forced implementation of the measures with a top-down teaching approach resulted in a high quantitative output of adoption of standards (the dimensions are kept rather well), but did not generate farmer understanding of processes which is necessary to modify and adapt techniques to achieve the maximum output in terms of soil conservation
- the limitation of conservation technologies to only one standardised measure as a panacea was extremely reductionist and could never have worked effectively. This was further proven through recent results of adaptive on-farm trials (Chuma & Hagmann, 1995) where the performance of most conservation techniques proves to be highly dependent on the site, soil and the farmers with their skills and resources. Therefore, generalised extension messages are not adequate to achieve an optimum output with the majority of farmers
- conservation works addressed rill erosion only. Rampant sheet erosion mainly caused by ploughing (Elwell, 1983) was not tackled at all until recently when research started on conservation tillage. Adoption of conservation tillage however, is still very low so far

A change is required to enable the development of flexible options for conservation (mechanical, agronomic and biological), which can be modified and adapted to the situation specific environment and optimised by farmers themselves. This requires farmers to understand bio-physical processes rather than being taught standardised techniques. Education instead of teaching is the only way to change the output from keeping standards to achieving the goal of conservation. This goes along with raising awareness for conservation and understanding that degradation affects first of all farmers themselves and not the outsiders. If conservation is limited to soil conservation this is often not the case. Therefore the approach should aim at land husbandry rather than focus on a single factor.

In terms of spreading of innovations among farmers major socio-organisational constraints in communities were identified during participatory on-farm research (Hagmann, 1993, Nyagumbo, 1995). It revealed that social conflicts have weakened local institutions which are essential for the spreading process. 'Leadership and Cooperation crisis', 'generation conflicts', 'communication barriers' as well as 'fear of innovations' were some of the major problems identified by farmers themselves. It turned out that the social environment must be favourable for an effective spreading of innovations. Therefore, besides technical innovations, socio-organisational developments and innovations must be considered and addressed in extension.

In terms of the technology transfer in extension the strict top-down approach of extension workers was revealed as a constraint for farmer innovation as they rely on extension workers to tell them what to do. Own initiatives are often blocked and farmer participation is very low. The high farmer to extension worker ratio (about 1000:1) also contributes to a low outreach of extension. On average they hardly reach more than 10% of the farmers (mainly master farmers) who benefit from extension. A change in the approach has to opt for reaching the majority of farmers of all gender, age, wealth and knowledge categories and has to encourage farmer initiative and participation in the sense of self-mobilisation.

The requirements to overcome the constraints of technical, socio-organisational and extension-related aspects are increased capacities of individual land users, individuals in institutions as well as local institutions and government institutions as a whole. Capacity is used here as a synonym for capability to solve one's problems and tasks. This requires creativity to find new patterns to solve new problems and not to apply the old problem-solving pattern to new problems (e.g. more contour ridges). Creativity in problem solving can only be enhanced through a deep understanding and analysis and through exposure to a variety of options. These elements have been built in an approach for participatory technology development and extension which has been tested in pilot areas and described in the following chapter. The learning process which led to the approach is described in Hagmann et. al. (1997a).

2 Concept, approach and tools

The approach is called 'Kuturaya', which means 'to try' and was created by farmers who were asked to provide a word for 'research' in the local language, Shona. Later the name became a synonym for the 'spirit of Kuturaya' and the 'school of Kuturaya', for learning and improving through experimentation. It is based on dialogue, farmers' own experimentation and strengthening of self-organisational capacities in rural communities. The approach combines research and innovation development, and extension in one process. A flexible model was developed in an iterative process based on experiences during four years of on-farm research and extension. It is not meant to be a blueprint for any given situation, but has to be modified and adapted to the local context when applied in a different environment. Many elements of the approach have been derived from the concept of 'participatory technology development' (Waters-Bayer, 1989; Haverkort, 1991).

2.1 Philosophy and conceptual model for Kuturaya

2.1.1 The goal of 'Kuturaya'

The goal is sustainable management of natural resources and food security in smallholder farming areas in Zimbabwe. It aims at developing and spreading of sustainable farming practices and at enabling rural communities to better handle their problems in a self-reliant way, without depending on external incentives. It addresses communities as a whole and individual families as units (men and women together).

2.1.2 The philosophical and developmental framework

Kuturaya is based on 'Training for Transformation' (TFT), This training programme was developed in Kenya in 1974, adapted to Zimbabwean conditions by Hope & Timmel (1984). It originates in the pedagogy of Freire (1973) and is built on conscientisation through participatory education, where learning is based on experience in the living world of the actor. Teaching therefore consists of dialogue via problem-posing, which means facilitation of communication flow and asking questions to help groups find the causes and the solutions themselves instead of teaching 'foreign' knowledge and realities. TFT provides concrete methods to implement Freire's approach and empowers local people to control their lives through active participation in their own development and sharing of ideas and knowledge. It stresses the importance of participation and co-operation in organisational development in order to build institutions which enable people to become self-reliant. It aims at strengthening people's confidence (e.g. slogans like: "nobody knows everything and nobody knows nothing") and integrates social analysis to help groups to find the root causes of problems (Hope & Timmel, 1984).

Freire's key principles form a philosophical framework which is relevant for any individual living in a society and can be applied in almost all situations in life. The strong acceptance of and agreement on these principles by various characters with different attitudes and in different mainstreams is its major strength. It manages to integrate and unite these often conflicting interests under one umbrella, the key principles. As described above, in Shona society socio-cultural change has weakened the social coherence and security which was based on traditional rules and regulations. Therefore, according to our experience, a new 'umbrella' as a reference base which can partly substitute the old security or negotiation rules is particularly important as the desire of social harmony is extremely strong and dominates most decisions of individuals. Without providing a platform to develop the new "umbrella" and new ways of negotiation within the communities, co-operation and leadership structures in rural communities will generally remain weak and often dominated by the unresolved social conflicts, which also adversely affects innovation development and extension.

Farmers are introduced to this framework right at the beginning of the process in awareness raising workshops with the whole community. Regular follow-ups are built in different stages of the process.

2.1.3 The core process: 'learning and development through experimentation'

The core of Kuturaya is a participatory community development process with a specific focus on strengthening local (peoples') institutions and experimenting with new ideas. It is people-centred as villagers analyse and define their visions, goals, needs, problems and potentials and the activities they want to carry out. The intervention from outside facilitates the process, raises awareness, contributes methodologies and inspires with potential technical options, but does not dominate and push people to carry out certain (from outsiders) pre-conceived activities. It is an open-ended development process where research and extension are support agencies and ideally participate in peoples' programme and not vice-versa. With regard to soil and water conservation, this might mean that other priorities are equally or even more important and conservation activities therefore come in only in the second or third year, after the priority problems have been tackled. The prioritisation of problems also depends on the awareness of causes and consequences of a certain problem. Here, the strengthening of analytical capacities through social analysis, learning and sharing is important. This refers to socio-organisational problems as well as to realise the impact of degradation on agriculture and the relative importance of soil and water conservation. The social learning creates a collective accountability and commitment, the key for local ownership of activities and projects.

2.1.4 Development of innovative techniques

Innovation development is based on the trial and error principle. Farmers are encouraged to experiment with ideas and techniques emanating from their own source of knowledge or from outside sources. Farmer experimentation is to stimulate the re-evaluation and appreciation of traditional knowledge, its combination with new techniques and a synthesis of the two. The knowledge and understanding of bio-physical factors gained through the experimentation process should result in building farmers confidence in their own capacities and knowledge to generate new solutions to problems. This should also increase their ability to choose the best options, to develop and adapt solutions appropriate to their situation-specific ecological, economic and socio-cultural conditions and circumstances. Problems identified during the experimentation process are the basis for a research agenda and resulting on-farm trials in which more focus is put on quantitative data to support the findings. If technical processes are not fully understood, farmers' ideas are taken to the research station for further research under controlled conditions.

2.1.5 Spreading of innovative techniques

Spreading is stimulated through the strengthening of the self-organisational capacities of rural communities and institutions. Improvements of communication structures, skills and modes is facilitated with the help of the TFT philosophy and tools in order to enable people to create an environment where they feel free to communicate and share their skills and experiences with all members of the community. Once this level of communication flow is reached in the communities, a high dynamic in farmer to farmer sharing and extension should result. In technical terms, not the adoption of new techniques as such are promoted, but the experimentation with technical options and indigenous knowledge is encouraged in order to achieve a conservation output. Experiences and results of the experiments are shared and compiled by farmers and extension as guidelines/training materials which focus on the understanding of the factors which make the techniques succeed or fail. Important tools are annual community reviews where the technical and socio-organisational progress is reviewed and evaluated and adaptations to the planning made.

The mainstay of 'Kuturaya' is encouragement of interactive participation and dialogue among all actors on the local level as partners, e.g. farmers and their institutions, extensionists and researchers. The aspect of human development is crucial as with an increased confidence and the appreciation of indigenous/farmer knowledge the value of farmers in society also increases, an important factor for rural development in general.

2.2 'Kuturaya' implementation model and tools

After the preparatory phase, 'Kuturaya' was implemented in a seasonal cycle. The implementation model consists of three major components which are closely interlinked: the process initiation, the seasonal cycle and the support system. In the pilot activities, research was represented by a research project within the extension department, extension workers were employed by the extension department and also by NGOs.

2.2.1 Preparatory phase

Assessment of agro-ecological, socio-economic and farming systems information is obligatory for whatever project is planned by outsiders like extension and NGOs. Of particular importance for 'Kuturaya', however, is the institutional survey to identify the local institutional basis of the intervention. Peoples' local institutions are to be the 'owners' of the intervention who are mandated by the community to be accountable and responsible. This survey (Murwira, 1991) consisted of interviews with community leaders as well as with a sample of the local communities. Roles, strengths and weaknesses of the existing local institutions were identified as well as peoples' preferences for institutions which should be selected as vehicle for implementation. A household needs assessment was carried out on the basis of different wealth ranks to differentiate the felt needs.

2.2.2 Process initiation

The process was initiated well before the start of the agricultural season at the end of the dry season. It has proven to be one of the most important phases to clarify the purpose of the envisaged process and to motivate people to become the drivers of the process. The first step was an awareness raising workshop in the community to addresses the strengthening of self-organisational capacities. All stakeholders at local level (villagers, local leaders and extension workers) were invited to share ideas. Both, the information of the institutional survey and the needs assessment survey were fed back to the community in order to stimulate the selection of the most appropriate institutions for project implementation. Training for Transformation principles were introduced in five interlinked sessions as a sequence to analyse and solve problems in communities. Perception of visions and goals in development, the analysis of root causes of problems, self-organisation, co-operation and leadership, and openness/criticism/sharing are topics which were being worked on in small groups, presented in the plenary and further concluded on the basis of codes (pictures, songs, proverbs etc. which symbolise real live situations.

For the problem analysis and the analysis of local institutions, elements of tools like goal-oriented planning and of Participatory Rural Appraisal (PRA: e.g. Venn diagrams) (Theis & Grady, 1991) were utilised. In the last session the link between the identified problems and solutions and the need for experimentation (trial and error) in order to find the best way to solve the problem was created. On the basis of the community's priority needs, problems and identified potential solutions, further activities like leadership training courses, 'field days for options' or 'look & learn-visits' (e.g. to research stations, innovative farmers etc.) as exposure to different technical options etc. were identified and planned. As answer to social incoherence, leadership courses where leaders were exposed to learning about group dynamics, self-reliance, leadership skills, analytical skills and self-realisation, and where they gained an insight understanding of development, were essential to improve the local institutions. In terms of technology, support to farmers was mainly offered in the field of soil and water conservation. For other aspects and priorities, villagers were helped to source for know-how from other agents.

The second step in the process initiation, in some cases after a 'look & learn-visit' where only a limited number of farmers selected by the community could take part, was the organisation of a community 'field day for options', where researchers and farmers who were on look & learn visits reported about techniques addressing peoples' priority problems. This aimed at raising awareness and to expose the whole community to technical options in order to inspire farmers with additional ideas so that they can choose certain techniques, experiment with them and combine or modify them with their own knowledge. Important learning tools were visual teaching aids like small models of the environment, which illustrated the processes underlying certain techniques (in our case soil and water conservation) and which helped farmers to discuss and understand processes and necessary action to be taken. During these field days the experimentation process was initiated and planned. Competitions among all individual farmers in the communities for 'the best ideas' (not only soil and water conservation) and among neighbouring communities for the highest number of farmers participating, trials and ideas were introduced. They stimulated the process of experimentation and the revival of farmer knowledge which could be seen in the high number of trials originating in their indigenous knowledge.

With individual competitions the danger of victimisation of the innovators was big. However, if they were combined with competitions between communities, the innovators were important for each community in order to win and thus they were fully respected and appreciated even if failures occurred. For organisation and judging of the competitions, farmers were stimulated to organise or elect a committee. For these elections, farmers were encouraged to consider the leadership qualities they defined as important in the first workshop.

2.2.3 Seasonal project cycle

After the process initiation at the onset of a cropping season experimentation by farmers started. Farmers chose the options and ideas they thought were most responsive to their individual problems and tested them. During the first year, as long as the process was still somewhat outside-driven, it was advisable to focus on a limited number of technologies in order to create the necessary entry point. The more confident farmers felt with the approach, the more ideas were being tried out. The major tool in the experimentation phase was the simple paired design. Farmers were encouraged to test every innovative idea in comparison to the conventional technique, side by side in one field. This allowed farmers to carry out a continuous qualitative assessment of the performance and therefore helped them to understand the factors which contribute to the differences, which in-turn enabled them to improve on these factors in future. This process of learning by experimenting increased their analytical capacities. Researchers, extension workers and farmers agreed on some fields where researchers could collect quantitative data on the paired design in check-plots and a regular monitoring took place.

In the middle of the season, before crop maturity, a farmer evaluation of the field performance of the different ideas and techniques was organised by farmers assisted by extension staff. Facilitation was done by extension workers. All farmers in the community were invited to go around the fields (close to the village) to see the 'trials'. Each individual farmer who ran interesting experiments presented his/her fields, ideas and findings to the group where it was further discussed and sometimes challenged. The objective was the effective sharing of knowledge among farmers, a rise in confidence due to the presentations and as a result of both, more farmer to farmer extension. For researchers and extensionists farmers' evaluation was of great interest as it revealed their knowledge and criteria, often not spoken out in extension meetings.

In another session, farmers screened the techniques and ideas which they saw in the field for further research, for promotion or for deleting. Ranking techniques were useful for that purpose. The technologies which were screened for further research were being put in more quantitative on-farm trials or fed back to on-station in the following season. The technologies which were classified as ready for promotion were further promoted as options in neighbouring areas. Joint formulation of fact sheets on these techniques by farmers, researchers and extensionists, which described and summarised the experiences gained was initiated. The judging of the competition was carried out by the committee of the neighbouring community who were advised in the criteria catalogue put up jointly by farmers, researchers and extensionists (e.g. no. of trials, innovativeness of the idea, trial management, quality of presentation etc.). In case there were too many fields to visit, only the best farmers in the competition were visited in the field. At the end of the field visits, further activities were outlined.

One to two months before the start of the next season, the community was encouraged to organise a feedback/review and planning workshop. If researchers had collected quantitative data, the results were fed back to farmers. The results of the whole process were reviewed and evaluated with regards to the initial objectives. This included the strengthening of self-organisational capacities. Based on this analysis farmers elaborated a plan for the activities of the next season. The testing strategy for the following season was clarified and if necessary, an exposure to other options was organised.

2.2.4 Support system

Research and extension are the major players in the support system. Other stakeholders within the communities (traditional and political leaders, farmers' clubs, etc.) and outside (health workers, verterinary services, rural credit institutions etc.) have a strong influence and must be supportive to make 'Kuturaya' a success.

The role of agricultural extension was predominantly focused on facilitation of the process, in particular in the first years until farmer leaders were trained and experienced to take over the facilitation role themselves. Facilitation means here to provide the methodology for the process, to facilitate communication and information flow and to provide the technical backup and options. Extension guided and supported the process without dominating farmers. Documentation of farmer knowledge and experience as well as production of guidelines and fact sheets with and, most important, for farmers were started. Once the scaling-up of Kuturaya takes place, well-illustrated fact sheets and guidelines which are handed out to farmers are the basis for training farmers in options and processes. Extension will also be the link to research in case farmers define problems where researchers are needed. According to the experience gained so far, the time requirements for the support role of extension is demanding initially (minimum of 5 days/year in the field for a group of two to three communities, approx. 50 - 80 households), but with time when farmer-leaders are being trained in facilitation they take over most of the tasks which releases the extension workers. In some communities, after three years, the process has become self-running, where extension and research take supportive roles only.

Agricultural research played the role of a demand-driven service institution. Farmer-oriented extension defined an outline of relevant topics which required research. Farmers and researchers together then put up a research agenda. Except for some basic research, which requires strictly controlled conditions, most research was carried out on-farm in an interactive way in order to find applicable solutions to farmers' problems. Researchers joined farmers in 'Kuturaya' and, provided farmers utilised simple paired trial designs and researchers brought up the necessary time for monitoring, they were able to collect quantitative data as was shown in the ConTill Project (Hagmann et. al, 1995; Chuma & Hagmann, 1995). The most interesting farmer ideas were taken back to the research station in order to look into the detailed processes. The nearest research station was developed into a 'thinktank' of options for 'look & learn visits'. Apart from the main research programme, as many technical options as possible were placed in 'look & see' trials to inspire farmers, to interact with them and to allow them to comment and give feedback to researchers in the discussions during farmers' visits. Farmers' appreciation of the exposure to options on the research station was evident when a group of farmers who were not involved in the pilot activities had heard about the activities and organised a visit to the research station, paying for the transport on their own. Based on a request of the Zimbabwe Farmers' Union more than 1000 farmers were hosted during field days at the station in 1995.

The role changes required from extension and research staff towards facilitators demands major changes in attitudes and behaviour. This change and the related building of capacities of the support institutions is a long-term venture and a specific intervention which has to go along with the field activities. Continual on the job training of field workers and an organisational development programme within the extension department contributed were geared towards these changes (Hagmann et al., 1996).

3 Impact of the participatory approach

The impact of the approaches taken in the two projects has been evaluated in terms of the development of innovative technologies, spreading of these technologies and the effects on the local institutions/social organisation. In all three cases the results are encouraging and should be further monitored during the coming years.

3.1 Development of innovative technologies for soil and water conservation (SWC)

As a result of both projects, several options for SWC were developed by farmers. Others were developed on research stations, offered to farmers for testing and adapting and further developed together with farmers. Most of the techniques are related to practices originating in traditional farming practices. The principles of traditional farming such as high soil cover, retaining water, minimum soil disturbance, intercropping, agroforestry, fertilisation with anthill material etc. were adapted to the present farming system with its environmental conditions.

In the ConTill Project the focus was more on farmer experimentation and research. The results of the adaptive trials on conservation tillage showed that it is possible to obtain valid quantitative data from farmer participatory research. Besides research, several innovations on the management of conservation tillage like appropriate planting methods on ridges, methods of incorporation of manure and anthill material, methods of combining weeding and tying of ridges were generated. The joint experimentation on conservation tillage brought about several innovations on draught power saving tillage implements (see Table 1). The fanja-juu was also developed jointly on the basis of farmers adaptations to the contour ridges. The spirit of "Kuturaya" and experimentation among farmers spread on other subjects linked to SWC, creating numerous ideas and farmer trials (see table 1). After two seasons each participating farmer had at least two trials on his/her fields. Very innovative farmers came up with up to 12 self-initiated trials which were analysed and shared with other farmers. Some of farmers' innovations have been taken to the research station in order to carry out some scientific experiments on them. As problems and needs differ from area to area, not all the options are being promoted for testing in all areas. Considering the time frame of only four years, the amount of innovations which were developed and/or adapted is high compared to the output of the formal research system.

INNOVATION/EXPERIMENT:	SOURCE OF IDEA	STATE OF DEV.
Implements:		
animal drawn disc ridger	ConTill Project	on the market
donkey-drawn toolbar (multiple purpose)	Farmers	on the market
a ripper tine mounted on the plough beam	ConTill Project	on the market
a planting device mounted on the plough beam	Farmers	on the market
animal drawn weed roller	ConTill Project	under testing
Soil and water conservation techniques:		
tied ridges/furrows	ConTill/Chiredzi	promotable
Basin tillage (widely spaced ridges/ semi-circular bunds)	ConTill/Chiredzi/Farmer	under testing
vetiver applications	ConTill/CARD	test & promote
methods for rill reclamation	Farmers/ConTill	promotable
the modified 'fanja-juu'	ConTill/Farmers	promotable
infiltration pits	Farmers	promotable
stone bunds	Farmers	under testing
subsurface irrigation for gardens	Chiredzi Res. Station	promotable
inverted bottles for irrigation in gardens	Farmers	promotable
plasic sheet to prevent rapid drainage (gardens)	Farmers	test & promote
mulching in gardens	Farmers	promotable
mulching in fields	ConTill Project	test & promote
Other agronomic and biological soil management methods		
innovative planting techniques	Farmers	promotable
various planting dates (various crops)	Farmers	under testing
various methods of making compost	ConTill/Farmers	test & promote
spreading of termitaria as fertiliser	Farmers	promote & test
various manure and fertiliser applications	Farmers	under testing
green manure with crotalaria sp.	Farmers	under testing
planting and use of hedgerows	ConTill/CARD	under testing
a relay cropping system	Farmers	test & promote
various intercropping combinations	Farmers	under testing
natural pesticides	Farmers	test & promote
raising of indigenous trees	Farmers	under testing
chicken manure as topdressing	Farmers	under testing

Table 1: Experimentation on techniques by farmers: ideas, source of ideas and state of development

3.2 Impact on spreading of conservation techniques

In one district, within one local administrative unit with a total of 1136 households, at least 80% of the households are practising soil and water conservation in one form or another. The range of technologies currently in practice include mulching, tied ridges, use of clay pipes and plastic sheets and inverted bottles for irrigation of gardens, infiltration pits, intercropping and rock catchment water harvesting. The adoption of the different techniques during the first three cropping seasons is shown in Table 2.

TECHNIQUE	ADOPTED BY NO OF FARMERS			
	92/93	93/94	94/95	SOURCE OF TECHNIQUE
Cropped fields				
Tied Ridges/Furrows	28	>100	>500	ConTill Pr./Chiredzi Res. Station
Infiltration Pits	20	289	>800	Farmer innovation (Mr. Phiri)
Fanja-juu	0	4	n.d.	ConTill Project
Mulching	2	3	n.d.	ConTill Project
Intercropping	~50	>450	n.d.	Indigenous farmer knowledge
Spreading of termitaria	78	>128	n.d.	Indigenous farmer knowledge
Tillage implements	0	96	n.d.	ConTill Project
Gardens				
Sub-surface irrig/garden ~50	68	n.d.		Chiredzi Research Station
Plastics/inverted bottles	10	>200	n.d.	Unknown/Farmer
Compost in gardens	4	14*	n.d.	Farmer/Contill Project
Mulching in gardens	85	>300	n.d.	Farmer

* groups out of a total of 37

Table 2: Adoption of SWC techniques in Chivi (Ward 21) in 1992/93, 93/94 and 94/95

Despite very limited animal draught power, tied ridges were constructed by farmers by hand through work parties. Infiltration pits have become very popular and have spread beyond the ward through farmer to farmer sharing whereas the adapted fanja-juu has only recently been introduced. In another area an NGO in collaboration with the Natural Resources Board is recommending this technique for testing and after only one season more than 100 farmers have dug this improved version of a farmer innovation. Mulching was not so popular in the fields, but more in gardens where 60% of the group members are practising it for water saving purposes. Spreading of anthill material as fertilising material is a traditional technique which was revived by motivating farmers to increase soil fertility. Testing of tillage implements has only started in 1994/95. The focus is draught power saving implements and the use of donkey power. There are many other innovations which are being tested on a small scale. In general it appears as if water conservation techniques for gardens are most popular as they contribute directly to a reduction of labour and to increased production through prolongation of water availability in the wells. As gardening is a major source of income generation in the area, these techniques are particularly important.

3.3 Impact on strengthening of local institutions and capacities

The impact on local institutions was assessed through surveys, informal discussions among various collaborating farmer groups and workshops. Some results are presented here.

Twenty farmer groups who were exposed to leadership training were surveyed in view of their leadership structures two years after the training. The leadership was classified as excellent by 1 group, as good by 13 groups, as satisfactory by 5 groups and as unsatisfactory by 1 group. Group members explained that they decided not to allow multiple leadership anymore. Leadership training in "Training for Transformation" (TFT) made them aware of the drawbacks due to single persons who occupied most of the leadership positions in various institutions. Therefore one person can now only occupy one post at a time and is elected democratically. They also stressed that their leaders of the various institutions now link more closely in order to share ideas and to encourage people through competitions. Ways to improve leadership were requested. Ten out of the twenty groups recommended sending the leaders on leadership courses within TFT, eight groups suggested including people with leadership qualities in the committee and two groups opted to elect a new committee.

In a workshop farmers were asked about their perception of the old extension approach compared with the new participatory approach in soil and water conservation techniques. Their comments are presented in table 3.

CHARACTERISTICS OF APPROACHES

Old approach
- Forceful methods were used
- Only few people could benefit (e.g. literate)
- Intercropping was forbidden
- Failed to address SWC convincingly
- We were told to do things without questioning
- Usefulness of cons. works never explained
- No dialogue between farmers and extensionists
- Little co-operation among farmers
- Extension agents treated our fields as theirs

New/participatory approach
- Everyone to benefit as all are free to attend meetings now
- There is dialogue
- Process is well explained (teaching by example)
- Farmers are the drivers now
- Intercropping is encouraged to boost yields
- Farmers are being treated as partners and equals
- No discrimination against poor or rich, educated or uneducated
- We are given a choice of options
- They pay attention to us and take time to find solutions to farmers' problems
- We are being encouraged to try out new things

The most important aspects of the new approach
- It helps farmers to work co-operatively
- Farmers practise SWC with enough knowledge of why they should do it
- Learning from others through exchange visits/ learning through sharing
- It helped farmers to develop the ability to encourage each other in farm. activities
- Encouragement to practise SWC through various options
- It is capable to mobilise large numbers of people with satisfaction
- The approach brings about desirable SWC techniques through participation
- Farmers are free to ask for advice
- Yields have increased through SWC techniques
- The dedication of modern extension agents/researchers
- It has brought development in the area
- It is very effective in conservation of trees, soil, water

Table 3: Farmer's perception of the different approaches and the characteristics of the participatory approach

In particular three major differences need to be highlighted. Firstly, farmers now feel that everybody can participate in soil and water conservation. In the past, Master farmer club members formed an elite who did not want non-members to participate in innovations. The second major difference is the process of dialogue together with the sound explanation of processes rather than confronting farmers with recommendations and even imposing them. Farmers also noted that they were encouraged to cooperate and share knowledge between themselves and the researchers. It appears that all these criteria should form the basis for any successful extension approach.

When farmers specified the institution which they favour for taking the initiative in bringing about the change, the extension worker was nominated to take the lead in the process, together with the farmers' club chairmen and the traditional leaders. The role of the government-introduced, modern institutions (Village development committee, counsellors, etc.) is not considered as very important in the process.

The participatory process has initiated a major drive towards improving leadership of local institutions. With leadership and co-operation being one of the major problems often mentioned by farmers these results indicate that institutional strengthening through Training for Transformation has a positive effect on the farmers' capacity to organise themselves and increase participation in agricultural development through club membership. After farmers decided to use farmer's clubs as a vehicle for participatory development in the Food Security Project and after leadership training, the number of clubs increased from 9 with 120 members in 1991 to 34 with 800 members in 1994. Similar increases were observed with garden groups. They increased from 10 with 250 members to 37 groups with 1073 members in 1994. A stronghold is the involvement of women who are dominant in terms of memberships. It is apparent that the high participation in the strengthened institution provides an ideal basis for promoting conservation techniques.

Besides the impacts described above, the interactive process between researchers and farmers has led to increased self-confidence and building up of knowledge based on experience. This manifested itself-initiated and organised field days to share knowledge with their neighbours and communities in which researchers and extensionists were invited as guests, not as showmasters, resulting in more farmer to farmer extension. In addition competitions were introduced for the best ideas and for the highest numbers of trials in the communities.

On the other hand, one has to admit that such a success requires considerable endurance, continuous stimulation, support and the will to move things. In some communities the process has been very complicated as too many leadership conflicts dominated and blocked an effective negotiation process. Therefore, for a large-scale implementation the success might not be uniform in all areas.

3.4 The future: institutionalising the approach into the extension service

The case study of the two projects has shown the potential of the participatory approaches to research and extension in soil and water conservation. The next step is to introduce these approaches into the governmental extension service on a large scale. This is a long term process and a major challenge as it requires a change of attitudes of extension staff towards farmers. Considerable progress has been made, but the results in terms of increased institutional capacities require more time to be realised (Hagmann et. al., 1998)

4 Conclusions and recommendations

The following conclusions and recommendations can be drawn from the work on participatory development and extension of soil and water conservation in Zimbabwe:
- the concept of the conventional transfer of (standardised) technology approach in research and extension can not generate the desired output as conditions for conservation technologies are too diverse and situation-specific to allow generalisation. Research and extension have to be output-(conservation) oriented instead of focusing on technology adoption.

- technology development and extension can not be seperated from the social context. Technologies are part of a socio-technical system in which social, political, economic and technical interests are interacting and have to be negotiated among the stakeholders. The role of outsiders is to facilitate the platform for these negotiations
- conservation extension and research has to focus on integrated land husbandry as single conservation measures can not provide an appropriate solution to land degradation, nor do they generate enough economic benefit. The most promising technologies have been generated by involving the farmer right from the start and by building upon farmers' knowledge in the experimentation process
- the new approach to technology development and extension requires new roles of extension and research as they are to support and strengthen farmers' experimental capacities. A role change from 'teacher' to 'facilitator' is a prerequisite.
- interactive farmer participation has been the key to success in our work. However, it revealed that participation is a process involving far more than applying some PRA tools by extension workers. It requires a change of attitudes in particular on the side of extension. This makes it dependent on personalities. It also implies a shift in paradigm which is a medium to long-term process. Therefore, for large-scale application of participatory approaches, extension should not expect miracles in a short time but rather focus on the learning process which enables extension workers to transform themselves and to understand the paradigm shift.
- to strengthen farmer participation in conservation and natural resource management, outsiders should shun being superior and paternalistic in the way they interact with farmers. They should not create new structures but work with local institutions and help to strengthen them so that they can develop the capacity to solve their own problems.

References

Alvord, E.D. (1926): The great hunger: the story of how an African Chieftaincy improved its farming methods under European guidance. Native Affairs Department Annual, Salisbury (now Harare)

Chuma, E. & Hagmann, J. (1995): Summary of results and experiences from on-station and on-farm testing and development of conservation tillage systems in semi-arid Masvingo. In: Twomlow S., Ellis-Jones J., Hagmann J., Loos H.: Soil and water conservation for smallholder farmers in semi-arid Zimbabwe. Proceedings of a technical workshop held 3-7 April 1995 in Masvingo. Belmont Press, Masvingo, Zimbabwe.

Elwell, H.A. (1983): The degrading soil and water resources of the communal areas. The Zimbabwean Science News **17** (9/10): 145-147

Elwell, H.A. (1991): A need for low input sustainable farming systems. The Zimbabwean Science News **25** (4/6): pp. 31-36

Elwell, H.A. (1993): Development and adoption of conservation tillage practices in Zimbabwe. In: FAO (1993): Soil tillage in Africa: needs and challenges. FAO Soils Bull. 69, Rome

Freire, P. (1973): Pädagogik der Unterdrückten. Reinbek, Rowohlt

Hagmann, J. (1993): Farmer participatory research in conservation tillage; approach, methods and experiences from an adaptive trial programme in Zimbabwe. In: Kronen, M. (ed.) (1993): Proceedings of the 4th annual scientific conference of the SADC Land and Water Management Programme, held in Windhoek, Oct. 11 to 15, 1993, Gaborone, Botswana, 217 - 236

Hagmann, J. (1996): Mechanical soil conservation with contour ridges: cure for, or cause of, rill erosion - which alternatives. In: Land Degradation & Development, Vol. 7, No. 2/1996, 145-160.

Hagmann, J., Chuma, E, Gundani, O. (1995): Integrating formal research into a participatory process. Information Centre for Low External Input Sustainable Agriculture (ILEIA) Newsletter Vol. 11/2

Hagmann, J., Chuma, E., Murwira, K. (1996): Improving the output of agricultural extension and research through participatory innovation development and extension. In: European Journal of Agricultural Education and Extension, Vol. 2, No. 3, 15-24.

Hagmann, J., Chuma, E., Murwira, K.(1997a): Kuturaya; participatory research, innovation and extension. In:

van Veldhuizen, L., Waters-Bayer, A., Ramirez, R., Johnson, D. & Thompson, J.: Farmers' research in practice: lessons from the field. IT publications, London, 153-173.

Hagmann, J., Chuma, E. & Connolly, M. & Murwira, K. (1998): Client-driven Change and Institutional Reform in Agricultural Extension: An Action Learning Experience From Zimbabwe. ODI (AGREN) Agricultural Research and Extension Network Paper No. 78.

Haverkort, B. (1991): Farmers' experiments and participatory technology development. In: Haverkort, B., Kamp, J. van der, Waters-Bayer, A. (eds.): Joining farmers' experiments: Experiences in participatory technology development. IT Publications, London, 3-16.

Hope, A. & Timmel, S. (1984): Training for Transformation; a handbook for community workers. Mambo Press, Gweru, Zimbabwe

Madondo, B.B.S. (1995): Agricultural transfer systems of the past and present. In: Twomlow S., Ellis-Jones J., Hagmann J., Loos H.: Soil and water conservation for smallholder farmers in semi-arid Zimbabwe. Proc. of a tech. workshop, 3-7 April 1995 in Masvingo; Belmont Press, Masvingo, Zimbabwe.

Murwira, K. (1991): Report on institutional survey in Ward 21 (Chomuruvati Area) in Chivi District, Masvingo Province, Zimbabwe. Intermed. Technol. Dev. Group (ITDG) Zimbabwe, Harare

Norton, A.J. (1987): Conservation tillage: what works. Paper presented to the Natural Resources Board Workshop on Conservation Tillage, 19 June 1987, 15pp., Institute for Agric. Eng. (IAE), Harare, Zimbabwe.

Norton, A. (1995): Soil and water conservation for smallholder farmers in Zimbabwe: past, present and future. In: Twomlow S., Ellis-Jones J., Hagmann J., Loos H.: Soil and water conservation for smallholder farmers in semi-arid Zimbabwe. Proc. of a tech. workshop, 3-7 April 1995, Belmont Press, Masvingo, Zimbabwe, 5-21.

Nyagumbo, I. (1995): Socio-cultural constraints to development projects in communal areas of Zimbabwe; a review of experiences from farmer participatory research in conservation tillage. Research report 14, Conservation Tillage Project, Inst. of Agric. Eng., Harare

Theis, J. & Grady, H.M. (1991): Participatory Rapid Appraisal for community development. A training manual based on experiences in the Middle East and North Africa. IIED, London.

Waters-Bayer, A. (1989): Participatory technology development in ecologically-oriented agriculture: some approaches and tools. AAU/ODI, London.

Addresses of authors:
J. Hagmann
Natural Resource Management Consultant
Talstraße 129
D-79194 Gundelfingen, Germany
E. Chuma
University of Zimbabwe
Institute of Environmental Studies
POB MP 167
Harare, Zimbabwe
K. Murwira
ITDG Zimbabwe
P.O. Box 1744
Harare, Zimbabwe

Women's Participation in Soil Conservation: Constraints and Opportunities. The Kenyan Experience

L.M.A. Omoro

Summary

This paper highlights the constraints faced by women participating in soil conservation. The paper discusses the interventions of the Ministry of Agriculture, Livestock Development & Marketing in Kenya in supporting women's participation. The various efforts by women to enhance their participation are also stated.

Keywords: Women, participation, conservation, group

1 Introduction

According to Young (1989), soil conservation entails maintaining soil fertility and reducing soil losses. The measures applied to achieve soil conservation address both crop and land husbandry. These measures are either through structural protection of land or the biological protection through management of vegetation. Examples of structural methods are terraces and water diversions ditches while biological ones include grass strips and crop cover. Availability of both labour and time is critical for effective implementation of these methods. The structural measures in particular, are strenuous and have great demands for labour. In order to be effective, the structural measures must be stabilized which usually is achieved through planting of vegetative materials to bind the soil particles together. The planted grass takes time to establish in order to serve effectively. At best, these two methods complement one another for effective soil conservation. If there is sufficient rainfall, the biological measures can in some cases be implemented on their own.

In the high rainfall areas with steep slopes, structural measures are necessary for effective soil conservation. Many farmers in these areas undertake appropriate measures by using people within a household, hired labour or work collectively in groups. In cases where no extra labour is available, little conservation is likely to be achieved. In Kenya, the participation of individual women in the implementation of the structural measures is often difficult to assess. However, for those women who become involved, the success rate is higher when they participate in groups.

2 Constraints to women's participation

Women are faced with many constraints which affect their participation. Some of these are discussed belows:

2.1 Gender, labour division and resources

The gender division of labour tends to assign many tasks to women. For example, in agriculture it is generally accepted that women provide 60-75% of the labour force (Boserup, 1990; Anderson et al., 1984). In Central province, women are involved in vegetable gardening, dairy and poultry farming to supply Nairobi City's population which is its major market outlet. The women are also involved in coffee picking and plucking of tea, which are the major cash crops. All these activities are time consuming to an extent that little time is available for them to engage in the construction of structural measures. While one would argue that such construction would raise farm productivity, the women find it extremely difficult to undertake additional responsibility when these measures are strenuous. Available evidence indicates that it is the men who perform these labour intensive and time consuming construction activities (Östberg, 1987).

The other constraint relating to time concerns women's involvement in household chores. In a survey on soil conservation work in Murang'a, Kenya, women's labour allocation was found to be as follows:

50% of their time is devoted to various tasks compared to 18% for men and 20% for children. Out of the 50%: 26% was spent on house chores, 20% on farm work, 2.7% on terracing and 1.3% on others; while the men's 18% were spent as: 4% on terracing, 11% on farm work, 2% domestic chores and 1% on others (MOA, 1993).

Thus, due to the numerous responsibilities, the women cannot effectively participate in the construction of structural conservation measures.

Lack of resources is another constraint that affect women's participation in soil conservation. In Kenya, 40% (World Bank, 1994) of the households are headed by women. Thus women undertake all aspects of agricultural production in those households along with other responsibilities. Some of these responsibilities include fulfillment of social obligations such as financial contributions to funerals and weddings. Such contributions reduce women's available resources and therefore, their inability to improve on their farm productivity. They are also less able to hire the much needed labour to construct structural measures.

2.2 Culture and societal norms

Cultural constraints that affect women's participation in soil conservation activities include:
(i) Communication barriers exist between the women and the extension personnel. Culturally women find it difficult to communicate easily with men, preferring instead to confide in their fellow women as the issues they discuss often go beyond farming such as social issues. Unfortunately the extension services are dominated by men. In Central Province of Kenya, 98% of the agents are men (Morna et al., 1990). Extension services are supposed to enhance management capabilities in the farms. This situation is particularly serious among the Masaais, Somalis and the Muslims. This problem deprives the women of the extension information they earnestly require.
(ii) Some societal norms inhibit women's full participation in certain aspects of soil conservation, such as tree planting. In some communities especially in the Western Province of Kenya, it is believed that if a woman plants a tree she would lay claim to that land on which the tree is grown. The problem is that women are not supposed to own land. It is also believed that if a woman plants a tree she would be barren or her husband would die. Such beliefs deter women from incorporating trees as biological soil conservation methods. Yet these are relatively easy to apply and are less labour intensive and time consuming.
(iii) A related problem is the practice in which the women have to consult their husbands before they implement any measures proposed by the extension personnel. This arises from traditional

requirement that the women consult their husbands on activities that affect the households. The problem with this practice is that the information provided by the extension personnel tend to be distorted as the women may not present them to the men as accurately as extension personnel would have done. The procedure/practice also takes time thus there is a delay in implementation.

2.3 Gender and farm produce

Some soil conservation measures such as tree planting have immediate and perceivable benefits. The women's access to such benefits (tree products of fuelwood) are at times subject to rules distinct from those that govern men's access in some societies in Kenya (KWAP, 1990). In Kakamega District, for example, acute fuelwood scarcity adversely affected women's obligations, particularly that of cooking. The irony in the area however, was that while the women experienced this problem, men usually had several trees on the farms for cash income which the women could not use for fuelwood. The women could not access this product because the trees were considered to be men's cash crops. Thus women cannot benefit from extension services, including information relating to tree planting.

2.4 Power and knowledge

Women's understanding of agricultural innovations and principles can be limited when they have low levels of education. Hornik (1988) suggests that education enables farmers to deal with and have access to external information and enables them to increase their allocative and innovative efficiencies. Despite the high literacy rates in Kenya estimated at 65% (Govt. of Kenya, 1986), the majority of illiterates are women. This means they cannot access printed materials relating to soil conservation. In addition, most of the extension materials and media such as radio programmes designed to communicate agricultural and other development information are aired in Kiswahili. This is a language which most women do not understand nor able to read. This affects the women's ability to understand the extension messages intended for them.

Another limitation in the training and provision of extension services to the women is that, historically, many women extension workers have been trained in home economics as opposed to general agriculture, particularly at the Diploma level adopted from the American model of training.

Consequently, the type of extension service provided has been characteristic of women-specific programmes relating to home economics rather than on information and skills necessary for improvement in agricultural production or soil conservation. Therefore, the extension service has overlooked the fact that the women are involved in farming in general.

3 Possibilities and Opportunities

Kenyan women have attempted to overcome the constraints they face in the implementation of soil conservation measures in several ways. These include:

3.1 Establishment of women groups

The women have in many cases formed groups to facilitate contacts with extension agents. These groups set aside time to wait for the extension agents to visit and address them on issues relating to enterprises they are involved in. Many women groups have established and managed tree nurseries and have been able to engage in tree planting ventures as a biological soil conservation measure. This has eliminated some of the cultural constraints cited above as they are able to access the numerous tree products.

Some financially capable women groups are able to assist their members to invest in other ventures or pay for construction of the structural soil conservation measures. In areas where successful soil conservation ventures have been recorded such as in the Kitui, Machakos, Makueni and Mwingi districts, regions that receive relatively low rainfall, water conservation is a very vital activity in crop production. In these districts, the women have organized themselves into self-help groups known as *Mwethya*. These groups rotate on every member's farm to undertake the physical excavations. Such groupings have enabled the women to overcome some of the constraints cited above due to:

(i) The labour problem that restricts the implementation of structural soil conservation measures is minimized because the group rotates on members' farms to implement the measures. However, there is a long time lapse before complete conservation can be achieved.
(ii) Elimination or less emphasis on socio-cultural issues that minimize women's participation. The decision to work and the discussions that are held with the extension agents are on a collective basis.
(iii) Women-headed households implement measures due to pressure that enhances the empowerment of the women to install structural measures on their farms without having to contact their husbands.

3.2 Indigenous conservation efforts

Over the years, women have undertaken certain activities which contributed to soil conservation although they may not have been *perceived* as soil conservation measures. In some cases, these measures were implemented with the objectives of saving both time and labour to accomplish farm activities. For instance, in the Siaya district, most women prefer to spreading seeds without any land preparation. After spreading the seeds, they hand-hoe the farms. In the process, the seeds are buried without having to be resown. This practice may not fulfill crop agronomic requirements, but it is a method that promotes soil conservation almost similar to minimum tillage.

In the high rainfall areas, on the other hand, farmers practice relay planting. The aim being to maximize on their land holdings for subsistence. Planting of vegetables, beans and maize is common among mature crops that are due for harvest. These practices ensure that crop cover is provided throughout the year and this achieves conservation.

Shifting cultivation is also a prevalent activity in the Nyanza and Coast Provinces of Kenya. In the semi-arid lands of the Eastern and North Eastern Provinces, there is selective use of wood resources. These practices ensure soil conservation because the crop cover is well managed. Ridging is also widely practiced for sweet potato growing by women. Ridging is a water conservation effort which reduces water run-off.

3.3 Opportunities for women in the National Soil and Water Conservation Programme

The National Soil and Water Conservation Programme (NSWCP) of the Ministry of Agriculture, Livestock Development & Marketing (MoALD & M) is implemented by the Soil & Water Conservation Branch (SWCB). The Branch applies a strategy that targets areas called catchments. This approach of extension services for soil conservation activities uses Participatory Rural Appraisal (PRA), whereby a multi-disciplinary team of experts mobilize, sensitize and discuss with the community the importance of soil conservation. A committee comprising a number of farmers is elected to represent the entire catchment. The Programme has set that at least 25% of the representation is by women in these committees.

The roles played by the catchment committees include: laying out the soil conservation structures for the farmers; encouraging individual farmers to install the recommended soil conservation measures and to exert peer group pressure among the farmers as they are often the first to implement the recommended structures. The soil conservation staff in turn plan and make specific recommendations on each farm based on the unique features observed during the discussions with the farmers including those farms under the management of women.

The strategy adopted in the catchment areas is flexible and in most cases, the agricultural staff discuss with the farmers the types of soil conservation measures they prefer to implement For most farms headed by women, the aged or where the head is sick, biological measures are normally recommended. These measures involve planting of grass strips along the contours to serve in controlling soil erosion. In situations where it is necessary to install structural soil conservation measures such as *fanya juu* terraces, retention ditches or cut-off drains, the catchment committee organizes the farmers into groups to undertake the tasks in those farms where the individual farmers are unable to.

The planting of grasses, particularly the *napier* along structural measures to stabilize the structures, have had the advantage of providing fodder to the livestock. In many farms within the agriculturally high potential areas, the women have found the biological conservation measures advantageous as it eases the management of the livestock since the feed is on farm. This has saved the women time for collection of animal feed and thus they are able to engage in other activities.

More significantly, the women are required to constitute 25% of the membership of these committees. This arrangement is intended to boost the influence of women in decision making of these committees.

4 Conclusions and recommendations

It is clear that attempts are being made to improve women's participation in agriculture. Much more can be done. Weidemann (1985) suggests that women's agricultural efficiency, rather than equity, should be increased. The women should be provided with information that can raise their production capacities. Such knowledge may empower the women to achieve more. Training and extension services need to be provided for women. The male extension workers can be made to work with the women by advocating for techniques that are feasible and consider cultural traditions.

Even though women have tried to solve their problems by forming groups, some improvements can be made on their performance and contribution to agricultural production. Some of these could emphasise biological rather than structural measures. Women have coped more easily with biological measures for soil conservation than the manually constructed structural measures.

References

Anderson, M.B., Cloud, K. and Austin, J.E. (eds) (1984): Gender Roles in Development Projects. Connecticut: Kumarin Press, West Hartford, 185-211.
Boserup, E. (1970): Women's Role in Economic Development. New York, St. Martins press; London George Allen and Urwin Ltd. 215-225.
Government of Kenya. Sixth Development Plan. (1986-1990): Government Printers, Nairobi, Kenya.
Hornik, R.C. (1988):. Development Communication. New York: Longman.

KWAP (1990): Sociological Issues in Agroforestry Extension. KWAP Lecture notes delivered to soil Conservation Agroforestry National Seminar, Kakamega, Kenya.

MOA/ Management Consultants and SIDA (1993): Socio-Economic Impact Assessment of Soil Cconservation Activities in Murang'a District. 1993. MOA/SIDA, Nairobi, Kenya.

Morna, C.K., Ephson, B., Quattara, S. and Topouzis, D. (1990): Women farmers emerge from the shadows. In: African Farmer. A key to the future #3.

Östberg, W. (1987): Ramblings on Soil Conservation. An Essay from Kenya. SIDA Stockholm.

The World Bank. 1994. Kenya Poverty Assessment, Report No. 13152-KE.

Weidemann, C.J. (1985): Extension systems and modern farmers in developing countries. Agriculture and Human Values, vol. 2 # 1, 56-59.

Young, A. (1989): Agroforestry for Soil Conservation. C.A.B. Int. U.K. Wallingford.

Address of author:
Loice M.A. Omoro
Ministry of Agriculture
Lnad Development Division
Soil and Water Conservation Branch
Box 30028
Nairobi, Kenya

Sustainable Land Use by Women as Agricultural Producers? The Case of Northern Burkina Faso

D. Kunze, H. Waibel & A. Runge-Metzger

Summary

The objective of a field study conducted in Bam Province of Burkina Faso from 1992 to 1994 was to identify the influence of women's access to land and participation in new, resource-conserving land management technologies on their agricultural production.

Based on farm surveys of 89 households women's land management practices and their contribution to total household production were analysed. Women's labour force invested in agricultural production on family and individual plots, women's labour employment in resource conservation and benefits from resource conservation on women's fields were assessed. Regarding land use arrangements the study shows that traditionally land allocation is based on collective, not individual use rights. Under this arrangement women in principle have the same access to land as other individual land users such as young unmarried men. Based on the results of the survey it appears that women receive land of less quality in terms of distance to home and less often plots with conservation works. There is however no discrimination concerning access to the most fertile lowland. Women's benefits from ameliorated family fields depend on the households sharing rules which are not yet fully understood.

In summary, results of the study suggest that the sustainability of land use systems is not so much a question of access to land as determined by either traditional or modern land tenure arrangements, but rather of intra-household land allocation rules.

Keywords: resource conservation, women, rock bunds, soil erosion, land use, sub-sahara

1 Introduction

Applying the concept of sustainable agriculture in the sub-sahelian region does not only require satisfying changing human needs but also rehabilitating the quality of the environment and conserving natural resources. A number of techniques in the fields of land reclamation, soil and water management, plant genetic resources, cropping practices and pest management are being proposed to contribute to sustainable land management (Reardon, 1995; Meerman et al., 1996).

Past experiences of soil and water conservation projects have shown that resource conservation cannot be addressed from a purely technical perspective without taking the social process and its actors into consideration (Marchal, 1986; Atampugre, 1993).

Women, who play a major role in western African agriculture, are often neglected in development activities (Savané, 1986). At the same time it is suspected that due to male labour migration, women will have to contribute increasingly to soil and water conservation activities (Monimart, 1989; Esser-Winckler and Sedogo, 1991; Howe, 1996).

Project intervention in this field is sometimes found to augment women's work load (Brasser and Vlaar, 1990). Traditional and modern land use rights are found to become a barrier to the full use of women's agricultural production capability (Monimart, 1989; Boserup, 1990; Birba, 1993). Time use, resources and responsibilities are unequally distributed between men and women (Goheen, 1988). It has been argued that women's social status forbids the mobilising of labour force, direct personal intervention and decision (Esser-Winckler and Sedogo, 1991; IFAD, 1992).

The introduction of resource conservation measures in Bam Province of northern Burkina Faso in the late 1980s as one step towards more sustainable land management has until today enjoyed a high degree of farmer participation. The question is raised if tenure arrangements influence women's land management practices including resource conservation.

This article explores several facets of women's role as agricultural producers and its possible implications for resource conservation measures. The allocation of land and labour, distribution of income, and investments in resource conservation within the context of gender-based responsibilities will be described in order to estimate the impact of women's role in sustainable agricultural production.

2 Theoretical considerations and hypotheses

Based on economic theory the household allocates labour in such a way as to maximise total household income. Labour can be allocated to the family field, individual fields and other activities including leisure and household work. In Figure 1 total family income (O_1) is composed of individual income, i.e. controlled by the individual household member (women) and income from family field controlled and distributed among other members by the household head. It is assumed that labour input on the family field is determined by the marginal product of women's labour on her individual field (w_1) thus resulting in labour input on the family field of l_1. For simplification, constant marginal returns to labour on women's field are postulated.

Land improvement on the family field such as rock bunds is expected to shift the production function and increases total family income from O_1 to O_2. Consequently optimal labour allocation within the family will shift towards a higher labour allocation to the family field. Women's labour input on the family field will increase to l_2. Considering women's indifference curve for labour (on her own field) and leisure we can show that the level of income which is directly controlled by women will decrease as indicated in the difference $W_1W'_1$ compared to $W_2W'_2$ after rock bund construction. Assuming a constant total labour input on individual and family fields, economic theory suggests that women become more dependent on the goodwill of their husband and household head regardless of existing land allocation rules.

The following questions were posed in this study :
- are women discriminated in land allocation in terms of quality and legal status of land?
- do women invest lower amounts of "external" inputs as compared to men?
- do women invest larger amounts of labour in agricultural production and resource conservation as compared to men?
- is women's share of households crop production and income important as compared to men's?
- do women benefit from family fields?
- does women's individual production decrease after resource conservation ?

A few of these questions will only be treated theoretically, because of missing before/after comparison. However some indicators can be derived which provide hints regarding implications of more sustainable land use practices for women's production.

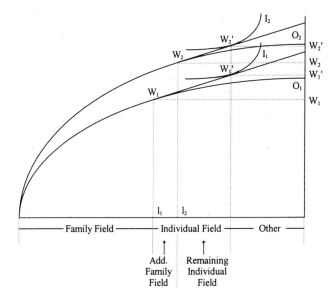

I_1, I_2	=	Preferences for labour and leisure
l_1	=	optimum labour input <u>before</u> innovation on family field
l_2	=	optimum labour input <u>after</u> innovation on family field
O_1	=	Total family income <u>before</u> innovation
O_2	=	Total family income <u>after</u> innovation
W_1, W_1'	=	Income from women's field <u>before</u> innovation
W_2, W_2'	=	Income from women's field <u>after</u> innovation
Assumptions:	-	constant marginal returns to labour on women's field
	-	constant total labour input to family and individual field

Figure 1: Allocation of women's labour time in rainy season with and without rock bunds

3 Methodology and source of data

This study is based on farm survey data from 1993/94 collected in the frame of PATECORE (Projet Aménagement des Terroirs et Conservation des Ressources dans le Plateau Central), a German-sponsored resource conservation project. The stratified random sample includes 89 households in 6 villages. In addition to the sedentary Mossi farmers in 5 villages, the sample included one village of Fulani households, who are traditional herders, but have now settled in the region.

Plant production income includes all harvest which was estimated by each individual cultivator in terms of local units. The threshed grain was weighed for each household. Farmers' estimates offer a reliable data base as comparative studies have provided evidence that this method produces valid production data compared to the standard crop-cut method (Murphy et al., 1991).

Crop gross margins are calculated as total crop value subtracting costs for seed input, fertiliser and pesticides. Time allocation in plant production was estimated per plot on 475 plots by each of the cultivators of a sub-sample of 36 households (10 Fulani households, 26 Mossi households). Labour input in resource conservation was analysed from project data of 30 construction sites.

Prices are based on the Kongoussi market, the main market of Bam Province, during the dry season for the subsistence part of harvest, the sold amount was valued at individual farm-gate prices. All monetary units in FCFA refer to the equivalent of 1 US$ = 294 FCFA before devaluation in January 1994.

Data processing was done using as SPSS statistics package. In case of qualitative or discrete variables Chi² as a non-parametric test of significance was used and Cramer's V as a chi²-based measure of association was computed in case of significance at the 0.01% level.

4 Results

This study wants to provide indicators for women's role in agricultural production with regard to three categories:
- access to land,
- allocation and distribution of inputs, especially labour, and
- the composition of total production and income.

During the course of analysis, family fields are compared to women's and men's individual fields. This approach differs from other studies which regard family plots as men's plots (Udry et al., 1995). Decisions referring to family fields are mostly taken by the household head, who is usually male. On the individual plots decisions are made by the respective user.

4.1 Access to land

Generally Mossi as well as sedentary Fulani women have the right to claim a piece of land for their own use from their husbands' household. In case of land scarcity at household level the claim will be going to the husband's larger family.

In accordance with local land tenure arrangements, women in principle have equal access to land as compared to other household members such as young unmarried men. Nevertheless they have little influence on the actual land allocation decision since it is usually the household head, but in any case a man, who distributes agricultural land. The quality and quantity of land at their disposal reflects their relation to the decision maker as a result of a bargaining process.

Quality of land has two dimensions, namely a locational dimension, in terms of the distance from the homestead and a soil quality dimension referring to soil type, slope and texture. An additional indicator of land quality is its ownership status.

Table 1 indicates that the majority of family fields are in close distance to the house thus facilitating accessibility and lowering transportation costs. On the other hand, women's fields are more likely to be away from the homestead despite the fact that they have to take care of post-harvest operations and other household work. Because of their distance to the homestead, women's fields are less likely to be managed intensively in terms of "external" input use. Young men's fields are mostly bush fields as their opportunity costs of time are lower.

Little difference can be observed with regard to soil quality as the majority of the plots are on sandy soils and only a small proportion are on lowland areas. No significant differences between user groups could be demonstrated in respect to land use rights referred to as "rented". The majority of the fields are found to be under hereditary possession meaning unlimited rights of usage. A survey in four provinces of Burkina Faso found this to be equally high with 85.7% (Ouedraogo et al., 1996).

Applying a Chi² test for the distribution of fields among the different users by quality criteria did not prove to be statistically independent for distance and soil quality parameters, but a low degree of association measured by Cramers' V implies only little dependency. The percentages given in Table 1 do not confirm the common perception that women have less access to the most fertile lowland.

	women's plots	men's plots	family plots	stat. testing
Distance:				Chi² ***, 0.21
Close to home	12.6	21.1	39.0	
Village fields	53.4	36.7	34.7	
Bush fields	34.0	42.2	26.3	
soil type:				Chi² ***, 0.16
Lowland	11.6	16.4	12.8	
Clay-sand	12.0	12.3	15.0	
Clay-loam-sand	2.6	6.7	11.9	
Sand	56.1	40.5	37.5	
Ferrallitic	17.9	24.1	22.9	
Rented[a]:	10.2	11.6	8.5	n.s.
	n=765	n=199	n=510	

stat. test: *** significant at the 0.01 level of Chi², Cramers' V; n.s. not significant.
[a] not in the sense of tenancy, without payment.
Source: Farm survey 1993/94.

Table 1: Distance, soil type and legal status of fields by type of field in %

In summary, results indicate that women are being discriminated in land allocation in respect to distance, but not to soil quality and legal status. Nevertheless, location theory suggests that when the transportation infrastructure is poor plots nearer to the homestead carry a considerable land rent making them more valid for melioration investments. The crucial question is whether this has any negative impact on the management of women's own fields and to what degree women participate in the additional benefits derived from the family fields.

4.2 Allocation of inputs for crop production including labour

Labour input in plant production is divided between family and individual plots. In a sub-group of 36 households and 475 fields labour input of 164,337 working hours in total is found to be distributed (Figure 2) between the different age and gender groups. It is shown that women almost reach the share of men in field work despite their regular household tasks.

Regarding labour input by type of field, it was found that family fields receive about 560 hours per hectare, which is roughly 20% more than women's fields (Figure 3). Men's fields receive only slightly less than family fields. Although labour input on women's field might be slightly underestimated because women sometimes receive help from their children, the difference nevertheless remains. Figure 3 also shows a slightly higher male than female labour input on family fields.

Individual fields are mainly cultivated by the user, supported by the help of the opposite sex. According to traditional labour division, men do the ploughing whereas women help with sowing, weeding and harvesting on men's plots. Women's labour input on men's individual plots is twice as high as men's support as indicated in Figure 3.

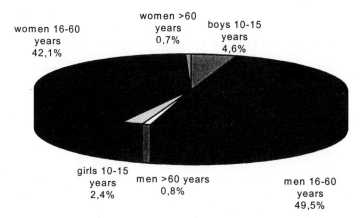

Source: farm survey 1993; total working hours of field work in 36 households.

Figure 2: Distribution of agricultural labour between age and gender

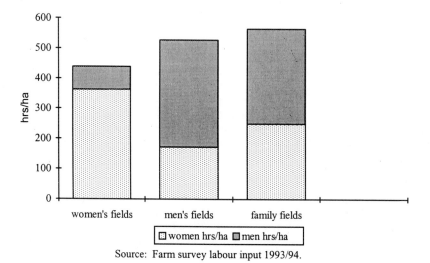

Source: Farm survey labour input 1993/94.

Figure 3: Labour input per hectare according to field user and workers gender 1993/94

Labour input in construction and maintenance of rock bunds usually takes place during the dry season. Data from 30 construction sites suggested significant differences in men's and women's labour participation according to the type of construction. While labour input at the construction of rock bunds, the measure with the lowest input, was equally distributed between men and women, men contributed a higher percentage of labour days for larger types of construction such as dikes and dams (Kunze, 1996). Labour time proved to be rather equally distributed between the two gender groups. This is contrary to the suspected sole responsibility of women by earlier authors (Esser-Winckler and Sedogo, 1991; IFAD, 1992).

As suggested by location theory it appears that the use of external inputs and other time consuming cultivation practices are lower on women's fields (Table 2). Seeding in rows or mulching with straw as well as applying inputs that require cash such as mineral fertiliser or pesticides is only practised on a small proportion of women's fields unlike on men's plots and especially family fields. Access to manure from the stable could be more difficult for women than men and bringing out manure is a time consuming activity. Resource conservation works are only very seldom built on women's fields, which confirms earlier results (Kunze, 1994). Compaoré (1994) found that women lack decision-making power regarding resource conservation practices.

	women's plots	men's plots	family plots	stat. testing
(%) of fields:				
Pesticides used	8.1	14.6	22.9	Chi² ***, 0,19
Seed rows	8.9	25.8	31.5	Chi² ***, 0.27
Miner. fertiliser	2.4	8.0	15.3	Chi² ***, 0.23
Manure	1.2	4.0	28.6	Chi² ***, 0.41
Straw mulch	0.1	1.0	3.3	--
Rock bunds	0.5	1.5	18.6	--
n=	765	199	510	

stat. test: *** significant at the 0.01 level of chi², Cramers'V, -- cells with less than 5 cases.
Source: Farm survey 1993/94

Table 2: Land management and crop production practices on women's, men's and family fields in 1993/94

4.3 Composition of household production and income

Plant production in the sub-sahelian region usually consists of growing cereals (sorghum, millet, maize, fonio (Digitaria exilis)) and cowpeas as the main staple food and minor crops as ingredients for the accompanying sauce. Cotton is grown for household use and market production but its area planted is declining.

	women's plots		men's plots		family plots	
		%		%		%
Households monetary crop income in FCFA[a]						
Mossi	22.100	16,5	13.800	10,3	98.100	73,2
Fulani	260	0,7	950	2,5	36.600	96,8
households cereal harvest[b] (kg)						
Mossi	230	10,6	210	9,7	1720	79,6
Fulani	10	1,5	18	2,7	632	95,8

[a] FCFA currency in January 1994 US$ 1 = FCFA 294 before devaluation.
[b] cereal harvest from sorghum, millet, maize.
Note: Number of observations is 1474 plots in 74 Mossi and 15 Fulani households.
Source: farm survey 1993/94.

Table 3: Distribution of plant production income according to ethnies and gender in 1993/94

The share of total household income in terms of total value of production which is derived from women's fields amounts to 16.5% while that from men's plots is approximately 10% for Mossi farmers. As Table 3 indicates, Fulani farmers' individual production is generally very low. In cereals Mossi women contributed 10.6% and men 9.7% with their individual production to total household production, for Fulani households individual shares were below 3% (Table 3). In cereal production gender differences are negligible, because men dominantly plant cereals on their individual fields, while women additionally plant groundnuts and bambara beans.

The difference in pattern and relative importance of agricultural production between Mossi and Fulani is explained by the fact that even sedentary Fulani have a lower preference in plant production than Mossi people. They rely on this activity only for urgent subsistence needs and are able to sell livestock in case of insufficient harvest or cash requirements.

5 Conclusions

In summary, women receive land of lower quality than man in terms of distance to their fields, use less external inputs and supposedly yield stimulation crop production techniques and yet have a higher share in the total household income derived from their individual fields. In addition women contribute about half of the total labour input in construction of rock bunds which dominantly takes place on family land controlled by the head of the household. This suggests that women can compensate additional tasks by apparently closing up any remaining efficiency gaps or further reducing their leisure time.

As claimed by some authors (Monimart, 1989; IFAD, 1992), although women do not benefit from resource conservation measures on their personal fields, they benefit from ameliorated family fields. This of course entirely depends on the households sharing rules.

On the other hand, this study showed that although women's role in agriculture is important in terms of labour input (not taking into account opportunity costs of labour in the dry season) and individual production, women's production nevertheless is not dramatically higher than that of men as in other African societies (Howe, 1996). Further data analysis has to be carried out to investigate whether tasks given to women in resource conservation causes adjustment in crop management in their personal fields that would lead to less resource-conserving practices.

Considering population growth, land will become even scarcer thus increasing the danger of more rapid resource depletion, which will increase the demand for compensatory measures such as rock bund construction. Women's lack of power in land allocation decisions poses a potential threat for the traditional institutional arrangements where women largely rely on the income from their individual fields. Enhancing the production activities where women have direct control, such as their individual or women's groups production, could be one step towards sustainable land use. Taking into account their contribution in resource conservation makes it mandatory to include women training programmes in order to improve their understanding of the relationships which drive sub-saharan agro-ecosystems.

Even though the study could not find women to be disadvantaged in terms of land use rights at present, changes in this field have to be followed closely in the future since land registration is on its way in most western African countries.

References

Atampugre, N. (1993): Behind the Lines of Stones; Oxfam Publ., Oxford.
Birba, T.A. (1993): Le droit foncier traditionnel; évolution et impact sur les mesures CES/AGF; CIEH/PATECORE; Burkina Faso.
Boserup, E. (1990): Economic and Demographic Relationships in Development, Baltimore and London.

Brasser, M.B., J.C.J. Vlaar (1990): Aménagement de Conservation des Eaux et des Sols par Digues Filtrantes, CIEH, Ouagadougou, Burkina Faso.

Comparoré, P. N. (1993): Les incidences du PATECORE sur la Situation Socio-Economique des Femmes dans la Province du Bam; Dept. de Géographie, Uni. de Ouagadougou, Burkina Faso.

Esser-Winckler, H., Sedogo, M. (1991): "Ressourcenschonende Bewirtschaftung auf dem Zentralplateau" in Burkina Faso, Band I, GTZ.

Goheen, M. (1988): Land and the Household Economy: Women Farmers of the Grassfields Today. In: Davison, J. (ed): Agriculture, Women and Land; the African Experience; Westview, Colorado

Howe, G. (1996): Ingegrated Community Driven Approaches; Africa Division, Project Management Dept., IFAD, Rome.

IFAD (1992): Soil and Water Conservation in Sub-Saharan Africa; Rome.

Kunze, D. (1994): Typologie des Systèmes d'Exploitation dans la Province du Bam/Burkina Faso, Rapport Final, PATECORE/Univ. Göttingen.

Kunze, D., Runge-Metzger, A. & Waibel, H. (1996): The Economics of Erosion Control Innovations in Northern Burkina Faso; Paper presented at the International Symposium: Food Security and Innovations; March 11--13, 1996, Hohenheim/Stuttgart, 57-71.

Marchal, J.-Y. (1986): Vingt ans de lutte antiérosive au nord du Burkina Faso, Cah. ORSTOM, sér. Pédol., vol XXII, no.2, 173-180.

Meerman, F., van de Ven, G.W.J., van Keulen, H. & Breman, H. (1996): Integrated crop management: an approach to sustainable agricultural development; Intern. J. o. Pest Managem. **42 (1)**, 13-24.

Monimart, M. (1989): Femmes du Sahel, Karthala et OCDE/Club du Sahel.

Murphy, J., Casley, D.J. & Curry, J.J. (1991): Farmers' Estimation as a Source of Production Data; World Bank Technical Paper Number 132, Africa Technical Dept. Series, Washington.

Ouedraogo, R.S., Sawadogo, J.P., Stamm, V. & Thiombiano, T. (1996): Tenure, agricultural practices and land productivity in Burkina Faso; Land Use Policy, Vol. 13, (3), 229-232.

Reardon, T. (1995): Sustainability Issues for Agricultural Research Strategies in the Semi-arid Tropics: Focus on the Sahel; Agricultural Systems, No. 48, 345-359.

Savané, M.A. (1986): The Effects of Social and Economic Changes on the Role and Status of Women in SSA, in: Moock, J.L.: Understanding Rural Households and Farming Systems, Boulder.

Udry, Ch., Hoddinott, J., Alderman, H. & Haddad, L. (1995): Gender differentials in farm productivity: implications for household efficiency and agricultural policy; Food Policy, Vol. 20, No. 5, 407-423, Great Britain.

Addresses of authors:
Dagmar Kunze
Am Reisenbrook 20d
D-22359 Hamburg, Germany
Hermann Waibel
Institut für Gartenbauökonomie
Universität Hannover
Herrenhäuser Str. 2
D-30419 Hannover, Germany
Artur Runge-Metzger
European Union
DG8
Brussels, Belgium

Rural Women and Conservation of Natural Resources: Traps and Opportunities

Swarn Lata Arya, J.S. Samra & S.P. Mittal

Summary

Soil and water conservation measures change the state of natural resources, especially that of water, soil and vegetation which in turn, have different consequences for women and men. Planners of soil and water conservation programs influence the position of women with their 'gender ideology' via assumptions which determine the structure and content of programs. Without understanding the mechanism in 'gender ideology', the use of natural resources, soil erosion and the position of women, even a friendly effort can result in increased pressure on farm women. This fact came into light during a study of integrated watershed management projects undertaken in a few villages located in the foothills of Shiwaliks in Northern India.

Keywords: Gender issues, natural resources, integrated watershed management, Shiwaliks, common property resources, rainwater harvesting, supplemental irrigation, Human Development Index, Gender Related Development Index, soil erosion, Hill Resource Management Society.

1 Introduction

The Shiwaliks are low mountains situated just below the Himalayas in North India, stretching west from Nepal to the Pakistan border. While they had thick lush green forests untill the middle of last century, today the Shiwaliks are amongst the most denuded hills in the country due to the geology coupled with excessive deforestation. It has been estimated that in these highly grazed Shiwalik hills, 6cm of top soil representing nearly 2400 years of local ecological history, can disappear in one year (Aggarwal et al., 1987). Alongside this process of deterioration of natural resources, has been the privatisation of Common Property Resources (CPR) which has favoured large farmers. CPRs account for as much as 25% or more of total income in these villages (Arya et. al. 1994). CPRs supply 85-100% of firewood and meet 60-80% of grazing needs of the landless and the poor.

As a result women and female children bear the main burden of this deterioration and decreasing access to CPR within poor rural households. As the main gatherers of fuel, fodder and water from faraway places (5 to 8 km), their work day has lengthened (Arya, 1995; Arya & Samra, 1995).

In all, six villages were selected for detailed study. These were Bunga, Dhamala, Sukhomajri, Gobindpur - Mandappa, Asrewali and Tibbi in Ambala district of Haryana state. In the first four villages, the watershed management program was implemented since the 1980's. The remaining two were the control villages, studied for the purpose of comparison.

2 Watershed management program

Watershed management program which attempt to conserve natural resources tend to optimize farm and forest production and therefore were supposed to help in reducing the drudgery of farm women.The projects focussed on rainwater harvesting by constructing small earthen dams and recycling stored water through a system of underground pipelines to the adjoining farm lands through gravity. Land-levelling and land - shaping operations were also carried out to conserve rainwater *in situ*. Simultaneously, improved agronomic practices were introduced to maximize the return. The cropping intensity in project villages tripled in some cases. The main crops of the area during winter are wheat (*Triticum aestivum* L.), gram (*Cicer arietinum* L.) arhar (*Cajanus cajan*) and during summer, maize (*Zea mays*), ground nut (*Arachis hypogea* L.) and bajra, (*Pennisetum typhoidium*). In addition to wheat, the irrigated crops being taken after the projects are paddy (*Oryza sativa*), sugar cane (*Saccharum officinarum* L.), and berseem fodder (*Trifolium alexandrinum*). There was a two-to four-fold increase in agricultural productivity as a result of supplemental irrigation (Arya and Samra, 1995). This in turn, helped in linking the economic interests of the villagers to the protection of adjoining forests and preventing their further degradation. Due to increased availability of fodder, both from fields as well as from forest, the dependence of villagers on forest was also reduced. To encourage cooperation and participation of villagers for protecting forest areas and maintenance of dams and pipelines and distributing irrigation water, Hill Resource Management Societies were formed in these villages.

3 Data collection and analyses

The data from six villages were collected by personal interviews for the agricultural years 1993-94 and 1994-95 with the help of a structured schedule. The information was collected on various activities particularly those in which women participated, and their contribution in these activities was worked out using simple tabular analyses. Human Development Index (HDI) and Gender Related Development Index (GDI) were computed as per the methods described by Kumar (1996). HDI was used to measure a society's average achievements in human capabilities whereas the GDI relates the male-female disparity to overall achievements in a society (Kumar, 1996). For computing GDI, the following three indices were calculated (i) equally distributed index of life expectancy (ii) the equally distributed index of educational attainment and (iii) the equally distributed index of income. The GDI is the average of these three indices and has a value ranging from 0 to 1.

4 Results and discussion

The average farm size in selected villages varied from 0.7 to 2.2 ha. The average number of animals per household varied from 3.8 to 17 in project villages whereas it was 74 and 29 in Aasrewali and Tibbi respectively (both non-project villages). Share of female population was 44 % in project and 47% in non-project villages. The female to male ratio was lower in both the project (0.79) and non-project villages (0.87) than the all-India average of 0.93. There were also gender differences in access to education.

4.1 Participation of women in different activities

The struggle of farm women is not only for food, but also for fodder, firewood and water. Deforestation over the years and possible subsequent lowering of the ground-water table forced the women to spend more time gathering fuelwood, and at the same time fetching drinking water from far-off places. The pattern of time expenditure on these activities changed after the watershed management project (Table 1). Although there was a reduction in time spent on grazing, collecting fuelwood and water by 264 % due to afforestation and rehabilitation of hills, yet the burden on farm women related to agriculture and animal husbandry works increased by 469% which further compelled them to pay less attention to self and child-care and social and leisure activities.

Particulars	Time spent		% Increase/ decrease
	Non-project Area	Project area	
Farm work	434	1078	(+) 148
Transporting fodder	218	495	(+) 127
Grazing animals	512	220	(-) 132
Animal husbandry (excluding grazing)	191	562	(+) 194
Collecting fuelwood and fetching water	661	285	(-) 132
Household chores	1360	1450	(+) 7
Child care and personal care	547	442	(-) 24
Miscellaneous (social & leisure time, knitting carpet, and cot weaving)	657	498	(-) 32
Total	4580	5030	(+) 11

Table 1 : Time spent by farm women on various activities (hrs/yr).

Activities	Time spent (hrs/yr)			
	Non-project area		Project area	
	Women	Men	Women	Men
Sowing	24.1	33.2	50	64
Transplanting	-	-	61	34
Fertilizer application	6	22.4	18	29
Irrigation	-	-	-	33
Weeding & other intercultural operations	60.4	71.2	128	116
Post harvest operations	72	121.6	160	184
	162	248.0	418	460

Table 2: Relative contribution of farm women and men in agriculture (per acre).

4.2 Relative contribution of men and women

Table 2 reveals that the female contribution to agricultural operations in project villages increased by 157 % whereas it was only 85% for males. Multiple-cropping systems and the introduction of fodder crops enhanced the involvement of rural women in agriculture. As fodder availabilities increased both from the fields and forests, the cattle migration decreased. The composition of animals also changed in favour of more productive cows and buffaloes which are mostly stall-fed in project villages. All these changes increased the female labour utilization by 38 % in project areas. Ratio of labour utilization in project compared with non-project villages was 0.96 and 1.39 for male and female labour respectively.

The time allocation pattern showed that the burden of ecological change has fallen disproportionately on females. The increased participation did not raise their access to and control over natural and domestic resources. An indepth analysis of earnings and spending carried out in Bunga village before and after the project period revealed that both earnings and spendings were collected and controlled by men only. Men usually indulged in enhanced liquor consumption and tended to overspend. Liquor consumption per household per annum increased from Rs.1710 in 1983-84 to Rs.4300 in 1993-94. Whenever women had any cash or kind it was used to fulfill family's basic needs, while men spent a part of their earnings on tobacco and liquor. Women were never involved in the activities of Hill Resource Management Societies from their inception in 1984-85. The consent of female members is always taken for granted.

GDI rank/villages	Project villages				Non-project villages		All India average
	Bunga	Dhamala	Sukho-majri	Gobindpur Mandappa	Aasrewali	Tibbi	
1. Gender related development index (GDI)	0.377	0.392	0.413	0.339	0.233	0.294	0.388
2. Equally distributed index of life expectancy	0.634	0.582	0.607	0.569	0.486	0.527	0.569
3. Equally distributed index of educational attainment	0.304	0.463	0.447	0.302	0.144	0.242	0.444
4. Equally distributed index of income	0.193	0.131	0.185	0.148	0.070	0.094	0.152
5. Human development index (HDI)	0.40	0.42	0.431	0.361	0.249	0.322	0.423
6. (HDI-GDI)/HDI(%)	5.98	7.76	4.18	6.09	6.42	8.69	8.2

Table 3: Gender related development index for sample villages.

4.3 Gender related development index (GDI)

The results revealed that values of both GDI and HDI were higher in project villages as compared with non-project villages (Table 3). The values of (HDI-GDI)/HDI show the percentage reduction of GDI from HDI. These values revealed that gender inequalities were more or less equally pronounced in some of the project as well as in non-project villages. It reinforces the fact that gender inequalities may not be correlated to income levels in any predictable manner. It also points to the anti-female bias and systematic deprivation of women in project areas.

In brief, there are five factors that have deprived women of the natural resources they need:
1. Resource depletion
2. Changing family structures
3. Introduction of new technologies
4. Development projects that fail to target women
5. Lack of access to crucial means of production such as land, credit and associated production technology.

In soil and water conservation programs, solutions are sought for soil erosion. Such solutions are the result of a series of decisions that are made at different levels and stages of a program. All decisions include certain assumptions based on the norms and values of planners. When these assumptions are incorrect, not only the success of a program is affected but also the land users, both men and women, in the area of interaction. To understand the effects of a program on women it is necessary to examine the decisions that are made during the course of a program and the assumptions on which these are based. Thus, before preparing any program on natural resource management, the following questions need to be answered.
1. What are the main natural resources involved in determining soil erosion?
2. a) Do women have access to and control over these natural resources?
   b) Does this differ from the access to and control over these natural resources than for men?
3. a) How do women use these natural resources?
   b) Does this use differ from that of men? If yes, why is that so?
   c) What is the effect of such differences in use on the occurence of soil erosion?
4. a) How are women affected by the deterioration of natural resources due to soil erosion?
   b) Are women affected differently than men ? If yes, why is this so?
5. a) What specific knowledge do women have about the management of natural resources?
   b) Using this knowledge how do women adapt to change in the state of natural resources?
6. a) Which stages and components of soil and water conservation programs can be discerned?
   b) What is the importance of technology in these programs?
   c) On what assumptions, which influence the position of women, are these programs based?
7. a) Is the interaction between women, soil erosion and soil and water conservation incorporated in the structure and content of these programs?
   b) What effect has this had or will it have on the position of women and on the nature and extent of soil erosion?

Indeed what is at issue today is a changing entire development model with its particular products and technological mix, its forms of exploitation of human and natural resources and conceptualisation of future strategies.

To find an alternative is of course not very easy. What is clear so far are its broad contours: that it needs to be transformational rather than welfare where development redistribution and ecology link in mutually regenerative ways. According to Aggarwal, (1989); we need to go beyond the demand for a mere redistribution of the "loaf" (or of entitlements to it) to a change in:
- the "loaf's" composition
- the technologies used to produce it,
- the processes by which a decision on its composition and its production technology are arrived at,
- the knowledge through which it is produced,
- the form by which nature's resources are appropriated,
- its very size

References

Aggarwal, A., Chopra, R. and Sharma, K. (1987): The State of India's Environment: The First Citizen's Report. Centre for Science and Environment, New Delhi, India

Aggarwal, B. (1989): Rural women, poverty and natural resources: Sustenance, sustainability and struggle for change. Economic and Political Weekly, October 28, WS46-WS65.

Arya, Swarn Lata, Agnihotri, Y. and Samra, J.S. (1994): Watershed management: Change in animal population structure, income and cattle migration, Shiwalik, India, Ambio **23**: 446-450.

Arya, Swarn Lata (1995): Soil and water conservation programme: Farm women burdened by development initiatives. Indian Farming **44**: 27-30.

Arya, Swarn Lata and Samra, J.S. (1995): Socio-economic implications and participatory appraisal of integrated watershed management project at Bunga. Technical Bulletin T-27/C-6, Central Soil and Water Conservation Research and Training Institute, Research Centre, Chandigarh, India.

Kumar, A.K. (1996): UNDP's Gender related development index. A computation for Indian States. Economic and Political Weekly. April 6, 887-895.

Address of authors:
Swarn Lata Arya
J.S. Samra
S.P. Mittal
Central Soil and Water Conservation Research and Training Institute
Research Centre
Sector 27-A
Madhya Marg
Chandigarh, 160019 India

Land Tenure and Land Management: Lessons Learnt from the Past, Challenges to be Met in the Future?

M. Kirk

Summary

After decades of neglect, the key role of land tenure for sustainable land use, environmental protection, socio-economic development and political stability is meanwhile undisputed. There is no doubt, that an unequal distribution of land and growing land scarcity affect young families, squatters, women, pastoralists or indigenous peoples. Their problems, however, are embedded in a more extensive interrelationship between land tenure and land management which create challenges for the future:

1. In Asia new threats are predictable through the dramatic conversion of land, land grabbing and new competition about its best use and through problems how to secure long-term investment and soil protection if no longer "access to land" but "access to income" is the future demand.
2. In Latin America the neo-liberal miracle to give the masses access to land via viable land markets did not occur. Accordingly the rebellion of marginal groups is entering a new, militant phase. If recent trends of rainforest conversion due to settlement into "open spaces" persist, ecological degradation, diminishing biodiversity and further global change are most likely.
3. The global land tenure crisis has already reached Africa, with increasing landlessness, insecure tenancy, eviction and extremely violent conflicts. Too many governments still ignore the role of customary law, of a minimum legal and regulatory framework and allow rent-seeking, corruption and land grabbing.
4. In transformation countries private ownership is not at all the silver bullet for access to credit, investment and resource preserving production.
5. In industrialized countries the presumption of an absolute right to produce food and fiber on private lands created an open-ended agricultural policy in which the state has become a captive of the sanctity of private land rights.

Keywords: Land tenure systems, land management, land policy, agrarian structure, tenancy agreement, land market, land conflict, land reform, autochthonous rights, indigenous peoples, developing countries, transforming economies

1 The land question reconsidered: old wine in new bottles?

Will land tenure problems really matter in the future? Are there not serious reasons such as scarce budgets, limited government implementation capacity as well as donor money and applied research that speak against concentrating on these specific problems?

Boserup (1965) wrote 30 years ago that population growth and land scarcity lead to intensification and changed land management. Moreover, she explained very clearly that this process is accompanied by a specification of property rights in terms of individualization and privatization. Therefore, theoretical concepts regarding the driving forces of change in land tenure and the orientation of this change are available to us.

With economic development, the contribution of agriculture to Gross Domestic Product and employment decreases. New leading sectors such as service industries or micro-electronics work nearly independently of land. Which relative importance will problems of land tenure and land management still have in industrialized countries in the future?

Even in agriculture (e.g., livestock production), technological progress causes increases in production to be less and less dependent upon land as a means of production. The dissemination of biotechnology in Asian agriculture will give an additional push in that regard and, therefore, winners and losers might be the same as in the first Green Revolution. Thus, it is above all technology which should win the race between food production and food demand. Social and economic institutions such as land tenure, must accordingly only be adapted in an optimal way.

Of the political debate of the FAO World Conference on Agrarian Reform in the 1970s, only a faint rumbling has remained. Technical terms underwent a miraculous metamorphosis from "land reform" over the broader concept of "agrarian reform", to the politically correct euphemism of "property rights".

The policy implications seem to be clear. The neo-classical demand to "get prices right" is doubtlessly valid for distorted land markets in order to provide incentives for long-term investment. In cases of badly defined property rights and a lack of access to land markets for millions of peasants, tenants, landless people, squatters or refugees, this policy advice comes up against limiting factors.

The call from 'New Institutional Economics' is to establish clearly defined property rights and to set up the legal framework. It is a necessary consequence for solving major land tenure problems, but it is not sufficient and not easy to implement, neither in Laotian shifting cultivation systems, nor in the *favelas* of Sao Paolo, nor in a landlord village in India, nor on Sahelian pastures, nor for former production co-operatives in East Germany.

Numerous interrelated serious problems and conflicts regarding land continue to exist:
- Resource degradation, which can be associated with the plundering of forests, land conversion, forced over-utilization of pastures and arable land or the decay of common property are progressing at an alarming rate.
- The way the state divests itself of land and the way it tries to satisfy land restitution claims are problems that have raised tempers in the recent transformation processes. Issues of sustainability and environmental protection have often been neglected in this debate.
- Land grabbing, land speculation, and environmentally destructive land use patterns of the new urban and old rural elite who benefit most from legal reforms, are a bone of contention worldwide.
- The empirical evidence, whether resource right regimes follow Boserup's predictions or whether they are an important constraint on sustainable agricultural intensification, is quite poor for large areas in the world (IFPRI, 1996).

Land tenure and land management still matter in very different scenarios so that a more detailed analysis related to the regions is urgently required.

2 Asia: will land tenure regimes cope with the ongoing rapid socio-economic change?

In Asia, land reforms are still a very controversial issue. Strictly egalitarian, redistributive but market-oriented reforms in Korea or Taiwan resulted in great increases in agricultural production and incomes. They are a cornerstone of the "East Asian Miracle" (World Bank, 1993). The other side of the coin in this success story is inadequate land management practices and massive environmental problems (Kuhnen, 1996): the bringing into line of land tenure with land use planning, especially urban planning, is still unsolved, the taxation of windfall profits from land speculation is insufficient, and preserving environmental goods such as clean water and landscapes for future generations have long been ignored, such as in rural Thailand (Panayotou ,1993).

In other regions such as in India or the Philippines, land reform has only experienced limited success and remains a ticking bomb. The reasons for this are weaknesses of governments in implementation and the countervailing pressure of influential landlords. Land reforms and decreasing farm sizes as an outcome of inheritance rules have only weakened the old peasant-landlord relationship without replacing it with new institutions which allocate land and settle conflicts.

Economists have already made their peace with sharecropping and bonded labor (Hayami and Otsuka, 1993). They will persist as imperfect, but functional arrangements for hundreds of millions of rural people. We already know much about short-term efficiency and equity issues of these institutions; we are still waiting for satisfactory answers to the environmental impact and the intergenerational efficiency of factor allocation within these controversial relationships. How can tenants be helped in improving their bargaining position? Or to put it the other way round, how can landlords be convinced that it is in their own interest to encourage soil-improving investments made by tenants? Without improvements of this kind, conflicts about land, which can only be postponed through migration, will be aggravated or they will be solved with weapons.

The concept of registered landownership as a result of land reform is not a country-wide panacea for optimal land use pattern, long-term investments and environmental protection (e.g. Laos) (Kirk, 1996). Whereas decision makers in African countries might learn from Asia about future challenges arising from tenancy arrangements, the Asian discussion on communal customary rights and decentralized, local co-operation for land use might also benefit from African experiences.

What is still required in many countries is the creation or the re-organisation of functional land markets and especially markets for tenancy rights to enable a smooth transfer of these rights according to owner preference. In most Asian countries, these markets are still very modest and imperfect, although land transfers are increasing rapidly as a result of the declining importance of agriculture and the phasing out of parts of the farming population.

Today, the reduction in farm size and new job alternatives have caused land to acquire a different meaning. While - as Kuhnen (1996) states - a generation ago the cry was "access to land" at the time of the land-to-the-tiller reforms, today the young generation wants "access to income" wherever it may come from. Remittances of migrant family members have already led to farm land of a poor soil quality being given up. This development, however, can be beneficial for the protection of endangered resources. Can direct transfers be a model for governments, NGOs or donors for environmentally sound land management?

With industrialization, the need for land for non-agricultural purposes is growing starkly. New and old land uses compete with each other: residential areas, industrial parks, mining and recreational areas, agricultural and forest land, nature reserves and water-protection areas. Are cost-effective, simple, flexible and easily accessible systems of land registration and land information available for supporting the land conversion?

To sum up: Land policy, land mangement and their linking with environmental policy are still neglected factors in the development equation of old, new and infant tigers and will have strong repercussions for them in future.

3 Latin-America: cemented unequal land distribution as a ticking social and environmental bomb?

The extraordinarily unequal bi-modal distribution of land, ongoing squatting and environmental destruction by marginalized users still persist after the failure of past land reforms (Thiesenhusen, 1995). Indigenous peoples in particular are affected by land conflicts. An increasing number of peasants are being forced to illegally cultivate marginal, fragile eco-systems so that investments in resource protection have been omitted. The result is degradation, erosion or permanent resource destruction.

So-called "open spaces" in the rain forest regions are utilized by governments as an outlet for new colonization, as in Brazil. This policy is a successful attempt at sneaking away from needed land reform and at postponing foreseeable social clashes. As a consequence, the rate of destruction and transformation into poor ranching pastures has been increasing steadily in recent years.

As past panaceas for the rural poor, such as "extension services", "tax reform" or "land reform", tend not to work by themselves, rural development was supposed to occur via the establishment of viable "land markets" in the past decade. Not surprisingly, it did not. New demands for agrarian reforms, as in Mexico, arise from the worldwide generalization of reforms towards open economies and free trade treaties (e.g. NAFTA) and more extended, complex interactions between agriculture, industry and services (Gordillo de Anda, 1993). However, some of them get bogged down from the beginning.

According to Thiesenhusen (1995), land markets became an excuse for policy makers not to invest in the rural poor. If land markets are to be supported, however, more titling and registration of farm property are required to allow its purchase and sale as well as for obtaining credit easily. The free functioning of the land market would also render the farming of large properties more uneconomic and cause *hacendados* to sell off to smaller proprietors.

These policies have not yet shown the salubrious results which they promised. All the more, it is questionable whether formal titling is the only way for *campesinos* to provide secure, negotiable property rights and to obtain credit, as more, for example, informal procedures practiced in Peru (Lastarria and Barnes, 1995). The neo-liberal miracle of diffusing the land question via land markets has not yet materialised.

Meanwhile, landless groups have raised their voices after years of quiescence. The Indian marches in Ecuador and Bolivia, the Chiapas rebellion in Mexico and the fights about the suspended land legislation in Brazil have all triggered off a more militant phase. There have been significant numbers of land invasions with a strong environmental impact. Much of the land leaches rapidly and, within a few years, peasants have had to move on to new land, thus destroying more of the forest. Alternatively, they have to migrate to *favelas* in the cities where defined rights of utilization or urban planning often do not exist. It would be easy to blame squatters for this environmental destruction, but that would be tantamount to blaming the victim (Thiesenhusen, 1995).

If recent trends persist in the future, ecological degradation and conflicts over land will increase, not only worsening the climate for foreign investment as was the case after the Chiapas revolution in Mexico. Genetic resources and biodiversity, which are completely undervalued at present, are going to be lost on a large scale. Negative climatic global changes are likely with follow-up costs attaining dimensions beyond one's imagination.

4 Africa: Sustainable land tenure and land management with or rather without the state?

The tenure crisis has already reached Africa, leading to increasing landlessness, insecure tenancy, evictions, and alarmingly violent local conflicts, even to civil wars which are rooted in conflicts over land (Bruce and Migot-Adholla, 1994; Kirk, 1998).

The core problem of this crisis seems to be above all a "tragedy of the state" and one of policy failure. Okoth-Ogendo (1997) stated for his country that "...Kenya has not been able, even after nearly fifty years, to arrive at a suitable property regime for the sustainable management of natural resources.". Shivji's (1997) statement as head of the land commission in neighbouring Tanzania is also discouraging: "...It is unlikely that the government on its own will move in the direction of democratising land tenure given vested interests within the state. A consistent and organised voice from civil society has to develop to take up the land issue."

After independence, almost all governments have completely failed to establish functioning land tenure systems for *all* citizens: for men and women, for the influential elders and for young innovative families or for agriculturists and for mobile pastoralists. Why is this so, and what are the consequences?

1) The complex interrelationship between autochthonous collective customary rights and statutory law is largely ignored. There was already a law <u>without</u> a central state. Why does the legislator and the land administration ignore the fact that land is still perceived both as a social space where people live and work, and that it is not perceived as a geographical space, measured by GIS and adjudicated, consolidated and registered as parcels? As long as this is not recognised, insecurity of access to and utilisation of land increases tremendously and lawlessness spreads (Münkner, 1995).

There is no doubt that it is not easy to make use of the collective character of autochthonous land tenure in national framework legislation. However, without integrating indigenous knowledge and institutional arrangements into this process, an investment in well-meant projects and programmes will not help in achieving development processes. Instead it can even be counter-productive, especially for women or livestock keepers (Kirk, 1998).

2) In most of the countries, the state has followed a hot-cold treatment between quasi-feudal, socialist and capitalistic experiments based on imported western blueprints of tenure concepts (for example, Benin, Ethiopia or even Kenya). Is it so astonishing that farmers do not invest in fruit trees, fencing, mulching or stump removal if there is always the risk that they may lose their land because of expropriation, resettlement, collectivisation or compulsory sale due to indebtedness or land consolidation without compensation? We should not complain that nationalised forests have been cut down, in many cases overnight, once the former control and command regime has collapsed, as in Ethiopia. There, we can see the final stage of the real "tragedy of the commons" created by the state, which ended in plundering (Bromley, 1989).

3) For international donors, it has been relatively easy to establish a reformed land legislation in the course of state divestiture or transformation. Unfortunately, the importance and the meaning of the necessary minimal legal and regulatory framework for supporting a consistent national land policy has often been underestimated by planners (Kirk and Adokpo-Migan, 1995). To what degree are land taxation, inheritance and family legislation created or reformed, and are they compatible with autochthonous rules? How can conflicts be solved when both spheres clash? In the same way, another question must be asked: is the whole western-inspired approach acceptable, is it applicable and cost-effective?

4) Despite the willingness to enforce the new law even in the remotest village, almost all African countries have failed miserably due to a lack of resources, appropriate institutions and qualified staff. The consequences are severe: only the new powerful elites with access to information have been able to make use of the "modern" instruments. This leads to numerous

conflicts and increasing insecurity about access to land, too little investment in soil and insecure tenancy.

Several African countries have tried to reverse this unfortunate development by systematically integrating local tenure institutions and autochthonous rules into the national legal system as in Botswana, Niger, Gambia and Senegal (Bruce et al., 1995). The results have so far been mixed. In general, only models developed by national experts together with the population in a participatory dialogue, as the Land Commission in Tanzania tried, can solve the increasing insecurity of land (United Republic of Tanzania, 1994). However, even then, new laws usually remain "dead letters" unless the machinery exists for their implementation. If not, resource degradation is likely to prevail and accelerate in Africa.

5 Countries in transformation: private land ownership as the silver bullet for sustainable land management?

In former socialist countries, state divestiture leads to a phase of institutional vacuum, since the empowerment of local users is difficult to implement, as experiences in eastern Europe as well as in Mozambique or in Laos show (World Bank, 1996; Effler, 1996; Myers, 1995; Kirk, 1996). Influential pressure groups and new Mafia-like structures try precisely to prevent this empowerment or try to use it to their advantage. What will finally succeed: the rule of law or the rule of the most powerful?

The question should be posed here as to whether or not privatisation and registration are the silver bullet for triggering off access to credit, investment and sustainable agricultural production. Are there not viable prospects for former production co-operatives and state farms for adopting a new form of autonomous, voluntary co-operation? What are the prospects for family farms of earning their living and producing in an environmentally sound way with only inheritable, long-term usufructuary rights?

6 Industrialised countries: about the sanctity of private property and imputed environmental costs

Private property is said to constitute the foundation of democracy, individual freedom and flourishing markets. In agriculture, this makes it possible to produce almost any commodity in almost any amount with numerous negative environmental effects (Bromley and Hodges, 1990). The presumption of an absolute right to produce food and fibre creates an open-ended agricultural policy in which the state and its treasury have become the captives of the sanctity of private rights to land. The property right regime in an industrialised agriculture heavily demands extensive financial obligations to avoid burdening the environment.

7 Conclusions

Land tenure and land management issues matter more than ever, although it is nearly impossible to draw general conclusions out of the complex and multifaceted material. Only a few lessons of the past have been learnt and have been exchanged between regions and continents. Pressing challenges are postponed or suppressed, as land tenure is one of the most sensitive issues within societies. If the solution of problems is further postponed to the future, the inefficient allocation of scarce resources, conflicts related to distribution, and the destruction of the environment will be aggravated and accelerated.

References

Boserup, E. (1965): The Conditions of Agricultural Growth, London: Allen & Unwin.
Bromley, D. (1989): Property Relations and Economic Development: The Other Land Reform, World Development 17, 867-877.
Bromley, D. & Hodge, I. (1990): Private property rights and presumptive policy entitlements: reconsidering the premises of rural policy, European Review of Agricultural Economics 17, 197-221.
Bruce, J. & Migot-Adholla, S. (eds.) (1994): Searching for Land Tenure Security in Africa, Dubuque, Iowa: Kendall/Hunt.
Bruce, J.; Freudenberger, M. & Tidiane Ngaido (1995): Old Wine in New Bottles: Creating New Institutions for Local Land Management, (GTZ-report), Eschborn, http://www.gtz.de/lamin/
Effler, D. (1996): National Land Policy and its Implications for Local Level Land Use. The Case of Mozambique, Proceedings of the 9th Conference of the International Soil Conservation Organisation, Bonn.
Gordillo de Anda, G. (1993): The Reform of Mexican Agriculture: Policies for the Construction of a New Paradigm, (Lecture delivered at the National University of Singapore) (mimeo).
Hayami, Y. & Otsuka, K. (1993): The Economics of Contract Choice, Oxford: Clarendon Press.
International Food Policy Research Institute (IFPRI) (1996): Property Rights and Collective Action. A Multi-Country Research Program, Washington, D.C.
Kirk, M. (1996): Land Tenure Development and Divestiture. In: Lao P.D.R. (GTZ-report), Eschborn, http://www.gtz.de/lamin/
Kirk, M. (1998): Land Tenure, Technological Change and Resource Use: Transformation Processes in African Agrarian Systems, Frankfurt a.M. etc. (forthcoming).
Kirk, M. & Adokpo-Migan, S. (1995): The Role of Land Tenure and Property Rights in Sustainable Resource Use: Studies on Benin In: DSE/GTZ (eds.) Market-based Instruments of Environmental Management in Developing Countries, Berlin, 43-60.
Kuhnen, F. (1996): Land Tenure in Asia, Hamburg: Dr. Kovac-Verlag.
Lastarria-Cornhiel, S. & Barnes, G. (1995): Assessment of the Praedial Property Registration System in Peru, (GTZ-report), Eschborn, http://www.gtz.de/lamin/
Münkner, H.-H. (1995): Synthesis of Current State and Trends in Land Tenure, Land Policy and Land Law in Africa, (GTZ-report), Eschborn, http://www.gtz.de/lamin/
Myers, G. (1995): Land Tenure Development in Mozambique, Implications for Economic Development, (GTZ-report), Eschborn, http://www.gtz.de/lamin/
Okoth-Ogendo, H.W.O. (1997): Land Tenure and Sustainable Land Management in Kenya. In: OSS/EDC (eds.), Land Tenure Issues in Natural Resources Management (Adddis Ababa 11-15 March, 1996), Paris, 86-100.
Panayotou, T. (1993): Green Markets. The Economics of Sustainable Development, San Francisco: ICS Press.
Shivji, Issa (1997): Land Tenure Problems and Reforms in Tanzania. In: OSS/ECA (eds.), Land Tenure Issues in Natural Resources Management (Addis Ababa 11-15 March, 1996), Paris, 101-118.
Thiesenhusen, W. (1995): Trends in Land Tenure Issues in Latin America: Experiences and Recommendations for Development Cooperation, (GTZ-report), Eschborn, http://www.gtz.de/lamin/
United Republic of Tanzania (1994): Report of the Presidential Commission of Inquiry into Land Matters, Vol. I: Land Policy and Land Tenure Structure, Uddevalla.
World Bank (1993): The East Asian Miracle, Economic Growth and Public Policy, Washington D.C..
World Bank (1996): World Development Report 1996, Washington D.C..

Address of author:
Michael Kirk
Institute for Co-operation in Developing Countries
Department of Economics
Philipps-Universität Marburg
Am Plan 2
D-35032 Marburg, Germany

Tenure Regimes and Land Use Systems in Africa: The Challenges of Sustainability

H.W.O. Okoth-Ogendo

Summary

This paper re-evaluates the characteristics of African land tenure regimes with a view to assessing their impact on land use decision-making in general, and the sustainable management of land resources in particular. This is done by examining what sustainable management of land resources requires, the structural and normative characteristics, the contextual framework that made African tenure regimes particularly amenable to that principle, the pressures that have all but destroyed these regimes over time, and what options are available to governments in the reform of such regimes.

1 Introduction

African land tenure regimes have been much maligned in the literature. For nearly a century, the conventional wisdom from Western (or western-trained) anthropologists, economists and land use "specialists" has been that these regimes are inherently incapable of providing a framework for efficient and productive management of land resources (Okoth-Ogendo) and Oluoch-Kosura, 1995). More recently, however, that position has been challenged through empirical investigations which indicate that no tenure regimes offers a panacea for all the land use ills of an agrarian system (Bruce and Migot-Adholla, 1994). It is now being accepted, albeit grudgingly, that the structural characteristics of a tenure regime, are not, per se, sufficient determinants of its efficacy (or lack of it) in the domain of land use.

2 The challenge of sustainability

It was the Brundtland Commission of 1987 which gave international currency to the view that the sustainable management of natural resources is fundamental to human survival (Brundtland, 1987). Since then, sustainability has become the organising idiom of development in the final decade of the 20th century (UN, 1993; Trzyna, 1995). But sustainability as a concept which addresses the utilisation of resources through time and space and within and across generations can only be understood in the context of the system of norms that inform given social organisations. Concerning land use and land management, therefore, sustainability is necessarily an incident inter alia, of the viability of social institutions and practices in the light of rapid changes in society. This means that the regimes that deal specifically with resource control and use in society must, as an absolute pre-requisite, evolve from and remain an integral part of social organisation. Until relatively recently, African land tenure regimes, though acknowledged to be fully integrated into social structures were, nonetheless, condemned as so chaotic, diffuse and dysfunctional, as to be totally incapable of providing a dynamic framework for the sustainable management of land.

3 The nature of African tenure regimes

The basis of that condemnation lay, firstly in the misdescription of the structural characteristics of African tenure regimes; secondly, in the assumption that tenure rights are secure only if they are exclusively defined; and thirdly, in the belief that African agrarian systems operated on an essentially consumptive rather than preservationist, view of resource use.

3.1 The problem of misdescription

The misdescription is clear in the assertion, so often made, especially by anthropologists and later by liberal economists, that African tenure regimes are „communal"; meaning, thereby, that all the incidents of property over land are vested in communities as owners. From that it is generally argued that community „ownership" being thus diffuse for want or impossibility of identifying specific rights, African tenure regimes are in reality open access systems (Demsetz, 1967). Thus they are thought to be systems in which individuals have mere privileges and no rights hence no corresponding obligations to take care of the resources. From that it was easy to arrive at the conclusion that those regimes would necessarily lead to social irresponsibility and resource plunder.

And yet there has always been evidence to the effect that while regarded as a community asset, land in Africa is not community owned (Bromley and Cernea, 1989) and indeed that the terminology of „ownership" itself is not particularly useful as a conceptual framework with which to analyse such regimes (Okoth-Ogendo, 1989). A more rigorous conceptual exposition of these regimes will reveal that tenure relations have always been and substantially remain an incident of the social organisation of particular territorial communities. For that reason tenure of land does not simply describe man-land relations; rather it defines the nature of reciprocal obligations which individuals owe to one another between and across generations, in virtue of access to land in a given society (Gluckman, 1965). The proper language of analysis in the study of African land tenure must therefore shift from the vocabulary of „who owns what interest in what land?" (Bentsi-Enchill, 1965), to „who has access to what interest in what land and how is the use of that land controlled?" (Okoth-Ogendo, 1989).

Thus reformulated, it will be found that African tenure regimes are by no means open access systems. For as I have argued elsewhere, access to and control of land are very complex phenomena. For while access is always:
- an attribute to membership in some unit of production,
- specific to a resource management or production function, or group of functions, and
- guaranteed through active participation in the processes of production and reproduction at particular levels of social organisations;

control itself is:
- an attribute of the sovereign status of a particular community,
- exercised at different levels of social organisation depending on the specific resource management or production function that is in issue,
- exercised solely for the purpose of guaranteeing legitimate access to land, and
- redistributive both in space and time, and between and across generations in response to changes in the agrarian system as a whole (Okoth-Ogendo, 1989)

Consequently, African tenure regimes have generally been particularly sensitive to resource availability and scarcity, responsive to changes in technologies of production, and particularly protective of transgenerational needs in society. Indeed so much is this the case that land, throughout Africa, has always been the object of great reverence, mythology or even deification (Fisiy, 1994). And the fact that indigenous medical practice and ritual artifacts still draw heavily on land based products, means that the preservation of bio-diversity is always high on the African land use agenda. The sustainable management of land resources is therefore not simply a principle; it is a requirement in African land relations.

3.2 The issue of security

The assumption that tenure regimes will confer security only if the rights they define are <u>exclusive</u>, is based on the doctrine that rational management of land is only possible under individual ownership. Since the categories of people who might gain membership into a territorial community, hence acquire access to its land resources, is never really closed, African tenure regimes were thought to be conflict prone, hence <u>inherently insecure</u>. A closer analysis of the literature indicates, however, that what most writers are concerned about is security for the individual <u>against, rather than within, society</u>! Without belabouring the point, it is now accepted that because access to land is tied, at all times, to social organisation, African tenure regimes offer a great deal more security to individuals and collectivities than the forces of the market in which private property regimes operate can afford. It is for this reason that the former have been known to absorb a great deal of external pressures designed to convert them into the latter. I have shown elsewhere (Okoth-Ogendo and Bilsborrow, 1992) that African tenure regimes are remarkably resilient in the light of sabotage from such influences as colonial expropriation, deliberate state intervention, population pressure, and exogenous technological intrusions.

What that level of security implies, is not simply the centrality of land and land-based resources in Africa's political economy. It also implies that a duty lies upon members of society severally and collectively, not to destroy; but rather to replenish the quality of those resources. And since community membership is accorded not only to the <u>living</u>, but also to the <u>dead</u> (the ancestors) and the <u>unborn</u>, access rights are, correspondingly, not only active <u>in presenti</u>, but also in mortis causa and in futuro. Rules of social organisation will not allow such rights to be extinguished, or otherwise adversely interfered with. This is one reason why the power of absolute disposal of rights and interests in land is unknown in indigenous law.

3.3 The allegation of consumptive orientation

The supposition that indigenous agrarian systems are essentially <u>consumptive</u> in orientation, rather than <u>preservationist</u>, is probably derived from the chronically subsistence nature of many production operations in Africa. But that simple description of the rather complex processes that enter into production processes in Africa, has now ben shown to be incorrect. Thriving "commercial" agricultural operations have been and continue to be sustained on a long term basis in many African countries on the basis of indigenous tenure regimes. Conversely, an analysis of tenure reform processes in some African countries, especially those designed to „privatise" landholding and use, suggests that no dramatic changes in production and productivity can be attributed exclusively to such changes (Bruce and Migot-Adholla, 1994).

What this supposition ignores is that for more than a century, indigenous agrarian systems have been under a perpetual state of siege. Under colonialism, for example, land was constantly being mined in the face of population growth coupled with the absence of a land frontier, productivity was always poor due to the fossilisation of indigenous technologies, and indigenous food crops were generally marginalised as a result of state preference of export crops over them. Nothing much has changed in the post-colonial period. Indeed, apart from perpetuating colonial agragrian policies, contemporary African governments, have, through resource plunder and mismanagement worsened the predicament of indigenous agragrian systems. The „African agrarian crisis" as this situation has come to be known (Cummins et. al., 1986), is therefore rooted in past policies and ideologies and contemporary developments and not a product of inherent defective tenure regimes.

What that condition points to is that the sustainable management of land resources has become that much harder in Africa. For apart from the need to address its effects on land tenure regimes, African countries must also face up to the challenges presented by the global technological revolution, democratic pressures for „good" governance, and rapid population growth, to mention but the primary ones.

4 Reform of African tenure regimes

The effect of past and current agrarian policies and programmes on African land tenure regimes have been recorded elsewhere. The general observation is that despite their resilience, it is clear that a great deal of normative confusion and structural decay have entered into the fabric of African tenure relations as a result of those policies and programmes. For example, while access rights remain a function of social organisation, control mechanisms and authorities are everywhere in disarray or simply dormant (Okoth-Ogendo and Bilsborrow, 1992). Similarly, the fact that alternative forms of livelihood are now available and often more attractive than land based pursuits, means that some of the pressures for the preservation of land, in time and space, which made Afri-can tenure regimes especially sensitive to issues of past, present and future quality, are no longer as acute as they may have been several decades ago (Turner et. al., 1993). Consequently reform of African tenure regimes has become an important concern of public policy throughout Africa.

The current orthodoxy about how best to do this, however says that indigenous property systems must be eradicated and replaced with those based on notions of private property (Okoth-Ogendo, 1995a). As a result, programmes involving the conversion of „ownership" structures are in operation or contemplation in many parts of Africa. The most comprehensive of such programmes has been in process in Kenya since 1954 and is set to continue well into the 21st century (Okoth-Ogendo, 1996). Similar experiments have also been tried in a number of countries including Uganda, Zambia, Malawi, Swaziland, Ghana and a number of French-speaking African countries.

Apart from the fact that these are not land reform, let alone tenure conversion measures, evidence from empirical studies now indicate that these programmes have introduced even greater confusion into African land relations. For as I have repeatedly asserted in respect of Kenya (Okoth-Ogendo, 1996) in addition to the absence of easy conceptual fit between community based property rights and private property, and the fact that the incidents of indigenous tenure regimes will not disappear quickly, many other socially disruptive processes and practices are bound to emerge in land relations as a result of these rather narrowly conceived, programmes (Okoth-Ogendo, 1986). That quite clearly should call into question the current orthodoxy about how best to reform land tenure regimes in Africa. For far from facilitating the sustainable management of land resources, reform, if limited to changes in the technical description of title, or in ownership characteristics, per se, as it now appears to be, may well put that normative objective in grave jeopardy. „Tenure" it is worth emphasising is much more than just „ownership", it includes the total context (structural and normative) in which access to land is obtained and exercised.

5 Concluding statement

Land in Africa is clearly an arena for the interaction of many social, cultural, economic and environmental factors all of which determine its utility as a unit of production and a context in which social values and processes evolve and operate. The role of land tenure in the sustainable management of land resources cannot, therefore, be fully assessed simply in terms of how and, to whom, property regimes allocate rights over the physical solum. Tenure regimes, whatever their essential characteristics, must form part of the complex set of principles and values which bind social groups together, before they can provide an adequate framework for sustainable land use. This is the direction which tenure or, better still, land reform processes and programmes in Africa ought to follow, particularly in the light of the very rapid and, sometimes, rather traumatic changes now sweeping across Africa.

References

Bentsi-Enchill, K. (1965): Do African Systems of Land Tenure Require a Special Terminology? Journal of African Law, Vol. 5 (2).

Bromley, D.W. and Cernea, M.M. (1989): The Management of Common Property Natural Resources World Bank Discussion Paper No. 57, Washington, D.C.

Bruce, J.W. and Migot-Adholla, S.E. (1994): Searching for Land Tenure Security in Africa Kendall/Hund, Iowa.

Brundland, Gro H. (1987): Our Common Future, Report of the World Commission on Environment and Development (Oxford University Press).

Cummins, S.K. et. al. (1986): African Agrarian Crisis: The Roots of Famine, Boulder, Colorado.

Demsetz, H. (1967): Towards a Theory of Property Rights, American Economic Review 57, 347-359.

Fisiy, C.A., (1994): The Death of a Myth System: Land Colonisation on the Slopes of Mount Oku, Cameroon. In: Bahema, R.J. (Ed), Land Tenure and Sustainable Land Use, KIT, Amsterdam Bulletin No. 332.

Gluckman, H. (1965): Ideas in Barotse Jurisprudence, Manchester University Press, London.

Okoth-Ogendo, H.W.O. (1996): Reforming Land Tenure in Africa: Conceptual, Methodological and Policy Issues, Paper for the Alistair Berkeley Seminar on Land Tenure and Tenurial Reform, London School of Economics and Political Science, London, May 29-31.

Okoth-Ogendo, H.W.O. and Oluoch-Kosura, W. (1995): Land Tenure and Agricultural Development in Kenya, Paper for the Ministry of Agriculture, Livestock Development and Marketing, Kenya, April

Okoth-Ogendo, H.W.O. (1995): Reform of Land Tenure and Resource Management: A Comment, Entwicklung Ländlicher Raum, Bonn, Germany

Okoth-Ogendo, H.W.O. and Bilsborrow, R.E. (1992): Population-driven Land Use in Changes in Developing Countries AMBIO Vol. 21(1), pp. 37-45.

Okoth-Ogendo, H.W.O. (1989): Some Issues of Theory in the Study of Tenure Relations in African Agriculture, AFRICA, Vol. 59.

Okoth-Ogendo, H.W.O. (1986): The Perils of Land "Tenure" Reform. In: Arntzen, J.W., et. Al. (Eds), Land Policy and Agriculture in Eastern and Southern Africa, United Nations University.

Trzyna, T.C. (Ed.) (1995): A Sustainable World: Defining and Measuring Sustainable Development, IUCN.

Turner, D.L., Hyden, G. and Kates, R.W. (1993): Population Growth and Agricultural Change in Africa, University Press of Florida.

UN (United Nations) (1993): Agenda 21: Programme of Action for Sustainable Development, New York.

Address of author:
H.W.O. Okoth-Ogendo
Department of Public Law, Faculty of Law
University of Nairobi
P.O. Box 30197
Nairobi, Kenya

Are There Land Tenure Constraints to the Conservation of Soil Fertility?
A Critical Discussion of Empirical Evidence from West Africa

V. Stamm

Summary
Economic theory claims that individual property rights guarantee an optimal allocation of the factors of production. Often, the conclusion is drawn that so-called customary forms of land management represent a major obstacle to the extension of sustainable agricultural methods and to the increase of productivity in rural areas. This point of view is challenged here.

My field work and a critical discussion of the (rather poor) empirical evidence show that
- no statistically significant relationship exists between form of access to land and factor allocation;
- rented plots are not less productive than owned ones;
- limiting factors in the use of soil conservation technologies are rather of an economic or social order: they just do not fit given farming systems.

As an explanation, it is hypothesized that customary land tenure systems offer a greater security of land use than often supposed. This assumption is supported by some further empirical findings.

Keywords: Land tenure, West Africa, soil fertility

1 Introduction

Confronted with continuing soil degradation and the hesitation or inability of farmers to apply soil conservation techniques, the question may be asked if this latter behaviour is not due to institutional constraints, such as imperfect land use rights. This position is held by orthodox economics, and it is largely shared by development organisations.

Surprisingly enough, empirical evidence for the underlying set of assumptions is, with regard to Sub-Saharan Africa, quite poor. "We simply do not know whether land rights are an important constraint on sustainable agricultural intensification in large areas of the developing world." (IFPRI 1996: 39-40).

It is the intention of this paper to examine some of the empirical data available from West Africa.

West African land tenure systems are characterised by a large variety of different, often overlapping, forms of rights. They may roughly be divided into
- pre-eminent rights of control, most often exercised by the most ancient of the indigenous groups;
- rights of management, held by more and more restricted social groups;

- rights of usage accorded to migrants, young families, women and, more generally, to all in need of land.

It is important to note that rights of access to land are quite flexible and adopt easily to changes in the economic or social environment. In doing so, they move on a continuum between more collective and more individualised forms, between non-commercial and commercial practices (see Basset and Crummey, 1993; LeRoy et al., 1996; Stamm, 1996a).

2 Tenure, factor allocation, and the productivity of land

A simple, but significant example of soil improvement shall open the debate, i.e. the use of organic and chemical fertilisers. The first may be interpreted as a form of labour-led intensification, since organic matter is generally not purchased but collected, stored and spread on the fields (household waste), or weeds and harvest residues are incorporated into the soil. The use of chemical fertilisers is an example of capital-led intensification. A survey of 1,175 plots in different regions of Burkina Faso is summarized in Tables 1, 2 and 3.

land right code	applied		not applied		total
	cases	%	cases	%	
1 quasi property	238	23.7	766	76.3	1004
2 short term use rights	0	0	6	100	6
3 long term use rights	39	24.1	123	75.9	162
total (missing cases: 3)	277	23.6	895	76.4	1172

(Source: Ouedraogo et al. 1996)

Table 1: Land rights and organic fertiliser

Land right code	applied		not applied		total
	cases	%	cases	%	
1 quasi property	39	3.9	966	96.1	1005
2 short term use rights	1	16.7	5	83.3	6
3 long term use rights	3	1.9	159	98.1	162
Total (missing cases: 2)	43	3.7	1130	96.3	1173

Table 2: Land rights and chemical fertiliser

In neither case was a significant relationship found between the use of fertilisers and the form of land rights. The low use made of chemical fertilisers is remarkable.

These results are confirmed by Neef (1997) in South Benin, in a quite different social context, marked by high frequency of short-term borrowing and more commercial forms of access to land than in Burkina Faso.

It should be noted that the findings from our simple example reflect the results of a more general analysis of factor inputs: the allocation of both labour and capital is independent of the form of land rights.

Thus Ouedraogo (1991) found that in the Western parts of Burkina Faso the quantities of labour or capital used per hectare did not differ as a function of land rights, neither in the case of cotton nor for corn or millet. Furthermore, the productivity of corn and millet fields is independent of the form of access to them.

In a country-wide sample covering four provinces in different zones of Burkina Faso, the use of capital goods or of agricultural inputs did not vary with changing forms of access to land

(Sawadogo, 1996). Biaou (1991), studying a Southern region of Benin, stated that the intensity of labour and capital did not correlate with land rights of the plot concerned and that, as a consequence, tenure did not influence the land productivity.

These regional and partial findings are largely confirmed by a cross-country evaluation using farm data from Kenya, Rwanda and Ghana. Place and Hazell (1993) concluded that "With few exceptions, land rights were not found to be a significant factor in determining whether or not farmers made land-improving investments or used yield-enhancing inputs." (see also Bruce and Migot-Adholla, 1994).

In our understanding, soil conservation implies a sustainable intensification of farming systems. It may be labour or capital-led, but the latter is more likely to produce the desired effects (Reardon et al., 1996; for labour intensive systems and their dynamics, see Stamm 1994 and Hahn, 1996). Since we could not identify significant relations between land tenure and intensity in the use made of labour or capital, probably neither form of intensification is severely restricted by factors related to land tenure.

Unfortunately, in our opinion, soil conservation is often not understood from a farming systems approach, but as being the result of the application of some particular techniques, most often construction of stone bunds and planting of trees. Though we do not really trust in this kind of sectoral interventions, we do not wish to disregard the effect of the cited methods as their application may be a beginning for further intensification.

The question may be asked, therefore, if *their* specific extension is limited because of land tenure arrangements, as often asserted in literature.

3 The case of some particular techniques of land amelioration

In a case study of 10 Ethiopian peasant associations, none of their speakers attributed the failure to adopt resource management techniques, especially agro-forestry, to land tenure constraints. All of them confirmed that the kind of use rights they had are a sufficient stimulant to engage in investments (Tolossa and Asfaw, 1995, 56-57). The limited acceptance of tree planting results from natural or technical constraints with regard to given farming systems :
- unfavourable climatic conditions
- damage to trees caused by cattle
- trees protect birds destroying the harvest
- trees in the field make farming more difficult.

The latter argument and the place occupied by trees - loss of arable land - is generally forwarded by Sahelian farmers.

Households surveyed in the Bam Province (Central Burkina Faso) expressed the view that they are free to improve the soil quality of their plots by all adequate measures, including the construction of stone bunds (Sawadogo, 1993), independent of their position of land owners or borrowers. Kunze (1994), working in the same region but using a larger sample, found that 40% of all land borrowers were authorised to ameliorate their plots, but only about 20% made use of this opportunity. The author did not observe a significant difference in the frequency of stone bunds on inherited fields or borrowed ones, the only exception being the very small number of plots given for a short period. She concluded that there is no reason to believe that land tenure constraints may limit the amelioration of borrowed plots (p. 67).

Going north in Burkina Faso to the Sahelian regions, Schmitt (1995, 104) stated that land borrowers (most often migrants) may farm the fields given to them as they like, including the right to invest, to plant trees or to leave them in fallow.

These examples could be multiplied (see Stamm, 1996c); they do somehow contrast to the commonly shared impression that land tenure issues form a severe and omnipresent obstacle to the application of erosion control measures.

Nevertheless, data do suggest that, even if not rendered impossible, some methods of soil conservation such as heavy construction or tree planting form a critical element in land use patterns (Sawadogo, 1996, Ouedraogo et al., 1996), since they may give rise to conflicts about land property. It is in this sense that Pélissier interpreted trees as symbolising ownership over land, which is certainly exaggerated. Ngaido, to cite only one example of a study dealing with such conflictual situations, stated that many tenant farmers in Western Niger planted trees, with or without the consent of the respective land owners. If the latter objected, negotiations took place, with a large variety of possible solutions, such as cutting the trees, selling them to the land owner, or sharing their benefits. Given this ambiguous situation, Ngaido's (1996, 257) recommendation to improve contractual tenure security is fully justified (for some tools, see Stamm, 1997 and 1998).

But one has always to keep in mind that almost everywhere in our region of study, inherited and relatively secure land rights are predominant, and that on these fields the application rate of the discussed soil conservation measures is quite low. There are more severe barriers to their extension than that of land tenure .

4 Discussion of results

The results obtained:
- common intensification patterns are not more difficult to attain because of local land tenure practices,
- insufficient acceptance of specific soil conservation techniques cannot be fully attributed to land tenure constraints,

may be explained by the combined influence of several factors to be sketched in this section.

4.1 The relative stability of land rights

The relationship between different right holders is generally not so conflictual as often supposed. In an extraordinary sample covering more than 400,000 hectares obtained during the pilot operations of the Plan Foncier Rural in the Ivory Coast (see Stamm, 1998), only 2 - 3 % of all plots were subject to some conflict regarding land rights (see also Cousins, 1996 and Stamm, 1996b)

4.2 The relative security of land rights

Where no massive state intervention occurs such as resettlement or land to the tiller policy, and where commercial interests of urban classes are not predominant, land rights even if not sanctioned by formal titles are quite secure. We found in Burkina Faso that the great majority of plots surveyed were used by virtue of quasi (hereditary) property rights, and even borrowed fields most often give rights to long term use.

Land right code	%	Total
Quasi property	85.7	1007
Short term use rights	0.5	6
Long term use rights	13.8	162
Total	100.0	1175

Table 3: Frequency of land rights

4.3 Limited possibilities of investment

The investment capacity of small farms is severely restricted by financial constraints, especially in the case of rain-fed agriculture, and by a lack of experienced and viable technologies of soil conservation. The first part of this hypothesis is confirmed by the insignificant use made of chemical fertilisers and its further reduction as a reaction to price increases in the course of adjustment programs. Concerning available technologies, stone bunds seem to be profitable on the farm level, but it is surprising to see that, after years of heavy extension and sponsoring, they are not generally adopted by the population, even not on their own fields where no tenure restrictions exist. This may be due to the often neglected opportunity costs of labour and to the logistics necessary for their construction (see Kunze 1994; Kunze et al. 1996; Ouedraogo et al. 1996).

As to agroforestry programs, they are still in a state of experimentation. After more than 10 years of research in Benin, agroforestry systems are not used by the farmers, unless urged by projects or against remuneration on experimental sites.

5 Conclusion

The evidence presented does not suggest that land tenure patterns in West Africa form a major obstacle to sustainable intensification of farming methods. However a few cases of soil conservation techniques turned out to be critical. Thus contractual arrangements between land managers and land users may be necessary to facilitate their adoption.

References

Basset, T.J. and Crummey, D.E. (1993): Land in African Agrarian Systems. Madison, Wisconsin
Biaou, G. (1991): Régime foncier et gestion des exploitations agricoles sur le Plateau Adja (Bénin). University of Abidjan (unpublished thesis)
Bruce, J.W. and Migot-Adholla, S.(ed.) (1994): Searching for land tenure security in Africa. Dubuque, Iowa
Cousins, B. (1996): Conflict management for multiple resource users in pastoralist and agro-pastoralist contexts. Third International Technical Consultation on Pastoral Development, Brussels, May 1996
Hahn, H.P. (1996): Wieviel Beratung benötigt ressourcenschonender Landbau in Westafrika? Eine Fallstudie bei den Kasena in Burkina Faso. In: Entwicklung und ländlicher Raum 3/1996: 17-19
IFPRI (International Food Policy Research Institute) (1996): Property Rights and Collective Action in Natural Resource Management. Multicountry Research Program MP-11. Washington
Kunze, D. (1994): Typologie des systèmes d'exploitation dans la province du Bam/Burkina Faso. University of Göttingen (unpublished report)
Kunze, D., Runge-Metzger, A. and Waibel, H. (1996): The Economics of Erosion Control. Innovations in Northern Burkina Faso. Contribution to the International Symposium Food Security and Innovations: Successes and Lessons Learned. University of Hohenheim
LeRoy, E., Karsenty, A. and Bertrand, A. (1996): La sécurisation foncière en Afrique. Paris
Neef, A. (1996): L'insécurité foncière au sud du Bénin: ses causes et ses effets. In: T. Bierschenk, P.-Y. Le Meur and M. von Oppen (Eds.), Institutions and Technologies for Rural Development in West Africa. Weikersheim, 321-331
Ngaido, T. (1996): Redefining the Boundaries of Control: Post-Colonial Tenure Policies and Dynamics of Social and Tenure Change in Western Niger. University of Wisconsin, Madison (unpublished PhD thesis)
Ouedraogo, R.S., Sawadogo, J.P., Stamm, V. and Thiombiano, T. (1996): Tenure, agricultural practices, and land productivity in Burkina Faso. Some recent empirical results. In: Land Use Policy **13**,3: 229-232
Ouedraogo, S. (1991): Influence des modes d'accès à la terre sur la productivité des exploitations agricoles: Le cas de la Zone Ouest du Burkina Faso. University of Abidjan (unpublished thesis)
Place, F. and Hazell, P. (1993): Productivity Effects of Indigenous Land Tenure Systems in Sub-Saharan

Africa. In: Amer. J. Agr. Econ. **75**: 10-19

Reardon, T., Kelly, V., Diagana, B., Dione, J., Crawford, E., Savadogo, K. and Boughton, D. (1996): Sustainable Capital-led Intensification in Sahel Agriculture: Addressing Structural Constraints after Macroeconomic Policy Reform. Contribution to the International Symposium Food Security and Innovations: Successes and Lessons Learned. University of Hohenheim

Sawadogo, J.P. (1993): Le droit foncier traditionnel et ses conséquences économiques à l'exemple de trois villages du Bam. University of Ouagadougou (unpublished MA thesis)

Sawadogo, J.P. (1996): Impact de la tenure foncière sur l'utilisation du capital, des intrants et sur la réalisation des améliorations foncières au Burkina Faso. Approche probabilistique. University of Ouagadougou (unpublished thesis)

Schmitt, G. (1995): Konkurrenz und Konflikte. Beziehungen zwischen Feldbauern und Viehhaltern in der Sahelregion von Burkina Faso. University of Frankfurt, Inst. für Hist. Ethnologie (unpublished MA thesis)

Stamm, V. (1994): Anbausysteme und Bodenrecht in Burkina Faso. In: Afrika Spectrum 4/1994: 247-264

Stamm, V. (1996a): Bodenordnung und ländliche Entwicklung in Afrika südlich der Sahara. In: Institut für Afrikakunde, R. Hofmeier (Ed.), Afrika-Jahrbuch 1995, Hamburg, 73-82

Stamm, V. (1996b): Landkonflikte in West-Afrika und Ansätze zu ihrer Lösung. In P. Meyns (Ed.), Staat und Gesellschaft in Afrika - Erosions- und Reformprozesse. Münster/ Hamburg, 536-545

Stamm, V. (1996c):. Zur Dynamik der westafrikanischen Bodenverfassung. Hamburger Beiträge zur Afrikakunde Vol. 49, Hamburg

Stamm, V. (1997): Approches de solution locale aux problèmes fonciers. In: T. Bierschenk, P.-Y. Le Meur and M. von Oppen (Eds.), Institutions and Technologies for Rural Development in West Africa. Weikersheim, 333-340

Stamm, V. (1998): Structures et politiques foncières en Afrique de l'Ouest. Paris

Tolossa, G. and Asfaw, Z. (1995): Land Tenure Structure and Development in Ethiopia. A Case Study of 10 Peasant Associations in Wara Jarso Woreda. Eschborn, GTZ

Address of author:
Volker Stamm
Robert-Schneider-Str. 15a
D-64289 Darmstadt, Germany

National Land Policy and Its Implications for Local Level Land Use: The Case of Mozambique

D. Effler

Summary

Land policy in Mozambique has changed dramatically during the last few decades and has direct impacts on land tenure and land use systems of rural populations. The consequences of Portuguese colonialism, socialist policy, structural adjustment programmes, civil war and millions of returning refugees still dominate change. Traditional rural communities have become the most vulnerable group of land users. Both traditional leaders and local government authorities are in charge of land allocation, backed by different land tenure systems. The resulting institutional vacuum at local level leads to an increase of land tenure insecurity and severe conflicts. Land use is short-term oriented and sustainable land use practices are rarely applied. Against this background, a model has been developed which aims at a cooperation between different interest groups to improve land tenure security and - as a consequence - to promote the application of more sustainable land use systems.

Keywords: Land tenure, traditional land use rights, land policy, land conflicts, land law, land reform, formal and informal land administration, conflict settlement, private-public-communal partnership.

1 Introduction

From 1994 to 1996, a study was conducted at national, regional and local levels in Mozambique. Recent development processes of land policy were studied including formal and informal land administration and their impacts on land management by traditional communities, and their relations with other land users or interest groups. During the research process a set of participatory methods was applied, complemented by structured and semi-structured interviews with key informants and farmers at all levels, institutional analysis, remote sensing techniques and personal observations.

2 Historical overview: The process of change in land tenure

The following time table illuminates some stages of recent Mozambican history. It highlights significant *changes of frame conditions* for land tenure and their direct impacts on rural populations in Manica province.

Pre-independence (before 1975)
Portuguese colonial government and administration grant land to Portuguese farmers, international enterprises and African "assimilados" as privately owned real estates.

Family farmers are displaced to less fertile areas (slopes, mountains, dry plains) and are forced to work on commercial farms or to produce cash-crops for them. Traditional communities have no formal land rights. As the colonial administration uses traditional leaders, informal land tenure systems are accepted in areas occupied by their communities.

Soon after independence (1975)
FRELIMO party establishes a socialist system. Land and natural resources become state-property.

Most commercial farms are converted into state-enterprises, some remain in the hands of private farmers. There is no significant change in access to land for traditional communities.

State enterprises and cooperatives become priviledged forms of land management, but also the private and family sectors are recognised. The creation of communal villages becomes the major objective for the rural areas.

Most traditional leaders are deprived of their power, informal land tenure systems are invalidated. Family farms are regarded as unproductive. Families are expected or forced to move into communal villages and to join cooperatives. Most of the family sector farmers lose their inherited or traditionally allocated land and receive new plots from village administrations.

Early 1980s
With South Africa's intervention in Mozambique by backing RENAMO, civil war prevails.

Millions of rural people are displaced from their land. In their areas of influence, RENAMO begins to reestablish traditional administration and land tenure structures.

Structural adjustment programmes of Mozambican government (late 1980s)
Government policy changes from state economy to market economy. The frame conditions for private enterprises are liberalised and national programmes are developed to attract foreign investors. A new constitution is passed in 1990. Land and natural resources remain property of the state. The 1979 land law with socialist objectives remains in force.

Most state enterprises fail economically. Their land use rights are re-allocated to other interest groups: communal villages, private investors and public servants. In many rural areas, ex-state enterprise land remains unutilised because of the war. Family sector farmers occupy many sites "illegally".

After the peace accord of Rome (1992)
Government decides on a change of national land policy. A national land commission is founded to review all land tenure related legislation and to propose new guidelines to the government. With peace, the process begins of return, repatriation and reintegration of more than 4.5 million refugees and displaced people.

For the first time in Mozambique's history, a government recognises the role of family farmers, analyses traditional structures and promises the development of the family sector. Most of the people return to their "areas of origin", complete communities reoccupy their ancient areas and reactivate traditional leadership structures and land tenure systems. Many communal villages are being abandoned. Conflicts arise between returning refugees and recent occupants of land. The gap grows between formal and informal land tenure and administration.

After the elections (1994)
The ruling FRELIMO party wins the first free and democratic elections. In 1995, the new government decides on a new land policy. A new land law is to be established at the end of 1996.

Family sector farmers now play a key role in future land use. They are expected to satisfy the major part of the nation's food crop needs. Informal land tenure and management systems shall be recognised by formal land administration. Land tenure security for all user groups is now one of the major targets of government policy.

3 Land tenure and land use in Manica Province

During the process of change, traditional communities and family sector farmers have become the most vulnerable group of land users in rural areas. In this context, land tenure insecurity can be regarded as an indicator for vulnerability. It influences mainly the decision-making process of a family and is therefore the most direct cause for the non-application of appropriate and sustainable land use practices. Formal land tenure is relevant for both the private and the family sector, whereas informal land tenure is only relevant for the family sector (Effler, 1996).

3.1 Application of the formal land tenure system and land tenure security

Land is property of the state, it cannot be sold, alienated or mortgaged (Moçambique 1975, art. 8; 1979, art. 1; 1990, art. 35). It is an exclusive right of the state to grant temporary land utilisation rights by issuing land use titles. Legal requirements on land administration by state institutions are high. Clearly defined procedures at different levels are described in the legal and administrative framework. Decisons on the allocation of land use rights are made at different government levels according to the size of the land applied for (Moçambique 1987, art. 4-10). Coordination rarely takes place (Alexander, 1994).

The private sector
Private sector enterprises obtain formal land tenure security through a title issued by the competent government authority. Control of the compliance of the private sector's activities with legal requirements does not take place. An institutional vacuum leads to an additional increase of de-facto land tenure security for title holders against the interests of the family sector. There are no known cases of withdrawal of a title in case of severe violations of regulations as foreseen in the legal framework. Private sector land users can be divided into two groups: title holders for the exploitation of a natural resource (e.g. timber) and title holders for a specific kind of land use (e.g. agriculture). In general, land use by the private sector can be regarded either as exploitive (first group), extractive or non-sustainable (both groups) or as non-productive or simply non-existant (second group).

The family sector
Family sector farmers can apply for a "family occupation certificate" which grants the utilisaton of a demarcated plot for subsistance production to a family. Other formal land use rights for family sector farmers are not recognised by the legal framework. With the exception of some better-off families around towns, there are no holders of such certificates. Families are either not aware of this regulation or cannot afford the costs of a registration process, or do not trust government institutions. The main impact of the formal land tenure system on land tenure security for family sector farmers consists in their extremely weak position against land use title holders in case of conflict (Myers, 1995). The issuing authority or its local representative is also responsible for conflict settlement. Therefore, conflicts are generally settled in favour of the private sector.

3.2 Application of informal land tenure systems and land tenure security

In most of the rural areas of Manica province, the application of informal land tenure systems is the decisive way for obtaining and securing access to land and natural resources for family sector farmers. They differ regionally, show a great variability at local levels and trace back to the ancestors of today's leading families. All land and natural resources are property of the ancestors of a leading family. The head of this family, the *régulo*, is regarded as the leader of the community and the administrator of the land. He is vested with a variety of tasks and exclusive rights concerning land and land use. He allocates land and resource use rights for a clearly defined piece of land to both members of the community and outsiders, settles conflicts and judges. He is responsible for the conservation of the fertility of the land for future generations. In general, *régulos* are respected by their communities and their decisions are definitive. The legitimacy of the man who exercises this position is supervised by a council of elders who, in case of doubt or conflict, can take over the *régulo's* duties and rights. The elders are respected members of the community whose knowledge and decisions are regarded as wise.

The replacement of many *régulos* by village secretaries during the socialist period did not lead to a significant change of the process of land allocation but to a complete reorganisation of local level land tenure. Today, in most of the rural areas of Manica province, communal villages are being dissolved, traditional forms of settlement are being revived and *régulos* are regaining power and influence. Conflicts arise between family sector farmers who have been allocated their land by either a *régulo* or a village secretary, especially between returning refugees and remaining residents.

The de facto informal land allocation process in communal villages is accepted by formal administration and benefits of formal land tenure security. Both the area of the communal village and the function of the village secretary are recognised by higher government and administration levels. Their rights are therefore protected against external interest groups. The informal land allocation process in traditional communities is not recognised by formal administration. Land settled and used by a traditional community is therefore often described as "unoccupied land" in land use title application forms by district administration. So, land tenure security is higher for remaining communal villages than for traditional communities.

Although informal land tenure systems are not officially recognised by formal authorities, there is one traditional rule that is regularly applied in case of conflicts by both formal and informal institutions: everybody has the right "to harvest what he or she has planted", whether the land has been allocated or occupied without any permission. On one side, this right leads to a certain short-term land tenure security for the most vulnerable groups of land users. On the other side, it leads to extremely short-term oriented land use practices. Most of the returning refugees have received soil conservation training in the refugee camps in Zimbabwe. But they do not apply any of their knowledge and are still used to rely on short-term survival strategies as do the remaining residents.

4 The new land policy and its aimed implications on land tenure and land use

Alleviation of poverty and improvement of living conditions for the population are overall purposes of the new government programme (Moçambique 1995a, art. 13). Land is regarded as a starting point for self-sustaining development. Access to land for all citizens and protection of areas occupied by the family sector are therefore fundamental principles of the new land policy. A guideline has been formulated to express the two columns of this policy (op.cit., art. 18): **"Save the rights of the Mozambican people on land and natural resources and promote investments and sustainable and appropriate utilisation of these resources"**

The new land policy contains some fundamental innovations and guiding principles which may have a significant influence on land tenure security for the family sector. Land and natural resources remain property of the state (op.cit., art. 17). Customary rights of rural populations shall be recognised and formalised (op.cit., art. 20) and traditional communities can be registered as communal holders of land utilisation titles (op.cit., art. 23-25). They shall get the right to negotiate directly with private interest groups about land utilisation, under supervision of local administrations (op.cit., art. 25). Mozambicans shall become active partners of (foreign) private enterprises (op.cit., arts. 17 and 25) and land utilisation rights shall become transferable (op.cit., art. 17). The efficiency and potentials of traditional communities in self-administration of their land is appreciated by the new land policy. It is expected that formal administration will be able to simplify administrative procedures and reduce costs by utilising these opportunities (op.cit., art. 21). The family sector, small scale and large scale agricultural enterprises are named as target groups. In the long term, it is expected that new (small scale) private enterprises will develop out of the family sector (op.cit., art. 29).

5 Future cooperation between the main interest groups in land use?

Sustainable land use development needs access to land and land tenure security for all user groups as well as formal and informal institutions that are capable of proper and just land administration. Based on discussions in the national land commission and referring to arts. 23-25 of the new land policy (Moçambique 1995b), a model for private-public-communal partnership has been developed. This model targets the creation of local level land tenure security for all groups, an increase of sustainable production systems as well as the improvement of institutional capacities at local levels. The following scenario describes the core elements of the model.

A local community, represented by its leadership and/or a special committee, is the exclusive holder of a land utilisation title for its traditionally occupied area. This area and the community are documented by formal cadastral and registry institutions and statutes define rights and duties of community members by "formalising the informal". A private investor negotiates with the community and local government or administration mediates and legalises. The community allocates a clearly demarcated area to the investor for a specific land utilisation type and a formally recognised contract regulates rights and duties of the partners as well as the procedures to be followed. Other issues such as marketing and transport facilities or direct production partnership could be included in such a contract. A "win-win relation" makes the three main actors in the rural areas beneficiaries of the private-public-communal partnership. The investor (*private*) benefits from a legal set of land use rights, recognised by the traditional community. The local administration (*public*) benefits from land use and other taxes paid by the investor, an improvement of rural development and a general increase of welfare in the region by increased agricultural productivity. The traditional community (*communal*) benefits from rural development projects to be financed by the investor as a form of rent and from a high degree of land tenure security.

This model is still a sketch. It has been developed to initiate further discussion on local level cooperation, improvement of land tenure security, achievement of sustainable forms of land use and initiation of self-sustaining rural development by utilising alternative sources of funding for local development projects and public budgets. Its feasability depends on the further development of legal systems and land policy in Mozambique and the will to cooperate with all groups.

References

Alexander, J. (1994): Land and political authority in post-war Mozambique: A view from Manica Province, LTC, Madison.

Effler, D. (1996): Bodenrecht, Bodenordnung und Landnutzungsplanung im Kontext der ländlichen Entwicklung in der Manica-Provinz, Mosambik, GTZ, Eschborn
Moçambique (1975): Constituição, INM, Maputo.
Moçambique (1979): Lei 6/79 (Lei de terras), INM, Maputo.
Moçambique (1987): Decreto 16/87 (Regulamento da lei de terras), INM, Maputo.
Moçambique (1990): Constituição, INM, Maputo.
Moçambique (1995a): Programa do Governo, BIP, Maputo.
Moçambique (1995b): Política nacional de terras, INM, Maputo.
Myers, G. (1995): Land tenure in Mozambique, GTZ, Eschborn

Address of author:
Dirk Effler
Projet Protection des Ressources Agro-Silvo-Pastorales
PASP/GTZ
B.P. 10814 Niamey, Niger
or
Grüner Weg 334
D-34369 Hofgeismar, Germany

Socio-Economic Aspects of Soil Conservation in Developing Countries: The Swaziland Case

H.M. Mushala & G. Peter

Summary

Soil conservation entails efficient land management for higher productivity leading to sustainable land use. Various factors influence efficient land management. Under subsistence agriculture, as practised in many developing countries, factors like land tenure, size of farm holdings, the use of farm inputs, availability of agricultural credit, availability and effectiveness of agricultural extension services, farmers' awareness of available technologies, farmers' abilities to afford and apply the technologies and the overall agricultural infrastructure contribute significantly to the achievement of sustainable land use.

Some studies have been undertaken in Swaziland, a small country in Southern Africa, to assess the impact of socio-economic factors on soil conservation and land degradation in particular. Early attempts at soil conservation were focused on physical structures on government aided projects but the use of grass strips has survived most of the experiments. There is need however to review some of the existing legislation on land use so as to ease the pressure on Swazi Nation Land, and provide some incentives for sustained land use. A re-examination of existing land tenure arrangements is necessary if cost-effective measures of soil conservation are to be implemented.

Keywords: Soil conservation, land tenure, land management, sustainable land use.

1 Introduction

Effective soil conservation in any farming system assumes proper land husbandry for the sustainable use of land to ensure the country's food security and enhance economic development. In many developing countries however, agricultural production is extended to fragile lands as a result of growing populations, and the poor performance of the non-agricultural sectors. For many farming communities poverty, land tenure constraints, absence of expert advisory services on soil conservation, and some selected cultural practices are among the factors that have accounted for limited success in soil erosion control. Coupled with governments' neglect of subsistence farming in the affected countries, soil erosion remains a major environmental and social problem especially in developing countries. Soil conservation determines land cover characteristics so as to counteract the magnitude of raindrop impact; maintains quality characteristics of soils so that they are least erodible; and reduces the steepness of slopes so that rainfall impact and velocity are curtailed. It is only through good land husbandry that soil conservation can be achieved. In some cases, as this paper shows, accelerated soil erosion can be a result of land use intensification or inadvertent change in land use imposed by socio-economic constraints under which land users have to operate. Usually these constraints overwhelm the individual land user to the extent that proper land husbandry is precluded (Grainger, 1990; Mushala, 1992).

This paper examines the socio-economic constraints pertaining to soil conservation in Swaziland as an example of conditions prevailing in many developing countries. The thesis upheld here is that many times the subsistence farmers are aware of the environmental implications of the production methods adopted, but fail to utilise the soil resources in a sustainable manner that conserves the environment because of limiting socio-economic conditions. This results in reduced soil productivity, less efficient production systems with poor returns which inhibit further investment in land management to maintain good husbandry. As the farmers remain in perpetual poverty the environment is continually degraded.

2 Swaziland: A background

Swaziland is a small country covering approximately 17,365 km^2. It is located between latitudes 25^{0}08' and 30^{0}44' East. Only a small portion of the country adjoins Mozambique but the rest of the country is surrounded by the Republic of South Africa.

The country is divided into four physiographic regions which account for more or less of the configuration and distribution of the natural resource base. The regions run almost parallel in a North-South direction and occur sequentially as the Highveld, Middleveld, Lowveld and Lubombo.

The climate is characterised by two distinct seasons. Summer occurs between October to March and winter occurs from April to September. Winter is usually dry and over eighty percent of the annual rainfall is received during summer. The Highveld receives the most precipitation with some areas getting a mean annual rainfall of about 1500 mm. The Lowveld receives the least precipitation with a mean annual rainfall of about 500 mm. The long term mean annual rainfall for the whole country is about 910 mm. The rainfall is effective in initiating erosion because it occurs after a prolonged dry season when vegetation cover is at its minimum.

Schulze and Smithen (1982) estimated rainfall erosivity in the country and established the erosivity value of 300-350 KJmm/m^2/hr for the Highveld and 250 KJmm/m^2/hr for the Lowveld and Lubombo Plateau. Kiggundu (1986), however, using the EI$_{30}$ index established the annual average erosivity ranging from 600-500 KJmm/m^2/hr.

Land use is mainly smallholder crop cultivation, estate farming and uncontrolled grazing. Most of the smallholder cultivation and uncontrolled communal grazing is under the Swazi Nation Land (SNL), while estate farming is on Individual Title Deed Lands (ITDL). The land tenure system has some implications on land husbandry which allows for extensive gully growth and development in selected areas.

In a study to establish the extent of erosion in the country, Watson (1986) observed that in an area of only 17,350 km^2, there are more than 2,500 gullies of varying sizes. For some parts of the country gully densities are as high as one major gully system for every 5 km^2. An earlier estimate by Spaargaren (1977) indicated that certain soils without conservation were losing 25-35 tons per hectare per year when tolerable soil losses were held from 3 to 5 tons per hectare per year. These estimates indicate clearly that soil erosion remains a socio-economic problem in the country unless appropriate measures are taken to control it. The following section identifies some of the soil conservation strategies adopted in Swaziland over time.

3 Historical background of soil conservation in Swaziland

The main studies on soil conservation in Swaziland include those by Reij (1984), Nsibandze (1987) and Mushala and Manyatsi (1991). Aspects of soil erosion in Swaziland can be traced back to 1909 when the country was partitioned between the indigenous Swazis and Europeans. About 640,000 hectares were divided into 31 Native Areas, while an equal amount of land was allocated to a few Europeans, and the remainder of the land was proclaimed Crown Land (Scott, 1951). This led to a population congestion in the Native Areas accompanied by high stocking rates. Initial measures to contain the problem included the Native Land Settlement of 1944 whose objective was to settle people living on European owned farms. This was followed in 1949 by the grass stripping programme which still exists and was launched and enforced by King Sobhuza II. By royal command all arable land was to be put under grass strips. The programme was extended to all subsistence farming plots apart from a few progressive farmers who were already implementing it voluntarily. However with increased pressure on land, some of the grass strips are gradually being ploughed and in some cases they are now too thin to be effective (Mushala and Manyatsi, 1991).

Land alienation increased pressure on Swazi Nation Land (SNL) and resulted in the deterioration of the resource base. Government initiatives during the 1949-1960 development plan period, introduced better land husbandry techniques and soil conservation was regarded as an initial step towards intensive farming (McDaniel, 1966). These initiatives, however, were only ad hoc and were not part of a protracted national land utilisation programme. A comprehenive programme would tend to combine aspects of integrated range management with destocking and anti-erosion programmes if soil conservation on rangelands was the expected outcome. However as Scott (1951) observed, the soil conservation programmes put much emphasis on anti-erosion measures and precluded pasture management strategies hence were not of lasting value.

Over time, resource base deterioration became so severe that gullies were spectacular landscapes on rangelands. It was necessary to have tractors plough deeply in order to break the developed badlands, and at the same time allow for rainwater infiltration before the areas were revegetated with grass. On arable land, farmers were required to construct graded drains and embankments to check gully erosion and were subsidised by the government (Reij, 1984).

These government soil conservation efforts were not very successful on Swazi Nation Land. For example, the established physical structures were not complemented by specific land use policies, and hence were not sustainable. At the same time the established grass strips were not properly maintained. Even some of the education programmes on soil conservation measures organised by the Natural Resources Board obtained reasonable results only with the Individual Tenure Farms.

The establishment of physical structures, such as storm water drains and the ploughing of rangelands, used costly machinery provided to the government through external aid. The machinery could not be maintained hence the programme was not sustainable. Even other soil conservation measures such as grassed waterways and terracing were equally capital-intensive and most subsistence farmers did not have adequate technical skills and financial resources to apply them. Besides, there was limited scientific data on soils and land capability in many of the affected areas such that the recommended ameliorative measures usually were not commensurate with the physical environment. In any case the entire programme did not attract participation by the local communities in terms of planning and implementation of activities. This is because the government imposed the programme on the people without prior consultation, but also because the projects relied too much on heavy equipment, which most farmers could not operate nor afford (Spaargaren, 1977). Given the small size of the holdings the ramifications of land tenure constraints were therefore quite obvious.

4 Socio-economic constraints to soil conservation

It has been observed that land tenure arrangements constrain favourable land use practices and therefore land degradation is initiated and perpetuated. This has been evaluated through the examination of the land tenure system in Swaziland relative to the farming systems. Swaziland is characterised by a dual land tenure system, i.e., land held under customary tenure (Swazi Nation Land) and land held under freehold tenure (Individual Tenure Farms). The evolution of the dual system is traced back to the early 1900's. Individual freehold tenure is heralded for better productivity, high income returns, better management and utilisation of modern agronomic techniques and its enabling environment for obtaining credit. The system, however, keeps most of the land in the hands of a few people thus creating landlessness and insecurity to the traditional family. The SNL, on the other hand, is characterised by small fragmented farm holdings, low productivity, insecurity of tenure and its inability to support the growing population. Land management under the system enhances land degradation (Mushala et. al., 1994). A similar observation has been made by Funnel (1991) who considered SNL farm holdings too small and fragmented. He also observes that there are very serious erosional problems on SNL resulting from overgrazing and that modern innovations such as fencing and credit are discouraged.

The existing land ownership systems in Swaziland not only concentrate farmers and pastoralists on small farm plots but pushes them into using marginal lands. Ninety percent of SNL falls under grazing land and only ten percent is under cultivation. A majority of the small holder farmers on SNL have small plots ranging from 1-5 ha of land with an average of 2.5 ha. The size of the plots limits improvements on them, and only allows limited agricultural operations such as mechanization. Thus farmers cannot be expected to produce sufficient food supplies with the poor and inefficient production technologies employed without damaging the environment. Worse still, most of the small holder farm plots are under continuous cultivation but their productivity declines gradually. As a result the subsistence farmers tend to overcultivate and overgraze their lands, thus increasing surface runoff rates on slopes and enhancing soil erosion. The small holder farmers on SNL have limitations in terms of adopting appropriate soil conservation measures to combat soil erosion. They own land communally and have only usufruct rights over the land they use. They do not have adequate incentives to invest in long-term soil conservation measures.

The situation in Swaziland is not unique because there are many scholars who are convinced that land degradation is widespread in Africa because it is unprofitable for farmers to invest in soil conservation (Stocking 1983). A similar opinion is held by Craswell (1985) who argues that

"The benefits of soil and water conservation are not captured by the individual because rights of land are either 'communal' so that the benefits are consumed by all, (hence) the individual has no incentive to incur private costs to produce only social benefits."

Ryszkowski (1993) quotes examples from Europe whereby insecurity of ownership tends to discourage adoption of land conservation measures. For example, squatters may be reluctant to plant trees because these do not provide sufficient immediate benefits to them. Also in some countries the squatters fear that by improving the land with trees they may prompt the resumption of land by the state or land owners.

In Swaziland every Swazi household is entitled to unlimited grazing area on communal areas. The tendency now is that individual farmers accumulate as much livestock as possible without due consideration to its impact on the environment. As a result, on most parts of SNL, stocking rates exceed the carrying capacity of the land leading to soil erosion.

An examination of the land use systems in Swaziland indicate that grazing land accounts for over 60% of the total land area, of which 77% is under SNL. These areas experience the most severe types of soil erosion. The remaining 23% of grazing land is under Individual Tenure Farms. These include private ranches established mainly in the Lowveld, in which the areas are fenced for controlled grazing, and some pastures irrigated for livestock fodder in winter when the natural veld is dry or degraded.

In a study undertaken by Mushala (1992), it was observed in the study area that there were changes in land-use patterns which, over time, saw an increase in the number of livestock as evidenced by increasing numbers of cattle tracks. As overgrazing took place due to depletion of vegetation cover, increased surface runoff led to the carving out of rills and eventually gullies developed along cattle tracks given weaknesses in the soil substratum. Land ownership under SNL tends to be restrictive in terms of soil conservation because the sizes of holdings are small and the agricultural extension service is not effective. In addition institutional and infrastructural support for small holder farming is non-existent. Production is inefficient because land husbandry is not at its best.

In addition to the inefficiency in production associated with SNL, the distribution of population also contributes to the problem. According to the 1986 National Population Census, the population of Swaziland is unevenly distributed. The overall population density was established at 43 persons per km^2. The population on SNL constitutes almost 66% of the total land and is spread over 52% of the total land area. Individual Tenure Farms occupy about 47% of the total land area but accommodate only 19% of the total population. This is an indication of a very high population density on SNL than on ITF and compounds the problem even further given that the country has a population growth of 3.6%. The problem of population growth is also associated with increases in bovine population, since livestock symbolizes wealth and social status in Swazi society.

The land tenure system in Swaziland therefore constrains soil conservation because the majority of the population is confined to SNL, where the sizes of farm holdings are relatively small and too fragmented to accommodate soil conservation techniques which require large pieces of land. Similarly the farm output is too small to generate enough resources for subsistence and a surplus for re-investment in soil conservation. Many families, therefore, depend on off-farm incomes. Also, the communal ownership of rangelands permits the over-exploitation of resources through livestock accumulation which results in overgrazing. In this way the "tragedy of the commons" prevails because the situation does not provide enough incentives for individuals to incur private costs to produce benefits for everyone. Finally, the land tenure system creates an inherent sense of insecurity because of the power accorded to the chiefs to allocate land and banish "undesirable" characters within their domain. Although rarely put into practice, the tendency instills some fear and inhibits individual innovativeness and entrepreneurship in land husbandry, including appropriate soil conservation technologies.

There are many examples elsewhere in the world where the adoption of appropriate technology have been able to increase yields and conserve soils. Harrison (1987) reports on the use of 15 k per ha of fertilizer which doubled cowpeas yields. Also,the use of contour ridge cultivation, pit cultivation, and improving soils by deep ploughing and use of larger amounts of fertilizer are known methods of good land husbandry (Dazhong, 1993). Most researchers now agree that technologies for soil conservation are available but are not easily implemented nor adopted by farmers (Craswell 1985; Ryszkowski 1993; Dasmann et al., 1973; Hurni, 1993).

It is contended that the greatest problems in the adoption of the technologies revolve around the social, political, tenurial and economic aspects of land use (Craswell, 1985). Hence there is need to assist farmers in designing erosion prevention measures, provision of the necessary resources or inputs for improving land husbandry or building conservation structures. This assistance is essential in developing countries because the stakeholders are subsistence farmers and resource poor.

Some researchers have also observed that where such farmers are not assisted, their agricultural activities inevitably lead to environmental degradation, because they are likely to over exploit their land, expand their activities into still more vulnerable areas, resort to primitive land use practices or abandon their efforts and migrate to town (Dasmann, 1973), Hurni (1993) and Dazhong (1993) share similar views.

Conditions in Swaziland are similar. The land tenure system constrains land use and there is need for reform so that more land is available to small holder farmers. The agricultural infrastructure favours private ownership of land thus excludes most stakeholders. Access to more land, the establishment of a land-use policy, and improvements in the delivery of agricultural extension service can revolutionalise land husbandry on small holder farms so that soil conservation is given high priority.

5 Conclusion

The socio-economic conditions of soil conservation in Swaziland apply in many developing countries especially Africa. Most of the agricultural production is subsistence. The resource-poor farmers are given much support by their governments in terms of establishing the basic infrastructure (including marketing, credit, inputs), adopting enabling land policies, or creating appropriate democratic institutions that allow full political participation of the small holder farmers. The tendency to overcultivate or overgraze and thus causing land degradation is perhaps out of desperation because there are no other options available. For example if the communal lands were given some price tag through land reform, it would create some incentives for improved land husbandry. It would allow some levels of competitiveness, innovativeness and inevitably lead to adoption of technologies that lead to soil conservation. At the same time government support would be required to establish effective extension programmes that create more farmers' awareness, provide the necessary expertise in land husbandry with the ultimate aim of improving productivity. Poverty can be considered as the main constraint towards better land management strategies and efforts made at alleviating poverty are likely to stimulate better soil conservation especially in small holder farming communities.

References

Craswell, E. T. (ed) (1985): Soil Erosion Management: Proceedings of A Workshop held at PCARRD. Los Banos. Phillipines.
Dasmann, R.F., Mitton, J.P. and Freeman, P.E. (1973): Ecological Principles for Economic Development. New York. John Wiley.
Dazhong, W. (1993): Soil Erosion and Conservation in China. In: Pimentel, D. (ed), World Soil Erosion Conservation. Cambridge Studies in Applied Ecology and Resource Management. Cambridge University Press, 63-86.
Funnel, D.C. (1991): Under the Shadow of Apartheid: Agrarian Transformation in Swaziland. Avenbury Academic Publishing Group, England.
Grainger, A. (1990): The Threatening Desert: Controlling Desertification. International Institute for Environment and Development Publications, Earthscan, London.
Harrison, P. (1987): The greening of Africa: An International Institute for Environment and Development - Earthscan Study. London. Paladin Grafton Books.
Hurni, H. (1993): Land Degradation, Famine and Land Resource Scenarios in Ethiopia. In: Pimentel, D. (ed), World Soil Erosion and Conservation, 27-62.
Kiggundu, L. (1986): Distribution of Rainfall Erosivity in Swaziland. SSRU Research Paper No. 22. University of Swaziland.
McDaniel, J. B. (1966): Some government measures to improve African Agriculture in Swaziland. Geographical Journal 13: 506 - 513.
Mushala, H. M. (1992): The Impact of Land Tenure Systems on Land Degradation: A Swaziland Case Study. Final Report to OSSREA. Addis Ababa.

Mushala, H. M., Kanduza, A., Simelane, N.O., Rwelamira, J. and Dlamini, N.F. (1994): Comparative and Multidimensional Analysis of Communal and Private Resources (Land) Tenure in Africa: The Case of Swaziland. A Study Commissioned by Agrarian Reform and Land Settlement Service. FAO. Rome.

Mushala, H. M. and Manyatsi, A. (1991): The History of Soil Conservation In SADC Countries: A Critical Appraisal. Paper presented at the Workshop on the History of Agrarian Reform. University of Swaziland.

Nsibandze, B. (1987): The History of Soil Conservation in Swaziland. Mbabane. Ministry of Agriculture and Cooperatives.

Reij, C. (1984): Back to the Grass Strips: The History of Soil Conservation in Swaziland. SSRU Research Report. University of Swaziland.

Ryszkowski, L. (1993): Soil Erosion and Conservation in Poland. In: Pimentel, D. (ed.), World Soil Erosion and Conservation. Cambridge University Press, 217-232.

Schulze, R.E. and Smithen, A.A. (1982): The Spatial distribution in Southern Africa of Rainfall Erosivity for use in USLE. Ministry of Agriculture and Cooperatives. Mbabane.

Scott, P. (1951): Land Policy and The Native Population of Swaziland. Geographical Journal **117**(4): 435-447.

Spaargaren, W. T. (1977): Estimated soil loss due to erosion. Mbabane. Ministry of Agriculture. Land Use Planning Division.

Stocking, M. (1983): Farming and Environmental Degradation in Zambia: The Human Dimension. School of Development Studies, University of East Anglia, Norwich.

Watson, A. (1986): Investigations of the causes and hydrological implications of gully erosion in Swaziland. W.M.S. Associates. Canada.

Addresses of authors:
H.M. Mushala
University of Transkei
Private Bag X1
UNITRA
Umtata 5117, South Africa
G. Peter
University of Swaziland
Private Bag 4
Kwaluseni, Swaziland

Impact of Land Use and Tenure Systems on Sustainable Use of Resources in Zimbabwe

M. Chasi

Summary

This paper reviews the impact of land use and tenure systems in the Communal Area Sector (CAS) with communal tenure and in the Large Scale Commercial Sector (LSCS), the freehold sector. The LSCS enterprise production relationship is a function of large farm size, land suitability and maximum profit margins but with under-utilisation of up to 80% of potential arable land. The LSCS sector is supported by powerful farmer institutions and deemed ecologically sustainable, but due to skewed land distribution, LSCS is considered politically unsustainable. The CAS exhibits over-utilisation, with population densities three times the estimated potential carrying capacity. The land tenure, land use practices and land administration systems are tolerant of continuous population absorption leading to high land use pressure, degradation and poverty, and rendering CAS ecologically unsustainable.

Keywords: Tenure, communal, freehold, land, institutions, sustainability.

1 Introduction

Land use systems in Zimbabwe are closely linked with tenurial systems of the agricultural sectors, notably the Large Scale Commercial and the Small Scale Commercial which are the freehold sectors, and the small holder communal area sector. There is a relationship between land tenure pattern and the nature and distribution of environmental resources. This is attributable to the historical legacy of inequitable land and natural resource access between the Large Scale Commercial Sector (LSCS) and the Communal Area Sector (CAS).

CAS land degradation has been attributed to the tenure system. In his comparative analysis of the Small Scale Commercial Sector (SSCS) farms and CAS in the same agro-ecological region, Ashworth (1993) concluded that the tenure system has contributed to better access to credit and more production improving technology resulting in better yields and outputs in SSCS than in the CAS. He found that the CAS agricultural production system and population pressure resulted in resource imbalances giving rise to degradation and lack of basic needs. This does not discount the multifaceted variables which account for the whole degradation processes on all land types.

2 Tenure systems of the agricultural sectors of Zimbabwe.

Zimbabwe has multifaceted tenure patterns corresponding to its four agricultural sectors: freehold, leasehold, state ownership and communal ownership. The LSCS freehold, and CAS communal tenure system which together account for 72% of Zimbabwe's soil, are discussed below.

ISBN 3-923381-42-5
© 1998 by CATENA VERLAG, 35447 Reiskirchen

2.1 The Large Scale Commercial Sector (LSCS) and the Small Scale Commercial Sector (SSCS): Freehold

Land distribution is skewed in favour of the 4,600 LSCS farmers occupying and controlling 30% of Zimbabwe's land. The mean farm size is 2,285 hectares in the sector. The mean farm size increases from 1,640 hectares in the wetter Agro-ecological Region I to 13,460 hectares in semi-arid, marginal Agro-ecological Region V (Central Statistical Office, 1994).

The freehold tenure allows for property subdivision as well as consolidation within the provisions of the Regional, Town and Country Planning Act of 1976. The rate of subdivisions is low since the prerequisite procedures were made stringent to discourage creation of 'economically and agriculturally unsustainable' land sizes. Property consolidation is simpler and fast becoming important in the constitution of nature conservancies in semi-arid and marginal zones with a potential for wildlife. Large properties are extended for wildlife management giving rise to ecologically sustainable production systems which are generally associated with exclusion of less profitable crop and beef production systems. Wildlife-based tourism has higher profit margins and reduced variable costs. Freehold title facilitates the enforcement of the Natural Resources Act, which is the main conservation legal instrument. Degradation in the LSCS is not as discernible as that occurring in CAS.

The sector has developed powerful institutions to sustain itself. Historically, the Commercial Farmers Union and affiliated LSCS institutions have successfully negotiated and lobbied on national and international fronts for the political and economic objectives of their constituency.

2.2 Communal area sector: Communal tenure system

The communal tenure rights are laid down in the cultural context within the provisions of the Communal Lands Act of 1982. Households have property rights for residential and arable land, with rights to subdivide, bequeath and inherit. Natural rangelands resources including grazing are held on a communal basis. Communal farmers have usufruct rights, they are unable to alienate or transfer ownership to another individual or community. CAS occupies 42% of Zimbabwe's land area, generally located in marginal semi-arid zones and supports one million farmers and 70% of the rural population. By 1983, 6.8% of CAS residents had no access to arable land and up to 50% of the farmers had less than 2.5 hectares of arable land (Central Statistical Office, 1984).

A research study by the Land Tenure Commission (LTC) set up in 1994 concluded that the communal land tenure system imposed economic constraints adversely affecting agricultural efficiency and sustainability under conditions of increasing population pressure. The constraints listed included: free right of access and administrative allocation of land; continual reduction of arable land sizes through subdivision; effects of the usufruct basis of tenure; the seemingly open access to communal natural resources and ineffective village level institutions (Zimbabwe Commission of Inquiry into Agricultural Land Tenure Systems, 1994).

Better accepted unwritten popular rules have developed parallel to and despite the official by-laws and the associated legal framework. Apparent is the conflict between the village based modern democratically elected institutions, the Village Development Committees (VIDCO) and the traditional institutions headed by village heads. Conflict between modern and traditional institutions in land administration weakened the enforcement mechanism, reducing the security of exclusion rights, resulting in land allocation conflicts. This poor management of communal resources resulted in over-utilisation, uncontrolled settlement and encroachment onto grazing land causing severe erosion, degradation, placing limitations on investments and productivity. In those cases where neither the traditional nor the elected leadership are in control, poor land administration and management have had adverse retrogressive consequences on the resource base.

3 Land use systems in the LSCS and CAS

A review of the LSCS diverse, high input, mechanised farming practices reflects an economically viable and ecological sustainable system. The LSCS properties indicate large farms with existence of dis-economies of scale noted by Mlambo and Zitsanza (1994) which were in conformity with the findings of Thutle et al. (1993), quoted in Chasi et al. (1994). Comparison of stocking rates on farms and estimated livestock carrying capacity indicate grazing utilization levels of 41% to 84% in LSCS, while the arable land utilisation levels varied between 15% and 33% of the potential arable land on farms (Chasi et al., 1994).

The dominant land use system in CAS is cropping supported by ox draught power. The cropping patterns do not match the suitability of the agro-ecological base. Maize, the staple food is cropped throughout the agro-ecological zones despite the semi-arid, marginal and risky nature of Agro-ecological Regions IV and V where 75% of the CAS are located. The farming system is based on low input levels but labour intensive employing 5.4 persons per hectare as compared to 1.6 persons in the LSCS. Unlike the LSCS; CAS farmers do not always follow crop rotation systems due to shortage of land.

Investment has a skewed and dualistic characteristic, a direct result of a deliberate past Government policy. Government provided infrastructure such as roads, electricity and water at lower levels in CAS accounting for 2% as compared to the LSCS at 98% of total investment. Formal credit is low in CAS (AFC, 1994). It is estimated that investment per hectare in the LSCS is three times higher than in the CAS. This compares with the average productivity ratio per hectare cropped between the two sectors especially for maize. It therefore follows that investment in conservation works is much lower in the CAS due to lack of financial resources.

In the CAS, livestock are kept for subsistence purposes with an input function as draught power and manure provision rated at 50% of the value. Land is overstocked by up to 200% of assessed carrying capacity leading to severe degradation of rangelands. At the same time the herd cannot adequately sustain and support the current farming system with draught power shortages at 30% to 50% of requirements in some communities (Chasi et al., 1994). Nationwide the sectoral distribution of poverty is estimated at 81% of the poor in the CAS and 51% among the LSCS workforce, (Ministry of Public Service, Labour and Social Welfare, 1996).

Population pressure increases are possible in CAS due to the land tenure system which permits continuous absorption of people despite long term adverse environmental effects. For example, the LSCS and the industrial sector offload their redundant labour arising from economic structural adjustment programmes into the communal sector. The socio-cultural attitude, legal and administrative framework is generally permissive of subdivision and fragmentation of arable land. Household arable blocks which were 8.6 hectares in the 1960s are now 2.5 hectares.

4 Conclusion

The differences between the CAS and LSCS tenure systems was used as a basis for differential treatments between the sectors by colonial regimes. The LSCS can now be regarded as a self-perpetuating exclusive sector, which is ecologically more sustainable than the CAS. Nonetheless it reflects inequitable and skewed land ownership patterns, investment and supportive institutions and legal framework. There is potential conflict arising due to inequity in landholding between sectors, rendering the LSCS unsustainable from a political perspective. The LSCS is not responsible for socio-economic constraints of other sectors but it should absorb its fair share of the potential sustainable carrying capacity on equity principles, to offset the shocks in the remaining sectors.

The evidence gathered directly by the LTC through open hearings, written and oral evidence from the indigenous CAS inhabitants including women's groups indicated that the majority felt that the communal tenure was appropriate, despite the high population pressure at three times the recommended carrying capacity, notwithstanding the constraints attributed to communal tenure. What is needed, the people felt, is to solve the land shortage through land redistribution and intensification of production systems. The LTC concluded the traditional land administration authority persisted despite legal provisions for its replacement by VIDCOs and Rural District Councils. Evidence suggested traditional leadership was preferred. Government intervention is required to institute appropriate policy, legal and administrative framework which are supportive of the traditional authority and capacity conducive to management of sustainable village livelihood systems. In formulating such a framework, CAS should lobby and negotiate for its own needs and ensure equity issues are redressed.

Multifaceted tenurial systems cannot operate independently within one country. Government policies should be in place to create a healthy interface which leads to the successful co-existence of the tenurial patterns to meet needs and aspirations of different communities. In the case of Zimbabwe, political decisions in terms of land redistribution have been called for through land taxes, land designation, and land ceilings in the LSCS. The goal is to create a sustainable and vibrant agriculture and industry across sectors.

References

AFC (1994): Agricultural Finance Cooperation, Annual Report. The AFC, Harare, Zimbabwe.

Ashworth, V. (1993): Towards Sustainable Small holder Agriculture, Ministry of Lands, Agriculture and Water Development Report, Harare.

Central Statistical Office (1994): Census 1992, Zimbabwe National Report, Central Statistical Office, Harare.

Chasi, M., Chinembiri, Mudiwa, C., Mudimu, G. and Johnson, P. (1994): Land Fragmentation, A Report prepared for the Commission of Inquiry into Land Tenure Systems in Zimbabwe, Government of Zimbabwe, Harare.

Mlambo & Zitsanza (1994): Economics of Scale, Capacity Utilisation and Productivity Measurement in Zimbabwean Commercial Agriculture: Invited Papers and Contributed Papers, XXII International Conference of Agricultural Economists Held in Harare, Zimbabwe 1994. Published by the International Association of Agricultural Economists.

Ministry of Public Service, Labour and Social Welfare (1996): 1995 "Poverty Assessment Study Survey Preliminary Report," Ministry of Public Service, Labour and Social Welfare, Harare.

Thirtle, C., Atkins, J., Bottomley, P., Gonese, N. and Govere (1993): Agricultural Productivity in Zimbabwe, 1970 to 1990. Economic Journal, Vol. 103.

Zimbabwe (1994): Commission of Inquiry into Agricultural Land Tenure Systems (Rukuni Report) Volume Two: Technical Report, Volume One: Main Report, Government Printer, Harare

Zimbabwe Commission of Inquiry into Agricultural Land Tenure Systems (1994): Rukuni Report, Volume Two: Technical Report, Volume One: Main Report. Government Printer, Harare.

Address of author:
Mutsa D. Chasi
Department of Agricultural Technical and Extension Service
Agritex Harare
P.O. Box CY 639
Harare, Zimbabwe

Assessing the Potential and Acceptability of Biological Soil Conservation Techniques for Maybar Area, Ethiopia

K. Goshu

Summary
After carefully assessing the conditions in the Maybar area and the economic and technological limitations of the farmers, the potential of biological conservation techniques was evaluated on the basis of their possible applicability. The selected techniques included row-planting of maize on contour ridge, vetch inter-row-planting in maize without ridges, relay planting of Teff in maize, and early planting of horsebeans. The effects of these techniques on production and on erosion control were examined for two consecutive years (1988-89) in two similar fields in comparison with farmer practices under their own management

Soil loss results suggest that erosion can be reduced to some extent by applying relatively easily applicable biological soil conservation techniques. The 1989 results of row planting of maize on contour ridges and vetch inter-row planting in maize without ridges (RI) showed a reduction in soil loss by about 49%, and 16%, respectively. Relay planting of Teff in maize reduced soil loss by 13%; and early planting of horsebeans reduced by 6.8% to 18.1%, compared to the farmer practices.

Row-planting of maize on contour ridges showed a reduction of about 7% in 1988 and 10% in 1989 in grain yield compared to the farmer method. Vetch inter row planting in maize without ridges also showed a reduction of about 10% in grain yield in 1989, though it showed an increase of about 16% in grain yield in 1988. Relay planting of Teff in maize showed about a 20% increase in grain and 9% in biomass, while early planting of horsebeans showed 2.3% to 8.2% increases in grain and 3.3% to 8.4% in biomass. Moreover, both the reductions in soil loss and the increases in crop yield observed in the experiments are far less than the variations between years due to climatic fluctuations. It was only the increase in supplementary fodder production (479%) from vetch inter-row planting in maize without ridges that was significant enough to be noticed by the farmers. It is more likely that the small beneficial effects of the relatively easily applicable biological soil conservation techniques are considered by the farmers as part of the variations due to the annual fluctuations rather than resulting from conservation techniques themselves. Hence farmers remain reluctant from applying these techniques.

Keywords: Applicability, biological soil conservation techniques, effectiveness

1 Introduction

The highlands of Ethiopia were and still are the sources of livelihood of most Ethiopians, mainly because of their favorable climatic conditions. "Most of the sedentary agriculture of the country has been going on for centuries, particularly in the northern part, which has been settled since more than 5000 years ago" (Hurni, 1985). Constable (1985) Estimated that the highlands of Ethiopia, which include those areas of above 1500 m asl and their associated valleys, cover 43% of the country's total area and accommodate 88% of the population, around 67% of the livestock and over 90% of the permanently cultivated area. However, this preferential concentration of people and their associated agricultural activities on the highlands for centuries, coupled with constantly increasing population and poor management practices, have gradually resulted in deforestation, overcultivation and overgrazing, which in turn led to inevitable accelerated soil erosion.

This study was undertaken in 1987 as part of the research programmes of the Soil Conservation Research Project (SCRP). Its aims were to assess the potential of biological soil conservation techniques for croplands in Maybar and the surrounding areas, and to supplement the then on-going soil conservation implementation programme with appropriate biological soil conservation techniques to be selected on the basis of the study.

The Maybar Research Unit of the SCRP is located in Dessie Zuria Wereda, at 39° 39'E and 10° 59'N, 14 km SSE of Dessie, the capital of South Wello Administrative Zone. The area is part of the central highlands of Ethiopia, which has been settled for centuries and gradually degraded by water erosion. The altitude ranges from 2500 to over 2800 m asl. There are two rainy seasons, the main rainy season (Kremt) from July to September, and the minor rainy season (Belg) from March to May. Both seasons are used for crop production.

2 Methodology

The study had two phases: preliminary assessment, and experimentation.

2.1 Preliminary assessment

This involved an examination of economic and technological limitations of applying biological soil conservation techniques. It was done in the 1987 Kremt through: (a) close observations of the agricultural activities, (b) extended discussions with randomly selected farmers and (c) consultations with experts and review of appropriate literature.

A pre-experiment evaluation of the conservation techniques was then made on the basis of the possible applicability of each technique in view of the economic and technological conditions of the farmers and relatively easily applicable biological conservation techniques were selected for experimentation. Due consideration was also given to widely grown crops of the area while selecting the techniques.

The biological soil conservation techniques selected for experimentation were:
- Alternative planting methods for maize and Teff:.
  maize: row-planting on contour ridges and vetch inter-row-planting without ridges
  Teff: relay planting with maize
- Alternative planting time for horsebeans:
  early planting

2.2 Experimentation

2.2.1 Site selection and runoff plots establishment

The experiments were carried out on the farmers' fields using runoff plots and under their own management. For this, two similar and representative fields of the cultivated lands, with about 10%

slope and soil type of deep Haplic Paeozems, were selected and arbitrarily designated as Field I (FI) and Field II (FII). Eight similar and parallel runoff plots with individual collection systems were constructed in each field. Each runoff plot had an area of 45 m² (3 m wide and 15 m long) and was enclosed with removable corrugated iron sheet borders to avoid flowing into or out of the runoff plots. Runoff and soil loss were collected in two tanks, one about 250 liter in capacity directly connected with an inlet tube from the collection ditch at the lower end of each runoff plot, and the other tank taking only 1/10 of the eventual overflow from the first tank through a slot-divisor.

2.2.2 Experimental design

Experimental design for maize: The experimental design used for maize was completely randomized with unequal replications i.e. 3 treatments by 3 or 2 replicates. The treatments were row-planting on contour ridges (RR), vetch inter-row-planting without ridges (RI) and broadcasting, farmers' practice as a control (C) (see Figure 1 for the layout).

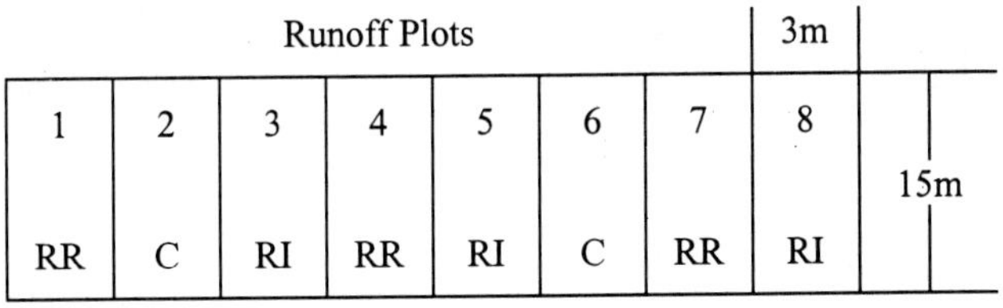

			Runoff Plots				3m	
1	2	3	4	5	6	7	8	
RR	C	RI	RR	RI	C	RR	RI	15m

Figure 1: Layout of three alternative planting methods of maize

Experimental design for horsebeans: The experimental design used for horsebeans was two treatments by four replicates. Early planting (EP) refers to the planting of horsebeans as soon as the Kremt rains start and the soil is wet enough for easy ploughing, while farmers' planting time (FP) relates to the planting of horsebeans at the time at which about half of the farmers in the area finish planting. Randomized layout was not used here because it precludes ploughing during the farmers' planting time as this would damage the seedlings of early planting in the neighbouring runoff plots (see Figure 2 for the layout).

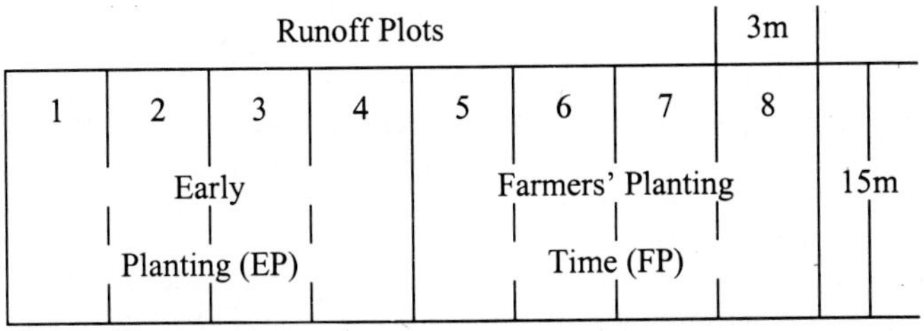

			Runoff Plots				3m	
1	2	3	4	5	6	7	8	
	Early Planting (EP)				Farmers' Planting Time (FP)			15m

Figure 2: Layout of two alternative planting times of horsebeans.

Experimental design for Teff: The experimental design used for Teff was also two treatments by four replicates. Relay planting of Teff (RPT) refers to the planting of Teff before the harvesting of the maize crop, that is to the planting of Teff in the maturing maize crop. Standard planting of Teff (SPT) relates to the planting of Teff after the harvesting of the maize crop as traditionally practised in the area. This involves removing all the residues from the field, ploughing and compacting the soil. For the same reasons given above, randomized layout was not used here too, hence a similar layout to that of Figure 2 was applied.

2.2.3 Data collection

Regardless of the different kinds of crops and interventions used in the experiments, the same procedures of data collection have been followed. Data on runoff, soil loss and crop yield were collected from each runoff plot. Runoff and soil loss data were collected on a storm basis. Measuring and sampling were made when any one of the direct tanks were more than 0.25 m full or when the daily rainfall was more than 12.5 mm, and there was some water in the tanks. To get the average depth of the runoff water in each tank, measurements were made at three points. Then the runoff water was drained saving a sample of one liter from the tanks when the water was muddy. Soil loss from each of the direct tanks was measured using a bucket and a balance and a 500g sample, if the collected soil in each tank was less than 500g, total soil, was saved in a plastic sack. These measurements were later used to determine the soil loss and the runoff for each runoff plot and for each treatment after the oven-dried weight of each soil sample was obtained.

Plant growth was measured weekly in an area of one square meter of each runoff plot. At harvest, grain and biomass were weighed separately for each runoff plot.

3 Results of the experiments and discussions

3.1 Alternative planting methods on maize

3.1.1 Runoff and soil loss

In general, the 1988 runoff and soil loss results were very low for all planting methods (Table 1) hence, the effect of each planting method in reducing soil loss was not fully observed in this particular experiment. Relatively higher runoff and soil loss results were observed in the 1989 experiment as compared to the 1988 results. Subsequently, the differences in runoff and soil loss among the three alternative planting methods (treatments) were also large. The highest monthly soil loss in all the three planting methods was in April, during which the height of maize plant was less than 20 cm and the maize crop cover presumably was insufficient, although the highest monthly rainfall during the experiment was in August. In fact, 88.1% of the soil loss in row-planting on contour ridges (RR), 69.5% in vetch inter-row-planting without ridges (RI) and 62.8% in broadcasting (C) of the entire experiment are results of a single storm that occurred on April 4. These also signify how damaging the single erosion events are in the area. On average, row-planting on contour ridges (RR) showed the lowest runoff and soil loss, reducing by 54.9% and 48.6% respectively as compared to broadcasting (C), and by 22% and 38.6% when compared to vetch inter-row-planting without ridges (RI). Vetch inter-row-planting without ridges (RI) also reduced runoff by 42.2% and soil loss by 16.2% when compared to broadcasting (C) (Table 1).

3.1.2 Yield

In 1988, vetch inter-row -planting without ridges (RI) showed the highest grain yield (2.24 t/ha) while broadcasting (C) showed the highest biomass yield (6.81 t/ha). Row-planting on contour ridges (RR) showed the lowest grain (1.79 t/ha) and biomass (6.10 t/ha) yields. In 1989, broadcasting (C) showed the highest yield of grain (6.07 t/ha) and biomass (16.16 t/ha) compared to the two interventions, which showed almost similar yield results (Table 1). However, the differences in both years were not-significant and they could possibly be due to the wide spacing (80 cm) between the rows used in the experiment.

Planting methods (Treatments)	Field and Year	Erosion		Yield (t/ha)		
		Runoff (mm)	Soil loss (t/ha)	Biomass	Grain	Supp. Fodder
Broadcasting (C)	(1988) FII	9.8	1.14	6.81	1.93	
Row-planting on contour ridges (RR)	FII	10.7	1.27	6.10	1.79	
Changes as % of C		9.2+	11.4+	10.4-	7.3-	
Vetch inter-row- planting without ridges (RI)	FII	10.4	1.19	6.58	2.24	
Changes as % of C		6.1+	4.4+	3.4-	16.1+	
Broadcasting (C)	(1989) FI	72.5	10.42	16.16	6.07	0.77
Row-planting on contour ridges (RR)	FI	32.7	5.36	14.40	5.44	0.40
Changes as % of C		54.9-	48.6-	10.9-	10.4-	48.1-
Vetch inter-row-planting without ridges (RI)	FI	41.9	8.73	14.22	5.41	4.46
Changes as % of C		42.2-	16.2-	12.0-	10.9-	479.2+

Key: + higher   - lower

Table 1: Erosion and crop yield results under the three alternative planting methods of maize.

In the Maybar area, maize plantation is widely used to supplement animal feeds. In this experiment, the amount of green feed produced with each of the three alternative planting methods, including the vetch biomass for RI, was also observed as supplementary fodder. The maize fodder yield was very low; less than 1 t/ha in all the three planting methods. But, when the vetch biomass is added to the supplementary fodder yield, vetch inter-row-planting without ridges (RI) showed the highest fodder yield (4.46 t/ha) (Table 1). This difference was very noticeable and significant enough to be realized by the farmers and to attract their interest.

3.2. Alternative planting methods on Teff

3.2.1. Runoff and soil loss

In both planting methods, both runoff and soil loss were greatest in August when the monthly rainfall was the highest. In fact, two storms, one on August 5 and another on August 13, accounted for 45.8% of the monthly soil loss in the relay planting (RPT) and 50% of the monthly soil loss in the standard planting of Teff (SPT). On average, relay planting of Teff (RPT) showed about 13% reduction in soil loss as compared to the standard planting (SPT) (Table 2).

3.2.2. Crop yield

The yield differences between the two alternative planting methods (treatments) were not-significant, though relay planting of Teff (RPT) showed slightly higher results both in biomass and in grain (Table 2).

Planting methods (Treatments)	Field and Year	Erosion		Yield (t/ha)	
		Runoff (mm)	Soil loss (t/ha)	Biomass	Grains
Standard planting of Teff (SPT)	(1988) FII	374.4	25.35	3.26	1.37
Relay planting of Teff (RPT)	FII	418.4	22.16	3.55	1.64
Changes as % of SPT		11.6+	12.6-	8.8+	19.7+

Table 2: Erosion and crop yield results under the two alternative planting methods of Teff.

3.3 Alternative planting times on horsebeans (broadbeans)

3.3.1 Runoff and soil loss

In 1988, higher soil losses were observed in July and August in both planting times of horsebeans. In both, the highest monthly soil loss occurred in July, soon after planting, though the highest monthly rainfall which also produced the highest monthly runoff was in August. In the 1989 experiment, both the runoff and soil loss results were much less than the results observed in the 1988 experiment. In both planting times, the highest monthly runoff and soil loss occurred in August in parallel to the highest monthly rainfall. On average, early planting (EP) reduced soil loss by 18.1% in 1988 and by 6.8% in 1989 as compared to farmer planting time (FP) (Table 3).

Planting times (Treatments)	Field and Year	Erosion		Yield (t/ha)	
		Runoff (mm)	Soil loss (t/ha)	Biomass	Grains
Farmers' planting time (FP)	(1988) FI	493.8	60.25	4.82	2.56
Early planting (EP)	FI	485.3	49.35	4.98	2.62
Changes as % of FP		1.7-	18.1-	3.3+	2.3+
Farmers' planting time (FP)	(1989) FII	150.1	10.36	4.90	3.04
Early planting (EP)	FII	142.9	9.66	5.31	3.29
Changes as % of FP		4.8-	6.8-	8.4+	8.2+

Table 3: Erosion and crop yield results under the two alternative planting times of horsebeans.

3.3.2. Crop yield

Biomass and grain yield results of horsebeans of each planting time are given in Table 3. In both years early planting (EP) showed higher yields though the differences were not-significant.

4 Conclusion

Controlling erosion in the croplands of the Maybar area through the application of biological soil conservation techniques is a difficult undertaking. Erosion hazards are more serious in the croplands, especially during ploughing and planting times, and single erosion events are very damaging. Because of very high economical and technological limitations, it is only the relatively easily applicable techniques that can be used by the farmers in the area. However, the beneficial effects of the relatively easily applicable biological soil conservation techniques, both in controlling erosion and improving production, are too small to attract farmers. In fact, it is more likely that the small beneficial effects of the relatively easily applicable techniques can be considered by the farmers as part of the variations in annual rainfall fluctuations rather than being the results of the conservation techniques. Furthermore, some of the relatively easily applicable techniques require extra labour for non-rewarding returns. Therefore, the potential of biological soil conservation techniques for controlling erosion in the croplands of the Maybar area and by implication in the highlands of Ethiopia is limited.

References

Constable, M. (1985): Ethiopian Highland Reclamation Study (EHRS) Summary (draft), EHRS working paper 24. LUPRD, MOA, Addis Ababa, Ethiopia. 34 pp.

Hurni, H. (1985):. Erosion-productivity-conservation systems in Ethiopia. Paper to the 4th International Soil Conservation Conference, Maracay, Venezuela. 20 pp.

Address of author:
Kassaye Goshu
Soil Conservation Research Programme (SCRP)
P.O. Box 10787
Addis Ababa, Ethiopia

Land Tenure Systems and Sustainable Land Use in Andhra Pradesh: Locating the Influencing Factors of Confrontation

B.J. Rao, R. Chennamaneni & E. Revathi

Summary

Land use pattern has an intricate relationship with the nature of land tenure systems in effect in a country. Different socio-, economic-, and cultural factors influence legal and social attempts for change. The authors attempt to explain the inter-connected influence of various factors over land tenure systems implementation aimed towards the development of the Adivasis Region. Effective handling of social, cultural, legal and demographic factors can reduce confrontation and increase necessary co-operative perception among the Adivasis people. Such policies and strategies can enhance sustainable land use in the long run.

Keywords: Land tenure systems, Adivasis and land relations, land tenure policies, sustainable land use

1 Introduction

Land use pattern has an intricate relationship with the nature of land tenure systems in effect in a country. Different socio-economic and cultural factors influence legal and social attempts for change. The failure of effective land tenure systems due to the above factors led to a constant confrontation between the Adivasi people and governmental institutions. In this paper we try to explain the inter-connected influence of various factors over the efficacy of land tenure system implementation. The effect of land dispossession on the Adivasis people as an outcome of land tenure systems is the main objective of the paper.

2 Methodology

The methodology is based on time series analyses, interviews and case studies. The following issues have been dealt with a) Origin, growth and application of land tenure systems in the Adivasi regions of Andhra Pradesh b) Implementation of protective Land Tenancy Laws in Adivasis Region between 1950-1995 c) Efficacy of the Protective Land Laws and Sustainable Land Use - Response from the Generations - Empirical Evidence d) Land Disposession, Restoration - Confrontational Factors e) Institutional Response to the Adivasi Land Claims and Issues of Sustainable Land Use.

3 The Problem

Adivasis Land Tenure Systems have been a constant unresolved problem in many Indian States. The existing structural inequalities have their roots in the problem of a Land Tenure System which was largely shaped by colonial governance. During the British colonial period, the local collective institutions of self-rule of the tribal communities (which were existing in nascent form) were totally ignored. Commenting on this aspect of the inducement of the institutional process, Dharma Kumar states that, "For India, the major institutional discontinuity of modern times was the colonial conquest. ... The transition of Independence, while horribly costly in terms of human deaths and dislocation, was in institutional aspects relatively smooth" (Dharma, 1993).

Thus, colonial conquest and modern jurisprudence have simultaneously made their entry into the tribal regions of India. The immediate implication was the introduction of an alien land revenue system in the tribal areas. The Land Revenue System was alien because it had not taken local institutions into consideration when the adjustment and entitlement to land resources began to take shape. Habib (1967) and Rothermund (1978) have aptly perceived this transition. Habib observes that "Initially neither the notions of landed property nor rental value applied to rural India. ... Peasants were aware of their right to till the soil which they had cleared and inherited, lords of various kinds claimed their share of the produce but these claims were based on public functions and not on private property - land transfers did take place and grants show an awareness of all the conditions of the control of land. ... But there was no land market" (Habib, 1967). Further, Rothermund states that the decisions of a revenue officer with regard to property presumed to be correct unless challenged in this way, they could at the most be subjected to the scrutiny of higher revenue authorities who were normally bound to back him, because theoretically they had to sanction everything he did and by admitting the validity of a complaint they were thus censuring themselves. The peasant had, therefore, no real chance to protest against a high assessment. If he defaulted, his land was attached and sold by the revenue administration. If his moneylender paid the revenue for him and he could not pay his debts, he was soon confronted with an exparte decree which the moneylender quickly obtained from one of the ubiquitious courts (Rothermund, 1978).

Thus, the net result was the ousting of the peasant from the land. Hence, large scale land alienation had taken place. The original tillers of the land were sidelined. The inability to pay the tax was the handy excuse for removal of peasants from the lands. Commenting on this, Justice Raymond West made an implicit observation. In his words, "the introduction of these doctrines had deprived the old proprietary castes and classes of their land and had replaced them with the moneylender who in keeping with his caste and profession was a mean coward on whom the Government could not relay at a time of crisis." West warned that the courts were getting into disrepute because they provided the money lender with powers which he had never had under indigenous rule (Rothermund, 1978).

The net result was the occurrence of large scale land alienation both in the plain areas as well as in tribal regions. Thus, designing an alien revenue system and its effects on India's land system was not only limited to the plain areas of India but expanded also soon to its tribal regions. The 'Revenue Augmentative Logic' of the colonial master played a disastrous role (Anderson and Huber, 1988). A number of social consequences, such as coining new sets of revenue, forest rules and forest formulation of the land auctioning policy by the Government, were the results (Guha and Gadgil, 1989).

Thus, modern state and governance in colonial and post-colonial periods have laid the beginnings of the selective enrichment of the social classes in India. Selective enrichment process both in the spheres of power and wealth was strengthened with the colonisation of India by the British. As a part of these processes, the prescription and application of laws conducive to its own rule had been imposed to regions which had previously not been an integral part of the mainstream hierarchial rule. "Legality" is a declared legal arrangement of the governing to offer justice to the so-called governed. It speaks the codes, acts and articulates the association between the state and civil society. Contrary to the medieval state's legal behaviour, modern state legality is believed to have been evolved on the principles of democracy, rule of law and constitutional accountability. The classic example of the subversion of the constitutional law and its adherence to the said ideas and the quality of lawlessness can be seen in the case of tribe and land relations in India. The replication of these broad realities are seen even in the Andhra Pradesh tribal protest.

4 All India scenario

Of the 220 to 300 million indigenous people of the world, 67.7 million live in India (1991 census), the country with the largest indigenous population in the world (see Table 1). Constituting 8% of the total population of the country, the indigenous people of India are popularly called "Adivasis", while the government has classified them as 'scheduled tribes'. They are spread over 26 states and Union Territories (UT's). They are not evenly distributed over the Indian land mass but are essentially found in pockets or regions across the country, mainly the mountainous and forested areas. The 6 regions of Adivasi concentrations are:
    I   Central Region
    II Western Region
    III North-Eastern Region
    IV North-Western Region
    V Southern Region
    VI Island Region

There are 527 communities ranging from the Great Andamanese who are only 33 in number as compared to the Gonds, Santals and Bhils who are 5 million, 4 million and 3.5 million, respectively. More than half (55%) of Adivasis are located in the Central Region consisting of Madhya Pradesh, Orissa, Bihar, Andhra Pradesh and West Bengal, while in the North Western Region (Himachal Pradesh and Uttar Pradesh) there are only 0.75%. The Island Region of Andaman and Nicobar and Lakshadweep have a Adivasi population of 0.11%. In terms of Adivasis ratios to non-Adivasis of the region, they are the highest in the North-Eastern Region and lowest in the Southern Region comprising of Karnataka, Kerala and Tamilnadu.

5 Land tenure and Adivasis in Andhra Pradesh

Alienation of tribal lands in Andhra Pradesh has been allowed to grow in proportion in the past decades. The main reasons have been a feeble restoration process and strong resistence by the powerful propertied sections either through force or manipulation of the laws. Consequently, the implementation of Land Transfer Regulation has neither remained effectively prohibitive in arresting the land transactions between the tribes to non-tribes in actual terms nor resulted in any significant degree of restoration of the lands to the tribes. The available data on the process of land restoration and the implementation of land transfer regulations reveal the intensity of the problem and the injustice faced by the tribes.

The Government of Andhra Pradesh issued the Andhra Pradesh (A.P.) (Scheduled Areas) Land Transfer Regulation, 1959, for protecting the tribal land rights. This was in force in the Scheduled Area of Andhra Pradesh. It was subsequently extended for application to the Scheduled Area of Telangana Region. This extension was done through A.P. Scheduled Area (Extension Amendment) Regulations, 1963. Subsequently, the amendment of the 1970 act was mooted mainly to plug the loopholes of the 1959 Regulation. The Land Transfer Regulation of 1970 declared that any transfer of immovable property situated in the agency tracts by a person, whether or not such a person is member of a Scheduled Tribe, was absolutely null and void, unless such transfer is made in favour of a person, who is a member of Scheduled Tribe or a Society, registered under the Co-operative Societies Act, 1964, which is composed solely of members of the Scheduled Tribes.

Region	Population	Tribal %	Total %
I Central Region			
1.Madhya Pradesh	15.399.034	23.27	22.73
2.Orissa	7.032.214	22.21	10.38
3.Bihar	6.516.914	7.66	9.77
4.Andhra Pradesh	4.299.481	6.31	6.20
5.West Bengal	3.808.760	5.60	5.62
Total Region	37.056.403		54.70
II Western Region			
6.Maharashtra	7.318.281	9.27	10.80
7.Gujarat	6.161.775	14.92	9.09
8.Rajasthan	5.474.881	12.44	8.08
9.Dadra & Nagar Haveli	109.380	78.99	0.16
10.Goa	376	0.03	-
11.Daman and Diu	11.724	11.54	0.01
Total Region	18.076.427		28.14
III North Eastern Region			
12.Meghalaya	1.517.927	85.53	2.24
13.Arunachal Pradesh	550.351	63.66	0.81
14.Nagaland	1.060.822	87.70	1.57
15.Manipur	632.173	34.41	0.93
16.Mizoram	653.565	94.75	0.97
17.Tripura	853.345	30.95	1.26
18.Sikkim	90.901	22.82	0.13
19.Assam	2.874.441	12.82	4.24
Total Region	8.233.525		12.15
IV North Western Region			
20.Himachal Pradesh	218.349	4.23	0.32
21.Uttar Pradesh	287.901	4.83	0.43
Total Region	506.250		0.75
V Southern Region			
22.Karnataka	1.915.691	4.26	2.83
23.Kerala	320.067	1.10	0.47
24.Tamilnadu	574.194	1.02	0.85
Total Region	2.810.852		4.15
VI Island Region			
26.Andaman and Nicobar	26.770	9.54	0.04
27.Lakshadweep	48.163	93.15	0.07
Total Region	74.933		0.11
All India Total	67.758.390	8.08	100

Source: Government of India. 1991 Census. New Delhi,1992. pp.12. Census was not held in Jammu and Kashmir, but a projected figure for total population has been included.

Table 1: Population in each State/Union Territories (UTs), percentage with respect to total population of the state/UTs and the percentage with respect to total tribal population of India.

This regulation had generated a considerable amount of land restoration activity in the tribal areas of the state up to 1979 - specially in the districts of Khammam, Warangal and Adilabad. This was the result of a few committed officials, who were the Special Deputy Collectors for Tribal Welfare (an Agency which was created by the government for land restoration in the tribal area of each district in the state). These officials traced out the whole details of land-ownership and prepared the lists of alienated tribes and issued suo moto notices to the non-tribal occupants. However, these notices were issued only to the non-tribal cultivators, specifically those who were declared as purchasers of "tribal owned land" after the year of 1963. The year 1963 was considered as the cut-off year for all sales and purchases of land transactions in these regions. Consequently, those who were said to have "purchased" the lands prior to 1963 were exempted from the purview of the Land Transfer Regulations. On this point, those landlords who were given 'Land Ownerships' over large chunks of forest lands long ago during the period of Nizam of Telangana (Telangana was a protectorate in British India), were not brought under this Regulation. In the case of coastal Andhra Pradesh, the cut-off year was 1970.

The official figures speak about the restored land as 81, 241 acres. This is seemingly a case of clear manipulation or miscalculation of numbers. The real extent of the land acreage restored to tribes is only 61, 599 acres. Further, it is startling that even the said restored land (61,599 or 81,241 acres) very rarely exist under the immediate control of the tribal owners. The reality is that the tribals were asked to take over the land restored by the restoring agencies, but they could not do so because of the extraneous influence of the non-tribal rich and other classes of immigrant peasantry. In the case of the poor peasantry, both tribal and non-tribal, the conflict was intense against the exploiting class of the immigrant rich peasantry operating in these areas. Thus, in the context of these realities, there were:

1. Inoperative or partially operative Land Transfer Regulations
2. Inertly surmounting contradictions among the immigrant peasantry and the native tribes
3. Formation of the new tribal groups (with the creation of Government Orders - G:O:'s in 1977 and the inclusion of Lambadas in the Scheduled Tribe category)

Apart from the failure to effectively implement Land Transfer Regulations and increasing contradictions between immigrant peasantry and native tribes, the polarisation between exploited classes within the tribal as well as non-tribal communities was accentuated in the process of the formation of new tribal groups (Rao, 1996). The influential classes are often supported by the various sections of political parties for political reasons. These factors resulting in increased landlessness, have sharpened the class and community conflicts in the tribal context of Andhra Pradesh.

The data shown in Table 2 clearly explains how the tribes-governance relations on the restoration of land rights exists in concrete terms. Factors that have contributed to the marginalisation of the Adivasis from their land holdings and assets are summarised below:

1. Introduction of an alien legal system, mainly protecting the interests of the encroachers of tribal lands.
2. Adoption of several 'weakening methods' of the enforcement of the protective law such as
    a) Passing contradictory government orders facilitating the land grabbers.
    b) Neglect of the manipulations by the revenue officials that disfigure the Adivasi land holdings.
    c) Resorting to administrative delay in favourable disposing of the land disputes.
    d) Encouraging 'legal litigation' between the Adivasis and non-Adivasis conflict formation.
    e) Defects and willful distortions in recording the actual possession of the land ownership by the survey-settlement machinery.
3. Application of force to suppress the genuine mobilisation of the Adivasis organised in favour of the protection of their land rights.

4. Strong trends of migration of the non-adivasi population has shaken the land tenure system, gradually weakening its objectives. Decade after decade, the waves of „incentive" migration due to the 'cheated land availability' in the areas not only resulted in demographic changes but also led to substantial formation of the confrontative situation between the 'mighty migrants' and 'native Adivasis'. The legitimate force i.e., the state, played the role of an invisible actor.

1. No. of Non-Tribal Occupations in Scheduled Villages detected which are prima facie violative of LTR. 57,150
2. Extent of Land involved in that is 245,589 acres
3. Cases in which enquiries were initiated under LTR. are 57,150
4. Extent of land involved in that is 245,581 acres
5. No. of cases disposed off are 48,234
6. Extent of land involved is 217,574 acres
7. No. of cases disposed off in favour of non-tribals is 23,702
8. Extent of land in that is 118,486 acres
9. No. of cases disposed in favour of Tribals is 24,532
10. Extent of land covered in that is 99,087 acres
11. No.of cases in which land was restored to tribals are 20,233
12. Extent of land covered (under Col. No. 11) 68,520 acres
13. No.of cases pending disposal are 2100
14. Extent of land covered (under Col. No. 13) is 7,653 acres

Table 2 : Details of land restored under land transfer regulation and protecting tenure system till end of June 1995 in A.P.

6 Conclusion

The above factors led to large extent of land alienation of tribal population and enormous pressure on the available forest and grazing land resources. The incidence of the loss of forest in the state rose to dangerous proportions. Loss of the forest as an effect of encroachments had given way to periodical recurrence of the floods in these areas. The effect of the floods is again not merely 'human dislocation' but a 'permanent loss of the land as a resource through continuous erosion'. Thus, continuing use of alien lands and subsequent defective implementation of legal acts, have promoted the increased incidence of confrontation. This confrontation, therefore, is largely due to the specific character of the land tenure system that has denied the rights of the rightful land owners. The implications are shifts and changes in land use pattern inviting both the distant emigrants as well as commercial cultivating castes who pose an eventual threat to the very basis of the Adivasi economy and society in Andhra Pradesh.

References

Anderson, R.S. and Huber, W. (1988): The Hour of the Fox, Vistaar, New Delhi. pp. 21-26.
Dharma, K. (1993): States and Civil Societies in Modern Asia, Economic and Political Weekly **42**, 2266-2269.
Guha, R. and Gadgil, N. (1989): Forestry and Social Conflict in British India, Past and Present **123**, 141-177.
Habib, I. (1967): Sixteenth Century Agrarian Conditions, Indian Economic and Social History Review **5**, 13-47

Rao, B. (1996): Adivasis in India - Characterisation of Transition and Development. In: T.V. Satyamurthy, (ed.), Region, Religion, Caste, Gender and Cultural Contemporary India, Oxford University Press. Delhi, 417-443.

Rothermund, D. (1978): Government, Landlord and Peasant in India: Agrarian Relations under British Rule: 1865 - 1935, Franz Steiner Verlag, Wiesbaden, 38-39.

Addresses of authors:
B. Janardhan Rao
E. Revathi
Department of Public Administration
Kakatiya University
Warangal - 506 000, A.P., India
Ramesh Chennamaneni
Department of Agricultural Economics and Social Sciences
Humboldt University
Luisenstraße 56
D-10099 Berlin, Germany

Effects of Land Tenure and Farming Systems on Soil Erosion in Northwestern Peru

H. Cotler

Summary

The relationship between population and environment is shown by the effects of land tenure and farming systems on soil erosion in an Andean watershed. In this region the land belongs to the peasant communities. The crops are cultivated on individual fields, while the pasturing takes place on communal grazing areas. Integrating soil landscape analysis with land use showed the limitaions and the degree of adaptation of the present farming systems and shed light on erosion dynamics. The structural instability index is significantly higher in the communal grazing areas than on the individual fields, thus the greatest erosion problems (rills and gullies) are related to communal land use. This is due to the lack of communal cohesion in the peasant communities that makes it difficult to exert control and resources accountability.

Keywords: Farming systems, land tenure, peasant communities, erosion, northwestern Peru.

1 Introduction

The relationship between population and environment has been studied in the way in which populaion growth influences changes in land use, and how this, in turn, may be related to environental degradation (Bilsborrow and Okoth-Ogendo, 1992; Jolly and Boyle, 1993). Many studies carried out in Latin America revealed links between population growth, changes in land uses, land tenure and environmental degradation (Stonich, 1989).

The land tenure and the pattern of fragmented land determine the possibility of conservation practices. Thus patterns of fragmented land ownership make it impossible to design and encour-ge farmers to implement conservation measures (Pimentel et al., 1993). However, common possesion of the land could cause problems in relation to land degradation (McCay and Acheson, 1990). The removal of vegetation may lead not only to rapid erosion but also to degradation of soil structure, as erosion by rainfall and runoff alternates with drying of the surface. Land degradation is thus a function of clearing and the mode of agriculture including the choice of crops. In Peru, 58% of the agricultural population is organized in peasant communities (Kervyn, 1990). Originally the peasant communities were formed by families linked by social, economic and cultural ancestral loops expressed in the communal ownership of the land, communal work and assistance. However, the demographic, social and economic changes of the last decades have provoked an ecological imbalance manifested as soil erosion. The main objective of the present study is to evaluate the effects of land tenure and farming systems on soil erosion in northwestern Peru.

2 Study area

This study was conducted in the Mangas watershed, north-western of Peru, on the occidental side of the Andean Cordillera, between 950 and 3000 m. The climate is a function of the south anticyclone, trade winds, orographic effects, the El Niño Southern Oscillation (ENSO) and characterized by dry winters and rainy summers. This basin is partially isolated during the rainy season due to dangerous road conditions. Economical and social isolation is reflected in its low population growth rate (0.5%) and in its high migration rate (9%). This has promoted agriculture as the principal economic activity in the region. As much as 80% of the population depends on agriculture for subsistence (Bernex and Revesz, 1988).

3 Materials and methods

Aerial photographs (1:20,000), topographic maps (1:50,000), geological maps (1:50,000) and field work were used to identify the geomorphological units and land use in the study area. Soil properties at each of these units were determined through 30 profile descriptions and sampled for further laboratory analyses. The combination of the geomorphological and soil maps resulted in a integrated soil landscape map. We analyzed the existing farming systems in the Mangas watershed; the crop pattern and livestock system of each peasant community between May and September. Information on land tenure was obtained from community records.

Soil erodibility was evaluated using an index proposed by Combeau and Monnier (1961). This index is based on structural characteristics that have proved useful in predicting erosion risks for a wide range of soils. The index evaluates soil structural instability, i.e. soil vulnerability to rainfall splash and consequently surface crusting. Structural instability is inversely proportional to stable soil structure and is obtained from the formula:

$$Is = \frac{\% \ maximum \ (clay + fine \ loam)}{\% \ mean \ coarse \ fragments - 0.9 \ (\% \ coarse \ sand)}$$

This index vary from 0 to 3, with 3 the highest structure instability. We compared soil erodibility under different land uses by analyzing samples from their surface horizons.

4 Results

4.1 Soils

The Mangas watershed presents a generalized steep and rough environment with a difference in elevation of 2050 m in only 32 km. The dominant landform consists of hillsides formed from andesitical material, with slopes up to 130%. The summits over andesitical material presents soils with a well-developed illuvial clay subsoil (Dystric Cambisols). Their slopes, formed also from andesitical material, presents moderately developed soils (Dystric Regosols). In such cases, mass land movements homogenize the profile physically and chemically. The andesitical downslopes give rise to linear and concentrated run-off (Dystric Regosols). The slopes over feldspathic sandstone generate soils with a moderately developed structure, with a slight illuviation (Dystric Vertisols). These soils are characterized by superficial cracks during the dry season. Slopes over clastic-volcanic rocks develop superficial soils (Dystric Leptosols) and their dejection cone has superficial soils (Dystric Regosols). On the other hand, the slopes over pyroclastic rocks are subject to continuous mass movement forming Dystric Cambisols. In narrow alluvial valleys young soils have formed with only incipient profile development (Dystric Fluvisols).

4.2 Farming systems

Five peasant communities exist in the study area (Table 1) where the peasants have adapted to demographics changes, the new techniques and market forces. Thus the right to cultivate specific parcels of land is given according to the demographic density increase. The lowest person/ha ratio is 1.4 at Suyupampa community and the highest is 30.4 in the Tacalpo community. The strong pressure on the land has intensified the fragmentation until it is now common to find plots as small as 50m².

Community	Nr. of Families	Population (1981)	Land surface (ha)					
			Irrigated	Non irrigated	Natural pastures	Forest	Others	Total
Suyupampa	614	4250	1200	1849	-	3050	64	6163
Cuyas-Cuchayo	571	3560	250	1028	170	2262	591	4301
Tacalpo	408	2762	91	-	-	1892	211	2194
Arreypite-Pingola	580	3476	1258	1119	-	8213	30	10620
Cujaca	267	1628	138	447	6069	44	115	6813

Table 1: Peasant communities characteristics in the Mangas watershed

The soil landscape characteristics determine the specialization of productive activities in this area. The principal farming systems in the study area are the traditional agroforestry system, two single-crops, both in individual parcels, and the grazing system of communal land. The first system is located on hillsides over andesitic, pyroclastic and on piedmont rocks. Slopes on these cultivated areas vary from 15% to 60%, therefore only manual tools are used by the peasants (sickle "misha" and manual plow "chaquitaclla"). In this traditional agroforestry system, the most usual crop associations are: corn-bean, corn-vetch, corn-broad bean, corn-banana, corn-cassava-coffee, corn-cassava-banana, banana-cassava and banana-coffee-bean. Intermixed with these crop associations are orange, lemon, lime, avocado and mango trees. With this land use, several plant strata are established in the same plot allowing a good soil cover. The two single-crop farming systems are rain fed cereals and irrigated potatoes. Cereals are cultivated on hillsides over andesitic rocks, with slopes ranging from 20 to 50%. Potatos are grown on summits and hillsides over andesitic rocks, with slopes ranging from 15% to 70%.

Animal husbandry is the most frequent complementary activity in the area. This system consist of two main and complementary forms, ruled by the rainy season. The first of these is characterized by gathering together, in common grazing areas, the cattle belonging to every community member. This happens from the beginning of the rainy season (October) until the beginning of the dry season (June-July). During July-September (dry season), cattle is placed on each farmer field which are already harvested or sown with pasture. The common grazing areas are either located on slopes over clastic-volcanic rocks and on slopes over feldspathic sandstone. In the first case, soils are very shallow and superficial with local outcrops of the parental rock. In the case of the feldspathic sandstone area, soils show superficial cracks. The first rain of the season usually falls on a very compact and hard soil as a result of the combined action of cattle trampling and sun desiccation. This situation promotes sheet erosion and a microrelief in the shape of cattle paths. Independently of the type of parental material, the highest index of soil erodibility was found on lands occupied by woodland or tree savanna which are used for communal grazing (Table 2).

Farming system	log 10 Is	Erosion forms
Cereals single crop without irrigation	1.72	Terracette, stem inclination, mass movements.
Association corn-bean-banana	1.17	Corn inclination, "scratches" on soil surface.
Association corn-cassava-coffee	1.13	Corn inclination, "scratches" on soil surface, local surface crusting.
Sugar cane crop	1.19	Diffuse run-off.
Grazing area (woodland)	2.10	Roots exposed, protrusion of thick elements (gravel, stones), paving formation, furrow, erosion gully.
Grazing area (tree savanna)	1.86	Roots exposed, protrusion of thick elements, polychromy of soil.

Table 2: Soil's structural instability according to the Combeau and Monnier index (1961) according to land use

Thus on these lands, the most intense and active gully formation process take place. In fact, the differences of the structural instability index among the production systems are statistical significant. On the individual parcels, soil erosion change as a function of the farming systems. Thus, cereals have an erosive character because they grow on steep slopes on clayey soils with strong plasticity. On the other hand, by means of this index and field observations, it is possible to conclude that agroforestry management (corn-beans and corn-cassava-coffee) produces the least soil erosion.

5 Discussions and conclusion

The integration between soil landscapes and land use analyses in the Mangas watershed reveals agroecological management according to the limitations and the potential of the soil landscapes. Thus, the cultural systems are superimposed on the soil landscapes, where the limitations besides the slope, are overcoat of chemical order. The extensive animal husbandry is restricted to two soil landscapes units with physical limitations for agriculture: the vertic character of the soils over feldspathic sandstone and the superficial soil's stones over clastic-volcanic rocks. Soil erosion is clearly more intense in the soil landscape units dedicated to the common grazing areas. This is reflected in the higher structural instability index and in the erosion phenomena. However, erosion problems are not caused exclusively by the natural conditions of the respective soil landscape units. The actual management of the grazing areas is characterized by the lack of restrictions established by the peasant communities. This has led to the absence of a rotation system that makes vegetation reconstitution practices difficult, and uncontrolled introduction of livestock which leads to overgrazing. As in others parts of Peru (Cotlear, 1989), the problem resides in the weakness of the communal organization which makes rules difficult to establish and enforce. In addition, this kind of communal grazing areas brings with it the cattles of many different owners, so it is not economically attractive for them to invest in these lands to maintain their productivity. In the peasant communities of this basin, many factors have contributed to the breakdown of communal-property mechanisms: pressure on the resource due to human population growth, economic changes and social and economic characteristics of the resources users who don't permit two main features of the common-property resource users (Feeny et al., 1990). The results of integrating soil landscape units and land use permits us to conclude that the greatest erosion problems are related to communal land use. Thus in the Mangas watershed, the controversy is not over land tenure but the socio-economic characteristics.

References

Bernex de, F & Revesz, B. (1988): Atlas Regional de Piura. Centro de Humanidades-Centro de Investigación en Geografía Aplicada, Pontificia Universidad Católica del Perú, Lima, Perú, 207pp.

Bilsborrow E.R. & Okoth Ogendo, O.H.W. (1992): Population-driven changes in land use in developing countries. Ambio **21**, 37-45.

Combeau, A. & Monnier, G. (1961): A method for the study of structural stability: application to tropical soils. African Soils **6**, 33-52.

Cotlear, D. (1989): Desarrollo campesino en los Andes. Instituto de Estudios Peruanos, Lima-Perú, 324p.

Feeny, D., Berkes, F., McCay, B. & Acheson, M. (1990): The tragedy of the commons:twenty-two years later. Human Ecology **18(1)**: 1-19.

Jolly, L.C. & Boyle, B. (1993) (eds.): Population and land use in developing countries. Report of a workshop. Comittee on Population. Commission on Behavioral and Social Sciences and Education National Research Council. National Academy Press, Washington, 159pp.

Kervyn, B. (1990): Les communautés paysannes au Pérou: un système de propriété dépassé? Agricultures et paysanneries en Amérique latine. Colloque International Toulouse (France), Université de Toulouse-Mirail, Atelier V, 301-308.

McCay, J.B & Acheson, M.J. (1990): Human Ecology of the Commons. In: McCay J.B & Acheson M.J. (eds.), The Question of the Commons. The University Of Arizona Press, pp. 1-34.

Pimentel, D., Allen, J., Beers, A., Guinand, L., Hawkins, A., Linder, R., Mclaughlin, P., Meer, B., Musonda, D., Perdue, D., Poisson, S., Salazar, R., Siebert, S. & Stoner, K. (1993): Soil erosion and agricultural productivity. In: Pimentel, D. (ed.): World Soil Erosion and Conservation, Cambridge University Press, 277-292.

Stonich, S.C. (1989): The dynamics of social processes and environmental destruction: a Central American case study. Popul. Dev. Rev. **15**: 269-296.

Address of author:
Helena Cotler
Instituto de Ecología
Universidad Nacional Autónoma de México
México D.F., México

Conclusions and Recommendations

Taking action that matters

H. Eger, E. Fleischhauer, A. Hebel & W.G. Sombroek (Editors)[1]

Under the topic „Toward Sustainable Land Use - Furthering Cooperation Between People and Institutions" the Conference brought together 900 scientists, government representatives, representatives of national and international organisations as well as NGO's and networks from 120 countries, discussing the topics of soil protection and soil conservation.

It was stated that the limited availability of soil resources for the production of food and renewable biotic resources caused by a steady growth in population and accelerated soil degradation can have a bigger negative impact on living conditions on earth than the human-induced greenhouse effect caused by mankind. Anthropogenic soil management and land use processes should be considered to be more destructive than climate change consequences - at least during the next decades. A comprehensive analysis of the driving variables for climate change and their effects on soil properties can hardly reveal the required short term indicators for the assumed long term changes in soil quality.

Land needs to be considered as a finite resource. Its allocation must aim to satisfy the needs of the various land users in the most equitable and sustainable way. Combating soil degradation and investing in the conservation of soil resources for future generations will be a major political task promoting sustainable development and nature protection. A global partnership is required to protect and restore the health of the Earth's terrestrial ecosystems.

What is required is a holistic approach for planning, development and management of land resources which methodically identifies human and environmental needs. The process needs to address cross sectoral issues such as responses to pressures on the land caused by poverty, unsustainable consumption and production systems, clarification and security of land rights and land tenure and land ownership reforms. Furthermore, this approach requires the integration of issues of water resources and biodiversity as they relate to land use. A mismanagement of land and water often leads to land degradation through erosion, flooding, waterlogging and salinity and the depletion of groundwater resources. Moreover, soil and water degradation through contamination by agricultural, urban and industrial effluents is of increasing importance in developed and developing countries.

For formulating and implementing policies it is essential to collect and process and disseminate timely and reliable information and to utilise modern land assessment and evaluation technologies to create sound scientific knowledge for proper decision support.

---

[1] The editors made extensive use of the specific conclusions/recommendations of the various ISCO working groups and of the ISCO pre-Conference brochure Precious Earth. They gratefully acknowledge the invaluable support of Hans Hurni, Galina Motuzova, Gisela Prasad, Michael Stocking, and of Kebede Tato and all chair persons of the working groups of the paper presentation sessions.

1 Managing a planning process for the use of land resources with all stakeholders

Improvements in sustainable land use and development need the consideration of the interests of all involved individuals and groups. Therefore, a multi-level stakeholder approach for the planning process is essential to obtain socially balanced results in which the economic and ecological objectives have been weighed up.

All stakeholders such as farmers/conservationists, owners/tenants, individuals/communities as well as administrators, planners and governments etc. in a problem setting must be identified and invited to take part in a broad participatory process to analyse problems, and express and evaluate their needs, interests and aims. Stakeholders must then negotiate options and priorities for action. This is a way of ensuring that action at the local level can be co-ordinated, and that alternative scenarios can be compared with a view to their potential for long-term improvement.

The multi-level stakeholder approach shall imply democratic - and to a certain extent - formalised procedures. They shall be based on a sound information basis which includes all data about the properties of the land, the land uses and their functions for the resilience of ecosystems.

Land use planning that considers all the functions of the land and that contains full participation of all stakeholders in the planning process is an important tool to implement the recommendations of UNCED's Agenda 21 in particular the section dealing with the concentration and management of resources for development.

2 Creating an enabling environment

A positive, "enabling" institutional environment at the national and the international level offers the potential for substantial support of sustainable land use by creating favourable conditions in which land users and communities can benefit by improving existing shortcomings. We have to think about the perceptions and decisions of stakeholders at the various communication levels such as households, communities, national societies and the interrelationships between them as interactive units.

National policies directly affect land users. Many national governments are increasingly attempting to integrate environmental, economic and social concerns into national planning processes. Declaration of a national policy on sustainable land use is an important measure that can help bring about necessary political, institutional and economic changes.

The emergence of enabling incentives marks a shift away from command and control approaches - which force land users to adopt or abandon a particular course of action - towards the creation of an environment which allows them to choose their own course of action.

Policy-enabling incentives acknowledge the need for a coherent natural resource policy in all policy aspects, co-ordinate economic and financial policies with environmental policy, co-ordinate strategies of agrarian and development policies with natural resource policy, create an institutional framework which supports natural resource policy, co-ordinate between different government institutions, guarantee regional autonomy and delegation of responsibility for natural resources to the communal and local levels, and enforce sustainable use of natural resources in local communities.

3 Creating a positive learning environment for sustainable land use

A common ground should be created to enable people to learn from others and by direct experience, through „participant observation", accepting non-standard sources of information, data and observations such as oral history, folk wisdom, intuition, emotions and feelings. Second, science would be seen as a set of provisional hypotheses, to be constructed, re-constructed and reformulated as additional knowledge becomes available. Third, in place of separate disciplines, multi- and inter-disciplinary institutions would be created under one roof to provide an institutional setting in which different forms of knowledge are equally acceptable. Fourth, the reward structures in education for both student and teacher would be changed, affecting exams, career development, prestige and pay.

We must consider how to foster openness to knowledge characterised by respect and understanding for different views rather than conflict and rejection of what is not part of one's own knowledge system. Let us start with the assumption that people must have good reasons why they should share their know-how, their understanding of the physical world, and their concepts of a sustainable society. That means to put local knowledge into the fore-front.

Local knowledge reflects natural and societal factors, and is embedded in social organisations as well as in cultural traditions and preferences. Land users' knowledge systems are dynamic; land users themselves continuously interact with the environment and make changes as they encounter new problems. Among the characteristics of local technical knowledge are low external input in materials, the low risk usually associated with the technologies at hand, and the fact that it is based on the preferences and skills of local society.

4 Enhancing action-oriented research

- **Introducing economic and ecological thinking and instruments in land use management**

Comprehensive economic analyses can increase the chances of success in promoting new technologies, if they clearly identify constraints which have been neglected in the past and which can be overcome by appropriate policy interventions.

Special attention shall be drawn to analysis of economic viability that means:
- Cost-benefit analyses by the stakeholders will provide insights into the profitability of land use types.
- Analysis of the economic environment will reveal impediments to changes in agricultural practices and land use patterns.
- Analysis of institutional constraints and imperfections will provide insights into other variables which influence stakeholders' decisions on land uses and investments.
- A policy analysis must be carried out based on the results of the previous three analytical steps. Appropriate policy instruments ranging from the macro-economic to the micro-economic level can then be selected.

Equally important is the awareness for ecological sustainability which is composed of the analyses of the
- soil functions showing whether these functions (production, regulation, cultural heritage, living space) are being maintained;
- functionality of ecosystems helping determine whether such ecosystem components as the water cycle, soil nutrient balance and microclimate will remain intact after the introduction of new land management technologies;
- biodiversity showing whether new land management technologies have negative impacts on fauna and flora;
- ecological resilience indicating the extent to which an ecosystem can tolerate depletion and/or accumulation of material without exceeding the capacity for natural regeneration and/or human activities which reverse damaging processes.

There is an increasing need for indicators for soil and land quality. At least three important indicators can be of value in indicating the health of soils and of landscapes in which they are found:

- One such indicator is the stability of plant production, in the form of crop and pasture yields, assessed from year to year.
- Visible signs of land degradation, as evidenced by e.g. the symptoms of excessive erosion and runoff, and/or declining biodiversity in natural and agricultural ecosystems, are another indicator.
- A third indicator is what farm families themselves perceive as changes.

- **Furthering the development of integrated technologies**

The basic principle of sustainable land management is to increase biomass production with technologies which make maximum use of solar energy, water and soil nutrients, and which do not have negative impacts on the environment.

When assessing techniques suitable for the local context, five broad issues should be considered:
- Analysis of productivity will show whether a given measure meets land user/household needs, does not take up too much space, and is adapted to available inputs.
- Analysis of security will show whether the measure minimises risks, leaves sufficient management flexibility, uses local resources, and reduces dependency.
- Analysis of continuity will give indication of soil quality, recycling of nutrients, prevention of soil degradation, maintenance of biomass and biodiversity, efficient use of water, and neutral off-site effects.
- Analysis of identity will be shown by integration into the land use systems and infrastructure, by strengthening of cultural systems, by consistency with policies, and by benefiting underprivileged groups.
- Analysis of adaptability will be demonstrated by spontaneous adoption, rapid success, flexibility in adaptation, and easy communication to other land users.

5 Developing suitable implementation approaches and establishing networks of observation systems

Any research on implementation approaches will have to consider the specific land use systems in place (pastoralist, agrarian, forestry, integrated), as well as the relationship between land use and settlement, infrastructure, industry, and mining. It must also include the impacts of agricultural mechanisation, industrialisation (pollution, construction) and climate change on soils and land, because these processes are likely to modify local conditions and land use systems. The objective of these approaches is to further social acceptability of which the following components shall be analysed
- social heterogeneity which will provide insights into different social groups as well as social conditions, e.g. poverty, equality, access to resources, including information, etc.
- demographic conditions which will examine such phenomena as migration, population growth, and ratios between people and resources (land, capital, etc.).
- social infrastructure which will shed light on the availability and the quality of various types of infrastructure such as schools, health care facilities, etc.
- norms and values which will indicate possible reasons for acceptance or rejection of new approaches in soil conservation and soil protection oriented land management.

Present-day problems arising from rapid change call for new methods which provide better reference points for assessing change. Environmental monitoring can only be carried out if suitable indicators are developed. Different types of indicators will be needed at the scientific, political and community levels. The objective will be to compare and appraise
- different technologies in different areas, and

- temporal change in socio-economic and biophysical conditions within an area.

From the institutional point of view, such reference points will require different monitoring networks. International networks should have observation systems in all major ecoregions of the world and focus on biophysical parameters as indicators of global change. National networks will have more refined systems which also take account of socio-cultural, economic and political parameters.

6 Furthering co-operation between people and institutions

- **Furthering co-operation between people and institutions in the local context**

There have been some revolutionary shifts in methods of communicating knowledge, such as participatory learning. If sustainability is to be more than a concept to which we merely pay lip-service, professional education must incorporate all sectors of society that interact with land use.

The ultimate goal is to make education and training a shared experience among equals. Training is needed in facilitation techniques, management and leadership skills, literacy and numeracy, public speaking and listening, and other skills related to learning and communication.

- **Harmonising international strategies and action plans for sustainable land use and development with national policies and land users' priorities**

The disparities that exist among countries in terms of economic status, natural resources, educational level, etc., should be reflected in the design of action plans.

Policy issues in sustainable land management include co-ordination of land titling, economic policy, nature conservation policy, and population policy. Therefore, national strategies for sustainable use of natural resources need to thoroughly harmonise, adapt, and integrate the different strategies and policies of governments and their ministries which are directly or indirectly linked to the use of natural resources by stakeholders.

Informal and formal institutions and organisations - from farmer groups, local NGO's and communities to ministries, government policies, and legislation - can only be sustained, if they are accepted and supported by their respective populations. This means that local knowledge systems, norms and values, must be respected. Negotiation processes among all stakeholders, which must be a part of good governance and administrative management, can be enhanced by better information and knowledge about land users' visions, options, and needs with respect to sustainable land management.

- **Co-ordinating global agreements and conventions**

Global conventions dealing with sustainable land management include the Convention to Combat Desertification, the Convention on Biological Diversity, and the Framework Convention on Climate Change. All three emphasise global solidarity in their ratification procedures, and they all initiate action programmes through a variety of means. In addition, Agenda 21, the Tropical Forestry Action Plan, and other international action programmes or regional frameworks for action are also concerned with promoting sustainable land management.

However, all these global initiatives and programmes display three chronic deficiencies:
- they are very far from the world of local land users;
- they have been poorly financed to date; and
- there is little co-ordination between their action plans at the local level.

Policy-makers continue to discuss the creation of a convention/code-of-conduct on sustainable land management. Justification for such a convention can be found in accelerating degradation of the world's land resources, and slow progress in promoting better management. Specific points such as nature reserves, natural world heritage sites, and wildlife preservation in natural habitats could also be included. Ecoregional approaches and basin-wide watershed development could be better co-ordinated under the auspices of such a convention. New and existing programmes could emphasise soil and water management and combine this with land use planning towards sustainable land management.

7 Rethinking UNCED '92

For enhancing sustainable land use in the framework of creating a sound economic and social development in all countries the assembly of ISCO recalls the demands noted in the UNCED's Agenda 21
- to review and develop policies to support the best possible use of land for the sustainable management of land resources
- to strengthen institutions and co-ordinating mechanisms for land resources
- to improve and strengthen planning, management and evaluation systems for land and land resources and
- to create mechanisms to facilitate the active involvement and participation of all concerned, particularly communities and people at the local level, in decision making on land use and land management

When the General Assembly of the United Nations will gather in a *Special Meeting* on Environment and Development in June 1997, the topics **Combating Soil Degradation** and **Promoting Sustainable Land Management** should be addressed in a widened sense for further international negotiations. The International Convention to Combat Desertification (CCD) will be enforced on 27 December 1996 which will enable the international community to tackle one important part of these problems.

ISCO '96 TOPICS

TOPIC 1
Soil conservation and sustainable land use -erosion, desertification and land use planning

- Management based on the natural watershed unit may be the most effective bases for natural resource conservation in general and water erosion control in particular.
- Consideration of on long-term at-site and off-site costs, benefits and environmental externalities of soil erosion must be a part of management strategies. Soil conservation programmes should be given greater priority in the larger context of environmental protection.
- There can be significant interactions between wind and water erosion at the same place. Nutrient loss and enrichment effects in wind erosion require better recognition.
- An understanding of processes involved in water and wind erosion through modelling can assist in matching these processes with appropriate and effective soil conservation measures. The purpose or objective of these models should be clearly defined. Such purposes can range from attempting to describe current knowledge of erosion processes, to the ranking of alternative management systems in level of capacity to reduce soil loss, or be a guide in soil conservation planning or erosion assessment. Modelling of wind erosion is possible, if adequate information on factors is available.
- There is an urgent need to develop georeferenced information on natural resources and socio-economic conditions, in order to monitor the change of land qualities over the time. It is therefore recommended to implement international methodologies such as Soils and Terrain Digital Databases (SOTER) and to link these databases with information on soil degradation and soil conservation following the expert system approach of GLASOD/WOCAT.
- There are no sound and easily measured indicators for soil organic matter (=SOM) quality and SOM-mediated fertility that could be used as a land quality indicator for sustainable land use, although there is general consensus that SOM is an important component of land quality. Further research is needed on a) the optimum level of SOM, b) the implications for nutrient balance of a climate change and a doubling of CO_2, c) the relationship between biomass production, organic matter production and long-term carbon storage.
- The building up of soil organic matter content can be considered to be a capital investment in soil as a national and community level resource. It also constitutes a substantial sequestering of human-induced atmospheric CO2, and therefore would diminish the hazards of global climate change. National governments that are signatories to the Framework Convention on Climate Change and that are considering the concept of "joint implementation" as a means to fulfil their requirements under the convention, may recognise carbon sequestration in degraded or natural low-activity soils as an attractive "win-win" proposition.
- The role of a land use planner in the promotion of sustainable land use is not one of decision-making but one of <u>facilitating</u> the process of stakeholders' agreement on wise use or non-use of the land of their interest, through a) the identification of likely future trends and needs, in consultation with the stakeholders, b) the provision of all necessary technical information on the basis of integration of data gathered by various disciplines, in an easily understood form, c) the itemising of the various viable land use options or alternatives, per identified agro-ecology and socio-economic land unit, d) the developing of procedures for a fully participatory approach in negotiated decision-making, at local and at national level, and at in-between levels (district planning offices and platforms), e) the disposition, through socio-psychologic training, to act as neutral mediator in platforms for conflict solving and decision-making on future land use, f) the monitoring and evaluation, together with the stakeholders, of land use practises that

are the result of the planning process, and to suggest modification of these practises where and when required. Participatory management based on socio-economic and gender issues of equitable sharing of benefits and responsibilities should be promoted.
- Recognising that land use planning is a dynamic process, the formal national and international land use planning <u>institutions</u> should be flexibly organised, with clearly defined mandates and lines of co-operation with each other, both vertically (from national to village level, and back) and horizontally (between disciplines).
- The integrated approach to land resources planning and management, as advocated in UNCED's Agenda 21 (Chapter 10), and elaborated in relevant FAO reports requires full-scale inter-institutional co-operation (government institutions and NGO's) instead of competition, development of goodwill of all stakeholders concerned through the functioning of platforms for conflict resolution and participatory decision-making, at all levels, and the sustained provision of basic financial resources, credit and marketing facilities, and technical support services for data gathering and processing, for rural extension services etc.
- Recognising that soils, or lands, are the exclusive patrimony of a country's inhabitants and a sovereignty-linked national resource, any international initiative to promote good and sustainable land husbandry world-wide may have to be restricted to the formulation of a "statement-of-principles". Individual countries may then be willing to adhere to such principles, on a non-legally binding basis, through the adoption of national land policies or charters that are modelled on such a statement-of-principles.

TOPIC 2
New forms of soil degradation

With growing populations and increasing industrialisation new forms of soil degradation have become increasingly important. Particularly soil degradation caused by air and water pollution with organic and inorganic pollutants is a matter of prime concern in industrialised as well as in developing countries. The following principles should guide soil protection and soil conservation measures:
- Diffuse and direct substance input into the ecosystem may in long term not exceed the possibilities to produce biomass and to filter and transform substances. Therefore unwelcome inputs of substances have to be minimised by all polluters such as industry, traffic and transport, agriculture and households at their source. Since the substance flows in many industrialised countries have already been put out of balance to a considerable degree, the measures to minimise substance inputs require planning with regard to areas and time.
- The consumption of soils for uses linked with a loss of the natural soil functions certainly has to be kept as low as possible. This also means that changes in further developments of the forms of land use have to be adapted to the natural site conditions. They trigger interlinking effects on ecosystems as to time and area which can be explained only to a very limited extent.

Threshold values for the assessment of soil strains and soil contamination have to be elaborated. It is necessary to agree internationally on simple, comparable and common soil investigation and evaluation methods. In many areas of the world agriculture and activities of the mining industry and hazardous waste sites and closed down industrial sites as relicts of modern life have affected the environment and often require soil and land rehabilitation measures. Although agrochemicals are essentials of modern agriculture and necessary for sufficient yields, improper handling of fertilisers and pesticides as well as of sewage sludge and composted municipal wastes may lead to contamination of the soil and water. Pesticide adsorption processes in soils and the concomitant formation of bound residues stress the necessity for an integrated pest management and of minimising pesticide use in agriculture.

There is no principal question about the beneficial effect of mineral fertilisers in supplementing the nutrient supply of agricultural soils and optimising crop yields and quality. However, since the buffering capacity of the soil system for soluble fertiliser salts is limited, a most careful matching of plant nutrient inputs to the outputs by crops is required. On less fertile and resilient soils, especially those in the tropics, the predominant use of NPK fertilisers tends to aggravate other deficiencies, particularly of micro-nutrients. Therefore, organic and mineral fertilisers should supplement each other, and the quantities supplied should always take into consideration the loading capacity of the whole environment.

In many developed countries livestock manure has to be used with more care for the ecosystems and especially for the water resources. Soils have a finite capacity for buffering and retaining excessive additions of plant nutrients from animal waste, and even the loading capacity for phosphorus eventually comes to an end. The resulting pollution of ground and surface waters with nitrate and phosphate, and of the air with ammonia, indicate that many existing animal husbandry systems cannot be sustained. Economical facts and constraints are no excuse for delaying the necessary adaptions. Policy makers should be made aware of the rising and eventually unaffordable external effects and costs.

The limitations for the use of sludge and composted organic municipal wastes are principally the same as for mineral fertilisers and for manure. Unfortunately, sewage sludge is a very effective sink for all kinds of environmental questionable contaminants, organic as well as inorganic, particularly of heavy metals. Only at very low rates of sludge and compost contamination, it is possible to maintain a long-term equilibrium between inputs and outputs of potential soil pollutants at the lowest possible level. On the other hand, recycling of suitable wastes from society into soils is the only possible way to reduce the unidirectional fluxes of plant nutrients from rural into the urban environment. Therefore, an essential prerequisite for future sludge and compost utilisation is that urban societies reduce the load of persistent contaminants in their waste waters by all possible means. This quality problem becomes even more relevant when the consequent recycling of appropriate organic wastes will result in compost quantities which have never been experienced before. Hence the principal desirability of closing plant nutrient circuits as much as we can, justifies all possible efforts to minimise the load of contaminants in municipal composts. Here again, however, there cannot be any compromise as far as the maintenance of soil multifunctionality is concerned. Solving this problem requires both government action and public education.

The sustainability of all measures considered to be beneficial for soil productivity have to be judged against the background of their long-term environmental effects. Existing soil quality standards are not always sufficiently based on the concept of sustainability, so that many soil "amendments" need to be reconsidered in the light of long term soil protection. "Prevention" should be the leading principle in soil protection, since "rehabilitation" is extremely difficult and often not possible at all, once symptoms of soil degradation can be observed.

Other issues which need much more attention are soil compaction and soil sealing. Soil compaction both in agriculture and in forestry is a world wide problem, which gets more important because of increasing use of machines, higher frequency of wheeling and increasing dynamic forces applied to the soil. Also surface sealing either by natural processes or by human activities, especially in densely populated areas, leads to a loss of soil functions in landscapes.

TOPIC 3
Influence of demographic, socio-economic and cultural factors on sustainable land use

- Rapid demographic change is a major threat to the environment. While projected future populations can be fed in aggregate, there remains many areas where severe stress will occur and where the demands of local people cannot possibly be met by the soil resources. Typically, such areas include conflict zones as in the Horn of Africa, sites where there has been rapid migration as in Montane Mainland South East Asia. Elsewhere, many commu-nities are learning to cope with demographic pressures and adapting their own technologies and responses to a changing environment. Society can adapt its land management to cope with demographic change. These are the people and situations from which we can learn ways in which sustainable land management may be practised.

- Policy and research responses to the influence of demographic change on land use must include: addressing inequitable and inappropriate land tenure systems; providing infra-structural, marketing and institutional support to projects, programmes and communities; targeting rehabilitation programmes to areas where special factors have made it impossible for communities to cope and adapt to demographic change.
- The engine for moving towards sustainable land management is the provision for people's social and economic needs. All examples of successful conservation projects have a primary focus on local society and livelihood security. Without these ingredients, no intervention or innovation will succeed. For land management practitioners, it therefore means they first have to listen and learn; secondly, they must match what they have to offer - technologies, education, professional advice, physical resources, money or whatever - with the socio-economic needs of the people, and finally, they must monitor and evaluate what happens.
- Policy and research responses to the influence of socio-economic factors must include: greater decentralisation and empowerment of local communities; promotion of a more participatory and incremental approach to interventions; support for analytical tools to understand the social and economic conditions; an understanding of the diverse societal and economic perspectives of people - such as gender differentiation and economic rationality at the household level. National governments should support financial and economic baseline studies of the implications of continued degradation, and use these to design appropriate policy responses in, for example, support to extension, research, marketing facilities and related programmes in healthcare, education and support services.
- Pollution brings special social and economic problems which need specific actions. Rural communities cannot shoulder the burden for all society of off-site impacts of land degradation. The careful design of incentive structures such as cost sharing of pollution control and of legal frameworks is called for in order to target action responses in the appropriate sections of societies - local people, farmers, men and women, young and old, poor and elite, rural and urban, landless and owners.
- Cultural factors such as respect for traditions, indigenous knowledge, religious and ceremonial needs condition many people's approach to the land. Enduring cultural traditions can be the motivating force for protecting the environment. They can be the catalyst for an integration of indigenous and formal scientific knowledge.
- Land tenure is fundamental to land management. A tenure regime should be clear, flexible and secure.
- Governments should be the facilitators of change, not the controllers. Soil, land and water resources are both public goods entrusted to society and private goods entrusted to the individual. Administrative, institutional and legal provisions must enable, not generalise.
- TRIGGERS. We must identify the triggering mechanisms for rapid and large-scale adoption of improved and managed practices. Triggers will undoubtedly live in social, cultural and economic factors. SECURITY. Long-term security of land use is a vital precondition to conservation, land husbandry and sustainable land management. INTERVENTIONS. They must be long-term and be monitored and evaluated.
- Our own assumptions, perceptions, biases and interests must be questioned continuously, and we should be receptive to new ideas.

TOPIC 4
From soil and water conservation to sustainable land management

- The diversity of potentially conflicting land management objectives, the multi-level decision making associated with various sectors of society and the proliferation of environmental regulations require that decision aids be designed to take these factors into account. Many tools, data bases, simulation models and quantitative understanding of indicators of sustainability which lend themselves to rational multi-criteria decision making are necessary for multiple objective decision making. High quality information, data bases and applicable models are crucial for decision making on productive and sustainable land use systems. Decision aids for informative land use planning must be user friendly and must provide choices not single solutions to sustainability problems.
- Multi-level stakeholder approach to sustainable land management appears to have been increasingly accepted as a tool not only for supporting local land uses and communities but also for research and extension. There is an urgent need to involve the farmers in technology development and adjustment so that their local experiences and knowledge can be fused into the research.
- Handling of information relevant to sustainable land use systems should be based on both conventional data processing and pre-processed data with subject oriented models. These should fit the integrated model where socio-economic data are also incorporated. The important phase will be the validation of the integrated models by means of expert evaluation and real values observations.
- It is recognised that there is a clear gap of data and sound soil and water conservation guidelines to assist poor farmers with options of technologies.
- In those situations where watershed is selected as entry point for resource management, there are clear problems of adoption of introduced conservation technologies due among other things to lack of people's participation, inconsistency and insufficiency of commitment on the part of the involved agencies, lack of co-operation among the stakeholders etc.
- Preventing degradation of productive land shall be given more priority than restoring already degraded land.
- There is need to make more use of information already on hand than to initiate more data collection in new projects.
- There is a need to adjust and adopt already known technologies for different bio-physical and socio-economic conditions.
- Institutions which finance research take account of the global need of sustainable land management and give guidance to their decision making bodies towards supporting more integrated approaches. The share between basic, disciplinary research on the one hand and integrated, multidisciplinary or transdisciplinary research on the other hand be half-half.
- Monitoring by means of sustainable development analysis of local conditions should guide any institutional approaches and activities supporting land users and their communities.
- There is need to bring together/to establish multi-disciplinary teams of natural and social sciences, decision/policy makers and the land users to jointly formulate scientifically based investigations, select indicators and principles for M&E (=monitoring and evaluation) and be responsible for the M&E, define basic and meaningful concepts of sustainable land use systems, serve as conceptual groups to advise sectoral departments and give guidance to institutions' policies.
- Researchers should not only develop sophisticated methods but also rough and quick methods that can be applied by non-researchers. Researchers should not conclude their tasks with scientific reports but be more actively involved in implementing the results.

- Long-term objective setting and commitment must be recognised by donors and govern-ments, if sustainable land management should be productive.
- Promotion of a suitable mix of technologies rather than any single technology be preferred.
- In semi-arid environments it is recognised that there is evident competition for crop residue as mulch, animal feed and fuel. In the short run therefore, trade-offs in the utilisation is recommended.
- It is recognised that keeping the environment and ecosystems intact is not always the first priority of the land users and at times the government as well. Rather, they expect tangible results from investments in land management. Therefore, considerations/preferences shall be given to technologies that show results in a short time.

TOPIC 5
Furthering co-operation between people and institutions

Monitoring refers to an information system, not just data collection. Data without an audience, an objective and analysis is a waste of time and effort. Monitoring in the context of sustainable land use has a wide range of audiences, all of whom are stakeholders in the development of more sustainable forms of land use.

1. How to deal with a situation of incompatibility of project length (short time frame) and the time it takes for impact to show itself (often much longer when it comes to some biophysical/socio-economic trends)?
2. How to deal with a dominant interest in quantitative data when often qualitative data is more relevant and revealing?

Investments in human capital is today recognised as the most cost-effective strategy in human-kind's endeavour to banish poverty. Successful human resource development programmes aim to enhance the ability of individuals so that they can have more options which will ultimately give them more control over their lives. A diversity of institutions at policy, research, implementation and local levels form a key focus for this strategy. Considerable gains have been made by analysing past and current human resources development practices revealing that there are institutional barriers which need to be examined with a sense of urgency.

The focus has to change from strengthening institutions to transforming institutions. In addition a considerable commitment has to be made to the establishment of innovative institutional mechanisms especially at the community level. The strategy for the future must focus first on institutional transformation and locate human resource development in this context of change. Organisation and human resource development needs to be encouraged in two broad areas, i.e. COMMUNITY BASED ORGANISATIONS in local level institutions. These consist of existing and traditional community organisations which are both formal and informal (e.g. user groups) and indigenous or introduced (e.g. co-operatives); EXTERNAL INSTITUTIONS which consist of those institutions which are not community based but all the same work with local communities such as youth extension department, research institutions, doctors, etc.

Knowledge creation becomes efficient through monitoring and adaptive management in a participatory approach. This needs mechanisms to capture this knowledge.

There is urgent need to research mechanisms of participatory approaches (in different socio-economic settings), specifically how to cut across boundaries between institutions and disciplines, how to ensure that the participatory process is on-going, and how to capture and disseminate useful knowledge.

Personal Records of the Editors

Prof. Dr. Dr. h.c. **Hans-Peter Blume**, born 1933 in Magdeburg, is agricultural scientist and holds a PhD from Kiel University and a *habilitation* from Hohenheim University, both in soil science. He serves as director of the Institute of Plant Nutrition and Soil Science and is Managing Director of the Ecoystem Research Centre of the Kiel University. His research focuses on the genesis, the ecology and the distribution pattern of soils in the temperate zones and on soils in cold and warm deserts as well as in urban-industrial agglomerations. He is editor of several books related to soil science and currently President of the German Soil Science Society.

Dr. **Helmut Eger**, born 1952 in Weißenburg/Bavaria, holds a PhD from Tübingen University in geoscience. Since 1992, he has been working for the Deutsche Gesellschaft für Technische Zusammenarbeit (GTZ) GmbH in the Division for multisectoral urban and rural development programmes at the Head Office in Eschborn. His work focuses on regional rural development and land management with main emphasis on land use planning, land information systems, remote sensing, decentralisation and rural appraisals. He has worked first as a lecturer on Geography, then as a coordinator, team leader and specialist adviser in various countries, mainly in Western Africa, South America and the Middle East.

Dr. **Eckehard Fleischhauer**, born 1934 in Freienwalde, studied agricultural sciences and general economics and holds a PhD from Gießen University in agricultural economics. From 1987 until 1996 he worked as Head of the Division of Soil Pollution and Soil Quality Targets for the German Federal Ministry for the Environment, Nature Conservation and Nuclear Safety in Bonn. His interest centers on ecosystem research, environmental monitoring and land use planning in rural and urban areas. Currently he has been working as consultant to the Ministry on the issues of soil quality evaluation and the harmonisation of soil investigation methods, particularly with Eastern European countries.

Dr. **Axel Hebel**, born 1961 in Herford, is geographer and holds a PhD from Hohenheim University in agricultural sciences/soil science. He worked in the Land and Water Development Division of the Food and Agriculture Organization of the UN in Rome, Italy, and was co-ordinator of a research project on the regeneration of ecosystems at the Centre for Environmental Research in Leipzig. Since 1996 he has been Technical Officer Soil Conservation in the German Federal Ministry for the Environment, Nature Conservation and Nuclear Safety in Bonn. His work focuses on soil fertility maintenance, land resources appraisal and sustainable land management.

Dr. **Kurt Georg Steiner**, born 1943 in Berlin, studied agricultural sciences at the Universities of Gießen and Hohenheim and holds a PhD in Phytopathology. As staff member of the Deutsche Gesellschaft für Technische Zusammenarbeit (GTZ) GmbH at GTZ headquarters in Eschborn he currently coordinates the Sustainable Land Use Systems Project. His professional experience covers various subjects such as plant pathology, plant protection smallholder farming systems, soil fertility maintenance, soil conservation and policy issues of sustainable land management. He worked for many years in West, Central and East Africa.

Chris Reij, born 1948 in Hilversum, is a human geographer with almost 20 years of experience in soil and water conservation and water harvesting, mainly in Africa, but also in South and Central Asia. He has undertaken policy formulation, technical support and evaluation missions for a wide range of donor agencies. As a senior staff member of the Natural Resources Management Unit at the Vrije Universiteit, Amsterdam, he is currently managing the second phase of a major Dutch-funded programme on Indigenous Soil and Water Conservation in Africa, linking researchers, field staff and farmer-innovators in experiments to improve the efficiency of traditional and modern SWC practices.

Authors' Index Volume I and Volume II

Abd El-Hafez, S.A.	597	Chartres, C.J	539	Frazier, B.E.	327
Abdel Aal, A.I.N.	509	Chasi, M	1519	Freibauer, A	637
Abend, S.	781	Chennamaneni, R	1125, 1531	Fryrear, D.W.	291
Abo Soliman, M.S.M.	597	Chittaranjan, S.	1281	Fugger, W.D.	615
Adhikari, R.N.	1281	Christiansson, C.	317	Fuguang, Xie	1225
Adolph, B.	1107	Chuma, E.	1187, 1447	Gabbard, D.S.	257
Agbenin, J.O.	435	Clark, R.	879	Gawander, J.S.	959
Akhtar, M.	465	Cornish, P.S.	1029	Gebski, M.	713
Alexander, I.	1239	Costa, E.S.	443	Giger, M.	1405
Almaraz, R.	153	Cotler, H.	1539	Giugliarini, L.	105
Aloui, T.	341	Csillag, J.	673	Glazunov, G.P.	301
Amelung, W.	217, 225	Cuevas, E.	415	Golev, M.	791
Ammosova, J.	791	Currle, J.	1389	Gomer, D.	1099
Armstrong, J.	1275	Da Silva, J.E.	637	Goshu, K.	1523
Arnalds, A.	919	Darling, W.	1217	Greeves, G.W.	539
Arsyad, S.	1275	Darolt, M.R.	1205	Griesbach, C.-J.	867
Arya, Swarn Lata	1479	De Graaff, F.	45	Grigorév, V.	805
Asmussen, P.	1023	De Haan, F.A.M	607	Grossman, R.B.	175
Aune, J.B.	37, 375, 1247	De Kimpe, C.R.	3	Guindo, D.	1261
Ayarza, M.	637	Dechen, S.C.F.	355	Gupta, S.K.	629, 745
Ayoub, A.	457	Defoer, T.	1083	Häfele, S.	1261
Babu, Ram	897	Delgado, F.	133	Hagmann, J.	1187, 1447
Badora, A.	681	Derpsch, R.	1179	Haigh, M.J.	767, 775
Badraoui, M.	503	Dhman, H.	333	Hämmann, M.	629
Bajracharya, R.M.	231	Dhyani, B.L.	897	Hannam, I.	945
Barber, R.G.	1311	Diese, N.	661	Hape, M.	73
Barros, J.	477	Diestel, H.	73	Haraldsen, T.K.	621
Bartels, G.	1093	Dilkova, R.	125	Hari, T.	745
Battenfeld, A.	485	Dilly, O.	29	Haro, G.O.	1047
Baumann, J.	573	Donovan, C.	1261	Hassanin, S.A.	597
Baumgartl, Th.	797	Dontsova, K.	581	Hebel, A.	1
Beji, M.A.	341	Dosch, F.	933	Hecker, J.-M.	73
Bekele, T.	1009	Douglas, M.	1435	Heitzer, A.	53
Bielders, C.	1287	Driessen, P.	477	Hellin, J.	1383
Bligh, K.	1217	Eckelmann, W.	169	Helming, K.	589
Blum, W.E.H.	755, 1099	Effler, D.	1505	Hermawan, A.	1161
Blume, H.P.	29, 239, 515, 781	Eger, H.	XIX, 819, 1001	Herweg, K.	1429
Böhm, P.	271	El Hakim, M.H.	509	Hilhorst, Th	1083
Boli Baboule, Z.	395	El-Mowelhi, N.M.	597	Hoffmann, R.	495
Bosshart, U.	403	El-Swaify, S.A.	63	Holzwarth, F.	unpaged
Breburda, J.	285, 485, 697	Eswaran, H.	153	Horn, R.	527
Breland, T.A.	1247	Ewing, S.	953	Hornetz, B.	97
Bruijnzeel, L.A.	1267	Farhat, A.	503	Hraško, J.	927
Bucher, E.H.	905	Fawcett, R.G.	1217	Huang, C.	257
Bujtás, K.	673	Felix-Henningsen, P.	121, 697	Hurni, H.	827, 1037
Bunderson, W.T.	1073	Ferrero, A.F.	557	Huszar, P.C.	889, 905
Bundt, M.	721	Ferro, M.	1205	Izac, A.-M.	1073
Busacca, A.J.	327	Filip, Z.K.	21	Jankauskas, B.	389
Calegari, A.	1205, 1239	Filipek, T.	681	Jankauskiene, G.	389
Castrignanò, A.	105	Fischer, G.	143	Janku, C.	113
Chalise, S.R.	1151	Flach, K.W.	217	Jansons, V.	621
		Fleischhauer, E.	XIX	Kabała, C.	705

Authors' Index

Kaihura, F.B.S.	375	
Kamar, M.J.	1057	
Kantè, S.	1083	
Kanyama-Phiri, G.	1073	
Karbo, N.	1375	
Karczewska, A.	705	
Karki, Sameer	1151	
Kelly, T.G.	1107	
Kerchev, G.	125	
Kercheva, M.	125	
Khan, H.R.	239	
Kilasara, M.	375	
Kimble, J.M.	175, 185, 231	
Kirk, M.	1485	
Kirsch, B.	797	
Klima, K.	279	
Kobza, J.	689	
Kollender-Szych, A.	485, 697	
Kolpakov, A.L.	813	
Körschens, M.	423	
Kotlyar, A.L.	89	
Kouri, L.	1099	
Kranz, B.	615	
Krebs, R.	745	
Kretzschmar, S.	721	
Krinari, G.A.	813	
Krone, F.	169	
Kroschel, J.	615	
Kruger, A.S.	965	
Kullaya, I.K.	375	
Kunze, D.	1469	
Kupriyanova-Ashina, F.G.	813	
Laajili Ghezal, L.	341	
Laflen, J.M.	257	
Lal, R.	175, 185, 231, 375	
Langyintou, A.S.	1375	
Larrea, S.	1383	
Lehmann, J.	1255	
Leinweber, P.	113	
Lentoror, E.I.	1047	
Lentz, R.D.	1233	
Leonard Little Finger	XVI	
Leschinskaya, I.B.	813	
Leslie, L.M.	307	
Leumann, C.D.	745	
Lilienfein, J.	637	
Liniger, H.	1037, 1167	
Little, D.	1217	
López, R.	133	
Losch, S.	933	
Luizão, F.J.	443	
Luizão, R.C.	443	
Lukács, A.	673	
Lyalko, V.I.	89	
Lyons, W.F.	307	
Maarleveld, M.	971	
Mahler, F.	1287	
Makota-Mahlangeni, C. V.	VII	
M'Hamdi, A.	503	
Malinda, D.K.	1217	
Manthrithilake, H.	879	
Marek, K.-H.	89	
Mascarenhas, J.	1117	
Massawe, A.	37	
Matzner, E.	661	
Mayer, F.A.	1303	
Mazzoncini, M.	105	
McCool, D.K.	327	
Mercik, S.	713	
Merkel, Angela	III	
Merzouk, A.	333, 503	
Michels, K.	1287	
Minae, S.	1073	
Mittal, S.P.	1479	
Mndeme, K.C.H.	1965	
Moll†, W.	573	
Montgomery, J.A.	327	
Moriya, K.	1179	
Morrás, H.	1211	
Mouazen, A.M.	549	
Moukhtar, M.M.	509	
Moyo, A.	363	
Mühlig-Versen, B.	1287	
Mukamuri, B.B.	1367	
Mukherjee, K.	1135	
Müller, J.	79	
Munro, R.K.	307	
Murwira, H.K.	1367	
Murwira, K.	1447	
Mushala, H.M.	1511	
Mwakalombe, B.	1247	
Mwambazi, T.N.	1247	
Němeček, J.	735	
Ndiaye, M.K.	1261	
Nébié, B.	1261	
Neményi, M.	549	
Németh, T.	673	
Nibbering, J.W.	45	
Nigrelli, G.	557	
Nitsch, M.	495	
Norton, D.	581	
Norton, L.D.	257	
Oberle, A.	83	
Okoth-Ogendo, H.W.O.	1493	
Omoro, L.M.A.	1463	
Oomen, A.	985	
Oppitz, S.	89	
Östberg, W.	1349	
Ouaar, M.	1099	
Oweis, T.	83	
Padmalah, M.	1281	
Paschen, H.	1099	
Patra, A.K.	645	
Pawitan, H.	1275	
Peter, G.	1511	
Pfeiffer, E.M.	193	
Pfisterer, U.	239	
Piccolo, G.	1211	
Podlešáková, E.	735	
Portillo, L.	495	
Prasad, S.N.	589	
Prasuhn, V.	161	
Pretty, J.	837	
Preu, C.	113	
Prinz, D.	83	
Prüeß, A.	727	
Purwanto, Edi	1267	
Qinglan Wu	781	
Raizada, A.	1281	
Rammelt, R.	745	
Ranković, N.	911	
Rao, B.J.	1531	
Rao, M.S. Rama Mohan	1281	
Reij, C.	819, 1349, 1413	
Reintam, L.	451	
Rerkasem, B.	1319	
Rerkasem, K.	1319	
Revathi, E.	1531	
Rexilius, L.	781	
Rishirumuhirwa, Th.	1197	
Robbins, C.W.	1233	
Römkens, M.J.M.	589	
Roose, E.	395, 1197	
Rose, C.W.	247	
Runge-Metzger, A.	1469	
Ryabokonenko, A.D.	89	
Saborió, G.	721	
Salcedo, I.H.	415	
Salviano, A.A.C.	369	
Samra, J.S.	263, 1145, 1479	
Sanders, D.	867, 1421	
Sauerborn, J.	615	
Scharpenseel, H.-W.	193	
Schaub, D.	161	
Schleuß, U.	781	
Schneichel, M.	1023	
Scholten, T.	121	
Scholz, R.W.	53	
Schröder, D.	565	
Schulte-Karring, H.	565	
Setiani, C.	1161	
Shao, Y.	307	
Shaxson, T.F.	11	
Shevchenko, V.N.	89	
Shi Xuezheng	285	

Short, M.	797	Vukelić, G.	911
Shuijin, Su	1225	Waibel, H.	1469
Sidorchuk, A.	805	Wang, A-Bih	383
Sikka, A.K.	1145	Wang, Lin Kai	1225
Singh, B.R.	375	Wang, Lixian	1017
Sinukaban, N.	1275	Warkentin, B.P.	3
Soane, B.D.	517	Weber, O.	53
Sojka, R.E.	1233	Wenger, K.	745
Sombatpanit, S.	1421	Werner, G.	573
Sommer, K.	713	Westerberg, L.-O.	317
Soudi, B.	503	White, R.	879
Spaan, W.P.	1295	Wilcke, W.	689, 721
Sparovek, G.	369, 431	Wopereis, M.C.S.	1261
Spranger, C.-D.	VIII	Wu, Chia-Chun	383
Dowdeswell, E.	XI	Wulf, S.	1255
Spricis, A.	621	Xie, G.	485
Squires, V.R.	209	Xing Tingyan	285
Srivastava, A.K.	1281	Yili, Li	1225
Stamm, V.	1499	Yohannes, G. M.	1359
Steer, A.	851	Yosko, I.	1093
Steiner, K.	819	Yu Dongsheng	285
Stelz, E.	1303	Zdruli, P.	153
Stocking, M.	355, 857, 879	Zech, W.	217, 225, 637, 697, 721, 1255
Stolbovoi, V.	143	Zekri, S.	341
Sudars, R.	621	Zhang, Xudong	217, 225
Süßmuth, Rita	I	Zihler, J.	629
Szabolcs†, I.	469	Zike, W.	1009
Szerszeń, L.	705	Zlatić, M.	911
Tengberg, A.	355	Zweifel, H.	991
Terytze, K.	79		
Theerawong, S.	1421		
Thippannavar, B.S.	1281		
Thomas, D.B.	1037, 1167		
Tian, Xiao	1225		
Tiessen, H.	415		
Tietje, O.	53		
Tiffen, M.	1333		
Toledo, C.S.	905		
Trott, H.	697		
Unwin, R.J.	1399		
Utermann, J.	169, 495		
v. Willert, F.	1255		
Vagstad, N.	621		
Van der Zee, S.E.A.T.M.	607		
van Dijk, K.J.	1295		
Van Ouwerkerk†, C.	517		
van Veldhuizen, L.	979		
Vieira, S.R.	369		
Vogel, H.	169		
Vogt, H.	1099		
Vogt, T.	1099		
von Lossau, A.	1047		
Von Wedemeyer, H.-Chr.	565		
Voplakal, K.	653		
Voulfson, L.D.	89		

GeoEcology textbook

M. Kutilek & D. Nielsen

Soil Hydrology

Textbook for students of soil science, agriculture, forestry, geoecology, hydrology, geomorphology and other related disciplines

370 pages / numerous figures, tables
ISBN 3-923381-26-3
list price: DM 59,-/US $ 39,-

We have intended to present an introduction to the physical interpretation of phenomena which govern hydrological events related to soil or the upper most mantle of the earth's crust. The text is based upon our teaching and research experience. The book can serve either as the first reading for future specialists in soil physics or soil hydrology. Or, it can be a source of basic information on soil hydrology for specialists in other branches, e.g. in agronomy, ecology, environmental and water management. (from Authors)

Contents (shortened)

1	Soils in Hydrology	6	Elementary Soil Hydrologic Processes
1.1	Soils	6.1	Principles of Solutions
1.2	Concepts of Soil Hydrology	6.2	Infiltration
2	Soil Porous System	6.3	Soil Water Redistribution and Drainage after Infiltration
2.1	Soil Porosity		
2.2	Classification of Pores	6.4	Evaporation from a Bare Soil
2.3	Methods of Porosity Measurement	6.5	Evapotranspiration
2.4	Soil Porous Systems	7	Estimating Soil Hydraulic Functions
2.5	Soil Specific Surface	7.1	Laboratory Methods
3	Soil Water	7.2	Field Methods
3.1	Soil Water Content	8	Field Soil Heterogeneity
3.2	Measurement of Soil Water Content	8.1	Variability of Soil Physical Properties
4	Soil Water Hydrostatics	8.2	Concept of Soil Heterogeneity
4.1	Interface Phenomena	8.3	Spatial Variability and Geostatistics
4.2	Soil Water Potential	8.4	Scaling
4.3	Soil Water Retention Curve	8.5	State-space Equations for Multiple Locations
5	Hydrodynamics of Soil Water	9	Transport of Solutes in Soils
5.1	Basic Concepts	9.1	Solute Interactions
5.2	Saturated Flow	9.2	Miscible Displacement in a Capillary
5.3	Unsaturated Flow in Rigid Soils	9.3	Miscible Displacement in Surrogate Porous Media
5.4	Two Phase Flow	9.4	One-dimensional Laboratory Observations
5.5	Flow in Non-rigid (Swelling) Soils	9.5	Theoretical Descriptions
5.6	Non-Isothermal Flow	9.6	Implications for Water and Solute Management

CATENA VERLAG GMBH   *GeoScience Publisher*

Advances in GeoEcology Formerly **CATENA SUPPLEMENTS**

Dan H. Yaalon & S. Berkowicz (Editors)

History of Soil Science
- International Perspectives -

Advances in GeoEcology 29
(follow-up series of CATENA SUPPLEMENTS)
438 pp   1997   DM 264,00 / US$ 176.-
ISBM 3-923381-40-9

The book presents a wideranging perspective on the history of soil science comprising a collection of 22 papers. Following an overview on the main paradigms, developments of the concepts of humus, horizons, classification of soil types and soil series usage are treated in specific chapters. Some selected topics in the history of soil chemistry and soil physics are treated in detail. A number of articles deal with regional aspects of soil science and the contribution of some outstanding personalities from the 18th to the 20th centuries. This is the first original history in soil science in the English language.

H.-P. Blume & S.M. Berkowicz (Editors)

Arid Ecosystems

Advances in GeoEcology 28
(follow-up series of CATENA SUPPLEMENTS)
229 pp   1995   DM 189,00 / US$ 126.-
ISBN 3-923381-37-9

K. Auerswald, H. Stanjek & J.M. Bigham (Editors)

Soils and Environment
Soil Processes from Mineral to Landscape Scale

Advances in GeoEcology 30
(follow-up series of CATENA SUPPLEMENTS)
422 pp,   1997   DM 189,00 / US$ 126.-
ISBN 3-923381-41-7

CATENA VERLAG GMBH   *GeoScience Publisher*